Sixth Edition

HUMAN HEREDITY

Principles and Issues

Michael R. Cummings
University of Illinois at Chicago

THOMSON
BROOKS/COLE

Australia • Canada • Mexico • Singapore • Spain
United Kingdom • United States

Biology Executive Editor: Nedah Rose
Development Editor: Marie Carigma-Sambilay
Assistant Editor: Christopher Delgado
Editorial Assistant: Rebecca Subity
Technology Project Manager: Keli Amann
Marketing Manager: Ann Caven
Advertising Project Manager: Linda Yip
Project Manager, Editorial Production: Teri Hyde
Print/Media Buyer: Tandra Jorgensen
Permissions Editor: Joohee Lee
Production Service: Graphic World Publishing Services

Text Designer: Laurie Janssen
Photo Researcher: Meyer Photo-Art
Copy Editor: Mary Ann Grobbel
Cover Designer: Larry Didona
Cover Image: Chromosomes © Lester V. Bergman/CORBIS; Magnifying glass with DNA © Andrew Brookes/ corbisstockmarket.com; Infants © Getty Images / V.C.L.; Aerial shot of double helix, Charles C. Benton
Cover Printer: Transcontinental Interglobe
Compositor: Graphic World, Inc.
Printer: Transcontinental Interglobe

Printed in Canada
1 2 3 4 5 6 7 06 05 04 03 02

For more information about our products, contact us at:
Thomson Learning Academic Resource Center
1-800-423-0563

For permission to use material from this text, contact us by:
Phone: 1-800-730-2214 **Fax:** 1-800-730-2215
Web: http://www.thomsonrights.com

Library of Congress Control Number: 2002102403

ISBN 0-534-39474-4

Brooks/Cole—Thomson Learning
511 Forest Lodge Road
Pacific Grove, CA 93950
USA

Asia
Thomson Learning
5 Shenton Way #01-01
UIC Building
Singapore 068808

Australia
Nelson Thomson Learning
102 Dodds Street
South Melbourne, Victoria 3205
Australia

Canada
Nelson Thomson Learning
1120 Birchmount Road
Toronto, Ontario M1K 5G4
Canada

Europe/Middle East/Africa
Thomson Learning
High Holborn House
50/51 Bedford Row
London WC1R 4LR
United Kingdom

Latin America
Thomson Learning
Seneca, 53
Colonia Polanco
11560 Mexico D.F.
Mexico

Spain
Paraninfo Thomson Learning
Calle/Magallanes, 25
28015 Madrid, Spain

To Colin and Maggie.

ABOUT THE AUTHOR

MICHAEL R. CUMMINGS received his Ph.D. in Biological Sciences from Northwestern University in 1968. His doctoral work, conducted in the laboratory of Dr. R. C. King, centered on ovarian development in *Drosophila melanogaster.* After a year on the faculty at Northwestern, he moved to a teaching and research position at the University of Illinois at Chicago. Here, he established a research program on the developmental genetics of *Drosophila* and began teaching courses in genetics, developmental genetics, and evolution. Currently an associate professor in the Department of Biological Sciences and in the Department of Molecular Genetics, he has also taught at Florida State University.

About fifteen years ago, Dr. Cummings developed a strong interest in scientific literacy. In addition to teaching genetics to biology majors, he organized and currently teaches a course in human genetics for non-majors, and participates in teaching general biology. He is now working to integrate the use of electronic resources such as the Internet and World Wide Web into the undergraduate teaching of genetics and general biology. His current research interests involve the role of the short arm/centromere region of human chromosome 21 in chromosomal aberrations. His laboratory is engaged in a collaborative effort to construct a physical map of this region of chromosome 21 to explore molecular mechanisms of chromosome interactions.

Dr. Cummings is the author and co-author of a number of widely used college textbooks, including *Biology: Science and Life, Concepts of Genetics,* and *Essentials of Genetics.* He has also written sections on genetics for the *McGraw-Hill Encyclopedia of Science and Technology,* and has published a newsletter on advances in human genetics for instructors and students.

He and his wife, Lee Ann, are parents of two adult children, Brendan and Kerry, and have two grandchildren, Colin and Maggie. He is an avid sailor, enjoys reading and collecting books (biography, history), music (baroque, opera, and urban electric blues), eating the fine cuisine at Al's Beef, and is a long-suffering Cubs fan.

Contents

CHAPTER 1

Genetics as a Human Endeavor 1

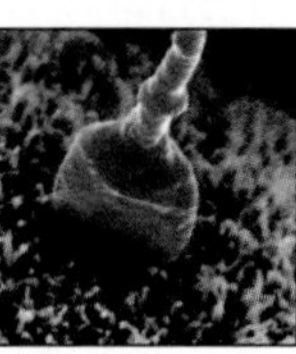
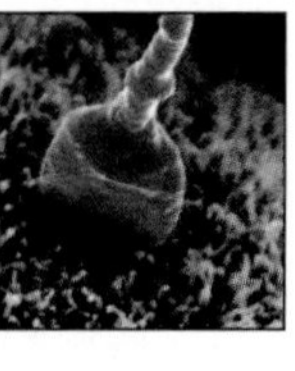

CHAPTER 2

Cells, Chromosomes, and Cell Division 16

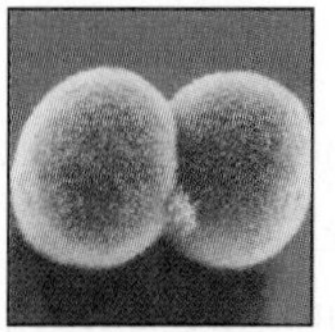

CHAPTER 3

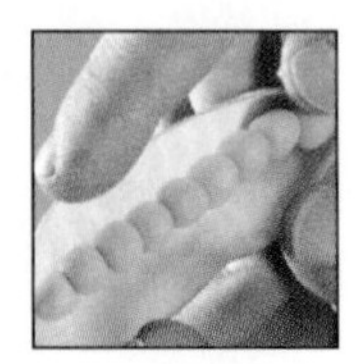

Transmission of Genes from Generation to Generation 47

CHAPTER 4

Pedigree Analysis in Human Genetics 74

CHAPTER 6

Cytogenetics: Karyotypes and Chromosome Aberrations 140

CHAPTER 7

Development and Sex Determination 174

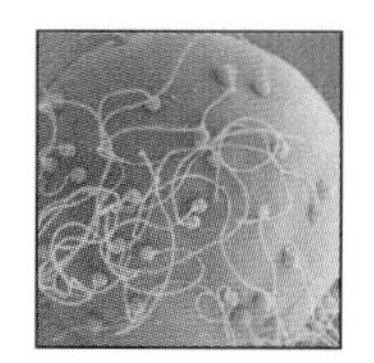

CHAPTER 8

DNA Structure and Chromosomal Organization 202

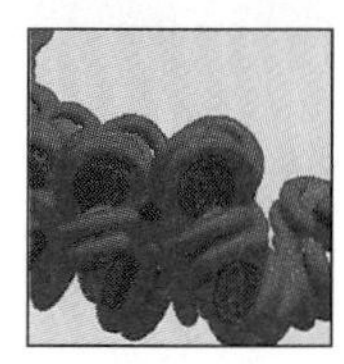

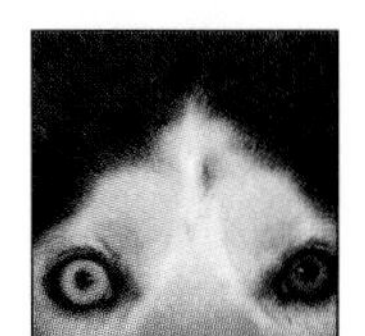

CHAPTER 12

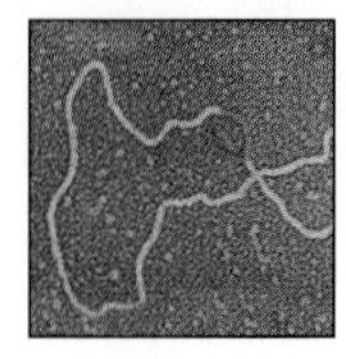

An Introduction to Cloning and Recombinant DNA 294

CHAPTER 13

Biotechnology and Genomics 316

CHAPTER 14

Genes and Cancer 342

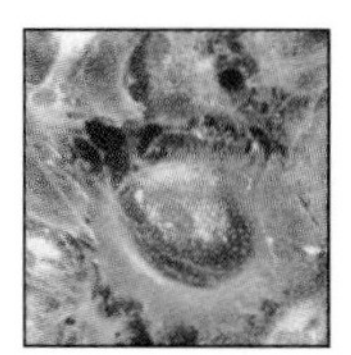

CHAPTER 15

Genetics of the Immune System 368

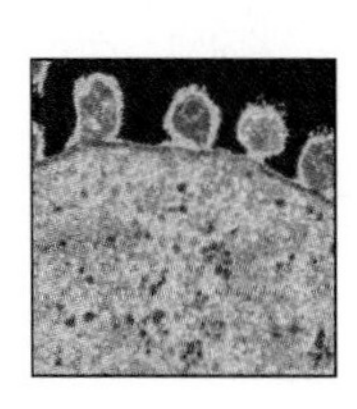

CHAPTER 16

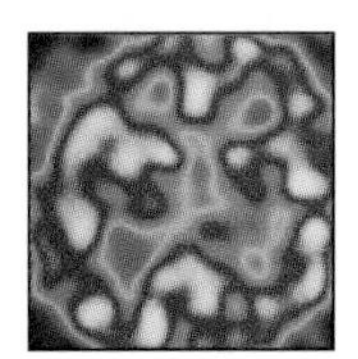

Genetics of Behavior 395

CHAPTER 17

Genes in Populations 423

CHAPTER 18

Human Diversity and Evolution 443

CHAPTER 19

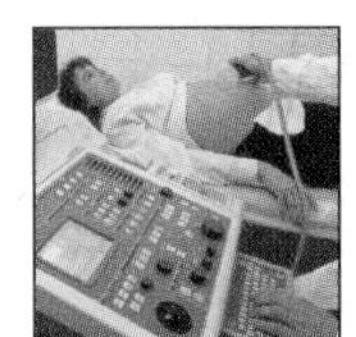

Genetic Testing, Gene Therapy, and Counseling 471

Preface

This edition of *Human Heredity* is being published at a time of transition in the field of human genetics, shortly after the draft sequence of the human genome was published. Two landmark papers published in 2001, one in *Science* and the other in *Nature*, give us a glimpse into the size, organization, and evolutionary history of our genome. In the preface to the fifth edition, I noted that the sequence of the human genome would be available while that edition was in use in the classroom. That prediction has, for the most part, been fulfilled. What is now apparent, even in the short time following completion of the draft sequence, is that data from the genome project are already having a significant impact on research and on the diagnosis and treatment of disease. Genome sequence information, coupled with advances in technology such as DNA chips, is causing a fundamental shift in the way human genetics research is done. These developments allow researchers to use automated processes to gather data on a scale far beyond what was imagined only a few years ago. This increase in information flow is rapidly advancing our knowledge of genetics, creating new applications, and impacting other fields of research such as cell biology and developmental biology. Many of these new advances are reflected in the organization of this edition, and in the content of some chapters.

Genetic research, genome databases, and new technology are transforming research, industry, agriculture, and our everyday lives, from the medical tests we take to the food we eat. To help undergraduate students in humanities, social sciences, business, engineering, and other disciplines understand these fast-moving developments, I have written this book to be used in a one-term introductory human genetics course aimed at the nonmajor. It is written for students who have little or no background in biology, chemistry, or mathematics, but have an interest in learning something about human genetics. Some descriptive chemistry is used after an appropriate introduction and definition of terms. In the same vein, math is used in several places, but no advanced math skills are required to calculate elementary probabilities or to calculate genotype and allele frequencies.

With the ever-increasing amount of information flowing from genome projects and the personal and societal choices generated by biotechnology, it is important for all of us to have an understanding of basic genetic concepts. To help students grasp what is important, the book is organized around a hierarchy of ideas, rather than a collection of facts. Each chapter deals with a small number of basic concepts, and uses these to organize examples and relevant facts. For example, Chapter 3 deals with Mendelian principles, and uses crosses in pea plants to explain basic concepts. The use of an experimental organism (in this case, peas) to illustrate the principles that govern the transmission of traits from generation to generation makes it easy to define each concept and provide examples that are clear cut and unambiguous. With a firm grounding in these principles, students will be able to apply them to the transmission of traits in humans, where the methods are indirect and observational rather than experimental.

Knowledge from genetics is rapidly being transferred to many other fields. The spread of this knowledge, the technology it generates, and the issues raised make it clear that difficult but informed decisions are required at many levels, from the personal to the political. The public, elected officials, and policy makers outside the scientific community need a working knowledge of genetics to help shape the course of research and its applications in our society. This text has been written to transmit the principles of genetics in a straightforward and accessible way, without unnecessary jargon, detail, or the use of anecdotal stories.

GOALS OF THE TEXT

I work with undergraduates on a daily basis, teaching courses in a biology curriculum to both majors and nonmajors. Over the last decade, students have helped shape this book in many ways. They have suggested how to best introduce topics, identified the most effective examples and analogies in explaining concepts, and more importantly, been forthright in clarifying what does not work in the classroom. As in the past, the organization and content for this edition have been developed by incorporating classroom-tested ideas refined by feedback from students and from faculty who have used the book at other institutions, and by comments and suggestions from reviewers. From the first edition, this book has been written to achieve several well-defined goals. This edition reinforces and extends these goals in the context of the Human Genome Project. The goals are as follows:

1. Present the principles of human genetics in a clear, concise manner that gives students a working knowledge of genetics. The premise in this approach is that a limited number of clearly presented, linked concepts is the best way to learn a complex subject such as genetics.
2. Communicate an understanding of the origin and amount of genetic diversity present in the human population, and how this diversity has been shaped by natural selection.
3. Examine the social, cultural, and ethical implications associated with the use of genetic technology.
4. Begin the discussion of concepts at a level that students can understand, and provide relevant examples that students can apply to themselves, their families, and their work environment.

To achieve these goals, emphasis has been placed on clear writing with the use of accompanying photographs and artwork that teach rather than merely illustrate the ideas under discussion. In addition, the book features up-to-date coverage and flexible organization. Once again in this edition, a conscious effort has been made to pare down and eliminate unnecessary terms and jargon and to present the material in a straightforward and engaging manner.

In general, the text consists of three sections: Chapters 1 through 7 cover cell division, transmission of traits from generation to generation, and development. Chapters 8 through 13 emphasize molecular genetics, recombinant DNA genomics, and biotechnology. These chapters cover gene action, mutation, cloning, the applications of genetic technology, the Human Genome Project, and the social, legal, and ethical issues related to genetics. Chapters 14 through 19 consider specialized topics, including cancer, the immune system, population genetics, as well as the social aspects of genetics including behavior, genetic screening, genetic testing, and genetic counseling.

Because courses in human genetics have a wide range of formats, the book is organized so it will be easy to use, no matter what order of topics an instructor chooses. After the section on transmission genetics, the chapters can be used in any order. Within each chapter, the outline lets the instructor and students easily identify central ideas.

FEATURES OF THE SIXTH EDITION

New Chapter

Chapter 13 has been rewritten and retitled to serve as an introduction to the rapidly growing fields of genomics and biotechnology. After an introduction to the uses of recombinant DNA technology in mapping genes, applications including DNA fingerprints, the preparation of edible vaccines, and the use of transgenic food crops are considered. The social impact of genetic technology on law and social policy is emphasized, and ethical questions about the use of this technology are raised. The chapter also examines some of the first applications of information flowing from the Human Genome Project in creating new fields of study, and how these new fields of genomics and bioinformatics are being used to study human genetic disorders. This chapter provides an important perspective on some of the impact that genetics has on our everyday lives.

New Tools for Systematic Learning of Genetics Concepts

The best way to learn genetics is by solving problems. Built into this edition are some learning tools designed to develop and sharpen students' analytical and problem-solving skills. These new "Web-Integrated *(genetics concepts)* Toolboxes" walk students through the process of understanding, analyzing, and working out problems in such areas as solving genetics problems (Chapter 3), solving pedigree problems (Chapter 4), analyzing karyotypes (Chapter 6); building a DNA molecule (Chapter 8); DNA fingerprinting (Chapter 13), and population genetics (Chapter 17). Each toolbox focuses on a concept that is crucial to understanding genetics. It first gives (on the printed page) a concise introduction to the concept, and maps out a "strategy" for understanding it based on actual hands-on work. Then it points students to the book's Web site, where they get a chance to carry out this concrete learning strategy using animations, exercises, and additional information.

New Organization

The order of chapters is the same as in the last edition. The reorganization and reordering of chapters that took place in the fourth edition has proven successful and has been continued here. Mendelian inheritance and quantitative inheritance are in one section, and the chapter on mutation precedes the chapters on recombinant DNA technology and its applications. In this edition, emphasis has been placed on ensuring that each chapter focuses on a small number of basic concepts. To facilitate this, material has been moved or rewritten to shift emphasis, and new sections have been added as needed.

A few examples of these changes will serve to illustrate how this focus was developed. In Chapter 13, the sections on prenatal genetic testing and gene therapy have been moved to Chapter 19, allowing more discussion of the applications of genomics and biotechnology in Chapter 13. This also places all the material on gene therapy, genetic testing, and screening in Chapter 19. Having these topics in a single chapter provides a foundation for discussion of the ethical and social issues raised by the use of these techniques. In Chapter 2, the sections on organelles have been linked to genetic disorders that affect organelle systems, emphasizing the need to understand the structure and function of cells, and how a single mutation can disrupt both the structure and function at the cellular level.

To help the student focus on the concepts presented, each chapter summary is now organized using the chapter outline. In the summary, the main points of the chapter are listed as headings, followed by a summary of the concepts and examples presented.

This focus on the chapter's main points is continued in the "Questions and Problems" at the end of each chapter. The questions and problems are organized into sections under the headings from the chapter outline. This allows students to relate the problems and questions to specific topics presented in the chapter, to focus on concepts they find difficult, and to work the problems that illustrate these topics.

Expanded Questions and Problems

Recognizing that many students have difficulty solving genetics problems, the end-of-chapter questions and problems have been revised and expanded. The revisions and additions have been contributed by adopters of the text who have used their experience and student input to redesign, rewrite, and add to the problem sets. The questions and problems use both an objective question and a problem format, and are arranged by topic and level of difficulty. Because some quantitative skills are necessary in human genetics, almost all chapters include some problems that require the students to organize the concepts in the chapter and use these concepts in reasoning to a conclusion. Answers to selected problems are provided in Appendix B. Answers to all questions and problems are available in the Instructor's Manual.

Case Studies

To make issues in human genetics relevant to situations that students may encounter outside the classroom, questions have been added to the case studies at the end of each chapter. This section contains scenarios and examples of genetic issues related to health, reproduction, personal decision making, public health, and ethics. Many of the case studies and the accompanying questions can be used for classroom discussions, student papers and presentations, and role playing.

New Topics

Many new topics have been introduced, and some topics have received greater emphasis. It is impossible to list them all, but included among them are pharmacogenetics; background radiation exposure, radiation as a source of mutation, and DNA repair mechanisms; cloning of mammals from somatic cells; mapping genes by positional cloning, preimplantation genetic testing, the use of DNA chips in genetics, biopharming, and the ethical issues surrounding the Human Genome Project; DNA repair mutations in cancer, the role of gatekeeper and caretaker genes in cancer, genomic instability and cancer, and behavior and cancer; the search for genes controlling manic depression and schizophrenia; and genetic evidence for the spread of humans across the world.

Guest Essays

This edition features a series of essays written by distinguished scientists, describing how they became interested in science, what they have chosen to study, and how their research relates to the larger context of human genetics. These essays are not just *about* scientists, they are written *by* scientists, giving nonmajors some insight into the lives, thoughts, and motives of biologists and geneticists.

Genetic Databases

To foster awareness of the vast array of databases dealing with genetics, the genetic disorders mentioned in the book are referenced with the indexing number assigned to them in the comprehensive catalog assembled by Victor McKusick and his colleagues. This catalog is available in book form as *Mendelian Inheritance in Man: Catalog of Human Genes and Genetic Disorders.* It is also available at several World Wide Web sites as *Online Mendelian Inheritance in Man (OMIM).* The online version (with daily updates) contains text, pictures, and videos along with references to the literature and links to other databases, including those related to the Human Genome Project. Students and an informed public need to be aware of the existence and relevance of such databases, and OMIM is used here in part to promote such awareness.

Students who wish to learn more about a particular genetic disorder can use OMIM to obtain detailed information about the disorder, its mode of inheritance, phenotype and clinical symptoms, mapping information, biochemical properties, the molecular nature of the disorder, and a bibliography of relevant papers. In the classroom, OMIM and its links to other databases are a valuable resource for student projects and presentations.

New Internet Activities

The World Wide Web (WWW) is an important and valuable resource in teaching human genetics, and the exercises included in this edition can be used to expand on concepts covered in the text, and provide detailed information about specific genetic disorders. They can also be used to introduce the social, legal, and ethical aspects of human genetics into the classroom and serve as a point of contact with support groups and testing services.

This edition contains updated and expanded end-of-chapter Internet activities for students. These activities use WWW resources to enhance the topics covered in the chapter, and are designed to generate interaction and thought rather than passive observation. Sites for these activities can be reached through the book's home page.

PEDAGOGICAL FEATURES

The basic organization within chapters, which has been successful as a teaching resource, has been continued in this edition. In some chapters, the content and features have been updated and revised to reflect current topics, engage student interest, and improve the pedagogy.

Opening Vignettes

Each chapter begins with a short prologue directly related to the main ideas of the chapter, often drawn from real life. Topics include genetic discoveries, such as the chromosomal basis of Down syndrome, the development of *in vitro* fertilization and the birth of Louise Brown, the first IVF baby. These vignettes are designed to promote student interest in the topics covered in the chapter and to demonstrate that laboratory research often has a direct impact on everyday life.

Chapter Outlines

At the beginning of each chapter, an outline provides an overview of the main concepts, secondary ideas, and examples. To help students grasp the central points, many of the headings have been rewritten as narratives or summaries of the ideas that follow. These outlines also serve as convenient starting points for students to review the material in the chapter. To make the outline more useful, it is used to organize the summary and the questions and problems at the end of each chapter. In this way, students can more readily and clearly relate examples and questions to specific topics in the chapter.

Concepts and Controversies

Within most chapters, students will find boxes that present ideas and applications related to the central concepts in the chapter. Some of these present interesting but tangential examples that should be of interest to the student, while others examine controversies that arise as genetic knowledge is transferred into technology and services.

Guest Essays

Scattered throughout the book are essays written by prominent scientists. The essays introduce the human side of scientists and summarize how they became interested in science, and how their work relates to larger issues in society.

Margin Glossary

A glossary in the page margins gives students immediate access to definitions of terms as they are introduced in the text. This format also allows definitions to be identified when students are studying or preparing for examinations. These definitions have been gathered into an alphabetical glossary at the back of the book. Because an understanding of the concepts of genetics depends on understanding the relevant terms, more than 350 terms are included in the glossary.

Sidebars

Throughout the book, sidebars are used to highlight applications of concepts, present the latest findings, and point out controversial ideas without interrupting the flow of the text.

END-OF-CHAPTER FEATURES

The end-of-chapter features have been revised and updated for this edition. The questions and problems have been expanded and reorganized to reflect the order of topics in the chapter. Questions have been added to the case studies to enhance their use in the classroom, and new Internet activities have been added.

Case Studies

As described earlier, this section contains case studies of individuals and families using various genetic services, large-scale issues such as radioactive pollution, and the impact of the Human Genome Project. Many of these can be used as the basis for classroom discussions, student presentations, and role playing.

Summary

Each chapter ends with a numbered summary that restates the major ideas covered in the chapter. Chapters begin with an outline and end with a summary using the same outline to emphasize major concepts and their applications. This helps students relate the conceptual framework of the chapter to the topics and examples presented in the chapter and, it is hoped, minimizes the chance they will attempt to learn by rote memorization.

Questions and Problems

The questions and problems at the end of each chapter are designed to test students' knowledge of the facts and their ability to reason from the facts to conclusions. This section has been reorganized, revised, and expanded. The questions and problems are presented under the headings from the chapter outline to enable the student to relate the problems to the topics presented in the chapter. Revisions, additional questions, and problems have been contributed by adopters of previous editions, by professionals in genetic research and health care services, and by students.

Internet Activities

Activities at the end of each chapter use Web sites to engage the student in activities related to the concepts discussed in the text. This section has been revised and expanded in this edition. Internet resources are now an essential part of teaching genetics and this section introduces students to the many databases, instructional sites, and support groups available to them.

For Further Reading

A list of readings is presented at the end of each chapter. These include reviews and general articles that are accessible to the nonscientist, as well as the key papers that describe discoveries covered in the chapter. The references have been updated just before publication to provide the most current coverage of the literature.

ANCILLARY MATERIALS

The ancillary materials that accompany this edition are designed to assist the instructor in preparing lectures and examinations, and to help keep instructors abreast of the latest developments in the field.

Instructor's Manual with Test Bank

An expanded and updated instructor's manual is available to help instructors in preparing class materials. It contains chapter outlines, chapter summaries, teaching/learning objectives, key terms, additional test questions, and discussion questions. It also contains answers to all end-of-chapter questions and problems.

ExamView

A computerized test bank, available on CD-ROM, can be customized and printed according to the instructor's preferences and needs. It contains approximately 800 test items.

Transparency Acetates

A set of 100 color transparencies featuring key figures—including drawings, charts, and diagrams from the text—is available to adopters.

Study Guide

A student study guide has been prepared by Nancy Shontz of Grand Valley State University. It is intended to enhance understanding of the text and course material. It includes chapter objectives and summaries, lists of terms, case worksheets (based on case studies in the text), discussion problems and questions, and other practice test items in multiple-choice, fill-in-the-blank, and modified true/false formats.

Gene Discovery Lab

A CD-ROM lab manual that provides a virtual laboratory experience for the student in doing experiments in molecular biology. It includes experiments that use nine of the most common molecular techniques in biology, an overview of the scientific method and experimental techniques, and Web links to provide access to data and other resources.

Current Perspectives in Genetics

A reader features approximately 40 articles in molecular, classical, and human genetics. Each article begins with a brief introduction and ends with critical thinking or discussion questions. Answers are provided in the back of the book.

Genetics on the Web: A Brief Guide to the Internet

A handy reference for students who need an introduction to research on the Internet. It includes a list of relevant URLs listed by genetics topics.

Instructor's Edition

A special edition of the text for instructors that features extra frontmatter, including a visual preface.

CONTACTING THE AUTHOR

I welcome questions and comments from faculty and students about the book or about human genetics. Please contact me at: cummings@uic.edu

ACKNOWLEDGMENTS

One of the rewards of teaching is seeing the glimmer of recognition in students' eyes when they grasp a concept or connect two seemingly unrelated facts. For me, writing a text is a way of providing students with a means of reaching that point. Looking back at the first edition, published in 1987, it is clear how far genetics and the resources for teaching it have advanced. Over the years, new editions of the book have evolved from the interaction of several forces. One of these is the ever-accelerating rate of progress in genetics, which has not only spawned new scientific disciplines such as genomics and bioinformatics, but has provided many new details about gene organization, function, and evolution. More important perhaps, the book has been refined as a teaching tool by many people, including reviewers, editors, instructors, and students. Hopefully, with each of these changes, it becomes a more useful instrument for knowledge that reflects the collective wisdom of those who have contributed to it. This edition, then, contains a summary of advances in genetics, and a distillation of what I have learned from those who have helped me become a better author and better teacher. To all of them, I offer my thanks.

To the reviewers who helped shape this edition, I extend my thanks and gratitude for their efforts and many suggestions. Their work and insights have enhanced the focus and presentation of the material.

James A. Brenneman
University of Evansville, IN

Michelle Geary
West Valley College, CA

Sandra L. Gilchrist
New College, FL

Donald J. Nash
Colorado State University, CO

David L. Parker
Northern Virginia Community College, VA

Bernard Possidente
Skidmore College, NY

Judith A. Sparlin
Clarion University of Pennsylvania, PA

Olivia Masih White
University of North Texas, TX

Robert Wiggers
Stephen F. Austin State University, TX

Michelle Murphy Whaley of the University of Notre Dame and Peter Follette took on the daunting task of revising and adding to the end-of-chapter questions and problems as well as writing the questions for the case studies. Writing questions that illustrate important concepts is one of the most difficult tasks in teaching. Through their work, these features of the text have been expanded and pedagogically improved. Michelle Geary of West Valley College revised and updated the Internet activities, linking them to material in the text.

At Brooks/Cole, the book has benefited from the creative input of many talented individuals whose efforts are always directed at finding ways to help students learn. My editor, Nina Horne, laid the foundation for this edition before turning her attention to family responsibilities. I have been fortunate to have the guidance of her successor, Nedah Rose, in the development of this edition. Her broad background in science publishing has provided me with perspective on how to communicate science to the nonspecialist. Ann Caven, Marketing Manager, contributed her marketing expertise, and her suggestions have raised the pedagogical level of the text. Once again, it has been my good fortune to work with Marie Carigma-Sambilay as my development editor. Her insights and suggestions are responsible for many of the new features in this edition. Her optimistic outlook and gentle prodding when needed helped keep this project on track. In addition, she handled most of the usual crises, problems, and delays, allowing me to devote my efforts to the preparation of this edition. Rebecca Subity organized and commissioned the reviews, effectively and concisely laying out the scope of this project. Christopher Delgado handled the ancillaries, coordinating the book's content with the resources of the publisher. Keli Amann researched, organized, and coordinated the media resources for the Web-Integrated Toolboxes, one of the new learning tools in this edition. In the transition to publication, Teri Hyde effectively brought together and coordinated the many processes in turning a manuscript into a printed book. At Graphic World, Mike McConnell calmly and professionally handled the sometimes chaotic flow of material in addition to supervising the copyediting, design, art, and photo program. The photo research team of Don and Joan Murie at Meyer Photo-Art were a pleasure to work with. They were always willing to go the extra mile to get the best possible illustrations. Stephen Rapley was the creative force behind the cover design. Thanks also to my copyeditor, Mary Ann Grobbel, who patiently taught me the difference between *since* and *because,* and corrected my many faults of usage, punctuation, and grammar.

I wish to express special thanks to my colleague, Suzanne McCutcheon, who gave generously of her time to help supervise the research in my lab during my sporadic absences to work on this edition. I also thank my undergraduate research students Jeff Kochikaran and Young Ahn and my graduate student Holly Dimitropoulos for their understanding and forbearance.

Michael R. Cummings

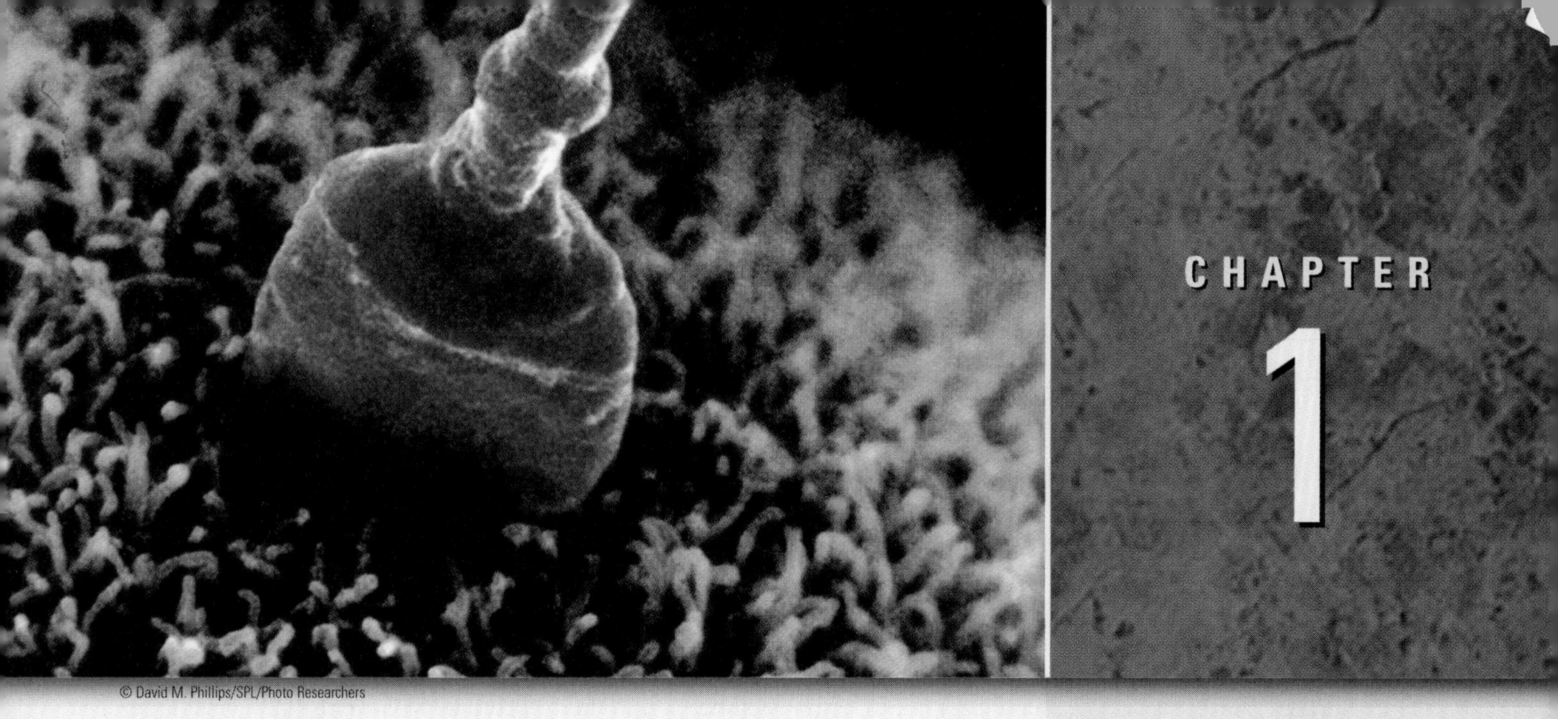
© David M. Phillips/SPL/Photo Researchers

CHAPTER 1

Genetics as a Human Endeavor

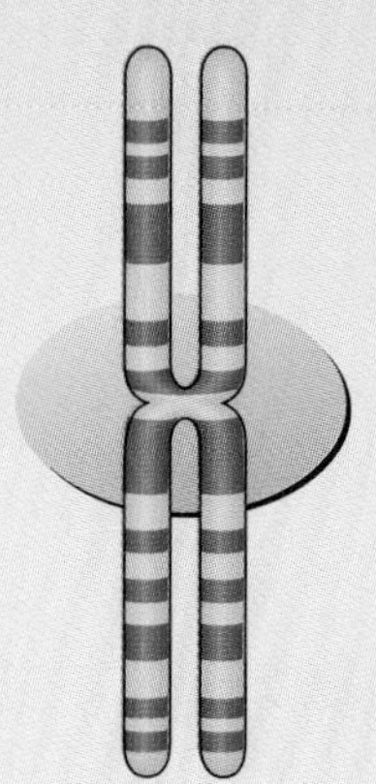

Chapter Outline

In December 1998, after much debate and over a determined opposition, the Icelandic Parliament (Althingi) passed a controversial bill allowing deCODE, a biotech company, to establish and operate a database that includes the medical records of all residents of Iceland. In addition, deCODE is compiling the genealogies of the approximately 800,000 Icelanders who lived since the colonization of the island in the ninth and tenth centuries. In combination with blood and tissue samples (for DNA extraction) provided by patients, the database is a powerful tool in the hunt for disease-causing genes. The new law grants the company the right to sell this information (and the DNA samples) to third parties, including the research labs of pharmaceutical companies, with the hope that once disease genes are identified, diagnostic tests and therapies will soon follow.

Why establish such a database in Iceland? Because Iceland has a small, genetically isolated population. The country was founded by a few hundred people in the ninth and tenth centuries and, until 50 years ago, was almost completely isolated from outside immigration. Plague (in the 1400s) and volcanoes (in the 1700s) decimated the population, further reducing genetic variation. The 270,000 inhabitants of Iceland share a remarkably similar set of genes, providing fertile ground for gene hunters seeking to identify disease genes.

Why the controversy? Opponents point out that the privacy provisions of the law are inadequate and may violate the ethical principle that health records must be kept confidential. Abuses and misunderstandings may affect employment, insurance, and even marriage. In addition, critics question whether a single company should have exclusive rights to medical information and whether the Icelandic population will derive health benefits from this arrangement. The argument is far from settled and may involve court action and censure by international medical societies.

With stories like those about Iceland becoming more common in the media, as we begin this book, we might pause and remember that genetics is more than a laboratory science; unlike some other areas of science, genetics and biotechnology have a direct impact on society.

WELCOME TO GENETICS

Genetics The scientific study of heredity.

As a first step in studying human genetics, we should ask what *is* genetics? **Genetics** is the science of heredity. Geneticists study how traits (like eye color and hair color) and diseases (like cystic fibrosis and sickle cell anemia) are passed from generation to generation. They also study the molecules that make up genes and gene products, and how genes are turned on and turned off. Population geneticists study why some genes are found more frequently in some populations than in others. Other geneticists work in industry to develop products for agricultural and pharmaceutical firms. This applied research, called biotechnology, has grown in recent years to become a multi-billion-dollar part of our economy.

In a sense, genetics is the key to all of biology, because genes control what cells look like and what they do, as well as how babies develop and how we reproduce. An understanding of how genes work is essential to our understanding of how life works.

In the chapters that follow, we will answer a series of questions about genes and genetics: How are genes passed from parents to their children? What are genes made of? Where are they located? How do they make gene products called proteins, and

Concepts and Controversies

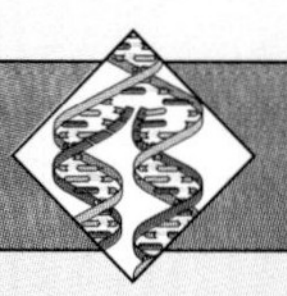

Genetic Disorders in Culture and Art

It is difficult to pinpoint when the inheritance of specific traits in humans was first recognized. Descriptions of heritable disorders often appear in myths and legends of many different cultures. In some ancient cultures, assigned social roles—from prophets and priests to kings and queens—were hereditary. The belief that certain traits were heritable helped shape the development of many cultures and social customs.

In some ancient societies, the birth of a deformed child was regarded as a sign of impending war or famine. Clay tablets excavated from Babylonian ruins record more than 60 types of birth defects, along with the dire consequences thought to accompany such births. Later societies, ranging from the Romans to eighteenth-century Europe, regarded malformed individuals (such as dwarfs) as curiosities rather than figures of impending doom, and they were highly prized by royalty as courtiers and entertainers.

Whether motivated by fear, curiosity, or an urge to record the many variations of the human form, artists have portrayed both famous and anonymous individuals with genetic disorders in paintings, sculptures, and other forms of the visual arts. These portrayals are often detailed, highly accurate, and easily recognizable today. In fact, across time, culture, and artistic medium, affected individuals in these portraits often resemble each other more closely than they do their siblings, peers, or family members. In some cases, the representations allow the disorder to be clearly diagnosed at a distance of several thousand years.

Throughout the book you will find fine-art representations of individuals with genetic disorders. These portraits represent the long-standing link between science and the arts in many cultures. They are not intended as a gallery of freaks or monsters, but as a reminder that being human encompasses a wide range of conditions. A more thorough discussion of genetic disorders in art is in *Genetics and Malformations in Art,* by J. Kunze and I. Nippert, published by Grosse Verläg, Berlin, 1986.

how do proteins help create the traits we can see and study? Because this book is about human genetics, we will use human genetic disorders as examples of inherited traits (see Concepts and Controversies: Genetic Disorders in Culture and Art). In addition, we will integrate the scientific aspects of genetics with the social, political, legal, and ethical implications of genetics and genetic technology.

Almost every day, the media reports on a story about advances in genetics and biotechnology. In many cases, as we will see, the technology is far ahead of public policy and laws. To make informed decisions about genetics and biotechnology in your daily life, you will need to understand something about what genes are and how they work. In the rest of this chapter, we will preview some of the basic concepts of human genetics and introduce some of the social issues inspired by genetic research. These concepts and issues will be explored in more detail in the chapters that follow.

WHAT ARE GENES, AND HOW DO THEY WORK?

Simply put, a gene is the basic structural and functional unit of genetics. In molecular terms, a gene is a string of chemical building blocks (nucleotides) in a **DNA** molecule (Figure 1.1). (DNA is shorthand for deoxyribonucleic acid.) There are four different nucleotides in DNA, and they store information in the form of a **genetic code.** The "letters" of the code (each nucleotide is a letter in the code) define the type and order of chemical subunits (amino acids) that make up gene products (polypeptides). When a gene is turned on, the information stored in the gene is decoded and used to make a molecule that folds into a three-dimensional shape. This molecule is known as a protein (Figure 1.2). The action of proteins produces the traits we see (like eye color or hair color) and are studied to see how they are passed from parents to children, generation after generation. We will cover these topics in Chapters 8 and 9.

DNA A helical molecule consisting of two strands of nucleotides that is the primary carrier of genetic information.

Genetic code The sequence of nucleotides that encodes the information for amino acids in a polypeptide chain.

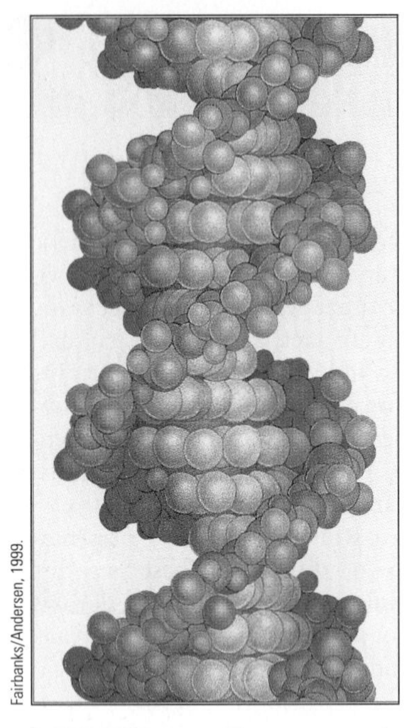

Fairbanks/Andersen, 1999.

FIGURE 1.1 The structure of DNA, discovered by James Watson and Francis Crick in 1952, marked the beginning of molecular genetics.

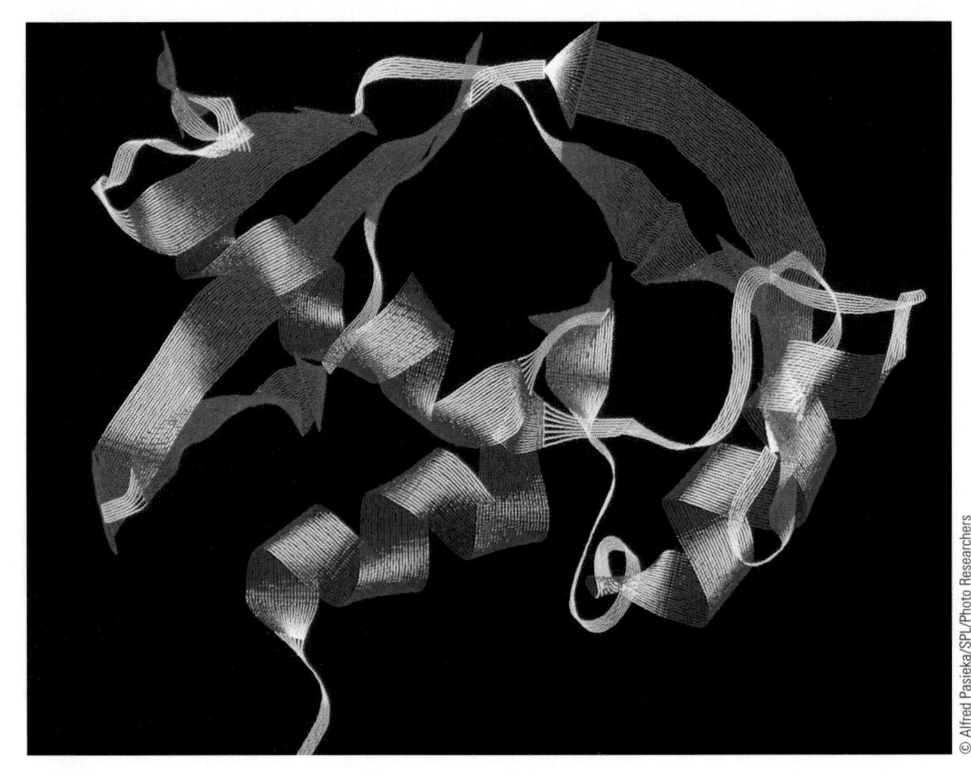

© Alfred Pasieka/SPL/Photo Researchers

FIGURE 1.2 A diagram showing how a polypeptide chain becomes folded into the three-dimensional structure of a protein.

Genes can also be defined by their properties. Genes are replicated (they are copied), they mutate (they can undergo change), they are expressed (they can be turned on and off), and they can recombine (they can move from one chromosome to another). In later chapters, we will explore these properties and see how they are involved in genetic diseases.

HOW ARE GENES TRANSMITTED FROM PARENTS TO OFFSPRING?

Thanks to the work of Gregor Mendel (Figure 1.3), a European monk, we understand how genes are passed from parents to offspring. At the time Mendel began his experiments, many people thought that the traits of the parents were blended together in the offspring. If a plant with red flowers was mated to one with white flowers, flower color should be blended in the next generation, and the offspring should have pink flowers. If this were true, eventually, there would only be pink flowers. Because plants with red and white flowers were produced in every generation, there were problems with this idea. Mendel's work provided the key to understanding how genes are passed from one generation to the next.

In the late nineteenth century, in what is now the Czech Republic, Mendel carried out a series of genetic experiments with pea plants. In his experiments, the parent plants each had a different, distinguishing characteristic called a **trait.** For example, he bred tall pea plants with short plants, or yellow-seeded plants with green-seeded ones, and kept careful records of the number of **progeny** with each trait. He followed the pattern of how the traits were passed through several generations. Based on his results, Mendel developed his ideas about how traits are inherited. Mendel showed that traits are passed from generation to generation through the inheritance of "factors." We now know that these factors are genes. He reasoned

Trait Any observable property of an organism.

Progeny Offspring, or descendants.

Portrait by Marcus Alan Vincent

FIGURE 1.3 Gregor Mendel, the Augustinian monk whose work on pea plants provided the foundation for genetics as a scientific discipline.

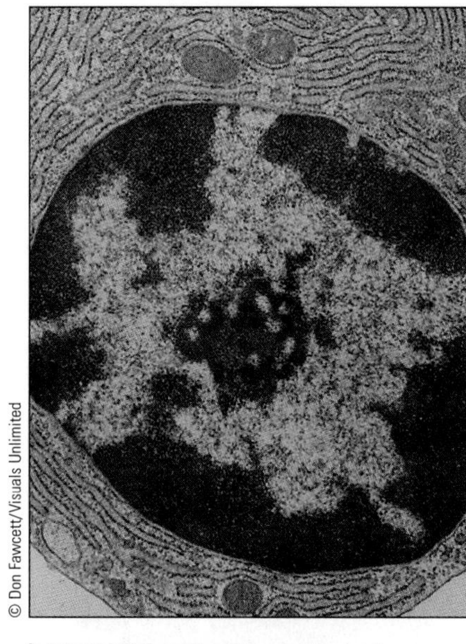

© Don Fawcett/Visuals Unlimited

FIGURE 1.4 The nucleus of a eukaryotic cell as seen in the electron microscope. The uncoiled chromosomes are represented by the dark clumps of chromatin throughout the nucleus. (Original magnification, ×15,500.)

that each parent carries a pair of genes for a given trait and that each parent contributes one gene to the traits shown in the offspring.

Mendel concluded that pairs of genes separate from each other during the formation of egg and sperm. When the egg and sperm fuse during fertilization to form a zygote, the genes from the mother and father become members of a new gene pair in the offspring. Later, researchers discovered that genes are located on chromosomes, made up of DNA molecules complexed with proteins. Chromosomes are found in the nucleus of **eukaryotic cells** (Figure 1.4). The separation of genes during formation of the sperm and egg and the reunion of genes at fertilization is explained by the behavior of chromosomes in a form of cell division called meiosis.

Eukaryotic cell A cell with a nuclear membrane surrounding the genetic material.

When Mendel published his work on the inheritance of traits in pea plants (discussed in Chapter 3), there was no well-accepted idea of how traits were transmitted from parents to offspring. Mendel's theories changed all this, and his work started the scientific study of heredity, which has expanded in many directions in the past 100 years.

HOW DO SCIENTISTS STUDY GENES?

Ideas that form the foundation of genetics were discovered by studying many different organisms, including bacteria, fungi, insects, and plants, as well as mammals including humans. Because the principles of genetics are universal, discoveries made using one organism (like bacteria) can be applied to other species, including humans. In spite of the fact that geneticists study many different species, they use a small number of basic approaches in their work.

Approaches to the Study of Genetics

The most basic approach in genetics, called **transmission genetics,** (Chapters 3 and 4) studies the pattern generated when traits are passed from generation to generation. Using experimental organisms, geneticists carry out mating experiments to analyze the transmission of traits (such as height, eye color, hemophilia, and so on) from parents

Transmission genetics The branch of genetics concerned with the mechanisms by which genes are transferred from parent to offspring.

FIGURE 1.5 A pedigree is a representation of the inheritance of a trait through several generations of a family. In this pedigree, males are symbolized by squares, females by circles. Filled symbols indicate those affected by the trait.

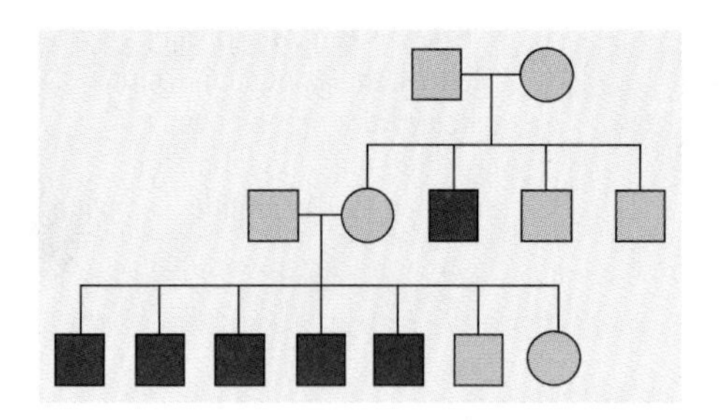

to offspring. The experimental results are analyzed to establish how a trait is inherited. Gregor Mendel did the first significant work in transmission genetics, using pea plants as his experimental organism. The methods he developed form the foundation of transmission genetics. In humans, however, experimental matings are not possible for several reasons, and another method, called pedigree analysis, is used. **Pedigree analysis** uses family history to reconstruct the path a trait follows through several generations of a family, and this information is used to determine its pattern of inheritance (Figure 1.5). Pedigrees are constructed from interviews, letters, diaries, photographs, and family records. As more of this information is converted to electronic forms that can be deleted with the push of a button, it is interesting to think what effect this will have on pedigree construction.

Pedigree analysis The construction of family trees and their use to follow the transmission of genetic traits in families. It is the basic method of studying the inheritance of traits in humans.

Cytogenetics The branch of genetics that studies the organization and arrangement of genes and chromosomes using the techniques of microscopy.

Karyotype A complete set of chromosomes from a cell that has been photographed during cell division and arranged in a standard sequence.

Molecular genetics The study of genetic events at the biochemical level.

Recombinant DNA technology Technique for joining DNA from two or more different organisms to produce hybrid, or recombined DNA molecules.

Cytogenetics is a branch of genetics that uses microscopes to study chromosomes (Chapter 6). At the beginning of the twentieth century, cytogenetics was used to propose the idea that genes are located on chromosomes. Cytogenetics is one of the most important investigative approaches in human genetics and is used to study chromosome abnormalities. In human genetics, cytogenetics is used to prepare **karyotypes,** a display of chromosomes in a standard arrangement (Figure 1.6). Chromosomes are arranged in a karyotype by size and other properties that will be described in Chapter 6. Cytogenetics is also used in locating genes on specific chromosomes and specific chromosome regions.

A third approach, **molecular genetics**, has had the greatest impact on human genetics over the last four or five decades. Molecular genetics uses **recombinant DNA technology** to identify, isolate, clone (produce multiple copies), and analyze genes. Cloned genes can be used to study how genes are organized and how they work; cloned genes can be transferred between organisms and between species. Gene therapy to treat human genetic disorders uses cloned genes. Molecular genetics is also used for prenatal diagnosis of genetic disorders and DNA fingerprinting. Advances in molecular genetics have generated much of the debate about the social, legal, and ethical aspects of genetics and biotechnology.

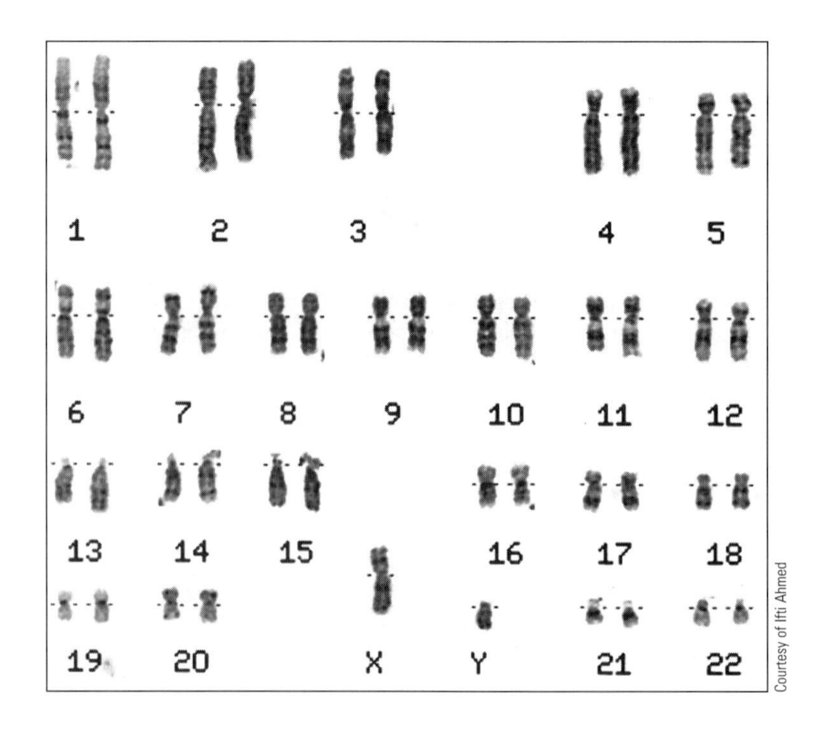

FIGURE 1.6 A karyotype arranges the chromosomes in a standard format so they can be analyzed for abnormalities. This karyotype is that of a normal male.

A fourth way to study genes is by looking at populations. Population geneticists are interested in the forces that change gene frequencies and drive the evolutionary process. Investigations in **population genetics** have defined how much genetic variation exists in populations and how forces such as migration, population size, and natural selection change this variation. The coupling of population genetics with recombinant DNA technology has helped us understand the evolutionary history of our species, and the patterns of migrations that distributed humans across the earth.

Population genetics The branch of genetics that studies inherited variation in populations of individuals and the forces that alter gene frequency.

Genetics Is Used in Basic and Applied Research

Because the principles of genetics have widespread uses, genetics is a discipline that crosses and recrosses the line between basic and applied research. Basic research is done by scientists in laboratory and field settings with the goal of understanding how something works or why it works the way it does. In basic research, there is no goal of solving a practical problem, or making a commercial product; knowledge itself is the goal. In turn, the results of basic research generate new ideas about how other systems work and stimulate more basic research. In this way, we are able to gain detailed information about the internal working of organisms. Among other things, basic research in genetics has provided us with precise knowledge about genes, how they work, and what happens when they fail to work properly.

Applied research is usually done with the goal of solving a practical problem or turning a discovery into a commercial product. Applied research uses methods of basic genetics research, including experimental matings, but also uses the latest methods of biotechnology to make a product. In agriculture, applied genetic research has helped increase crop yields, lower the fat content of pork by selective breeding, and make farm plants more resistant to disease. In medicine, new diagnostic tests, the synthesis of customized proteins for treating disease, and the production of vaccines are just a few examples of applied genetic research.

The use of some applied research is controversial and is generating debate about the merits and risks of applied research. Occasionally, the public rejects some products. For example, there are serious controversies about the environmental release of genetically modified organisms, the sale of food that has been modified by recombinant DNA technology (◗ Figure 1.7), the use of recombinant DNA-derived growth hormone in milk production, and the irradiation of food.

Courtesy of Calgene

◗ **FIGURE 1.7** Transgenic tomatoes, genetically modified by recombinant DNA techniques to slow softening, were withdrawn from the marketplace partly because of consumer resistance to genetically modified food.

An understanding of the basic concepts of genetics will help us make informed decisions about the use of biotechnology in our lives, ranging from the food we eat, the diagnostic tests we elect to have performed, and even the breeding of our pets. This course will provide you with knowledge about the basic concepts that can be used as a framework to make these informed decisions.

WHAT IS THE HISTORICAL IMPACT OF GENETICS ON SOCIETY?

Genetics and biotechnology not only affect our personal lives but also raise larger questions about ethics, social policy, and law. You may be surprised to learn that genetics has had a significant impact on law and social policy in the past. Knowledge about this era may help us avoid mistakes and pitfalls as we consider new issues raised by genetics.

Genetics Directly Affected Social Policy

Eugenics The attempt to improve the human species by selective breeding.

Hereditarianism The idea that human traits are determined solely by genetic inheritance, ignoring the contribution of the environment.

After the publication of *The Origin of Species* by Charles Darwin, his cousin Francis Galton (▶ Figure 1.8) thought that natural selection could be used to improve the human species. He founded **eugenics**, a method he thought could improve the intellectual, economic, and social level of mankind. Bypassing legal and ethical considerations, Galton proposed that people with desirable traits such as leadership or musical ability should be encouraged to have large families, whereas those with undesirable traits should be discouraged from reproducing. One flaw in Galton's reasoning was that he believed that traits were handed down without any environmental influence. The idea that all human traits are genetically determined is known as **hereditarianism.**

Eugenics took hold in the United States, and eugenicists worked to promote selective breeding in the human population (▶ Figure 1.9) and to control reproduction of those labeled as genetically defective. From about 1905 through 1933, eugenics was a powerful and influential force in passing laws regulating social policy.

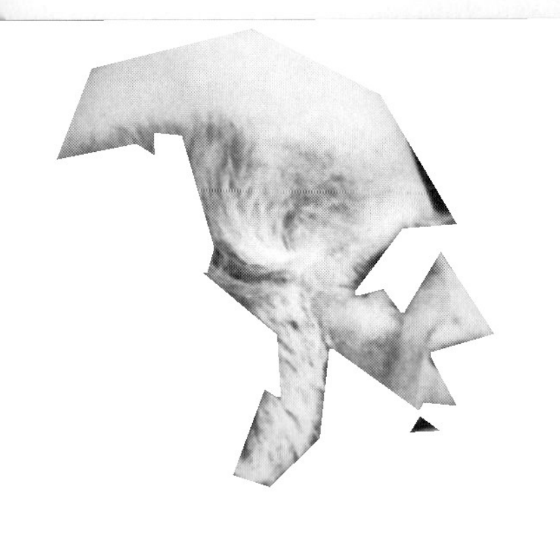

© Corbis

▶ **FIGURE 1.8** Sir Francis Galton, cousin of Charles Darwin and the founder of the eugenics movement.

Eugenics Helped Change Immigration Laws

In the early decades of the twentieth century, the high levels of unemployment, poverty, and crime among immigrants from southern and eastern Europe were taken as evidence that these people were genetically inferior. Based on testimony by eugenics experts, Congress passed the Immigration Act of 1924, which was signed into law by President Coolidge. This law, based on faulty eugenic assumptions, effectively closed the door to America for millions of people in southern and eastern Europe by setting low entry quotas for immigrants from that region. The Chinese Exclusion Acts of 1882 and 1902 previously restricted immigration from Asia. In addition, a 1907 agreement between the U.S. government and the Japanese government restricted the immigration of Japanese citizens. In the early decades of the twentieth century, there was little immigration from Africa, and lawmakers saw no need to regulate entry from this continent.

Immigration laws based on faulty eugenics were in effect for just over 40 years and were finally corrected in 1965, when a new immigration law was passed. Under the new law, immigrants are admitted to the United States by order of application, rather than by country of origin.

Eugenics Helped Restrict Reproductive Rights

In addition to influencing immigration policy, the eugenics movement in the United States was closely associated with laws requiring sterilization of individuals regarded as genetically inferior. A committee of eugenicists convened to study the

American Philosophical Association

FIGURE 1.9 In the early part of the century, eugenics exhibits were a common feature at fairs and similar events. Such exhibits served to educate the public about genetics and the benefits of eugenics as public policy. These exhibits often included contests to find the eugenically perfect family.

problem concluded that up to 10% of the U.S. population should be isolated from the gene pool by being institutionalized or sterilized. Laws requiring sterilization for genetic defectives and those convicted of certain crimes were passed in many states. It may surprise you to learn that the U.S. Supreme Court in 1927 (*Buck v. Bell*) upheld the right of the state of Virginia (and all other states) to use forced sterilization for eugenic reasons. This ruling, which has never been overturned, includes the following statement:

> It is better for all the world, if instead of waiting to execute degenerate offspring for crime, or to let them starve for their imbecility, society can prevent those who are manifestly unfit from continuing their kind. The principle that sustains compulsory vaccination is broad enough to cover cutting the fallopian tubes.

Soon after this decision, Carrie Buck, the plaintiff in the suit, was sterilized. By about 1930, some 24 states passed sterilization laws, and by the mid-1930s, about 20,000 sterilizations had been carried out. In the 1960s, some states began to repeal these laws, but they are still on the books in several states, although federal regulations restrict their use.

Eugenics Was Associated with the Nazi Movement

In Germany, eugenics (known as *Rassenhygiene*) fused with genetics and the political philosophy of the Nazi movement (see Concepts and Controversies: Genetics, Eugenics, and Nazi Germany). This relationship evolved into the systematic killing of those defined as socially defective, including the physically deformed, the retarded, and the mentally ill. Later this rationale was expanded in an attempt to eradicate entire ethnic groups such as the Gypsies and Jews. The association of eugenics with the government of Nazi Germany led to the decline of the eugenics movement in the United States by the mid-1930s.

With advances in genetics and biotechnology, we are once again facing a period when new social policies and laws are being formulated, based on what we know (or think we know) about genetics. As we contemplate the possibility of cloning

Concepts and Controversies

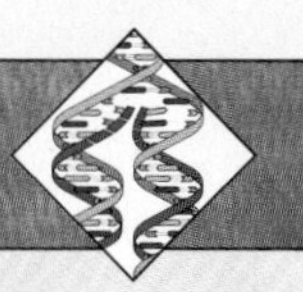

Genetics, Eugenics, and Nazi Germany

In the first decades of the twentieth century, eugenics advocates in Germany were concerned with preservation of racial "purity," as were their colleagues in other countries, including the United States and England. By 1927, many states in the United States had enacted laws that prohibited marriage of "social misfits" and made sterilization compulsory for the "genetically unfit" and for those found guilty of certain crimes. In Germany, the laws of the Weimar government prohibited sterilization, and there were no laws restricting marriage on eugenic grounds. As a result, several leading eugenicists became associated with the National Socialist Party (Nazis), which advocated forced sterilization and other eugenic measures to preserve the purity of the Aryan "race."

Adolf Hitler and the Nazi Party came to power in January 1933. By July of that year a sterilization law was in effect. Under the law, those regarded as having lives not worth living, including the feebleminded, epileptics, the deformed, those having hereditary forms of blindness or deafness, and alcoholics, were to be sterilized.

By the end of 1933, the law was amended to include mercy killing (*Gnadentod*) of newborns who were incurably ill with hereditary disorders or birth defects. This program was gradually expanded to include children up to 3 or 4 years of age, then adolescents, and finally, all institutionalized children, including juvenile delinquents and Jewish children. More than two dozen institutions in Germany, Austria, and Poland were assigned to carry out this program. Children were usually killed by poison or starvation.

In 1939, the program was extended to include mentally retarded and mentally defective adults and adults with certain genetic disorders. This program began by killing adults in psychiatric hospitals. As increasing numbers were marked for death, gas chambers were installed at several institutions to kill people more efficiently, and crematoria were used to dispose of the bodies. This practice spread from mental hospitals to include defective individuals in concentration camps and then to whole groups of people in concentration camps, most of whom were Jews, Gypsies, Communists, homosexuals, or political opponents of the government.

humans, or modifying our genetic makeup by gene therapy, we would do well to remember the mistakes of the past, admit that our knowledge is incomplete, and formulate policy and laws wisely.

WHEN DID HUMAN GENETICS GET STARTED?

In the first part of the twentieth century, most geneticists avoided studying human genetics because of its association with eugenics. Immediately after World War II, however, serious research in human genetics began to focus on the identification of Mendelian traits and the use of mathematical formulas to study genes in different human populations. During this period, human genetics emerged as a separate branch of genetics.

In 1949, a group of researchers led by Linus Pauling at Cal Tech discovered that a genetic disorder, sickle cell anemia, was caused by a defective hemoglobin molecule (Figure 1.10). This discovery helped found the field of human molecular genetics. Human cytogenetics began in 1956 when J. H. Tijo and A. Levan found that humans have 46 chromosomes, settling years of debate over how many chromosomes are present in human cells. Shortly thereafter, in 1959, Down syndrome was identified as the first chromosomal disorder.

The revolution in molecular genetics that began with the discovery of the structure of DNA in 1953 had an immediate impact on human genetics. In the years following, the molecular basis of several human genetic disorders was explained.

With the development of recombinant DNA technology in the 1970s, human genetics has made rapid strides. In the span of about 20 years, we have learned how to predict the sex of unborn children, to diagnose genetic disorders prenatally, and to manufacture gene products to treat some genetic diseases. The Human Genome

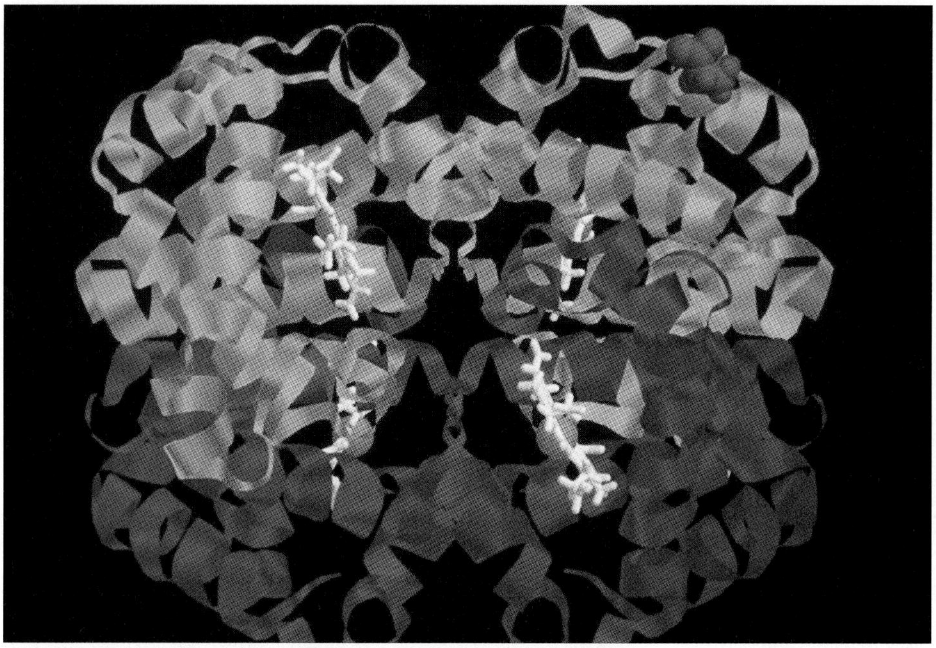

FIGURE 1.10 Hemoglobin is an oxygen-transporting protein found in red blood cells.

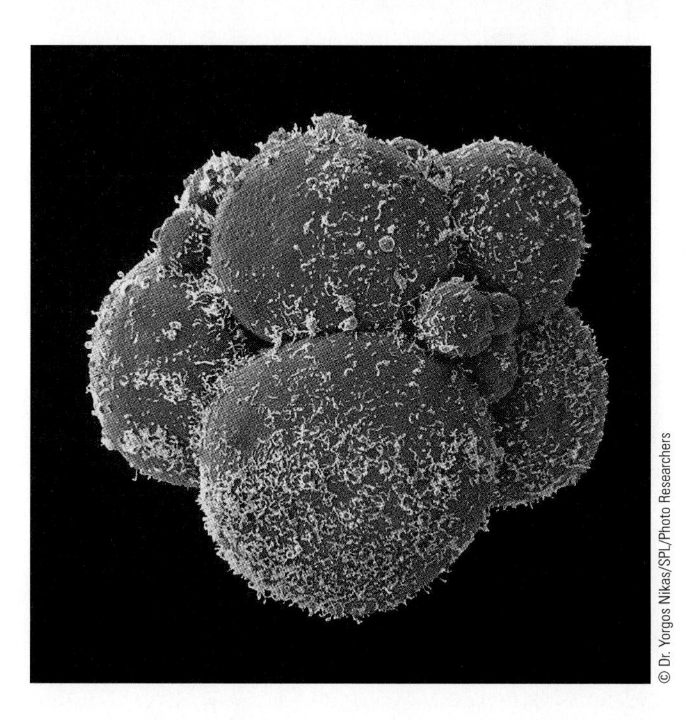

FIGURE 1.11 Human embryo, shortly after fertilization in the laboratory. Embryos at this stage of development can be analyzed for genetic disorders before implantation into the uterus of the egg donor or that of another, surrogate mother.

Project, started in 1991, is an international effort to map all the genetic information carried by humans at its most elemental level: the 3 billion nucleotides in our DNA. As of this writing, the project has completed a first draft of the sequence and is now cataloging the 35,000 or so genes identified to date. Genetic technology has made it possible to produce human embryos by fusion of sperm and eggs in a laboratory dish (Figure 1.11) and to transfer the developing embryo to the womb of a surrogate mother. Embryos can also be frozen for transfer to a womb at a later time. In a small number of cases, genetic defects are now corrected by inserting normal genes that act in place of mutant genes, a technique called gene therapy. We can even insert human genes into animals, creating new types of organisms, which produce human proteins. These proteins are purified and used to treat human diseases such as emphysema.

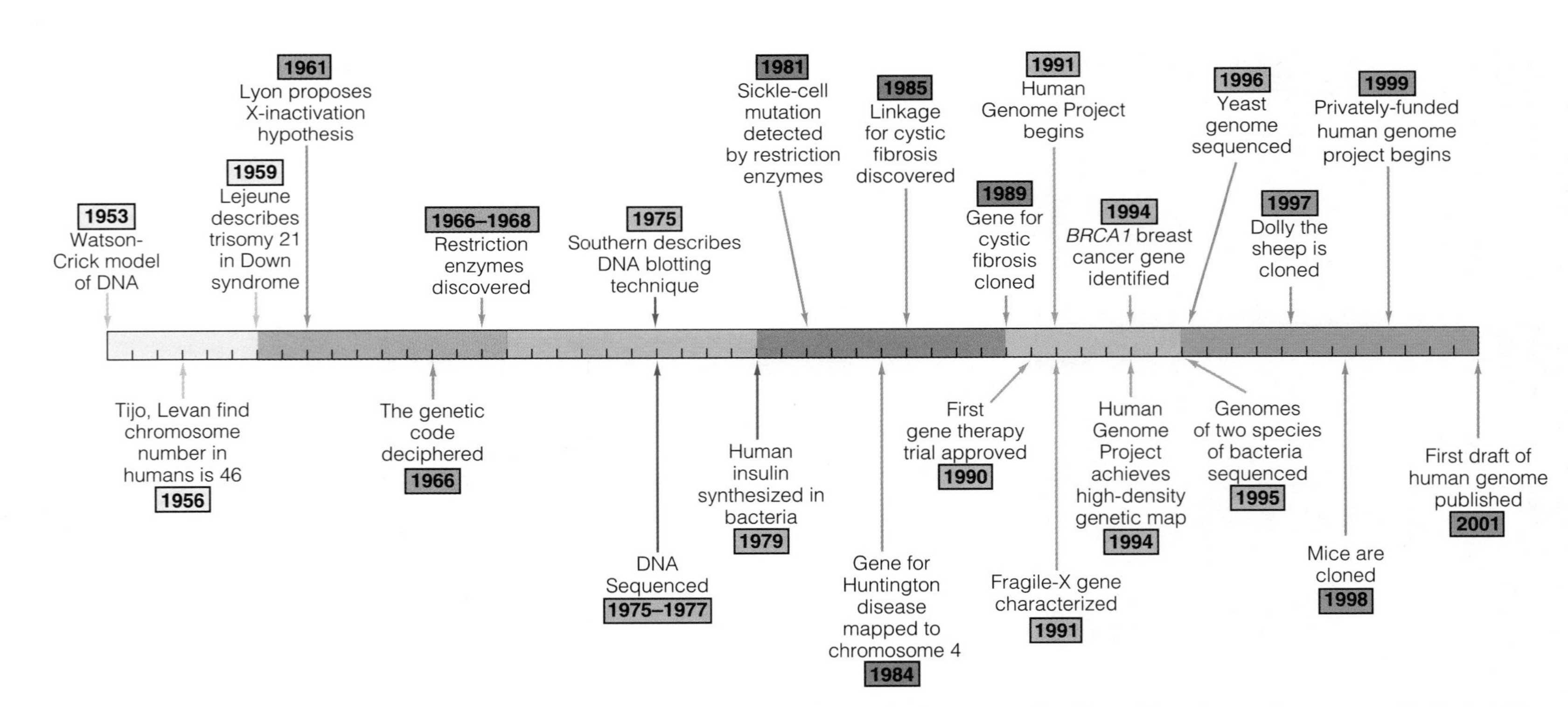

FIGURE 1.12 A timeline showing the major advances in genetics with a major impact on human genetics. Most of the advances have occurred in the last 50 years, making human genetics a relatively young science. (All of the events on this timeline are described in this text.)

WHERE IS HUMAN GENETICS GOING IN THE FUTURE?

Interest in heredity and the transmission of traits from parents to offspring can be traced back thousands of years. Evidence indicates that animals and plants were domesticated by preliterate cultures between 10,000 and 12,000 years ago. The heritable traits of domesticated species were manipulated to produce strains tailored to human needs, a process known as selective breeding. Although it is effective, selective breeding is a process of trial and error. An understanding of how traits are passed from generation to generation was not achieved until late in the nineteenth century, when the science of genetics began.

Almost all of what we know about genetics as a scientific discipline has been learned in the last 100 years. The amount of scientific knowledge is thought to double about every 10 years. In human genetics, knowledge appears to be doubling even more rapidly (◗ Figure 1.12). If this is the case, the next few decades will see an explosion of information and technology related to human genetics.

More importantly, we are beginning to apply genetic knowledge in ways that will affect us all, and this trend will probably accelerate in the coming years. One of the most visible applications of genetic knowledge is the Human Genome Project. This project, involving thousands of scientists, has generated knowledge about complex genetic disorders such as heart disease and mental disorders and will provide insight into cellular processes that may help unravel the causes of cancer.

Developments in biotechnology have produced fundamental changes in the diagnosis of genetic disorders and infectious diseases. Prenatal diagnosis for dozens of genetic diseases such as sickle cell anemia and cystic fibrosis is now possible. Biotechnology has been instrumental in the development of genetically engineered plants and animals. Researchers are experimenting with genetically modified food plants that carry a vaccine, so that eating vegetables will protect against infectious diseases. In the pharmaceutical industry, genetically engineered bacteria and farm animals are being used to synthesize human gene products for therapeutic purposes. Human insulin was the first such product, with over two dozen other recombinant drugs now available.

The use of all this genetic knowledge raises new ethical questions that we must face and answer in the near future. As a student of human genetics, you have elected to become involved in the search for answers to these important questions.

Summary

Welcome to Genetics

1. Genetics is the science of heredity. In a sense, genetics is the key to all of biology, because genes control what cells look like and what they do. Understanding how genes work is essential to our understanding of how life works.

What Are Genes and How Do They Work?

2. The gene is the basic structural and functional unit of genetics. It is a string of chemical building blocks (nucleotides) in a DNA molecule. When a gene is turned on, the information stored in the gene is decoded and used to make a molecule that folds into a three-dimensional shape. This molecule is known as a protein (Figure 1.2). The action of proteins produces the traits we see (like eye color or hair color).

How Are Genes Transmitted from Parent to Offspring?

3. From his experiments on pea plants, Mendel concluded that pairs of genes separate from each other during the formation of egg and sperm. When the egg and sperm fuse during fertilization to form a zygote, the genes from the mother and father become members of a new gene pair in the offspring. The separation of genes during formation of the sperm and egg and the reunion of genes at fertilization is explained by the behavior of chromosomes in a form of cell division called meiosis.

How Do Scientists Study Genes?

4. Genes are studied using several different methods. Transmission genetics studies how traits are passed from generation to generation. Cytogenetics studies chromosome structure and the location of genes on

chromosomes. Molecular geneticists study the molecular makeup of genes and gene products and the function of genes. Population genetics focuses on the dynamics of populations and their interaction with the environment that results in changing gene frequencies over several generations.

What Is the Historical Impact of Genetics on Society?

5. Eugenics was an attempt to improve the human race by using the principles of genetics. In the early years of the twentieth century, eugenics was a powerful force in shaping laws and public policy in the United States. This use of genetics was based on the mistaken assumption that genes alone determined human behavior and disorders and neglected the role of the environment. Eugenics fell into disfavor when it became part of the social programs of the Nazis in Germany.

When Did Human Genetics Get Started?

6. Although scientists studied human genetics early in the twentieth century, human genetics began to grow in the years after 1945, with studies in population genetics and biochemical genetics and the discovery of the correct number of chromosomes carried in human cells. The use of molecular techniques to study human genetics, culminating in the Human Genome Project has made human genetics one of the fastest growing areas of science.

Where Is Human Genetics Going in the Future?

7. With the completion of the Human Genome Project, the ability to manipulate human reproduction, and the ability to transfer genes, we are on the threshold of being able to genetically modify humans. The ethical use of this information and these technologies will require participation by a broad cross section of society.

Internet Activities

The following activities use the resources of the World Wide Web to enhance the topics covered in this chapter. To investigate the topics described below, log on to the book's home page (www.brookscole.com/biology_d) and follow the "Student Resources" link to your subject and text.

1. *Learning Styles.* You can learn more from your studies in any subject if you know something about your personal learning preferences. At the *Active Learning Site* (http://www.active-learning-site.com/vark.htm) you may take a simple, informal assessment of your learning style. After completing the VARK learning style inventory, explore the tips for using your preferred style(s) to enhance learning. How might you strengthen your ability to learn using the other styles?
2. *How to Study Biology.* The University of Texas maintains a Web site that provides suggestions on how to approach the study of biology, including genetics. At the *How to Study Biology (and Succeed)* Web site (http://www2.tltc.ttu.edu/dini/BIOL1403/Regular/howtostudybiology.html) check out the general study suggestions for biology courses. How does low-quality study differ from higher-quality study according to these suggestions? Try developing a concept map, as outlined at this Web site, for some of the topics being covered in your genetics course.
3. *Genetics as a Contemporary Field of Research.* Genetics is one of the most active research fields in biology today. Go to the Web site for the Genetics Society of America (http://www.faseb.org/genetics/gsa/gsamenu.htm) and browse the information on the journal, meetings, and awards. Using the link to the "Careers Brochure" (http://www.faseb.org/genetics/gsa/careers/bro-menu.htm), read what a number of prominent geneticists have to say about their careers. How has genetics been used to study dog behavior or played a role in reuniting kidnapped children in Argentina with their families?
4. *The Ongoing Eugenics Debate.* For a history of the eugenics movement in the United States, take a look at the "Eugenics Slide Show" (http://ns2.d20.co.edu/kadets/lundberg/ethics/). Although the eugenics movement in the United States declined by the mid-1930s, there are those who argue that eugenicists are alive and active among us. Check out the "Eugenics" page (http://www.beloit.edu/~biology/genethics/eugenics.html) for links to several points of view on this issue.

For Further Reading

Adams, M. B. (1990). *The wellborn science: Eugenics in Germany, France, Brazil, and Russia.* New York: Oxford University Press.

Ahmed, I. (Ed.). (1992). *Biotechnology: A hope or threat?* London: St. Martin's Press.

Anderson, W. F., & Dircumakos, E. G. (1981, July). Genetic engineering in mammalian cells. *Sci. Am., 245,* 106–121.

Barnum, S. R. (1998). *Biotechnology: An introduction.* Belmont, CA: Wadsworth.

Bishop, J. E., & Waldholz, M. (1990). *Genome.* New York: Simon & Schuster.

Bud, R. (1993). *The use of life: A history of biotechnology.* New York: Cambridge University Press.

Corcos, A. (1984). Reproductive hereditary beliefs of the Hindus, based on their sacred books. *J. Hered., 75,* 152–154.

Davis, B. D. (Ed.). (1991). *The genetic revolution: Scientific prospects and public perceptions.* Baltimore: Johns Hopkins Press.

Dunn, L. C. (1962). Cross currents in the history of human genetics. *Am. J. Hum. Genet., 14,* 1–13.

Dunn, L. C. (1965). *A short history of genetics.* New York: McGraw-Hill.

Fairbanks, D. J., & Andersen, W. R. (1999). *Genetics: The continuity of life.* Pacific Grove, CA: Brooks/Cole.

Glick, B. J., & Pasternak, J. J. 1998. *Molecular biotechnology: Principles and applications of recombinant DNA* (2nd ed.). Washington, DC: ASM Press.

Hopwood, D. A. (1981, September). The genetic programming of industrial microorganisms. *Sci. Am., 245,* 91–102.

Kevles, D. J. (1985). *In the name of eugenics: Genetics and the use of human heredity.* New York: Knopf.

Kevles, D. J., & Hood, L. (Eds.). (1992). *The code of codes: Scientific and social issues in the Human Genome Project.* Cambridge, MA: Harvard University Press.

Kuhl, S. (1994). *The Nazi connection: Eugenics, American racism, and the German national socialism.* New York: Oxford University Press.

Larson, E. J. (1995). *Sex, race and science: Eugenics in the deep South.* Baltimore: Johns Hopkins University Press.

Lifton, R. J. (1986). *The Nazi doctors: Medical killing and the psychology of genocide.* New York: Knopf.

McKusick, V. J. (1975). The growth and development of human genetics as a clinical discipline. *Am. J. Hum. Genet., 27,* 261–273.

Muller-Hill, B. (1988). *Murderous science: Elimination by scientific selection of Jews, Gypsies, and others, Germany, 1933–1945.* Oxford, England: Oxford University Press.

Proctor, R. N. (1988). *Racial hygiene: Medicine under the Nazis.* Cambridge, MA: Harvard University Press.

Rafter, N. H. (1988). *White trash.* Boston: Northeastern University Press.

Reilly, P. R. (1991). *The surgical solution: A history of involuntary sterilization in the United States.* Baltimore: Johns Hopkins University Press.

Shreve, J. (1998, May). The code breaker. *Discover,* 45–51.

Silver, L. 1997. *Remaking Eden: Cloning and beyond in a brave new world.* New York: Avon Books.

Stern, C., & Sherwood, E. (1966). *The origin of genetics: A Mendel sourcebook.* San Francisco: Freeman.

Stubbe, H. (1972). *History of genetics: From prehistoric times to the rediscovery of Mendel's laws.* Cambridge, MA: MIT Press.

Tijo, H. J., & Levan, A. (1956). The chromosome number of man. *Hereditas, 42,* 1–6.

Torrey, J. G. (1985). The development of plant biotechnology. *Am. Sci., 73,* 354–363.

Weiss, S. F. (1988). *Race hygiene and national efficiency: The eugenics of William Schallmeyer.* Berkeley: University of California Press.

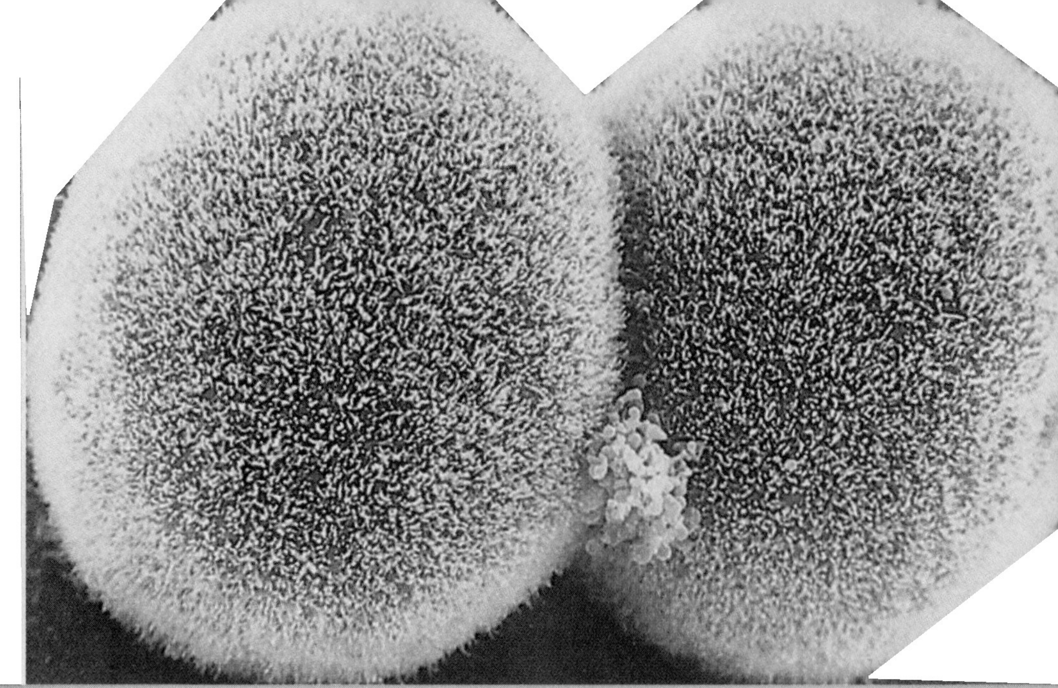

© David M. Phillips/Visuals Unlimited

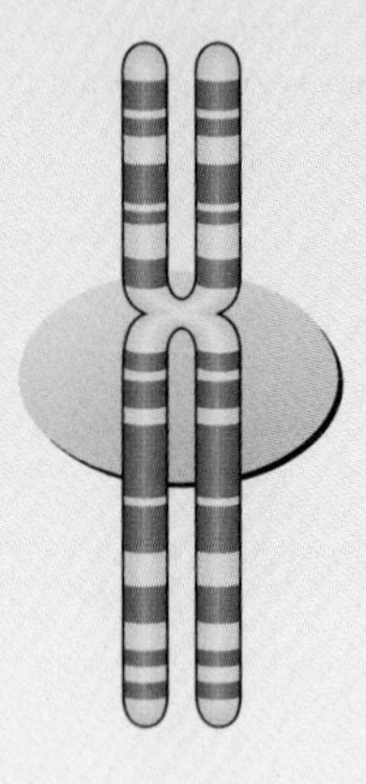

Cells, Chromosomes, and Cell Division

Chapter Outline

Emerging from the womb, a newborn human represents the culmination of a series of genetically programmed events that began some 38 weeks earlier. At birth an infant contains trillions of cells derived from the fertilized egg by cell division, under the control of genetic information contributed by each parent. The infant's body contains several hundred different cell types. Each has a distinctive organization, often associated with highly specialized functions.

Despite their apparent variation in size and shape, all cells in each human carry the same set of genetic information and share a basic architecture. Each is surrounded by a plasma membrane, and each possesses a membrane-enclosed nucleus for all or part of its life cycle. In addition, cells possess internal structures known as organelles. Because the shape, structure, and internal components of each cell are determined by genetic information, a large part of human genetics involves the study of cells. Many genetic diseases are expressed at the cellular level. For example, a mutant gene in sickle cell anemia causes an alteration in the shape of red blood cells, and most of the resulting physical symptoms are a direct result of this altered shape.

As cells grow and divide, the genetic information they carry must be faithfully copied and distributed to their progeny by structures within the nucleus known as chromosomes. The process of chromosome distribution and cell division is known as mitosis. The sperm and egg that fuse to form the fertilized egg or zygote are the products of a special form of cell division known as meiosis. Mistakes in either of these processes can have serious genetic consequences.

Because cells are the building blocks of the body and because the reproduction and transmission of genetic traits are mediated by cells, we begin with an outline of their basic structural features. We also consider cell division, emphasizing the replication and distribution of chromosomes. Human reproduction is accomplished via specialized cells known as gametes, and knowledge of how genetic traits are distributed during gamete formation is essential to understanding heredity.

THERE ARE TWO MAIN CELL TYPES: PROKARYOTIC AND EUKARYOTIC

At the cellular level, organisms can be classified as **prokaryotes** or **eukaryotes**. Prokaryotic cells are enclosed by a plasma membrane but do not have a membrane-bound nucleus. Instead, the genetic information carried by the cell (in the form of a DNA molecule) is in direct contact with the cytoplasm. In addition, prokaryotic cells do not have any membrane-contained organelles. Organisms such as bacteria and blue-green algae are prokaryotes.

Prokaryotes Organisms without a nucleus.

Eukaryotes Organisms that have a nuclear membrane surrounding the genetic material and have other membrane-bound organelles in the cytoplasm.

Eukaryotic cells have a plasma membrane and a membrane-bound nucleus. The cytoplasm is divided into a number of compartments by internal membrane systems and membranous organelles. Organisms such as protozoa, fungi, plants, and animals (humans) are eukaryotes.

CELL STRUCTURE REFLECTS FUNCTION

In the following discussion, we will review some of the basic aspects of eukaryotic cell structure. Although eukaryotic cells differ widely in their size, shape, functions, and life cycle, they are fundamentally similar to one another—they all have a plasma

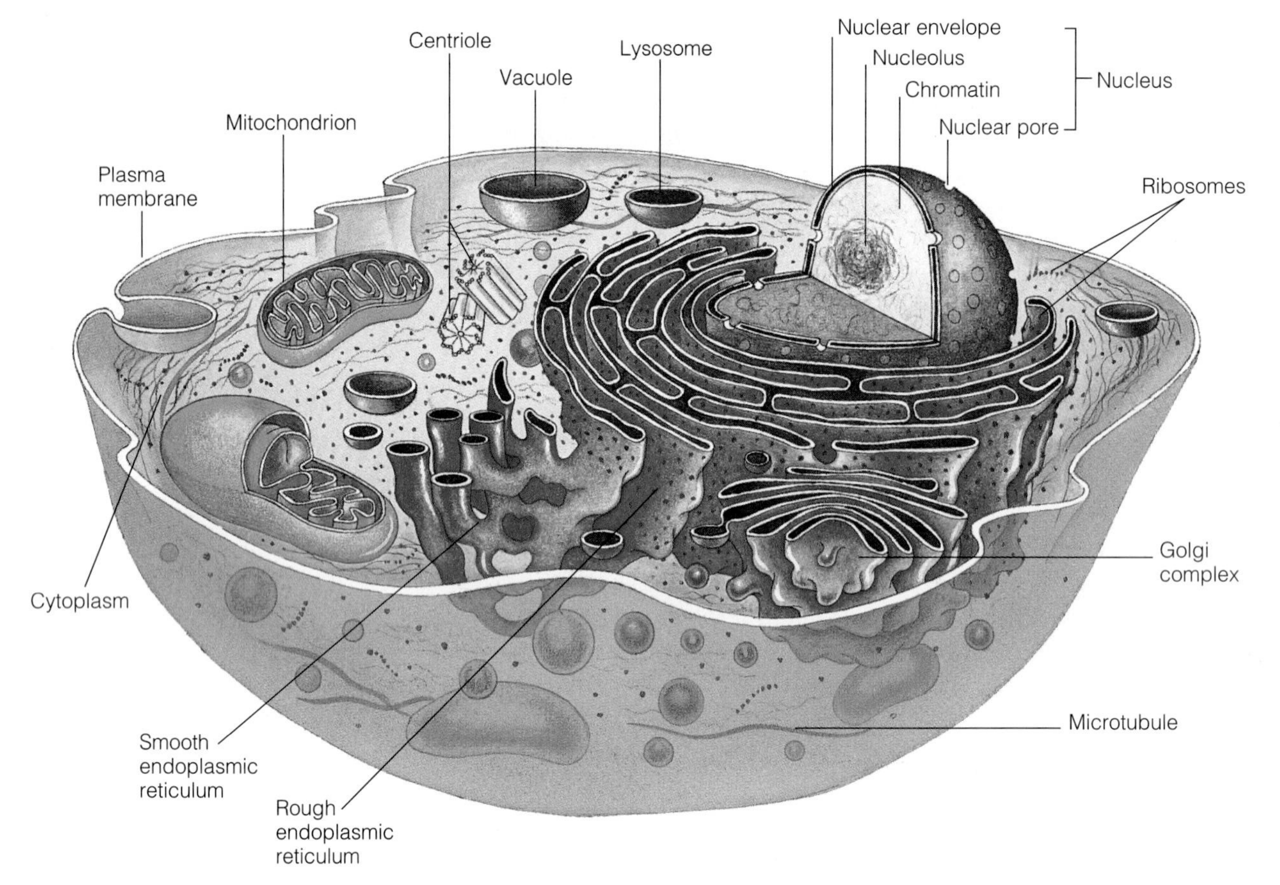

FIGURE 2.1 A generalized eukaryotic cell showing the organization and distribution of organelles as they would appear in the transmission electron microscope. The type, number, and distribution of organelles found in cells is related to cell function.

membrane, cytoplasm, membranous organelles, and a membrane-bound nucleus. An idealized eukaryotic cell is shown in Figure 2.1. A cell's structure and function are under genetic control, and many genetic disorders cause changes in cellular structure and/or function.

The Plasma Membrane and the Cytoplasm Are Two Cellular Domains

A double-layered plasma membrane separates the cell from the external environment. This membrane is a dynamic and active component of cell function, and controls the exchange of materials with the environment outside the cell (Figure 2.2). Gases, water, and some small molecules pass through the membrane easily, but large molecules are transported by energy-requiring systems. Molecules in and on the plasma membrane give the cells a form of molecular identity. The type and number of these molecules are genetically controlled and are responsible for many important properties of cells, including blood type and compatibility in organ transplants. Several genetic disorders, including cystic fibrosis (MIM/OMIM 219700), are associated with the plasma membrane. (See Chapter 4 for an explanation of MIM/OMIM numbers and the catalog of human genetic disorders.)

The plasma membrane encloses the cytoplasm, which is a complex mixture of **molecules** and structural components. Within the cytoplasm, a three-dimensional network of protein filaments and tubules form the **cytoskeleton**, which gives a cell its shape and helps anchor cellular structures. The cytoplasm also contains a number of specialized structures known collectively as **organelles**.

Molecules Structures composed of two or more atoms held together by chemical bonds.

Cytoskeleton A system of protein microfilaments and microtubules that allows a cell to have a characteristic shape.

Organelles Cytoplasmic structures that have a specialized function.

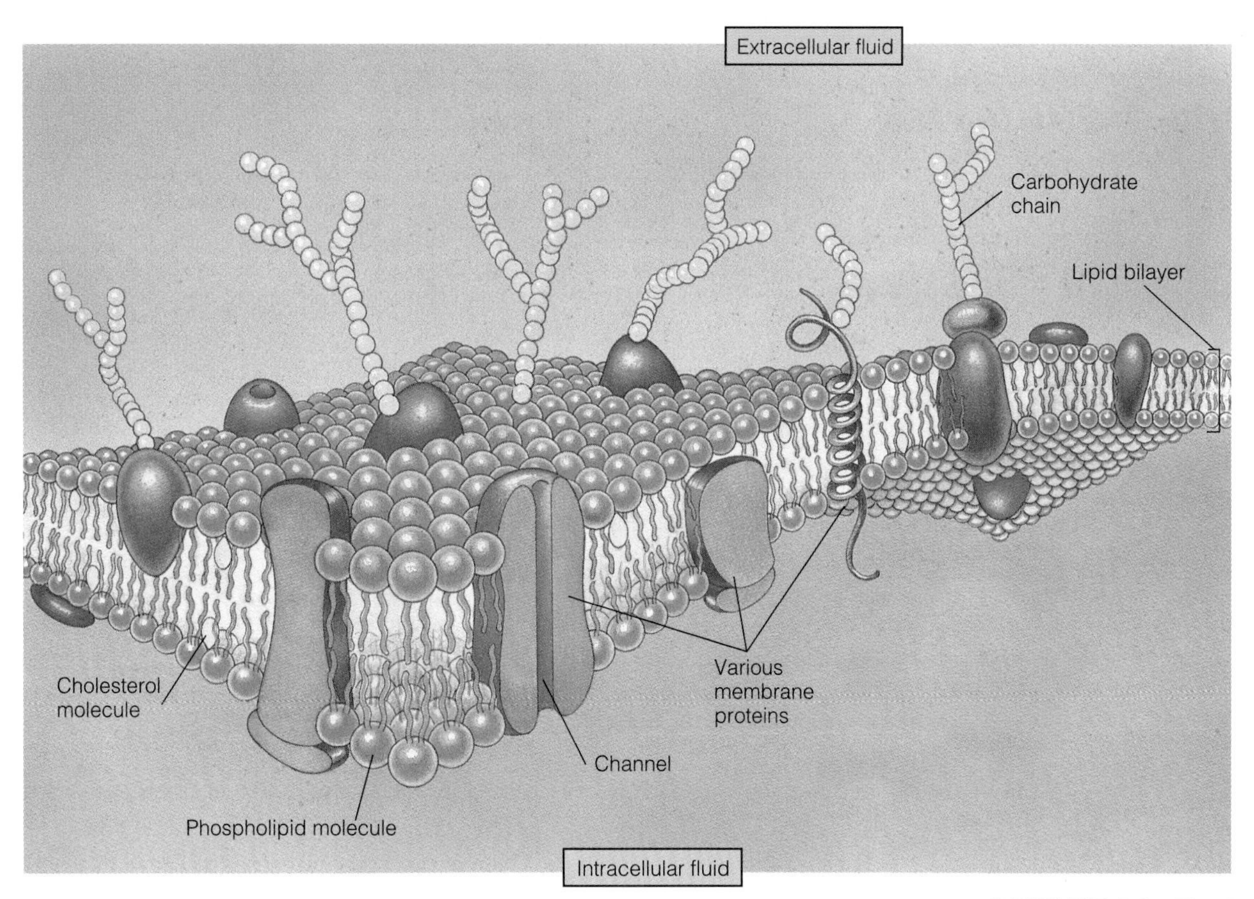

FIGURE 2.2 The plasma membrane. Proteins are embedded in a double layer of lipids. Short carbohydrate chains are attached to some proteins on the outer surface of the membrane.

Organelles Are Specialized Structures in the Cytoplasm

The cytoplasm in a eukaryotic cell has an organization that is related to the cell's function and often differs from one cell type to another. Cytoplasmic organelles are formed from membranes and divide the cell into a number of functional compartments. Table 2.1 summarizes the major organelles and their functions. We will review some of them here.

Endoplasmic Reticulum The **endoplasmic reticulum (ER)** is a network of membranes that form channels in the cytoplasm. There is a lot of ER in cells that make large amounts of protein (Figure 2.3). The space inside the ER, called the lumen, is a separate compartment in the cytoplasm. In the ER lumen, proteins are folded, chemically modified, and prepared for transport to other locations within the cell or for secretion from the cell.

Endoplasmic reticulum (ER) A system of cytoplasmic membranes arranged into sheets and channels that function in synthesizing and transporting gene products.

The outer surface of the ER is often covered with **ribosomes**, another cytoplasmic component (Figure 2.3). Ribosomes are the most numerous cellular structures and can be found free in the cytoplasm; they can also be attached to the outer surface of the ER or to the outside of the nuclear membrane. Under an electron microscope, ribosomes appear as spherical structures but are actually composed of two subunits. Ribosomes are the site of protein synthesis. (The process of protein synthesis is discussed in Chapter 9.)

Ribosomes Cytoplasmic particles composed of two subunits that are the site of protein synthesis.

Golgi Apparatus Animal cells contain clusters of flattened membrane sacs, called the **Golgi apparatus** (Figure 2.4). The Golgi receive proteins from the ER and distribute them to their destinations inside and outside the cell. Abnormalities of the Golgi are responsible for Menkes syndrome (MIM/OMIM

Golgi apparatus Membranous organelles composed of a series of flattened sacs. They sort, modify, and package proteins synthesized in the ER.

Table 2.1 Overview of Cell Organelles

Organelle	Structure	Function
Nucleus	Round or oval body; surrounded by nuclear envelope.	Contains the genetic information necessary to control cell structure and function. DNA contains heredity information.
Nucleolus	Round or oval body in the nucleus consisting of DNA and RNA.	Produces ribosomal RNA.
Endoplasmic reticulum	Network of membranous tubules in the cytoplasm of the cell. Smooth endoplasmic reticulum contains no ribosomes. Rough endoplasmic reticulum is studded with ribosomes.	Smooth endoplasmic reticulum (SER) is involved in producing phospholipids and has many different functions in different cells. Rough endoplasmic reticulum (RER) is the site of the synthesis of lysosomal enzymes and proteins for extracellular use.
Ribosomes	Small particles found in the cytoplasm; made of RNA and protein.	Aid in the production of proteins on the RER and ribosome complexes (polysomes).
Golgi apparatus	Series of flattened sacs usually located near the nucleus.	Sorts, chemically modifies, and packages proteins produced on the RER.
Secretory vesicles	Membrane-bound vesicles containing proteins produced by the RER and repackaged by the Golgi apparatus; contain protein hormones or enzymes.	Store protein hormones or enzymes in the cytoplasm awaiting a signal for release.
Lysosome	Membrane-bound structure containing digestive enzymes.	Combines with food vacuoles and digests materials engulfed by cells.
Mitochondria	Round, oval, or elongated structures with a double membrane. The inner membrane is thrown into folds.	Complete the breakdown of glucose, producing NADH and ATP.
Cytoskeleton	Network of microtubles and microfilaments in the cytoplasm.	Gives the cell internal support, helps transport molecules and some organelles inside the cell, and binds to enzymes of metabolic pathways.

309400) and other genetic disorders. The Golgi apparatus is also a source of membranes for other organelles, including lysosomes.

Lysosomes Membrane-enclosed organelles that contain digestive enzymes.

Lysosomes The **lysosomes** are membrane-enclosed vesicles that contain digestive enzymes. The enzymes originate in the ER and are transported to the Golgi, where they are packaged into vesicles that bud off the Golgi to form the lysosomes (Figure 2.4). Lysosomes degrade a wide range of materials, including proteins, fats, carbohydrates, and viruses that might enter the cell. These organelles are important in cellular maintenance, and genetic disorders that disrupt or halt lysosome function are often fatal. For example, Tay-Sachs disease (MIM/OMIM 272800) and Pompe disease (MIM/OMIM 232300) affect lysosomal function. The outcomes of these disorders include severe mental retardation, blindness, and death by the age of 3 or 4 years. These disorders serve to reinforce the point made earlier that the functioning of the organism can be explained by events that occur within its cells.

Mitochondria (singular: mitochondrion) Membrane-bound organelles present in the cytoplasm of all eukaryotic cells that are the sites of energy production within the cells.

Mitochondria Energy transformation takes place in **mitochondria** (Figure 2.5). Mitochondria carry their own genetic information in the form of circular DNA molecules. Mutations in mitochondrial DNA can cause a number of genetic disorders, including Kearns-Sayre syndrome (MIM/OMIM 530000) and MELAS syndrome (MIM/OMIM 535000). These and other genetic disorders affecting mitochondria are discussed in Chapter 4.

FIGURE 2.3 (a) Three-dimensional representation of the endoplasmic reticulum (ER), showing the relationship between the smooth and rough ER. (b) An electron micrograph of ribosome-studded rough ER.

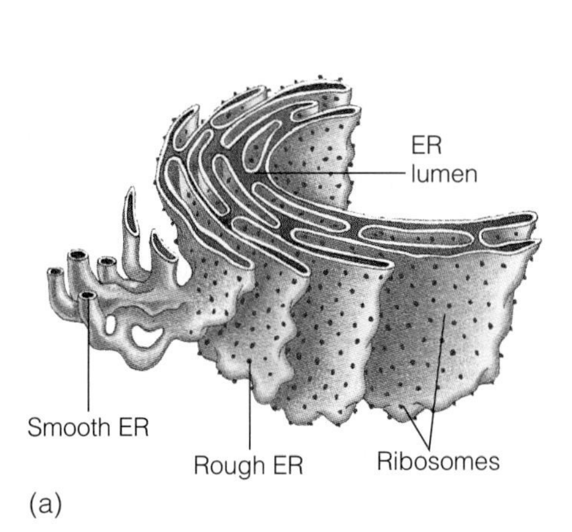

(a)

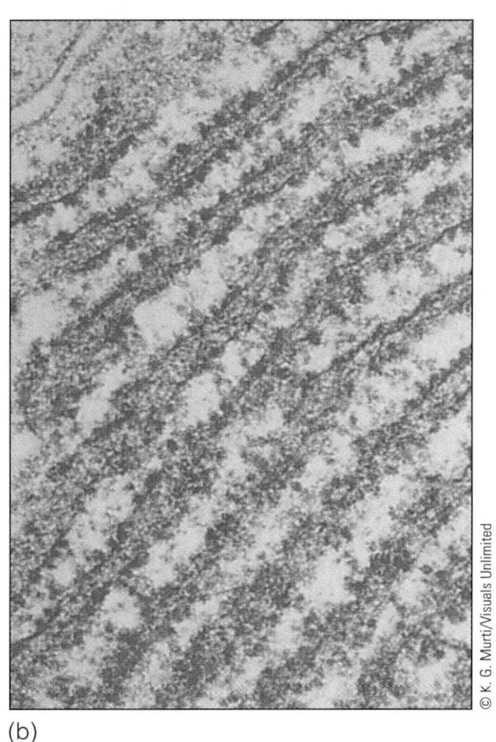

(b)

© K. G. Murti/Visuals Unlimited

SIDEBAR

A Fatal Membrane Flaw

Cystic fibrosis is a genetic disorder that leads to an early death. Affected individuals have thick, sticky secretions of the pancreas and lungs. Diagnosis is often made by finding elevated levels of chloride ions in sweat. According to folklore, midwives would lick the forehead of newborns. If the sweat was salty, they predicted that the infant would die a premature death. Despite intensive therapy and drug treatments, the average survival of persons with this disorder is only about 25 years. Cystic fibrosis is caused by a functional defect in a membrane protein that controls the movement of chloride ions across the plasma membrane. In normal cells, this protein functions as a channel controlling the flow of chloride, but in cystic fibrosis, the channel is unable to open. This causes chloride ions to accumulate inside the cell. To balance the chloride ions, the cells absorb excess sodium. In secretory glands, this leads to decreases in fluid production, resulting in blockage of flow from the pancreas and the accumulation of thick mucus in the lungs. The symptoms and premature death associated with this disorder point out the important role of membranes in controlling cell function.

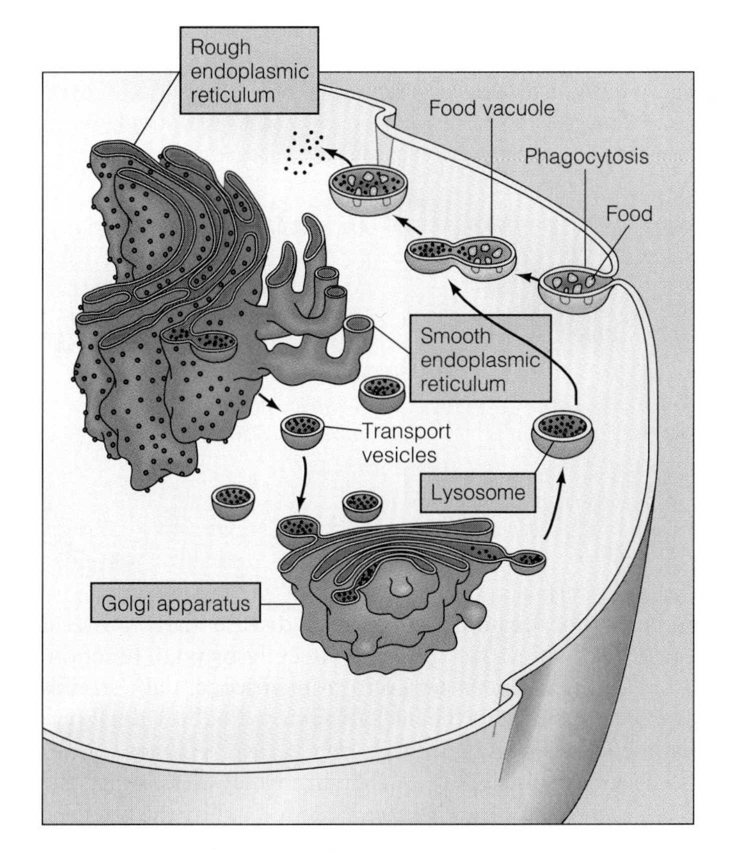

FIGURE 2.4 The relationship between the Golgi complex and lysosomes. Digestive enzymes are synthesized in the ER and move to the Golgi in transport vesicles. In the Golgi, the enzymes are modified and packaged. Lysosomes pinch off the end of the Golgi membrane. In the cytoplasm, lysosomes fuse with and digest the contents of vesicles that are internalized from the plasma membrane.

Nucleus The largest and most prominent cellular organelle is the **nucleus** (Figure 2.6a). The nucleus is enclosed by a double membrane that is covered with pores that allow direct communication between the nucleus and cytoplasm (Figure 2.6b). Within the nucleus, dense regions known as **nucleoli (singular: nucleolus)**

Nucleus The membrane-bounded organelle in eukaryotic cells that contains the chromosomes.

Guest Essay: Research and Applications

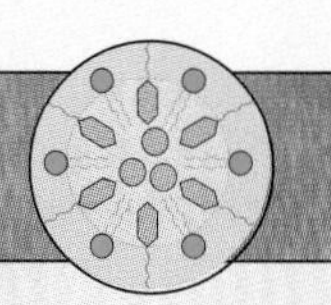

Exploring Membranes

Anne Walter

While growing up, it seemed to me that a biologist could be only a physician or a quintessential naturalist with sensible shoes, binoculars, and field notebook. Yet even though I loved biology, neither of these futures felt right. Fortunately, two wonderful teachers, Miss Hill and Miss Strosneider, taught biology in the Washington, DC, public schools. With a paper chromatography experiment to extract plant pigments and a lab on enzymes that I remember to this day, these two remarkable teachers were the first to show me that the basis for much of biology was the precise and intricate interactions of specialized molecules. This was amazing! My interest must have showed, because I was encouraged to compete for an American Heart Association research opportunity that resulted in a summer at George Washington University working on lipid metabolism. Little did I expect lipids to be in my future.

My experiences in college helped me decide that I was interested in physiology and I started graduate studies with Maryanne Hughes, who wanted to understand how seagulls can drink seawater. These birds have several special adaptations, including a gland that secretes an incredibly salty solution after they've had a salty drink. I had learned that because water permeability across cell membranes is high, cells equilibrate rapidly with their external medium. So why didn't the seagull salt glands rapidly lose water and shrink? I concluded that there must be something unusual about their membrane lipids and went on to study at Duke University in a department that specializes in membranes. My research problem was to define the permeability properties of the lipid bilayer in the absence of protein. My results confirmed that hydrophobic solutes like CO_2, ethanol, and aspirin all penetrate the membrane rapidly, whereas hydrophilic molecules, such as glucose or amino acids, penetrate very slowly. Part of my research was to test our methods thoroughly to ensure that our values reflected the true permeabilities, allowing us to be very confident in our conclusion that the membrane behaves like a hydrocarbon that is very constrained . . . like an oil, but not like an oil. One potential application of this research may be a phospholipid "sponge" for cleaning up hydrocarbon pollutants. If lipid bilayers can be a sponge to soak up other molecules, they can also release molecules. This idea is being used by the cosmetics and pharmaceutical industries to develop phospholipid dispersions as safe, slow-release systems.

My current research asks whether the behavior of transmembrane proteins is affected by their environment. With Neal Rote's research group at Wright State University, I have helped explore the possibility that the autoimmune disease called "antiphospholipid antibody syndrome" might be caused by antibodies against phosphatidylserine that react when this lipid is exposed on the outside of the cell during platelet activation and possibly during placenta formation. I never would have guessed that expertise with membranes and lipids would be important in trying to figure out a disease process whose main symptoms are blood clotting disorders and poor placental development.

Lipid bilayers are essential to all living cells. In fact, it has been suggested that the "primordial soup" contained lipids that spontaneously formed closed vesicles and bilayer surfaces that both protected and concentrated the protoenzymes as one of the first steps in the origin of living cells. Discovering the molecular basis for these properties is a puzzle that is turning out to be quite exciting to put together.

ANNE WALTER is a professor in the Biology Department of St. Olaf College in Northfield, Minnesota. She received a B.A. in biology in 1973 from Grinnell College in Iowa and a Ph.D. in physiology and pharmacology in 1981 from Duke University, North Carolina.

(Figure 2.6a) synthesize the ribosomes. Under an electron microscope, dark strands and clumps of **chromatin** are seen throughout the nucleus (Figure 2.6c). As a cell prepares to divide, the chromatin condenses and coils to form the **chromosomes.**

The nucleus contains the genetic information that ultimately determines the structure and shape of the cell and the range of functions carried out by the cell. The genetic information is composed of DNA and organized into units called **genes.** DNA and its associated proteins are organized into chromosomes.

Because chromosomes carry genetic information and transmit genetic information from generation to generation, they occupy a central position in human genetics. The correct number of chromosomes in humans (46) was not determined until 1956. In the 1970s, advances in chromosome staining and new

Nucleolus (plural: nucleoli) A nuclear region that functions in the synthesis of ribosomes.

Chromatin The component material of chromosomes, visible as clumps or threads in nuclei under a microscope.

Chromosomes The threadlike structures in the nucleus that carry genetic information.

Genes The fundamental units of heredity.

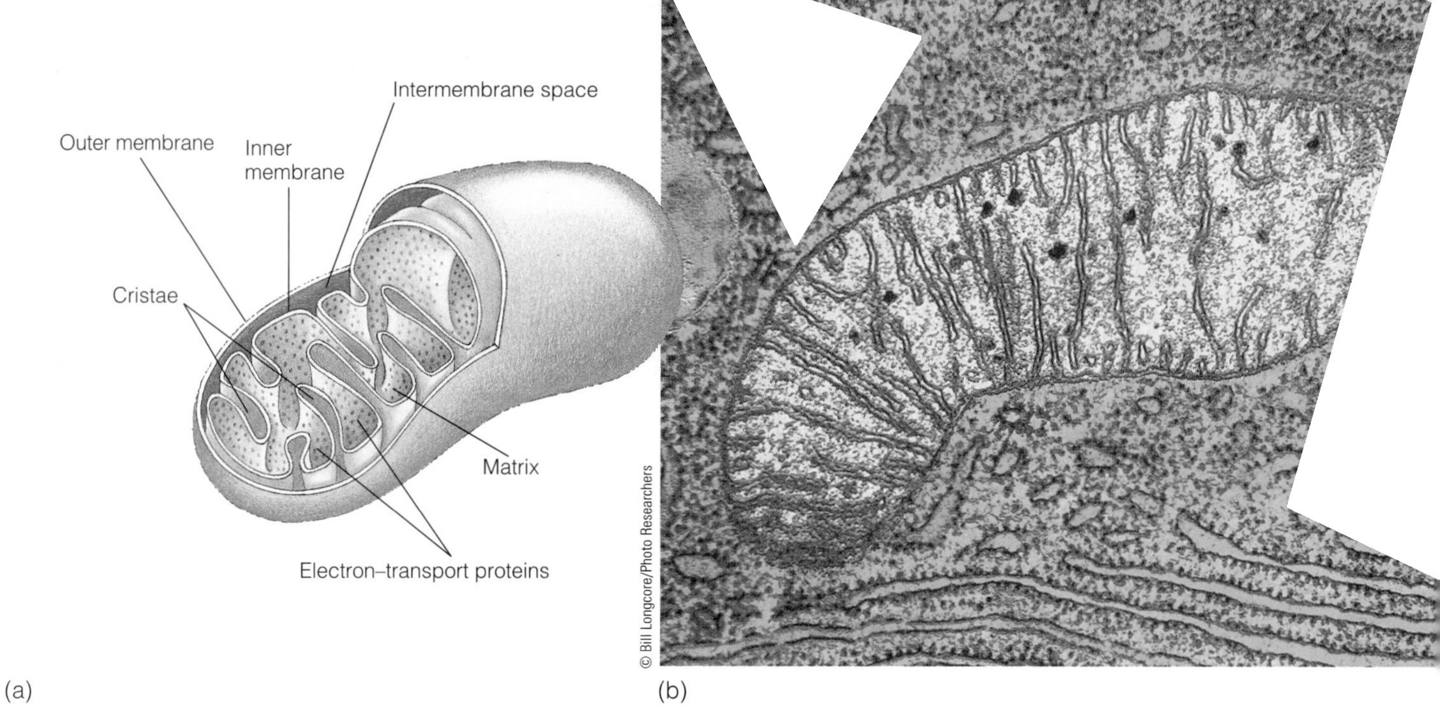

FIGURE 2.5 The mitochondrion is a cell organelle involved in energy transformation. (a) The infolded inner membrane forms two compartments where chemical reactions transfer energy from one form to another. (b) A transmission electron micrograph of a mitochondrion.

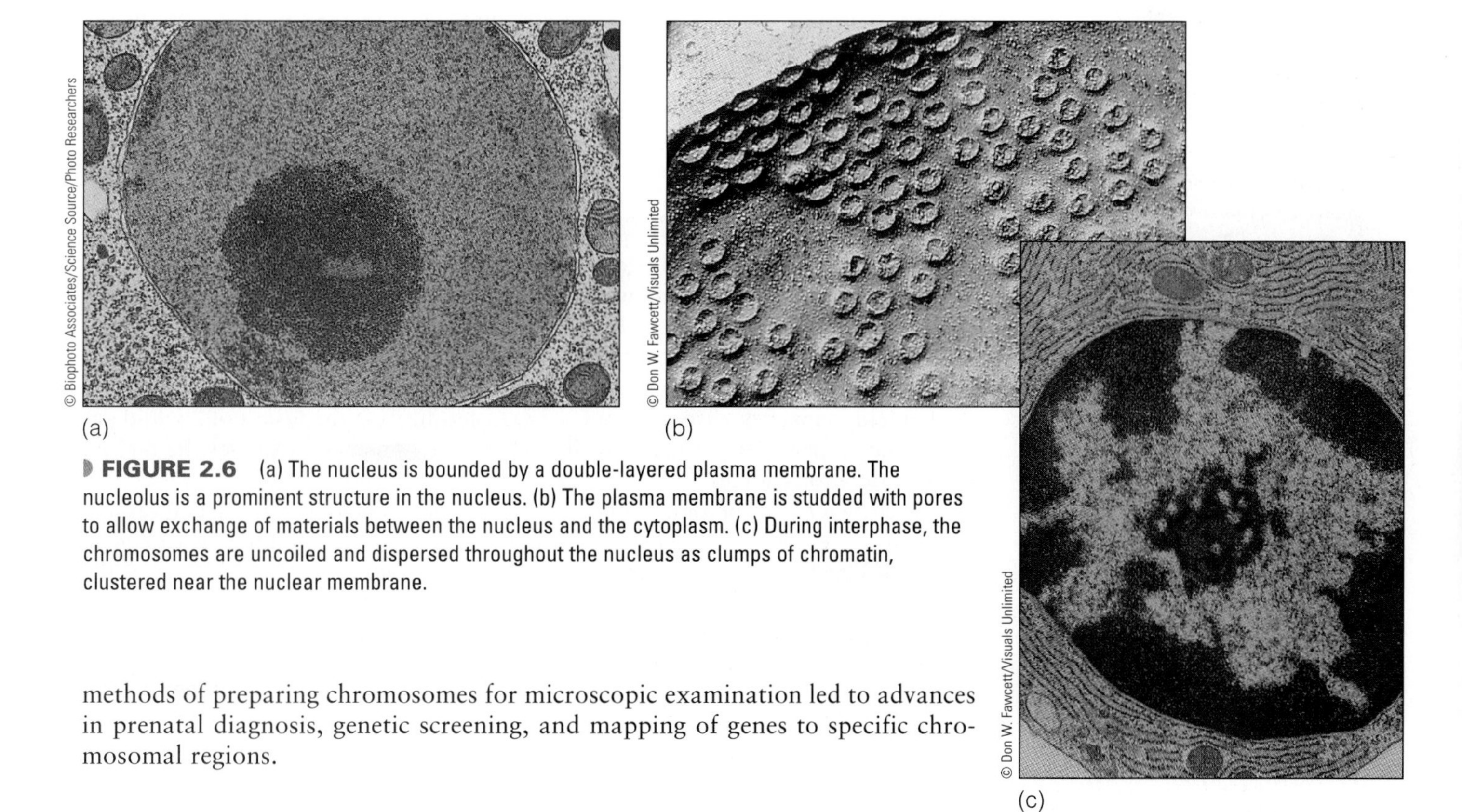

FIGURE 2.6 (a) The nucleus is bounded by a double-layered plasma membrane. The nucleolus is a prominent structure in the nucleus. (b) The plasma membrane is studded with pores to allow exchange of materials between the nucleus and the cytoplasm. (c) During interphase, the chromosomes are uncoiled and dispersed throughout the nucleus as clumps of chromatin, clustered near the nuclear membrane.

methods of preparing chromosomes for microscopic examination led to advances in prenatal diagnosis, genetic screening, and mapping of genes to specific chromosomal regions.

Chromosome Number and Chromosome Pairs

Chromosomes are not usually visible in the nuclei of nondividing cells because they are uncoiled to form the threads and clumps of chromatin found throughout the nucleus. As a cell prepares to divide, the thin strands of chromatin coil and shorten

Table 2.2 Chromosome Number in Selected Organisms

Organism	Diploid Number (2*n*)	Haploid Number (*n*)
Human (*Homo sapiens*)	46	23
Chimpanzee (*Pan troglodytes*)	48	24
Gorilla (*Gorilla gorilla*)	48	24
Dog (*Canis familiaris*)	78	39
Chicken (*Gallus domesticus*)	78	39
Frog (*Rana pipiens*)	26	13
Housefly (*Musca domestica*)	12	6
Onion (*Allium cepa*)	16	8
Corn (*Zea mays*)	20	10
Tobacco (*Nicotiana tabacum*)	48	24
House mouse (*Mus musculus*)	40	20
Fruit fly (*Drosophila melanogaster*)	8	4
Nematode (*Caenorhabditis elegans*)	12	6

Homologs Members of a chromosomal pair.

Diploid The condition in which each chromosome is represented twice as a member of a homologous pair.

Haploid The condition in which each chromosome is represented once in an unpaired condition.

Zygote The diploid cell resulting from the union of a male haploid gamete and a female haploid gamete.

Centromere A region of a chromosome to which fibers attach during cell division. The location of a centromere gives a chromosome its characteristic shape.

Metacentric chromosomes Chromosomes that have a centrally placed centromere.

Submetacentric chromosome A chromosome whose centromere is placed closer to one end than the other.

Acrocentric chromosome A chromosome whose centromere is placed very close to, but not at, one end.

Sex chromosomes In humans, the X and Y chromosomes that are involved in sex determination.

Autosomes Chromosomes other than the sex chromosomes.

to form structures recognizable as chromosomes. The number of chromosomes in the nucleus is characteristic for a given species: cells of the fruit fly *Drosophila melanogaster* have 8 chromosomes, corn plants have 20 chromosomes, and humans have 46 chromosomes. The chromosome numbers for several species of plants and animals are given in Table 2.2. The chromosomes of humans and most other eukaryotic organisms occur in pairs. One member of each chromosome pair is derived from the female parent and the other from the male parent. Members of a chromosome pair are known as **homologs.**

Cells with pairs of homologous chromosomes are called **diploid** cells, and the number of chromosomes is known as the diploid, or 2*n*, number of chromosomes. In humans, the diploid number of chromosomes is 46. Certain cells, such as eggs and sperm (gametes), which contain only one copy of each chromosome, are called **haploid** cells. These cells have a haploid, or *n*, number of chromosomes. In humans, the haploid number of chromosomes is 23. At fertilization the fusion of haploid gametes and their nuclei produces a cell, known as a **zygote,** that carries the diploid number of chromosomes.

As chromosomes become visible just before cell division, certain structural features can be recognized. Each chromosome contains a specialized region known as the **centromere,** which divides the chromosome into two arms. The location of the centromere is characteristic for a given chromosome (Figure 2.7). Chromosomes with centromeres at or near the middle have arms of equal length and are known as **metacentric chromosomes** (Figure 2.8). If the centromere is located away from the center, the arms are unequal in length, and the chromosome is called a **submetacentric chromosome.** If the centromere is located very close to one end, the chromosome is called an **acrocentric chromosome.**

Human males and females (and other animal species) have one pair of chromosomes that are not completely homologous. Members of this pair are involved in sex determination and are known as **sex chromosomes.** There are two types of sex chromosomes, X and Y. Females have two homologous X chromosomes, and males have a nonhomologous pair, consisting of one X and one Y chromosome. Chromosomes other than sex chromosomes are called **autosomes.**

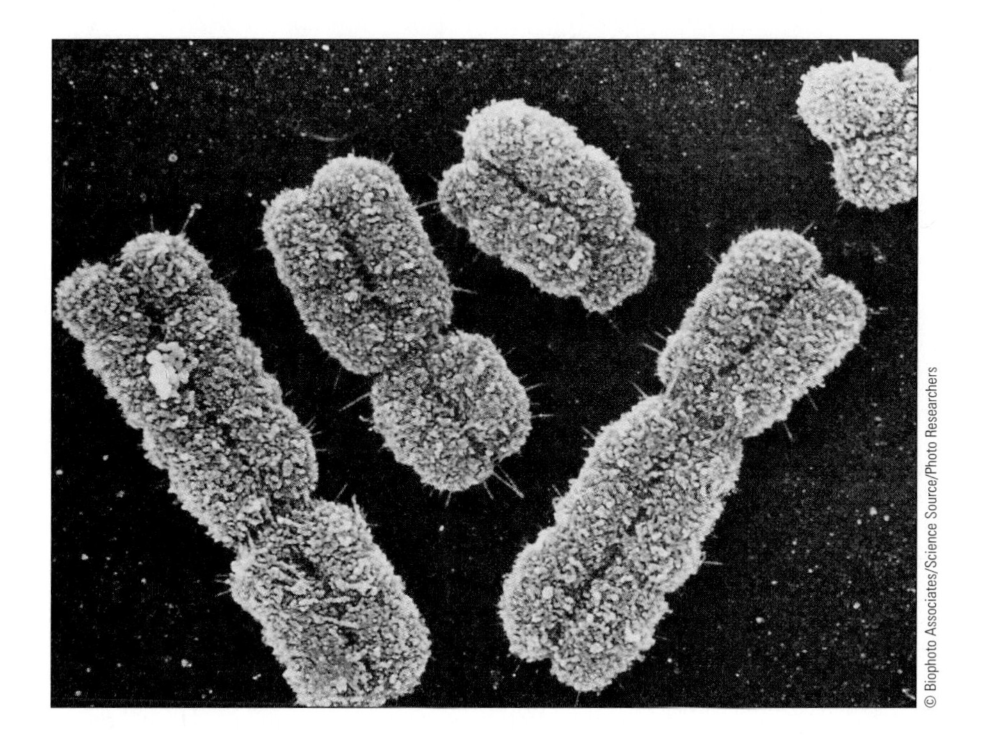

FIGURE 2.7 Human chromosomes as seen at the metaphase of mitosis in the scanning electron microscope. The replicated chromosomes appear as double structures, consisting of sister chromatids joined by a single centromere.

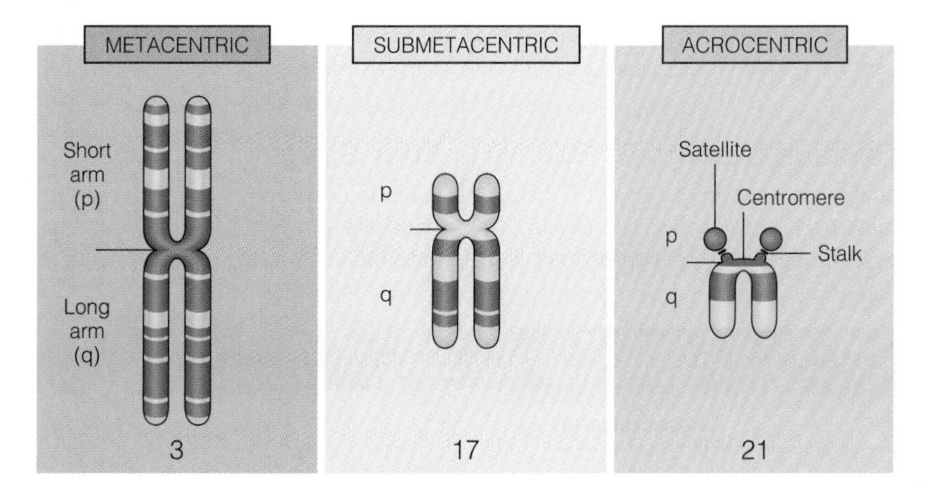

FIGURE 2.8 Human metaphase chromosomes are identified by size, centromere location, and banding pattern. The relative size, centromere locations, and banding patterns for three representative human chromosomes are shown. Chromosome 3 is one of the largest human chromosomes, and because the centromere is centrally located, is a metacentric chromosome. Chromosome 17 is a submetacentric chromosome because the centromere divides the chromosome into two arms of unequal size. Chromosome 21 has a centromere placed very close to one end and is called an acrocentric chromosome. In humans, the short arm of each chromosome is called the p arm, and the long arm is called the q arm.

THE CELL CYCLE DESCRIBES THE LIFE HISTORY OF A CELL

Many cells in the body alternate between two states: division and nondivision. The interval between cell divisions can vary from minutes (in embryonic cells) to months or even years in some cells of adults. The sequence of events from one division to another is called the **cell cycle**. A cycle consists of three phases, **interphase, mitosis,** and **cytokinesis** (Figure 2.9). Interphase, the time between cell divisions, is the first major part of the cell cycle. The period of division consists of two phases, mitosis (division of the chromosomes) and cytokinesis (division of the cytoplasm).

Cell cycle The sequence of events that takes place between successive mitotic divisions.

Interphase The period of time in the cell cycle between mitotic divisions.

Mitosis Form of cell division that produces two cells, each of which has the same complement of chromosomes as the parent cell.

Cytokinesis The process of cytoplasmic division that accompanies cell division.

Interphase Has Three Stages

A good place to begin a discussion of the cell cycle is with a cell that has just finished division. After a cell divides, the resulting daughter cells are about one-half the size of the parental cell. Before they can divide again, they must undergo a period of growth and synthesis. These events take place during the three stages of interphase: G1, S, and G2.

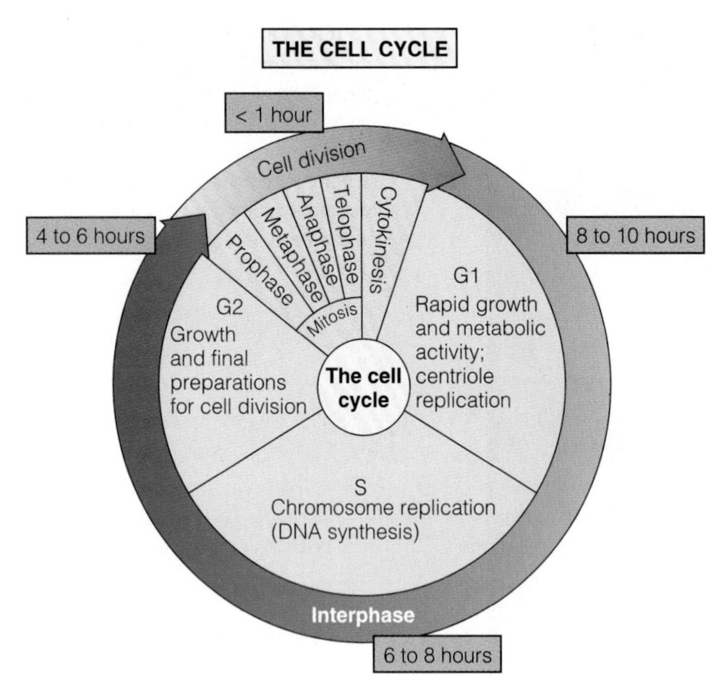

FIGURE 2.9 The cell cycle has three stages: interphase, mitosis, and cytokinesis. Interphase has three components: G1, S, and G2. Times shown for the stages are representative for cells grown in the laboratory.

The G1 stage begins immediately after division and is a period when many cytoplasmic components including organelles, membranes, and ribosomes are synthesized. The synthetic activity in G1 almost doubles the cell size and replaces components lost in the previous division. G1 is followed by the S (synthesis) phase, during which a duplicate copy of each chromosome is made. A second period of cellular growth, known as the G2 phase, takes place before the cell is ready to begin a new round of division. By the end of G2, the cell is ready to divide. In cells grown in the laboratory, the time spent in the three stages of interphase (G1, S, and G2) varies from 18 to 24 hours. The events of mitosis usually take less than 1 hour, so cells spend most of their time in interphase. Table 2.3 summarizes the phases of the cell cycle.

The life history and cell cycles vary for different cell types. Some cells, like those in bone marrow, pass through the cell cycle continuously and divide regularly to form blood cells. At the other extreme, some cell types (certain cells in the nervous system) become permanently arrested in G1 and never divide. In between are cell types that are arrested in G1 but can divide under certain circumstances (white blood cells).

When cells escape from the controls that are part of the cell cycle, they can become cancerous (see Concepts and Controversies: Sea Urchins, Cyclins, and Cancer).

Cell Division by Mitosis Occurs in Four Stages

When a cell reaches the end of the G2 stage, it is ready to undergo division, the second major part of the cell cycle. During this period, two important steps are completed. A complete set of chromosomes is distributed to each daughter cell (mitosis),

Table 2.3 **Phases of the Cell Cycle**

Phase	Characteristics
Interphase	
G1 (gap 1)	Stage begins immediately after mitosis. RNA, protein, and other molecules are synthesized.
S (synthesis)	DNA is replicated. Chromosomes become double stranded.
G2 (gap 2)	Mitochondria divide. Precursors of spindle fibers are synthesized.
Mitosis	
Prophase	Chromosomes condense. Nuclear envelope disappears. Centrioles divide and migrate to opposite poles of the dividing cell. Spindle fibers form and attach to chromosomes.
Metaphase	Chromosomes line up on the midline of the dividing cell.
Anaphase	Chromosomes begin to separate.
Telophase	Chromosomes migrate or are pulled to opposite poles. New nuclear envelope forms. Chromosomes uncoil.
Cytokinesis	Cleavage furrow forms and deepens. Cytoplasm divides.

Concepts and Controversies

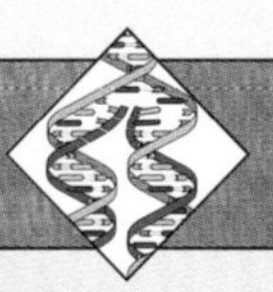

Sea Urchins, Cyclins, and Cancer

Advances in human genetics and cancer research sometimes come from unexpected directions. One such story has its beginning at the Marine Biological Laboratories at Woods Hole, Massachusetts, in 1982. There, a group of young scientists led by Tim Hunt gathered for the summer to study biochemical changes that take place after fertilization in sea urchin eggs. They fertilized a batch of sea urchin eggs and, at 10-minute intervals, analyzed the newly made proteins during the first 2 to 3 hours of development. The fertilized egg first divides at about 1 hour and again about 2 hours after fertilization, resulting in a four-cell embryo.

Several new proteins appeared almost immediately after fertilization, including one that was continuously synthesized, but then destroyed just before each round of cell division. Because of its cyclic behavior, this protein was called cyclin. Work with newly fertilized clam eggs revealed that this species also has cyclins that disappear just before mitosis. Because of their pattern of synthesis and destruction, Hunt and his colleagues concluded that cyclins might be involved in controlling cell division.

Subsequent work showed that cyclins are present in the cells of many organisms and act as important switches in controlling cell division. Sea urchins have only one cyclin, but humans and other mammals have as many as 8 to 12 different cyclins, each of which controls one or more steps in cell division. What does all this have to do with cancer? It turns out that some nondividing cells are arrested in the G1 phase. The mechanism that determines whether cells move through the cycle operates in G1. A critical switch point commits a cell to enter the S phase, G2, and mitosis or causes the cell to leave the cycle and become nondividing. The nature of this switch point, one of the central regulatory mechanisms in all of biology, is slowly being revealed by research in genetics and cell biology. The synthesis and action of cyclins generate the chemical signals that are part of this switch point. At the G1 control point, a cyclin combines with another protein, causing a cascade of events that moves the cell from G1 into S.

Cancer cells have disabled this signal and can divide continuously. Mutations in genes that control the synthesis or action of cyclins are important in the transition of a normal cell into a cancer cell. This important discovery is built on a foundation of work done on sea urchin embryos. Because eukaryotic cells share many properties, work done on yeast, sea urchin eggs, or clam embryos can be used to understand and predict events in normal human cells and in cells that have undergone mutations and become cancerous. For his work on cyclins, Tim Hunt shared the 2001 Nobel Prize for Physiology or Medicine with two other scientists who also worked on cell division.

---

and the cytoplasm is distributed more or less equally to the two daughter cells (cytokinesis). The division of the cytoplasm is accomplished by splitting the cell into two parts, each of which receives enough cytoplasm, organelles, and plasma membrane to form the new cells. Although cytoplasmic division can be somewhat imprecise and still be operational, the division and distribution of the chromosomes must be accurate and unfailing for the cell to function properly.

The chromosomes and the genetic information they contain are replicated during the S stage of interphase, and a complete set of chromosomes is distributed to each of the daughter cells during mitosis. The net result is two daughter cells. In humans, each daughter cell receives a set of 46 chromosomes derived from a single parental cell with 46 replicated chromosomes. Although the distribution of chromosomes in cell division is usually precise, errors in this process do occur. These mistakes often have serious genetic consequences and are discussed in detail in Chapter 6.

Although mitosis is a continuous process, for the sake of discussion it is divided into four phases: prophase, metaphase, anaphase, and telophase (▶ Figure 2.10). These phases are accompanied by changes in chromosome organization as described in the following sections.

Prophase At the beginning of **prophase,** the chromosomes coil, thicken, shorten, and become recognizable as distinct structures under a microscope (Figure 2.10b). At first, the chromosomes appear as long, thin, intertwined threads. As prophase continues, the chromosomes become shorter and thicker. In human cells, 46 such

Prophase A stage in mitosis during which the chromosomes become visible and split longitudinally except at the centromere.

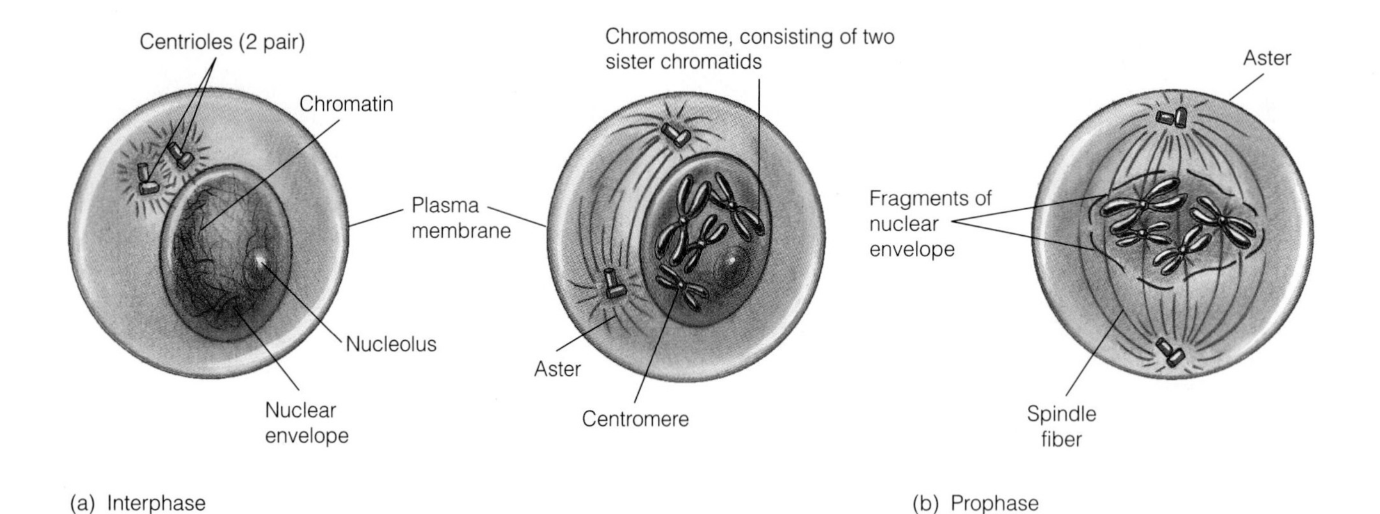

(a) Interphase

(b) Prophase

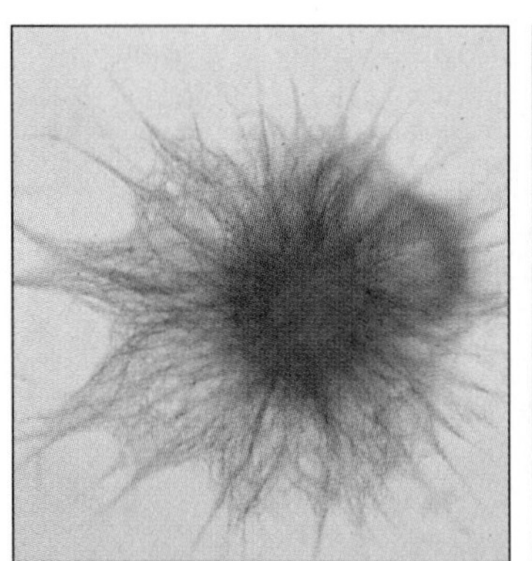

Interphase.

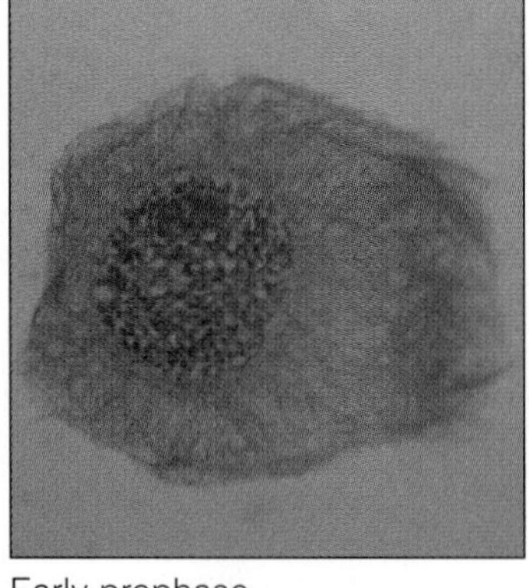

Early prophase.

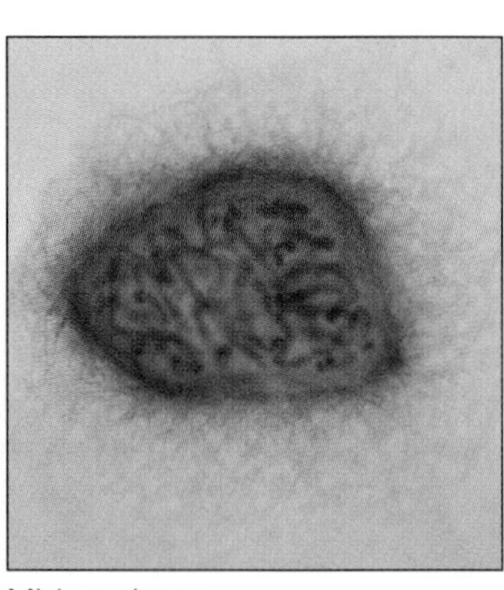

Midprophase.

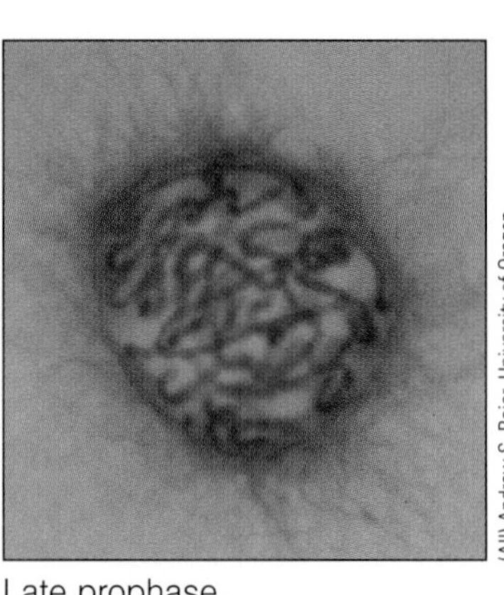

Late prophase.

(All) Andrew S. Bajer, University of Oregon

FIGURE 2.10 Stages of mitosis. (a) During interphase replication of chromosomes takes place. (b) In prophase the chromosomes coil and become visible as threadlike structures. In late prophase they are visible as a double structure, consisting of sister chromatids joined by a single centromere. At the end of prophase, the nuclear membrane breaks down. (c) In metaphase, chromosomes become aligned at the equator of the cell. (d) In anaphase the centromeres divide, converting the sister chromatids into chromosomes, which move toward opposite sides of the cell. (e) At telophase the chromosomes uncoil, the nuclear membrane re-forms, and the cytoplasm divides.

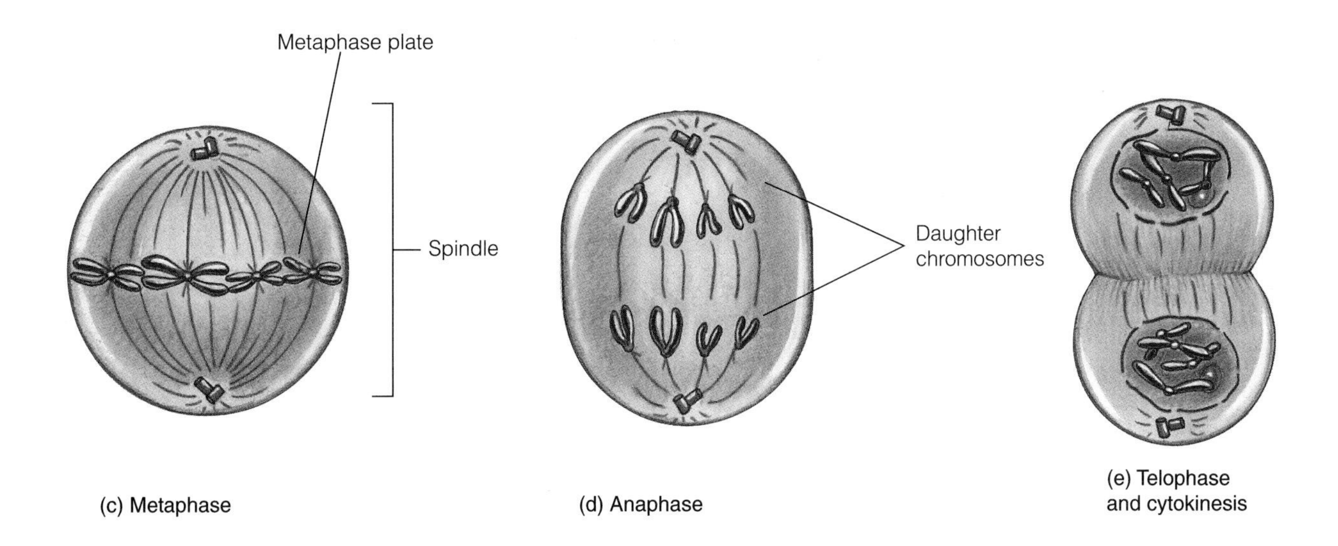

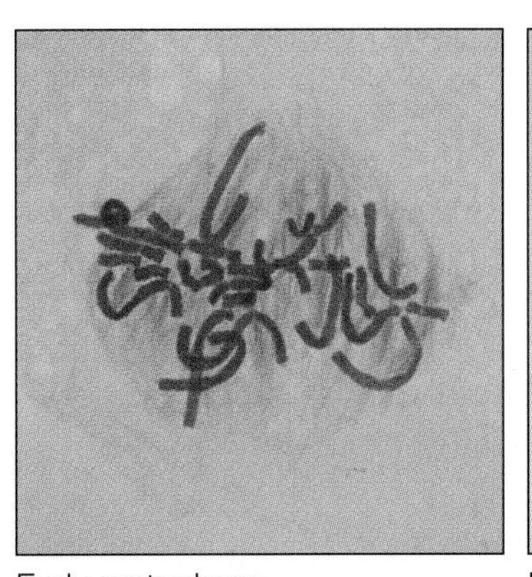
Early metaphase.

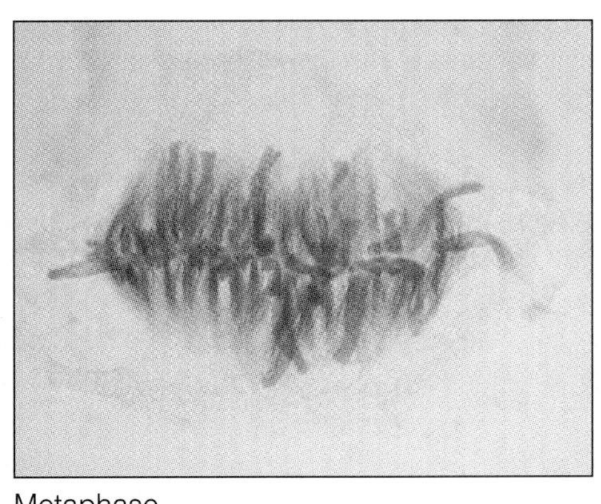
Metaphase.

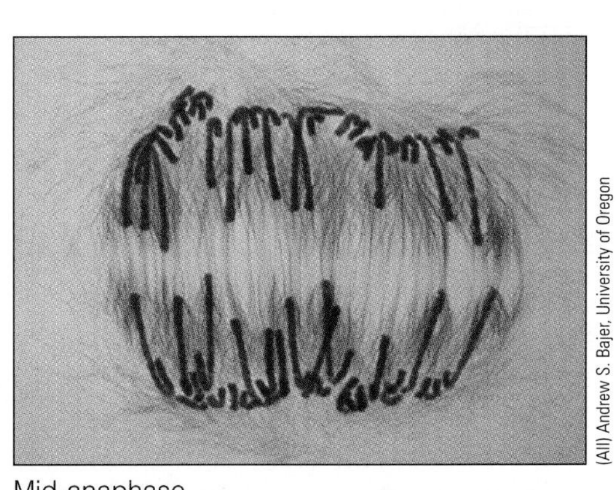
Mid-anaphase.

(All) Andrew S. Bajer, University of Oregon

Late anaphase.

Telophase.

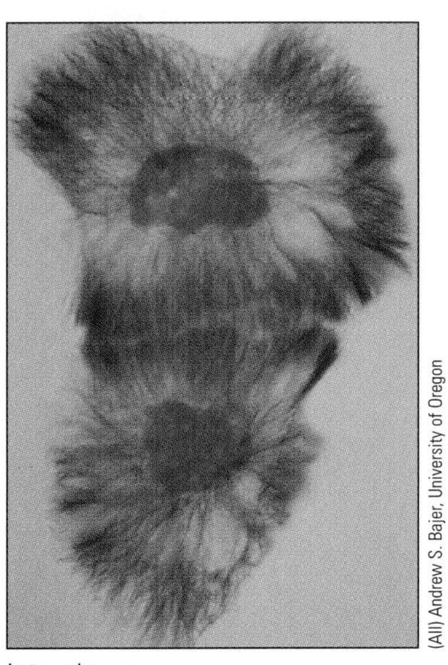
Interphase.

(All) Andrew S. Bajer, University of Oregon

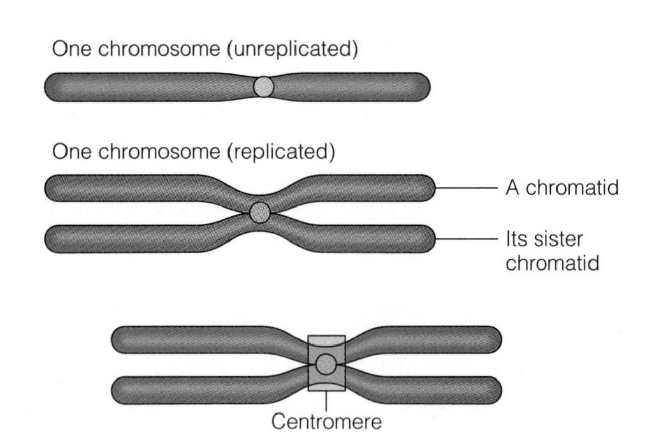

FIGURE 2.11 Chromosomes replicate during the S phase. While attached to the centromere, the replicated chromosomes are called sister chromatids.

structures can be seen. The condensation of chromosomes at the beginning of prophase serves an important function. In a shortened and contracted form, the chromosomes untangle from each other and move freely during mitosis. Near the end of prophase, each chromosome consists of two longitudinal strands known as **chromatids.** The chromatids are separate structures, held together by the centromere. Two chromatids joined by a centromere are known as **sister chromatids** (Figure 2.11). Near the end of prophase, the nucleolus disappears, and the nuclear membrane breaks down. At this time, a collection of specialized tubules known as spindle fibers begins to form in the cytoplasm. When fully organized, the fibers stretch from centriole to centriole, forming an axis (Figure 2.10b).

Chromatid One of the strands of a duplicated chromosome, joined by a single centromere to its sister chromatid.

Sister chromatids Two chromatids joined by a common centromere. Each chromatid carries identical genetic information.

Metaphase A stage in mitosis during which the chromosomes move and become arranged near the middle of the cell.

Metaphase The **metaphase** stage begins when the nuclear membrane disappears completely, and the chromosomes are free in the cytoplasm. At metaphase, each chromosome consists of two chromatids attached to a centromere, giving the chromosomes an X-shaped appearance. The chromosomes move to the middle, or equator, of the cell, where spindle fibers attach to the centromeres (Figure 2.10c). At this stage there are 46 centromeres, each attached to two sister chromatids.

FIGURE 2.12 Roberts syndrome is a genetic disorder caused by malfunction of centromeres during mitosis. In this painting by Goya (1746–1828), the child on the mother's lap lacks limb development, which is characteristic of this syndrome.

Anaphase At the beginning of anaphase, the centromeres divide, converting each sister chromatid into a chromosome (Figure 2.10d). An abnormality of centromere function during prenatal development is responsible for Roberts syndrome (MIM/OMIM 268300) (Figure 2.12). After centromere division, the two chromosomes that were originally sister chromatids are genetically and structurally identical to each other. Spindle fibers attached to the centromeres begin to shorten, and the chromosomes migrate toward opposite ends of the cell. As the spindle fibers shorten, the centromeres move first, and the chromosome arms trail behind, making the chromosomes appear V-shaped or J-shaped.

By the end of anaphase, there is a complete set of chromosomes at each end of the cell. Although anaphase is the briefest stage of mitosis, it is essential for ensuring that each daughter cell receives a complete and identical set of 46 chromosomes.

Anaphase A stage in mitosis during which the centromeres split and the daughter chromosomes begin to separate.

Telophase The last stage of mitosis, during which division of the cytoplasm occurs, the chromosomes of the daughter cells disperse, and the nucleus re-forms.

Telophase Two events take place during telophase, the final stage of mitosis: formation of nuclei and division of the cytoplasm (Figure 2.10e). As the chromosomes reach opposite ends of the cell, the spindle fibers break down. Membrane buds from the ER form a new nuclear membrane. Inside the new nucleus, the chromosomes uncoil, lengthen, and form threads and clumps of chromatin.

Cytokinesis, the division of the cytoplasm, begins with the formation of the cell furrow, a constriction of the cell membrane that forms at the equator of the cell (Figure 2.13). The constriction gradually tightens and divides the cell in two. The major features of mitosis are summarized in Table 2.4.

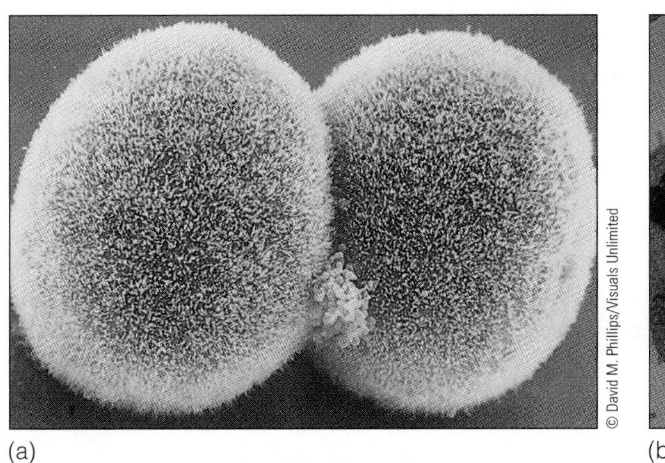

(a)

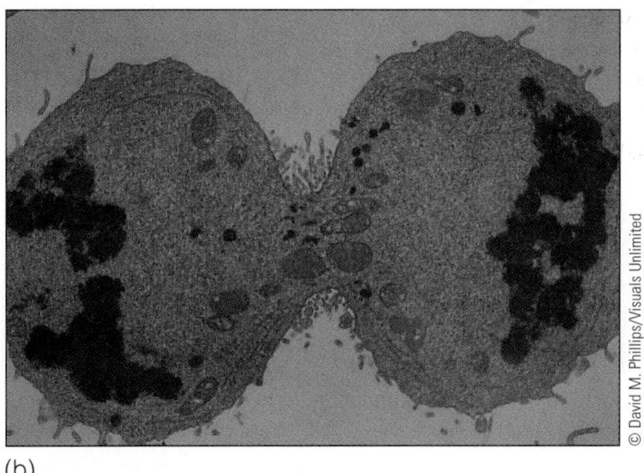

(b)

FIGURE 2.13 Cytokinesis. (a) A scanning electron micrograph of cleavage as seen from the outside of the cell. (b) A transmission electron micrograph of cytokinesis in a cross section of a dividing cell.

Table 2.4 Summary of Mitosis

Stage	Characteristics
Interphase	Replication of chromosomes takes place.
Prophase	Chromosomes become visible as threadlike structures. As they continue to condense, they are seen as double structures, with sister chromatids joined at a single centromere.
Metaphase	Chromosomes become aligned at equator of cell.
Anaphase	Centromeres divide, and chromosomes move toward opposite poles.
Telophase	Chromosomes uncoil, nuclear membrane forms, and cytoplasm divides.

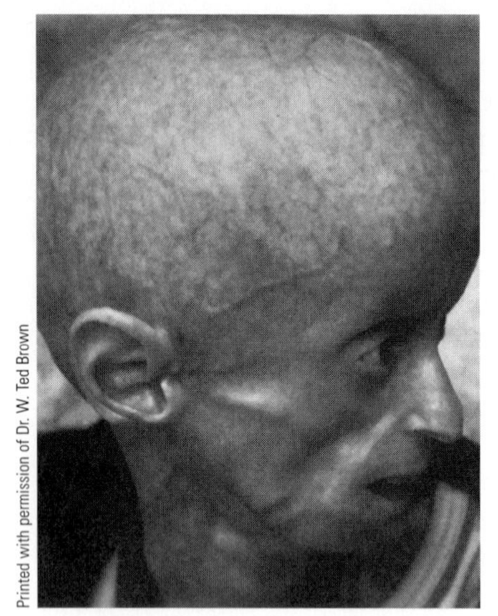

FIGURE 2.14 A 13-year-old girl who has progeria.

MITOSIS IS ESSENTIAL FOR GROWTH AND CELL REPLACEMENT

Mitosis is an essential process in humans and all multicellular organisms. In humans—throughout adult life—some cells retain their capacity to divide, whereas others do not divide after adulthood is reached. Cells in the bone marrow continually move through the cell cycle and produce about 2 million red blood cells each second. Skin cells divide to replace dead cells that are continually sloughed off the surface of the body. Skin cells also divide during wound healing to repair tissue damaged in an injury. By contrast, most cells in the nervous system remain permanently in G1 and cannot divide. As a result, when nerves are damaged or destroyed, they cannot be replaced. For this reason many injuries to the spinal cord result in permanent paralysis.

The mechanism that determines whether cells are cycling or noncycling operates in the G1 phase of the cell cycle. The general features of this regulation are known and are discussed in Chapter 14.

Cells grown in the laboratory undergo a characteristic number of mitotic divisions. Once this number, known as the Hayflick limit, is reached, the cells die. Cells from human embryos have a limit of about 50 divisions, a capacity that includes all divisions necessary to produce an adult and cell replacement during a lifetime. Cells from adults and elderly individuals can divide only about 10 to 30 times before dying.

Human cells (and those of many other multicellular organisms) contain an internal mechanism that determines the maximum number of divisions, which in turn determines the ultimate life span of the individual. This process is under genetic control; several mutations that affect the aging process are known. One of these is a rare genetic disorder known as progeria (MIM/OMIM 176670), in which affected individuals age rapidly. Seven- or 8-year-old children who have this disease physically and mentally resemble individuals of 70 or 80 years (Figure 2.14). Affected individuals usually die of old age in their teens. Werner syndrome (MIM/OMIM 277700) is another genetic disorder associated with premature aging. In this case, the disease process begins between the ages of 15 and 20 years, and affected individuals die of age-related problems by 45 to 50 years.

CELL DIVISION BY MEIOSIS: THE BASIS OF SEX

Meiosis The process of cell division during which one cycle of chromosomal replication is followed by two successive cell divisions to produce four haploid cells.

The genetic information we inherit is contained in two cells, the sperm and the egg. These cells are produced by a form of cell division called **meiosis** (Figure 2.15). Recall that in mitosis, each daughter cell receives a diploid set of 46 chromosomes. In meiosis, members of a chromosome pair are separated from each other, and so each cell receives a haploid set of 23 chromosomes. These haploid cells form the gametes (sperm and egg). Union of the two gametes in fertilization restores the chromosome number to 46 and provides a full set of genetic information to the fertilized egg (zygote).

The distribution of chromosomes in meiosis is an exact process. Each gamete contains one member of each chromosome pair, not a random selection of 23 of the 46 chromosomes. The two rounds of division in meiosis accomplish the precise reduction in the chromosome number.

Cells in the testis and ovary called germ cells give rise to gametes. These cells are diploid and initially divide by mitosis. Some of the daughter cells (but no other cells in the body) undergo meiosis. In meiosis, diploid ($2n$) cells undergo one round of chromosomal replication followed by two divisions to pro-

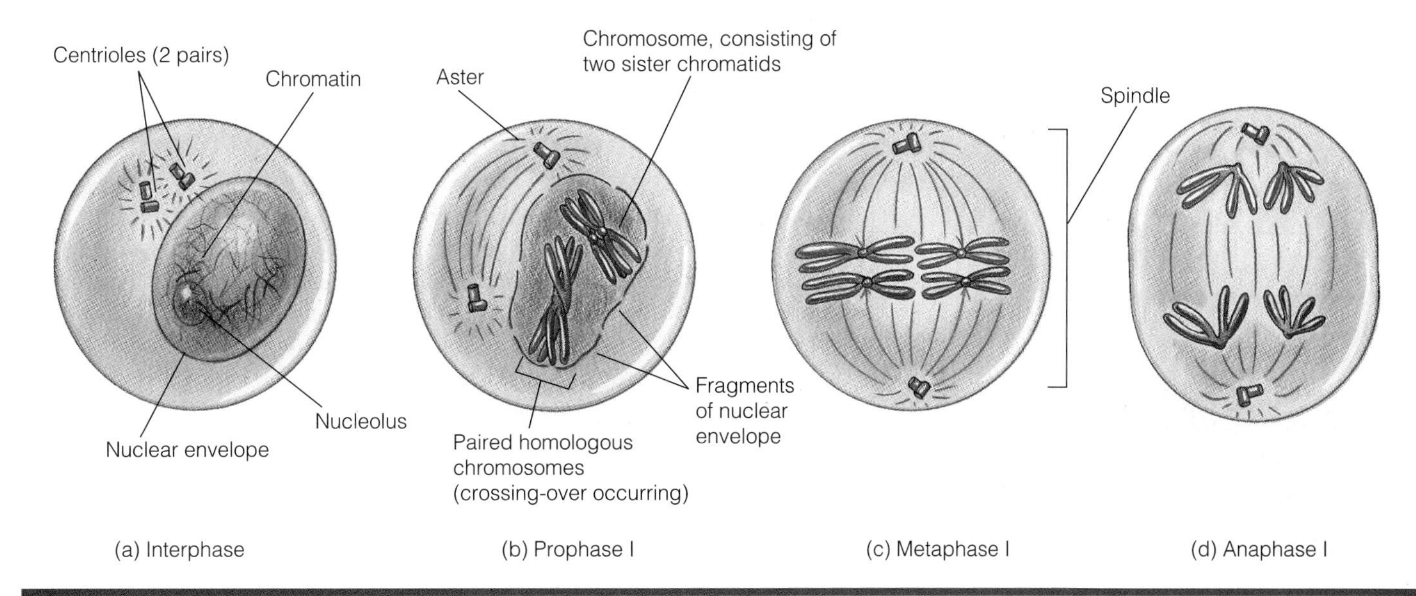

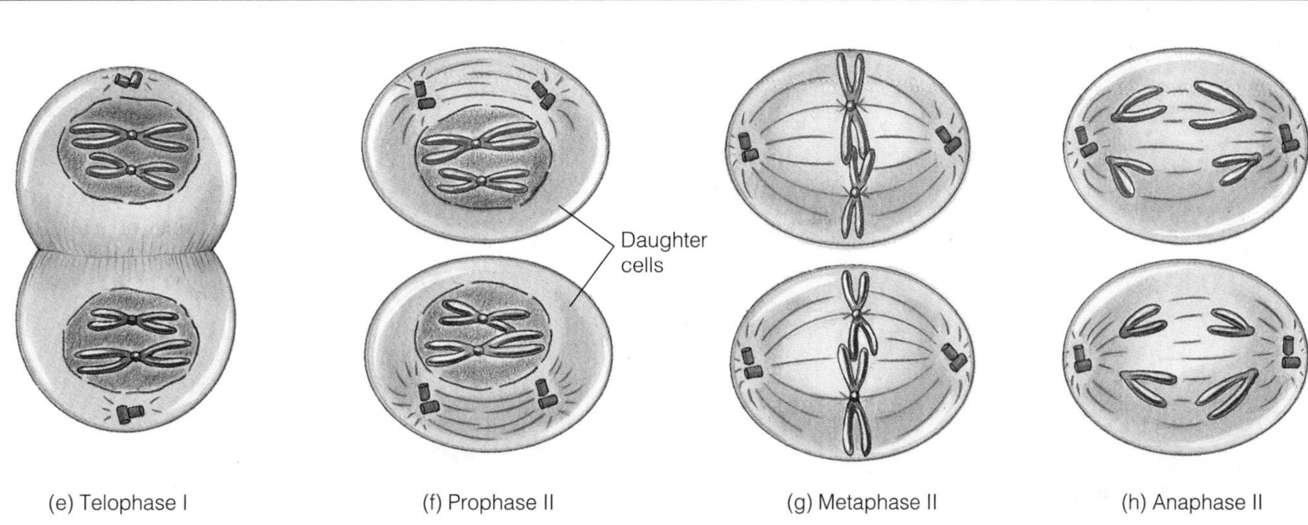

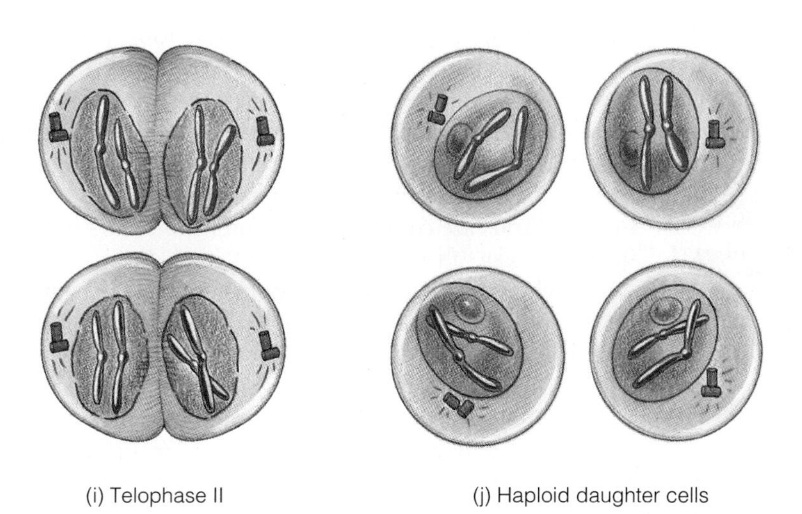

FIGURE 2.15 The stages of meiosis. Chromosomes become visible as threadlike structures early in prophase I (b). Once visible, homologous chromosomes pair and split longitudinally, except at centromeres. While paired, homologous chromosomes undergo crossing-over. In metaphase I (c) the chromosomes become arranged at the cell's equator. In anaphase I (d) members of all chromosome pairs separate from each other and migrate to opposite sides of the cell. The result is the formation of two new cells (e, f). In prophase II (f) the nuclear membrane breaks down, and the chromosomes align at the metaphase plate (g). In anaphase II (h) the centromeres split, converting the sister chromatids into chromosomes, and the newly formed chromosomes move apart. After telophase (i) the result is four haploid daughter cells (j), each containing one copy of each chromosome.

duce four cells. Each cell has the haploid (n) number of chromosomes. The two divisions are referred to as meiosis I and meiosis II, respectively.

Meiosis I Reduces the Chromosome Number

Before cells enter meiosis, the chromosomes are replicated during interphase (Figure 2.15a). In the first part of prophase I, the chromosomes coil, condense, and become visible under a microscope (Figure 2.15b). As the chromosomes coil, the nucleoli and nuclear membrane disappear, and the spindle becomes organized. Each chromosome physically associates with its homologue in a process known as **synapsis.** Chromosomal pairing usually begins at one or more points along the chromosome and proceeds until the chromosomes are aligned.

Synapsis The pairing of homologous chromosomes during prophase I of meiosis.

Chiasmata The crossing of nonsister chromatid strands seen in the first meiotic prophase. Chiasmata represent the structural evidence for crossing-over.

Crossing-over The process of exchanging parts between homologous chromosomes during meiosis, which produces new combinations of genetic information.

Next, the sister chromatids of each chromosome become visible so that each chromosome consists of two sister chromatids joined by a single centromere (Figure 2.15b). Also during this stage, the chromatids of homologous chromosomes form cross-shaped or X-shaped regions known as **chiasmata** (singular: chiasma) (Figure 2.15b). These structures are the points at which physical exchange of chromosome parts between chromatids can take place. This event, known as **crossing-over,** is of great genetic significance and is discussed later.

In metaphase I (Figure 2.15c), the members of a chromosomal pair line up at the equator of the cell. In anaphase I, members of each chromosomal pair separate from each other and move toward opposite sides of the dividing cell (Figure 2.15d). Each daughter cell receives one member of each chromosomal pair. Cytokinesis (division of the cytoplasm) occurs in telophase I, producing two haploid cells (Figure 2.15e). Two major events take place during meiosis I: *reduction,* the random separation of maternal and paternal chromatids, and *crossing-over,* the physical exchange of segments between homologous chromosomes.

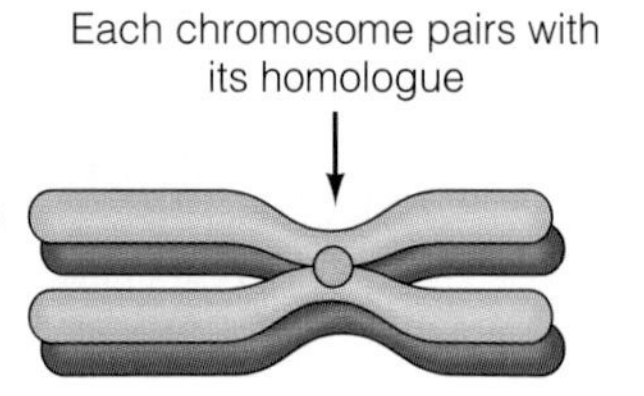

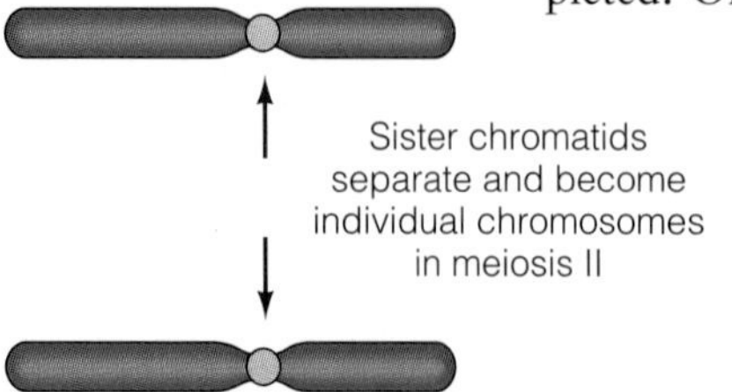

FIGURE 2.16 Summary of chromosome movements in meiosis. Homologous chromosomes appear and pair in prophase I. At metaphase I members of a homologous pair align at the equator of the cell and separate from each other in anaphase I. In meiosis II, the centromeres split, and sister chromatids are converted into individual chromosomes. Each of the resulting haploid cells has one set of chromosomes.

Meiosis II Begins with Haploid Cells

In prophase II, the chromosomes coil and condense and a spindle forms (Figure 2.15f). Each unpaired chromosome consists of two sister chromatids joined by a centromere. At metaphase II (Figure 2.15g), the 23 unpaired chromosomes line up with spindle fibers attached to the centromeres. Anaphase II (Figure 2.15h) begins when the centromere of each chromosome divides for the first time. The 46 chromatids are converted to chromosomes and move to opposite poles of the cell.

In telophase II, the chromosomes uncoil, the nuclear membrane re-forms, and division of the cytoplasm takes place (Figure 2.15i). The process of meiosis is now completed. One diploid cell with 46 chromosomes has undergone one round of chromosomal replication and two rounds of division to produce four haploid cells, each of which contains one copy of each chromosome (Figure 2.15j).

The movement of chromosomes is summarized in Figure 2.16, and the events of meiosis are presented in Table 2.5. Figure 2.17 compares the events of mitosis and meiosis.

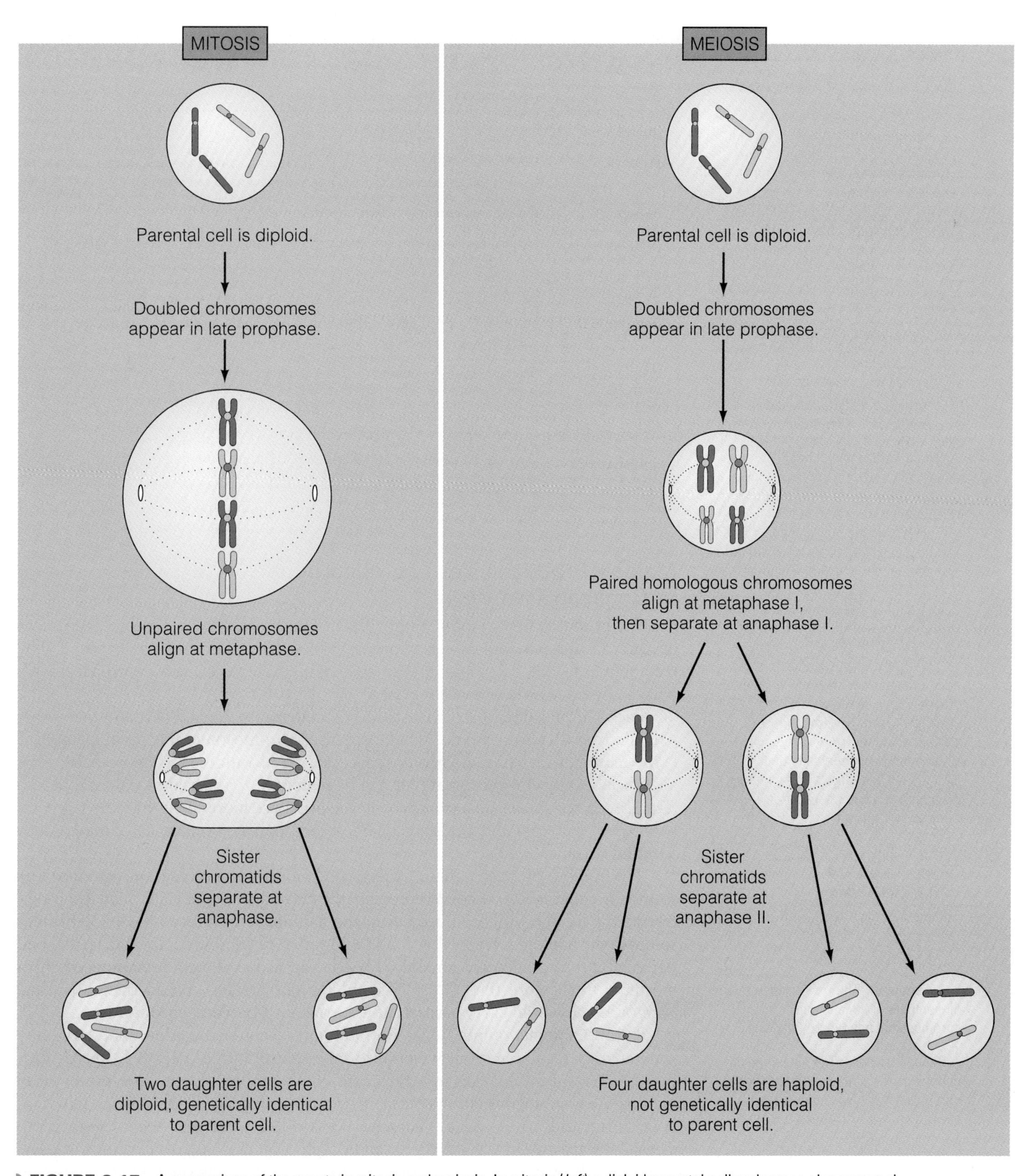

FIGURE 2.17 A comparison of the events in mitosis and meiosis. In mitosis (*left*) a diploid parental cell undergoes chromosomal replication and then enters prophase. The chromosomes appear doubled during late prophase, and unpaired chromosomes align at the middle (equator) of the cell during metaphase. In anaphase, the centromeres split, converting the sister chromatids into chromosomes. The result is two daughter cells, each of which is genetically identical to the parental cell. In meiosis I (*right*) the parental diploid cell undergoes chromosome replication and then enters prophase. Homologous chromosomes pair, and each chromosome appears doubled, except at the centromeres. Paired homologues align at the equator of the cell during metaphase, and members of a chromosome pair separate during anaphase. In meiosis II, the unpaired chromosomes in each cell align at the equator of the cell. During anaphase II the centromeres split, and one copy of each chromosome is distributed to daughter cells. The result is four haploid daughter cells, which are not genetically equivalent to the parental cell.

SIDEBAR

Cell Division and Spinal Cord Injuries

Many highly differentiated cells, like those of the nervous system, do not divide. They are sidetracked from the cell cycle into an inactive state called G0 (G-zero). The result is that injuries to nervous tissue, such as the spinal cord, cause permanent loss of cell function and paralysis. For years, scientists have worked to learn how to stimulate growth of spinal cord cells so that injuries can be repaired. Past efforts have met with failure, but recent work suggests that it may soon be possible for nerves in the spinal cord to reconnect to their proper targets and to restore function in nerve cells that are damaged, but not cut. In one approach, researchers have shown that severed spinal cords of young rats can be reconstructed by transplanting the corresponding section of spinal cord from rat embryos. When the rats reached adulthood, most of the sensory function and movement were restored.

Other researchers have isolated a growth factor found only in the central nervous system that causes cell growth from the ends of severed spinal cords—whereas related growth factors have no effect. Whether such growth can result in reconnection of nerves to their proper muscle targets and whether function can be restored are unresolved questions. These advances may represent the turning point in understanding how to manipulate cell growth to repair spinal cord injuries.

Table 2.5 Summary of Meiosis

Stage	Characteristics
Interphase I	Chromosome replication takes place.
Prophase I	Chromosomes become visible, homologous chromosomes pair, and sister chromatids become visible. Recombination takes place.
Metaphase I	Paired chromosomes align at equator of cell.
Anaphase I	Homologous chromosomes separate. Members of each chromosome pair move to opposite poles.
Telophase I	Cytoplasm divides, producing two cells.
Interphase II	Following a brief pause, chromosomes uncoil slightly. This is not a real interphase as such.
Prophase II	Chromosomes re-coil.
Metaphase II	Unpaired chromosomes become aligned at equator of cell.
Anaphase II	Centromeres split. Daughter chromosomes pull apart.
Telophase II	Chromosomes uncoil, nuclear membrane reforms, cytoplasm divides, and meiosis is complete.

Meiosis Produces New Combinations of Genes in Two Ways

Meiosis produces new combinations of genes in two ways: by random **assortment** of maternal and paternal chromosomes, and the alleles of genes they contain and by **recombination,** the exchange of chromosome parts between sister chromatids. We will briefly discuss these two processes.

Each chromosomal pair consists of one maternally derived (M) and one paternally derived (P) chromosome. When the paired chromosomes line up in metaphase I, the maternal or paternal chromosomes line up at random (▶ Figure 2.18). The alignment of any chromosomal pair can be maternal to paternal or paternal to maternal. As a result, the cells produced in meiosis I are more likely to receive a combination of maternal and paternal chromosomes than a complete set of maternal or paternal chromosomes.

The number of maternal and paternal chromosome combinations produced by meiosis is equal to 2^n, where 2 represents the chromosomes in each pair, and n represents the number of chromosomes in the haploid set. Humans have 23 chromosomes in the haploid set, so 2^{23} or 8,388,608 different combinations of maternal and paternal chromosomes are possible. If each parent has 2^{23} combinations of chromosomes, then more than 7×10^{13} combinations are possible in their offspring, each of which would carry a different assortment of parental chromosomes.

This astronomic number does not take into account the additional variability generated by recombination, which also takes place during meiosis I. Recombination is a second mechanism by which new combinations of genes are produced in meiosis. As homologous chromosomes pair during prophase I, the arms of two nonsister chromatids can overlap. (This overlap is called a chiasma; see ▶ Figure 2.19.) These overlaps are sites where there can be a physical exchange of chromosome segments and the genes they carry. For example, if a pair of homologous chromosomes carries different forms of a gene (for example, *A* and *a* or *B* and *b*), crossing-over reshuffles this genetic information and creates new chromosomal combinations (▶ Figure 2.20).

Without crossing-over, the arrangement of alleles on a particular chromosome would remain coupled together indefinitely. Crossing-over allows new and perhaps

Assortment The random distribution of members of homologous chromosomal pairs during meiosis.

Recombination The exchange of genetic material between homologous chromosomes. Also known as crossing over.

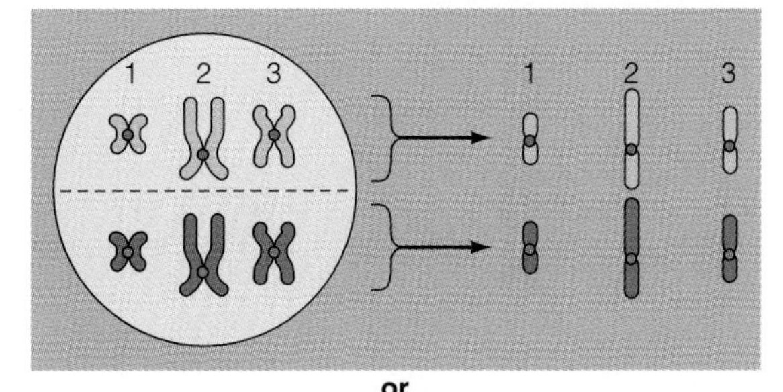

or

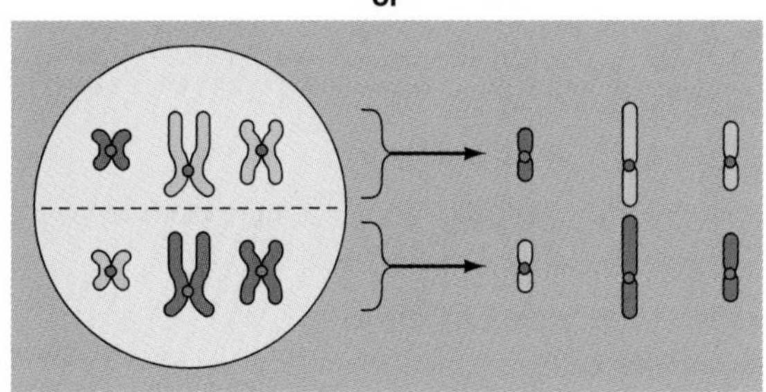

or

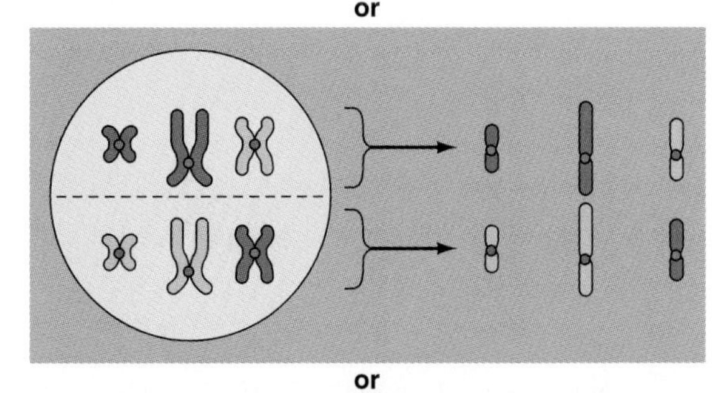

or

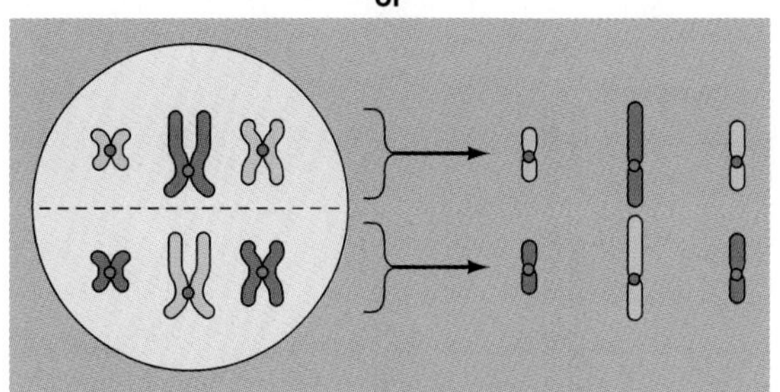

FIGURE 2.18 The orientation of members of a chromosome pair at meiosis is random. Here three chromosomes (1, 2, and 3) have four possible alignments (maternal members of each chromosome pair are light blue; paternal members are dark blue). There are eight possible combinations of maternal and paternal chromosomes in the resulting haploid cells.

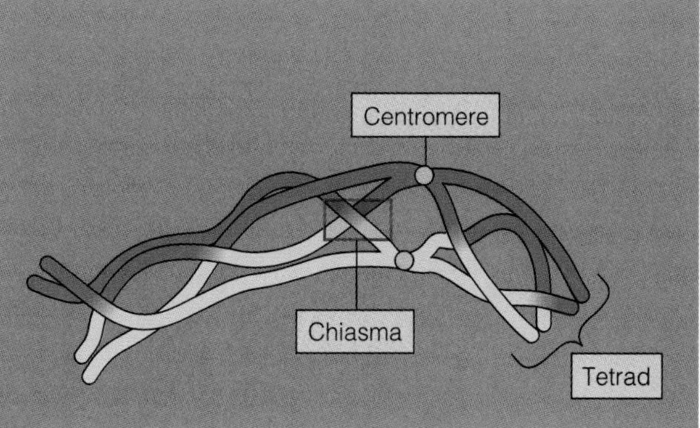

FIGURE 2.19 Crossing-over is the physical exchange of chromosome parts between homologous chromosomes. The X-shaped regions are the site of crossing-over.

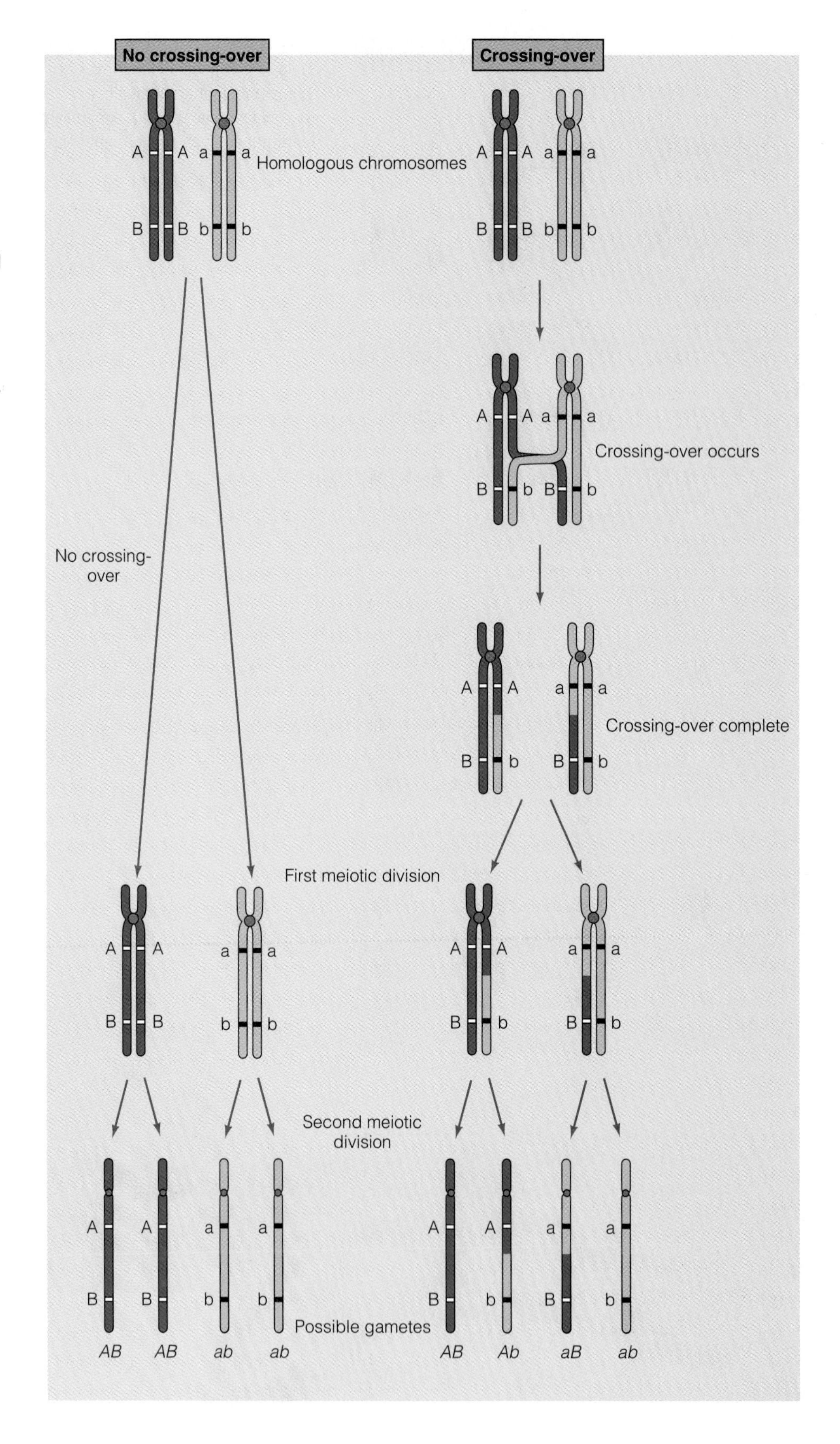

FIGURE 2.20 Crossing-over increases genetic variation by combining genes from both parents on the same chromatid. At left is shown the combination of maternal and paternal genes when no crossing-over occurs. At right, new combinations (Ab, aB) are produced by crossing-over, increasing genetic variability in the haploid cells that will form gametes.

advantageous combinations of alleles to be produced. When the variability generated by crossing-over is added to that produced by random chromosomal combinations, the number of different genetic combinations that a couple can produce in their offspring has been estimated at 80^{23}. Obviously, the offspring of a couple represents only a very small fraction of all these possible gamete combinations. For this reason, it is almost impossible for any two children (aside from identical twins) to be genetically identical.

SPERMATOGENESIS AND OOGENESIS ARE PROCESSES THAT FORM GAMETES

In males, the production of sperm, known as spermatogenesis, occurs in the testis. Cells called **spermatogonia** line the tubules of the testis and divide by mitosis from puberty until death, producing daughter cells called primary spermatocytes (Figure 2.21). The spermatocytes undergo meiosis, and the four haploid cells that result are known as **spermatids.** Each spermatid undergoes a period of development into mature sperm. During this period, the haploid nucleus (sperm carry 22 autosomes and an X or a Y chromosome) becomes condensed and forms the head of the sperm. In the cytoplasm, a neck and whiplike tail develop, and most of the remaining cytoplasm is lost. The entire process of spermatogenesis takes about 48 days: 16 for

Spermatogonia Mitotically active cells in the gonads of males that give rise to primary spermatocytes.

Spermatids The four haploid cells produced by meiotic division of a primary spermatocyte.

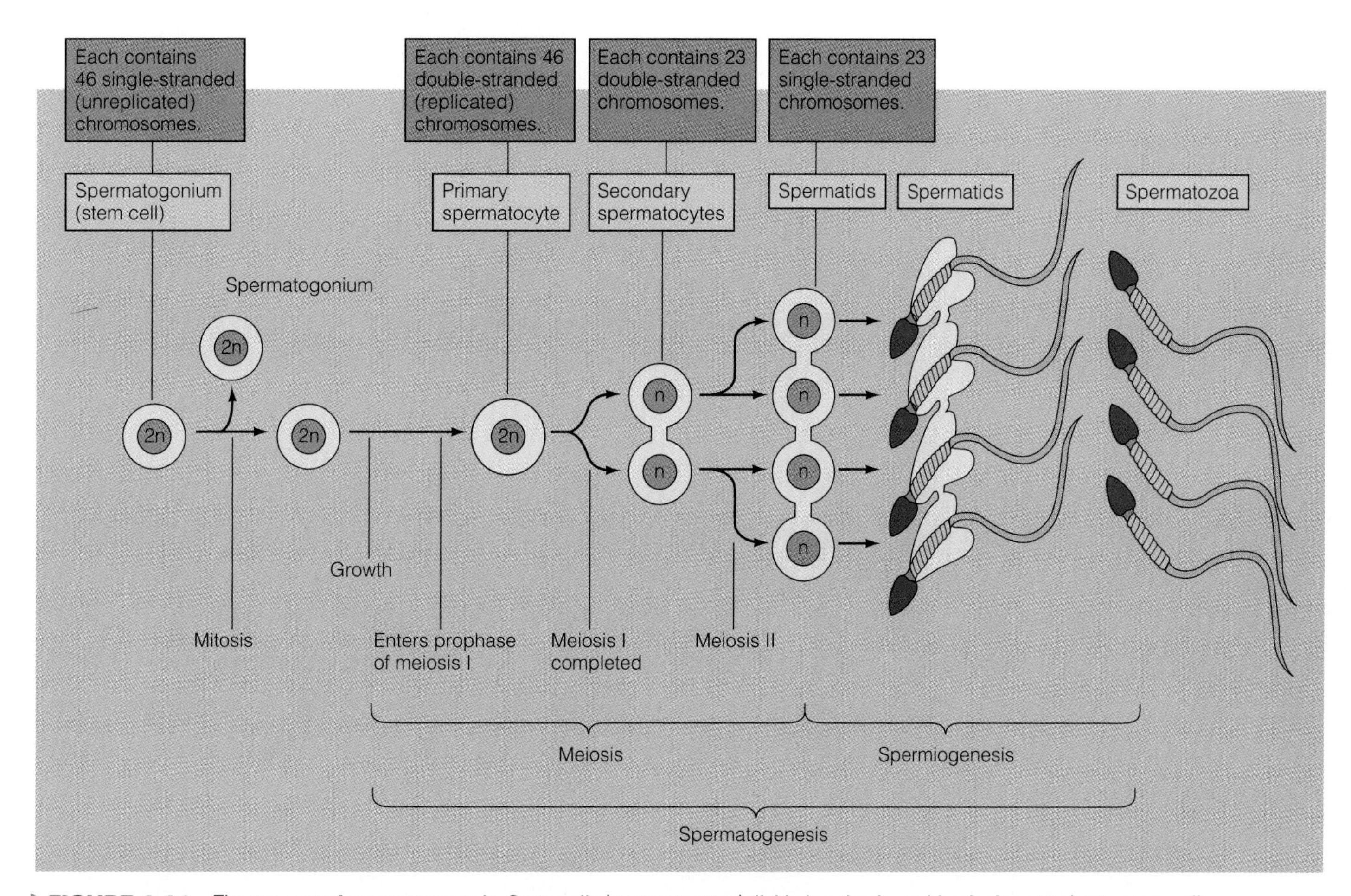

FIGURE 2.21 The process of spermatogenesis. Germ cells (spermatocytes) divide by mitosis, and beginning at puberty some cells produced in this way enter meiosis as primary spermatocytes. After meiosis I, the secondary spermatocytes contain 23 double-stranded chromosomes. After meiosis II, the haploid spermatids contain 23 single-stranded chromosomes. Spermatids undergo a series of developmental changes (spermiogenesis) and are converted into mature spermatozoa.

FIGURE 2.22 The process of oogenesis. Germ cells (oogonia) divide by mitosis, and some cells enter meiosis as primary oocytes during embryonic development. The primary oocytes arrest in prophase I. At puberty, one oocyte per menstrual cycle continues meiosis, and the product of the first meiotic division, the secondary oocyte, is released from the ovary. Production of the secondary oocyte is accompanied by unequal cytoplasmic cleavage, producing the secondary oocyte and a polar body. The secondary oocyte is arrested in meiosis II until fertilization. Penetration of the sperm stimulates completion of the second meiotic division, producing the ovum and the second polar body.

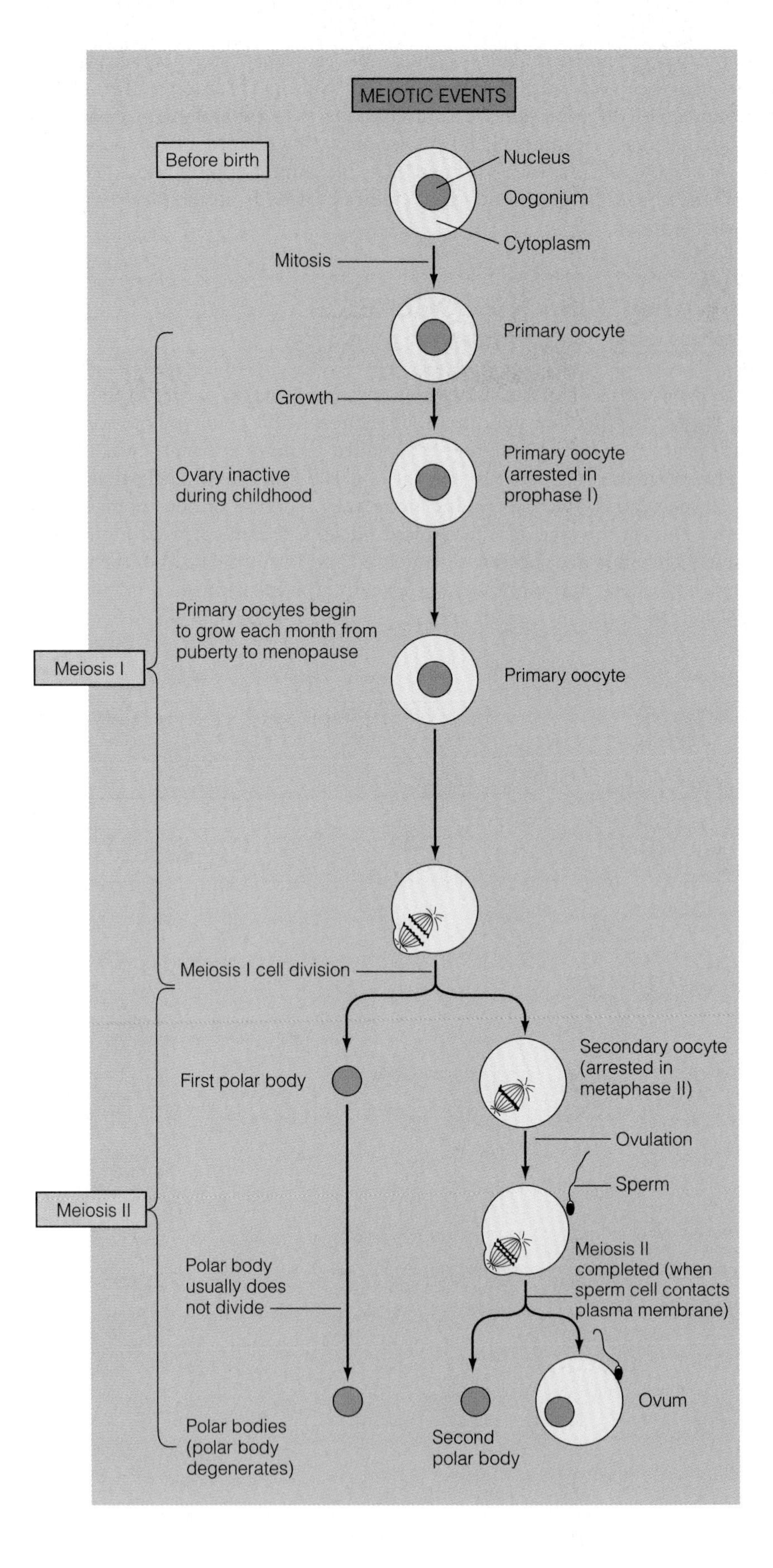

Table 2.6 A Comparison of the Duration of Meiosis in Males and Females

Spermatogenesis		Oogenesis	
Begins at Puberty		**Begins During Embryogenesis**	
Spermatogonium ↓		Oogonium ↓	Forms at 2 to 3 months after conception
Primary spermatocyte ↓	16 days	Primary oocyte ↓	Forms at 2 to 3 month of gestation. Remains in meiosis I until ovulation, 12 to 50 years after formation.
Secondary spermatocyte ↓	16 days	Secondary oocyte ↓	
Spermatid ↓	16 days	Ootid	Less than 1 day, when fertilization occurs
Mature sperm		Mature egg-zygote	
Total time	48 days	Total time	12 to 50 years

meiosis I, 16 for meiosis II, and 16 to convert the spermatid into the mature sperm. The tubules within the testis contain many spermatocytes, and large numbers of sperm are always in production. A single ejaculate may contain 200 to 300 million sperm, and over a lifetime a male produces billions of sperm.

In females the production of gametes, known as oogenesis, takes place in the ovary. Cells in the ovary, called **oogonia,** divide by mitosis to form primary oocytes that undergo meiosis (Figure 2.22). The cytoplasmic cleavage in meiosis I does not produce cells of equal size. One cell, destined to become the oocyte, receives about 95% of the cytoplasm and is known as the **secondary oocyte.** In the second meiotic division, the same disproportionate cleavage results in one cell retaining most of the cytoplasm. The large cell becomes the functional gamete (the **ootid**), and the non-functional, smaller cells are known as **polar bodies.** Thus, in females, only one of the four cells produced by meiosis becomes a gamete. All oocytes contain 22 autosomes and an X chromosome.

Oogonia Mitotically active cells that produce primary oocytes.

Secondary oocyte The large cell produced by the first meiotic division.

Ootid The haploid cell produced by meiosis that becomes the functional gamete.

Polar bodies Cells produced in the first or second division in female meiosis that contain little cytoplasm and will not function as gametes.

The timing of meiosis and gamete formation in the human female is different than in the male (Table 2.6). Oogonia begin mitotic divisions early in embryonic development and finish by 7 to 8 weeks after fertilization. Because no more mitotic divisions take place, the female is born with all of the primary oocytes she will ever possess. All the primary oocytes begin meiosis during embryonic development and then stop. They remain in meiosis I until the female undergoes puberty. At puberty, usually one oocyte per menstrual cycle completes the first meiotic division, is released from the ovary, and moves down the oviduct. If the egg is fertilized, it quickly completes meiosis II, producing a diploid zygote. Unfertilized eggs are sloughed off during menstruation, along with uterine tissue. Each month until menopause, another oocyte completes meiosis I and is released from the ovary. Altogether, a female releases about 450 oocytes during the reproductive phase of her life.

In females, then, meiosis takes years to complete. It begins with prophase I, while she is still an embryo, and continues to the completion of meiosis II following fertilization. Depending on the time of ovulation, meiosis can take from 12 to 50 years in human females.

Summary

There Are Two Main Cell Types: Prokaryotic and Eukaryotic

1. Prokaryotic cells (bacteria, blue-green algae) do not have nuclei or internal membrane systems. Eukaryotic cells (humans, other animals, and plants) have a membrane-enclosed nucleus and internal membrane systems that divide the cell into regions with different functions.

Cell Structure Reflects Function

2. The cell is the basic unit of structure and function in all organisms, including humans. Because genes control the number, size, shape, and function of cells, the study of cell structure helps us understand how genetic disorders disrupt cellular processes. The nucleus contains the genetic information that controls the structure and function of the cell. This genetic information is carried in chromosomes. The number, size, and shape of chromosomes are species-specific traits. In humans, 46 chromosomes—the $2n$, or diploid, number—is present in most cells, whereas specialized cells known as gametes contain half of that number—the n, or haploid number—of chromosomes.

The Cell Cycle Describes the Life History of a Cell

3. At some point in their life, cells pass through the cell cycle, a period of nondivision (interphase) that alternates with division of the nucleus (mitosis) and division of the cytoplasm (cytokinesis). Cells must contain a complete set of genetic information. This is ensured by replication of each chromosome and by the distribution of a complete chromosomal set in the process of mitosis. Mitosis (division) is one part of the cell cycle. During interphase (nondivision), a duplicate copy of each chromosome is made. The process of mitosis is divided into four stages: prophase, metaphase, anaphase, and telophase. In mitosis, one diploid cell divides to form two diploid cells. Each cell has an exact copy of the genetic information contained in the parental cell.

Mitosis Is Essential for Growth and Cell Replacement

4. Human cells are genetically programmed to divide about 50 times. This limit allows growth to adulthood and repairs such as wound healing. Alterations in this program can lead to genetic disorders of premature aging or to cancer.

Cell Division by Meiosis: The Basis of Sex

5. Meiosis is a form of cell division that produces haploid cells containing only the paternal or maternal copy of each chromosome. In meiosis, homologous chromosomes pair with each other. At this time, each chromosome consists of two sister chromatids joined by a common centromere. Nonsister chromatids can exchange chromosome segments, an event known as crossing-over. In metaphase I, pairs of homologous chromosomes align at the equator of the cell. In anaphase I, the homologues separate. In meiosis II, the unpaired chromosomes align at the equator of the cell. In anaphase II, the centromeres divide, and the daughter chromosomes move to opposite poles. The four cells produced in meiosis contain the haploid number (23 in humans) of chromosomes.

Spermatogenesis and Oogenesis Are Processes That Form Gametes

6. Spermatids, produced in male meiosis, undergo structural changes to convert them to functional sperm. In female meiosis, division of the cytoplasm is unequal, leading to the formation of one functional gamete and three smaller cells known as polar bodies.

Case Studies

CASE 1

It is May 1989 and the scene is a crowded research laboratory with beakers, flasks, and pipettes covering the lab bench. People and equipment take up every possible space. One researcher, Joe, passes a friend staring into a microscope. Another student wears gloves while she puts precisely measured portions of various liquids into tiny test tubes. As he walks, Joe glances at the DNA sequence results he is carrying. Something is wrong. There it is, a three base pair deletion—a type of genetic mutation in a DNA sequence. The genetic information required to describe one amino acid in a protein chain is missing, as if one bead had fallen from a precious necklace. Instead of sitting down, Joe rushes to tell his supervisor Dr. Tsui (pronounced "Choy") that he has found a three base pair deletion in a person with cystic fibrosis (CF), but he does not see this same deletion in a normal person's genes. CF is a fatal disease that kills about one out of every 2,000 Caucasians (mostly children).

Dr. Tsui examines the findings and is impressed but wants more evidence to prove that the result is real. He has had false hopes before, so he is not going to celebrate until they check this out carefully. Maybe the difference between the two gene sequences is just a normal variation among individuals.

Five months later, Dr. Tsui and his team identify a "signature" pattern of DNA on either side of the base-pair deletion, and using that as a marker, they compare the genes of 100 normal people with the DNA sequence from 100 CF patients. By September 1989, they are sure they have identified the CF gene.

After several more years, Tsui and his team discover that the DNA sequence with the mutation encodes the information for a protein called CFTR (cystic fibrosis transmembrane conductance regulator), a part of the plasma membrane in cells that make mucus. This protein regulates a channel for chloride ions. Proteins are made of long chains of amino acids. The CFTR protein has 1,480 amino acids. Most children with CF are missing one single amino acid in their CFTR. Because of this, their mucus becomes too thick, causing all the other symptoms of CF. Thanks to Tsui's research, scientists now have a much better idea of how the disease works. We can now easily predict when a couple is at risk for having a child with CF. With increasing understanding, scientists may also be able to devise improved treatments for children born with this disease.

CF is the most common genetic disease among those of European ancestry. Children who have CF are born with it. Half of them will die before they are 25 and few make it past 30. It affects all parts of the body that secrete mucus: the lungs, stomach, nose, and mouth. The mucus of children with CF is so thick sometimes they cannot breathe. Why do 1 in 25 Caucasians carry the mutation for CF? Tsui and others think that people who carry it may also have resistance to diarrhea-like diseases.

1. Dr. Tsui's research team discovered the gene for cystic fibrosis. What medical advances can be made after a gene is cloned?
2. Why do you think a three base pair deletion in the CF gene can cause such severe effects in CF patients? Relate your answer to the CFTR protein function and the cell membrane.

CASE 2

Jim, a 37-year-old construction worker, and Sally, a 42-year-old business executive, were eagerly preparing for the birth of their first child. They, like more and more couples, chose to wait to have children until they were older and more financially stable. Sally had an uneventful pregnancy with prenatal blood tests and an ultrasound indicating that the baby looked great and everything seemed "normal." Then, a few hours after Ashley was born, they were told she was born with Down syndrome. In shock and disbelief, the couple was devastated. How could something so right go so wrong? What did they do to deserve a disabled child?

It has long been recognized that the risk of having a child with trisomy 21 increases with maternal age. For example, the risk of having a child with Down syndrome when the mother is 30 years old is 1 in 1,000 and at age 40 is 9 in 1,000. Well-defined and distinctive phenotypic features characterize Down syndrome, which is the most frequent form of mental retardation caused by a chromosomal aberration. Most individuals (95%) with Down syndrome or trisomy 21 have three copies of chromosome 21. Errors in meiosis that lead to trisomy 21 are almost always of maternal origin; only about 5% occur during spermatogenesis. It has been estimated that meiosis I errors account for 76% to 80% of maternal meiotic errors. In about 5% of patients, one copy is translocated to another chromosome, most often chromosome 14 or 21. No one is at fault when a child is born with Down syndrome, but the chances of it occurring increase with advanced maternal age.

Individuals with Down syndrome often have specific major congenital malformations such as those of the heart (30% to 40% in some studies). Individuals with Down syndrome have an increased incidence (10 to 20 times higher) of leukemia than the normal population. Ninety percent of all Down syndrome patients have significant hearing loss. The frequency of trisomy 21 in the population is 1 in 650 to 1,000 live births.

1. With your knowledge of meiosis in females and egg formation, why would maternal age be associated with an increase in errors that lead to Down syndrome?
2. What prenatal tests could have been done to detect Down syndrome before birth? Should they have been done?
3. Down syndrome is characterized by mental retardation. Can individuals with Down syndrome go to school or hold a job?
4. Should people with mental disabilities be integrated into the community? Why or why not?

Questions and Problems

Cell Structure and Function

1. Which statement is *not* true about eukaryotic cells?
   a. Human cells are eukaryotic.
   b. Eukaryotic cells have a membrane-bound nucleus.
   c. Eukaryotic cells have membrane-bound organelles.
   d. Insect cells are eukaryotic.
   e. Bacterial cells are eukaryotic.
2. Assign a function(s) to the following cellular structures:
   a. plasma membrane
   b. mitochondrion
   c. nucleus
   d. ribosome
3. How many autosomes are present in a body cell of a human being? In a gamete?
4. Define the following terms:
   a. chromosome
   b. chromatin
5. Human haploid gametes (sperm and eggs) contain:
   a. 46 chromosomes, 46 chromatids
   b. 46 chromosomes, 23 chromatids
   c. 23 chromosomes, 46 chromatids
   d. 23 chromosomes, 23 chromatids
6. What is meant by the term *homologous chromosomes*?

The Cell Cycle and Mitosis

7. What are sister chromatids?
8. Draw the cell cycle. What is meant by the term "cycle" in the cell cycle? What is happening at the S phase and the M phase?
9. In the cell cycle, at which stages do *two* chromatids make up *one* chromosome?
   a. beginning of mitosis
   b. end of G1
   c. beginning of S
   d. end of mitosis
   e. at the beginning of G2
10. Does the cell cycle refer to mitosis as well as meiosis?
11. It is possible that an alternative mechanism for generating germ cells could have evolved. Consider meiosis in a germ cell precursor (a cell that is diploid but will go on to make gametes). If the S phase were skipped, which meiotic division (meiosis I or meiosis II) would no longer be required?
12. Identify the stages of mitosis, and describe the important events that occur during each stage.
13. Why is cell furrowing important in cell division? If cytokinesis did not occur, what would be the end result?
14. A cell from a human female has just undergone mitosis. For unknown reasons, the centromere of chromosome 7 failed to divide. Describe the chromosomal contents of the daughter cells.
15. During which phases of the mitotic cycle would the terms *chromosome* and *chromatid* refer to identical structures?
16. Describe the critical events of mitosis that are responsible for ensuring that each daughter cell receives a full set of chromosomes from the parent cell.
17. Mitosis occurs daily in a human being. What type of cells do humans need to produce in large quantities on a daily basis?
18. Speculate on how the Hayflick limit may lead to genetic disorders such as progeria or Werner syndrome. How is this related to cell division?
19. How can errors in the cell cycle lead to cancer in humans?

Cell Division by Meiosis

20. List the differences between mitosis and meiosis in the following chart:

	Mitosis	Meiosis
Number of daughter cells produced		
Number of chromosomes per daughter cell		
Do chromosomes pair? (Y/N)		
Does crossing-over occur? (Y/N)		
Can the daughter cells divide again? (Y/N)		
Do the chromosomes replicate before division? (Y/N)		
Type of cell produced		

21. In the following diagram, designate each daughter cell as diploid (2N) or haploid (N)

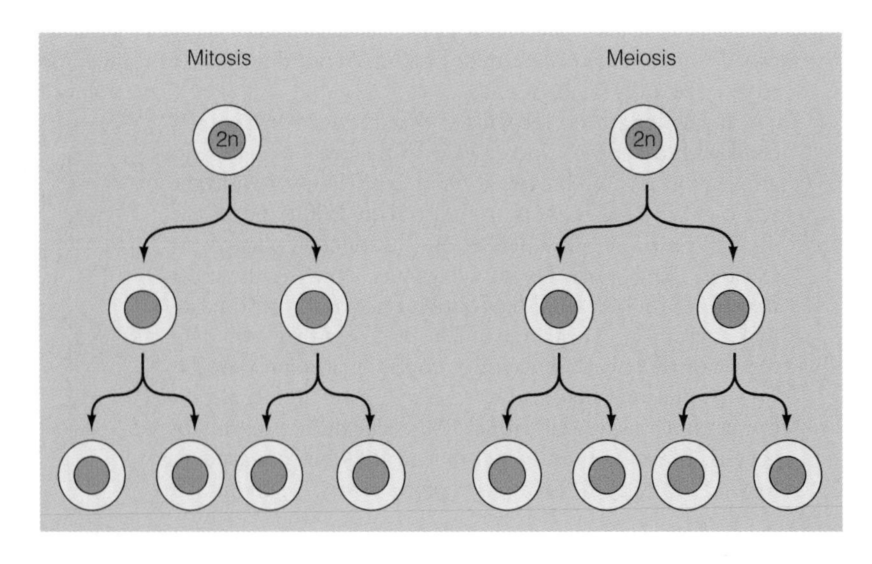

22. Which of the following statements is *not* true when comparing mitosis and meiosis?
    a. Twice the number of cells are produced in meiosis versus mitosis.
    b. Meiosis is involved in the production of gametes, unlike mitosis.
    c. Crossing-over occurs in meiosis I, not in meiosis II or mitosis.
    d. Meiosis and mitosis both produce cells that are genetically identical.
    e. In both mitosis and meiosis, the parental cell is diploid.
23. Match the phase of cell division with the diagrams below. In these cells, $2n = 4$.
    a. anaphase of meiosis I
    b. interphase of mitosis
    c. metaphase of mitosis
    d. metaphase of meiosis I
    e. metaphase of meiosis II

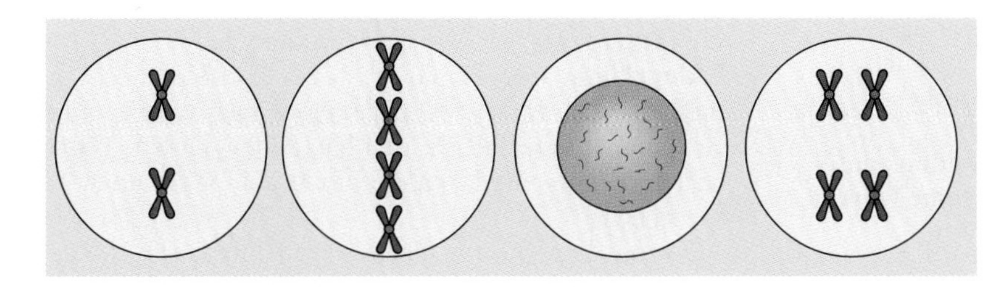

24. A cell has a diploid number of 6 ($2n = 6$).
    a. Draw the cell in metaphase of meiosis I.
    b. Draw the cell in metaphase of mitosis.
    c. How many chromosomes are present in a daughter cell after meiosis I?
    d. How many chromatids are present in a daughter cell after meiosis II?
    e. How many chromosomes are present in a daughter cell after mitosis?
    f. How many tetrads are visible in the cell in metaphase of meiosis I?
25. A cell ($2n = 4$) has undergone cell division. Daughter cells have the following chromosome content. Has this cell undergone mitosis, meiosis I, or meiosis II?

a.

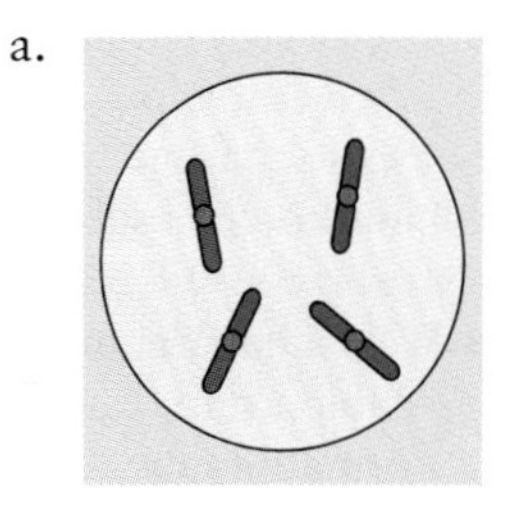

b.

c.

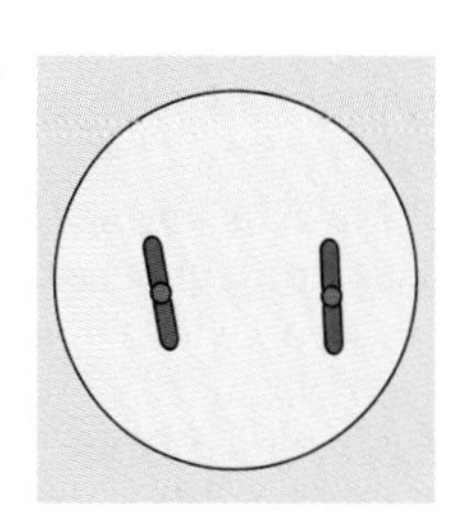

26. We are following the progress of human chromosome 1 during meiosis. At the end of synapsis, how many chromosomes, chromatids, and centromeres are present to ensure that chromosome 1 faithfully traverses meiosis?
27. What is the physical structure that is associated with crossing-over?
28. Compare meiotic anaphase I with meiotic anaphase II. Which meiotic anaphase is more similar to the mitotic anaphase?
29. Discuss and compare the products of meiosis in human females and males. How many functional gametes are produced from the daughter cells in each sex?
30. Provide two reasons why meiosis leads to genetic variation in diploid organisms.
31. A human female is conceived on April 1, 1949, and is born on January 1, 1950. Onset of puberty occurs on January 1, 1962. She conceives a child on July 1, 1994. How long did it take for the ovum that was fertilized on July 1, 1994, to complete meiosis?

Internet Activities

The following activities use the resources of the World Wide Web to enhance the topics covered in this chapter. To investigate the topics described below, log on to the book's home page at: www.brookscole.com/biology_d, and follow the "Student Resources" link to your subject and text.

1. *Structure and Function of the Nucleus.* The *Cell Biology Topics* Web site maintained by the University of Texas (http://cellbio.utmb.edu/cellbio_menu.htm) presents basic information about cell biology arranged by organelle system.
   a. Choose the "Nucleus" link (http://cellbio.utmb.edu/cellbio/nucleus.htm) and explore the numerous structures within the nucleus. Which structures and molecules always remain within the nucleus? Which ones are transported out of the nucleus? Take a closer look at the "nuclear pore" (http://cellbio.utmb.edu/cellbio/nuclear_envelope.htm#Pore). What types of molecules must be entering and exiting through these pores? How might the pores regulate the movement of molecules into and out of the nucleus?
   b. Within the "Nucleus" topic (http://cellbio.utmb.edu/cellbio/nucleus.htm), choose the "chromosome" link (http://cellbio.utmb.edu/cellbio/nucleus2.htm) to compare heterochromatin and euchromatin, the two different forms of DNA in the nucleus. How are these two forms different structurally? How might the structural differences contribute to the functional differences—for example, the transcriptional inactivity of heterochromatin?
2. *Diversity of Cell Types.* The cellular world is almost unimaginably diverse, and modern technology has not only permitted new ways of viewing this diversity, it has made it possible to share this information worldwide. At the *Molecular Expressions Photo Gallery* (http://micromagnet.fsu.edu/micro/gallery.html#top), check out any of the "Galleries" on the contents page to view a variety of cells, organisms, cellular structures, and (occasionally) everyday objects photographed using a variety of different photomicrographic techniques. For an overview of different types of cellular structure (with colorful line drawings but no photomicrographs), follow the "Cell and Virus Structure" link (http://micro.magnet.fsu.edu/cells/index.html).
3. *Mitosis Overview.* The "Mitosis" link at the *Molecular Expressions Photo Gallery* (http://micro.magnet.fsu.edu/micro/gallery/mitosis/mitosis.html) has both photomicrographs and an interactive tutorial for reviewing the phases of mitosis.
4. *Cell Size—and More Mitosis.* At the *Cells Alive!* Web site (http://www.cellsalive.com), follow the "Cell Biology" link and compare the sizes of different cells at the "How Big is a . . . ?" page. How does the size of a virus compare to the size of a bacterium? How big is a bacterium compared to a red blood cell, or a human sperm cell?

   Further Exploration: There is also a mitosis tutorial at the *Cells Alive!* Web site (http://www.cellsalive.com). How does the mitosis tutorial at this site compare to the mitosis overview at the *Molecular Expressions Photo Gallery*? Do you think one site would be more useful for some learning styles (see Chapter 1's Internet activities for information on learning styles) but not for others? What are the strong points (and the weak points) of each presentation?

For Further Reading

Bretscher, M. (1985). Molecules of the cell membrane. *Sci. Am., 253,* 100–109.

Cross, P. C. (1993). *Cell and tissue ultrastructure: A functional perspective.* Upper Saddle River, NJ: Prentice Hall.

Crow, E. W., & Crow J. F. (2002). 100 years ago: Walter Sutton and the chromosome theory of heredity. *Genetics, 160,* 1–4.

Gosden, R., Krapez, J., & Briggs, D. (1997). Growth and development of the mammalian oocyte. *Bioessays, 10,* 875–882.

Kessel, R., & Shih, C. (1974). *Scanning electron microscopy in biology: A students atlas of biological organization.* New York: Springer-Verlag.

Russell, P. (1998). Checkpoints on the road to mitosis. *Trends Biochem. Sci., 10,* 399–402.

Therman, E., & Susman, M. (1993). *Human chromosomes: Structure, behavior, effects* (3rd ed.). New York: Springer-Verlag.

Travis, J. (1996). What's in the vault? *Sci. News, 150,* 56–57.

Wagner, R. P, Maguire, M. P., & Stallings, R. L. (1993). *Chromosomes: A synthesis.* New York: Wiley-Liss.

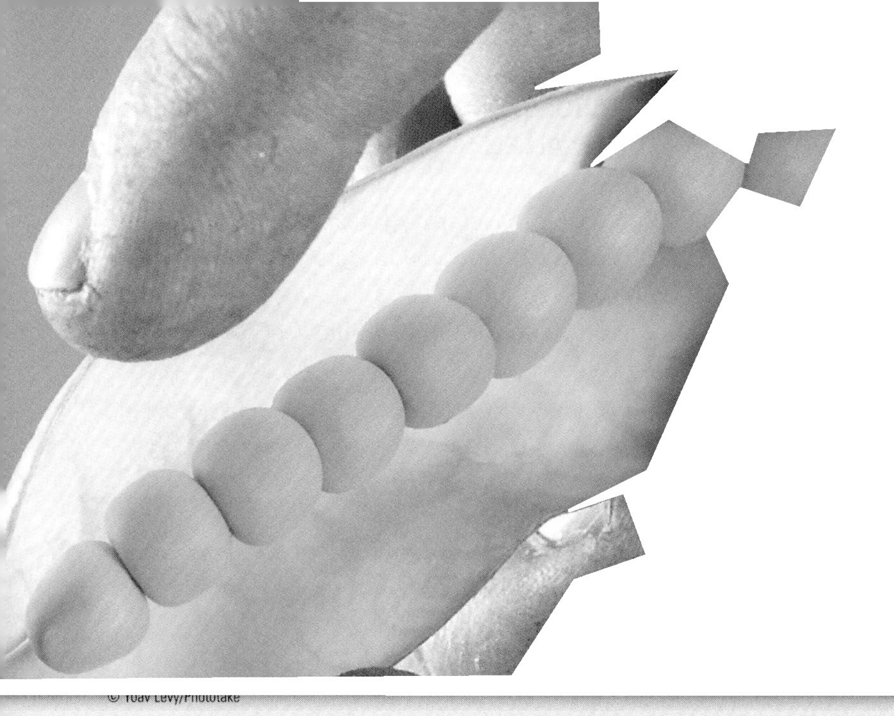

© Yoav Levy/Phototake

Transmission of Genes from Generation to Generation

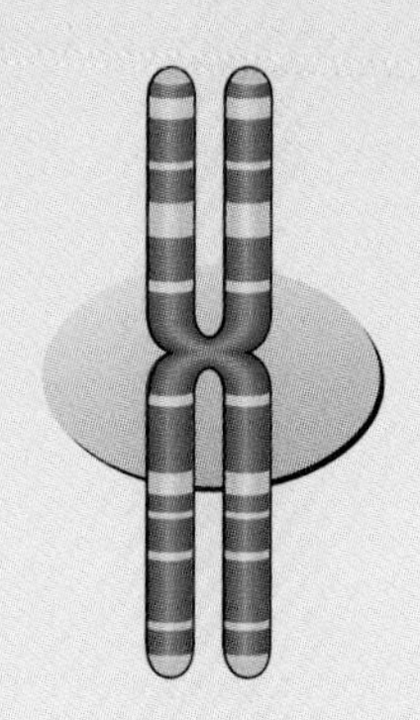

Chapter Outline

Charles Darwin was not the only member of his family to make significant contributions to our understanding of genetics and evolution. Darwin's cousin, Francis Galton, was concerned that Darwin's account of evolution did not explain how small variations in appearance were generated and set out to experimentally test Darwin's idea about the way traits are inherited.

Darwin originally subscribed to the generally accepted idea of the time that parental traits are blended in the offspring. According to the blending idea, if a plant with red flowers is crossed to a plant with white flowers, the offspring should have pink flowers (a blend of red and white). But Darwin realized that if traits were blended generation after generation, variation would be reduced, not increased. To get around this problem, Darwin decided to support a second idea about the way traits are inherited. This idea was really the ancient theory of pangenesis presented in a slightly different form. According to this concept, instructions to form structures of the body are contained in particles called "gemmules." These particles, formed in various parts of the body, move through the blood to the reproductive organs and are transmitted from there to the offspring. Gemmules from each parent make a contribution to the traits expressed in the offspring.

Galton decided to put the idea of gemmules to a test. He used rabbits that had different coat colors and transfused blood between them. He thought that a transfusion should mix the gemmules from the two rabbits, changing the coat color in the offspring. Blood from black-coated rabbits was transfused into rabbits with white coats, and these white rabbits were bred to each other. If the gemmules from the black rabbits mixed with the gemmules in the white rabbits, then crossing the transfused white rabbits with each other should produce at least some offspring with gray coats (a blend of black and white).

The results of Galton's experiments did not support the idea of pangenesis. No mixing of fur color occurred when transfused rabbits were interbred. He presented his results to the Royal Society, the highest scientific body in England, on March 30, 1871, and said: "The conclusion from this large series of experiments is not to be avoided, that the doctrine of pangenesis, pure and simple, as I have interpreted it, is incorrect." The report showed that traits were not transmitted by pangenesis, but it left unanswered the question of how traits are inherited.

Galton went on to make other significant contributions to the study of inheritance. He set up the mathematical basis for studying traits controlled by several genes and pointed out the importance of twin studies in human genetics.

Unfortunately, it appears that neither Galton nor Darwin read the work of Gregor Mendel on the inheritance of traits in the garden pea published in 1866.

This work, titled "Experiments in Plant Hybrids," is one of the most important scientific papers ever published. Here Mendel departs from the usual reporting of observations and takes the additional step of fitting his experimental results into a conceptual framework that can be used to explain the mechanism of heredity in any organism, not just the garden pea. In this chapter, we reconstruct the experiments of Mendel and show how he moved from recording his results to drawing conclusions about the principles of heredity.

HEREDITY: HOW ARE TRAITS INHERITED?

Johann Gregor Mendel was born in 1822 in Hynice, Moravia, a region now part of the Czech Republic. He showed great promise as a student, but poverty prevented him from attending a university. At the age of 21, he entered the Augustinian monastery at Brno as a way of continuing his interests in natural history. After completing his monastic studies, Mendel enrolled at the University of Vienna in the fall of 1851. In his first year he took courses in physics, mathematics, chemistry, and the natural sciences. In his botany courses, Mendel encountered the new concepts that all organisms are composed of cells and that cells are the fundamental unit of all living things. The cell theory raised several new questions and many controversial issues. One was the question of whether both parents contribute equally to the traits of the offspring. Because the female gametes in most plants and animals are so much larger than those of the male, this was a logical and much debated question. Related to this was the question of whether the traits present in the offspring result from blending the traits of the parents. If so, did the female contribute 40%, 50%, 60%, or more of the traits? In 1854 Mendel returned to Brno to teach physics and began a series of experiments that were to resolve these questions.

MENDEL'S EXPERIMENTAL DESIGN RESOLVED MANY UNANSWERED QUESTIONS

Mendel's success in explaining the mechanisms of inheritance was the result of carefully planned experiments. Having determined what he wanted to investigate, Mendel set about choosing an organism for his experiments. Near the beginning of his landmark paper on inheritance, Mendel wrote:

> The value and validity of any experiment are determined by the suitability of the means as well as by the way they are applied. In the present case as well, it cannot be unimportant which plant species were chosen for the experiments and how these were carried out. Selection of the plant group for experiments of this kind must be made with the greatest possible care if one does not want to jeopardize all possibility of success from the very outset.

Then he listed the properties that an experimental organism should have. First, it should have a number of different traits that can be studied. Second, the plant should be self-fertilizing and have a flower structure that minimizes accidental pollination. Third, the offspring of self-fertilized plants should be fully fertile so that further crosses can be made.

He paid particular attention to pea plants and their relatives because the flower structure of these species minimizes accidental pollination. He found that 34 varieties of pea plants with different traits were available from seed dealers. The plant had a relatively short growth period, could be grown in the ground or in pots in the greenhouse, and could be self-fertilized or artificially fertilized by hand (Figure 3.1).

Mendel then tested all 34 varieties of peas for 2 years to ensure that the traits they carried were true-breeding, that is, that self-fertilization gave rise to the same traits in

SIDEBAR

Mendel and Test Anxiety

Mendel entered the Augustinian monastery in 1843 and took the name Gregor. While studying at the monastery, he served as a teacher at the local technical high school. In the summer of 1850 he decided to take the examinations that would allow him to have a permanent appointment as a teacher. The exam was in three parts. Mendel passed the first two parts, but failed one of the sections in the third part. In the fall of 1851, he enrolled at the University of Vienna to study natural science (the section of the exam he flunked). He finished his studies in the fall of 1853, returned to the monastery, and again taught at a local high school.

In 1855 he applied to take the teacher's examination again. The test was held in May of 1856, and Mendel became ill while answering the first question on the first essay examination. He left and never took another examination. As a schoolboy and again as a student at the monastery, Mendel had also suffered bouts of illness, all associated with times of stress.

In an analysis of Mendel's illnesses made in the early 1960s, a physician concluded that Mendel had a psychological condition that today would probably be called "test anxiety." If you are feeling stressed at exam time, take some small measure of comfort in knowing that it was probably worse for Mendel.

▶ **FIGURE 3.1** The study of the way traits such as flower color in pea plants and pod shape are passed from generation to generation provided the material for Mendel's work on heredity.

▶ **FIGURE 3.2** The monastery garden where Mendel carried out his experiments in plant genetics.

all offspring, generation after generation. From these, he selected 22 varieties to plant every year for the next 8 years for his experiments (▶ Figure 3.2). For his experiments, he studied seven characters that affected the seeds, pods, flowers, and stems of the plant (Table 3.1). Each character was represented by two distinct forms: plant height by tall and short plants, seed shape by wrinkled and smooth peas, and so forth.

To avoid errors caused by small sample sizes, he planned experiments on a large scale. Over the next 8 years, Mendel used some 28,000 pea plants in his experiments. In his work, he kept track of each character separately over several generations. He began by studying one pair of traits at a time and repeated his experiments for each trait to confirm his results. Using his training in physics and mathematics, Mendel analyzed his data according to the principles of probability and statistics. His methodical and thorough approach to his work and his lack of preconceived notions were the secrets of his success.

Table 3.1 Traits Selected for Study by Mendel

Structure Studied	Dominant	Recessive
SEEDS		
Shape	Smooth	Wrinkled
Color	Yellow	Green
Seed coat color	Gray	White
PODS		
Shape	Full	Constricted
Color	Green	Yellow
FLOWERS		
Placement	Axial (along stems)	Terminal (top of stems)
STEMS		
Length	Long	Short

CROSSING PEA PLANTS: MENDEL'S STUDY OF SINGLE TRAITS

To show how Mendel developed his ideas about inheritance, we will first describe some of his experiments and the results he obtained. Then we will follow the reasoning Mendel used in reaching his conclusions and outline some of the further experiments that confirmed his ideas.

In his first set of experiments, Mendel studied the inheritance of seed shape. Because this involved only one character (seed shape), he called it a **monohybrid cross**. He took plants with smooth seeds and crossed them to plants with wrinkled seeds. In making this cross, flowers from one variety were fertilized using pollen from the other variety. In this first experiment, Mendel performed 60 fertilizations on 15 plants. The seeds that formed as a result of these fertilizations were all smooth. This result was true whether the pollen was from a plant with smooth peas or a plant with wrinkled peas. The next year, Mendel planted the smooth seeds from this cross. When the plants matured, the flowers were self-fertilized, and 7,324 seeds were collected. Of these, 5,474 were smooth and 1,850 were wrinkled.

Monohybrid cross A cross between individuals that differ with respect to a single gene pair.

P1 : Smooth × wrinkled
F1 : All Smooth
F2 : 5,474 Smooth and 1,850 wrinkled

In his experiments, Mendel called the parental generation P1 and the offspring he termed the F1 (first filial) generation. The second generation, produced by self-fertilizing the F1 plants, was called the F2 (or second filial) generation. His experiments with seed shape are summarized in Figure 3.3.

Results and Conclusions from Mendel's First Series of Crosses

The results from experiments with all seven characters were similar to those Mendel observed with smooth and wrinkled seeds (Figure 3.4). In all crosses, the following results were obtained:

- The F1 offspring showed only one of the two parental traits and always the same trait.
- In all crosses, it did not matter which plant the pollen came from. The results were always the same.

SIDEBAR

Why Wrinkled Peas Are Wrinkled

Wrinkled peas were one variety used by Mendel in his experiments. At the time, nothing was known about how peas became wrinkled or smooth. All Mendel needed to know was that a factor controls seed shape and that it had two forms—a dominant one for smooth shape and a recessive one for wrinkled shape.

Recently, scientists have discovered how peas become wrinkled, providing a connection between a gene and its phenotype. While the pea is developing, starch is synthesized and stored as a food source. Starch is a large, branched molecule made up of sugar molecules, and the ability to form branches is controlled by a gene. Normally, starch molecules are highly branched structures. This allows more sugar to be stored in each molecule. In peas that have the wrinkled genotype, the branching gene is inactive. Thus, the developing pea converts sugar into starch very slowly, and excess sugar accumulates. The excess sugar causes the pea to take up large amounts of water, and the seed swells. In a final stage of development, water is lost from the seed. In homozygous wrinkled peas, more water is lost than in the smooth seeds, causing the outer shell of the pea to become wrinkled.

Mendel's contribution was to show that a specific factor controlled a trait and that a given gene could have different forms. Now we know that genes exert their effect on phenotype through the production of a gene product.

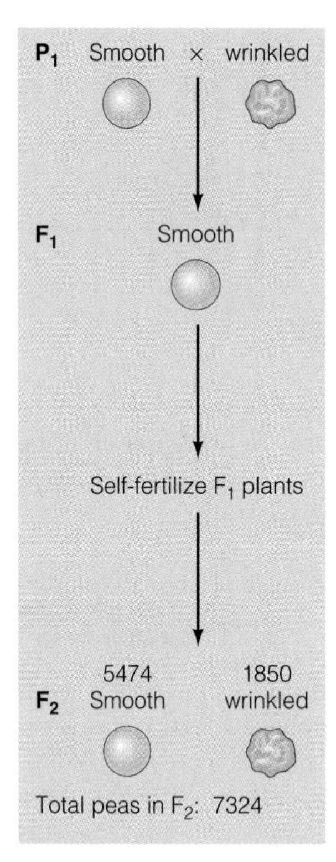

FIGURE 3.3 One of Mendel's crosses. Pure-breeding varieties of peas (smooth and wrinkled) were used as the P1 generation. All the offspring in the F1 generation had smooth seeds. Self-fertilization of F1 plants gave rise to both smooth and wrinkled progeny in the F2 generation. About three-fourths of the offspring were smooth and about one-fourth were wrinkled.

Trait Studied	Results in F_2	
Seed shape	5474 smooth	1850 wrinkled
Seed color	6022 yellow	2001 green
Seed coat color	705 gray	224 white
Pod shape	882 inflated	299 constricted
Pod color	428 green	152 yellow
Flower position	651 along stem	207 at tip
Stem length	787 tall	277 dwarf

FIGURE 3.4 Results of Mendel's monohybrid crosses in peas. The numbers represent the F2 plants showing a given trait. On average, three-fourths of the offspring showed one trait, and one-fourth showed the other (a 3:1 ratio).

- The trait not shown in the F1 offspring reappeared in about 25% of the F2 offspring.

The results from these crosses were Mendel's first discoveries. His experiments showed that traits remained unchanged as they passed from parent to offspring. Traits did not blend together in any of the offspring. Although traits might not be expressed, they remained unchanged from generation to generation. This convinced him that inheritance did not work by blending the traits of the parents in the offspring. Instead, he concluded that traits were inherited as if they were separate units that did not blend together.

In his experiments, Mendel made reciprocal crosses, so that the variety used as a male parent in one set of experiments was used as the female parent in the next set of crosses. In all cases, it did not matter whether the male or female plant had smooth or wrinkled seeds; the results were the same. From these experiments he concluded that each parent makes an equal contribution to the genetic makeup of the offspring.

Based on the results of his crosses with each of the seven characters, Mendel came to several conclusions:

- **Genes** (Mendel called them factors) that determine traits can be hidden or unexpressed. For example, in a cross between plants with smooth seeds and plants with wrinkled seeds, all the F1 seeds were smooth. When these seeds were grown and self-fertilized, the next generation of plants (the F2) had some wrinkled seeds. This means that the F1 seeds contained a gene for wrinkled that was present but not expressed. He called the trait that is not expressed in the F1 but is expressed in the F2 plants a **recessive trait.** The trait present in F1 plants he called the **dominant trait.** Mendel called this phenomenon dominance.
- By comparing the offspring of the P1 plants with smooth seeds and the F1 plants with smooth seeds, Mendel concluded that despite their identical appearances, the P1 and F1 plants must be genetically different. When P1 plants are self-fertilized, all the plants in the next generation have only smooth seeds. But when F1 plants are self-fertilized, plants in the F2 have both smooth and wrinkled seeds. Mendel realized that it was important to make a distinction between the appearance of an organism and its genetic constitution. We now use the term **phenotype** to describe the trait we can see and the term **genotype** to describe the genetic makeup of an organism. In our example, the P1 and F1 plants that have smooth seeds have identical phenotypes, but must have different genotypes.
- The results of these self-fertilization experiments show that the F1 plants must carry genes for smooth and wrinkled traits because both types of seeds are present in the F2 generation. The question is, how many genes for seed shape are carried in the F1 plants? Mendel had already reasoned that the male and female parent contributed equally to the traits of the offspring. The simplest explanation is that each F1 plant carried two genes, one for smooth that was expressed, and one for wrinkled that remained unexpressed (see Concepts and Controversies: Ockham's Razor). By extension of this reasoning, each P1 and F2 plant must also contain two genes for seed shape. To symbolize genes, uppercase letters are used to represent genes with a dominant pattern of inheritance, and lowercase letters are used to represent those with a recessive pattern of inheritance (S = smooth, and s = wrinkled). Using this shorthand, we can reconstruct the genotypes and phenotypes of the P1 and F1 shown in Figure 3.5.

Genes The fundamental units of heredity.

Recessive trait The trait unexpressed in the F1 but reexpressed in some members of the F2 generation.

Dominant trait The trait expressed in the F1 (or heterozygous) condition.

Phenotype The observable properties of an organism.

Genotype The specific genetic constitution of an organism.

How a Single Trait Is Inherited: The Principle of Segregation

If genes exist in pairs, there must be some way to prevent them from doubling in each succeeding generation. (If each parent has two genes for a given trait, why doesn't the offspring have four?) Mendel reasoned that members of a gene pair must separate or segregate from each other during gamete formation. As a result, each gamete receives only one of the two genes that control a given trait. The separation of members of a gene pair during gamete formation is called the principle of **segregation,** or Mendel's First Law.

Segregation The separation of members of a gene pair from each other during gamete formation.

Figure 3.6 diagrams the separation of a gene pair so that only one member of a pair is included in each gamete. In our example, each member of the F1 generation can make two kinds of gametes in equal proportions (S gametes and s gametes). At fertilization, the random combination of these gametes produces the genotypic combinations shown in the Punnett square (a method for analyzing genetic crosses devised by R. C. Punnett). The F2 has a genotypic ratio of 1 SS:2 Ss:1 ss and a phenotypic ratio of 3 smooth:1 wrinkled (dominant to recessive).

▶ **FIGURE 3.5** The phenotypes and genotypes of the parents and offspring in Mendel's cross involving seed shape.

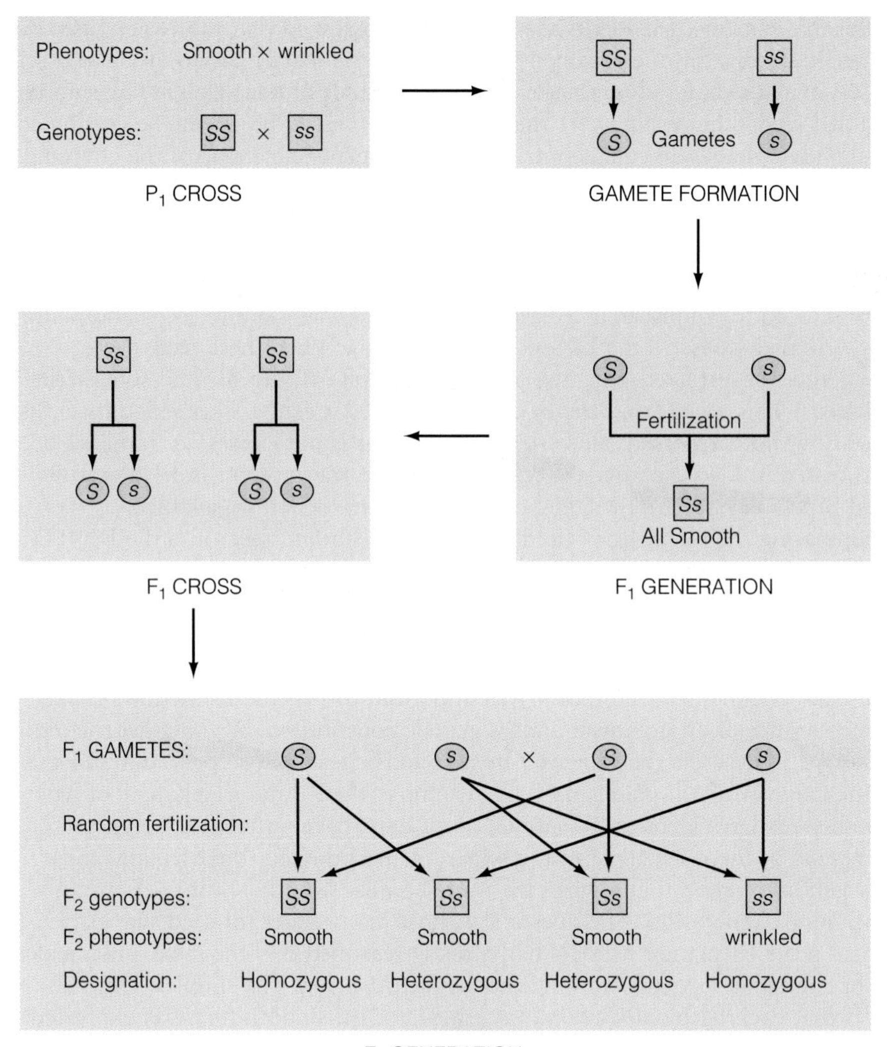

Mendel's reasoning allows us to predict the genotypes of the F2 generation. One fourth of the F2 plants should carry only the genes for smooth seeds (*SS*), and, when self-fertilized, all the plants will have smooth seeds. Half of the F2 plants should carry genes for both smooth and wrinkled (*Ss*) and give rise to smooth and wrinkled plants in a 3:1 ratio when self-fertilized to produce an F3 generation (▶ Figure 3.7). Finally, one fourth of the F2 plants should be homozygous for wrinkled (*ss*) and have all wrinkled progeny if self-fertilized. In fact, Mendel fertilized a number of plants from the F2 generation and five succeeding generations to confirm these predictions.

Mendel carried out his experiments before the discovery of mitosis and meiosis and before the discovery of chromosomes. As we discuss in a later section, his conclusions about how traits are inherited are, in fact, descriptions of the way chromosomes behave in meiosis. Seen in this light, his discoveries are all the more remarkable.

Today, we call Mendel's factors genes and refer to the alternate forms of a gene as **alleles**. In the example we have been discussing, the gene for seed shape (*S*) has two alleles, smooth (*S*) and wrinkled (*s*). Individuals that carry identical alleles of a given gene (*SS* or *ss*) are **homozygous** for the gene in question. Similarly, when two different alleles are present in a gene pair (*Ss*), the individual has a **heterozygous** genotype. The *SS* homozygotes and the *Ss* heterozygotes show the dominant smooth phenotype (because *S* is dominant to *s*), and *ss* homozygotes show the recessive, wrinkled phenotype.

Allele One of the possible alternative forms of a gene, usually distinguished from other alleles by its phenotypic effects.

Homozygous Having identical alleles for one or more genes.

Heterozygous Carrying two different alleles for one or more genes.

Concepts and Controversies

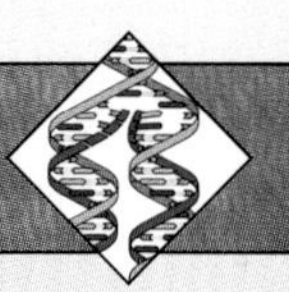

Ockham's Razor

When Mendel proposed the simplest explanation for the number of factors contained in the F1 plants in his monohybrid crosses, he was using a principle of scientific reasoning known as parsimony, or Ockham's razor.

William of Ockham (also spelled Occam) was a Franciscan monk and scholastic philosopher who lived from about 1300 to 1349. He had a strong interest in the study of thought processes and in logical methods. He is the author of the maxim known as Ockham's razor: "Pluralites non est pondera sine necessitate," which translates from the Latin as "Entities must not be multiplied without necessity." In the study of philosophy and theology of the Middle Ages, this was taken to mean that when constructing an argument, you should never go beyond the simplest argument unless it is necessary. Although Ockham was not the first to use this approach, he used this tool of logic so well and so often to dissect the arguments of his opponents that it became known as Ockham's razor.

The principle was transferred to scientific hypotheses in the 15th century. Galileo used the principle of parsimony to argue that because his model of the solar system was the simplest, it was probably correct (he was right). In modern terms, the phrase is taken to mean that in proposing a mechanism or hypothesis, use the least number of steps possible. The simplest mechanism is not necessarily correct, but it is usually the easiest to prove or disprove by doing experiments and the most likely to produce scientific progress.

For a given trait, Mendel concluded that both parents contribute an equal number of factors to the offspring. In this case, the simplest assumption is that each parent contributed one such factor and that the F1 offspring contained two such factors. Further experiments proved this conclusion correct.

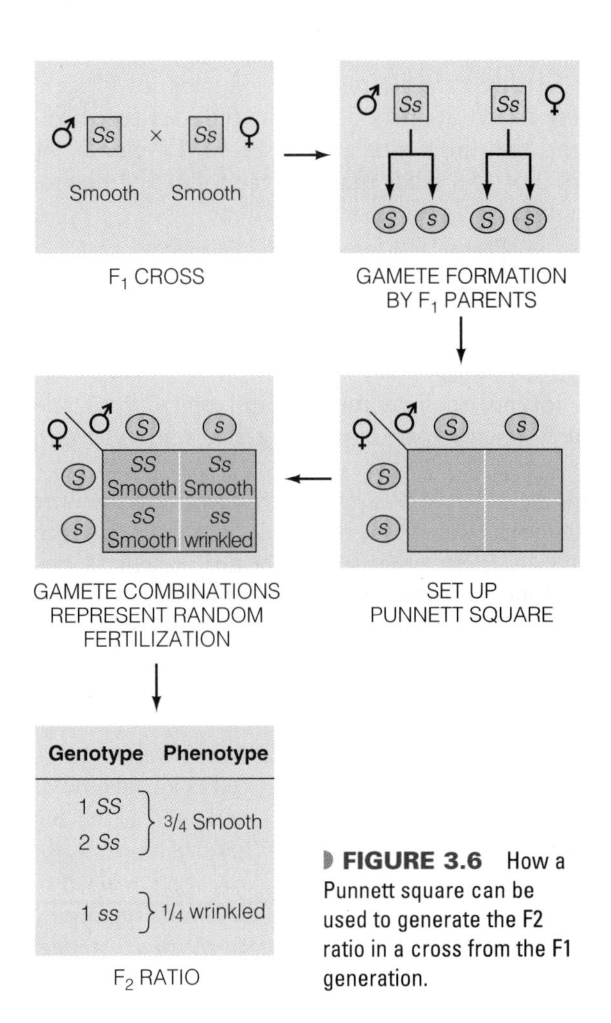

FIGURE 3.6 How a Punnett square can be used to generate the F2 ratio in a cross from the F1 generation.

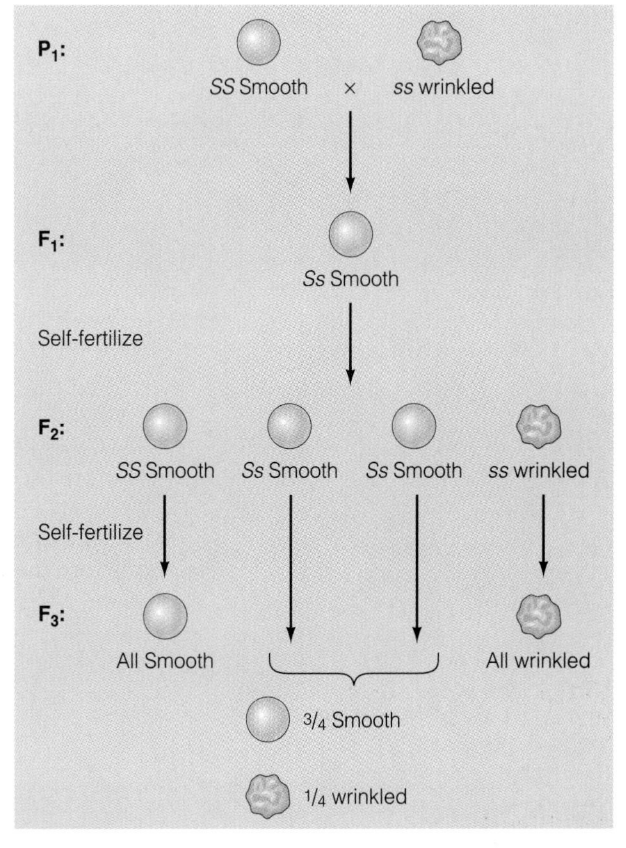

FIGURE 3.7 Self-crossing the F2 plants demonstrates that there are two different genotypes among the plants with smooth peas in the F2 generation.

MORE CROSSES WITH PEA PLANTS: THE PRINCIPLE OF INDEPENDENT ASSORTMENT

Mendel realized the need to extend his studies on the inheritance from monohybrid crosses to more complex situations. He wrote:

> In the experiments discussed above, plants were used which differed in only one essential trait. The next task consisted in investigating whether the law of development thus found would also apply to a pair of differing traits.

For this work, he selected seed shape and seed color as traits to be studied, because, as he put it, "Experiments with seed traits lead most easily and assuredly to success." A cross that involves two sets of characters is called a dihybrid cross.

Crosses Involving Two Traits

As in the first set of crosses, we will analyze the actual experiments of Mendel, outline his results, and summarize the conclusions he drew from them. From previous crosses, Mendel knew that for seeds, smooth is dominant to wrinkled and yellow is dominant to green. In our reconstruction of these experiments we will use the following symbols: smooth (*S*), wrinkled (*s*), yellow (*Y*), and green (*y*).

Analyzing the Results and Drawing Conclusions

Mendel selected true-breeding plants with smooth, yellow seeds and crossed them with true-breeding plants with wrinkled, green seeds (Figure 3.8). In the F1 plants, the seeds were all smooth and yellow, confirming that smooth and yellow are dominant traits. Then, Mendel self-fertilized the F1 plants, producing an F2 generation. These F2 plants produced four types of seeds, often found together in a single pod. From 15 plants, he counted a total of 556 seeds that had the following phenotypic distributions:

315 smooth and yellow
108 smooth and green
101 wrinkled and yellow
32 wrinkled and green

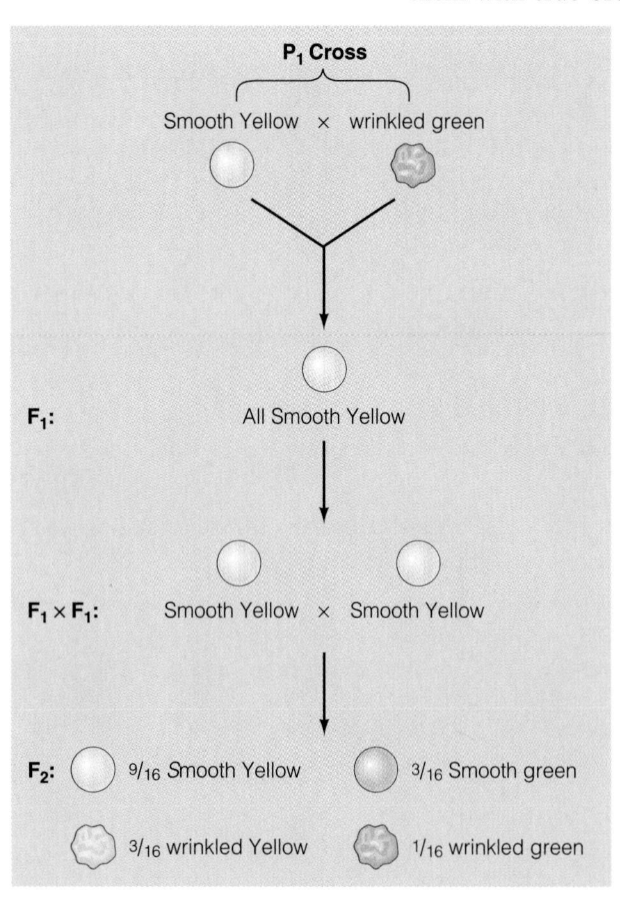

FIGURE 3.8 The phenotypic distribution in a dihybrid cross. Plants in the F2 generation show the parental phenotypes and two new phenotypic combinations.

The F2 phenotypes include the parental phenotypes and also two new phenotypes (smooth green and wrinkled yellow).

To determine how the two genes in a dihybrid cross were inherited, Mendel first analyzed the results of the F2 for each trait separately, as if the other trait were not present (Figure 3.9). If we look at seed shape (smooth or wrinkled) and ignore seed color, we expect three-fourths smooth and one-fourth wrinkled offspring in the F2. Analyzing the results, we find that the total number of smooth offspring is 315 + 108 = 423. The total number of wrinkled seeds is 101 + 32 = 133. The proportion of smooth to wrinkled seeds (423:133) is very close to a 3:1 ratio. Similarly, if we consider only seed color (yellow or green), there are 416 yellow seeds (315 + 101) and 140 green seeds (108 + 32) in the F2 generation. These results are also close to a 3:1 ratio.

Once he established a 3:1 ratio for each trait separately (consistent with the principle of segregation), then Mendel considered the inheritance of both traits simultaneously.

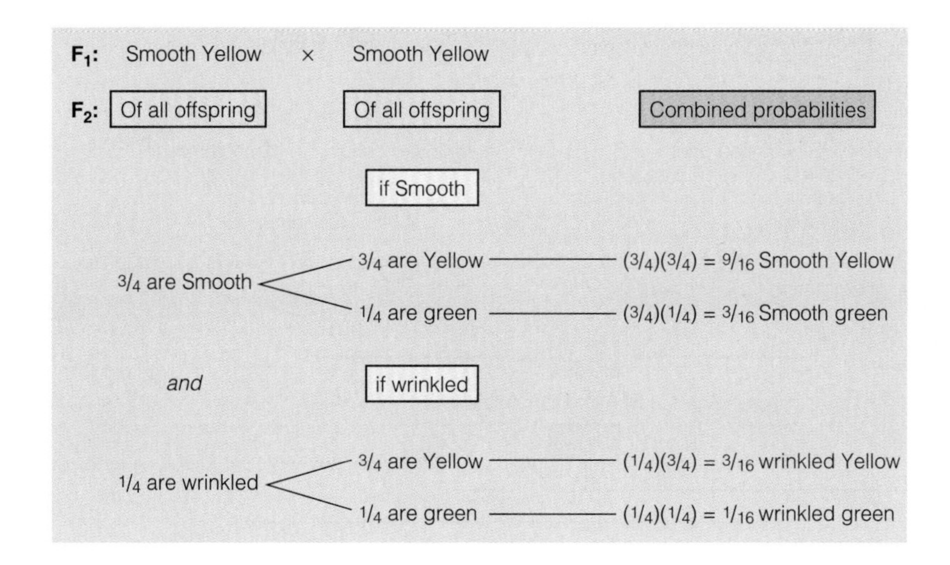

FIGURE 3.9 Analysis of a dihybrid cross for the separate inheritance of each trait.

The Inheritance of Two Traits: The Principle of Independent Assortment

Before we discuss what is meant by independent assortment, let's see how the phenotypes and genotypes of the F1 and F2 were generated. The F1 plants with smooth yellow seeds were heterozygous for both seed shape and seed color. This means that the genotype of the F1 plant must have been *SsYy*, and the *S* and *Y* alleles were dominant to *s* and *y*. Mendel had already concluded that members of a gene pair separate or segregate from each other during gamete formation. In this case, the segregation of the *S* and *s* alleles occurred independently from the segregation of the *Y* and *y* alleles (Figure 3.10).

Because each gene pair segregated independently, the gametes formed by the F1 plants contained all combinations of these alleles in equal proportions: *SY*, *Sy*, *sY*, and *sy*. If fertilizations involving the four types of male and female gametes occurred at random (as expected), 16 possible combinations would result (Figure 3.10).

An inspection of the 16 combinations in the Punnett square demonstrates the following:

- Nine have at least one copy of the dominant alleles *S* and *Y*.
- Three have at least one copy of the dominant allele *S* and are homozygous for *yy*.
- Three have at least one copy of the dominant allele *Y* and are homozygous for *ss*.
- One combination is homozygous for *ss* and *yy*.

In other words, the 16 combinations of fertilization events (genotypes) fall into four phenotypic classes:

smooth and yellow
smooth and green
wrinkled and yellow
wrinkled and green

These combinations correspond to the number of phenotypic classes in the F2 generation and to the proportions of seeds in each class (Figure 3.10). For example, 315 of 556 seeds were smooth and yellow, corresponding to about nine-sixteenths of the total number of offspring; 108 of 556 seeds were smooth and green, corresponding to about three-sixteenths of the offspring; and so forth. This distribution of offspring in the F2 corresponds to a phenotypic ratio of 9:3:3:1 (see Concepts and Controversies: Evaluating Results—The Chi Square Test).

FIGURE 3.10 Punnett square of the dihybrid cross shown in Figure 3.9. There are two combinations of dominant and recessive traits that result in heterozygous F1 individuals. One is a cross between smooth, yellow and wrinkled, green. The other is a cross between wrinkled, yellow and smooth, green.

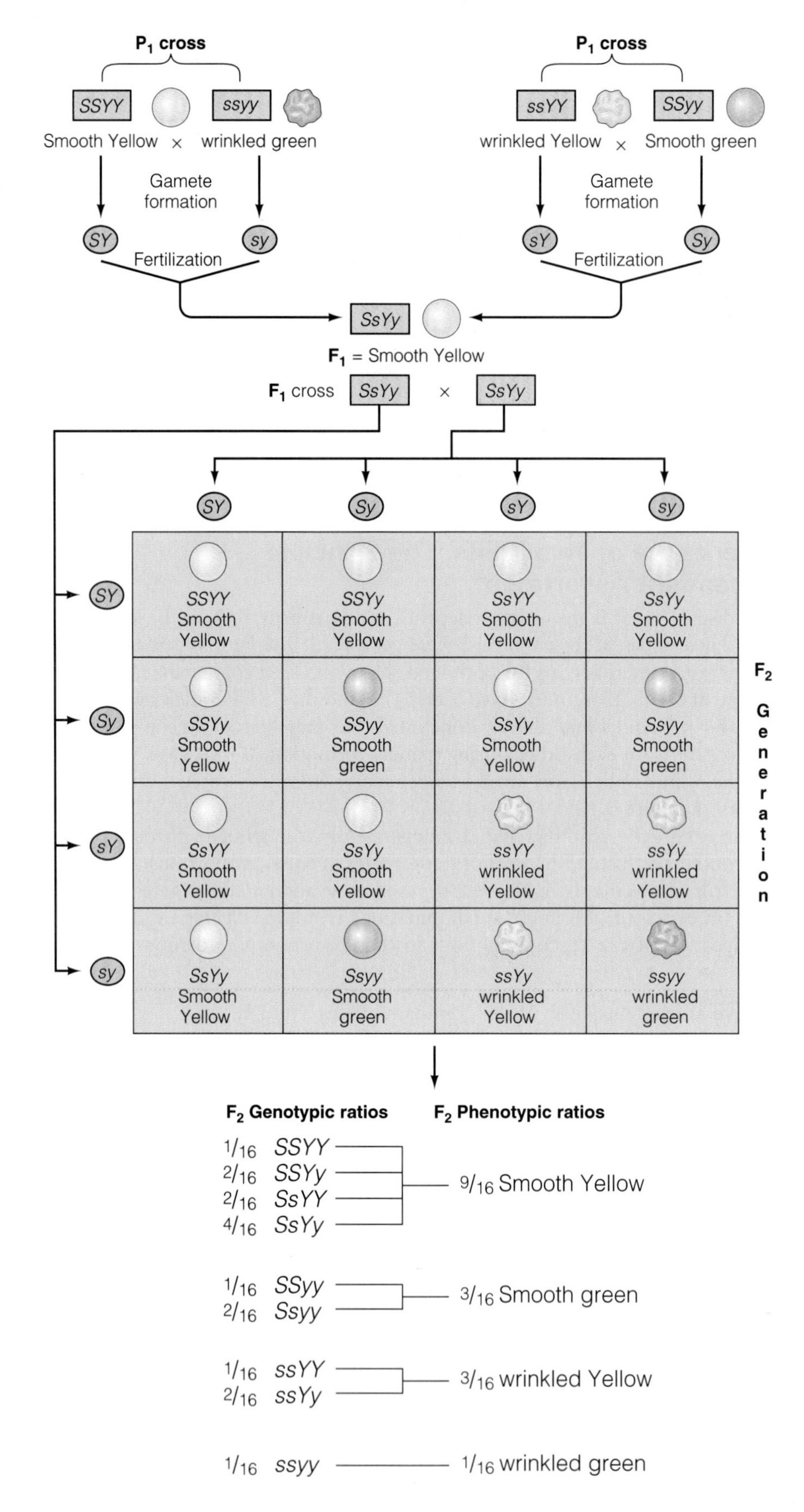

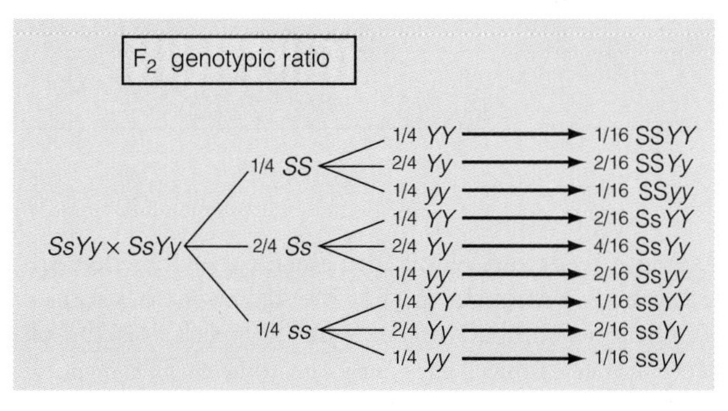

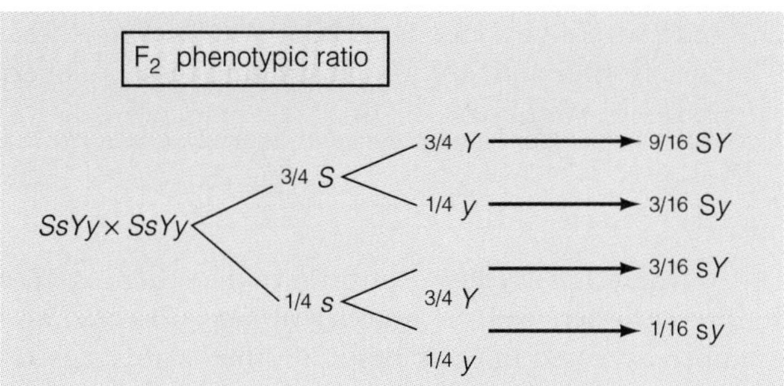

FIGURE 3.11 The phenotypic and genotypic ratios of a dihybrid cross can be derived using a branched-line method instead of a Punnett square.

SIDEBAR

The Branched-Line Method

Instead of using a Punnett square to determine the distribution and frequency of phenotypes and genotypes in the F2, we can also use a branch diagram, or branched-line method, which is based on probability. In the dihybrid F2, the probability that a plant will be smooth is three in four, and the chance that it will be wrinkled is one in four. Because each trait is inherited independently, each smooth plant has a three-in-four chance of being yellow and a one-in-four chance of being green. The same is true for each wrinkled plant. Figure 3.11 shows how these probabilities combine to give the phenotypic ratios.

The results of this cross can be explained by assuming (as Mendel did) that during gamete formation, alleles of one gene pair segregate into gametes independently of the alleles belonging to other gene pairs, resulting in the production of gametes containing all combinations of alleles. This second fundamental principle of genetics outlined by Mendel is called the principle of **independent assortment**, or Mendel's Second Law.

Independent assortment The random distribution of genes into gametes during meiosis.

After 10 years of work involving thousands of pea plants, Mendel presented his results in 1865 at the February and March meetings of his local Natural Science Society. The text of his lectures was published the following year in the *Proceedings* of the society. Although his work was mentioned in several bibliographies and copies of the journal were widely circulated, the significance of Mendel's findings was unappreciated. Even Charles Darwin, in his search for a mechanism to explain the role of heredity in natural selection, failed to realize the significance of Mendel's work. Darwin's own copy of the *Proceedings* has many notes scribbled in the margin of the paper next to Mendel's, but not one pencil mark anywhere in Mendel's paper.

Finally, in 1900 three scientists, each working independently on the mechanism of heredity, confirmed Mendel's findings and brought his paper to widespread attention. These events stimulated a great interest in the study of what is now called **genetics.** Unfortunately, Mendel died in 1884—unaware that he had founded an entire scientific discipline.

Genetics The scientific study of heredity.

MEIOSIS EXPLAINS MENDEL'S RESULTS: GENES ARE ON CHROMOSOMES

When Mendel was experimenting with pea plants, the behavior of chromosomes in mitosis and meiosis was unknown. By 1900, however, the details of mitosis and meiosis were well known. As scientists confirmed that the fundamentals of Mendelian inheritance operated in many organisms, it soon became apparent that genes and chromosomes had much in common (Table 3.4). Both chromosomes and genes occur in pairs. In meiosis, members of a chromosome pair separate from each other, and members of a gene pair separate from each other during gamete formation. Finally, the fusion of gametes during fertilization restores

Concepts and Controversies

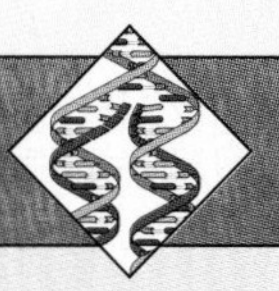

Evaluating Results—The Chi-Square Test

One of Mendel's innovations was the application of mathematics and combinatorial theory to biological research. This allowed him to predict the genotypic and phenotypic ratios in his crosses and to follow the inheritance of several traits simultaneously. If the cross involved two alleles of a gene (e.g., *A* and *a*), the expected outcome was an F2 phenotypic ratio of 3 *A*:1 *a* and a genotypic ratio of 1 *AA*:2 *Aa*:1 *aa*.

What Mendel was unable to analyze mathematically was how well the actual outcome of the cross fulfilled his predictions. He apparently realized this problem and compensated for it by conducting his experiments on a large scale, counting substantial numbers of individuals in each experiment to reduce the chance of error. Shortly after the turn of the twentieth century, an English scientist named Karl Pearson developed a statistical test to determine whether the observed distribution of individuals in phenotypic categories is as predicted or occurs by chance. This simple test, regarded as one of the fundamental advances in statistics, is a valuable tool in genetic research. The method is known as the chi-square (χ^2) test (pronounced "kye square"). In use, this test requires several steps:

1. Record the observed numbers of organisms in each phenotypic class.
2. Calculate the expected values for each phenotypic class based on the predicted ratios.
3. If O is the observed number of organisms in a phenotypic class, or category, and *E* is the expected number, calculate the difference *d* in each category by subtraction: $(O - E) = d$ (Table 3.2).
4. For each phenotypic category, square the difference *d*, and divide by the number expected (*E*) in that phenotypic class.
5. Add all the numbers in step 4 to get the χ^2 value.

If there are no differences between the observed and the expected ratios, the value for χ^2 will be zero. The value of χ^2 increases with the size of the difference between the observed and the expected classes. The formula can be expressed in the general form

$$\chi^2 = \Sigma \frac{d^2}{E}$$

Using this formula, we can do what Mendel could not—analyze his data for the dihybrid cross involving wrinkled and smooth seeds and yellow and green cotyledons that produced a 9:3:3:1 ratio. In the F2, Mendel counted a total of 556 peas. The number in each phenotypic class is the observed number (Table 3.2). Using the total of 556 peas, we can calculate that the expected number in each class for a 9:3:3:1 ratio would be 313:104:104:35 (9/16 of 556 is 313, 3/16 of 556 is 104, and so on). Substituting these numbers into the formula, we obtain

$$\chi^2 = \frac{2^2}{313} + \frac{4^2}{104} + \frac{3^2}{104} + \frac{3^2}{35} = 0.371$$

The χ^2 value is very low, confirming that there is very little difference between the number of peas observed and the number expected in each class. In other words, the results are close enough to the expectation that we need not reject them.

The question remains, however, how much deviation is permitted from the expected numbers before we decide that the observations do not fit our expectation that a 9:3:3:1 ratio will be fulfilled. To decide this we must have a way of interpreting the χ^2 value. We need to convert this value into a probability and to ask: What is the probability that the calculated χ^2 value is acceptable? In making this calculation, first we must establish something called degrees of freedom, *df*, which is one less than the number of phenotypic classes, *n*. In the dihybrid cross, we expect four phenotypic classes, so the degrees of freedom are as follows:

$$df = n - 1$$
$$df = 4 - 1$$
$$df = 3$$

Next we can calculate the probability of obtaining the given set of results by consulting a probability chart (Table 3.3). First, find the line corresponding to a *df* value of 3. Look across on this line for the number corresponding to the χ^2 value. The calculated value is 0.37, which is between the columns headed 0.95 and 0.90. This means that we can expect a deviation of this magnitude at least 90% of the time when we do this experiment. In other words, we can be confident that our expectations of a 9:3:3:1 ratio are correct. In general, a *p* value of less than .05 means that the observations do not fit the expected distribution into phenotypic classes and that the expectation needs to be reexamined. The acceptable range of values is

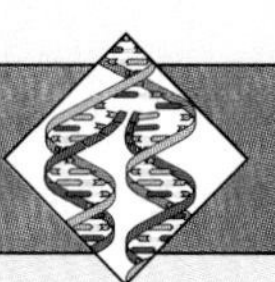

Table 3.2 Chi-Square Analysis of Mendel's Data

Seed Shape	Cotyledon Color	Observed Numbers	Expected Numbers (Based on a 9:3:3:1 Ratio)	Difference (d) ($O - E$)
Smooth	Yellow	315	313	+2
Smooth	Green	108	104	+4
Wrinkled	Yellow	101	104	−3
Wrinkled	Green	32	35	−3

Table 3.3 Probability Values for Chi-Square Analysis

	Probabilities								
df	0.95	0.90	0.70	0.50	0.30	0.20	0.10	0.05	0.01
1	.004	.016	.15	.46	1.07	1.64	2.71	3.84	6.64
2	.10	.21	.71	1.39	2.41	3.22	4.61	5.99	9.21
3	.35	.58	1.42	2.37	3.67	4.64	6.25	7.82	11.35
4	.71	1.06	2.20	3.36	4.88	5.99	7.78	9.49	13.28
5	1.15	1.61	3.00	4.35	6.06	7.29	9.24	11.07	15.09
6	1.64	2.20	3.83	5.35	7.23	8.56	10.65	12.59	16.81
7	2.17	2.83	4.67	6.35	8.38	9.80	12.02	14.07	18.48
8	2.73	3.49	5.53	7.34	9.52	11.03	13.36	15.51	20.09
9	3.33	4.17	6.39	8.34	10.66	12.24	14.68	16.92	21.67
10	3.94	4.87	7.27	9.34	11.78	13.44	15.99	18.31	23.21
	← Acceptable →								Unacceptable

Note. From *Statistical Tables for Biological, Agricultural and Medical Research* (6th ed.), Table IV, by R. Fisher and F. Yates, Edinburgh: Longman Essex, 1963.

indicated by a line in Table 3.3. The use of $p = .05$ as the border for acceptability has been arbitrarily set.

In the case of Mendel's data, there is very little difference between the observed and expected results (Table 3.2). Several writers have commented that Mendel's results fit the expectations too closely and that perhaps he adjusted his results to fit a preconceived standard.

In human genetics, the χ^2 method is very valuable and has wide applications. It is used in deciding modes of inheritance (autosomal or sex-linked), deciding whether the pattern of inheritance shown by two genes indicates that they are on the same chromosome, and deciding whether marriage patterns have produced genetically divergent groups in a population.

Table 3.4 Genes, Chromosomes, and Meiosis	
Genes	**Chromosomes**
Occur in pairs (alleles)	Occur in pairs (homologues)
Members of a gene pair separate from each other during meiosis	Members of a pair of homologues separate from each other during meiosis
Members of one gene pair independently assort from other gene pairs during meiosis	Members of one chromosome pair independently assort from other chromosome pairs during meiosis

Locus The position occupied by a gene on a chromosome.

the diploid number of chromosomes and two copies of each gene to the zygote, producing the genotypes of the next generation.

In 1903 Walter Sutton and Theodore Boveri each noted these similarities and independently proposed the idea that genes are carried on chromosomes. This chromosome theory of inheritance has been confirmed by many experiments in the following decades and is one of the foundations of modern genetics. Each gene occupies a specific site (called a **locus**) on a chromosome and each chromosome carries many genes. In humans it is estimated that 30,000 to 40,000 genes are carried on the 24 different chromosomes. ▶ Figure 3.12 shows how genes and chromosomes move through meiosis.

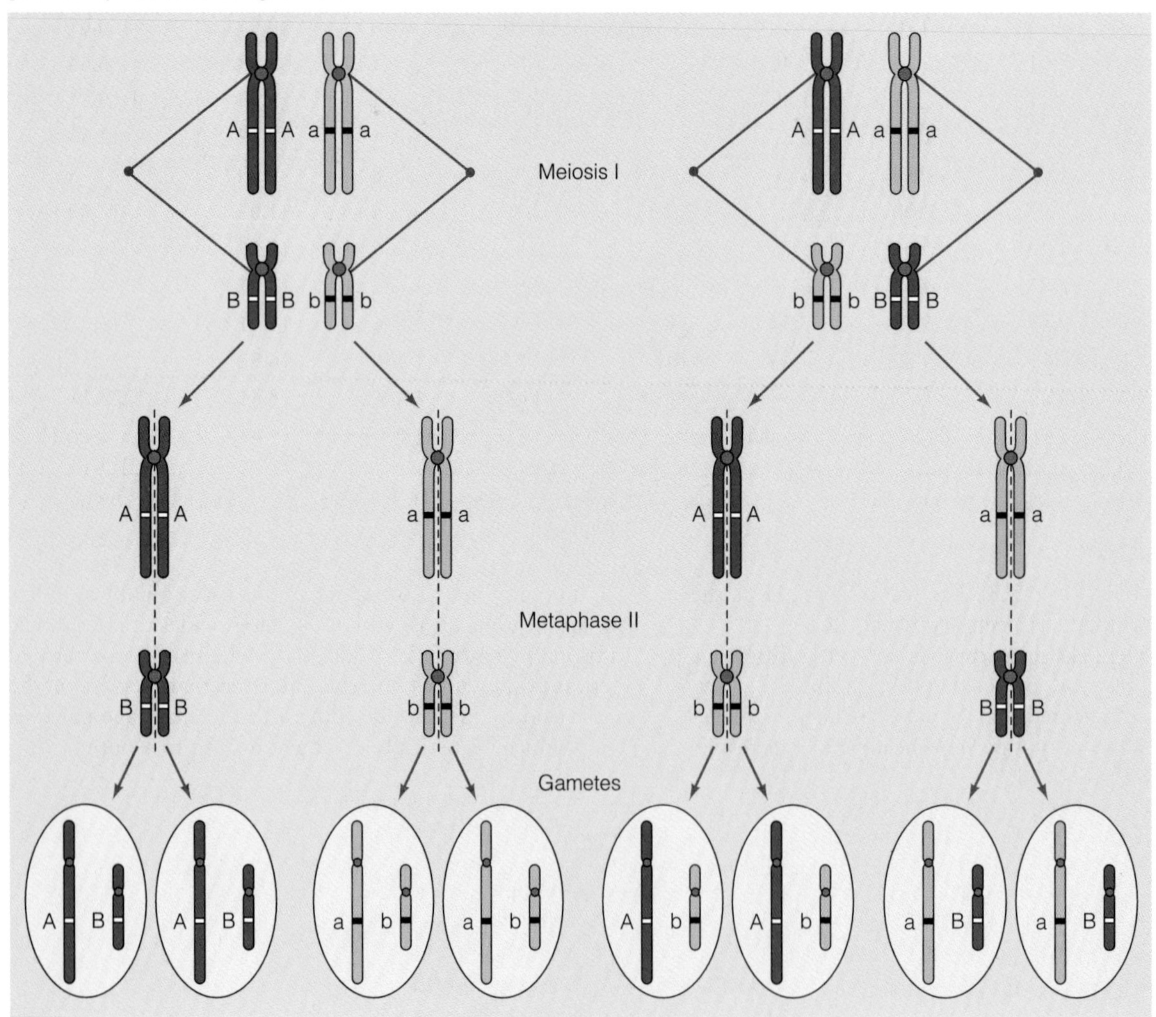

▶ **FIGURE 3.12** Mendel's observations about segregation and independent assortment are explained by the behavior of chromosomes during meiosis. The arrangement of chromosomes at metaphase I is random. As a result, all allelic combinations of the two genes are produced in the gametes.

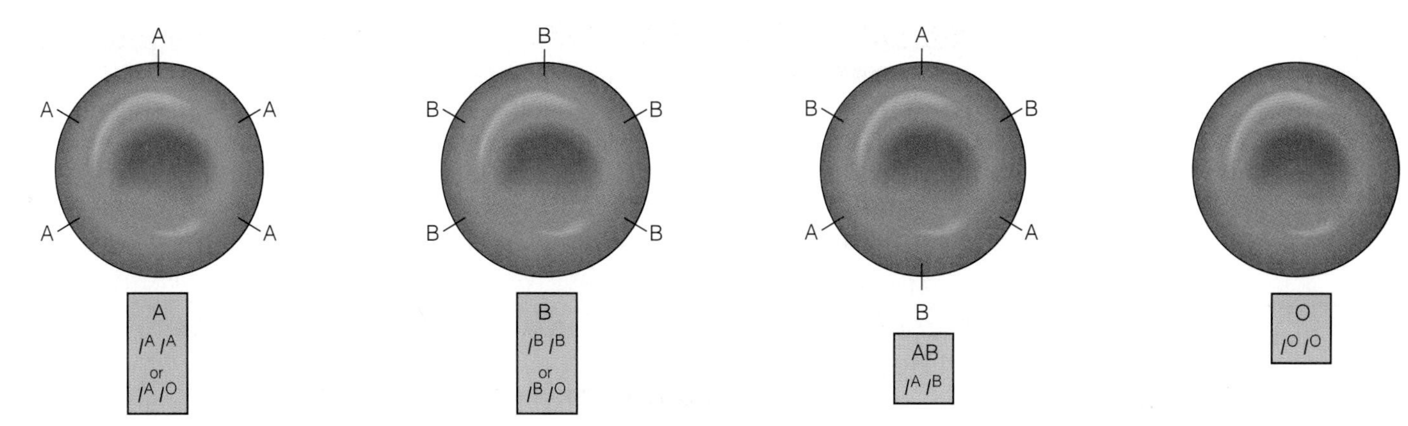

FIGURE 3.13 Each allele of codominant genes is fully expressed in the heterozygote. Type A blood has A antigens on the cell surface, and type B has B antigens on the surface. In type AB, both the A and the B antigen are present on the cell surface. The *A* and *B* alleles of the *I* gene are codominant. In type O blood, no antigen is present. The *O* allele is recessive to both the *A* and the *B* allele.

MANY GENES HAVE MORE THAN TWO ALLELES

So far, we have discussed only genes with two alleles. Because alleles represent different forms of a gene, there is no reason why a gene has to have only two alleles. In fact, many genes have more than two alleles. Any *individual* can carry only two alleles of a gene, but in a population, many different alleles of a gene can be present. In humans, the gene that determines ABO blood groups is an example of a gene with **multiple alleles.** ABO blood types are determined by molecules (proteins with attached sugars) on the surface of human red blood cells. These molecules provide the cell with an identity tag recognized by the body's immune system.

Multiple alleles Genes that have more than two alleles.

There is one gene (I) for ABO blood types, with three alleles, I^A, I^B, and I^O. The *A* and *B* alleles control the formation of slightly different forms of the molecule (called antigens) on the surface of blood cells. Those homozygous for the *A* allele (*AA*) have the A antigen on cells and have blood type A. Those homozygous for the *B* allele (*BB*) carry the B antigen and are type B. The third allele, (O), does not make any antigen, and individuals homozygous for this allele, I^O, carry neither the A nor the B antigen on their cells. The O allele is recessive to both the *A* and *B* alleles. Because there are three alleles, there are six possible genotypes (Figure 3.13).

VARIATIONS ON A THEME BY MENDEL

After Mendel's work became widely known, geneticists turned up cases in which the phenotypes of the F1 offspring did not resemble those of the parents. In some cases, the offspring had a phenotype intermediate to that of the parents or a phenotype in which the traits of both parents were expressed. These findings led to a debate as to whether these cases could be explained by Mendelian inheritance or whether there might be another, separate mechanism of inheritance that did not follow the laws of segregation and independent assortment.

Eventually, experiments with several different organisms showed that these cases were not exceptions to Mendelian inheritance. In this section, we will discuss some of these variations and show that although the phenotypes may not follow the predicted Mendelian ratios for dominance, they do follow the predicted ratios for genotypes.

Table 3.5

ABO Blood Types	
Genotypes	**Phenotypes**
I^AI^A, I^AI^O	Type A
I^BI^B, I^BI^O	Type B
I^AI^B	Type AB
I^OI^O	Type O

Codominance Full phenotypic expression of both members of a gene pair in the heterozygous condition.

Incomplete dominance Expression of a phenotype that is intermediate between those of the parents.

Codominant Alleles Are Fully Expressed in Heterozygotes

In **codominance**, heterozygous individuals fully express both alleles. In the ABO blood type, *AB* heterozygotes have both the *A* and *B* antigens on their cell membranes and are blood type AB. In this case, neither allele is dominant to the other. Because each allele is fully expressed, they are said to be codominant (Figure 3.13). As a result, the three alleles of the ABO system have six genotypic combinations that contribute to four phenotypes (Table 3.5).

Incomplete Dominance Has a Distinctive Phenotype in Heterozygotes

In a situation with **incomplete dominance**, the phenotype of the heterozygote is intermediate to those of the homozygous parents. For example, in snapdragons (▸ Figure 3.14), if a true-breeding variety with red flowers is crossed with one that produces white flowers, the F1 all have pink flowers. The phenotype of the F1 is different than either parent and is intermediate to the parental phenotypes. When the F1 plants are self-fertilized, they produce an F2 with red, pink, and white flowers in a 1:2:1 ratio.

The results of this cross can be explained using Mendelian principles. Each genotype in this cross has a distinct phenotype, and the phenotypic ratio of 1 red:2 pink:1 white is the same as the genotypic ratio of 1 *RR*:2 *Rr*:1 *rr*. Two copies of the *R* allele are needed to produce red flowers. One copy results in pink flowers. As a result, the *R* allele is incompletely dominant over the *r* allele. Because the *r* allele produces no color, the absence of *R* alleles produces white flowers.

MENDELIAN INHERITANCE IN HUMANS

After 1900, segregation and independent assortment were studied in a wide range of organisms. Although some believed that inheritance of traits in humans might be an exception to Mendelian principles, the first Mendelian trait (a hand

▸ **FIGURE 3.14** Incomplete dominance in snapdragon flower color. Red-flowered snapdragons crossed with white-flowered snapdragons produce offspring that have pink flowers in the F1. In heterozygotes, the allele for red flowers is incompletely dominant over the allele for white.

deformity called brachydactyly, MIM/OMIM 112500) was identified in 1905. Since then, over 5,000 such traits have been described.

Segregation and Independent Assortment of Human Traits

To illustrate that segregation and independent assortment apply to human traits, we will follow the inheritance of a recessive trait called albinism (MIM/OMIM 203100). Individuals homozygous (*aa*) for albinism cannot make pigment and have very pale, white skin; white hair; and colorless eyes (Figure 3.15). The dominant allele (*A*) controls normal pigmentation.

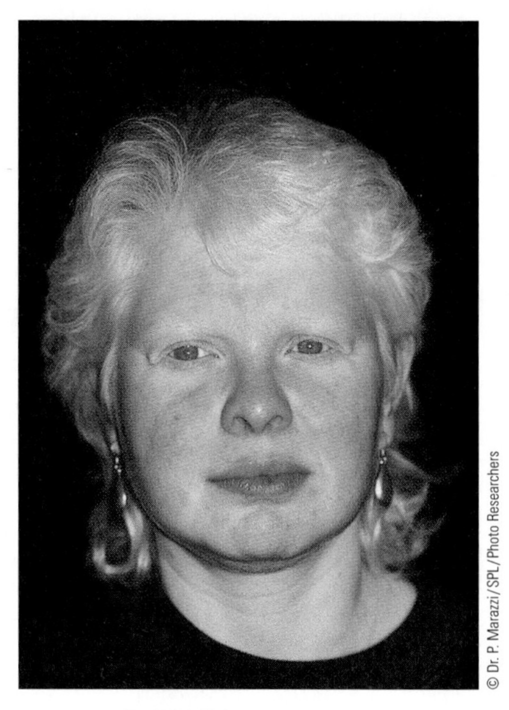

FIGURE 3.15 Albino individuals lack pigment in the skin, hair, and eyes.

In this example, both parents have normal pigmentation but are heterozygous for the recessive allele (*a*) (Figure 3.16). In each parent, the dominant and recessive alleles separate or segregate from each other at the time of gamete formation. Because each parent can produce two different types of gametes (one type with *A* and another with *a*), there are four possible combinations of these gametes at fertilization. If enough offspring are produced, these four possible types of fertilization events will show a predicted phenotypic ratio of 3 pigmented:1 albino offspring and a genotypic ratio of 1*AA* : 2*Aa* : 1*aa* (Figure 3.16). In other words, segregation of alleles during gamete formation produces the same outcome in both pea plants and humans. This does not mean that there will be one albino child and three normally pigmented children in every such family with four children. It does mean that in a mating between heterozygotes, each child has a 25% chance of being albino and a 75% chance of having normal pigmentation.

The simultaneous inheritance of two traits in humans follows the Mendelian principle of independent assortment (Figure 3.17). To illustrate, let's examine a family in which each parent is heterozygous for albinism (*Aa*) and heterozygous for another recessive trait, hereditary deafness (MIM/OMIM 220290). For hearing, the normal allele (*D*) is dominant and is expressed in the homozygous dominant (*DD*) or heterozygous genotype (*Dd*). During gamete formation, the alleles for skin color and the alleles for hearing assort into gametes independently. As a result, each parent produces equal proportions of four different types of gametes. If each parent produces four types of gametes, there are 16 possible combinations of these gametes at fertilization (four types of gametes in all possible combinations), resulting in four different phenotypic classes (Figure 3.17). An examination of the possible genotypes shows that there is a 1 in 16 chance that a child would be both deaf and an albino.

Pedigree construction Use of family history to determine how a trait is inherited and to determine risk factors for family members.

Pedigree chart A diagram listing the members and ancestral relationships in a family; used in the study of human heredity.

In pea plants and other organisms, such as *Drosophila*, genetic analysis is done using experimental crosses. In humans, experimental crosses are not possible, and geneticists must base their work on matings that have already taken place. In human genetics, the study of a trait begins with a family history, as outlined in the following section.

FIGURE 3.16 The segregation of albinism, a recessive trait in humans. As in pea plants, alleles of a gene pair separate from each other during gamete formation.

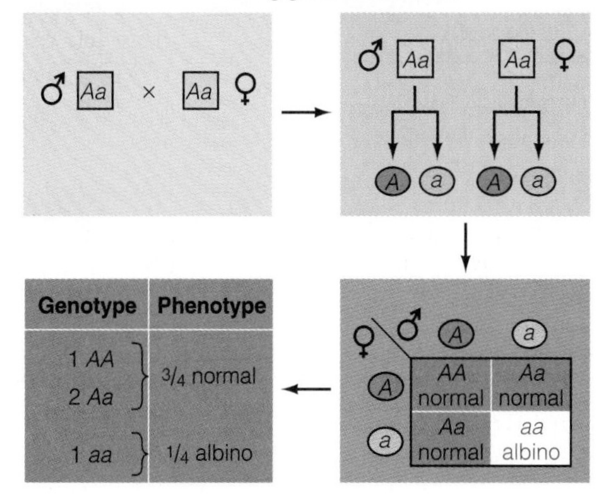

Pedigree Construction in Human Genetics

A basic method of genetic analysis in humans is the construction of a family history to follow the inheritance of a trait. This method is called **pedigree construction.** A **pedigree chart** is the orderly presentation of family information in the form of an easily readable chart. From such a family tree, the inheritance of a trait can be followed through several generations. Using Mendelian principles, the information in the pedigree can be

FIGURE 3.17 Independent assortment for two traits in humans follows the same pattern of inheritance as traits in pea plants.

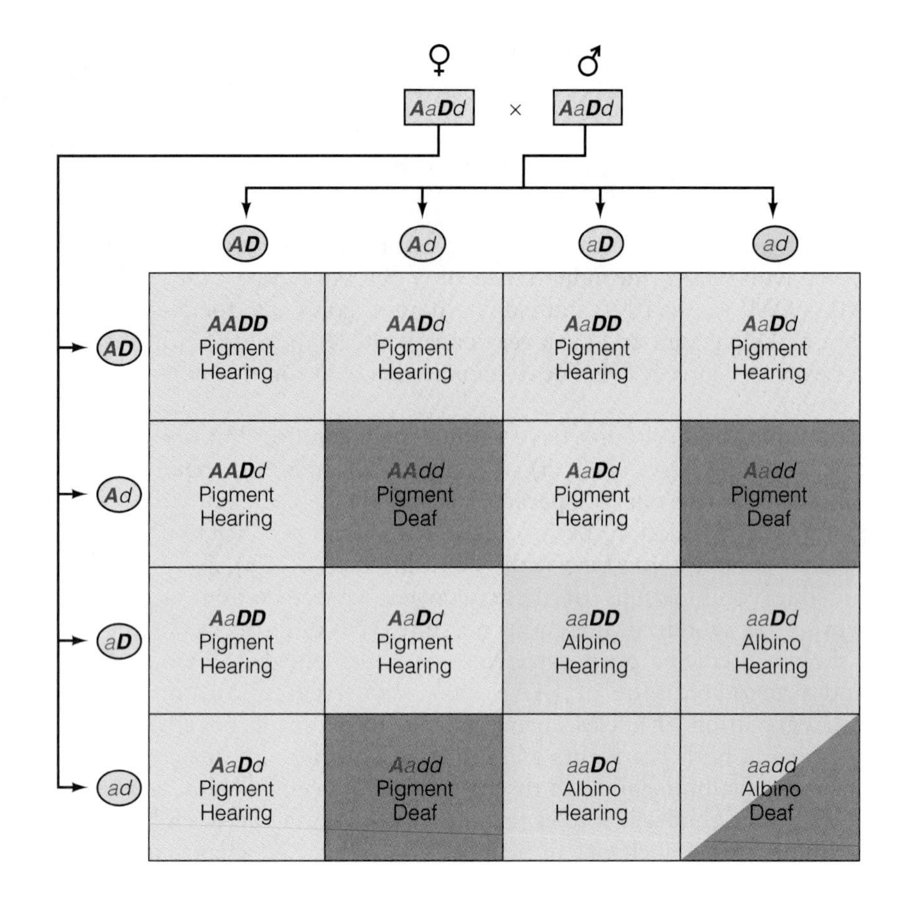

analyzed to determine whether the trait has a dominant or recessive pattern of inheritance and whether the gene in question is located on an autosome or a sex chromosome.

Pedigree construction uses a standardized set of symbols, many of which are borrowed from genealogy. In constructing a pedigree, males are represented by squares (□) and females by circles (○). An individual who exhibits a trait in question is represented by a filled symbol (■ or ●). Heterozygotes, when known, are indicated by half-filled symbols (◧ or ◐).

The relationships between individuals in a pedigree are shown as a series of lines. A horizontal line between two symbols represents a mating (□–○). A double horizontal line is used to symbolize matings between brother and sister or between close relatives. These are known as consanguineous matings. The offspring, listed from left to right in birth order, are connected to each other by a horizontal line (○○□) and to the parents by a vertical line:

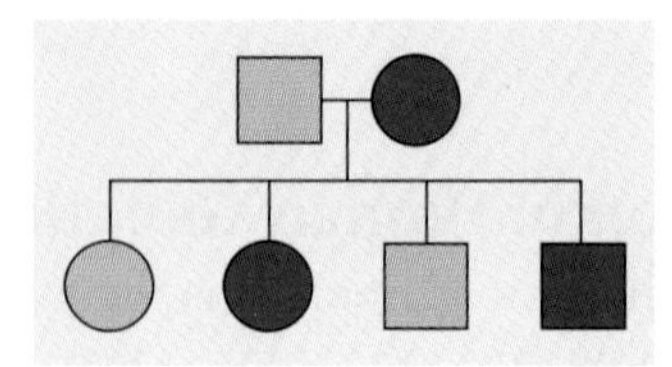

If the children are twins, they are identified as identical (monozygotic) twins ( ), which arise from a single egg, or nonidentical (dizygotic) twins ( ), resulting from the fertilization of two eggs.

Web-Integrated Toolbox

Solving Genetics Problems

Solving genetics problems requires a systematic approach and the development of some analytic skills. Although many problems can seem overwhelming at first glance, they can all be worked by following a few simple steps.

1. Read the problem carefully and determine what information is provided and what information is asked for. For example, does the problem give genotypes and ask for phenotypes? Are you given information about the parents and asked about the offspring? Sort out and organize the information provided, keeping in mind what is asked for.
2. Translate the words and terms of the problem into symbols. Use letters of the alphabet for alleles when assigning genotypes; use single words or short phrases for phenotypes. Keep track of generations by using symbols such as P1, F1, and F2.
3. Solve the problem by applying the concepts of Mendelian genetics, such as segregation or independent assortment. Use a Punnett square if necessary to keep track of all combinations of gametes and all possible fertilization events.

The best way to learn genetics is by solving problems. To help you develop your problem-solving skills, the book's Web site contains a link to exercises that will provide step-by-step guidance through the process of setting up and solving genetics problems, illustrating the principles of Mendelian inheritance.

By recreating the thought processes used by geneticists in analyzing problems, you will learn how to analyze problems and how to apply genetic concepts to solve them. Working problems will also help you become aware of what information you know and what you need to clarify or learn. The skills gained in solving these problems will also improve your overall abilities to think analytically about problems or situations you encounter in other aspects of life.

A numbering system is used in pedigree construction. Each generation is indicated by a roman numeral, and within a generation, each individual is identified by an arabic number:

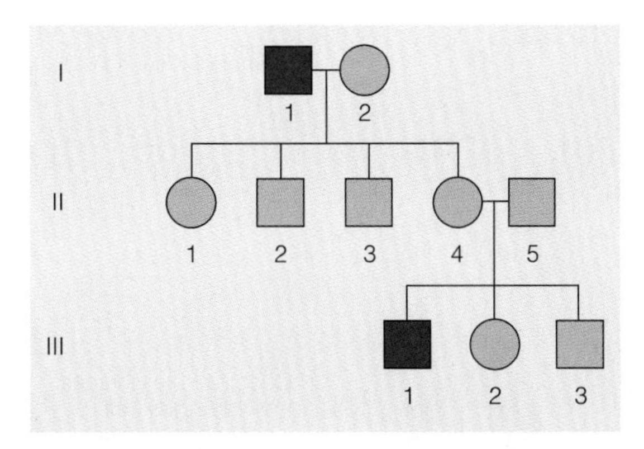

Pedigrees are often constructed after a family member afflicted with a genetic trait has been identified. This individual, known as the **proband,** is indicated on the pedigree by an arrow and the letter P:

Proband First affected family member who seeks medical attention.

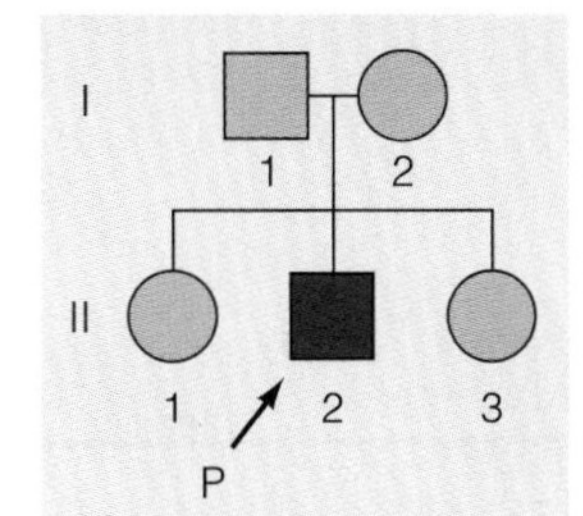

Because pedigree construction is essentially a family history, details about earlier generations may be uncertain. If the sex of someone is unknown or unimportant, this is indicated by a diamond shape: (◇). In some cases, spouses in a pedigree are omitted if they are not essential to the inheritance of the trait. If there is doubt that a family member possessed the trait in question, this is indicated by a question mark above the symbol. Many of the symbols and terminology used in constructing pedigrees are presented in ◗ Figure 3.18. A completed pedigree is a form of symbolic communication used by clinicians and researchers in human genetics (◗ Figure 3.19). It contains information that can establish the pattern of inheritance for a trait, can identify those at risk of developing or transmitting the trait, and is a resource for establishing biological relationships within a family. In Chapter 4, we will see how pedigree analysis is used to establish the genotypes of individuals and to predict the chances of having children affected with a genetic disorder.

◗ **FIGURE 3.18** Symbols used in pedigree analysis.

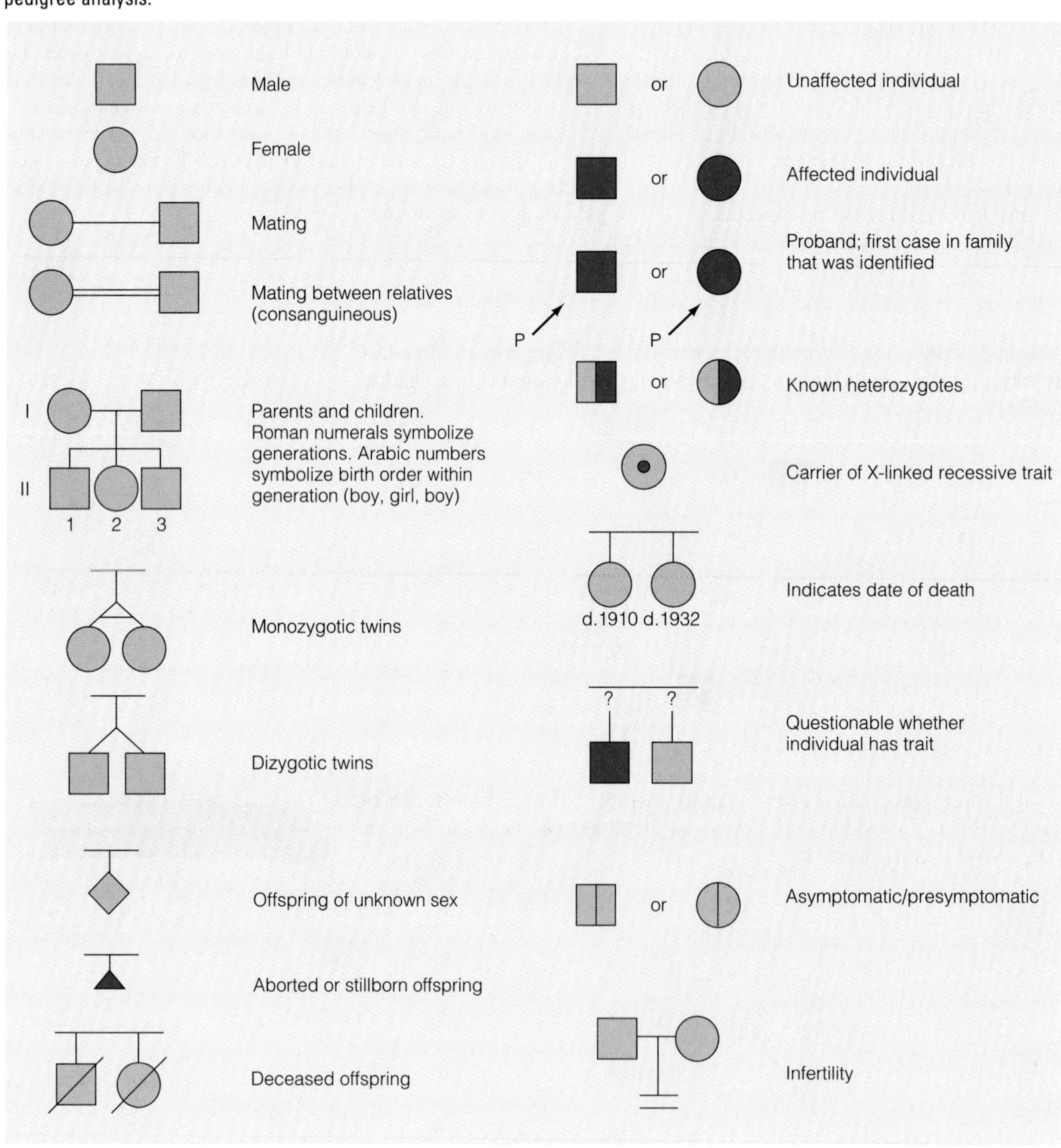

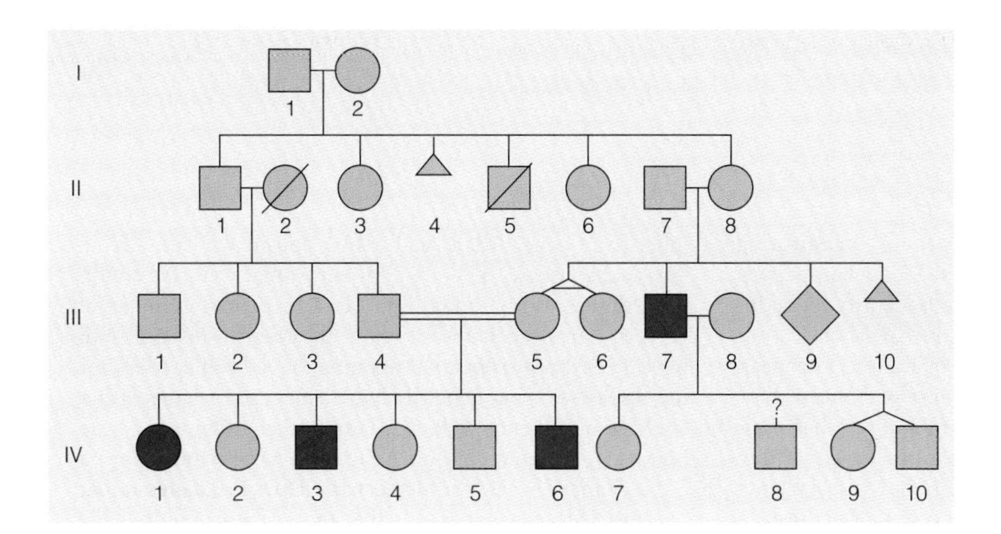

FIGURE 3.19 A pedigree chart showing the inheritance of a trait in several generations of a family. This pedigree and all those in this book use the standardized set of symbols adopted in 1995 by the American Society of Human Genetics.

Summary

Heredity: How Does It Work?

1. In the centuries before Gregor Mendel experimented with the inheritance of traits in the garden pea, several competing theories attempted to explain how traits were passed from generation to generation, but none were completely successful.

Mendel's Experimental Design Resolved Many Unanswered Questions

2. Mendel carefully selected an organism to study, kept careful records, and studied the inheritance of traits for several generations. In his decade-long series of experiments, Mendel established the foundation for the science of genetics.

Crossing Pea Plants: The Principle of Segregation

3. Mendel studied crosses in the garden pea that involved one pair of alleles and demonstrated that the phenotypes associated with these traits are controlled by pairs of factors, now known as genes. These factors separate or segregate from each other during gamete formation and exhibit dominant/recessive relationships.

More Crosses With Pea Plants: The Principle of Independent Assortment

4. In later experiments, Mendel discovered that members of one gene pair separate or segregate independently of other gene pairs. This principle of independent assortment leads to the formation of all possible combinations of gametes with equal probability in a cross between two individuals.

Meiosis Explains Mendel's Results: Genes Are on Chromosomes

5. Segregation and independent assortment of genes results from the behavior of chromosomes in meiosis. At the turn of the twentieth century, it became apparent that genes are located on chromosomes.

Many Genes Have More Than Two Alleles

6. Although any individual can carry only two alleles of a gene, many genes have multiple alleles, carried by members of a population.

Variations on a Theme by Mendel

7. Variations in phenotypes can change the expected outcome of crosses, but the genotypic ratios do not change. Codominant alleles are both expressed in the phenotype, whereas in incomplete dominance, the heterozygote has a phenotype intermediate to that of the parents.

Mendelian Inheritance in Humans

8. Because genes for human genetic disorders exhibit segregation and independent assortment, the inheritance of certain human traits is predictable, making it possible to provide genetic counseling to those at risk of having children affected with genetic disorders.

Pedigree Construction in Human Genetics

9. Instead of direct experimental crosses, traits in humans are traced by constructing pedigrees that follow a trait through several generations.

Case Studies

CASE 1

Pedigree analysis is a fundamental tool for investigating whether a trait is following a traditional Mendelian pattern of inheritance. It can also be used to help identify individuals within a family that may be at risk for the trait.

Adam and Sarah, a young couple of Eastern European Jewish ancestry, went to a genetic counselor because they were planning a family and wanted to know what their chances were for having a child with a genetic condition. The genetic counselor took a detailed family history from both of them and discovered several traits in their respective families.

Sarah's maternal family history is suggestive of an inherited form of breast and ovarian cancer with an autosomal dominant pattern of cancer from her grandmother to mother because of the young ages at which they were diagnosed with their cancers. If an altered gene that predisposed to breast and ovarian cancer were in Sarah's family, she, her sister, and any of her own future children could be at risk for inheriting this gene. The counselor told her that genetic testing is available that may help determine if an altered gene is in her family.

Adam's paternal family history has a very strong pattern of early-onset heart disease. An autosomal dominant condition, known as familial hypercholesterolemia, may be responsible for the number of individuals in the family who have died from heart attacks. Like hereditary breast and ovarian cancer, there is genetic testing available to see if Adam carries this altered gene. Testing may give the couple more information on the chances that their children could inherit the gene. Adam had a first cousin who died from Tay-Sachs disease (TSD), a fatal autosomal recessive condition, that is more commonly found in people who are of Eastern European Jewish descent. In order for his cousin to have TSD, both of his parents must have been carriers for the disease-causing gene. If that is the case, Adam's father could be a carrier as well. If Adam's father and mother were both carriers, each child of Adam and Sarah would have a 25% chance of being afflicted with TSD. If Adam's father has the TSD gene, it is possible that Adam inherited the gene. Because Sarah is also of Eastern European ancestry, she could be a carrier of this gene, although no one in her family has been affected with TSD. A simple blood test performed on both Sarah and Adam could determine if they are carriers of the TSD gene.

1. If Sarah carries the mutant cancer gene and Adam carries the mutant heart disease gene, what is the chance that they will have a child that is free of both diseases? Are these good odds?
2. Would you want to know the results of the cancer, heart disease, and TSD tests if you were Sarah and Adam? Is it their responsibility as good potential parents to find out this kind of information before they decide to have a child?
3. Would you decide to have a child if the test results said that you carry the cancer gene? The heart disease gene? The TSD gene? The heart disease and the TSD gene?

Questions and Problems

Mendel

1. Explain the difference between the following terms:
   a. Gene versus allele versus locus
   b. Genotype versus phenotype
   c. Dominant versus recessive
   d. Complete dominance versus incomplete dominance versus codominance
2. Of the following, which are phenotypes, and which are genotypes?
   a. *Aa*
   b. Tall plants
   c. *BB*
   d. Abnormal cell shape
   e. *AaBb*
3. Define Mendel's Law of Segregation
4. Define Mendel's Law of Independent Assortment

Mendel's Study of Single Traits—Monohybrid Crosses

5. Suppose that organisms have the genotypes listed below. What types of gametes will these organisms produce, and in what proportions?
   a. *Aa*
   b. *AA*
   c. *aa*
6. Given the following matings, what are the predicted genotypic ratios of the offspring?
   a. $Aa \times aa$
   b. $Aa \times Aa$
   c. $AA \times Aa$
7. Brown eyes (*B*) are fully dominant over blue eyes (*b*).
   a. A 3:1 phenotypic ratio of F1 progeny indicates that the parents are of what genotype?
   b. A 1:1 phenotypic ratio of F1 progeny indicates that the parents are of what genotype?

8. An unspecified character controlled by a single gene is examined in pea plants. Only two phenotypic states exist for this trait. One phenotypic state is completely dominant to the other. A heterozygous plant is self-crossed. What proportion of the progeny plants exhibiting the dominant phenotype is homozygous?
9. Sickle cell anemia (SCA) is a human genetic disorder caused by a recessive allele. A couple plans to marry and wants to know the probability that they will have an affected child. With your knowledge of Mendelian inheritance, what can you tell them if (1) both are normal, but each has one affected parent and the other parent has no family history of SCA; and (2) the man is affected by the disorder, but the woman has no family history of SCA?
10. If you are informed that being right- or left-handed is heritable and that a right-handed couple is expecting a child, can you conclude that the child will be right-handed?
11. Stem length in pea plants is controlled by a single gene. Consider the cross of a true-breeding, long-stemmed variety to a true-breeding, short-stemmed variety in which long stems are completely dominant.
    a. 120 F1 plants are examined. How many plants are expected to be long stemmed? Short stemmed?
    b. Assign genotypes to both P1 varieties and to all phenotypes listed in (a).
    c. A long-stemmed F1 plant is self-crossed. Of 300 F2 plants, how many should be long stemmed? Short stemmed?
    d. For the F2 plants mentioned in (c), what is the expected genotypic ratio?

The Inheritance of Two Traits—Dihybrid Crosses

12. Organisms have the genotypes listed below. What types of gametes will these organisms produce and in what proportions?
    a. *Aabb*
    b. *AABb*
    c. *AaBb*
13. Given the following matings, what are the predicted phenotypic ratios of the offspring?
    a. *AABb* × *Aabb*
    b. *AaBb* × *aabb*
    c. *AaBb* × *AaBb*
14. A woman is heterozygous for two genes. How many different types of gametes can she produce, and in what proportions?
15. Use the Punnett square and then the forked-line method to determine the possible genotypes and phenotypes of the F1 offspring from the following cross to show that the two methods achieve the same results.

    P1: *AaBb* × *AaBb*

16. Two traits are simultaneously examined in a cross of two pure-breeding pea-plant varieties. Pod shape can be either swollen or pinched. Seed color can be either green or yellow. A plant with the traits swollen and green is crossed with a plant with the traits pinched and yellow, and a resulting F1 plant is self-crossed. A total of 640 F2 progeny are phenotypically categorized as follows:

    swollen, yellow 360
    swollen, green 120
    pinched, yellow 120
    pinched, green 40

    a. What is the phenotypic ratio observed for pod shape? Seed color?
    b. What is the phenotypic ratio observed for both traits considered together?
    c. What is the dominance relationship for pod shape? Seed color?
    d. Deduce the genotypes of the P1 and F1 generations.
17. Consider the following cross in pea plants, in which smooth seed shape is dominant to wrinkled and yellow seed color is dominant to green. A plant with smooth, yellow seeds is crossed to a plant with wrinkled, green seeds. The peas produced by the offspring are all smooth and yellow. What are the genotypes of the parents? What are the genotypes of the offspring?
18. Consider another cross in pea plants involving the genes for seed color and shape. As before, yellow is dominant to green, and smooth is dominant to wrinkled. A plant with smooth, yellow seeds is crossed to a plant with wrinkled, green seeds. The peas produced by the offspring are as follows: one-fourth are smooth, yellow; one-fourth are smooth, green; one-fourth are wrinkled, yellow; and one-fourth are wrinkled, green.
    a. What is the genotype of the smooth, yellow parent?
    b. What are the genotypes of the four classes of offspring?
19. Determine the possible genotypes of the parents shown below by analyzing the phenotypes of their children. In this case, we will assume that brown eyes (*B*) is dominant to blue (*b*) and that right-handedness (*R*) is dominant to left-handedness (*r*).
    a. Parents: brown eyes, right-handed × brown eyes, right-handed
       Offspring: 3/4 brown eyes, right-handed
       1/4 blue eyes, right-handed
    b. Parents: brown eyes, right-handed × blue eyes, right-handed
       Offspring: 6/16 blue eyes, right-handed
       2/16 blue eyes, left-handed
       6/16 brown eyes, right-handed
       2/16 brown eyes, left-handed
    c. Parents: brown eyes, right-handed × blue eyes, left-handed
       Offspring: 1/4 brown eyes, right-handed
       1/4 brown eyes, left-handed
       1/4 blue eyes, right-handed
       1/4 blue eyes, left-handed
20. Think about this one carefully. Albinism and hair color are governed by different genes. A recessively inherited form of albinism causes affected individuals to lack pigment in their skin, hair, and eyes. In hair color, red hair is inherited as a recessive trait, and brown hair is inherited as a dominant trait. An albino woman whose

parents both have red hair has two children with a man who is normally pigmented and has brown hair. The brown-haired partner has one parent who has red hair. The first child is normally pigmented and has brown hair. The second child is albino.

a. What is the hair color (phenotype) of the albino parent?
b. What is the genotype of the albino parent for hair color?
c. What is the genotype of the brown-haired parent with respect to hair color? Skin pigmentation?
d. What is the genotype of the first child with respect to hair color and skin pigmentation?
e. What are the possible genotypes of the second child for hair color? What is the phenotype of the second child for hair color? Can you explain this?

21. Consider the following cross:

P1: *AABBCCDDEE* × *aabbccddee*
F1: *AaBbCcDdEe* (self cross to get F2)

What is the chance of getting an *AaBBccDdee* individual in the F2 generation?

22. In the following trihybrid cross, determine the chance that an individual could be phenotypically A, b, C in the F1 generation.

P1: *AaBbCc* × *AabbCC*

23. In pea plants, long stems are dominant to short stems, purple flowers are dominant to white, and round seeds are dominant to wrinkled. Each trait is determined by a single, different gene. A plant that is heterozygous at all three loci is self-crossed, and 2,048 progeny are examined. How many of these plants would you expect to be long stemmed with purple flowers, producing wrinkled seeds?

Incomplete Dominance and Codominance

24. A character of snapdragons amenable to genetic analysis is flower color. Imagine that a true-breeding red-flowered variety is crossed to a pure line having white flowers. The progeny are exclusively pink flowered. Diagram this cross, including genotypes for all P1 and F1 phenotypes. What is the mode of inheritance? Let F = red, and f = white.

25. In peas, straight stems (S) are dominant to gnarled (s), and round seeds (R) are dominant to wrinkled (r). The following cross (a test cross) is performed: *SsRr* × *ssrr.* Determine the expected phenotypes of the progeny and what fraction of the progeny should exhibit each phenotype.

26. Pea plants usually have white or red flowers. A strange pea-plant variant is found that has pink flowers. A self-cross of this plant yields the following phenotypes:

red flowers, 30
pink flowers, 62
white flowers, 33

What are the genotypes of the parents? What is the genotype of the progeny with red flowers?

27. A plant geneticist is examining the mode of inheritance of flower color in two closely related species of exotic plants. Analysis of one species has resulted in the identification of two pure-breeding lines—one produces a distinct red flower and the other produces no color at all; however, he cannot be sure. A cross of these varieties produces all pink-flowered progeny. The second species exhibits similar pure-breeding varieties; that is, one variety produces red flowers, and the other produces an albino flower. A cross of these two varieties, however, produces orange-flowered progeny exclusively. Analyze the mode of inheritance of flower color in these two plant species.

28. What are the possible genotypes for the following blood types?

a. type A
b. type B
c. type O
d. type AB

29. A man with blood type A and a woman with blood type B have three children: A daughter with type AB, and two sons, one with type B and one with type O blood. What are the genotypes of the parents?

30. What is the chance that a man with type AB blood and a woman with type A blood whose mother is type O can produce a child that is:

a. type A
b. type AB
c. type O
d. type B

31. A hypothetical human trait is controlled by a single gene. Four alleles of this gene have been identified: *a*, *b*, *c*, and *d*. Alleles *a*, *b*, and *c* are all codominant; allele *d* is recessive to all other alleles.

a. How many phenotypes are possible?
b. How many genotypes are possible?

Meiosis Explains Mendel's Results

32. Discuss the pertinent features of meiosis that provide a physical correlate to Mendel's abstract genetic laws of random segregation and independent assortment.

33. The following diagram shows a hypothetical diploid cell. The recessive allele for albinism is represented by a, and d represents the recessive allele for deafness. The normal alleles for these conditions are represented by A and D, respectively.

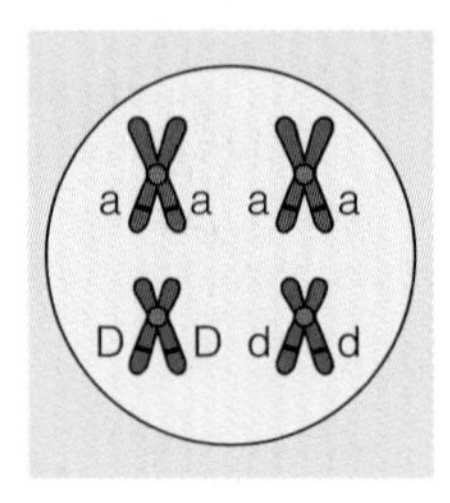

a. According to the principle of segregation, what is segregating in this cell?

b. According to Mendel's principle of independent assortment, what is independently assorting in this cell?
c. How many chromatids are in this cell?
d. How many tetrads are in this cell?
e. Write the genotype of the individual from whom this cell was taken.
f. What is the phenotype of this individual?
g. What stage of cell division is represented by this cell (prophase, metaphase, anaphase, or telophase of meiosis I, meiosis II, or mitosis)?
h. After meiosis II is complete, how many chromatids and chromosomes will be present in one of the four progeny cells?

Medelian Inheritance in Humans—Pedigree Construction

34. Define the following pedigree symbols:

a. 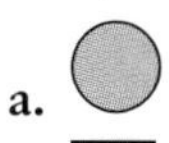

b. 

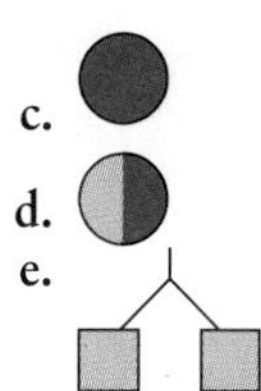

35. Draw the following simple pedigree. A man and a woman have three children, a daughter, then two sons. The daughter is married to a man and has monozygotic (identical) twin girls. The youngest son in generation II is affected with albinism.
36. Construct a pedigree given the following information. A woman, Mary, is 16 weeks' pregnant and was referred for genetic counseling because of advanced maternal age. Mary has one daughter, Sarah, who is 5 years old. Mary has three older sisters and four younger brothers. The two oldest sisters are married and each has one son. All her brothers are married, but none has any children. Mary's parents are both alive, and she has two maternal uncles and three paternal aunts. Mary's husband, John, has two brothers, one older and one younger, neither of whom is married. John's mother is alive, but his father is deceased.

Internet Activities

The following activities use the resources of the World Wide Web to enhance the topics covered in this chapter. To investigate the topics described here, log onto the book's homepage at: www.brookscole.com/biology_d, and follow the *Student Resources* link to your subject and text.

1. *Mendelian genetics/ plant genetics.* Gregor Mendel crossed pea plants to investigate the results of hybridization experiments. Now you give it a try! At the CUNY Brooklyn Mendelian Genetics site (http://www.brooklyn.cuny.edu/bc/ahp/MGInv/MGI.Inv.html), read the Introduction (http://www.brooklyn.cuny.edu/bc/ahp/MGInv/MGI.Intro.html) carefully (some of the steps are a little tricky) and then click on the *Plant Hybridization* link at the site to choose and perform some crosses of your own.
2. *Mendel's discoveries in his own words.* At the *MendelWeb* Web site (http://www.netspace.org/MendelWeb/) you can read Mendel's original paper in English and German. In addition to Mendel's original text, this site also has links to essays and commentary on Mendel's works and writings, and on the state of knowledge about heredity before Mendel.
3. *Meet Gregor Mendel?* Check out Professor John Blamire's fictionalized account of Mendel's life (http://www.brooklyn.cuny.edu/bc/ahp/MBG/MBG.main.html).

For Further Reading

Corcos, A. F., & Monaghan, F. (1985). Role of de Vries in the recovery of Mendel's work. I. Was de Vries really an independent discoverer of Mendel? *J. Heredity, 76,* 187–190.

Dahl, H. (1993). Things Mendel never dreamed of. *Med. J. Aust., 158,* 247–252.

Dunn, L. C. (1965). *A short history of genetics.* New York: McGraw-Hill.

Edwards, A. W. (1986). Are Mendel's results really too close? *Biol. Rev. Cambridge Philos. Soc., 61,* 295–312.

Gasking, E. B. (1959). Why was Mendel's work ignored? *J. Hist. Ideas, 20,* 62–84.

George, W. (1975). *Gregor Mendel and heredity.* London: Priory Press.

Hartl, D., & Orel, V. (1992). What did Gregor Mendel think he discovered? *Genetics, 131,* 245–253.

Heim, W. G. (1991). What is a recessive allele? *Amer. Biol. Teacher, 53,* 94–97.

Monaghan, F. V., & Corcos, A. F. (1985). Mendel, the empiricist. *J. Heredity, 76,* 49–54.

Orel, V. (1973). The scientific milieu in Brno during the era of Mendel's research. *J. Heredity, 64,* 314–318.

Orel, V. (1984). *Mendel.* New York: Oxford University Press.

Pilgrim, I. (1986). A solution to the too-good-to-be-true paradox and Gregor Mendel. *J. Heredity, 77,* 218–220.

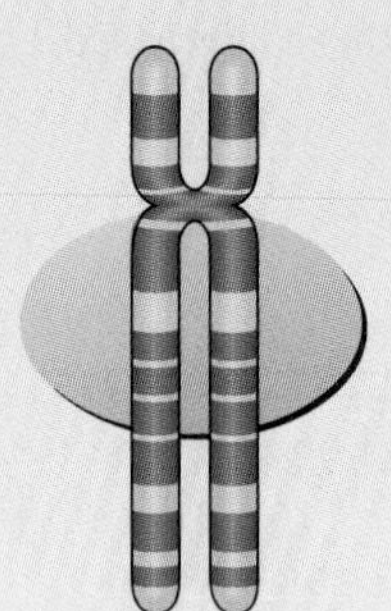

Pedigree Analysis in Human Genetics

Chapter Outline

Was Abraham Lincoln, the 16th president of the United States, affected with a genetic disorder? Several writers have speculated that Lincoln had Marfan syndrome. The evidence offered in support of this idea is based on Lincoln's physical appearance and the report of Marfan syndrome in a distant relative.

Photographs, written descriptions, and medical reports are available to provide ample information about Lincoln's physical appearance. He was 6 ft. 4 in. tall and thin, weighing between 160 and 180 lbs. for most of his adult life. He had long arms and legs with large, narrow hands and feet. Contemporary descriptions of his appearance indicate that he was stoop-shouldered, loose jointed, and walked with a shuffling gait. In addition, he wore eyeglasses to correct a visual problem.

Lincoln's appearance and ocular problems are suggestive of Marfan syndrome, a genetic condition that affects the connective tissue of the body, resulting in a tall, thin individual often with a shifted lens in the eye, blood vessel defects, and loose joints.

In 1960 a man diagnosed with Marfan syndrome was found to have ancestors in common with Lincoln (the common ancestor was Lincoln's great-great-grandfather). Added to the information about Lincoln's physical appearance, the family history suggests to some observers that Lincoln had Marfan syndrome.

Others strongly disagree with this speculation, arguing that the length of Lincoln's extremities and the proportions of his body were well within the normal limits for tall, thin individuals. In addition, although Lincoln had visual problems, an examination of his eyeglasses indicates that he was farsighted, whereas those with the classical form of Marfan syndrome are nearsighted. Lastly, Lincoln showed no outward signs of problems with major blood vessels, such as the aorta.

Did Lincoln really have Marfan syndrome, and should we care? Interest in this issue probably grows from our curiosity about the lives of historic figures and our fascination with the intimate details of the lives of the famous and the infamous. At the present there is no solid evidence to suggest that Lincoln had Marfan syndrome, although molecular testing on bone and hair fragments (discussed later) has been proposed. For now we are left only with speculation and inferential reasoning.

The more important question for this chapter is how can we tell when someone is affected with a genetic disorder? Lincoln's family history shows only one documented case of Marfan syndrome in nine generations. Is this enough to decide that he was also affected with this genetic condition?

In this chapter we show that the principles of inheritance discovered by Mendel in peas also apply to humans. Even though the methods employed in human genetics differ significantly from those used in experimental organisms, the rules for the inheritance of traits in humans are the same as those for pea plants.

STUDYING THE INHERITANCE OF TRAITS IN HUMANS

Mendel used pea plants for two primary reasons. First, they can be crossed in any combination. Second, each cross is likely to produce large numbers of offspring, an important factor in understanding how a trait is inherited. If you were picking an organism for genetic studies, humans would not be a good choice. With pea plants, it is easy to carry out crosses between plants with purple flowers and plants with white flowers and to repeat this cross as often as necessary. For obvious reasons, experimental matings in humans are not possible. You can't ask all albino humans to mate with homozygous normally pigmented individuals and have their progeny interbreed to produce an F2. For the most part, human geneticists base their work on matings that have already taken place, regardless of whether such matings would be the most genetically informative.

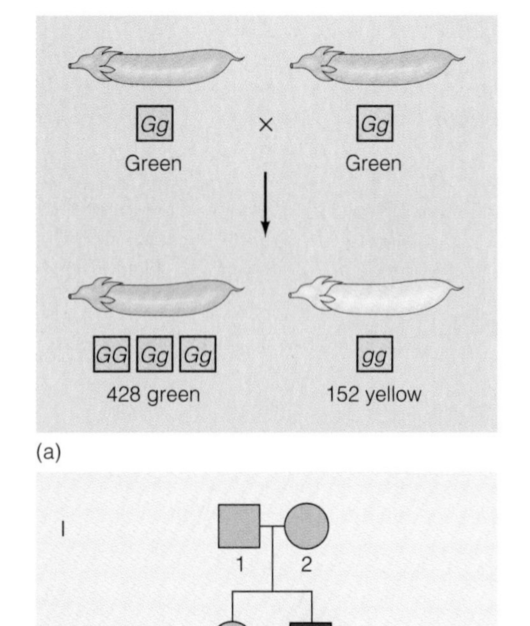

FIGURE 4.1 Inheritance in pea plants and humans. (a) In pea plants, a cross between two heterozygotes provides enough offspring in each phenotypic class to allow the pattern of inheritance to be determined. (b) Humans have relatively few offspring, often making it difficult to interpret how a trait is inherited.

Compared with the progeny that can be counted in a single cross with peas, humans produce very few offspring, and these usually represent only a small fraction of the possible genetic combinations. If two heterozygous pea plants are crossed ($Aa \times Aa$), about three-fourths of the offspring will express the dominant phenotype, and the recessive phenotype will be expressed in the remaining one-fourth of the progeny (Figure 4.1). Mendel was able to count hundreds and sometimes thousands of offspring from such a cross, to record progeny in all expected phenotypic classes, and to clearly establish a phenotypic ratio of 3:1 for recessive traits. As a parallel, consider two humans, each of whom is phenotypically normal. Suppose this couple has two children, one an unaffected daughter, and the other a son affected with a genetic disorder. The ratio of phenotypes in this case is 1:1. This makes it difficult to decide whether the trait is carried on an autosome or a sex chromosome, whether it is a dominant or recessive trait, and whether it is controlled by a single gene or by two or more genes.

This example reminds us that the basic method of genetic analysis in humans is observational rather than experimental and requires reconstructing events that have already taken place rather than designing experiments to test a hypothesis directly. As outlined in the last chapter, one of the first steps in studying a human trait is pedigree construction. Then, the information in the pedigree is used to determine how a trait is inherited. This chapter focuses on the analysis of pedigrees and their use in human genetics.

PEDIGREE ANALYSIS IS A BASIC METHOD IN HUMAN GENETICS

A pedigree is an orderly presentation of family information, using standardized symbols. Once a pedigree has been constructed, the principles of Mendelian inheritance are used to determine how the trait in question is inherited. Analysis of the pedigree can determine whether the trait has a dominant or recessive pattern of inheritance and whether the gene in question is located on an autosome or a sex chromosome. In addition, the information in the pedigree can be used in other ways, and we will discuss those applications later in this chapter.

Collection of pedigree information is not always straightforward. Knowledge about distant relatives is often incomplete, and recollections about medical conditions can be blurred by the passage of time. Older family members are sometimes

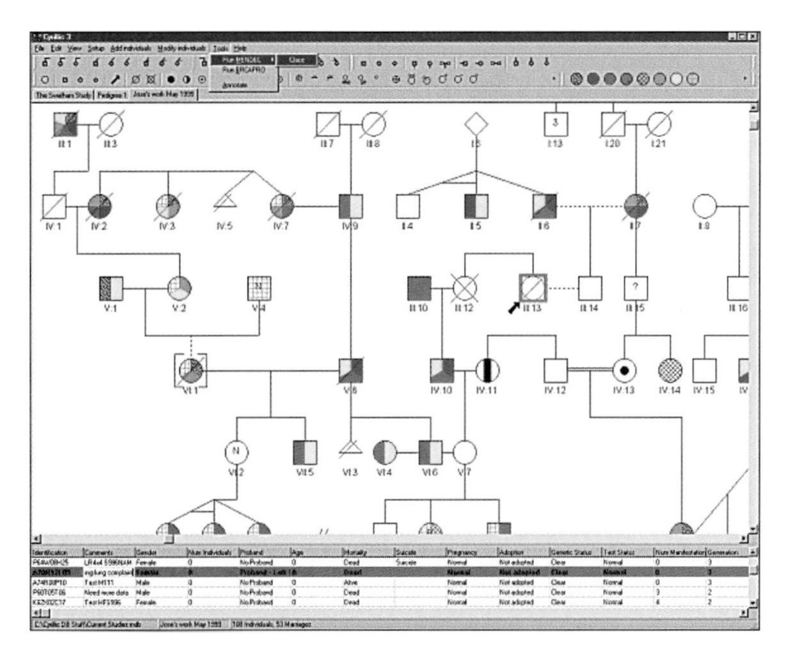

FIGURE 4.2 Software programs such as Cyrillic can be used to prepare pedigrees, store information, and analyze pedigrees. (Images by kind permission of FamilyGenetix Ltd. © 2002, FamilyGenetix Ltd. All rights reserved.)

reluctant to discuss relatives who had abnormalities or who were placed in institutions. As a result, gathering information for pedigree construction can be a challenge for the geneticist. In addition, organizing and storing the pedigree information for several generations in a large family can be a difficult task. The collection, storage, and analysis of pedigree information can now be done using software such as Cyrillic (Figure 4.2). These programs give on-screen displays of pedigrees and genetic information that can be used to analyze patterns of inheritance.

Once a pedigree has been constructed, the information in the pedigree is analyzed to determine how the trait is inherited. The patterns of inheritance we consider in this chapter include

- autosomal recessive
- autosomal dominant
- X-linked dominant
- X-linked recessive
- Y-linked
- mitochondrial inheritance

As outlined previously, pedigrees can be difficult to construct. For several reasons, they can also be difficult to analyze. In analyzing a pedigree, a geneticist first uses principles of Mendelian inheritance to rule out patterns of inheritance that are inconsistent with the pedigree. For example, only males carry a Y chromosome. If a trait is controlled by a gene carried on the Y, then only males will be affected. If the pedigree shows affected females, then Y-linked inheritance can be ruled out. Analysis of the pedigree is complete only when all possible patterns of inheritance have been considered. If all other possible ways of inheritance have been ruled out and only one pattern of inheritance is supported by the information in the pedigree, it is accepted as the pattern of inheritance for the trait being examined.

However, it may turn out that the pedigree does not provide enough information to rule out other possible patterns of inheritance. For example, analysis of a pedigree may indicate that a trait can be inherited in an autosomal dominant or an X-linked dominant fashion. If this is so, the pedigree is examined to determine if one manner of transmission is more likely. Then, the most likely type of inheritance is used as the basis for further work. If no information is available to select the most likely pattern, the geneticist is forced to conclude that the trait can be explained by autosomal dominant or X-linked dominant inheritance, and that more work is necessary to identify the pattern of inheritance. This may require adding more family members to the pedigree or the analysis of pedigrees from other families with the same trait.

As a further complication, some genetic disorders have more than one pattern of inheritance. Ehlers-Danlos syndrome (◗ Figure 4.3) (MIM/OMIM 130000 and other numbers), characterized by loose joints and easily stretched skin, can be inherited as an autosomal dominant, autosomal recessive, or X-linked recessive trait. In other cases, a trait can have a single pattern of inheritance but be caused by mutation in any of several genes. Porphyria (MIM/OMIM 176200 and other numbers), a metabolic disorder associated with abnormal behavior, is inherited as an autosomal dominant trait. However, it can be caused by mutation in genes on chromosomes 1, 9, 11, and 14.

It is important for several reasons to establish how a trait is inherited. If the pattern of inheritance can be established, it can be used to predict genetic risk in several situations, including

- pregnancy outcome
- adult onset disorders
- recurrence risks in future offspring

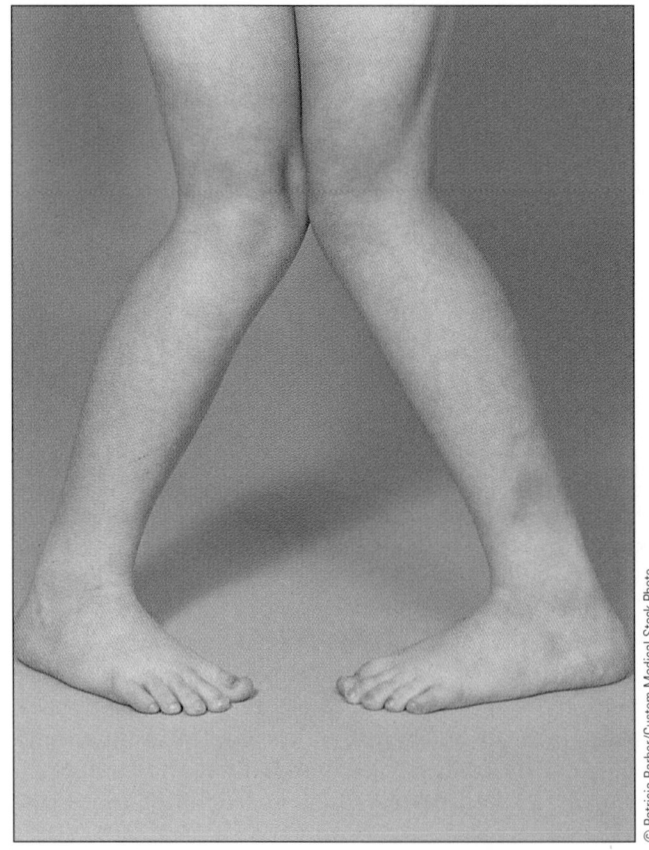

© Patricia Barber/Custom Medical Stock Photo

◗ **FIGURE 4.3** Ehlers-Danlos syndrome. This disorder can be inherited as an autosomal dominant, autosomal recessive, or X-linked recessive trait. People who have the common autosomal dominant form have loose joints and highly elastic skin, which can be stretched by several inches but returns to its normal position when released.

THERE IS A CATALOG OF HUMAN GENETIC TRAITS

In this chapter we deal with the six possible patterns of inheritance listed previously and use Mendelian principles to analyze pedigrees for these traits. For simplicity, we will limit our discussion to traits controlled by a single gene. Near the end of the chapter, we consider factors that can influence gene expression. In the next chapter, we discuss traits that are controlled by two or more genes.

To keep track of genetic disorders and the genes that control them, Victor McKusick, a geneticist at Johns Hopkins University, and his colleagues have compiled a catalog of human genetic traits. The catalog is published in book form as *Mendelian Inheritance in Man: Catalogs of Human Genes and Genetic Disorders* (MIM). The catalog is also available on the World Wide Web as "Online Mendelian Inheritance in Man" (OMIM). The online version contains text, pictures, references, and links to other databases (◗ Figure 4.4). Each trait is assigned a catalog number (called the MIM or OMIM number). In this chapter and throughout the book, the MIM/OMIM number for each trait discussed is listed. You can use the MIM/OMIM number to obtain more information about these traits. Access to OMIM is available through the book's home page.

PEDIGREE ANALYSIS FOR AUTOSOMAL RECESSIVE TRAITS

Although human families are relatively small, analysis of affected and unaffected members over several generations usually provides enough information to determine whether a trait has a recessive pattern of inheritance and is carried on an au-

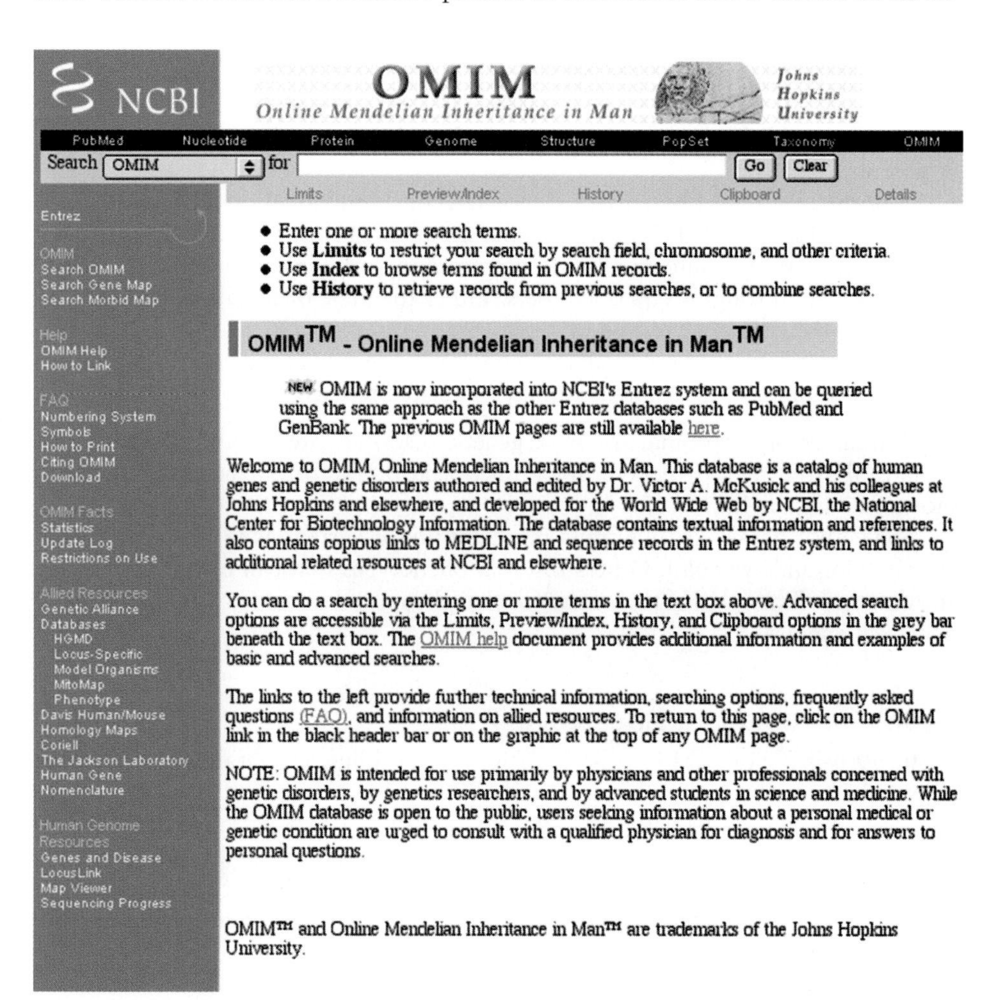

◗ **FIGURE 4.4** Online Mendelian Inheritance in Man (OMIM) is an online database that contains information about human genetic disorders. (Courtesy of Online Mendelian Inheritance in Man, National Center for Biotechnology Information.)

Concepts and Controversies

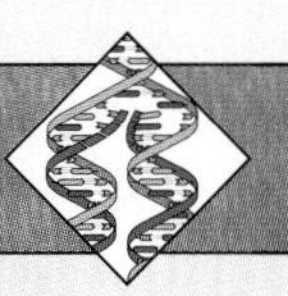

Was Noah an Albino?

The biblical character Noah, along with the ark and its animals, is among the most recognizable figures in the Book of Genesis. His birth is recorded in a single sentence, and although the story of how the ark was built and survived a great flood is told later, there is no mention of Noah's physical appearance. But other sources contain references to Noah that are consistent with the idea that Noah was one of the first albinos mentioned in recorded history.

The birth of Noah is recorded in several sources, including the Book of Enoch the Prophet, written about 200 B.C. This book, quoted several times in the New Testament, was regarded as lost until 1773, when an Ethiopian version of the text was discovered. In describing the birth of Noah, the text relates that his "flesh was white as snow, and red as a rose; the hair of whose head was white like wool, and long, and whose eyes were beautiful."

A reconstructed fragment of one of the Dead Sea Scrolls describes Noah as an abnormal child born to normal parents. This fragment of the scroll also provides some insight into the pedigree of Noah's family, as does the Book of Jubilees. According to these sources, Noah's father (Lamech) and his mother (Betenos) were first cousins. Lamech was the son of Methuselah, and Lamech's wife was a daughter of Methuselah's sister. This is important because marriage between close relatives is sometimes involved in pedigrees of autosomal recessive traits, such as albinism.

If this interpretation of ancient texts is correct, Noah's albinism is the result of a consanguineous marriage, and not only is he one of the earliest albinos on record but his grandfather Methuselah and Methuselah's sister are the first recorded heterozygous carriers of a recessive genetic trait.

tosome or a sex chromosome. Recessive traits carried on autosomes have several characteristics that can be established by pedigree analysis. Some of these are listed here:

- For rare or relatively rare traits, most affected individuals have unaffected parents.
- All of the children of two affected (homozygous) individuals are affected.
- The risk of an affected child from a mating of two heterozygotes is 25%.
- Because the trait is autosomal, it is expressed in both males and females, who are affected in roughly equal numbers. Either the male or female parent can transmit the trait.
- In pedigrees involving rare traits, the unaffected (heterozygous) parents of an affected (homozygous) individual may be related to each other.

A number of autosomal recessive genetic disorders are listed in Table 4.1. A pedigree illustrating a pattern of inheritance typical of autosomal recessive genes is shown in Figure 4.5.

Some autosomal recessive traits represent minor variations in phenotype, such as hair color and eye color. Others result in phenotypes that can be life threatening or even fatal. Examples of these more severe phenotypes include cystic fibrosis and sickle cell anemia.

Cystic Fibrosis Is a Recessive Trait

Cystic fibrosis A fatal recessive genetic disorder associated with abnormal secretions of the exocrine glands.

Cystic fibrosis (CF) (MIM/OMIM 219700) is a disabling and fatal genetic disorder inherited as an autosomal recessive trait affecting the glands that produce mucus, digestive enzymes, and sweat. Because the sweat glands are defective, they release excessive amounts of salt, and the disease is often diagnosed by analyzing the amount of salt in sweat. According to folklore, midwives would lick the forehead of newborns. If the sweat tasted too salty, they predicted that the child would die prematurely of lung congestion.

Table 4.1 Some Autosomal Recessive Traits

Trait	Phenotype	MIM/OMIM Number
Albinism	Absence of pigment in skin, eyes, hair	203100
Ataxia telangiectasia	Progressive degeneration of nervous system	208900
Bloom syndrome	Dwarfism; skin rash; increased cancer rate	210900
Cystic fibrosis	Mucous production that blocks ducts of certain glands, lung passages; often fatal by early adulthood	219700
Fanconi anemia	Slow growth; heart defects; high rate of leukemia	227650
Galactosemia	Accumulation of galactose in liver; mental retardation	230400
Phenylketonuria	Excess accumulation of phenylalanine in blood; mental retardation	261600
Sickle cell anemia	Abnormal hemoglobin, blood vessel blockage; early death	141900
Thalassemia	Improper hemoglobin production; symptoms range from mild to fatal	141900 / 141800
Xeroderma pigmentosum	Lack of DNA repair enzymes, sensitivity to UV light; skin cancer; early death	278700
Tay–Sachs disease	Improper metabolism of gangliosides in nerve cells; early death	272800

This disease has far-reaching effects because the affected glands perform a number of vital functions. CF causes the production of thick mucus, which clogs the ducts that carry digestive enzymes from the pancreas to the small intestine, reducing the effectiveness of digestion. As a result, affected children often suffer from malnutrition in spite of an increased appetite and increased food intake. Eventually, cysts form in the pancreas and the gland degenerates into a fibrous structure, giving rise to the name of the disease. Because CF causes the produc-

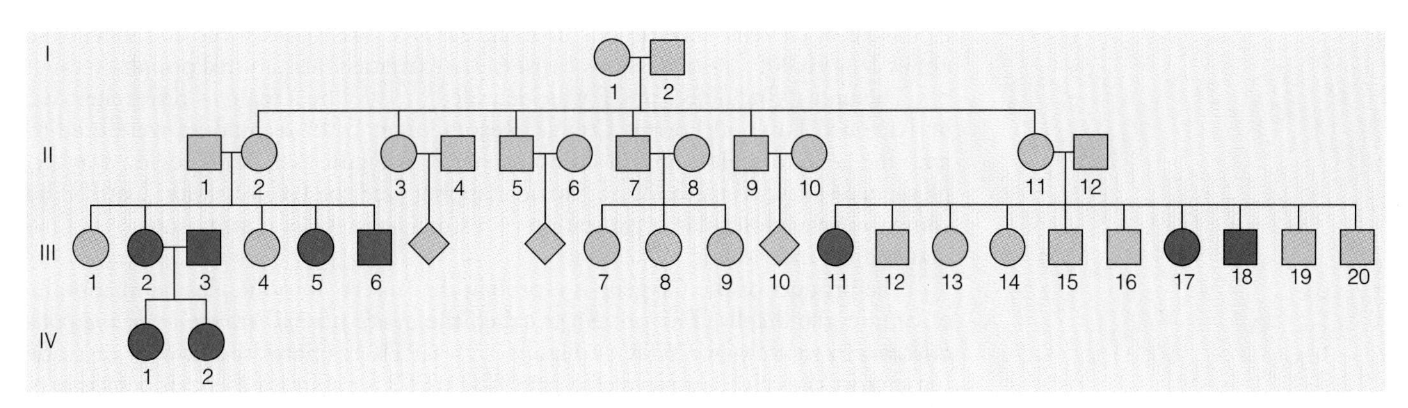

FIGURE 4.5 A pedigree for an autosomal recessive trait. This pedigree has many of the characteristics associated with an autosomal recessive pattern of inheritance. Most affected individuals have normal parents, about one-fourth of the children in large affected families show the trait, both sexes are affected in roughly equal numbers, and affected parents produce only affected children.

Web-Integrated Toolbox

Solving Pedigree Problems

Constructing and analyzing pedigrees is one of the most basic tools in human genetics. Pedigrees are used to determine whether a trait is inherited in a Mendelian pattern and can be used to identify family members who may be at risk for a deleterious trait or to predict the probability of having a child with this trait.

Pedigree analysis presents a different set of challenges than construction and is what we will concentrate on here. Analysis of multigenerational pedigrees requires different skills than those used to solve problems of Mendelian inheritance. For example, humans have far fewer offspring than peas or other organisms used in experimental genetics, leaving the geneticist to draw conclusions based on a small number of offspring. In addition, some disorders can be inherited in more than one way. For example, Ehlers-Danlos syndrome can be inherited as an autosomal dominant, autosomal recessive, or X-linked recessive trait.

The first task in pedigree analysis is to determine how the trait is inherited. Is it passed down as an autosomal dominant or autosomal recessive trait? Is the gene on the X or Y chromosome? Could a mitochondrial gene be responsible for this condition? Analysis proceeds by ruling out patterns of inheritance that do not fit with observations in the pedigree. For example, if affected females are present, a gene on the Y chromosome can be ruled out, because males pass this chromosome only to other males.

All patterns of inheritance that fail to be ruled out must be considered as possible ways the trait can be transmitted. Often, only one pattern of inheritance fails to be ruled out and can be accepted as the explanation for how the trait is passed from generation to generation. However, in some cases, two or more patterns remain after all others have been ruled out, making pedigree analysis a somewhat frustrating exercise to those who want a tidy solution to every problem.

The book's Web site contains a link to exercises that will help you develop the skills needed to analyze pedigrees. These exercises explain how and where to begin analyzing a pedigree and lay out the logic in a step-by-step process. Working out pedigree problems will help you see patterns of inheritance in pedigrees and complement the skills learned in solving Mendelian genetics problems.

tion of thick mucus that blocks airways in the lungs, most patients with cystic fibrosis develop obstructive lung diseases and infections that lead to premature death (◗ Figure 4.6).

Almost all cases of CF are children of phenotypically normal, heterozygous parents. CF is relatively common in some populations but rare in others (◗ Figure 4.7). Among the U.S. white population, CF has a frequency of 1 in 2,000 births, and 1 in 22 members of this group are heterozygous carriers. The disease is less common among the U.S. black population and has a frequency of 1 in 17,000 to 1 in 19,000. Among U.S. citizens with origins in Asia, CF is a rare disease whose frequency is about 1 in 90,000. Heterozygous carriers are extremely rare in this population.

The underlying defect in CF was identified in 1989 by a team of researchers led by Lap-chee Tsui and Francis Collins. Recombinant DNA techniques were used to map the gene to region q31 of chromosome 7 (◗ Figure 4.8). This region was explored using several methods of genetic mapping, and the CF gene was identified by comparing the molecular organization of a small segment of chromosome 7 in normal and CF individuals.

The product of the CF gene is a protein that inserts into the plasma membrane of certain gland cells. The protein is called the *cystic fibrosis transmembrane conductance regulator,* or CFTR (◗ Figure 4.9). CFTR regulates the flow of chloride ions across the cell's plasma membrane. The CFTR protein is absent or defective in affected individuals. Because fluids move across plasma membranes in response to the movement of ions, a defective CFTR protein causes less fluid to be added to the secretions of glands. The thickened secretions produce the characteristic symptoms of CF. It is hoped that further studies of the structure of the CF gene and the function of CFTR will lead to the development of new methods for treating this deadly disease.

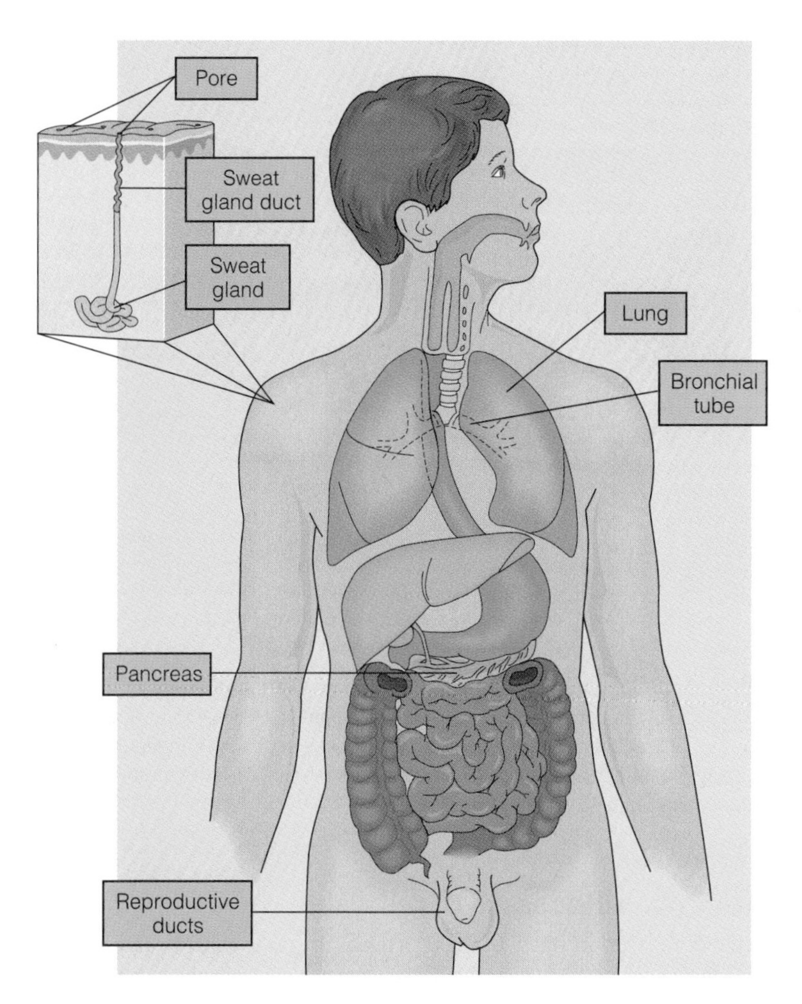

FIGURE 4.6 Organ systems affected by cystic fibrosis. Sweat glands in affected individuals secrete excessive amounts of salt. Thick mucus blocks the transport of digestive enzymes in the pancreas. The trapped digestive enzymes gradually break down the pancreas. The lack of digestive enzymes results in poor nutrition and slow growth. Cystic fibrosis affects both the upper respiratory tract (the nose and sinuses) and the lungs. Thick, sticky mucus clogs the bronchial tubes and the lungs, making breathing difficult. It also slows the removal of viruses and bacteria from the respiratory system, resulting in lung infections. In males, mucus blocks the ducts that carry sperm, and only about 2% to 3% of affected males are fertile. In women who have cystic fibrosis, thick mucus plugs the entrance to the uterus, lowering fertility.

FIGURE 4.7 About 1 in 25 Americans of European descent, 1 in 46 Hispanics, 1 in 60 to 65 African-Americans, and 1 in 150 Asian-Americans is a carrier for cystic fibrosis. A crowd such as this may contain a carrier.

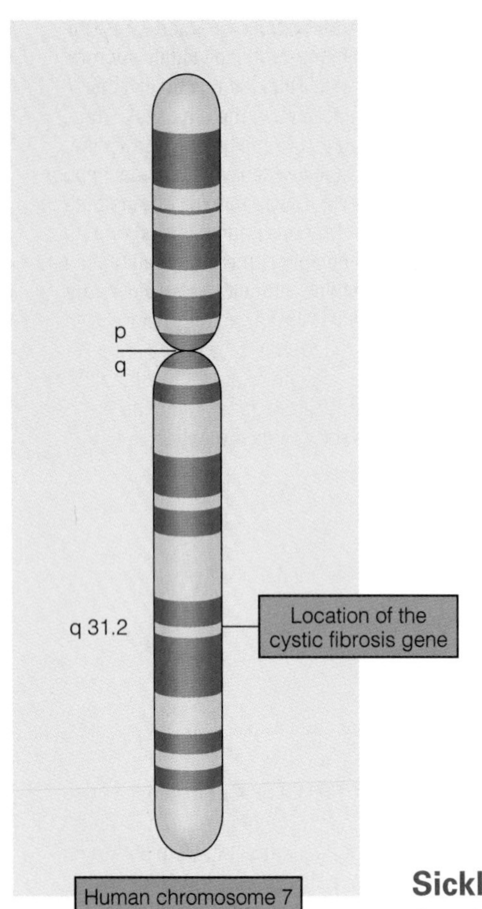

FIGURE 4.8 Human chromosome 7. The gene for cystic fibrosis (CF) maps to region 7q31.2-31.3, about two-thirds of the way down the long arm of the chromosome.

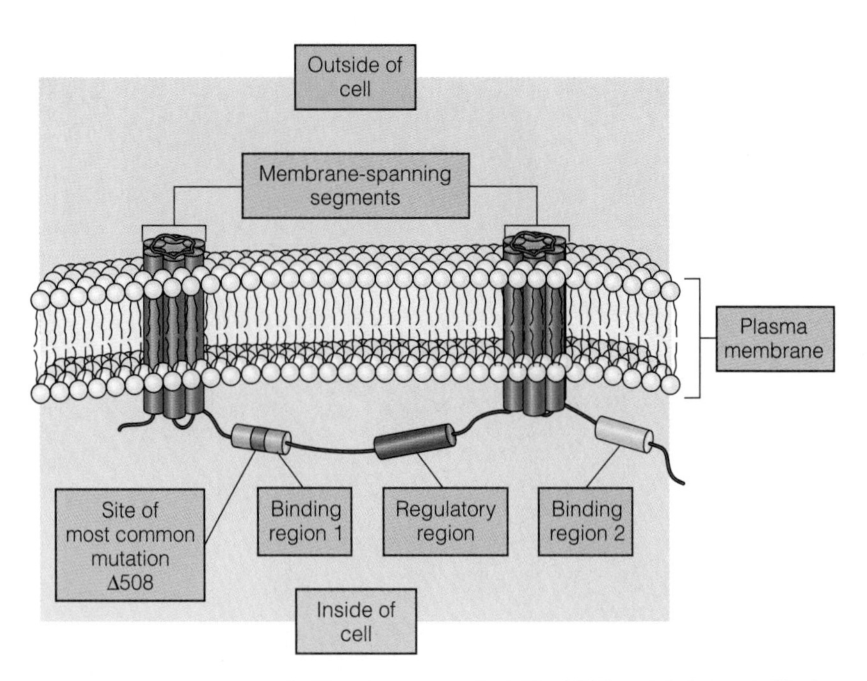

FIGURE 4.9 The cystic fibrosis gene product. The CFTR protein is located in the plasma membrane of the cell and regulates the movement of chloride ions across the cell membrane. The regulatory region controls the activity of the CFTR molecule in response to signals from inside the cell. In most cases (about 70%), the protein is defective in binding region 1.

Sickle cell anemia A recessive genetic disorder associated with an abnormal type of hemoglobin, a blood transport protein.

Sickle Cell Anemia Is a Recessive Trait

Americans whose ancestors lived in parts of West Africa, in the lowlands around the Mediterranean Sea, or in parts of the Indian subcontinent have a high frequency of **sickle cell anemia** (MIM/OMIM 141900). Individuals with this recessive genetic disorder produce an abnormal type of hemoglobin, a protein found in red blood cells. Normally, this protein transports oxygen from the lungs to the tissues of the body. Each red cell contains millions of hemoglobin molecules.

In sickle cell anemia, abnormal hemoglobin molecules polymerize to form rods (Figure 4.10). The formation of rods causes the red blood cells to become crescent- or sickle-shaped (Figure 4.11). The deformed red blood cells are fragile and break open as they circulate through the body. New blood cells are not produced fast enough to replace those that are lost, and the oxygen-carrying capacity of the blood is reduced, causing anemia. Individuals with sickle cell anemia tire easily and often develop heart failure because of the increased load on the circulatory system. The deformed blood cells also clog small blood vessels and capillaries, further reducing oxygen transport and sometimes initiating a sickling crisis. As oxygen levels fall in the circulatory system, more and more red blood cells become sickled, causing intense pain as blood vessels become blocked. In some affected areas of the body, ulcers and sores appear on the body surface. Blockage of the blood vessels in the brain leads to stroke and can result in partial paralysis.

Because of the number of organ systems affected and the severity of the effects, untreated sickle cell anemia can be lethal. Some affected individuals die in childhood or adolescence, but aggressive medical treatment allows survival into adulthood. As in CF, most affected individuals are children of phenotypically normal, heterozygous parents.

The high frequency of sickle cell anemia in certain populations is related to the frequency of malaria. Sickle cell heterozygotes are more resistant to malarial infection than homozygous normal individuals. The high frequency of this mutation in the U.S. black population is a genetic relic of African origins. In U.S. blacks, sickle cell anemia occurs with a frequency of 1 in every 500 births, and approximately

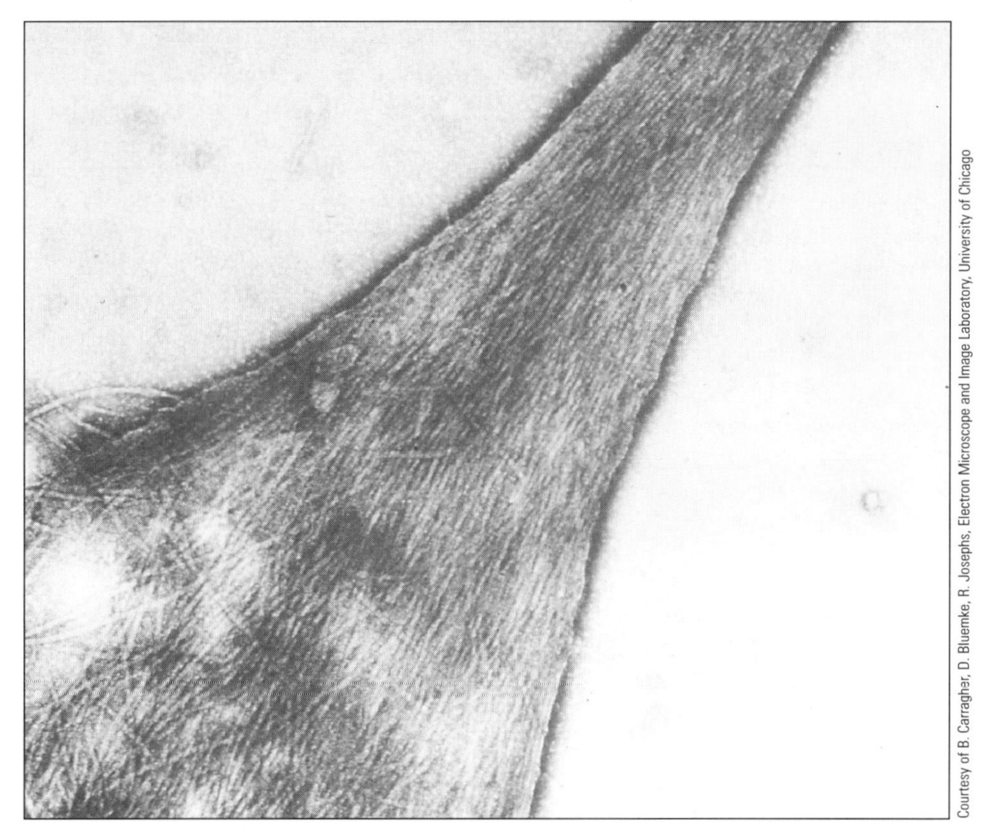

FIGURE 4.10 Hemoglobin molecules aggregate in sickle cell anemia. The mutant hemoglobin molecules in red blood cells stack together to form rodlike structures. The formation of these aggregates in the cytoplasm causes the red blood cells to deform and become elongated or sickle-shaped.

1 in every 12 individuals is heterozygous. The same is true for U.S. residents whose ancestral origins are in lowland regions of Italy, Sicily, Cyprus, Greece, and the Middle East. This hemoglobin defect has a double effect: It causes sickle cell anemia but also confers resistance to malaria. The molecular basis of this disease is well known and is discussed in later chapters.

PEDIGREE ANALYSIS FOR AUTOSOMAL DOMINANT TRAITS

In autosomal dominant disorders, heterozygotes and those with a homozygous dominant genotype express the abnormal phenotype. Unaffected individuals carry two recessive alleles. Careful analysis of pedigrees is necessary

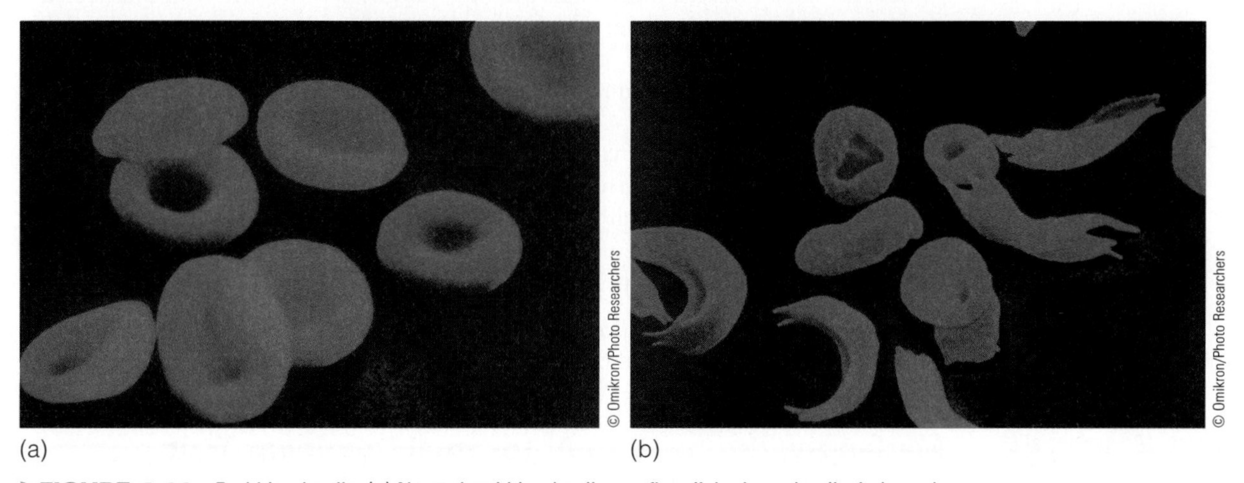

FIGURE 4.11 Red blood cells. (a) Normal red blood cells are flat, disk-shaped cells, indented in the middle on both sides. (b) In sickle cell anemia, the cells become elongated and fragile.

SIDEBAR

Gas Therapy for Sickle Cell Anemia

In sickle cell anemia (a recessive disorder), the mutant oxygen-carrying hemoglobin molecules polymerize, and the red blood cells become sickle-shaped. One approach to treating this disorder is to somehow increase the ability of the mutant hemoglobin molecules (HbS) to take up oxygen. Recently, researchers discovered that if red blood cells from individuals with sickle cell anemia are exposed to low concentrations of nitric oxide, they are able to carry up to 15% more oxygen. In further studies, volunteers with sickle cell anemia breathed air containing 80 ppm nitric oxide for 45 minutes. Red blood cells from these individuals showed an increase in oxygen-carrying capacity up to an hour after breathing the nitric oxide. Red blood cells from normal individuals showed no change in oxygen-carrying capacity. These results suggest that low concentrations of nitric oxide may be useful in the treatment of sickle cell anemia.

Table 4.2 **Some Autosomal Dominant Traits**

Trait	Phenotype	MIM/OMIM Number
Achondroplasia	Dwarfism associated with defects in growth regions of long bones	100800
Brachydactyly	Malformed hands with shortened fingers	112500
Camptodactyly	Stiff, permanently bent little fingers	114200
Crouzon syndrome	Defective development of mid-face region, protruding eyes, hook nose	123500
Ehlers–Danlos syndrome	Connective tissue disorder, elastic skin, loose joints	130000
Familial hypercholesterolemia	Elevated levels of cholesterol; predisposes to plaque formation, cardiac disease; may be most prevalent genetic disease	144010
Adult polycystic kidney disease	Formation of cysts in kidneys; leads to hypertension, kidney failure	173900
Huntington disease	Progressive degeneration of nervous system; dementia; early death	143100
Hypercalcemia	Elevated levels of calcium in blood serum	143880
Marfan syndrome	Connective tissue defect; death by aortic rupture	154700
Nail-patella syndrome	Absence of nails, kneecaps	161200
Porphyria	Inability to metabolize porphyrins; episodes of mental derangement	176200

to determine whether a trait is caused by a dominant allele. Dominant traits have a distinctive pattern of inheritance:

- Every affected individual should have at least one affected parent. Exceptions can occur in cases in which the gene has a high mutation rate. (Mutation is the sudden appearance of a heritable trait that was not transmitted by the biological parents.)
- Because most affected individuals are heterozygotes who mate with unaffected (homozygous recessive) individuals, there is a 50% chance of transmitting the trait to each child.
- Because the trait is autosomal, the numbers of affected males and females are roughly equal.
- Two affected individuals may have unaffected children, again because most affected individuals are heterozygous. In contrast, two individuals affected with an autosomal recessive trait have only affected children.
- In homozygous dominant individuals, the phenotype is often more severe than the heterozygous phenotype.

A number of autosomal dominant traits are listed in Table 4.2. The pedigree in Figure 4.12 is typical of the pattern found in autosomal dominant conditions.

Marfan Syndrome Is an Autosomal Dominant Trait

Marfan syndrome An autosomal dominant genetic disorder that affects the skeletal system, the cardiovascular system, and the eyes.

Marfan syndrome (MIM/OMIM 154700) is an autosomal dominant disorder that affects the skeletal system, the eyes, and the cardiovascular system. Affected indi-

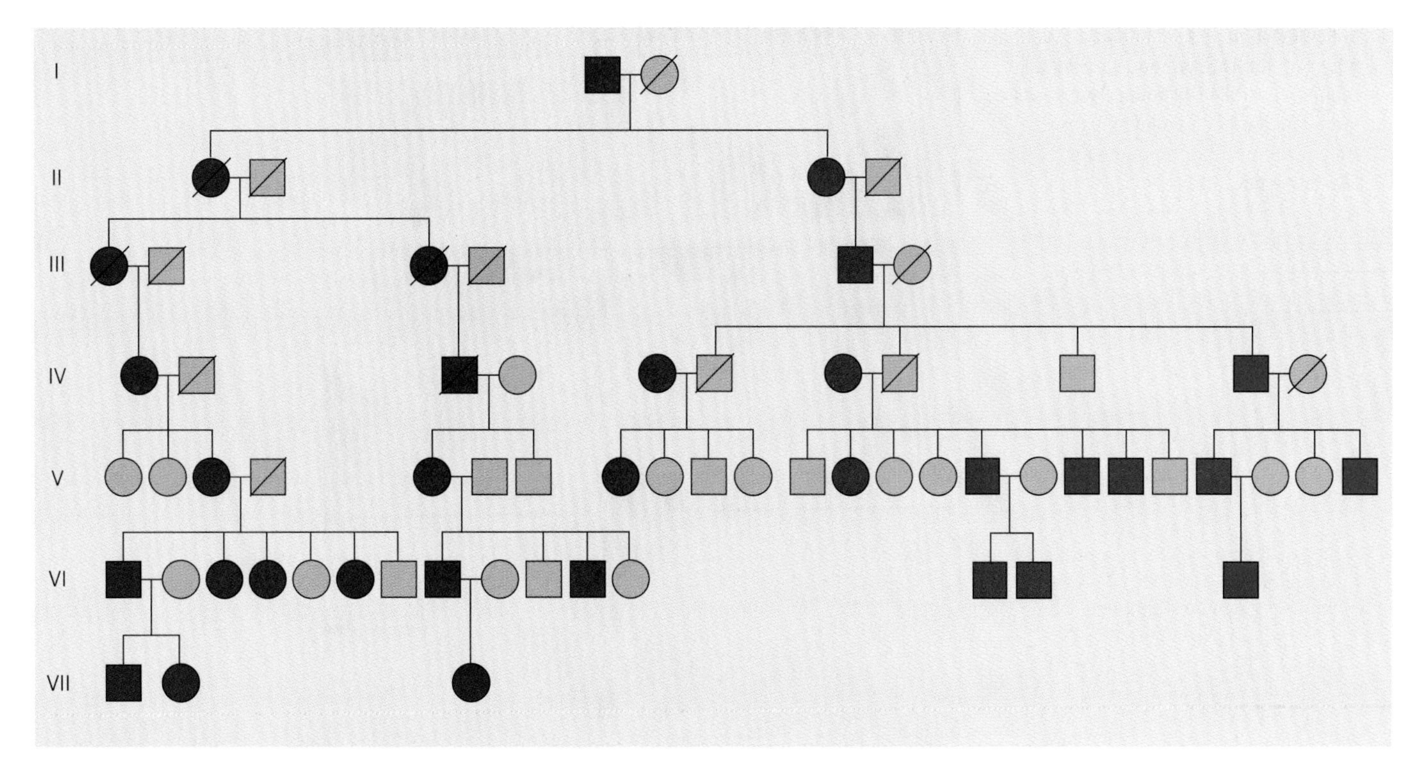

◗ **FIGURE 4.12** A pedigree for an autosomal dominant trait. This pedigree shows many of the characteristics of autosomal dominant inheritance. Affected individuals have at least one affected parent, about one-half of the children who have one affected parent are affected, both sexes are affected with roughly equal frequency, and affected parents can have unaffected children.

viduals tend to be tall and thin and have long arms and legs and long, thin fingers. Because of their height and long limbs, those with Marfan syndrome often excel in sports like basketball and volleyball, although nearsightedness and defects in the lens of the eye are also common (◗ Figure 4.13).

The most dangerous effects of Marfan syndrome are on the cardiovascular system, especially the aorta. The aorta is the main blood-carrying vessel in the body. As it leaves the heart, the aorta arches back and downward, feeding blood to all major organ systems. Marfan syndrome weakens the connective tissue around the base of the aorta, causing it to enlarge and eventually to split open (◗ Figure 4.14). In cases in which the enlargement can be detected, it can be repaired by surgery.

The gene that causes Marfan syndrome is located on chromosome 15. The normal product of the gene is a protein called fibrillin, which is part of connective tissue. The disorder affects males and females with equal frequency and is found in all ethnic groups at a frequency of about 1 in 10,000 individuals. About 25% of affected individuals appear in families with no previous history of Marfan syndrome, indicating that this gene undergoes mutation at a high rate.

As discussed in the chapter introduction, it has been suggested that Abraham Lincoln had Marfan syndrome. Bone fragments and hair from Lincoln's body are preserved at the National Medical Library in Washington, D.C. A group of research scientists has requested permission to analyze these samples using recombinant DNA techniques to decide whether Lincoln did, in fact, have Marfan syndrome. A committee appointed to review the request has agreed that this material can be tested, but it has recommended that testing be delayed until more is known about the fibrillin gene.

FIGURE 4.13 Flo Hyman was a star on the U.S. women's volleyball team in the 1984 Olympics. Two years later, she died in a volleyball game from a ruptured aorta caused by Marfan syndrome.

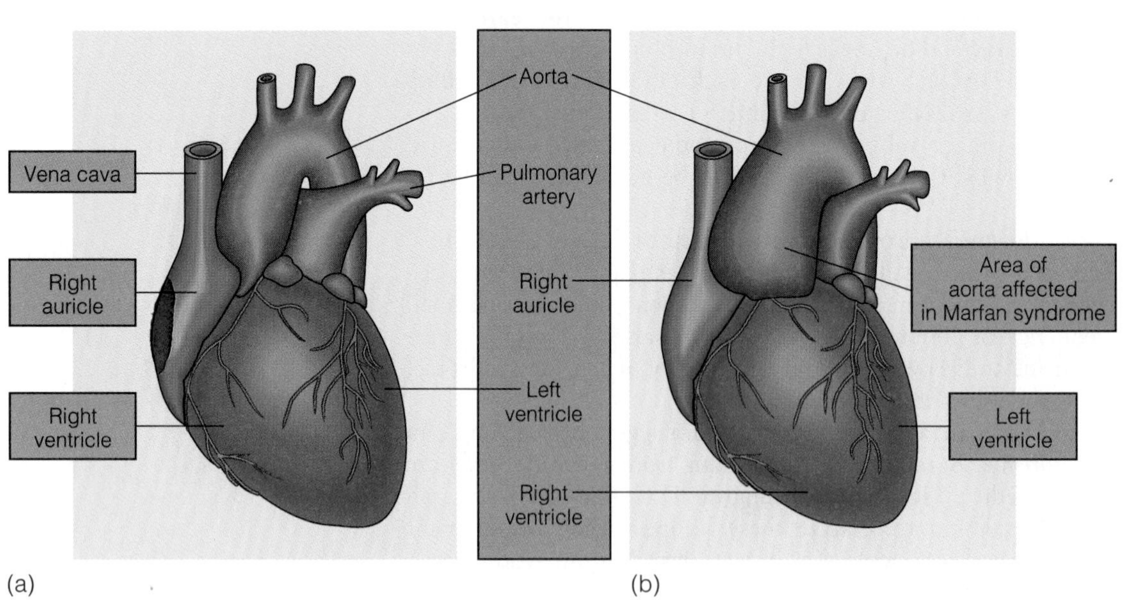

FIGURE 4.14 The heart and its major blood vessels. Oxygen-rich blood is pumped from the lungs to the left side of the heart. From there, blood is pumped through the aorta to all parts of the body.

SEX-LINKED INHERITANCE INVOLVES GENES ON THE X AND Y CHROMOSOMES

As noted in Chapter 2, females have two X chromosomes, and males have an X and a Y chromosome. These two chromosomes are very different in size and appearance. The X chromosome is a medium-sized, metacentric chromosome with a well-defined banding pattern. The Y chromosome is a much smaller, acrocentric chromosome, only about 25% as large as the X, with a variable banding pattern (Figure 4.15). At meiosis, the X and Y chromosomes pair only at a small region at the tip of the short arms. The absence of pairing along most of the chromosome suggests that the great majority of genes on the X chromosome are not represented on the Y.

This lack of genetic equivalence between the X and Y chromosomes is responsible for a pattern of transmission known as sex-linked inheritance. For genes on the X chromosome, the pattern is **X-linked**, and for genes on the Y chromosome, the pattern is **Y-linked.** Females carry two copies of all X-linked genes and can be heterozygous or homozygous for any of them. Males, on the other hand, have only one copy of the X chromosome. Because the Y chromosome does not carry copies of most X-linked genes, a male cannot be heterozygous for genes on the X chromosome.

This explains why males are affected by X-linked recessive genetic disorders more often than females. The term **hemizygous** is used for genes present in a single dose on the X chromosome in males. Because males cannot be heterozygous for these genes, recessive traits on the X chromosome are expressed in males. (Traits controlled by genes on the X chromosome are defined as dominant or recessive by their phenotype in females.)

X-linked The pattern of inheritance that results from genes located on the X chromosome.

Y-linked The pattern of inheritance that results from genes located only on the Y chromosome.

Hemizygous A gene present on the X chromosome that is expressed in males in both the recessive and dominant condition.

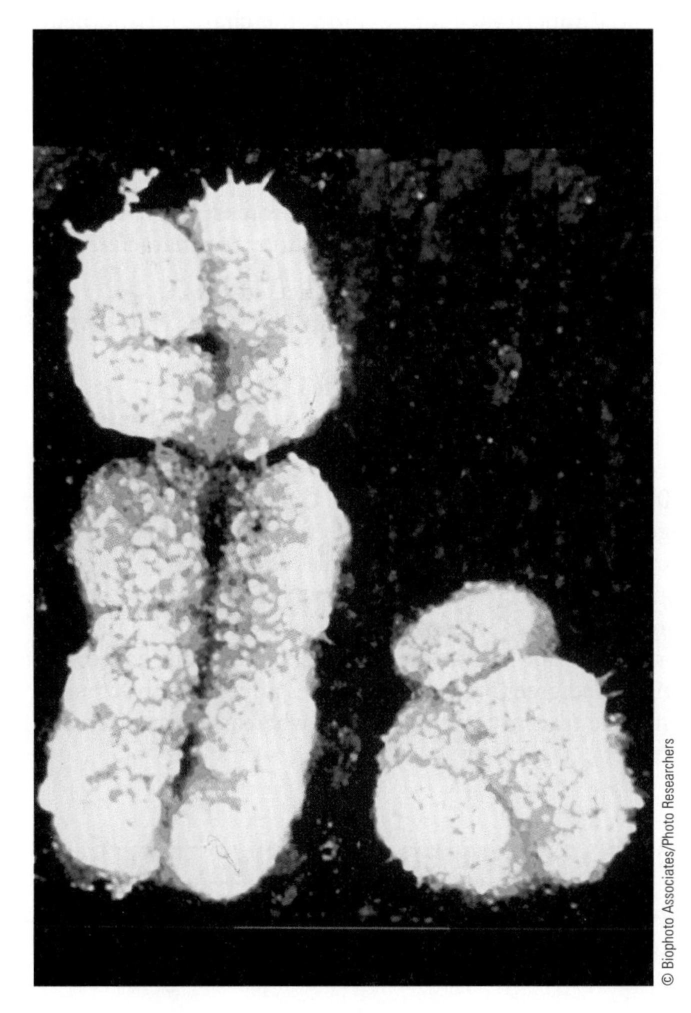

FIGURE 4.15 The human X chromosome (left) and the Y chromosome (right). This false-color scanning electron micrograph shows the differences between these chromosomes.

FIGURE 4.16 Distribution of sex chromosomes by parents. All children receive an X chromosome from their mothers. Fathers pass their X chromosome to all daughters and a Y chromosome to all their sons. The sex chromosome content of the sperm determines the sex of the child.

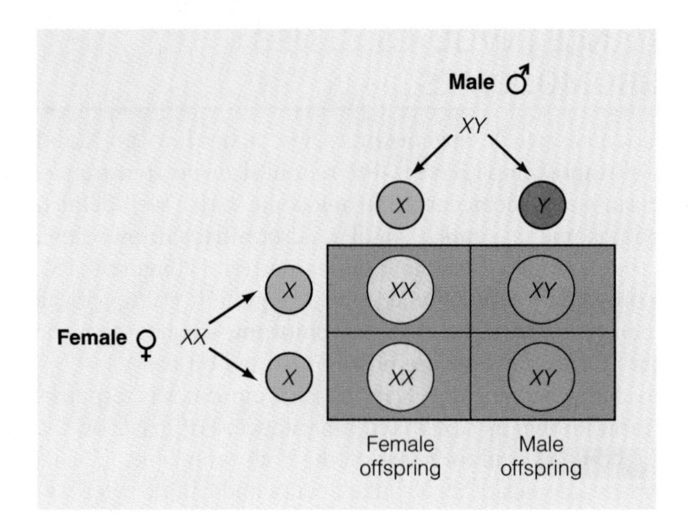

Males transmit their X chromosome to all daughters and their Y chromosome to all sons. Females randomly pass on one or the other X chromosome to all daughters and to all sons (Figure 4.16). As a result, the X and Y chromosomes have a distinctive pattern of inheritance. If a trait is X-linked, a male passes it to all of his daughters (who may be heterozygous or homozygous for the condition). If their mother is heterozygous for an X-linked recessive trait, sons have a 50% chance of receiving the recessive allele. In the following sections, we consider examples of sex-linked inheritance and explore the characteristic pedigrees in detail.

PEDIGREE ANALYSIS FOR X-LINKED DOMINANT TRAITS

Only a small number of dominant traits map to the X chromosome. Dominant X-linked traits have a distinctive pattern of transmission with three characteristics:

- Affected males produce all affected daughters and no affected sons.
- A heterozygous affected female will transmit the trait to half of her children, and males and females are equally affected.
- On average, twice as many females as males are affected.

Hypophosphatemia An X-linked dominant disorder. Those affected have low phosphate levels in blood and skeletal deformities.

As expected, a homozygous female will transmit the trait to all of her offspring. A pedigree for an X-linked dominant form of phosphate deficiency, **hypophosphatemia** (MIM/OMIM 307800) is shown in Figure 4.17. This disorder causes a type of rickets, or bowleggedness. To determine whether a trait is X-linked dominant or autosomal dominant, the children of affected males should be carefully analyzed. In X-linked dominant traits, affected males transmit the trait only to daughters. In autosomal dominant conditions, heterozygous affected males pass the trait to both daughters and sons, and about half of the daughters and about half of the sons are affected. As seen in the pedigree (Figure 4.17), males affected with X-linked dominant traits only transmit the trait to their daughters, whereas affected females have affected sons and daughters.

PEDIGREE ANALYSIS FOR X-LINKED RECESSIVE TRAITS

Males transmit an X chromosome to all daughters and not to their sons, and females transmit an X chromosome to all children. In addition, males are hemizygous for all genes on the X chromosome and express all X-linked genes. These

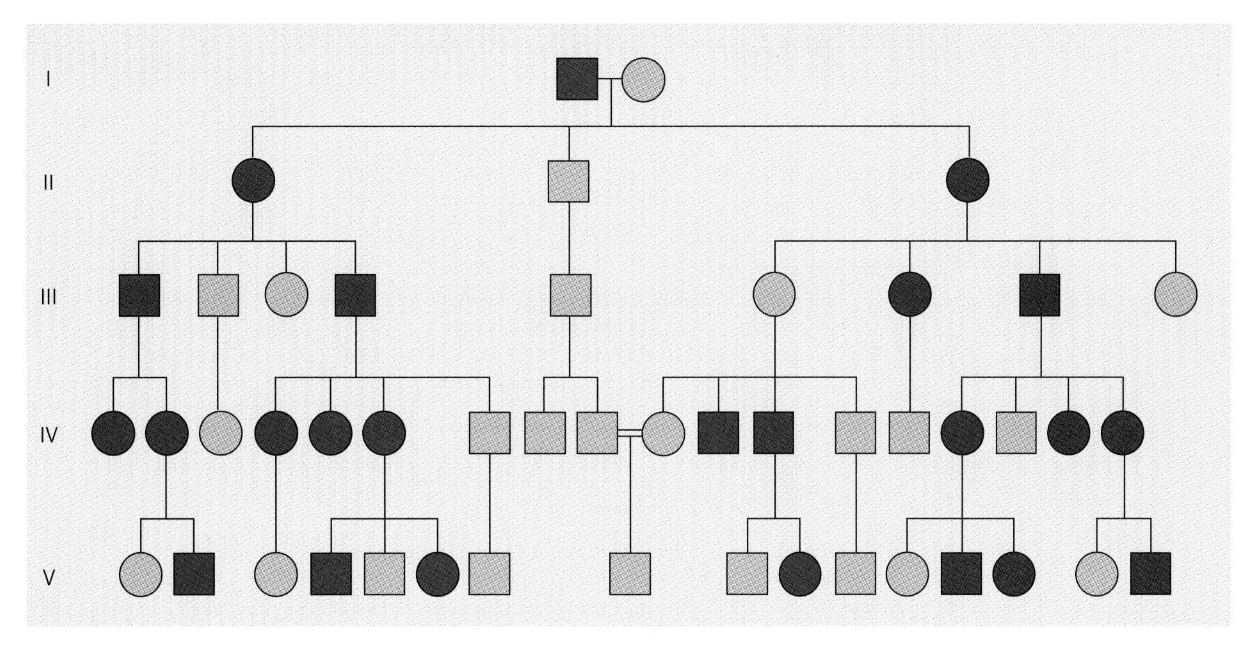

FIGURE 4.17 A pedigree for hypophosphatemia, an X-linked dominant trait. This pedigree shows characteristics of X-linked dominant traits. Affected males produce all affected daughters and no affected sons; affected females transmit the trait to roughly half their children, with males and females equally affected; and twice as many females as males are affected with the trait.

two factors produce a distinctive pattern of inheritance for X-linked recessive traits. This pattern can be summarized as follows:

- Hemizygous males and homozygous females are affected.
- Phenotypic expression is much more common in males than in females, and in the case of rare alleles, males are almost exclusively affected.
- Affected males get the mutant allele from their mothers and transmit it to all of their daughters but not to any sons.
- Daughters of affected males are usually heterozygous and therefore unaffected. Sons of heterozygous females have a 50% chance of receiving the recessive gene.

Some X-linked recessive conditions are listed in Table 4.3. A pedigree for an X-linked recessive trait is shown in Figure 4.18.

Color Blindness Is an X-Linked Recessive Trait

X-linked **color blindness** is actually a collection of several abnormalities of color vision. The most common form of color blindness, known as red-green blindness, affects about 8% of the male population in the United States. Red blindness (MIM/OMIM 303900) is characterized by the inability to see red as a distinct color (Figure 4.19). Green blindness (MIM/OMIM 303800) is the inability to see green and other colors in the middle of the visual spectrum (Figure 4.20). Both red blindness and green blindness are inherited as X-linked recessive traits. A rare form of blue color blindness (MIM/OMIM 190900) is inherited as an autosomal dominant condition that maps to chromosome 7.

Color blindness Defective color vision caused by reduction or absence of visual pigments. There are three forms: red, green, and blue blindness.

These three genes for color blindness encode different forms of opsins, proteins found in the cone cells of the retina (Figure 4.21). Normally, opsins bind to visual pigments in the red-, green-, or blue-cone cells, making the visual pigment/opsin complex sensitive to light of a given wavelength. If the red-opsin gene product is defective or absent, function of the red cones is impaired, and red color blindness results. Similarly, defects in the green or blue opsins produce green and blue blindness.

Table 4.3 Some X-Linked Recessive Traits

Trait	Phenotype	MIM/OMIM Number
Adrenoleukodystrophy	Atrophy of adrenal glands; mental deterioration; death 1 to 5 years after onset	300100
Color blindness		
Green blindness	Insensitivity to green light; 60 to 75% of color blindness cases	303800
Red blindness	Insensitivity to red light; 25 to 40% of color blindness cases	303900
Fabry disease	Metabolic defect caused by lack of enzyme alpha-galactosidase A; progressive cardiac and renal problems; early death	301500
Glucose-6-phosphate dehydrogenase deficiency	Benign condition that can produce severe, even fatal anemia in presence of certain foods, drugs	305900
Hemophilia A	Inability to form blood clots; caused by lack of clotting factor VIII	306700
Hemophilia B	"Christmas disease"; clotting defect cause by lack of factor IX	306900
Ichthyosis	Skin disorder causing large, dark scales on extremities, trunk	308100
Lesch–Nyhan syndrome	Metabolic defect caused by lack of enzyme hypoxanthine-guanine phosphoribosyl transferase (HGPRT); causes mental retardation, self-mutilation, early death	308000
Muscular dystrophy	Duchenne-type, progressive; fatal condition accompanied by muscle wasting	310200

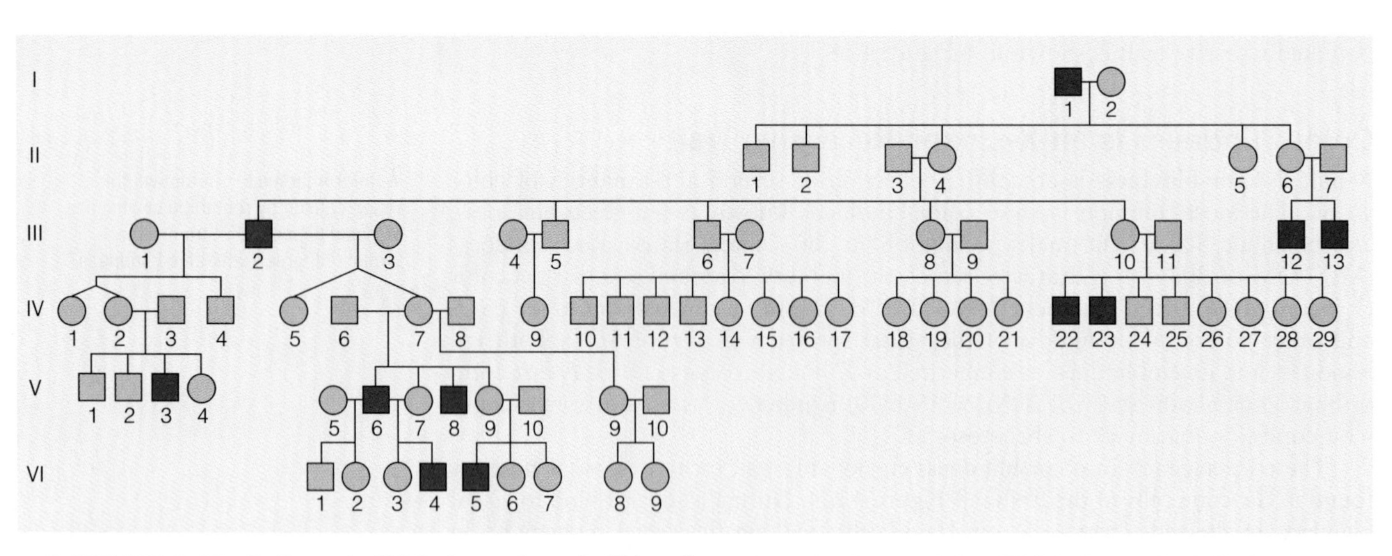

FIGURE 4.18 Pedigrees for X-linked recessive traits. This pedigree shows the characteristics of X-linked recessive traits: Hemizygous males are affected and transmit the trait to all daughters, who become heterozygous carriers, and phenotypic expression is much more common in males than in females.

Muscular Dystrophy Is an X-Linked Recessive Trait

Although often thought of as a single disorder, **muscular dystrophy** is a group of diseases with common features, including progressive weakness and wasting of muscle tissue. There are autosomal and X-linked forms of muscular dystrophy. Duchenne muscular dystrophy (DMD, MIM/OMIM 310200), an X-linked recessive disorder, is the most common form of muscular dystrophy. In the United States, DMD affects 1 in 3,500 males and usually has an onset between 1 and 6 years of age. Progressive muscle

Muscular dystrophy A group of genetic diseases associated with progressive degeneration of muscles. Two of these, Duchenne and Becker muscular dystrophy, are inherited as X-linked, allelic, recessive traits.

(both) © Eastcott/Momatiuk Photo Researchers

FIGURE 4.19 People who are colorblind see colors differen (a) Those who have normal vision see the red leaves. (b) Someone w is red-green colorblind sees the leaves as gray.

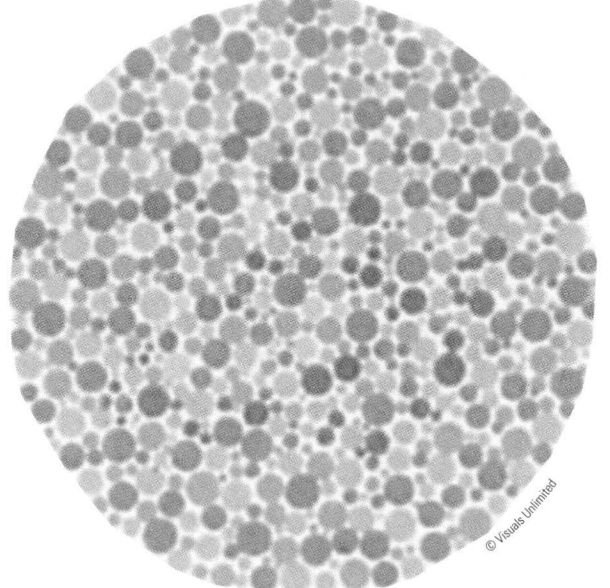

© Visuals Unlimited

FIGURE 4.20 People who have normal color vision see the number 29 in the chart; however, those who are colorblind cannot see any number.

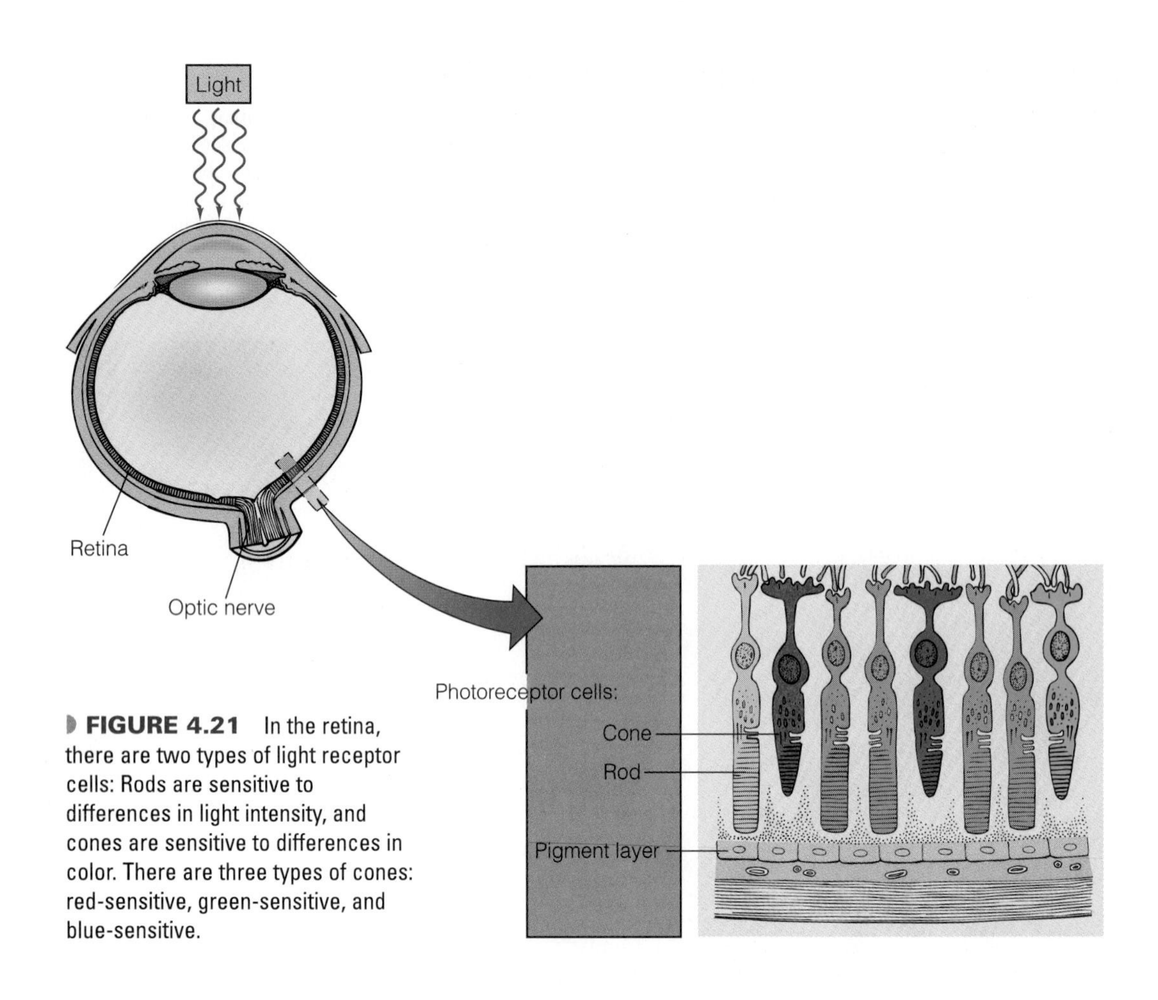

▶ **FIGURE 4.21** In the retina, there are two types of light receptor cells: Rods are sensitive to differences in light intensity, and cones are sensitive to differences in color. There are three types of cones: red-sensitive, green-sensitive, and blue-sensitive.

weakness is one of the first signs of DMD, and affected individuals use a distinctive set of maneuvers in rising from the prone position (▶ Figure 4.22). The disease progresses rapidly, and by 12 years of age, affected individuals are usually confined to wheelchairs because of muscle degeneration. Death usually occurs by the age of 20 years due to respiratory infection or cardiac failure.

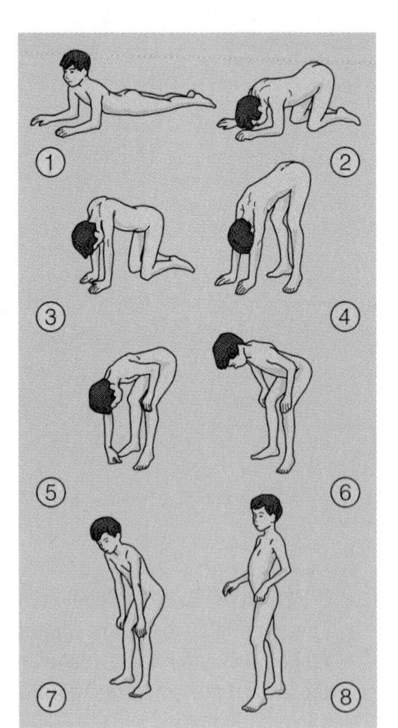

▶ **FIGURE 4.22** A sign of Duchenne muscular dystrophy. Children who have muscular dystrophy use a characteristic set of movements when rising from the prone position. Once the legs are pulled under the body, children use their arms to push the torso into an upright position.

The DMD gene is located at Xp21, a region in the middle of the short arm of the X chromosome, and encodes a protein called *dystrophin*. Normal forms of dystrophin attach to the cytoplasmic side of the plasma membrane in muscle cells and stabilize the membrane during the mechanical strains of muscle contraction (▶ Figure 4.23). When dystrophin is absent or defective, the plasma membranes gradually break down, causing the death of muscle tissue.

Most individuals with DMD have no detectable amounts of dystrophin in their muscle tissue. However, those affected with another form of muscular dystrophy, Becker muscular dystrophy (BMD, MIM/OMIM 310200), make a shortened dystrophin that is partially functional. As a result, those with BMD have a later age of onset, milder symptoms, and longer survival compared with DMD. DMD and BMD represent different mutant alleles of the same gene. The *DMD* gene has been isolated and cloned using recombinant

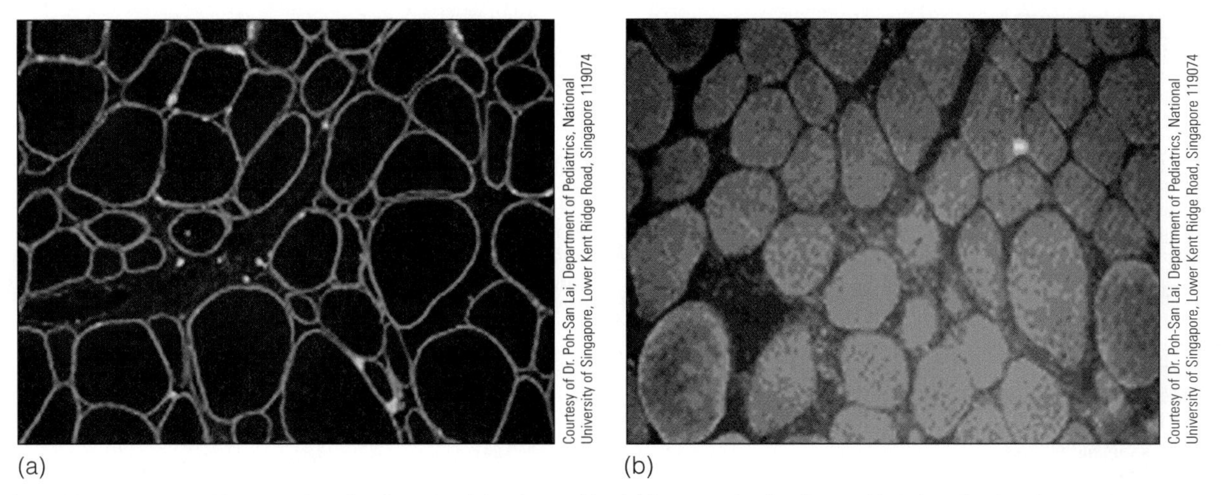

FIGURE 4.23 Photographs of cells stained for dystrophin. (a) In normal cells, dystrophin is localized to the membrane surrounding muscle cells. (b) In muscular dystrophy, no dystrophin is present.

DNA techniques. Future work on the structure and function of dystrophin will hopefully lead to the development of an effective treatment for muscular dystrophy.

Y-LINKED INHERITANCE INVOLVES TRANSMISSION FROM MALE TO MALE

Genes carried on the Y chromosome are said to be Y-linked. Because only males have Y chromosomes, these traits are present only in males and are passed directly from father to son. In addition, every Y-linked trait should be expressed, because males are hemizygous for all genes on the Y chromosome. Although the distinctive pattern of inheritance should make such genes easy to identify, only about three dozen Y-linked traits have been discovered. These include a gene for a protein found in the cell nucleus that may regulate gene expression. Another gene mapped to the Y chromosome, testis-determining factor (TDF/SRY, MIM/OMIM 480000), is involved in determining maleness in developing embryos. The *TDF/SRY* gene and early human development are discussed in Chapter 7. Some of the genes mapped to the Y chromosome are listed in Table 4.4. Figure 4.24 shows a pedigree for Y-linked inheritance.

Table 4.4 Some of the Genes Mapped to the Y Chromosome

Gene	Product	MIM/OMIM Number
ANT3 ADP/ATP translocase	Enzyme that moves ADP into, ATP out of mitochondria	403000
CSF2RA	Cell surface receptor for growth factor	425000
MIC2	Cell surface receptor	450000
TDF/SRY	Protein involved in early stages of testis differentiation	480000
H-Y antigen	Plasma membrane protein	426000
ZFY	DNA binding protein that may regulate gene expression	490000

SIDEBAR

Hemophilia, HIV, and AIDS

People with hemophilia who used blood and blood components to control bleeding episodes in the early 1980s were exposed to HIV, the virus that causes AIDS. This occurred because some blood donors unknowingly had HIV infection and contaminated the blood supply. The result was that many people, including more than half of the hemophilia patients in the United States, developed HIV infection. Most of the blood contamination took place before the cause of AIDS was discovered and before a test to identify HIV-infected blood was instituted. It has been estimated that nearly 10,000 individuals who have hemophilia are infected with HIV.

Fortunately, blood donor screening and new clotting products have virtually eliminated the risk of HIV transmission through blood products. As of January 1991, there have been no reports that anyone who exclusively received heat-treated, donor-screened products is infected with HIV.

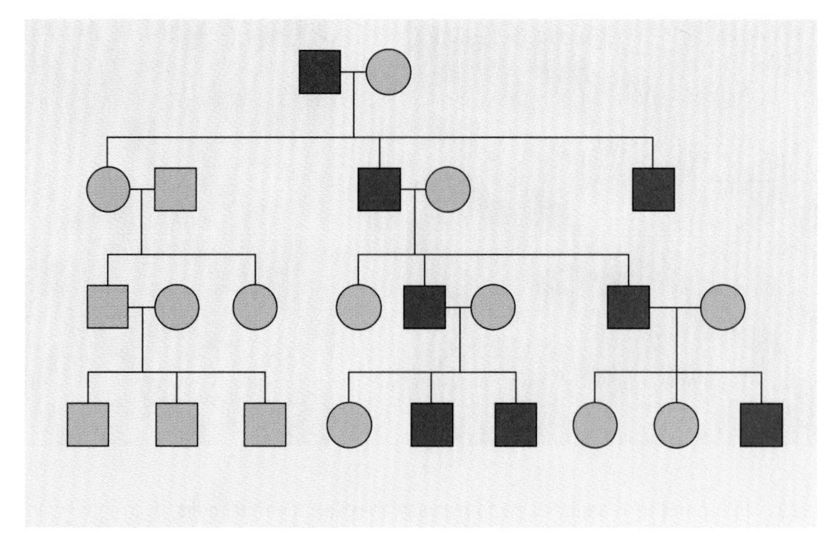

FIGURE 4.24 A pedigree for a Y-linked trait.

MITOCHONDRIAL GENES ARE TRANSMITTED FROM MOTHER TO OFFSPRING

Mitochondria are cytoplasmic organelles that convert energy from food molecules into ATP, a molecule that powers many cellular functions. Billions of years ago, ancestors of mitochondria were probably free-living bacteria that formed a symbiotic relationship with primitive eukaryotes. As an evolutionary relic of their free-living ancestry, mitochondria carry DNA molecules that encode information for some 37 mitochondrial genes. Most mitochondria carry five to ten of these DNA molecules, and each cell can contain from several hundred to over a thousand mitochondria. (Red blood cells are an exception; they have none.)

Mitochondria are transmitted from mothers to all their offspring through the cytoplasm of the egg. (Sperm lose all cytoplasm during maturation.) As a result, genetic disorders resulting from mutations in mitochondrial genes are maternally inherited. Both males and females can be affected by these disorders, but only females transmit mitochondria and any mutant genes they carry, producing a distinctive pattern of inheritance (Figure 4.25).

Genetic disorders in mitochondrial DNA are associated with defects in energy conversion. Tissues with the highest energy requirements are most affected. These include the nervous system, muscle, liver, and kidneys. Some of the disorders associated with mutations in mitochondria genes are listed in Table 4.5.

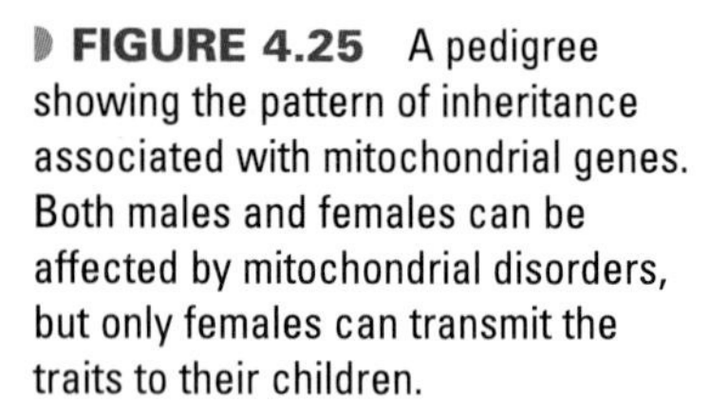

FIGURE 4.25 A pedigree showing the pattern of inheritance associated with mitochondrial genes. Both males and females can be affected by mitochondrial disorders, but only females can transmit the traits to their children.

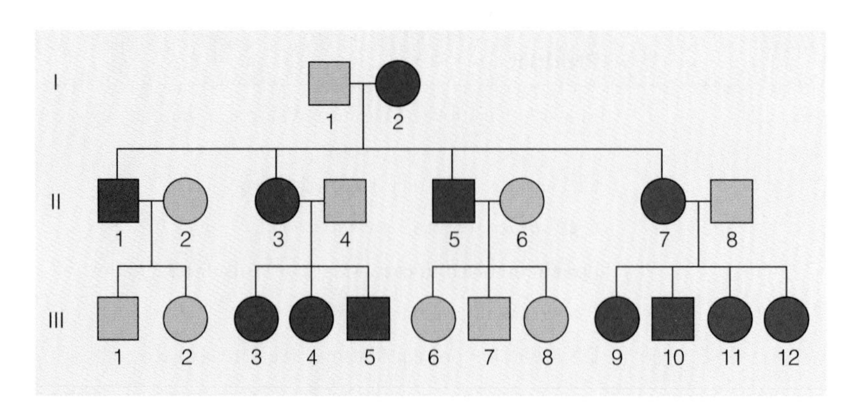

Concepts and Controversies

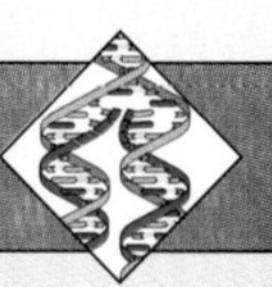

Hemophilia and History

Hemophilia, an X-linked recessive disorder, is characterized by defects in the mechanism of blood clotting. This form of hemophilia, called hemophilia A, occurs with a frequency of 1 in 10,000 males. Because only homozygous recessive females can have hemophilia, the frequency in females is much lower, on the order of 1 in 100 million.

Pedigree analysis indicates that Queen Victoria of England (the granddaughter of King George III) carried this gene. Because she passed the mutant allele on to several of her children (one affected male, two carrier daughters, and one possible carrier daughter), it is likely that the mutation occurred in the X chromosome she received from one of her parents. Although this mutation spread through the royal houses of Europe, the present royal family of England is free of hemophilia because it is descended from Edward VII, an unaffected son of Victoria.

Perhaps the most important case of hemophilia among Victoria's offspring involved the royal family of Russia. Victoria's granddaughter Alix, a carrier, married Czar Nicholas II of Russia. She gave birth to four daughters and then a son, Alexis, who had hemophilia. Frustrated by the failure of the medical community to cure Alexis, the royal couple turned to a series of spiritualists, including the monk Rasputin. While under Rasputin's care, Alexis recovered from several episodes of bleeding, and Rasputin became a powerful adviser to the royal family. Some historians have argued that the czar's preoccupation with Alexis's health and the insidious influence of Rasputin contributed to the revolution that overthrew the throne. Other historians point out that Nicholas II was a weak czar and that revolution was inevitable, but it is interesting to speculate that much of Russian history in the twentieth century turns on a mutation carried by an English queen at the beginning of the century.

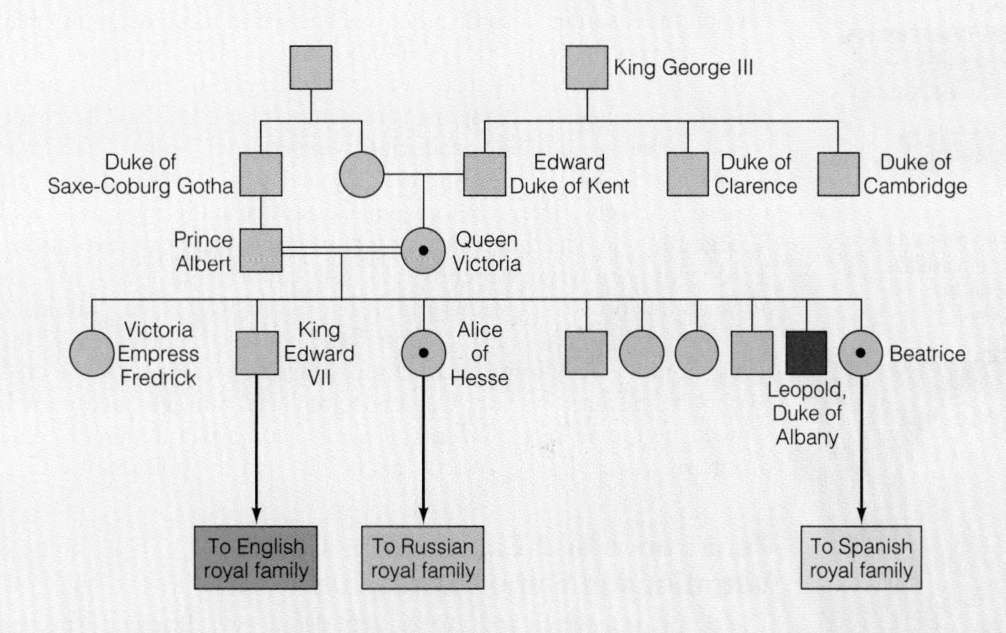

Table 4.5 Some Mitochondrial Traits

Trait	Phenotype	MIM/OMIM Number
Kearns–Sayre syndrome	Short stature; retinal degeneration	530000
Leber optic atrophy (LHON)	Loss of vision in center of visual field; adult onset	535000
MELAS syndrome	Episodes of vomiting, seizures, and stroke-like episodes	540000
MERRF syndrome	Deficiencies in the enzyme complexes associated with energy transfer	545000
Oncocytoma	Benign tumors of the kidney	553000

PEDIGREE ANALYSIS AND VARIATIONS IN GENE EXPRESSION

Many genes have regular and consistent patterns of expression, but others produce a wide range of phenotypes. In some cases, a mutant genotype may be present but remain unexpressed, resulting in a normal phenotype. Variation in phenotypic expression is caused by a number of factors, including interactions with other genes in the genotype and interactions between genes and the environment.

Age and Gene Expression

Although a large number of genes act before birth or early in development, the phenotypic expression of some genetic disorders is delayed until adulthood. One of the best known examples is **Huntington disease** (HD) (MIM/OMIM 143100), an autosomal dominant disorder. The phenotype of HD is first expressed between the ages of 30 and 50 years. Affected individuals undergo progressive degeneration of the nervous system, causing mental deterioration and uncontrolled, jerky movements of the head and limbs. The disease progresses slowly, and death occurs some 5 to 15 years after onset. This disorder is particularly insidious because onset usually occurs after the affected person has started a family. Because most affected individuals are heterozygotes, each child of an affected parent has a 50% chance of developing the disease. The gene for HD has been identified and cloned using recombinant DNA techniques. This makes it possible to test family members and identify those who will develop the disorder. This disorder is discussed in more detail in Chapter 16.

Huntington disease A dominant genetic disorder characterized by involuntary movements of the limbs, mental deterioration, and death within 15 years of onset. Symptoms appear between 30 and 50 years of age.

Porphyria A genetic disorder inherited as a dominant trait that leads to intermittent attacks of pain and dementia. Symptoms first appear in adulthood.

Penetrance The probability that a disease phenotype will appear when a disease-related genotype is present.

Expressivity The range of phenotypes resulting from a given genotype.

Camptodactyly A dominant human genetic trait that is expressed as immobile, bent, little fingers.

Porphyria (MIM/OMIM 176200), an autosomal dominant disorder, is also expressed later in life. This disease is caused by the inability to correctly metabolize porphyrin, a chemical component of hemoglobin. As blood levels of porphyrin increase, some is excreted, producing wine-colored urine. The elevated levels also cause episodes of seizures, intense physical pain, dementia, and psychosis. These symptoms rarely appear before puberty and usually appear in middle age. King George III, who was the British king during the American revolution, may have suffered from porphyria (Figure 4.26). He had a major attack in 1788 at the age of 50 years. He became delirious and suffered convulsions. His physical condition soon improved, but he remained irrational and confused. Early in 1789 his mental functions spontaneously improved, although his physicians took the credit. Later, after two more episodes, his son, George IV, replaced him. He died years later, blind and senile. The movie *The Madness of King George* is a fictionalized account of how porphyria affected King George, his family, and the politics of Great Britain.

FIGURE 4.26 King George III of Great Britain (1738–1820) was probably afflicted with porphyria, a genetic disorder that appears in adulthood and affects behavior.

Penetrance and Expressivity Cause Variations in Gene Expression

The terms *penetrance* and *expressivity* define two different aspects of variation in gene expression. **Penetrance** is the probability that a disease phenotype will appear when a disease-related genotype is present. If all individuals who carry the gene for a dominant disorder have the mutant phenotype, the gene is said to have 100% penetrance. If only 25% of those who carry the mutant gene show the mutant phenotype, the penetrance is 25%. Both genetic and environmental factors can affect penetrance. **Expressivity** refers to the range of phenotypic variation present. The following example shows the relationship between penetrance and expressivity.

The autosomal dominant trait **camptodactyly** (MIM/OMIM 114200) is caused by the improper attachment of muscles to bones in the little finger. The result is an immobile, bent, little finger. In some people, both little fingers are bent; in others, only one finger is affected; and in a small percentage of cases, neither finger is affected, even though a mutant allele is present (Figure 4.27). Because the trait is dominant, all heterozygotes and homozygotes should be af-

Guest Essay: Research and Applications

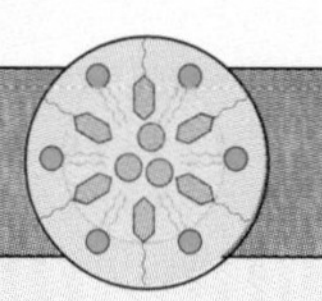

Nothing in Biology Makes Sense Without Evolution

Michael Rose

The best course I took in high school, maybe the best course I've ever taken, was comparative vertebrate anatomy. I loved slowly taking apart the bodies, learning the names of each part, carefully removing material to see little holes and tiny nerves. But before the gore came the scientific theory: fossils, Lamarck, Darwin, Mendel. The theory animated the corpses, made them meaningful, gave them some sense. At the time, I was a big science fiction fan, a genre in which every story had to have a portentous meaning. Evolution is the deep meaning behind life, and without it biology becomes a lot of details.

When I started my doctoral studies I was given the task of showing experimentally that evolution could make sense of biological aging, the irrevocable deterioration undergone by the adults of almost all animal species. At first I was entirely dismayed. Aging, throughout human history, has been a mystery. The people who talk or write most about it are charlatans, quacks, and hustlers. I feared that my career would be aborted by the combination of my failure to accomplish the task and my spattering with the mud of aging quackery. So far, my fears have proven erroneous.

Consider two dramatic genetic diseases, progeria and Huntington disease. It is now thought that each of these diseases is caused by a specific mutation at a single copy of a normal human gene, one gene for each disease. Progeria strikes children between 5 and 10 years and is extremely rare. Huntington disease, on the other hand, strikes primarily after the age of 30 and affects tens of thousands of individuals around the world.

Progeria strikes children and prevents their reproduction. In so doing, it also completely precludes the transmission of the progeria gene into the next generation. With 100% success, natural selection screens out new mutations for progeria. With Huntington disease, the patient may already have children before he shows any symptoms. Natural selection fails to screen the Huntington mutation out of the population. It fails because the gene has effects primarily at later ages, not earlier ages. The key is that the force of natural selection falls with adult age.

Genes that have effects on the health or development of young animals are sharply scrutinized by natural selection. Genes that have effects only at later ages have very little effect on reproduction and so are unimportant for natural selection. This fundamental idea was first proposed by evolutionary biologists in the 1930s and 1940s. My role has been to test this idea. One way to do so is to reverse the normal evolutionary situation, in which mostly early reproduction counts. Instead, we can prevent reproduction until experimental animals are much older and then keep the offspring produced at later ages to start the next generation. When I did this with lab populations of the common fruit fly *Drosophila,* they evolved increased life span over dozens of generations. This happened because we artificially strengthened natural selection at later ages. Fruit flies that couldn't survive to reproduce in their middle age were selected against. Thus life span was increased by natural selection, altering the genetics of aging and so increasing life span.

Flies that live longer are interesting beasts. They reproduce less when young, even when they have opportunity to do so. They resist stresses better, including starvation and desiccation. They move around more when they are older, whether walking or flying. They can reproduce more at later ages. Our longer lived flies are superior organisms—"superflies"!

Whenever I present these superflies to general audiences, the people in the audience want to know if I will ever be able to do the same things for them. Literally, the answer is no. But there is the possibility of learning more about the genetics of postponed aging in fruit flies, in the hope that we can apply our findings to humans. If this is ever done, it will radically transform the human life cycle, as aging becomes something that can be controlled, instead of merely endured.

Michael R. Rose is a professor of evolutionary biology at the University of California, Irvine. He received B.S. and M.S. degrees from Queen's University, Canada, and a Ph.D. from the University of Sussex, England. He is a member of the editorial boards of Experimental Gerontology, Journal of Anti-Aging Medicine, *and* Aging Research Reviews. *He has received research prizes from the American Society of Naturalists in 1992 and the World Congress of Gerontology in 1997.*

fected on both hands. The pedigree shows that one individual (III-4) is not affected even though he passed the trait to his offspring.

The pedigree in Figure 4.27 shows that nine people must carry the dominant allele for camptodactyly, but phenotypic expression is seen only in eight, giving 8/9, or 88%, penetrance. This is only an estimate because II-1, II-2, and III-1 produced no offspring and could also carry the dominant gene with no

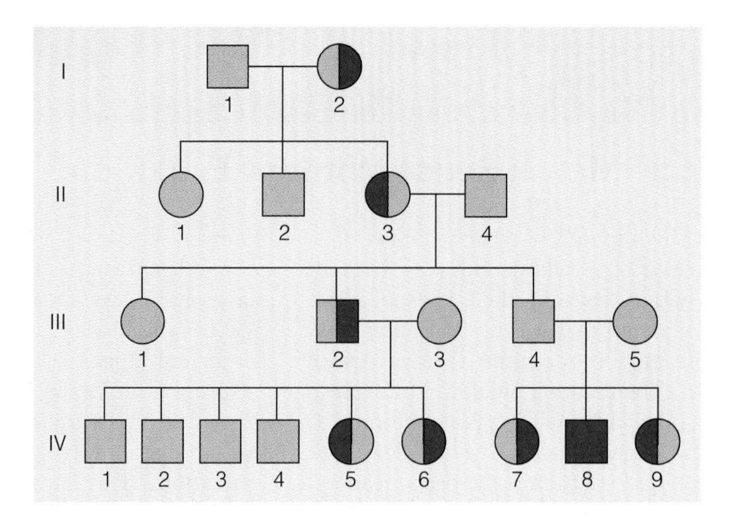

FIGURE 4.27 Penetrance and expressivity. This pedigree shows the transmission of camptodactyly in a family. Those who have two affected hands are shown as fully shaded symbols. Those affected only in the left hand are indicated by shading the left half of the symbol, and those affected only in the right hand have the right half of the symbol shaded. Symbols with light shading indicate unaffected family members.

penetrance. Many more samples from other pedigrees would be necessary to establish a reliable figure for penetrance of this gene.

Expressivity defines the *degree* of expression for a given trait. In the pedigree for camptodactyly, some individuals are affected on the left hand and others on the right hand; in one case both hands are affected; in another, neither hand is affected. This variable gene expression results from interactions with other genes and with nongenetic factors in the environment.

The variations in gene expression that we have discussed are all the result of the relationship between a gene and the mechanisms that produce the gene's phenotype. The inheritance of these genes follows the predictable pattern worked out by Mendel for traits in the pea plant, but expression can be complicated by factors that include temperature and age.

GENES ON THE SAME CHROMOSOME ARE LINKED

Linkage A condition in which two or more genes do not show independent assortment. Rather, they tend to be inherited together. Such genes are located on the same chromosome. By measuring the degree of recombination between linked genes, the distance between them can be determined.

There are about 35,000 genes distributed on the 24 human chromosomes (22 autosomes and the X and the Y chromosome). Obviously, this means that each chromosome carries many genes. Genes on the same chromosome are said to show **linkage,** because they tend to be inherited together. Julia Bell and J.B.S. Haldane discovered the first case of linkage in humans in 1936. By pedigree analysis, they demonstrated that the genes for hemophilia (MIM/OMIM 306700) and color blindness (MIM/OMIM 303800) are both on the X chromosome and are therefore linked.

Figure 4.28 shows a pedigree showing linkage between the gene for the ABO blood type (the *I* locus, MIM/OMIM 110300) and a condition called nail-patella syndrome (MIM/OMIM 161200). Nail-patella syndrome is an autosomal dominant disorder associated with deformities in the nails and kneecaps. In this pedigree, the *B* allele (I^B) and the allele for nail-patella syndrome occur together in individual I-2. In generations II and III, these two alleles tend to be inherited together. Individuals with type B blood tend to have nail-patella syndrome. In generation II, the *B* allele and the nail patella allele show linkage in individuals II-1, II-2, II-5, II-6, and II-14. In generation III, the two alleles are linked in individuals III-1 and III-2.

Although linked genes tend to be inherited together, they separate from each other in some offspring because of crossing-over between members of a chromosomal pair. As shown in the pedigree, type B blood and nail-patella syndrome are not always inherited together. Examination of the pedigree shows that 2 of the 16 individuals have inherited *one* of the two alleles but not both (individuals II-8 and III-3). This separation of the two alleles is the result of recombination between the two genes. Crossing-over and recombination were discussed in

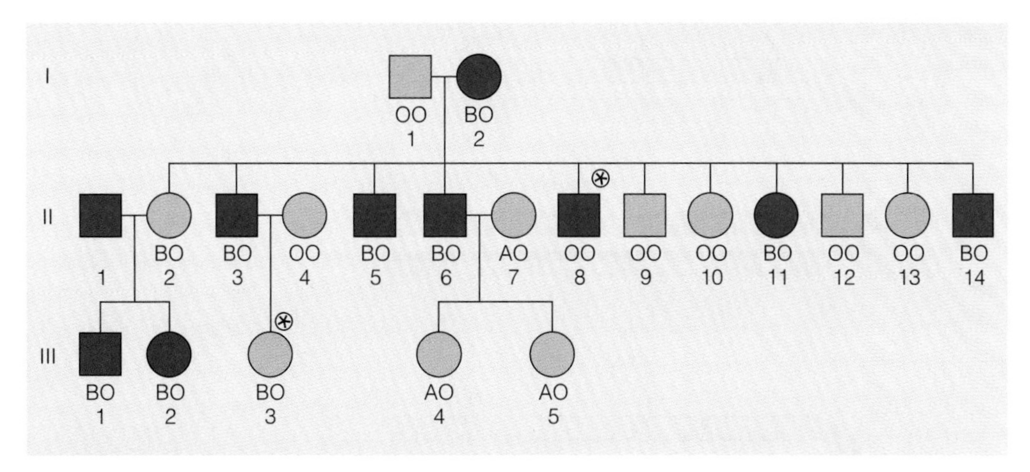

FIGURE 4.28 Linkage between nail-patella syndrome and the ABO blood type locus. Darkly shaded symbols in this pedigree represent those who have nail-patella syndrome, an autosomal dominant trait. Genotypes for the ABO locus are shown below each symbol. Nail-patella syndrome and the *B* allele are present in I-2; they tend to be inherited together in this family, and they are identified as linked genes. Individuals marked with an asterisk (II-8 and III-3) inherited the nail-patella allele or the *B* allele alone. This separation of the two alleles occurred by recombination.

Chapter 2. Recombination between linked genes can be used to establish the order and distance between genes on a chromosome.

Recombination Frequencies Are Used to Make Genetic Maps

Working with the fruit fly, *Drosophila*, Alfred Sturtevant realized that if two genes are close together on the same chromosome, crossing-over is rare; if they are far apart, crossing-over is common. He concluded that the amount of crossing-over between genes could be used to determine the order and distance between genes on a chromosome and produce a genetic map.

In a **genetic map**, genes are arranged in a linear order, and the distance between any two genes is measured by how frequently crossing-over takes place between them (Figure 4.29). In other words, genetic distance is measured by the frequency of recombination between loci on the same chromosome. The units are expressed as a percentage of recombination, where 1 map unit is equal to a frequency of 1% recombination. (This unit is also known as a centimorgan, or cM.)

Genetic map The arrangement and distance between genes on a chromosome deduced from studies of genetic recombination.

Remember, it is possible to construct genetic maps only because linked genes separate from each other by crossing-over during meiosis. How often this happens depends on the distance between the genes. The farther apart two genes are on a chromosome, the more likely it is that a crossover will separate them. Conversely, the closer together two genes are on a chromosome the less frequently they will be separated by a crossover (Figure 4.30).

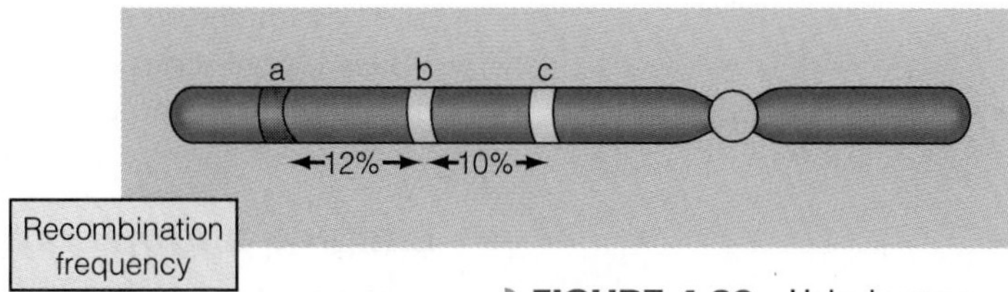

FIGURE 4.29 Linked genes are carried on the same chromosome. Recombinational frequencies can be used to construct genetic maps, giving the order and distance between genes.

Let's use this information to calculate the distance between the gene for nail-patella syndrome and the gene for the ABO blood group by measuring how often recombination occurs between them. In the pedigree in Figure 4.28, the nail-patella gene and the gene for the B blood type have separated in 2 of the 16 individuals (II-8 and III-3). From the frequency of recombination in this pedigree (2/16 = 0.125, or 12.5% recombination), the distance between the gene for the ABO locus and the gene controlling nail-patella syndrome can be calculated as 12.5 map units.

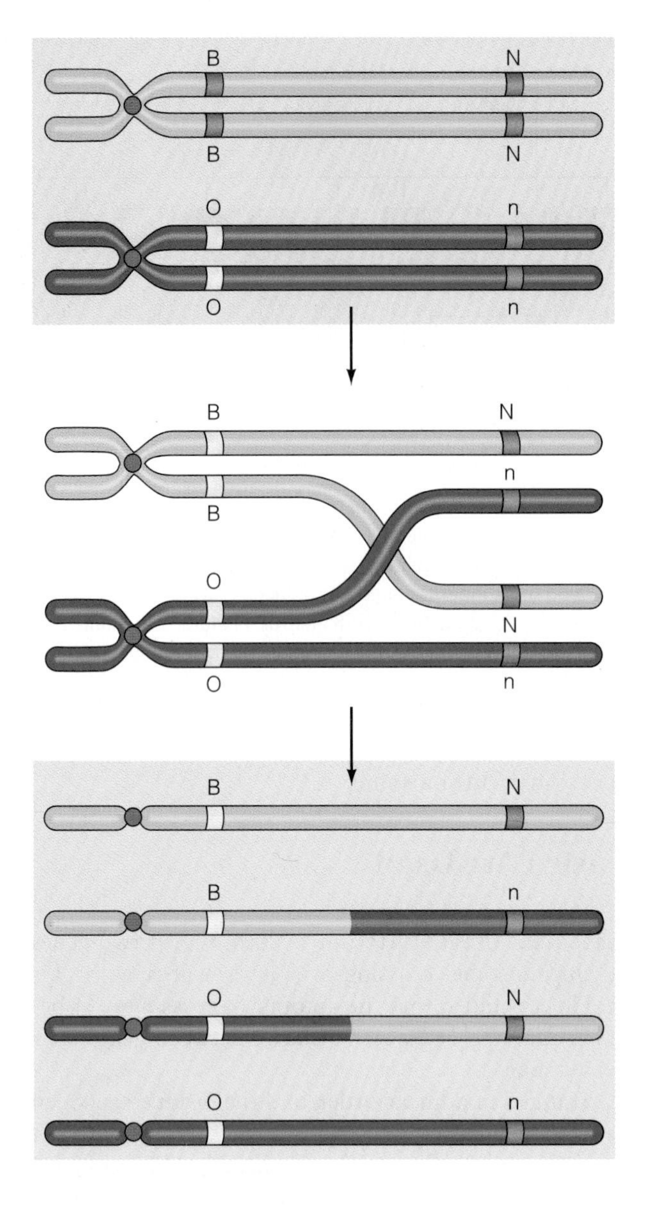

FIGURE 4.30 Crossing-over between homologous chromosomes during meiosis involves the exchange of chromosomal parts. In this case, crossing-over between the genes for blood type (alleles *B, O,* of gene *I*) and nail-patella syndrome (*N*) produces new allelic combinations. The frequency of crossing-over is proportional to the distance between the genes, allowing construction of a genetic map for this chromosomal region.

For accuracy, either a much larger pedigree or many more small pedigrees need to be examined to determine the extent of recombination between these two genes. When a large series of families is combined in an analysis, the map distance between the ABO locus and the nail-patella locus is about 10 map units.

Linkage and Recombination Can Be Measured by Lod Scores

In human genetics, it is difficult to establish linkage and measure the distance between genes because such studies require large pedigrees with many offspring covering several generations in families with two genetic disorders (the two genes being analyzed for linkage). In most cases, pedigrees do not include more than three generations—the grandparents, parents, and children. In addition, families with two genetic disorders that might map to the same chromosome are very rare. To get around these problems, geneticists use a statistical technique known as the **lod method** to determine whether two genes are linked and to measure the distance between them.

Lod method A probability technique used to determine whether genes are linked.

Lod scores can be calculated using software programs (like LINKMAP). First, an observed frequency of recombination between two genes is derived from pedigree studies. Then, the software is used to calculate two probabilities: the probability that the observed results would have been obtained if the two genes were linked,

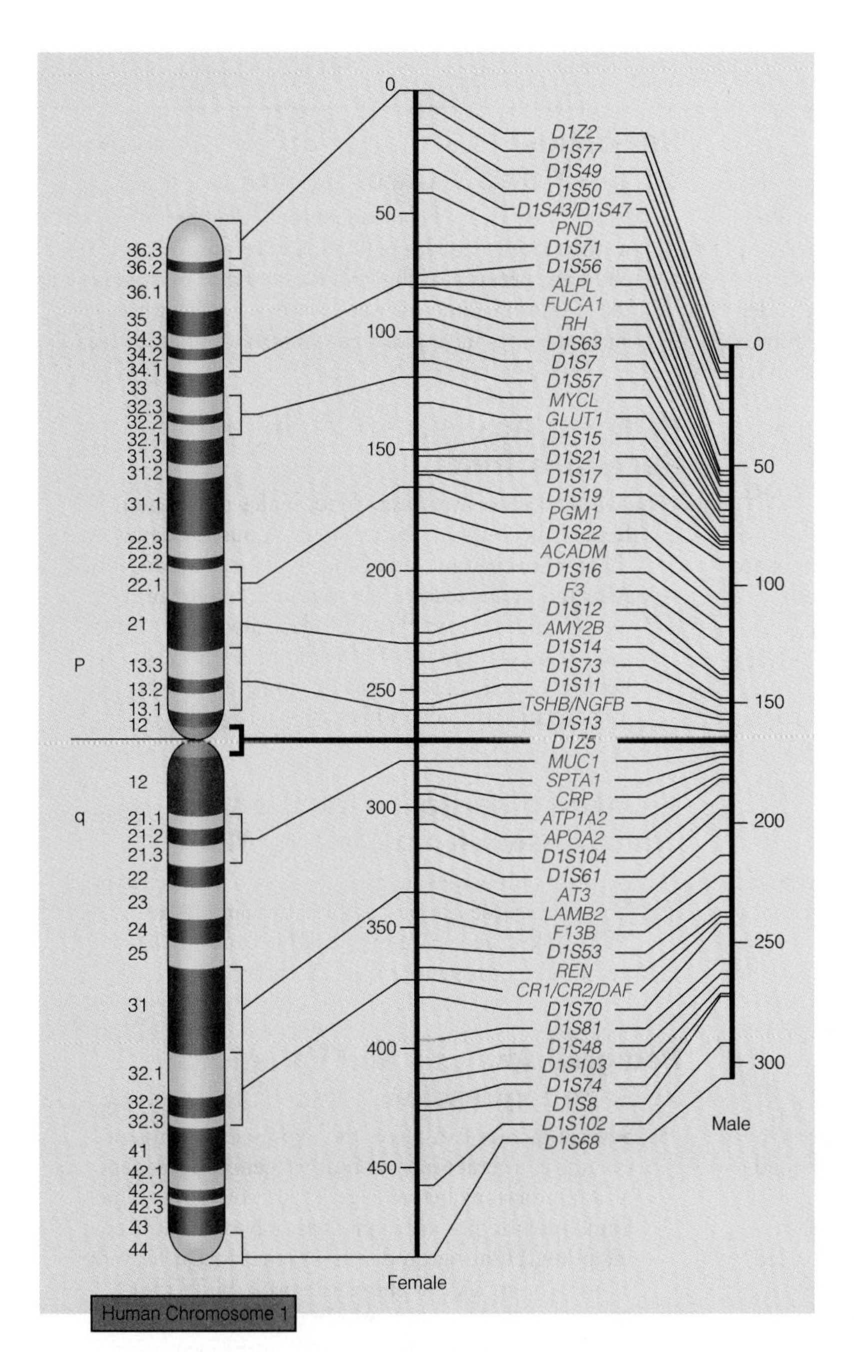

FIGURE 4.31 Genetic map of human chromosome 1. At the left is a drawing of the chromosome. The two vertical lines at the right represent genetic maps derived from studies of recombination in males and females. Between the genetic maps are the order and location of 58 loci, some of which are genes (*red*) and others (*blue*) that are genetic markers detected using recombinant DNA techniques. The map in females is about 500 cM long, and in males it is just over 300 cM. This is a result of differences in the frequency of crossing-over in males and females. This map provides a framework for locating genes on the chromosome as part of the Human Genome Project.

and the probability that the results would have been obtained even if the two genes were not linked. The results are expressed as the $\log_{10}$ of the ratio of the two probabilities, or **lod score** (*lod* stands for the log of the odds). A lod score of 3 means that the odds are 1,000 to 1 in favor of linkage; a score of 4 means that the odds are 10,000 to 1 in favor of linkage. Most geneticists agree that two genes are considered to show linkage when the lod score is 3 or higher.

Lod score The ratio of the probability that two loci are linked to the probability that they are not linked, expressed as a $\log_{10}$. Scores of 3 or more are taken as establishing linkage.

Using pedigree analysis and lod scores, genetic maps have been constructed for all of the human chromosomes. A genetic map for a human chromosome is shown in Figure 4.31. The map is drawn next to one of the chromosomes itself. The connecting lines show the sites on the actual chromosome where some of the genes are located.

Other methods of mapping use recombinant DNA techniques to map human chromosomes. These methods, along with linkage mapping, are part of the Human Genome Project, an international effort to map all of the genes in the human genome. These methods and the project itself are discussed in Chapter 13.

Summary

Studying the Inheritance of Traits in Humans

1. The inheritance of single gene traits in humans is often called Mendelian inheritance because of the pattern of segregation within families. These traits produce phenotypic ratios similar to those observed by Mendel in the pea plant. Although the results of studies in peas and humans may be similar, the methods are somewhat different.

Pedigree Construction Is a Basic Method in Human Genetics

2. Instead of direct experimental crosses, human traits are traced by constructing pedigrees that follow a trait through several generations of a family.

There Is a Catalog of Human Genetic Traits

3. As genetic traits are identified, they are described, cataloged, and numbered in a database called "Online Mendelian Inheritance in Man" (OMIM). This online resource is updated on a daily basis and contains information about all known human genetic traits.

Pedigree Analysis for Autosomal Recessive Traits

4. Information in the pedigree is used to determine how a trait is inherited. These patterns include autosomal dominant, autosomal recessive, X-linked dominant, X-linked recessive, Y-linked, and mitochondrial. Autosomal recessive traits have several characteristics: For rare traits, most affected individuals have unaffected parents; all children of affected parents are affected; the risk of an affected child with heterozygous parents is 25%.

Pedigree Analysis for Autosomal Dominant Traits

5. Dominant traits have several characteristics: Except in traits with high mutation rates, every affected individual has at least one affected parent; because most affected individuals are heterozygous, and have unaffected mates, each child has a 50% risk of being affected; because of heterozygosity, two affected individuals can have an unaffected child.

Sex-Linked Inheritance Involves Genes on the X and Y Chromosomes

6. Males pass an X chromosome to all daughters, but not to their sons. Females pass an X chromosome to all their children. Because of this and the fact that most genes on the X chromosome are not on the Y, genes on the sex chromosomes have a distinct pattern of inheritance.

Pedigree Analysis for X-Linked Recessive Traits

7. X-linked recessive traits affect males more than females because males are hemizygous for genes on the X chromosome. In X-linked recessive inheritance: Affected males receive the mutant allele from their mother and transmit it to all daughters, but not to their sons; daughters of affected males are usually heterozygous; sons of heterozygous females have a 50% chance of being affected.

Y-Linked Inheritance Involves Transmission from Male to Male

8. Because only males have Y chromosomes, genes on the Y chromosome are passed directly from father to son. All Y-linked genes are expressed because males are hemizygous for genes on the Y chromosome.

Pedigree Analysis and Variations in Gene Expression

9. Several factors can affect the expression of a gene, including interactions with other genes in the genotype and interactions between genes and the environment. Some phenotypes are expressed only in adulthood, including Huntington disease. Penetrance affects the expression of a gene and is the probability that a disease phenotype will appear when the disease-producing genotype is present. Another variation in gene expression is expressivity, which is the range of phenotypic variation associated with a given genotype. These variations can affect pedigree analysis.

Genes on the Same Chromosome Are Linked

10. The order and distance between two genes on the same chromosome can be estimated from the frequency of recombination between them. Lod scores can be used to estimate linkage and recombination distances. A lod score of 3 or 4 is generally regarded as evidence that two genes are linked.

Case Studies

CASE 1

Florence is an active 44-year-old elementary school teacher who was examined by her doctor because she was experiencing long periods of severe headaches and nausea. She told her physician that her energy level had been dramatically reduced the last few months, and her arms and legs felt like they "weighed 100 pounds each," particularly after she worked out in the gym. Her doctor did a complete physical and noticed that she did have reduced strength in her arms and legs and that her left eyelid was droopier than her right eyelid. He referred her to an ophthalmologist who discovered that she had an unusual pigment accumulation on her retina, which had not yet affected her vision. She then visited a clinical geneticist who examined the mitochondria in her muscles. She was diagnosed with a mitochondrial genetic disorder known as Kearns-Sayre syndrome.

Mitochondria are responsible for the conversion of food molecules into energy to meet the cell's energy needs. In mitochondrial disorders, these biochemical processes are abnormal, and energy production is reduced. Muscle tends to be particularly affected because it requires a lot of energy, but other tissues such as the brain may also be involved. Under the microscope, the mitochondria in muscle from people with mitochondrial disorders look abnormal, and they often accumulate around the edges of muscle fibers. This gives a particular staining pattern, known as a "ragged red" appearance, and this is usually how mitochondrial disorders are diagnosed.

Mitochondrial disorders affect people in many ways. The most common problem is a combination of mild muscle weakness in the arms and legs together with droopy eyelids and difficulty in moving the eyes. Some people do not have problems with their eye muscles, but have arm and leg weakness that gets worse after exertion. This weakness may be associated with nausea and headache. Sometimes muscle weakness is obvious in small babies if the illness is severe, and they may have difficulties in feeding and swallowing. Other parts of the body may be involved, including the electrical conduction system of the heart. Most mitochondrial disorders are mildly disabling, particularly in people who have eye muscle weakness and limb weakness. The age at which the first symptoms develop is variable, ranging from early childhood to late adult life.

About 20% of those with mitochondrial disorders have similarly affected relatives. Because only mothers transmit this disorder, it was suspected that some of these conditions are caused by a mutation in the genetic information carried by mitochondria. Mitochondria have their own genes, separate from the genes in the chromosomes of the nucleus. Only mothers pass mitochondria and their genes on to children, whereas the nuclear genes come from both parents. In about one third of people with mitochondrial disorders, substantial chunks of the mitochondrial genes are deleted. Most of these individuals do not have affected relatives and it seems likely that the deletions arise either during development of the egg or during very early development of the embryo. Deletions are particularly common in people with eye muscle weakness and the Kearns-Sayre syndrome.

1. Why would mitochondria have their own genomes?
2. How would mitochondria be passed from mother to offspring during egg formation? Why doesn't the father pass mitochondria?

CASE 2

The Smiths had just given birth to their second child and were eagerly waiting to take the newborn home. At that moment, their obstetrician arrived in the hospital room with some news about their daughter's newborn screening tests. The physician told them that the state's mandatory newborn screening test had detected an abnormally high level of phenylalanine in the blood of their daughter. The Smiths asked if this was just a fleeting effect, like newborn jaundice, that would "go away" in a few days. When they were told that this was unlikely, they were even more confused. Their daughter was born looking perfectly "normal" and the pregnancy had progressed without any complications. Mrs. Smith even had a normal amniocentesis early in the pregnancy. The physician asked a genetic counselor to come to their room to explain their daughter's newly diagnosed condition.

The counselor began her discussion with the Smiths by taking a family history from each of them. She explained that phenylketonuria (PKU) is a genetic condition that results when an individual inherits an altered gene from each parent. The counselor wanted to make this point early in the session in case either parent was blaming himself or herself or the other spouse for their daughter's condition. She explained that PKU is characterized by an increased concentration of phenylalanine in blood and urine and that mental retardation can be part of this condition if it is not treated at an early age. Therefore, it is important that they treat their daughter as soon as possible. To prevent the development of mental retardation, diagnosis must be done and dietary therapy must be started before the child is 30 days old. Dietary therapy is accomplished by consuming a special diet in which the bulk of protein is substituted for an artificial amino acid mixture

Case Studies (continued)

low in phenylalanine. The diet must start in the first month of life and should be continued indefinitely to be maximally effective.

PKU is one of several diseases known as the hyperphenylalaninemias. Hyperphenylalaninemias occur in 1 in 10,000 births. Classic PKU accounts for two thirds of them. It is an autosomal recessive disease, widely distributed among whites and Asians. It is rare in blacks. Heterozygous carriers do not show symptoms but may have slightly increased phenylalanine concentrations. If untreated, children with classic PKU can experience progressive impairment of cerebral function, seizures, and hyperactivity. EEG abnormalities; mousy odor of the skin, hair, and urine; tendency to hypopigmentation; and eczema complete the clinical picture.

1. Why did amniocentesis not detect the PKU? What conditions can amniocentesis detect?
2. Assume you are the genetic counselor. How would you counsel the parents to help them cope with their situation if one or both were blaming themselves for their child's condition?
3. What foods contain phenylalanine? How disruptive do you think the diet therapy will be to everyday life?

Questions and Problems

Pedigree Analysis

1. What are the reasons that pedigree charts are used?
2. Pedigree analysis permits all of the following, except:
   a. an orderly presentation of family information
   b. the determination of whether a trait is genetic
   c. the determination of whether a trait is dominant or recessive
   d. an understanding of which gene is involved in a heritable disorder
   e. the determination whether a trait is sex-linked or autosomal
3. Using the pedigree provided below, answer the following questions.
   a. Is the proband male or female?
   b. Is the grandfather of the proband affected?
   c. How many siblings does the proband have and where is he/she in the birth order?

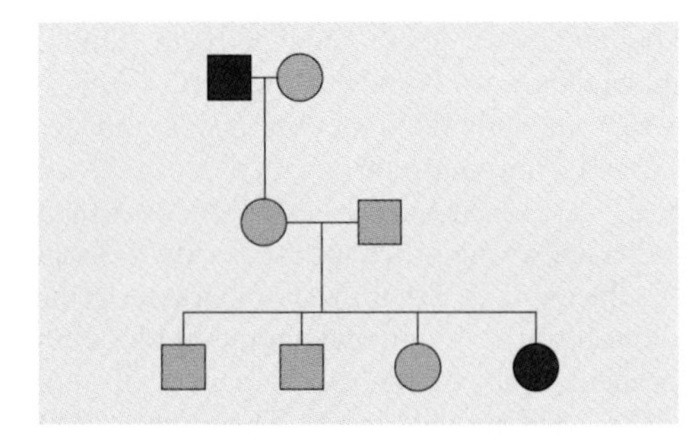

4. What does OMIM stand for? What kind of information is in this database?

Pedigree Analysis for Autosomal Recessive and Dominant Traits

5. What pattern of inheritance is suggested by the following pedigree?

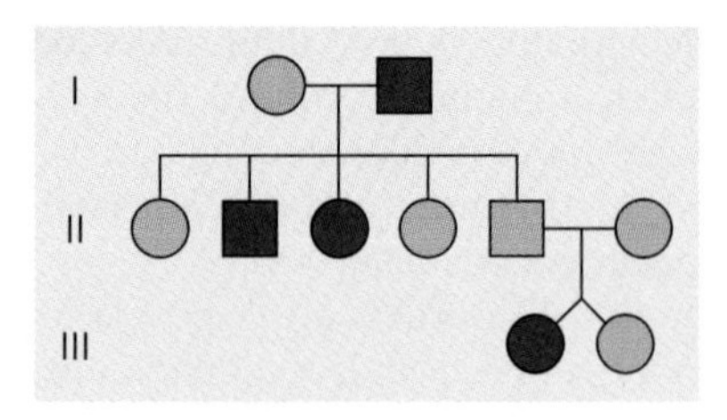

6. Does the indicated individual (III-5) show the trait in question?

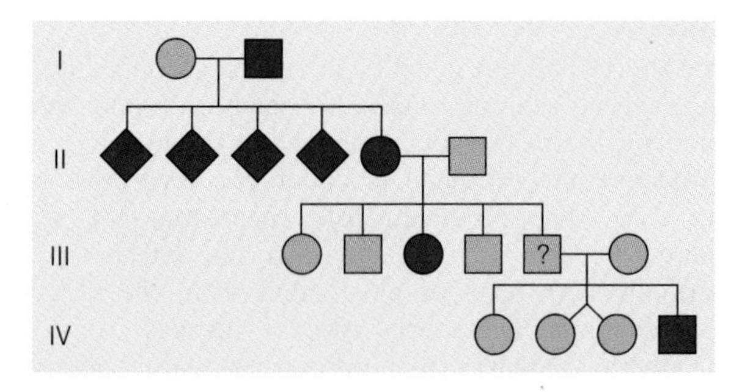

7. Use this information to respond to the following problems: (1) The proband (affected individual that led to the construction of the pedigree) exhibits the trait. (2) Neither her husband nor her only sibling, an older brother, exhibits the trait. (3) The proband has five children by her current husband. The oldest is a boy, followed by a girl, then another boy, and then identical twin girls. Only the second oldest fails

to exhibit the trait. (4) Both parents of the proband show the trait.

a. Construct a pedigree of the trait in this family.
b. Determine how the trait is inherited (go step by step to examine each possible pattern of inheritance).
c. Can you deduce the genotype of the proband's husband for this trait?

8. In the following pedigree, assume that the father of the proband is homozygous for a rare trait. What pattern of inheritance is consistent with this pedigree? In particular, explain the phenotype of the proband.

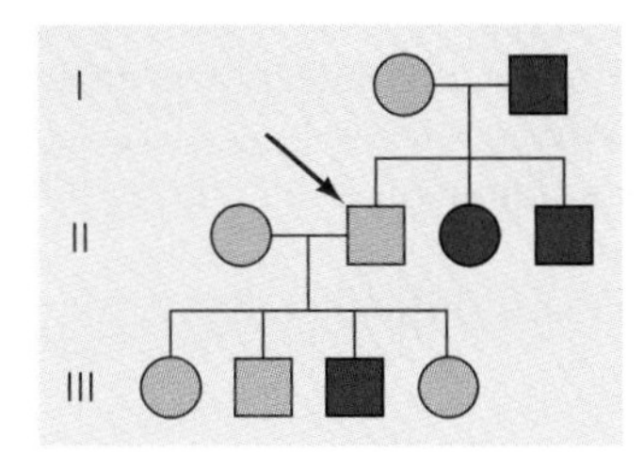

9. Using the following pedigree, deduce a compatible pattern of inheritance. Identify the genotype of the individual in question.

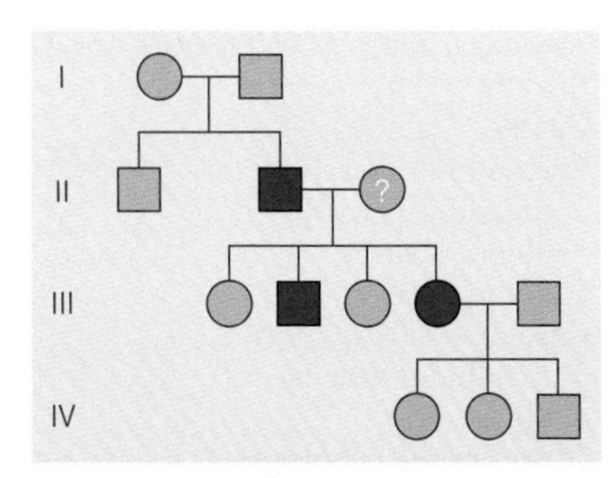

10. A proband female suffering from an unidentified disease seeks the advice of a genetic counselor prior to starting a family. Based on the following data, the counselor constructs a pedigree encompassing three generations: (1) The maternal grandfather of the proband suffers from the disease. (2) The mother of the proband is unaffected and is the youngest of five children, the three oldest being male. (3) The proband has an affected older sister, but the youngest siblings are unaffected twins (boy and girl). (4) All individuals suffering from the disease have been revealed. Duplicate the counselor's feat.
11. Describe the primary gene/protein defect and the resulting phenotype for the following diseases:
    a. cystic fibrosis
    b. sickle cell anemia
    c. Marfan syndrome
12. List and describe two other diseases inherited in the following fashion:
    a. autosomal dominant
    b. autosomal recessive
13. The father of 12 children begins to show symptoms of neurofibromatosis.
    a. What is the probability that Sam, the man's second oldest son (II-2), will suffer from the disease if he lives a normal life span? (Sam's mother and her ancestors do not have the disease.)
    b. Can you infer anything about the presence of the disease in Sam's paternal grandparents?
14. Huntington disease is a rare, fatal disease that usually develops in the fourth or fifth decade of life. It is caused by a single autosomal dominant allele. A phenotypically normal man in his twenties, who has a two-year-old son of his own, learns that his father has developed Huntington disease. What is the probability that he himself will develop the disease? What is the chance that his young son might eventually develop the disease?

Pedigree Analysis for Sex-Linked Traits

15. The X and Y chromosomes are structurally and genetically distinct. However, they do pair during meiosis at a small region near the tips of their short arms, indicating that the chromosomes are homologous in this region. If a gene lies in this region, will its pattern of transmission be more like a sex-linked gene or an autosomal gene? Why?
16. What is the chance that a colorblind man and a carrier woman will produce:
    a. a colorblind son?
    b. a colorblind daughter?
17. A young boy is colorblind. His one brother and five sisters are not. The boy has three maternal uncles and four maternal aunts. None of his uncles' children or grandchildren are colorblind. One of the maternal aunts married a colorblind man, and half of her children, both male and female, are colorblind. The other aunts married men who have normal color vision. All their daughters have normal vision, but half of their sons are colorblind.
    a. Which of the boy's four grandparents transmitted the gene for colorblindness?
    b. Are any of the boy's aunts or uncles colorblind?
    c. Are either of the boy's parents colorblind?
18. Describe the phenotype and primary gene/protein defect of the X-linked recessive disease muscular dystrophy.
19. In the beginning of this chapter, it was stated that a couple, both phenotypically normal, have two children, one unaffected daughter and one affected son. The phenotype ratio is 1:1. This makes it difficult to determine if the trait is autosomal or X-linked. With your knowledge of genetics, what are the genotypes of the parents and children in the autosomal case? In the X-linked case?
20. The following is a pedigree for a common genetic trait. Analyze the pedigree to determine whether the trait is inherited as:
    a. autosomal dominant
    b. autosomal recessive

c. X-linked dominant
d. X-linked recessive
e. Y-linked

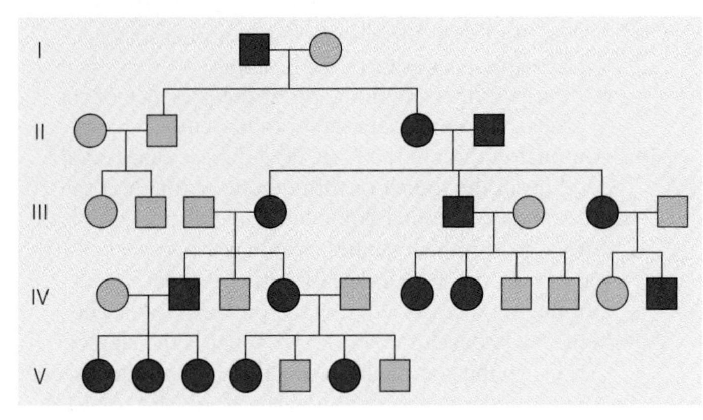

21. As a genetic counselor investigating a genetic disorder in a family, you are able to collect a four-generation pedigree that details the inheritance of the disorder in question. Analyze the information in the pedigree to determine whether the trait is inherited as:
    a. autosomal dominant
    b. autosomal recessive
    c. X-linked dominant
    d. X-linked recessive
    e. Y-linked

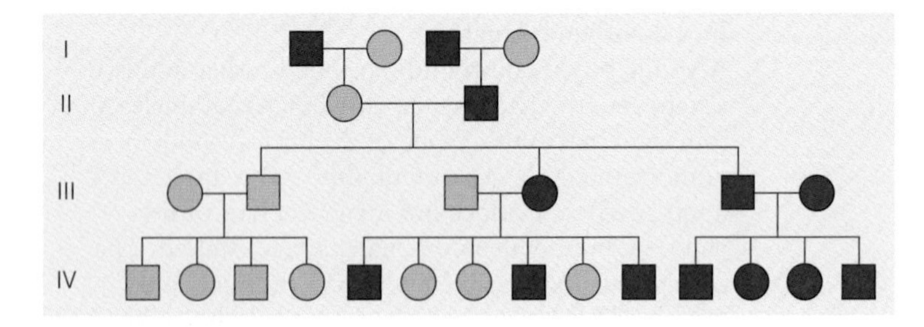

22. In the eighteenth century a young boy suffered from a skin condition known as ichthyosis hystrix gravior. The phenotype of this disorder includes thickening of skin and the formation of loose spines that are periodically sloughed off. This man married and had six sons, all of whom had this same condition. He also had several daughters, all of whom were unaffected. In all succeeding generations, this condition was passed on from father to son. What can you theorize about the location of the gene that causes ichthyosis hystrix gravior?

Mitochondrial Inheritance

23. What are the unique features of mitochondria that are not present in other cellular organelles in human cells?
24. How is mitochondrial DNA transmitted?

Variations in Gene Expression

25. Define penetrance and expressivity.
26. Suppose that space explorers discover an alien species possessing the same genetic principles as apply to humans. Although all 19 aliens analyzed to date carry a gene for a third eye, only 15 display this phenotype. What is the penetrance of the third-eye gene in this population?
27. A genetic disorder characterized by falling asleep in genetics lectures is known to be 20% penetrant. All 90 students in a genetics class are homozygous for this gene. Theoretically, how many of the 90 students will fall asleep during the next lecture?
28. Explain how camptodactyly is an example of expressivity.
29. How can dominant lethal alleles survive in a population?
30. Why are humans difficult subjects for genetic analysis?

Genes on the Same Chromosome Are Linked

31. Consider the autosomal recessive conditions albinism (*A*/*a*) and hereditary deafness (*D*/*d*).
    a. Given the following data (assume no crossing-over occurred), in which cases are the genes linked?
       Case 1:
       9 nonalbino, hearing
       3 nonalbino, deaf
       3 albino, hearing
       1 albino, deaf
       Case 2:
       3 nonalbino, hearing
       1 albino, deaf
    b. What are the genotypes of the parents?
    c. Which alleles are linked to each other (for example, *A* linked to *D* or to *d*)?
32. In case 2 of question 31, what would be the expected genotypes of the progeny if *A* was linked to *d* and *a* was linked to *D*?
33. A recombination experiment is carried out in the fruit fly *Drosophila* to measure the distance between two genes on chromosome 2. The results indicate that there is 58% recombination between the two loci. Are these genes linked?
34. The frequency of recombination between gene *A* and gene *B* is 8%; between gene *B* and gene *C*, the frequency is 12%; and between gene *A* and gene *C* the frequency is 4%. From this information, deduce the gene order.

Internet Activities

The following activities use the resources of the World Wide Web to enhance the topics covered in this chapter. To investigate the topics described below, log on to the book's homepage (www.brookscole.com/biology_d) and follow the *Student Resources* link to your subject and text.

1. *A Database of Human Genetic Disorders and Traits.* The Internet site *On-line Mendelian Inheritance in Man,* or OMIM (http://www.ncbi.nlm.hih.gov/Omim/), is an on-line database of human genetic disorders and genetically controlled traits that is updated daily. For any given disorder or trait, information including symptoms, mode of inheritance, molecular genetics, diagnosis, therapies, and more is given.
   a. At the OMIM site, enter the database and select a genetic disorder mentioned in the text, such as cystic fibrosis or Marfan syndrome. How is this trait transmitted from generation to generation? Is it an autosomal or a sex-linked trait? Dominant or recessive? What is known about the molecular genetics or the location of this trait?
   b. OMIM also contains information about human traits that are not considered to be diseases, such as eye color, handedness, uncontrolled sneezing (the achoo syndrome), alteration of taste sensations, earlobe creases, ear wax, and many more. Some of these traits may or may not be genetically controlled—the scientific jury is still out. Access the OMIM database and read about handedness. Is this a genetically controlled trait or not?
2. *Genetic Disorders and Support.* Information about many genetic disorders and support groups and organizations for persons with genetic disorders is available on the World Wide Web. If you have a personal or scientific interest in a particular disorder, try to find information about the disorder, treatments, or parent groups through Web search engines such as Google (www.google.com). In addition, this text's homepage has a link to a listing of genetic support groups that can be written to in order to obtain more information about a specific genetic disorder.
3. *Would You Want to Know If You Carried the Gene for a Disorder?* Not all dominant genetic disorders are obvious in early life, and of course an individual may be a carrier for a recessive disorder without displaying the characteristics of the trait. At *Do You Really Want to Know If You Have a Disease Gene?* (http://nasw.org/users/robinhenig/gene_testing.htm) journalist and author Robin Henig explores this question, which we will return to in Chapter 19, Genetic Screening and Counseling.

Further Exploration: For a simple version of the genetics of left-handedness, in addition to a look at what life is like for a southpaw, check out *Lorin's Lefthandness Site* (http://duke.usask.ca/ ~ elias/left/).

For Further Reading

Christensen, D. (2000). No news: nitric oxide may help treat sickle cell anemia. *Sci. News, 157,* 78–79.

Cochrane, G., & Ewald, P. (1999). High risk defenses. *Nat. Hist., 108,* 40–43.

Eaton, W., & Hofrichter, J. (1995). The biophysics of sickle cell hydroxyurea therapy. *Science, 268,* 1142–1143.

Embury, S. H. (1986). The clinical pathology of sickle cell disease. *Ann. Rev. Med., 37,* 361–376.

Francomano, C. A., Le, P. L., & Pyeritz, R. E. (1988). Molecular genetic studies in achondroplasia. *Basic Life Sci., 48,* 53–58.

Huntington's Disease Collaborative Research Group. (1993). A novel gene containing a trinucleotide repeat that is expanded and unstable on Huntington's disease chromosomes. *Cell, 72,* 971–983.

Kline, R. (2001). Whose blood is it anyway? *Sci. Am., 284,* 42–49.

Macalpine, I., & Hunter, R. (1969, July). Porphyria and King George III. *Sci. Am., 221,* 38–46.

Peltonen, L., & Kainulainen, K. (1992). Elucidation of the gene defect in Marfan syndrome. Success by two complementary research strategies. *FEBS Letters, 307,* 116–121.

Potts, D. M., & Potts, W. T. W. 1995. *Queen Victoria's gene: Hemophilia and the royal family.* Gloucestershire, England: Alan Sutton Publishing.

Rommens, J. M., Iannuzzi, M. C., Bat-Sheva, K., Drumm, M. L., Melmer, G., Dean, M., Rozmahel, R., Cole, J. L., Kennedy, D., Hidaka, N., Zsiga, M., Buchwald, M., Riordan, J. R., Tsui, L. C., & Collins, F. S. (1989). Identification of the cystic fibrosis gene: Chromosome walking and jumping. *Science, 245,* 1059–1065.

Seppa, N. (2000). Gene mutation for color blindness found. *Sci. News, 158,* 63.

Smithies, O. (1993). Animal models of human genetic diseases. *Trends Genet., 9,* 112–116.

Stanbury, J. B., Wyngaarden, J. B., & Fredrickson, D. S. (1983). *The metabolic basis of inherited disease* (5th ed.). New York: McGraw-Hill.

Stern, C. (1973). *Principles of human genetics* (3rd ed.). San Francisco: Freeman.

Welsh, M., & Smith, A. (1995). Cystic fibrosis. *Sci. Am., 273,* 52–59.

Worton, R. (1995). Muscular dystrophies: Diseases of the dystrophin-glycoprotein complex. *Science, 270,* 755–756.

Yoshida, A. (1982). Biochemical genetics of the human blood group ABO system. *Am. J. Hum. Genet., 34,* 1–14.

© Mark Burnett/David R. Frazier Photolibrary, Inc.

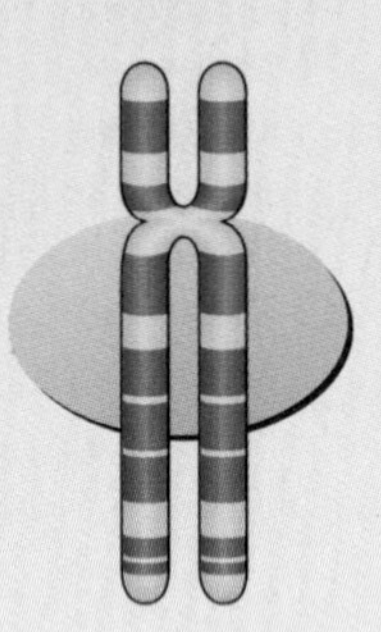

Polygenes and Multifactorial Inheritance

Chapter Outline

In 1713, a new king was crowned in Prussia. He immediately began one of the largest military buildups of the eighteenth century. In the space of 20 years, King Frederick William I, ruler of fewer than 2 million citizens, enlarged his army from around 38,000 men to just under 100,000 troops. Compare this to the neighboring kingdom of Austria, which had a population of 20 million and an army of just under 100,000 men, and you will understand why Frederick William was regarded as a military monomaniac. The crowning glory of this military machine was his personal guard, known as the Potsdam Grenadier Guards. This unit was composed of the tallest men obtainable. Frederick William was obsessed with having giants in this guard, and his recruiters used bribery, kidnapping, and smuggling to fill the ranks of this unit. It is said that members of the guard could lock arms while marching on either side of the king's carriage. Many members were close to 7 feet tall. Although someone 7 feet tall is not much of a novelty in the NBA, in eighteenth-century Prussia anyone taller than about 5 feet 4 inches was above average height.

King Frederick William was also rather miserly, and because this recruiting was costing him millions, he decided it would be more economical simply to breed giants to serve in his elite unit. To accomplish this, he ordered that every tall man in the kingdom was to marry a tall, robust woman, expecting that the offspring would all be giants. Unfortunately this idea was a frustrating failure. Not only was it slow, but most of the children were shorter than their parents. While continuing this breeding program, the king reverted to kidnapping and bounties, and he also let it be known that the best way for foreign governments to gain his favor was to send giants to be members of his guard. This human breeding experiment continued until shortly after King Frederick William's death in 1740, when his son, Frederick the Great, disbanded the Potsdam Guards.

SOME TRAITS ARE CONTROLLED BY TWO OR MORE GENES

What exactly went wrong with King Frederick William's experiment in human genetics? Selecting the tallest men and women as parents should result in tall children. After all, when Mendel intercrossed true-breeding tall pea plants, the offspring were all tall. Even when heterozygous tall pea plants are crossed, three-fourths of the offspring are tall.

The problem with this comparison is that a single gene pair controls height in pea plants, whereas several gene pairs determine height in humans. Traits determined by two or more gene pairs are called **polygenic traits.** The tall and short phenotypes in pea plants are two distinct phenotypes and show **discontinuous variation.** In measuring height in humans, it is difficult to set up only two phenotypes. Instead, height in humans is an example of **continuous variation.** Unlike Mendel's pea plants, people are not either 18 in. or 84 in. tall; they fall into a

Polygenic traits Phenotypes that depend on the action of a number of genes.

Discontinuous variation Phenotypes that fall into two or more distinct, nonoverlapping classes.

Continuous variation A distribution of phenotypic characters that is distributed from one extreme to another in an overlapping, or continuous, fashion.

series of overlapping phenotypic classes. Two or more separate gene pairs often control traits with a gradation of phenotypes.

Understanding the distinction between discontinuous and continuous traits was an important advance in genetics. It is based on accepting the idea that genes interact with each other and with the environment. Traits that are controlled by two or more genes and that also show strong interaction with the environment are called **multifactorial traits.**

Multifactorial traits Traits that result from the interaction of one or more environmental factors and two or more genes.

In this chapter, we examine traits controlled by genes at two or more loci (polygenic traits) and then discuss traits controlled by two or more genes and nongenetic factors, such as the environment (multifactorial traits). In multifactorial inheritance, the degree of genetic effects on a trait can be estimated by measuring heritability. We consider this concept and the use of twins as a means of measuring the heritability of a trait. In the last part of the chapter, we examine a number of human polygenic traits, some of which have been the subject of political and social controversy.

POLYGENES AND VARIATIONS IN PHENOTYPE

Mendel was not the only scientist in the late nineteenth century that experimented with the inheritance of traits. In one series of experiments, Josef Kölreuter crossed tall and dwarf tobacco plants. The F1 plants were all intermediate in height to the parents. When self-crossed, the F1 produced an F2 that contained plants of many different heights. Some of the F2 were as tall or as short as the P1, but most of the F2

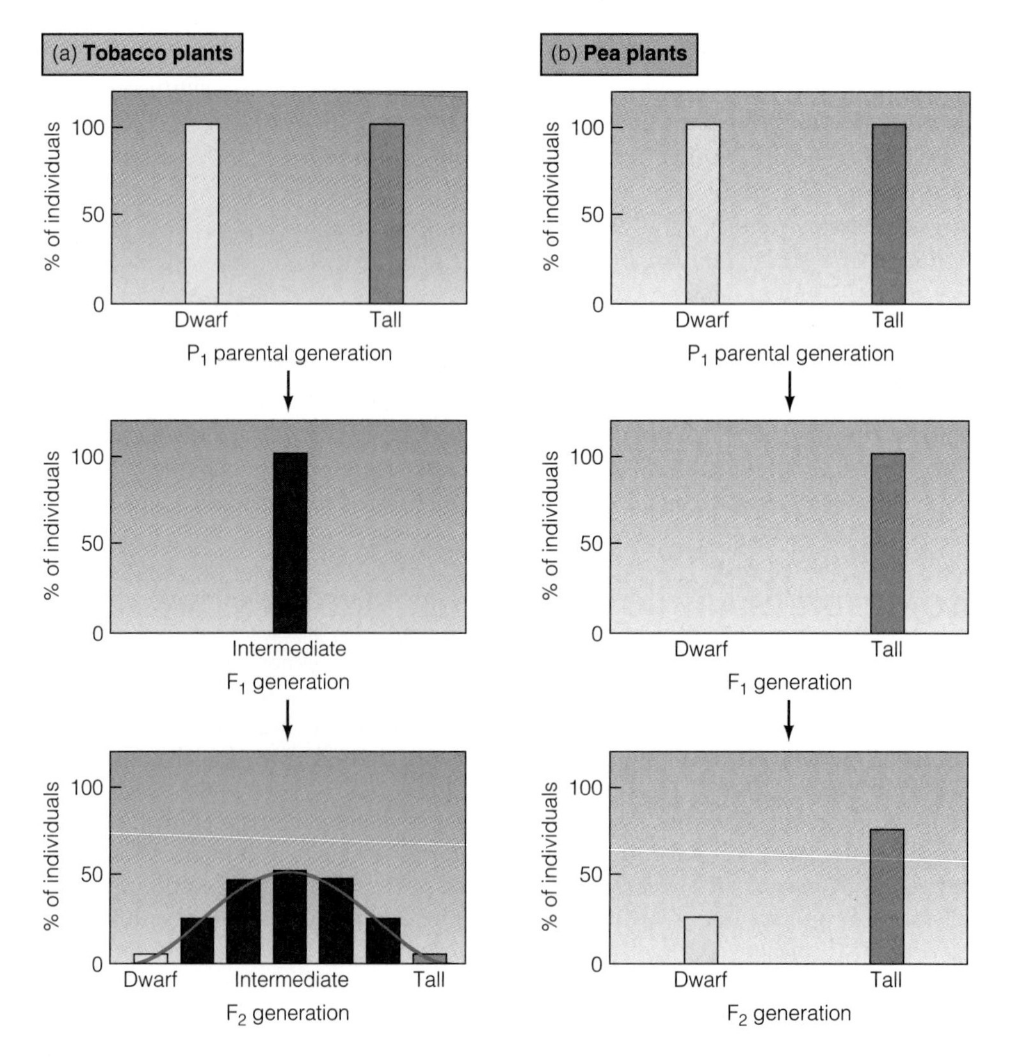

FIGURE 5.1 A comparison of traits that have continuous and discontinuous phenotypes. (a) Histograms show the percentage of plants that have different heights in crosses between tall and dwarf strains of tobacco plants carried to the F2 generation. The F1 generation is intermediate to the parents in height, and the F2 shows a range of phenotypes from dwarf to tall. Most plants have a height intermediate to those of the P1 generation. (b) Histograms show the percentage of plants that have different heights in crosses between tall and dwarf strains of the pea plant. The F1 generation has the tall phenotype, and the F2 has two distinct phenotypic classes: 75% of the offspring are tall, and 25% are dwarf. The differences between tobacco plants and pea plants are explained by the fact that height in tobacco plants is controlled by two or more gene pairs, whereas height in peas is controlled by a single gene.

were intermediate in height when compared with the parents (◗ Figure 5.1) and showed continuous variation in the distribution of phenotypes. Mendel's results with pea plants, however, produced evidence of discontinuous variation (Figure 5.1).

Shortly after the turn of the twentieth century, it was found that traits in different plants and animals show continuous phenotypic variation. In each of these cases, the offspring had a phenotype that seemed to be a blend of the parental traits. Geneticists debated whether continuous variation was consistent with the inheritance of Mendelian factors or whether this apparent blending of traits signaled the existence of another mechanism of inheritance. This argument is important to human genetics because many human traits and many genetic disorders show continuous variation.

In the years immediately after the rediscovery of Mendel's work, only a few traits were known in humans, and single genes control these traits. At the time, interest in human genetics was largely centered on discovering whether "social" traits, such as alcoholism, feeblemindedness, and criminal behavior, were inherited. Some geneticists constructed pedigrees and simply assumed that single genes controlled these traits. Other geneticists pointed out that these traits did not show the phenotypic ratios observed in experimental organisms and concluded that Mendelian inheritance might not hold true humans. In fact, the biomathematician Karl Pearson, who studied polygenic traits in humans, is reported to have said: "There is no truth in Mendelism at all."

Between 1910 and 1930, the controversy over continuous variation was resolved. Experimental crosses with corn and tobacco plants demonstrated that continuous phenotypic variation could be explained by Mendelian inheritance. These findings revealed that traits determined by a number of alleles, each of which makes a small contribution to the phenotype, show a continuous distribution of phenotypes in the F2 generation. This is true even though the inheritance of each gene follows the rules of Mendelian inheritance. This distribution of phenotypes follows a bell-shaped curve. The phenotypes include a small number of individuals who have extreme phenotypes (very short or very tall, for example). Most individuals, however, have phenotypes between these extremes; their distribution follows what statisticians call a normal curve (◗ Figure 5.2). This pattern of inheritance, known as polygenic or quantitative inheritance, is additive because it is controlled by two or more genes, and each allele adds a small but equal amount to the phenotype.

The continuous distribution of phenotypes in polygenic inheritance has several distinguishing characteristics:

- Traits are usually quantified by measurement rather than by counting.
- Two or more genes contribute to the phenotype. Each gene contributes in an additive way to the phenotype. The effect of individual alleles may be small, and some alleles may make no contribution.

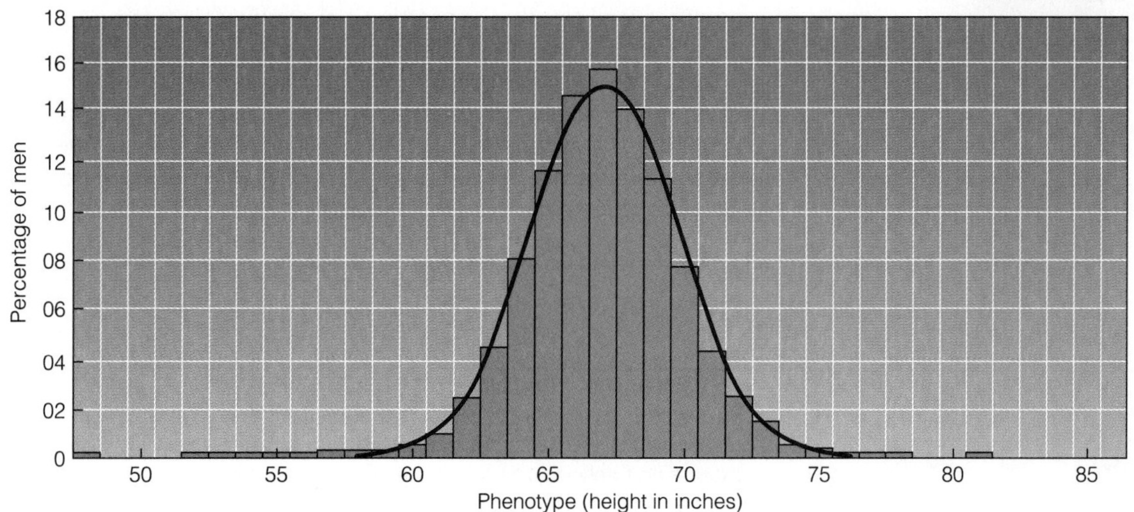

◗ **FIGURE 5.2** A bell-shaped or "normal" curve shows the distribution of phenotypes for traits controlled by two or more genes. In a normal curve, few individuals are at the extremes of the phenotype, and most individuals are clustered around the average value. In this case, the phenotype is height measured in a population of human males.

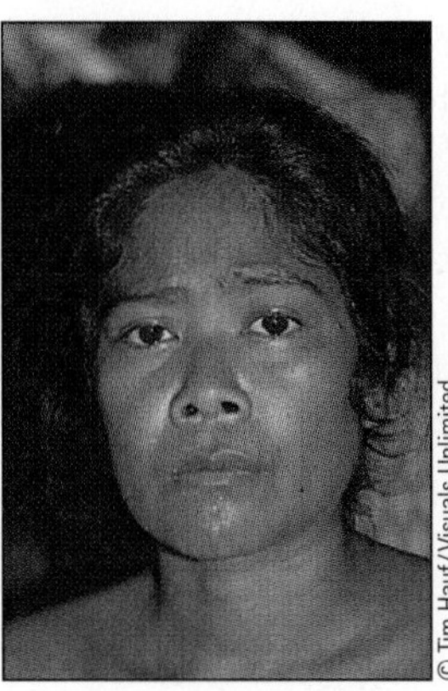

FIGURE 5.3 Skin color is a polygenic trait controlled by three or four genes, producing a wide range of phenotypes. Environmental factors (exposure to the sun and weather) also contribute to the phenotypic variation, making this a multifactorial trait.

- Phenotypic expression of polygenic inheritance varies across a wide range. This variation is best analyzed in populations rather than in individuals (Figure 5.3).

Polygenic inheritance is an important concept in human genetics. Traits such as height, weight, skin color, and intelligence are under polygenic control. In addition, congenital malformations (such as neural tube defects, cleft palate, and clubfoot) as well as genetic disorders (such as diabetes, hypertension, and behavioral disorders) are polygenic or multifactorial traits.

The distribution of phenotypes and F2 ratios in traits involving two, three, and four genes is shown in Figure 5.4. If two genes control a trait, there are five phenotypic classes in the F2, each of which has four, three, two, one, or zero dominant alleles. The F2 ratio of 1:4:6:4:1 results from the number of genotypic combinations that produce each phenotype. At one extreme is the homozygous dominant (*AABB*) genotype with four dominant alleles; at the other extreme is the homozygous recessive (*aabb*) genotype with 0 dominant alleles. The largest class (6/16) has 6 genotypic combinations of two dominant alleles (*AaBb, Aabb, aaBB,* etc.). As the number of genes that controls the trait increases and the amount of phenotypic variation increases because of environmental factors, the phenotypes blend together, generating a continuous distribution.

As the number of loci that controls a trait increases, the number of phenotypic classes increases. As the number of classes increases, there is less phenotypic difference between each class. This means that there is a greater chance for environmental factors to override the small difference between classes. For example, exposure to sunlight can alter skin color and obscure genotypic differences.

The results of phenotypic measurements in polygenic inheritance are usually expressed in frequency diagrams. Figure 5.5 shows a frequency distribution for a polygenic trait.

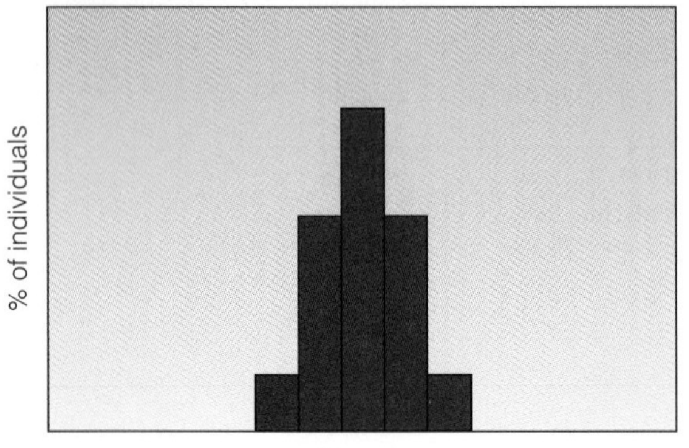

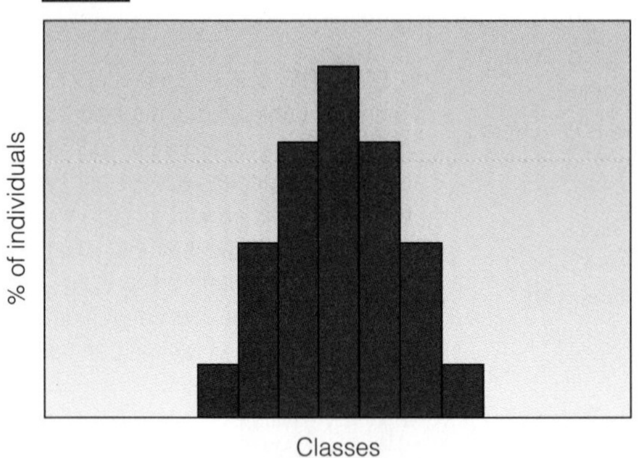

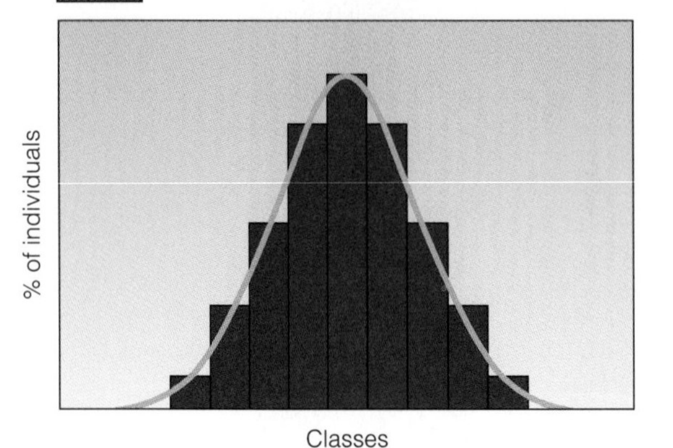

FIGURE 5.4 The number of phenotypic classes in the F2 generation increases as the number of genes controlling the trait increases. This relationship allows geneticists to estimate the number of genes involved in expressing a polygenic trait. As the number of phenotypic classes increases, the distribution of phenotypes becomes a normal curve.

The Additive Model for Polygenic Inheritance

To explain how polygenes contribute to a trait and how genotypes contribute to variation in phenotypic expression, let's consider a model for polygenic inheritance. To simplify the analysis, we will examine the model under the following conditions:

- The trait is controlled by several genes. We will assume three genes, each of which has two alleles (*A,a, B,b, C,c*).
- Each of the dominant alleles makes an equal contribution to the phenotype, and the recessive alleles make no contribution.
- The effect of each active (dominant) allele on the phenotype is small and additive.
- The genes controlling the trait are not linked; they assort independently.

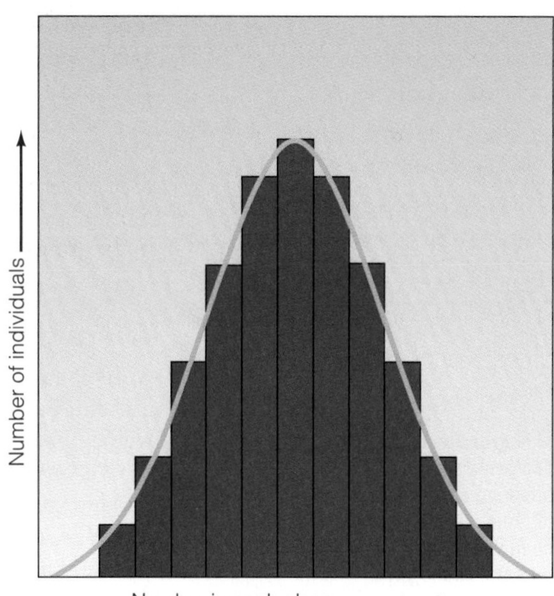

FIGURE 5.5 The results of phenotypic measurements in polygenic inheritance expressed as a frequency diagram, resulting from an interaction of polygenes with environmental factors.

This model will be applied to the inheritance of height. In this model, we assume that each dominant allele adds 3 in. to a base height of 5 ft. The recessive alleles *a, b,* and *c* add nothing to the base height. An individual who has the genotype *aabbcc* is 5 ft. tall, and someone who has the genotype *AABBCC* is 6 ft. 6 in. tall. If the environment acts equally on all genotypes, those who carry three of the six possible dominant alleles would represent the average height (5 ft. 9 in.).

Figure 5.6a shows the genotype frequencies and their phenotypes. With three alleles, there are seven phenotypic classes, with six, five, four, three, two, one, and zero dominant alleles. In real-life situations, the action of environmental factors blurs the distinction between the phenotypic classes, producing the continuous variation in height that is actually observed (Figure 5.6b). The most frequent phenotype

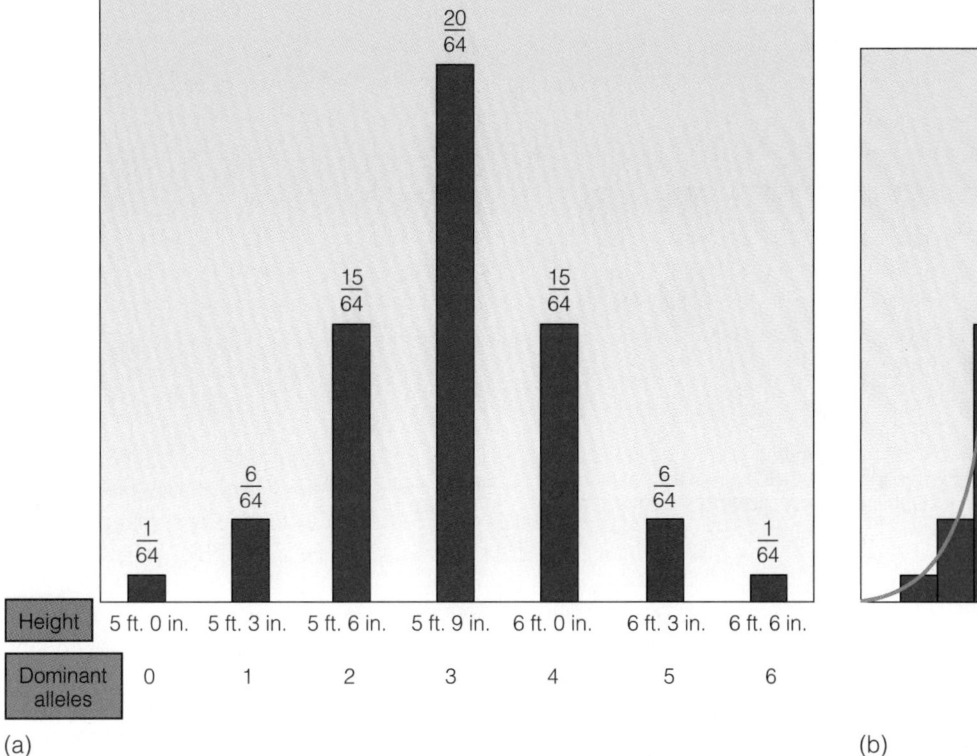

(a)

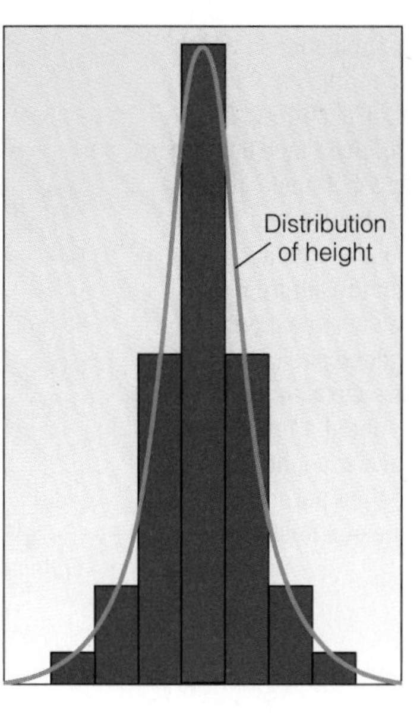

(b)

FIGURE 5.6 (a) The distribution of height in a three-gene model, where each dominant allele adds three inches above a base (homozygous recessive *aabbcc*) level of 5 ft. In this model, the phenotypic extremes are represented by 5 ft. and 6 ft. 6 in. (b) In reality, the interaction of genotypes with the environment produces a continuous distribution of phenotypes (height).

(one with three dominant alleles in the genotype) produces the average height of 5 ft. 9 in. At either end of the phenotypic range, the frequencies decline, and only 1/64 of the individuals will be as short as 5 ft. or as tall as 6 ft. 6 in.

In this example and in polygenic traits generally, the genotype represents the genetic potential for height. Full expression of the genotype depends on the environment. Poor nutrition during childhood can prevent people from reaching their potential heights. On the other hand, optimal nutrition from birth to adulthood cannot make someone taller than the genotype dictates.

Averaging Out the Phenotype: Regression to the Mean

A distinguishing characteristic of polygenic traits is that most offspring of individuals with one of the extreme phenotypes (tall × tall, for example) will have a less extreme phenotype. This phenomenon is known as **regression to the mean.**

Regression to the mean In a polygenic system, the tendency of offspring of parents who have extreme differences in phenotype to exhibit a phenotype that is the average of the two parental phenotypes.

As an example, let's look at King Frederick William's attempt to breed giants for his elite guard unit. We will assume that this breeding program used individuals at least 5 ft. 9 in. tall. For simplicity, we'll also assume that the dominant alleles *A*, *B*, and *C* each add 3 in. above a base height of 5 ft. 9 in. and the recessive alleles *a*, *b*, and *c* add nothing above the base height. An individual with the genotype *aabbcc* would be 5 ft. 9 in. tall, and an individual with the genotype *AABBCC* would be 7 ft. 3 in. tall.

Suppose that a 6 ft. 9 in. member of the guard who carries the genotype *AabCC* mates with a 6 ft. 3 in. woman with the genotype *AaBbcc*. The possible genotypic and phenotypic outcomes are diagrammed in Figure 5.7. In this case, there are four paternal and four maternal gamete combinations, and 16 possible types of fertilizations with five phenotypic classes.

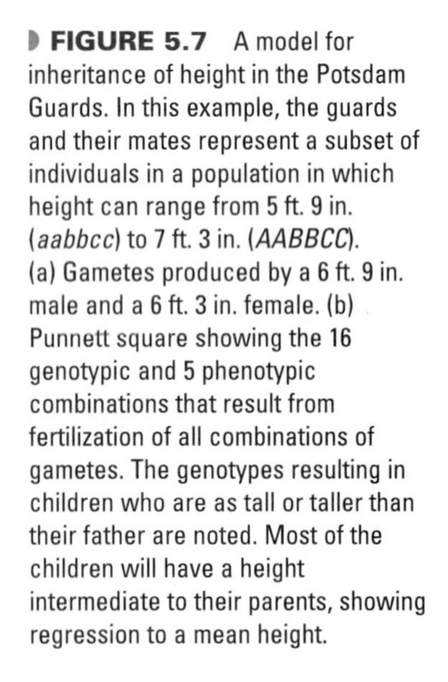

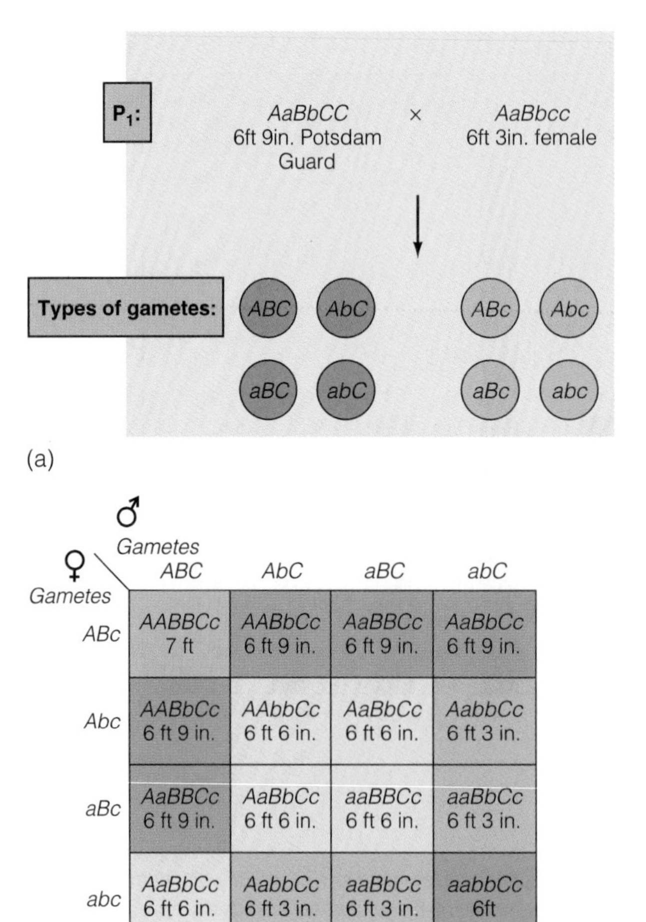

FIGURE 5.7 A model for inheritance of height in the Potsdam Guards. In this example, the guards and their mates represent a subset of individuals in a population in which height can range from 5 ft. 9 in. (*aabbcc*) to 7 ft. 3 in. (*AABBCC*). (a) Gametes produced by a 6 ft. 9 in. male and a 6 ft. 3 in. female. (b) Punnett square showing the 16 genotypic and 5 phenotypic combinations that result from fertilization of all combinations of gametes. The genotypes resulting in children who are as tall or taller than their father are noted. Most of the children will have a height intermediate to their parents, showing regression to a mean height.

The phenotypic classes and ratio of possible offspring are diagrammed in ◗ Figure 5.8. As the king discovered (after waiting about 18 years for them to grow up), most of the children tended toward the average height (6 ft. 6 in.) between the two parents. In fact, 11 of the 16 genotypic combinations will result in children shorter than their father. In succeeding generations, further regression to the mean will occur. To make matters worse, many of the Potsdam Grenadier Guards were tall because of endocrine malfunctions and did not have the genotypes to produce tall offspring under any circumstances.

Regression to the mean is brought about by dominance and additive effects along with gene interaction and environmental effects. If these other factors were included in our crosses and if the genotypes included individuals of average height (say, 5 ft. 4 in.), the regression toward the mean would be even more pronounced.

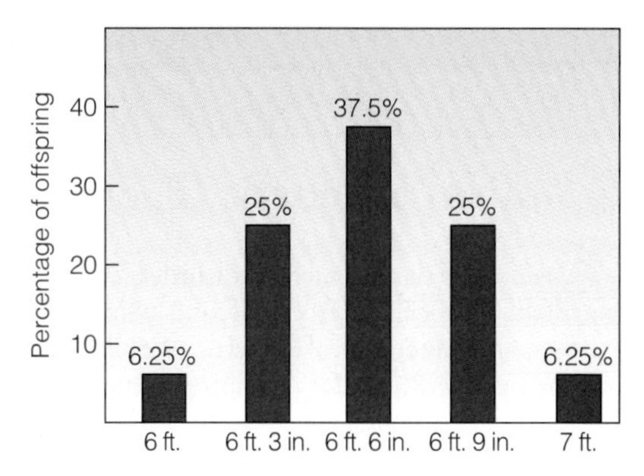

◗ **FIGURE 5.8** Frequency distribution of phenotypes from the possible offspring in Figure 5.7. Height of the offspring shows regression to the mean.

POLYGENES AND THE ENVIRONMENT: MULTIFACTORIAL TRAITS

In considering the interaction of polygenes and the environment, we begin by reviewing some basic concepts. The genotype represents the genetic constitution of an individual. It is fixed at the moment of fertilization and, barring mutation, is unchanging. The phenotype is the sum of the observable characteristics. It is variable and undergoes continuous change throughout the life of the organism. The environment of a gene includes all other genes in the genotype; their effects and interactions; and all nongenetic factors, whether physical or social, that can interact with the genotype (see Concepts and Controversies: Is Autism a Genetic Disorder?).

In assessing the interaction between the genotype and the environment, as in all science, you have to ask the right question. Suppose the question is, "How much of a given phenotype in an individual is caused by heredity and how much by environment?" Because each individual has a unique genotype and has been exposed to a unique set of environmental conditions, it is impossible to evaluate quantitatively the phenotype's genetic and environmental components. Thus, for a given individual, the question as posed cannot be answered. However, in a later section we see that if the question is changed, it is possible to estimate the genotypic contribution to a phenotype.

Threshold Effects and the Expression of Multifactorial Traits

Although the degree of interaction between a genotype and the environment can be difficult to estimate, family studies indicate that such interactions do occur. Some multifactorial traits do not show a continuous distribution of phenotypes; individuals are either affected or not. Congenital birth defects, such as clubfoot or cleft palate, are examples of traits that are distributed discontinuously but are, in fact, multifactorial.

A model can help explain the phenotypic distribution of such traits. In this model, a liability for a genetic disorder is distributed among individuals in a bell-shaped curve. This liability is in the form of genotypes, but only a limited number of genotypes express the phenotype (◗ Figure 5.9). The predisposition is caused by a number of genes, each of which contributes to the liability in an additive way. Individuals with a liability above a certain genetic threshold will develop the genetic disorder if exposed to the proper environmental conditions (Figure 5.9). In other words, environmental conditions are most likely to have the greatest impact on genetically predisposed individuals.

◗ **FIGURE 5.9** A model to explain the discontinuous distribution of some multifactorial traits. In this model, liability for a genetic disorder is distributed among individuals in a normal curve. This liability is caused by a number of genes, each acting additively. Only those individuals who have a genetic liability above a certain threshold are affected if exposed to certain environmental conditions.

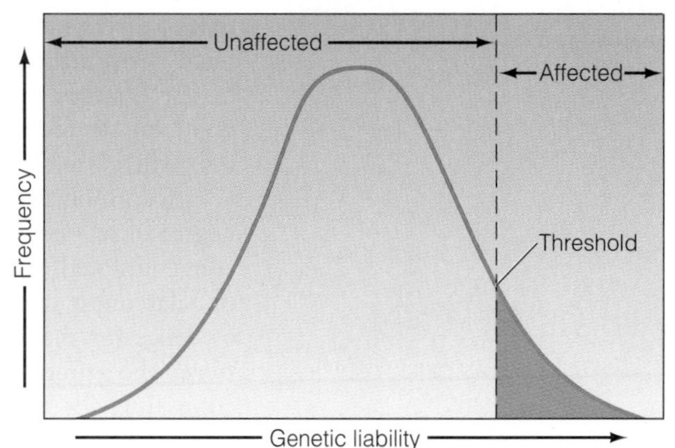

Concepts and Controversies

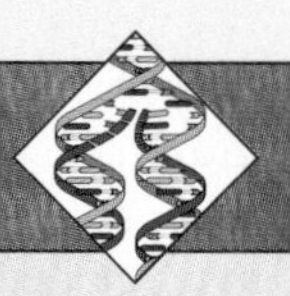

Is Autism a Genetic Disorder?

Autism is a developmental disorder characterized by impairment in social interactions and communication and by narrow and stereotypical patterns of abilities. As depicted in the movie *The Rain Man*, symptoms can include aversion to human contact, language difficulties that show up as bizarre speech patterns, difficulty in understanding what others think, and repetitive body movements. These characteristics seem to be associated with malfunctions of the central nervous system. Autistic individuals have changes in brain anatomy and biochemistry. Symptoms usually appear before the age of 30 months in affected individuals. As outlined in Chapter 4, the information from pedigree construction is used to determine whether a trait is genetically determined and to determine its mode of inheritance. Although these steps seem simple and clear-cut, in practice the decisions are often more difficult. To illustrate these difficulties, we will briefly consider two questions of current interest in human genetics: Is autism a genetic disorder, and if so, how is this trait inherited?

Autism affects about 4 of every 10,000 individuals in the general population. There is a much higher frequency of autism in pairs of identical twins than in nonidentical twins, and siblings of an autistic child are 75 times more likely to be autistic than are members of the general population. These observations indicate that autism has a strong genetic component. Many teams of researchers are working to identify the chromosome regions and the genes involved in autism. In an early study, a team of researchers from UCLA and the University of Utah studied the incidence and inheritance of autism, using almost all the families in the state of Utah as a study group. In 187 families, there was a single autistic child, and in 20 families, there were multiple cases. In the multiple case families, simple recessive or dominant Mendelian inheritance does not easily explain the pattern of transmission.

The accuracy of pedigree studies can be affected by several factors. Autism is a behavioral trait, and the phenotype is not always clearly defined. In addition, there may be a number of different diseases that all produce a similar set of symptoms. Some cases may have symptoms that are too mild to be diagnosed, whereas very severe cases may result in prenatal death, all of which can skew the pedigrees.

To resolve these problems, a group of 21 different institutions formed the International Molecular Genetic Study of Autism Consortium to use recombinant DNA technology to search for autism genes.

Using pedigree analysis and molecular markers, this team identified loci on several chromosomes that may contain such genes. A study of 153 families identified a region on the long arm of chromosome 7 that contains a susceptibility gene. Other studies have turned up genes on the long arm of chromosome 2 and the short arm of chromosome 16. Using a combination of DNA samples from affected families and sequence data from the Human Genome Project, researchers are working to identify genes on these chromosomes. These results, as well as those from twin studies, are most consistent with a polygenic mode of inheritance for autism or a susceptibility to autism. It is estimated that between 5 and 20 genes may contribute to autism. Once these genes have been identified, the role of environmental factors in triggering autism will need to be evaluated.

The threshold model is useful in explaining the frequency of certain disorders and congenital malformations. Evidence for a threshold in any given disorder is indirect and comes mainly from family studies. The frequency of the disorder among relatives of affected individuals is compared to the frequency of the disorder in the general population. In a family, first-degree relatives (parents-children) have one-half of their genes in common, second-degree relatives (grandparents-grandchildren) have one-fourth of their genes in common, and third-degree relatives (uncles-nieces or uncles-nephews, aunts-nieces or aunts-nephews) have one-eighth of their genes in common. As the degree of relatedness declines, so does the probability that individuals will share the same combination of alleles at multiple loci.

According to the threshold model, the risk for a disorder should decrease as the degree of relatedness decreases. The distribution of family risk frequencies for some congenital malformations, shown in Table 5.1, shows a declining risk as the degree of relatedness declines. The multifactorial threshold model provides only indirect evidence for the effect of genotype on traits and for the degree of interaction between the genotype and the environment. The model is helpful, however, in genetic counseling for predicting recurrence risks in families that have certain congenital malformations and multifactorial disorders.

Table 5.1 Familial Risks for Multifactorial Threshold Traits

Multifactorial Trait	Risk Relative to General Population: MZ Twins	First-Degree Relatives	Second-Degree Relatives	Third-Degree Relatives
Club foot	300×	25×	5×	2×
Cleft lip	400×	40×	7×	3×
Congenital hip dislocation (females only)	200×	25×	3×	2×
Congenital pyloric stenosis (males only)	80×	10×	5×	1.5×

Estimating the Interaction Between Genotype and Environment

How can we measure the interaction between the genotype and the environment? To do this, we must first examine the total variation in phenotype of a population, rather than looking at individual members of the population. Phenotypic variation is derived from two sources: (1) the presence of different genotypes in members of the population and (2) the presence of different environments in which all the genotypes have been expressed. Assessing the role of these factors in producing phenotypic variability in a population is embodied in the concept known as **heritability.**

Heritability An expression of how much of the observed variation in a phenotype is due to differences in genotype.

HERITABILITY MEASURES THE GENETIC CONTRIBUTION TO PHENOTYPIC VARIATION

Phenotypic variation that is caused by different genotypes is known as **genetic variance.** If there is phenotypic variation among individuals with the same genotype, this variation is known as **environmental variance.**

Genetic variance The phenotypic variance of a trait in a population that is attributed to genotypic differences.

Environmental variance The phenotypic variance of a trait in a population that is attributed to differences in the environment.

The heritability of a trait, symbolized by H, is that fraction of the total phenotypic variance caused by *genetic* differences. Heritability is always a variable, and it is not possible to obtain an absolute heritability value for any given trait. The value depends on several factors, including which population is being measured and the amount of environmental variation that is present at the time of measurement. Remember, heritability is a phenomenon observed in populations, not in individuals. In other words, heritability is a statistical value (expressed as a percent) that defines the genetic contribution to the trait being analyzed in a population of related individuals (see later discussion).

In general, if heritability is high (it can range up to 100% when $H = 1.0$), the phenotypic variation is largely genetic, and the environmental contribution is low. If the heritability value is low (it can be as low as zero when $H = 0.0$), there is little genetic contribution to the observed phenotypic variation, and the environmental contribution is high.

Heritability Estimates Are Based on Known Levels of Genetic Relatedness

Heritability is calculated from observations made among relatives because we know the fraction of genes shared by related individuals. As described in a previous section, parents and children share one-half their genes, grandparents and grandchildren share one-fourth their genes, and so forth. These relationships are expressed as **correlation coefficients.** The half-set of genes received by a child from its parent cor-

Correlation coefficients Measures of the degree to which variables vary together.

FIGURE 5.10 The dermatoglyphic patterns on the fingers and palms are laid down early in fetal development.

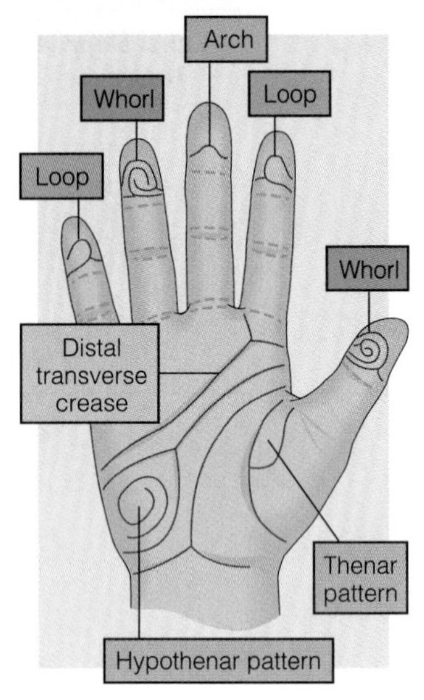

responds to a correlation coefficient of 0.5. The genetic relatedness of identical twins is 100% and is expressed as a correlation coefficient of 1.0. Unless a mother and father are related by descent, they should be genetically unrelated, and the correlation coefficient for this relationship is 0.0.

Using the degree of genetic relatedness among population members expressed as a correlation coefficient and using the measured phenotypic variation expressed in quantitative units (inches, pounds, etc.), a heritability value can be calculated for a given phenotype in the population under study. Values are expressed as a percentage. If for example, the heritability value is 0.72, this means that 72% of the phenotypic variability seen in the population is caused by genetic contributions to the trait.

Using Fingerprints to Estimate Heritability

Because of interactions between genes and the environment, it is difficult to find multifactorial traits that can be used to measure heritability. However, fingerprint ridges are one multifactorial trait that has been used to successfully measure heritability.

Fingerprint patterns are laid down in the first 3 months of embryonic development (weeks 6 to 13). They are controlled by several genes and can be influenced by the environment only during this short period. Everyone, including identical twins, has a unique set of fingerprints. Even though identical twins share the same set of genes and occupy the same uterus simultaneously, each lives in a slightly different subenvironment. These subtle environmental factors are enough to create different fingerprint patterns.

Dermatoglyphics The study of the skin ridges on the fingers, palms, toes, and soles.

Fingerprints are really rows of skin cells called dermal ridges. As they develop, these ridges are laid down in distinctive patterns (similar patterns are formed by the ridges on the palms, toes, and soles). Analysis of these patterns is known as **dermatoglyphics** (translated literally, the term means "skin writing"). The lines in the palms, known as flexion creases (the heart, life, and head lines used in palm reading by fortune-tellers), are formed at the same time as the dermal ridges (Figure 5.10).

Fingerprint patterns are classified by their shape and by ridge counts. The three shapes are loops, whorls, and arches (Figure 5.11). Ridge counts are the feature of fingerprints most useful to the study of phenotypic variance and heritability. They are easily and objectively measured and, once established, are not subject to social and environmental factors (Figure 5.11).

Using correlation coefficients, Sarah Holt studied total ridge counts (TRC) in 825 British males and 825 British females (Table 5.2). The results provide no information about the number or nature of the genes involved, but the almost total agreement between the observed and expected values indicates that TRC is almost totally under genetic control and that environmental factors play only a minor role.

Analysis of ridge counts in parents and their children can be used to estimate the heritability of ridge counts as $H = 0.95$. A heritability value of 0.95 means that most of the phenotypic variation in fingerprint ridges is genetic and is transmitted from parent to offspring. In fact, what this means is that 95% of the phenotypic variation seen in ridge counts is caused by differences in genotype. The small amount of nongenetic variation helps explain why identical twins have different fingerprint patterns.

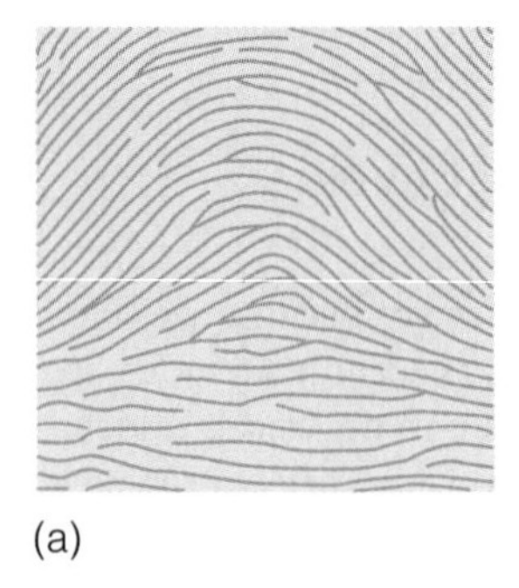
(a)

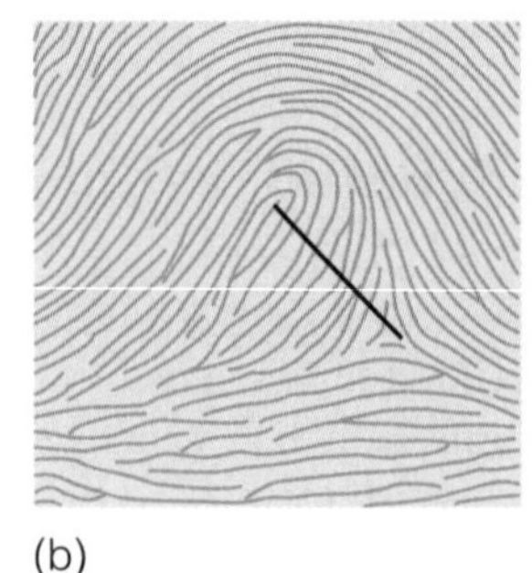
(b)

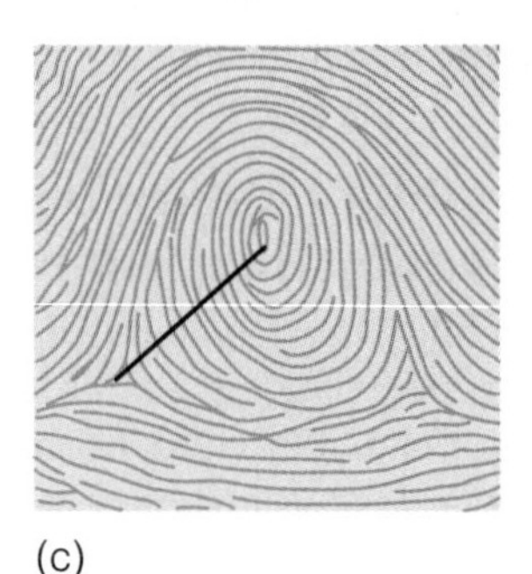
(c)

FIGURE 5.11 The three basic patterns of fingerprints. (a) Arch, (b) Loop, and (c) Whorl. The triangular areas in (b) and (c) where ridge patterns diverge are called triradii. Ridge counts are made from prints of loops and whorls by superimposing a line from the triradius to the center of the print and counting the number of ridges that cross the line.

Table 5.2 Correlations Between Relatives for Total Ridge Count (TRC)

Relationship	Number of Pairs	Observed Correlation Coefficient	Expected Correlation Coefficient Between Relatives	Heritability
Mother–child	405	0.48 ± 0.04	0.50	0.96
Father–child	405	0.49 ± 0.04	0.50	0.98
Husband–wife	200	0.05 ± 0.07	0.00	–
Sibling–sibling	642	0.50 ± 0.04	0.50	1.0
Monozygotic twins	80	0.95 ± 0.01	1.00	0.95
Dizygotic twins	92	0.49 ± 0.08	0.50	0.98

Note: From Quantitative genetics of fingerprint patterns, by S. B. Holt, (1961). Br. Med. Bull., *17, 247–250.*

TWIN STUDIES AND MULTIFACTORIAL TRAITS

Using correlation coefficients to measure the degree of observed phenotypic variability provides an estimate of heritability. This method, however, has one main problem. The closer the genetic relationship, the more likely it is that the relatives share a common environment. In other words, how can we tell whether parents and children have similar phenotypes because they have one-half of their genes in common or because they share a similar environment? Is there a way we can separate the effects of genotype on phenotypic variation from the effects of the environment?

To solve this problem, human geneticists look for situations in which genetic and environmental influences are clearly separated. One way to do this is to study twins (◗ Figure 5.12). Identical twins share the same genotype. If identical twins are

© Mark C. Burnett/Photo Researchers

◗ **FIGURE 5.12** Identical twins (monozygotic twins) have the same sex and share a single genotype.

separated at birth and raised in different environments, the genotype is constant, and the environment is different. To reverse the situation, geneticists compare traits in unrelated adopted children with those of natural children in the same family. In this situation, there is a constant environment and maximum genotypic differences. As a result, twin studies and adoption studies are important tools in measuring heritability in humans.

The Biology of Twins: Monozygotic and Dizygotic Twins

Monozygotic (MZ) Twins derived from a single fertilization involving one egg and one sperm; such twins are genetically identical.

Dizygotic (DZ) Twins derived from two separate and nearly simultaneous fertilizations, each involving one egg and one sperm. Such twins share, on average, 50% of their genes.

Sir Francis Galton (a cousin of Charles Darwin) pointed out the value of studying twins to obtain information about the effects of environment on heredity. Before examining the results of twin studies, we will look briefly at the biology of twinning.

There are two types of twins, **monozygotic** (**MZ**) (identical) and **dizygotic** (**DZ**) (fraternal). Monozygotic twins originate from a single egg fertilized by a single sperm. During mitosis at an early stage of development, two separate embryos are formed. Additional splitting is also possible (see Concepts and Controversies: Twins, Quintuplets, and Armadillos). This separation may occur at the first mitotic division of the zygote, and each of the two cells forms a separate but genetically identical em-

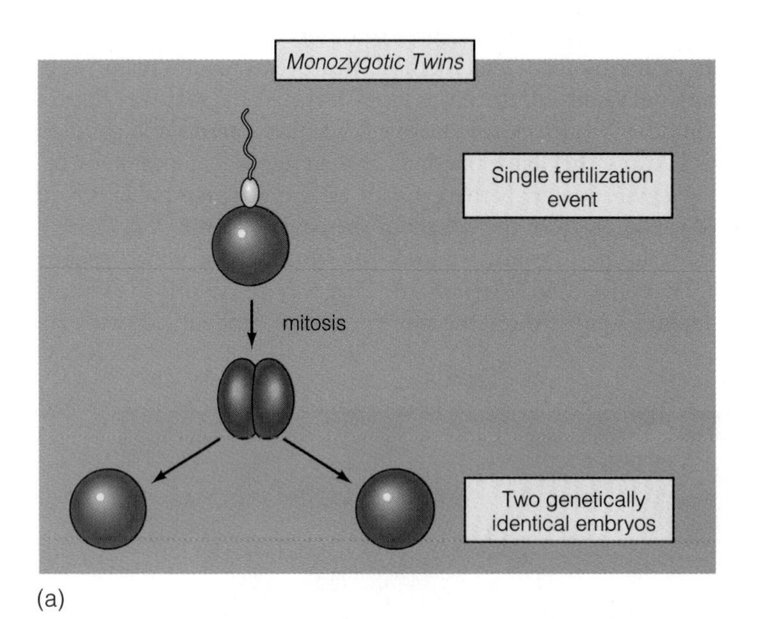

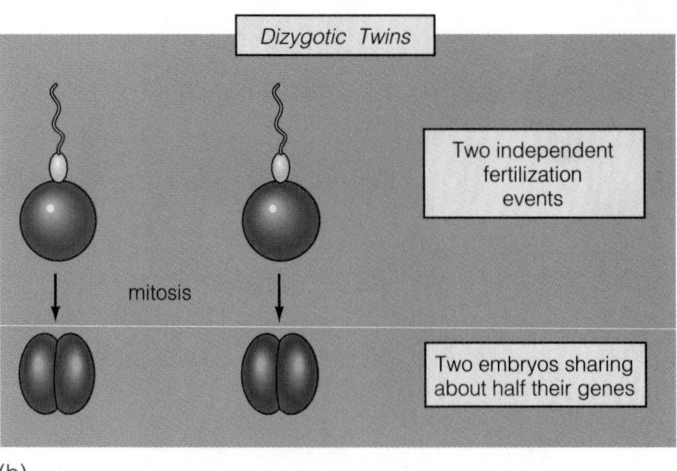

FIGURE 5.13 (a) Monozygotic (MZ) twins result from the fertilization of a single egg by a single sperm. After one or more mitotic divisions, the embryo splits in two and forms two genetically identical individuals. (b) Dizygotic (DZ) twins result from the independent fertilization of two eggs by two sperm during the same ovulatory cycle. Although these two embryos simultaneously share the same uterine environment, they share only about half of their genes.

Concepts and Controversies

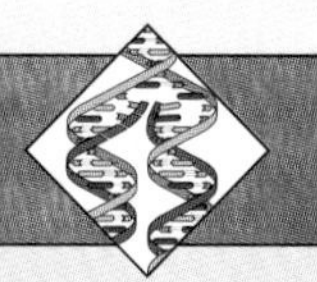

Twins, Quintuplets, and Armadillos

Monozygotic (MZ) twins are genetically identical because of the way in which they are formed. The process of embryo splitting that gives rise to MZ twins can be considered a form of human asexual reproduction. In fact, another mammal, the nine-banded armadillo, produces litters of genetically identical, same-sex offspring that arise by embryo splitting. In armadillo reproduction a single fertilized egg splits in two, and daughter embryos can split again, resulting in litters of two to six genetically identical offspring.

Multiple births in humans occur rarely. About 1 in 7,500 births are triplets, and 1 in 658,000 births are quadruplets. In many cases, both embryo splitting and multiple fertilizations are responsible for naturally occurring multiple births. Triplets may arise by fertilization of two eggs, in which one of them undergoes embryo splitting. The use of hormones to enhance fertility has slightly increased the frequency of multiple births. These drugs work by inducing the production of multiple eggs in a single menstrual cycle. The resulting fertilizations have resulted in multiple births ranging from twins to septuplets.

Embryo splitting in naturally occurring births was documented in the Dionne quintuplets born in May 1934. This was the first case in which all five members of a set of quintuplets survived. Blood tests and physical similarities indicate that these quintuplets arose from a single fertilization followed by several embryo splits.

From this, it seems that MZ twins, armadillos, and the Dionne quintuplets have something in common. They all arise by embryo splitting.

bryo. It may also occur at any time up to the first 2 weeks of development (Figure 5.13). Because they arise from a single fertilization event, MZ twins share the same genotype, have the same sex, and carry the same genetic markers, such as blood types. Dizygotic twins originate from two separate fertilization events: Two eggs, ovulated in the same menstrual cycle, are fertilized independently by two different sperm. DZ twins are no more related than other pairs of siblings, have half of their genes in common, can differ in sex, and may have different genetic markers, such as blood types.

For heritability studies, it is essential to know whether a pair of twins is MZ or DZ. Comparison of many traits using a variety of tests can identify twins as MZ. If one or more traits do not match, it means that the twins are DZ. Among the traits tested are blood groups, sex, eye color, hair color, fingerprints, palm and sole prints, DNA fingerprinting, and analysis of DNA molecular markers.

Concordance and Twins

A simple method for evaluating any phenotypic differences between twins is the use of traits that can be scored as present or absent rather than measured quantitatively. Twins show **concordance** if both have a trait and are discordant if only one twin has the trait. As noted, MZ twins have 100% of their genes in common, whereas DZ twins, on average, have 50% of their genes in common. For a genetically determined trait, the correlation in MZ twins should be higher than that in DZ twins. If the trait is completely controlled by genes, concordance should be 1.0 in MZ twins and close to 0.5 in DZ twins.

Concordance Agreement between traits exhibited by both twins.

In evaluating results, it is the degree of difference between concordance in MZ twins versus DZ twins that is important. The greater the difference, the greater the heritability. Concordance values in twins for several traits are listed in Table 5.3. The concordance value for cleft lip in MZ twins is higher than that for DZ twins (42% versus 5%). Although this difference suggests a genetic component to this trait, the value is so far below 100% that environmental factors are obviously important in the majority of cases. As this example shows, concordance values must be interpreted cautiously.

Table 5.3 Concordance Values in Monozygotic (MZ) and Dizygotic (DZ) Twins

	Concordance Values (%)	
Trait	**MZ**	**DZ**
Blood types	100	66
Eye color	99	28
Mental retardation	97	37
Hair color	89	22
Down syndrome	89	7
Handedness (left or right)	79	77
Epilepsy	72	15
Diabetes	65	18
Tuberculosis	56	22
Cleft lip	42	5

Concordance values can be converted to heritability values through a number of statistical methods. Some calculated heritability values are listed in the right column of Table 5.5. Remember that heritability is a relative value, valid only for the population measured and only under the environmental conditions in effect at the time of measurement. Heritability measurements made within one population cannot be compared with heritability measurements for the same trait in another population because the two groups differ in genotypes and environmental variables in unknown ways.

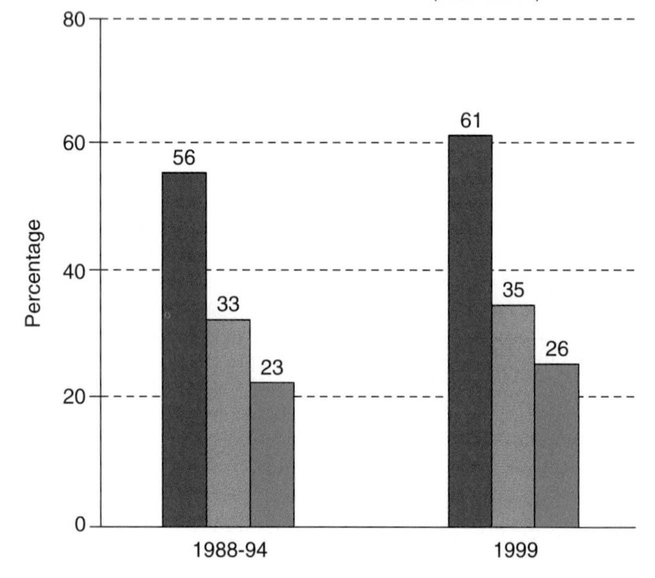

FIGURE 5.14 Age-adjusted prevalence of overweight and obesity among U.S. adults age 20 years and older. The 1999 survey shows that 61% of the adult population is overweight or obese. Comparison with the 1988-94 survey shows that only 56% of the population was overweight or obese at that time.

*Age-adjusted by the direct method to the year 2000 U.S. Bureau of the Census estimates using the age groups 20-29, 30-39, 40-49, 50-59, 60-69, 70-79, and 80 years and over.

Source: Centers for Disease Control and Prevention (http://www.cdc.gov/nchs/images/obsefig1.gif)

Using Twins to Study Obesity

Heritability estimates derived from studies of twins and adopted children are indirect ways of studying multifactorial traits. These studies are based on correlations rather than direct demonstration of cause and effect and are also subject to a number of uncertainties, as shown in the following sections.

Obesity is a trait that is said to "run" in families. Obesity is also a national health problem. A Federal study has estimated that up to 61% of the adult U.S. population is either overweight or obese (Figure 5.14). These individuals are at risk for diseases such as high blood pressure, elevated levels of cholesterol in the blood, coronary artery disease, and adult-onset diabetes.

Twin studies have been used to estimate the heritability of obesity. Let's examine one such study that used 1,974 pairs of MZ twins and 2,097 pairs of DZ twins who were born between 1917 and 1927 and who served in the armed forces. Weights and heights measured at induction into the armed forces were compared with similar measurements taken 25 years later. Obesity was measured by body mass index (BMI = weight in kilograms divided by height in meters squared) and by reference to tables listing ideal

Table 5.4 Concordance Values for Obesity in Twins

% Overweight	% Concordant at Military Induction		% Concordant 25 Years Later	
	MZ	DZ	MZ	DZ
15	61	31	68	49
20	57	27	60	40
25	46	24	54	26
30	51	19	47	16
40	44	0	36	6

Note: From A twin study of human obesity, by A. J. Stunkard, T. T. Foch, and Z. Hrubec, (1986). JAMA, *256, 51–54.*

weight–height relationships. Table 5.4 shows the concordance values from this study, for five levels of obesity. At military induction, concordance for MZ twins was much higher than that for DZ twins at all five levels of obesity. Twenty-five years later, the concordance for MZ twins was still much higher than that for DZ twins. The heritability values derived from these studies are shown in Table 5.5. The results show high values of heritability for obesity, suggesting that this condition has a strong genetic component.

Twin Studies Are Indirect: Potential Problems

Like all twin studies, the results from this study show a correlation, and do not establish a cause and effect relationship. Let's consider some potential problems with this type of study that can affect the conclusions and the interpretation of the results. First, this study included only men who had passed a preinduction physical that eliminated individuals who had marked obesity. This means that at the start, the twin population in this study was biased toward less obese individuals. Because obese individuals were screened out, this study would tend to underestimate the contribution of heredity to obesity. Second, the study did not include women, children, or men excluded from military service for other causes. This screening limited the researchers' ability to generalize the conclusions. Last, the study did not attempt to directly study the role of environmental factors, such as diet in obesity.

Even with these limitations, however, the heritability values are high enough to indicate that obesity is under strong genetic control. Although other, similar studies

Table 5.5 Heritability Estimates for Obesity in Twins (from Several Studies)

Condition	Heritability
Obesity in children	0.77–0.88
Obesity in adults (weight at age 45)	0.64
Obesity in adults (body mass index at age 20)	0.80
Obesity in adults (weight at induction into armed forces)	0.77
Obesity in twins reared together or apart	
Men	0.70
Women	0.66

have their own limitations, the conclusions reached in well-designed twin studies with a large sample size are useful in outlining the role of heritability in human multifactorial traits.

Other Approaches to Studying the Heritability of Obesity

Perhaps the most effective way of separating the effects of genes and environment and of controlling external factors is to study identical and fraternal twins reared under different environmental conditions. In a study of body mass index (BMI) in identical and fraternal twins reared together or apart, heritability values for BMI were calculated as 0.70 for men and 0.66 for women (Table 5.5). This indicates that 66% to 70% of the phenotypic variation (obesity) observed is due to differences in genotype, linking the condition to genetic factors.

Another method of assessing the role of genes and the environment in obesity is to study adopted unrelated children, who bring maximum genotypic differences into a family. In this case, the adopted child with an unrelated genotype is subjected to the same environment as family members with related genotypes. Studies comparing obesity in adopted children with obesity in the biological and adopted parents confirm the role of genetic factors as important influences on obesity and assign a minor role to the family environment. One such study compared obesity in 3,580 adoptees and their brothers and sisters, who were reared separately. The results indicate that obese adoptees tend to have obese siblings, even though they were raised in different environments. This finding reinforces the conclusion from the study of twins in the military that heredity plays an important role in obesity. The analysis also indicates that polygenic inheritance can account for body mass, ranging from thin to obese. These results are consistent with other studies showing that 75% of the phenotypic variation in obesity is explained by genetic factors.

Genetic Clues to Obesity

Heritability estimates are a form of genetic analysis performed at the phenotypic level. These values cannot be used to tell us anything about how many genes control the trait being studied, whether such genes are inherited in a dominant, recessive, or sex-linked fashion, or how such genes act to produce the phenotype. Several

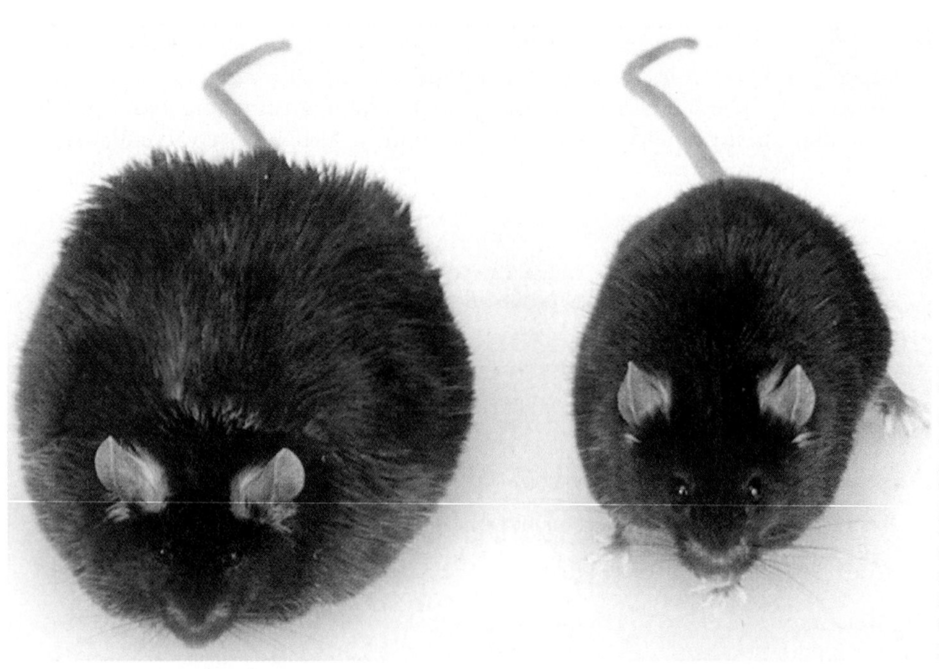

John Sholtis, The Rockefeller University, New York

FIGURE 5.15 The obese (*ob*) mouse mutant, shown on the left (a normal mouse is on the right), has provided many clues about the way weight is controlled in humans.

methods are available to identify genes that contribute to complex traits in humans (such as obesity). One of these methods is the use of animal model systems (such as the rat or mouse) to identify the contributing genes and to study their modes of action. Once genes have been identified in other mammals, recombinant DNA technology can be used to see if the same or related genes are present in humans and whether they work in the same way in mice and men.

Recent breakthroughs in understanding how genes regulate body weight have come from studies in mice. Several mouse genes that control body weight have been identified, isolated, cloned, and analyzed. The mouse mutants obese (*ob*) and diabetes (*db*) are both obese (◗ Figure 5.15). Using recombinant DNA techniques, these genes and their human equivalents have been isolated, mapped, and studied in detail. The *ob* gene encodes the weight-controlling hormone, **leptin** (from the Greek word for "thin"), which is produced by fat cells. In mice, the hormone is released from fat cells and travels through the blood to the brain, where cells of the hypothalamus have cell surface receptors for leptin. These receptors are encoded by the *db* gene. Binding of leptin to cells of the hypothalamus activates the leptin receptor and initiates a response that involves changes in gene expression. This response may involve the production of an appetite-suppressing hormone, glucagon-like protein 1 (GLP-1). In normal individuals the control system apparently regulates the amount of food converted into fat or muscle mass. This hormone may also alter the rate of energy consumption, helping to maintain weight within a relatively narrow range (◗ Figure 5.16). Control of body weight in mice may involve other genes, such as tubby (*tb*), which also has an obese phenotype.

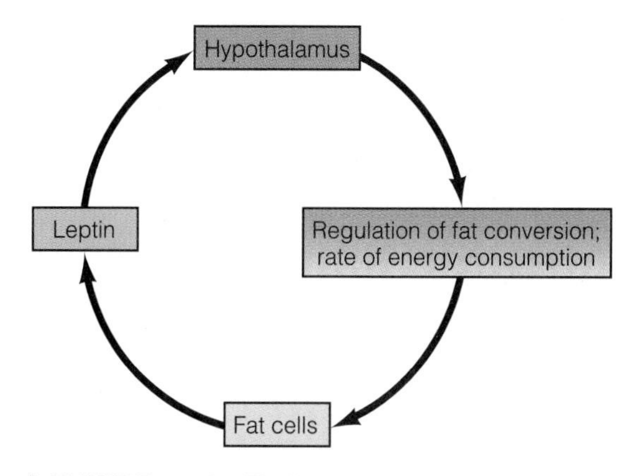

◗ **FIGURE 5.16** The hormone leptin is produced in fat cells, moves through the blood, and binds to receptors in the hypothalamus. Binding presumably activates a control mechanism (still unknown) that controls weight by regulating the conversion of food energy into fat and the rate of energy consumption.

Leptin A hormone produced by fat cells that signals the brain and ovary. As fat levels become depleted, secretion of leptin slows and eventually stops.

The human gene for leptin (MIM/OMIM 164160), which is equivalent to the mouse *ob* gene, maps to chromosome 7q31.1. The gene for the leptin receptor (MIM/OMIM 601007), which is equivalent to the mouse *db* gene, maps to chromosome 1p31. Studies have shown that most obese people do not have mutations in the *ob* gene, but do overproduce leptin. In these individuals, the defect seems to be in the receptor or in the control systems activated by the receptor. An understanding of how leptin and its receptor work to regulate weight will hopefully lead to the design of new drugs to treat obesity. Because Americans spend more than $30 billion dollars in weight control products every year, there is a large potential market for leptin and drugs that affect the leptin receptor. A biotech company has paid $20 million for a license to develop drugs based on leptin, and these products may be on the market in a few years.

Although the gene for leptin may regulate some aspects of body weight, in humans, there is little evidence that a single gene controls a complex phenotype like body weight. There are very few examples of a specific genotype that is directly related to obesity. To search for human genes controlling body weight, several groups have initiated genome-wide searches using large, multigenerational families and molecular markers to search for linkage between the molecular markers and obesity. Using 92 families selected for obesity, one group has found positive evidence for linkage between obesity and one or more genes located on the long arm of chromosome 20. Other groups, using similar methods, have identified loci on chromosome 2 and the long arm of chromosome 11 that may control body weight. Further work on these regions may identify specific genes involved in obesity and provide insight into the molecular events that regulate body weight.

SIDEBAR

Leptin and Female Athletes

Leptin, like some other hormones, may have multiple effects. In addition to signaling the hypothalamus about body fat levels, leptin may also signal the ovary. The obese strain of mice in which leptin was discovered has a mutated leptin gene, and these mice do not make any leptin. Females of this strain are infertile and do not ovulate. Injection of leptin into these females leads to ovulation and normal levels of fertility. The discovery that leptin receptors are found in the ovary indicates that leptin may act directly to control ovulation. In mammalian females, including humans, ovulation stops when body fat falls below a certain level. Many female athletes, such as marathon runners, have low levels of body fat and stop menstruating. Because leptin is produced by fat cells, females who have low levels of body fat may have lowered their leptin levels to the point at which ovulation and menstruation cease.

A SURVEY OF SOME MULTIFACTORIAL TRAITS

Much of the phenotypic diversity in humans results from interactions between genes and environmental factors. Many of these traits involve a number of genes that, taken one by one, have only a small effect on the phenotype. Each of these genes

Table 5.6

Risk Factors for Cardiovascular Disease
Heredity (history of cardiovascular disease before age 55 in family members)
Being male
Hypertension
High blood cholesterol (high LDL and/or low HDL)
Smoking
Obesity
Lack of exercise
Stress

can interact with the environment, producing a wide range of phenotypes and often obscuring the genetic components associated with a disease.

Some progress has been made in defining the genetic components of the multifactorial traits discussed later. New methods of screening for polygenes developed in the last few years and the results of the Human Genome Project may eventually explain how genetic and environmental factors contribute to these and other complex traits.

Cardiovascular Disease Has Genetic and Environmental Components

Two significant factors lead to cardiovascular disease: **hypertension** (MIM/OMIM 145500, high blood pressure) and **atherosclerosis** (MIM/OMIM 143890, deposition of plaque on artery walls). Both traits have significant environmental contributions (Table 5.6) but also have well-established genetic components.

Hypertension Elevated blood pressure, consistently above 140/90 mm Hg.

Atherosclerosis Arterial disease associated with deposition of plaques on inner surface of blood vessels.

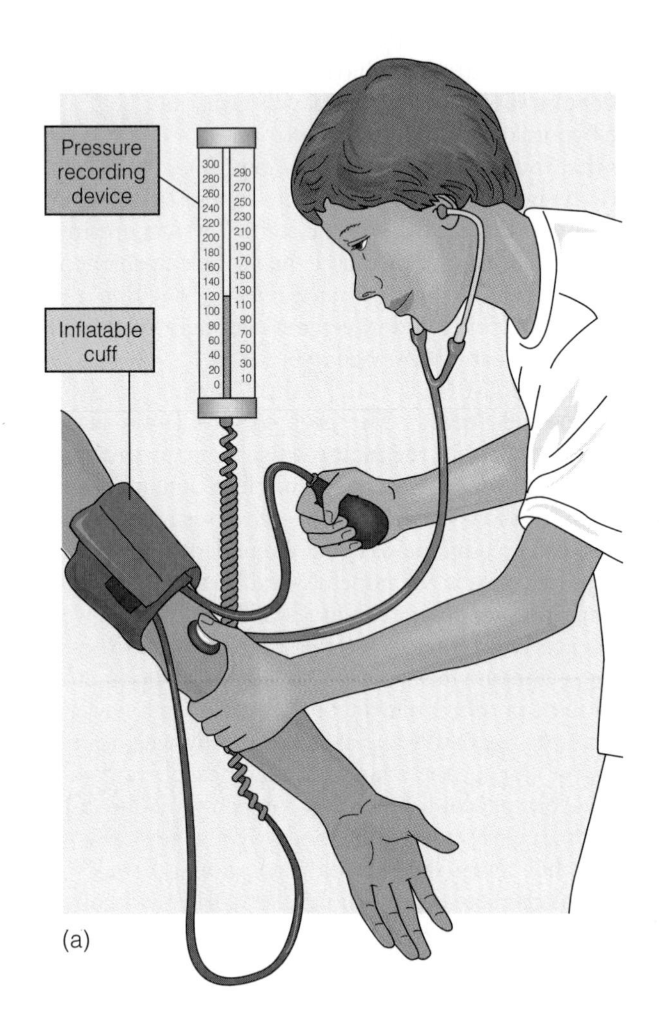

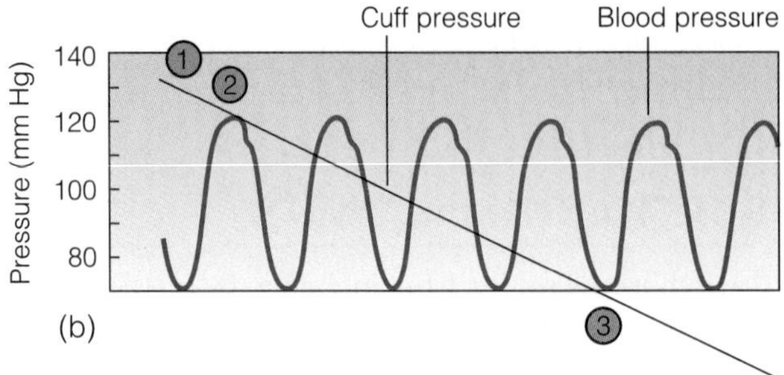

FIGURE 5.17 Blood pressure readings. (a) A blood pressure cuff is used to determine blood pressure. As shown in (b), blood pressure rises and falls as the left ventricle of the heart contracts and relaxes. To measure blood pressure, air is pumped into the cuff until it stops blood flow into the lower arm, and no sound can be heard (1). Air is gradually let out of the cuff, and when blood is heard flowing past the cuff, the pressure measured at this point is taken as the systolic pressure—the upper number in a blood pressure reading (2). As pressure in the cuff is gradually released, the artery becomes fully open, and no sound is heard. This is the diastolic pressure—the lower number in a blood pressure reading (3).

Hypertension occurs when blood pressure (◗ Figure 5.17) is consistently above 140/90 mm Hg (140 is the pressure generated when the heart ventricles contract, and 90 is the pressure when the ventricles are relaxed). At least 10 genes are involved in controlling blood pressure. Most appear to work by controlling the amount of salt and water reabsorbed into the blood by the kidney. One of these is the gene for angiotensinogen (AGT). AGT (MIM/OMIM 106150) is a protein made in the liver and secreted into the blood. In its active form, AGT controls salt and water retention, which in turn controls blood pressure. Some variants of this protein have been linked to a predisposition to hypertension, which is a silent killer because no obvious symptoms appear in early stages of the disease. Some 10% to 20% of the adult population of the United States suffers from hypertension, making it a serious health problem.

Not all forms of hypertension are caused by genes acting in the kidneys. A rare form of hypertension, familial primary pulmonary hypertension (FPPH) inherited as an autosomal dominant trait (MIM/OMIM 178600), has been identified as a mutation in a cell surface receptor gene. Affected individuals have an overproduction of endothelial cells in lung blood vessels, resulting in elevated pressure in the pulmonary arteries.

Atherosclerosis is the result of an imbalance between dietary intake and the synthesis, utilization, and breakdown of lipids, especially cholesterol. This imbalance can lead to blockage of blood vessels and the development of cardiovascular disease (◗ Figure 5.18). The most serious consequences come from blockages in the brain and heart, which cause strokes and heart attacks.

Cholesterol is needed to synthesize plasma membranes and steroid hormones (such as estrogen and testosterone) and is a component of bile. The liver synthesizes cholesterol, and it is not a necessary part of the diet. However, the accumulation of excess cholesterol in arterial plaques is an important factor in cardiovascular disease.

Large lipid molecules, such as cholesterol, are not soluble in blood plasma and are wrapped in a coat of proteins and phospholipids for transport. The coat and its contents are known as **lipoproteins** (◗ Figure 5.19). Lipoproteins are classified by their size and density. Cholesterol can be carried by two types of lipoproteins: low-density lipoprotein (LDL) and high-density lipoprotein (HDL). LDLs are about 45% cholesterol, and HDLs are about 20% cholesterol. LDLs transport cholesterol from the liver to tissues in the body for utilization and breakdown. HDLs transport cholesterol to the liver. The risk of atherosclerosis is related to the HDL/total cholesterol ratio. Higher HDL levels mean lower risk.

Lipoproteins Particles that have protein and phospholipid coats that transport cholesterol and other lipids in the bloodstream.

Genetic studies of cardiovascular disease are directed at finding genes that predispose to disease, identifying environmental factors (such as diet and exercise) that

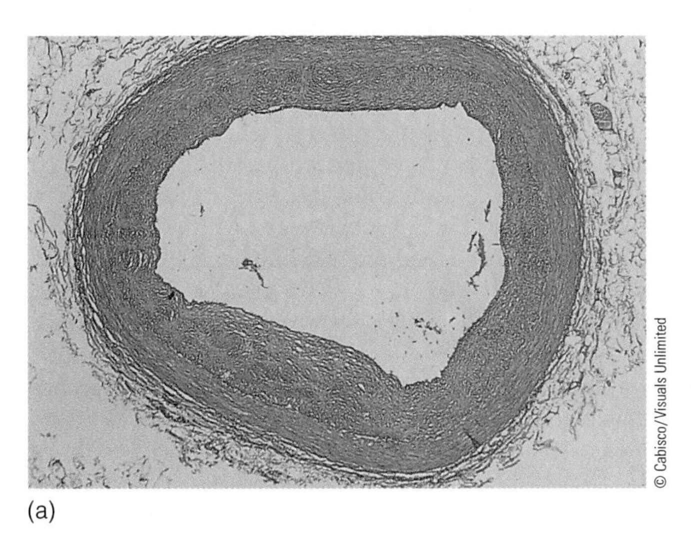

(a)

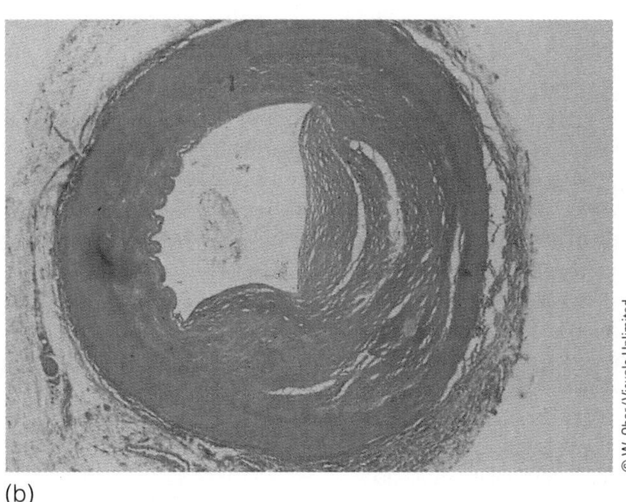

(b)

◗ **FIGURE 5.18** (a) A cross section of a normal artery. (b) A cross section of an artery partially blocked by atherosclerotic plaque. As excess cholesterol accumulates in the body, it accumulates in plaques, leading to cardiovascular disease.

affect disease progression and detecting individuals at risk for cardiovascular disease. Many genes, including those that encode apolipoproteins (the part of the lipoprotein that attaches to cell receptors) control cholesterol levels in the body. Other genes encode cell receptors that bind to and internalize lipoproteins and enzymes that degrade lipoproteins inside the cell. Mutations in one of the receptor genes are discussed in the next section.

Familial hypercholesteremia (FH) Autosomal dominant disorder with defective or absent LDL receptors. Affected individuals are at increased risk for cardiovascular disease.

The autosomal dominant disease **familial hypercholesterolemia (FH)** (MIM/OMIM 143890) is caused by defects in cell surface receptors that regulate the uptake of LDLs. Affected heterozygotes have elevated cholesterol levels in their blood serum and usually develop coronary artery disease between the ages of 40 to 50. Two general classes of mutants are known: defective receptors and absent receptors. Several types of defective receptors are known, including one that cannot recognize and bind to LDLs, another that has reduced ability to bind LDLs, and a third that recognizes and binds LDL but is unable to move the LDL into the cell. Altogether, more than 150 mutations of the gene on chromosome 19 that encodes the LDL receptor have been identified. Each of these mutations results in elevated levels of LDL-derived cholesterol in the blood serum and the deposition of plaques on the inner surface of arteries, leading to premature heart disease.

The frequency of heterozygotes for familial hypercholesterolemia in European, Japanese, and U.S. populations is about 1 in 500, although in regions of Quebec the frequency is 1 in 122. The highest reported frequency is in a South African community, where the frequency is 1 in 71. The average frequency of 1 in 500 makes this disorder one of the most common mutations in our species, and it is one of the major causes of cardiovascular disease.

Mutations in other genes that control lipid metabolism are also involved in cardiovascular disease. A gene for atherosclerosis susceptibility (*ATHS*) has been mapped to chromosome 19 near the locus for the LDL receptor. *ATHS* (MIM/OMIM 108725) is associated with an increase in LDL levels in the blood, a decrease in HDL levels, and a threefold increase in heart attack risk.

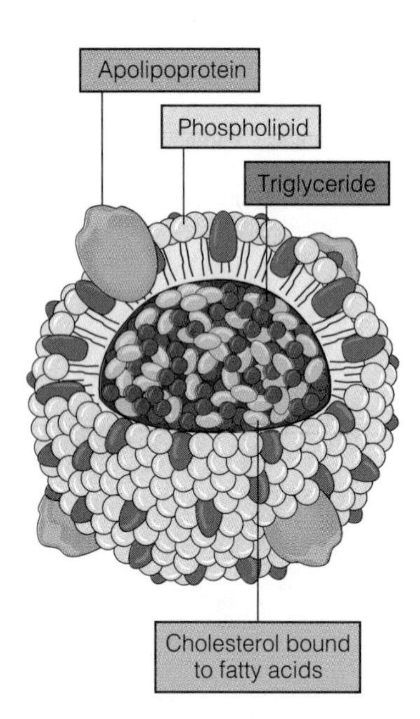

FIGURE 5.19 Lipoproteins carry cholesterol and other lipids through the blood enclosed in a protein and phospholipid coating. There are several types of lipoproteins that have different proportions of lipids. Low-density lipoproteins (LDLs) are composed of about 45% cholesterol, and high-density lipoproteins (HDLs) are made of about 20% cholesterol. High levels of LDLs and a low level of HDLs are risk factors for cardiovascular disease.

Skin Color Is a Multifactorial Trait

Charles and Gertrude Davenport, leading figures in the American eugenics movement, first tested the idea that some human traits are multifactorial. Between 1910 and 1914, they collected information on skin color in black–white marriages in Bermuda and in the Caribbean. Skin color is controlled by genetic and environmental factors, making it a multifactorial trait. Exposure to the sun can darken skin color and obscure genotypic differences. The Davenports were aware of this and measured skin color under the upper arms of their subjects.

To measure skin color, they used a top equipped with a disk containing colored sectors of various sizes (Figure 5.20). The color sectors were black, white, red, and yellow. When the top was spun, the colored sectors blended together and produced a color that could be matched to someone's skin color by changing how much of each colored sector was visible. When the skin color was matched, the Davenports measured the size of the black sector and assigned the individual to one of five categories, numbered 0 to 4.

The results of the Davenports' study illustrate several properties of polygenic traits. The offspring (F1) have skin color values intermediate between those of their parents. In the F2 generation, a small number of children were as white as one grandparent, a small number were as black as the other grandparent, and most had skin color between these two extremes (Figure 5.21). Because F2 individuals could be grouped into five phenotypic classes, the Davenports hypothesized that two gene pairs control skin color (see the distribution of genotypes in Figure 5.4a). Each F2 phenotypic class represented a genotype produced by the segregation and assortment of two gene pairs. To help explain their results, let's suppose that these genes are *A* and *B*, respectively. Class 0 has the lightest skin color and represents the genotype

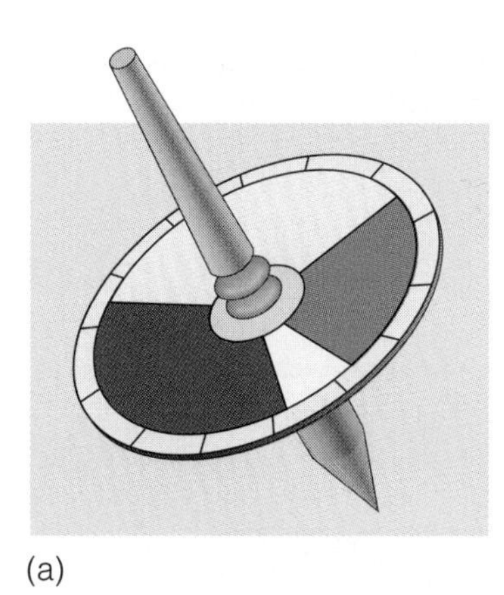

(a)

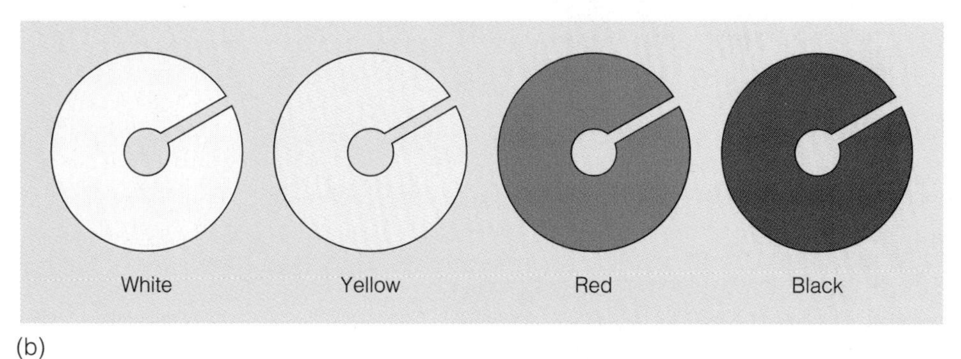

(b)

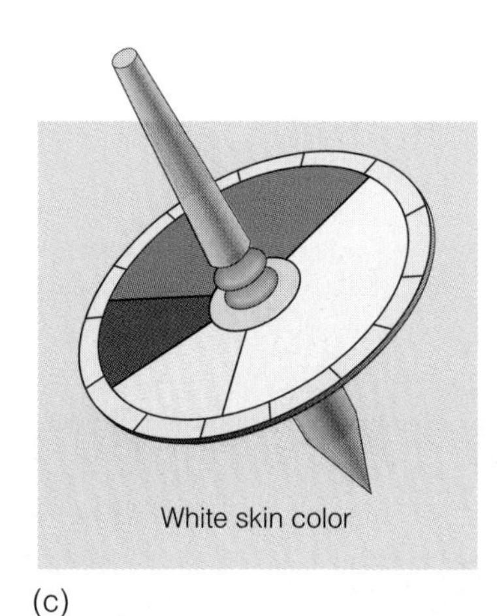

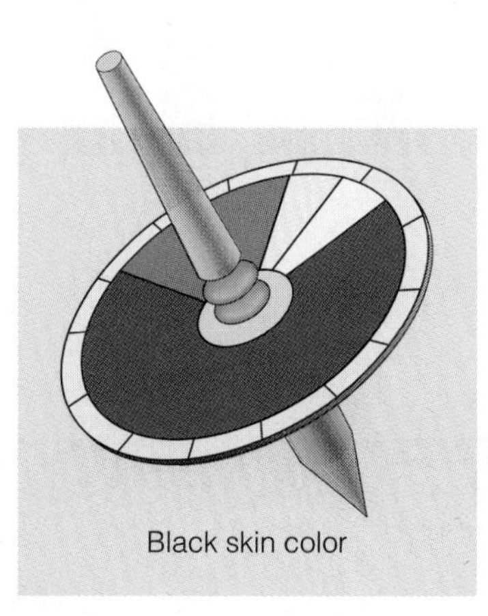

(c)

FIGURE 5.20 (a) The top used by the Davenports to measure skin color with overlaid disks. The color produced when the top is spun can be changed by changing the size of the sectors in the colored disks stacked together on the top. (b) The disks used to measure skin color. Disks of different amounts of black and white can be mounted on the top to produce a color that matches that of the person being studied. (c) Arrangement of sectors that produce skin colors characteristic of those persons studied by the Davenports.

aabb, Class 1 has the genotype *Aabb* or *aaBb,* and so forth, up to Class 5, which has the darkest skin color and represents the homozygous dominant (*AABB*) genotype.

Later work by other investigators, using sophisticated instruments that measure the reflection of light from the skin surface (a reflectometer), has shown that skin color is actually controlled by more than two gene pairs. The data are most consistent with a model that involves three or four genes (Figure 5.22). We will discuss the genetics of skin color in more detail in a discussion of human races in Chapter 18.

Intelligence and Intelligence Quotient (IQ): Are They Related?

The idea that intelligence is a distinct phenotype that can be quantitatively measured arose in the late eighteenth and early nineteenth centuries. Early on in the study of intelligence, phrenologists believed that the brain is composed of a series of compartments. Each compartment controls a single function, such as musical ability, courage, or intelligence. Because each area of the brain has an assigned function,

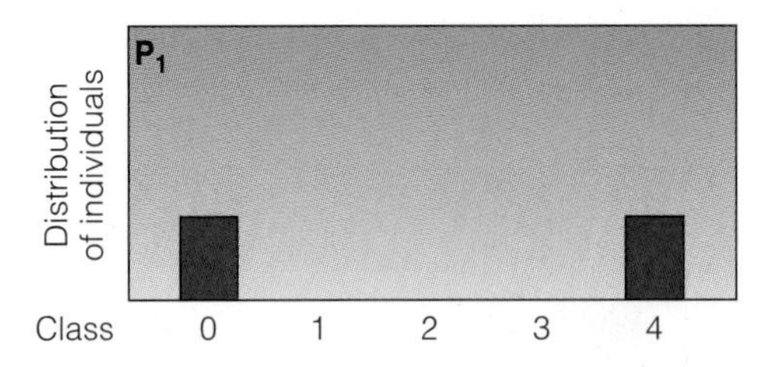

Black sector
(% of disk) 6–11 12–25 26–40 41–55 56–78
(a)

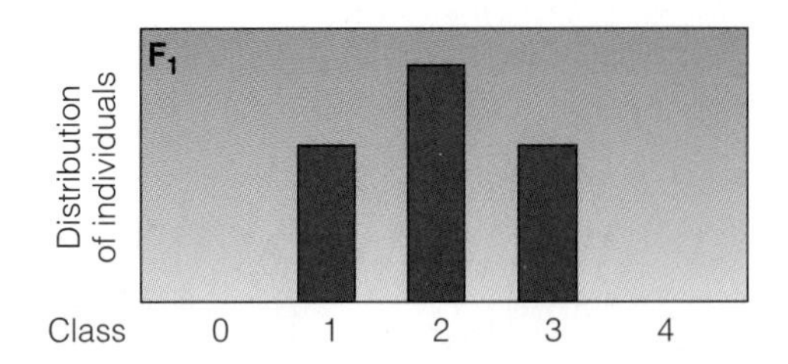

Black sector
(% of disk) 6–11 12–25 26–40 41–55 56–78
(b)

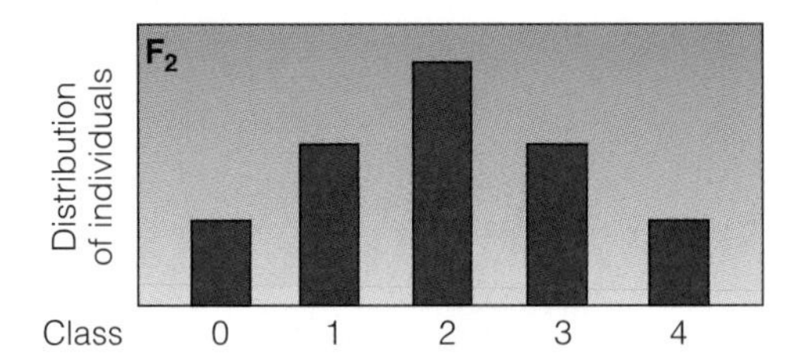

Black sector
(% of disk) 6–11 12–25 26–40 41–55 56–78
(c)

FIGURE 5.21 Frequency diagrams of skin colors obtained by the Davenports. (a) Skin color distribution in the parents falls into two discontinuous classes. (b) Color values of seven children from the parents in (a) are intermediate to those of their parents. (c) Skin colors of 32 children of the parents in (b). Color values range from one phenotypic extreme to the other, and most are clustered around a mean value. This normal distribution of phenotypes is characteristic of a polygenic trait.

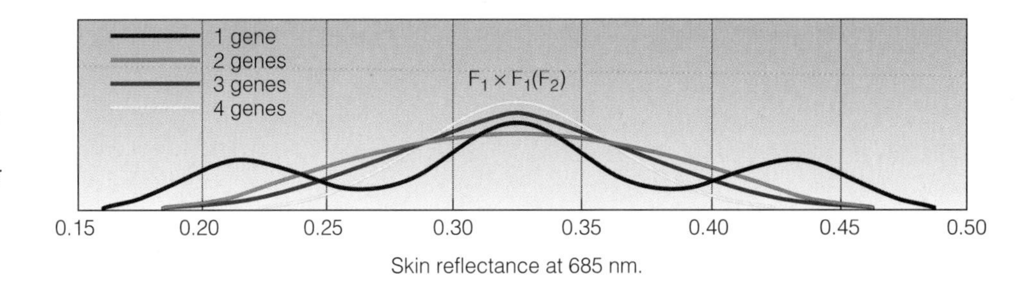

FIGURE 5.22 Distribution of skin color as measured by a reflectometer at a wavelength of 685 nm. The results are shown for an additive model of skin color, with environmental effects, for one to four gene pairs. Distributions observed in several populations indicate that three or four gene pairs control skin color.

phrenologists believed that by examining the shape and size of a particular region of the skull (Figure 5.23) they could determine how much intelligence, courage, and so forth that the individual possessed. In the late nineteenth century, some scientists believed that intelligence was directly related to overall brain size. This means of quantitatively measuring intelligence, called craniometry (the measurement of brain size by weight or volume), became the dominant means of assessing intelligence. Large brain size was associated with high intelligence, and small brain size with lower intelligence.

At the turn of the twentieth century, Alfred Binet, a French psychologist used psychological rather than physical methods to measuring intelligence. To identify children who needed special education, he developed a graded series of tasks related

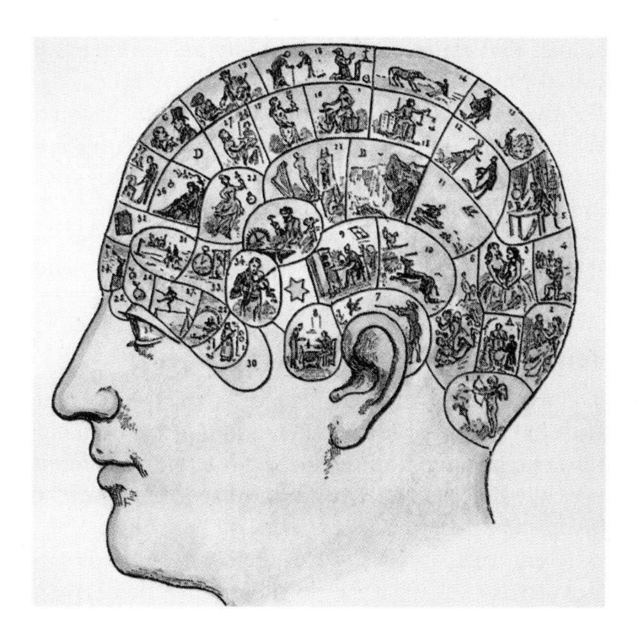

FIGURE 5.23 Phrenology model showing areas of the head overlying brain regions that control different traits. Intelligence was estimated by measuring the area of the skull overlying the region of the brain thought to control this trait.

to basic mental processes, such as comprehension, direction (sorting), and correction. Each child began by performing the simplest tasks and progressing in sequence until the tasks became too difficult. The age assigned for the last task performed became the child's mental age, and the intellectual age was calculated by subtracting the mental age from the chronological age. Binet's test became the basis for the Stanford-Binet intelligence tests in use today.

Wilhelm Stern, another psychologist, divided mental age by chronological age, and the number became known as the **intelligence quotient (IQ).** If a child of 7 years (chronological age) was able to successfully perform tasks for a 7-year-old but could not do tasks for an 8-year-old, a mental age of 7 would be assigned. To determine the IQ for this child, divide mental age by chronological age: Mental age = 7 divided by chronological age = 7 = 1.0 The quotient is multiplied by 100 to eliminate the decimal point ($1.0 \times 100 = 100$), and we obtain an IQ of 100 for this child.

Intelligence quotient (IQ) A score derived from standardized tests that is calculated by dividing the individual's mental age (determined by the test) by his or her chronological age and multiplying the quotient by 100.

The substitution of psychological for physical methods to measure intelligence does not change the underlying assumption that intelligence is a biological property that can be quantitatively expressed as a single number. In fact, if anything, the use of IQ tests by governments and educational institutions has strengthened the assumption that IQ measures a fundamental genetically determined physiological or biochemical property of the brain related to intelligence. The question is whether psychological methods, such as IQ tests, measure intelligence any more accurately than the discredited physical methods of phrenology or craniometry.

To determine whether IQ tests or any other method accurately measure intelligence, we must first define intelligence so that it can be objectively measured, in the same way we measure height, weight, or fingerprint ridge counts. Properties such as abstract reasoning, mathematical skills, verbal expression, ability to diagnose and to solve problems, and creativity are often cited as important components of intelligence. Unfortunately, there is no evidence that any of these properties are directly measured by an IQ test, and there is no objective way to quantify such components of intelligence.

Dozens of distinct neural processes may be involved in mental functions, many of which may contribute to intelligence. In spite of the rapid advances in neurobiology, little or no information is available about the biological basis of any of these processes. Without knowledge of the mechanisms involved, it is difficult to quantitatively measure the outcome of these mechanisms.

The values obtained in IQ measurements, however, do have significant heritable components. The evidence that IQ has genetic components comes from two areas:

studies that estimate IQ heritability and comparison of IQs in groups of individuals raised together (unrelated individuals, parents and children, siblings, and MZ and DZ twins) and individuals raised separately (unrelated individuals, siblings, and MZ twins). Heritability estimates for IQ range from 0.6 to 0.8. The high correlation observed for MZ twins raised together indicates that genetics plays a significant role in determining IQ (Figure 5.24). But rearing MZ twins apart or raising siblings in different environments significantly reduces the correlation and provides evidence that the environment has a substantial role in determining IQ.

The Controversy about IQ and Race

The assumption that intelligence is solely determined by biological factors coupled with the misuse or misunderstanding of the limits of heritability estimates have led to the conclusion that differences in IQ among different racial and ethnic groups are genetically determined. On standardized IQ tests, blacks score an average of 15 points lower than the average score of 100 by whites. These differences are consistent across different tests, and the scores themselves do not seem to be a serious issue. The controversy is over what causes these differences. Are such differences genetic in origin, do they reflect environmental differences, or are both factors at work? If both, to what degree does inheritance contribute to the differences? The debate about these questions has been renewed by the book *The Bell Curve* by Herrnstein and Murray.

The results of heritability studies have been used to support the argument that intelligence is mainly innate and inherited, citing heritability values of 0.8 for intelligence. In most cases, however, the reasoning used to support this argument misuses heritability. Recall that a measured heritability of 0.8 (for example) means that 80% of the phenotypic variation observed is due to genetic differences *within that* population. Heritability differences between two populations cannot be compared because heritability measures only variation within a population at the time of measurement and cannot be used to estimate genetic variation between populations. In other words, we cannot use heritability differences between groups to conclude that there are genetic differences between those groups.

It is quite evident that both genetic and environmental factors make important contributions to intelligence. Clearly, the relative amount each contributes cannot be

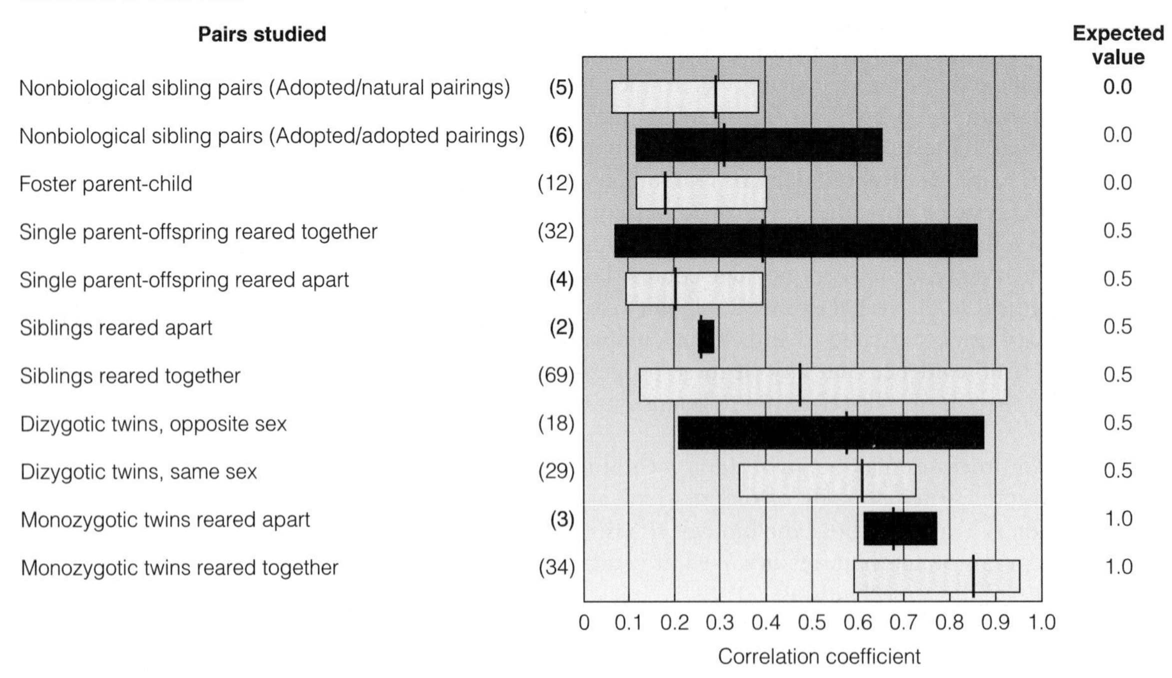

FIGURE 5.24 A graphical representation of correlations in IQ measurements in different sets of individuals. The expected correlation coefficients are determined by the degree of genetic relatedness in each set of individuals. The vertical line represents the median correlation coefficient in each case.

measured accurately at this time. Several points about this debate should be kept in mind. First, it is clear that IQ test scores cannot be equated with intelligence. Second, IQ scores are not fixed and can be changed significantly by training in problem solving and, in fact like many other phenotypes, change somewhat throughout the life of an individual. Variation in IQ scores is quite wide, and measurements in one racial or ethnic group overlap those of other groups, making comparisons more difficult.

The problem in discussing the differences in IQ scores between groups arises when the quantitative differences in scores are converted into qualitative judgments used to rank groups as superior or inferior. Genetics, like all sciences, progresses by formulating hypotheses that can be rigorously and objectively tested but is often misused for ideological ends.

Searching for Genes that Control Intelligence

As discussed in the section on obesity, heritability studies cannot provide information about the number, location, or identity of genes involved in intelligence. To learn about the genes themselves, scientists are using an expanded definition of intelligence that goes beyond IQ and recognizes that many genes are involved in normal cognitive and intellectual function, defined as **general cognitive ability.**

General cognitive ability Characteristics that include verbal and spatial abilities, memory, speed of perception, and reasoning.

To identify genes associated with intelligence, animal model systems such as the fruit fly *Drosophila* and the mouse have been used to identify single genes that control aspects of learning, memory, and spatial perception. After these genes have been mapped and characterized, the equivalent genes in humans are being identified, mapped, and isolated.

A second approach uses techniques of molecular genetics and recombinant DNA to identify genes that affect specific polygenic traits, such as reading ability and IQ. These polygenes are also known as **quantitative trait loci** (QTLs) (and are discussed in more detail in Chapter 16). The search for QTLs has uncovered genes associated with reading disability (developmental dyslexia) on chromosome 6 (MIM/OMIM 600202) and chromosome 15. More recently, a gene associated with general cognitive ability has been identified on chromosome 6. Using approaches that combine behavioral genetics, twin studies, and molecular genetics, more genes controlling aspects of cognitive ability are likely to be identified in the near future. As results from the Human Genome Project are analyzed, it will become easier to define the number and actions of genes involved in higher mental processes and provide insight into the genetics of intelligence. Once such genes are identified and isolated, fears may be raised about genetic discrimination against individuals who have certain genotypes and the possibility of using gene therapy to enhance intelligence. These issues and others related to the Human Genome Project are discussed in Chapter 13.

Quantitative trait loci (QTLs) Two or more genes that act on a single polygenic trait.

Summary

Some Traits Are Controlled by Two or More Genes

1. In this chapter we considered genes that affect traits additively or quantitatively. The pattern of inheritance that controls metric characters (those that can be measured quantitatively) is called polygenic inheritance because two or more genes are usually involved.

Polygenes and Variations in Phenotype

2. The distribution of polygenic traits through the population follows a bell-shaped, or normal, curve. Parents whose phenotypes are near the extremes of this curve usually have children whose phenotypes are less extreme than the parents' and are closer to the population mean. This phenomenon, known as regression to the mean, is characteristic of systems in which the phenotype is produced by the additive action of many genes.

Polygenes and the Environment: Multifactorial Traits

3. Variations in the expression of polygenic traits are often due to the action of environmental factors. Polygenic traits with a strong environmental

component are called multifactorial traits. The impact of environment on genotype can cause genetically susceptible individuals to exhibit a trait discontinuously even though there is an underlying continuous distribution of genotypes for the trait.

Heritability Measures the Genetic Contribution to Phenotypic Variation

4. The degree of phenotypic variation produced by a genotype in a given population can be estimated by calculating the heritability of a trait. Heritability is estimated by observing the amount of variation among relatives having a known fraction of genes in common. MZ twins have 100% of their genes in common and when raised in separate environments, provide an estimate of the degree of environmental influence on gene expression. Heritability is a variable, which is validly calculated only for the population under study and the environmental condition in effect at the time of the study.

Twin Studies and Multifactorial Traits

5. In twin studies, the degree of concordance for a trait is compared in MZ and DZ twins reared together or apart. MZ twins result from the splitting of an embryo produced by a single fertilization, whereas DZ twins are the products of multiple fertilization. Although twin studies can be useful in determining whether a trait is inherited, they cannot provide any information about the mode of inheritance or the number of genes involved.

A Survey of Some Multifactorial Traits

6. Many human traits are controlled by polygenes, including skin color, intelligence, and aspects of behavior. The genetics of these traits has often been misused and misrepresented for ideological or political ends. New genetic approaches using recombinant DNA methods are helping identify genes involved in these traits.

CASE 1

Sue and Tim were referred for genetic counseling after they inquired about the risk of having a child with a cleft lip. Tim was born with a mild cleft lip that was surgically repaired. He expressed concern that his future children could be at risk for a more severe form of clefting. Sue was in her 12th week of pregnancy, and both were anxious about this pregnancy because Sue had a difficult time conceiving. The couple stated that they would not consider terminating the pregnancy for any reason, but wanted to be prepared for the possibility of having a child with a birth defect. The genetic counselor took a three-generation family history from both Sue and Tim and found that Tim was the only person to have had a cleft lip. Sue's family history showed no cases of cleft lip. Tim and Sue had several misconceptions about how clefting occurs and the genetic counselor spent time explaining how cleft lips occur and some of the known causes of this birth defect. Outlined below is a summary of the counselor's discussion with this couple.

- Fathers, as well as mothers, can pass on genes that cause clefting.
- Some clefts are caused by environmental factors, which means the condition didn't come from the father or the mother.
- One child in 33 is born with some sort of birth defect. One in 700 is born with a cleft-related birth defect.
- Most clefts occur in boys; however, a girl can be born with a cleft.
- If a person (male or female) is born with a cleft, the chances of that person having a child with a cleft, given no other obvious factor, is 7 in 100.
- Some clefts are related to identifiable syndromes. Of those, some are autosomal dominant. A person with an autosomal dominant gene has a 50% probability of passing the gene to an offspring.
- Many clefts run in families even when there does not seem to be any identifiable syndrome present.
- Clefting seems to be related to ethnicity, occurring most often among Asians, Latinos, and Native Americans (1:500); next most often among persons of European ethnicity (1:700); and least often among persons of African origins (1:1,000).
- A cleft condition develops during the fourth to the eighth week of pregnancy. After that critical period, nothing the mother does can cause a cleft. Sometimes a cleft develops even before the mother is aware that she is pregnant.
- Women who smoke are twice as likely to give birth to a child with a cleft. Women who ingest large quantities of vitamin A or low quantities of folic acid are more likely to have children with a cleft.
- In about 70% of cases, the fetal face is clearly visible using ultrasound. Facial disorders have been detected at the 15th gestational week of pregnancy. Ultrasound can be precise and reliable when diagnosing fetal craniofacial conditions.

Case Studies (continued)

1. After hearing this information, should Sue and Tim feel that their chances of having a child with a cleft lip are increased over that of the general population? If yes, how much of an increase?
2. Can cleft lip be surgically corrected?
3. If the child showed a cleft lip through ultrasound analysis, and the parents then started blaming each other (because Sue is a smoker, and Tim was born with the defect), how would you counsel them?

CASE 2

Louise was an active 27-year-old gymnastic instructor at a local YMCA. She was recently accepted into law school and worked part time in the evenings at the local Gap clothing store. She was very busy and therefore thought nothing of her increasing fatigue until it started to affect her daily activities. She also complained of occasional dizziness, difficulty hearing, constipation, and problems with controlling her bladder. Her general practitioner conducted a series of blood tests to determine what was wrong. All tests showed she was in perfect health but her symptoms were becoming progressively worse. Over a period of months, she was forced to quit teaching gymnastics and was soon confined to a wheelchair. Her doctor finally referred her for a genetic evaluation. The clinical geneticist immediately recognized Louise's symptoms as signs of multiple sclerosis (MS). The geneticist explained that fatigue, almost to the point of being disabling, is the most common symptom of MS. Some medications could help her fatigue but their effect may not last. The best cure for fatigue is to listen to her body and just rest whenever possible.

The geneticist warned her of other possible symptoms of MS, including: (1) numbness, which is most likely a direct result of the destruction of myelin in the nerves; (2) tingling or pins-and-needles sensation in her feet (the same sensation you get when your foot "falls asleep"), for which not much can be done other than to massage the affected area and rest; (3) tremors, usually of the hands, also due to myelin destruction; (4) muscle spasms (sustained or temporary muscle contractions) in the arms, legs, abdomen, back—just about anywhere; (5) depression and mood swings; (6) memory problems; (7) loss of balance with or without dizziness; and (8) bowel and bladder problems. MS is the most common autoimmune disease involving the nervous system, affecting approximately 250,000 individuals in the United States. The cause of MS is unknown; however, some studies have suggested that the risk to a first-degree relative (sibling, parent, or child) of a patient with MS is at least 15 times that for a member of the general population. Unfortunately, no definite genetic pattern is discernible.

1. What could explain why a first-degree relative of an MS patient is 15 times more likely to have MS than the general population? Does this mean that MS has a genetic component?
2. What is an autoimmune disease? What kinds of genes could be involved in this process?
3. What kind of medical research questions should be asked to study MS and possible treatments?

Questions and Problems

Some Traits Are Controlled by Two or More Genes

1. Describe why continuous variation is common in humans, and provide examples of such traits.
2. The text outlines some of the problems Frederick William I encountered in his attempt to breed tall Potsdam Guards.
   a. Why were the results he obtained so different from those obtained by Mendel with short and tall pea plants?
   b. Why were most of the children shorter than their tall parents?
3. What role might environment have played in causing Frederick William's problems, especially at a time when nutrition varied greatly from town to town and from family to family?
4. Do you think Frederick William's experiment would have worked better if he had ordered brother-sister marriages within tall families instead of just choosing the tallest individuals from throughout the country?
5. As it turned out, one of the tallest Potsdam Guards had an unquenchable attraction to short women. During his tenure as guard he had numerous clandestine affairs. In each case children resulted. Subsequently, some of the children, who had no way of knowing that they were related, married and had children of their own. Assume that two pairs of genes determine height. The genotype of the 7-foot-tall Potsdam Guard was *A'A'B'B'*, and the genotype of all of his 5-ft. clandestine lovers was *AABB*, where an *A'*

or *B'* allele adds 6 inches to the base height of 5 ft. conferred by the *AABB* genotype.
   - **a.** What were the genotypes and phenotypes of all of the F1 children?
   - **b.** Diagram the cross between the F1 offspring, and give all possible genotypes and phenotypes of the F2 progeny.

Polygenes and Variations in Phenotype

6. Describe why there is a fundamental difference between the expression of a trait that is determined by polygenes and the expression of a trait that is determined monogenetically.

Polygenes and the Environment: Multifactorial Traits

7. A clubfoot is a common congenital birth defect in the Smith family. This defect is caused by a number of genes but appears to be phenotypically distributed in a noncontinuous fashion. Geneticists use the multifactorial threshold model to explain the occurrence of this defect. Explain this model. Explain predisposition to the defect in an individual who has a genotypic liability above the threshold versus an individual who has a liability below the threshold.

Heritability Measures the Genetic Contribution to Phenotypic Variation

8. Define genetic variance.
9. Define environmental variance.
10. How is heritability related to genetic and environmental variance?
11. Why are relatives used in the calculation of heritability?
12. If there is no genetic variation within a population for a given trait, what is the heritability for the trait in the population?

Twin Studies and Multifactorial Traits

13. Can conjoined (Siamese) twins be dizygotic twins given the theory that conjoined twins arise from incomplete division of the embryo?
14. Dizygotic twins:
   - **a.** are as closely related as monozygotic twins.
   - **b.** are as closely related as nontwin siblings.
   - **c.** share 100% of their genetic material.
   - **d.** share 25% of their genetic material.
   - **e.** none of the above.
15. Why are monozygotic twins who are reared apart so useful in the calculation of heritability?
16. Monozygotic (MZ) twins have a concordance value of 44% for a specific trait, whereas dizygotic twins have a concordance value of less than 5% for the same trait. What could explain why the value for MZ twins is significantly less than 100%?
17. If monozygotic twins show complete concordance for a trait whether they are reared together or apart, what does this suggest about the heritability of the trait?
18. In this chapter, you read about an obesity study in which MZ and DZ twins who served in the armed forces were studied at induction into the military and 25 years later. Results indicated that obesity has a strong genetic component.
   - **a.** What are the problems with this study?
   - **b.** Design a better study to test whether obesity has a genetic component.
19. What does the *ob* gene code for? How does it work? Is this just a gene found in animals, or do humans have it also?
20. What is the importance of the comparison of traits between adopted and natural children in determining heritability?

A Survey of Some Multifactorial Traits

21. Is cardiovascular disease (hypertension, atherosclerosis, and familial hypercholesterolemia) genetic?
22. Discuss the difficulties in attempting to determine whether intelligence is genetically based.
23. At the age of 9 years, your genetics instructor was able to perform the mental tasks of an 11-year-old. According to Wilhelm Stern's method, calculate his or her IQ.
24. Suppose that a team of researchers analyzes the heritability of SAT scores and assigns a heritability of 0.75 for this skill. This team also determines that a certain ethnic group has a heritability value that is 0.12 lower when compared to that of other ethnic groups. The group draws the conclusion that there must be a genetic explanation for the differences in scores. Why is this an invalid conclusion?

Internet Activities

The following activities use the resources of the World Wide Web to enhance the topics covered in this chapter. To investigate the topics described below, log on to the book's home page (www.brookscole.com/biology_d) and follow the *Student Resources* link to your subject and text.

1. *Twin Studies.* Access the Finnish Twin Cohort Study (http://kate.pc.helsinki.fi/twinhome.html) to view information on a long-term twin study project. The project overview mentions several different on-going studies. What are some of the traits that are being looked at as part of this project? Why are twin studies so helpful in the analysis of complex, multifactorial traits? This Web site also has a link to the International Society for Twin Studies (http://www.ists.qimr.edu.au/). At the ITST Web site, open and read the link to the "Declaration of Rights and Statement of Needs of Twins and Higher Order Multiples." Why do you think the authors of this statement saw the need for such a document?
2. *A Multifaceted Look at a Multifactorial Trait.* Body weight is an example of a multifactorial trait that results from the interaction between multiple genes and the environment. At the same time, the issue of body weight has taken on different connotations in different societies. The PBS series *Frontline* has a Web site that addresses both the question of what makes people "fat" (http://www.pbs.org/wgbh/pages/frontline/shows/fat/) and how fatness is perceived by our society.
   a. At the Web site, click on "What the Experts Said in This Report" regarding body weight, especially those comments regarding the various causes—cultural, societal, and genetic—of fatness in people. Does the fact that body weight is largely caused by genetic factors affect your personal opinion about and/or reaction to obesity? Certain genes have been found that may play a role in human obesity. Would you support the prospect of correcting mutant genes in humans to prevent or reverse obesity? Why or why not?
   b. Follow the link to the chapter of Richard Klein's book *Eat Fat* (http://www.pbs.org/wgbh/pages/frontline/shows/fat/readings/). Klein discusses how fatness and even the word *fat* once had very positive connotations; however, this is generally untrue today, especially in the industrialized Western world. Based on your reading, why do you think that popular perceptions of fat have changed so much over time? Do you think these perceptions might ever revert to conform with the old value system?

For Further Reading

Aldhous, P. (1992). The promise and pitfalls of molecular genetics. *Science, 257,* 164–165.

Balmor, M. G. (1970). *The biology of twinning in man.* Oxford, England: Clarendon.

Baringa, M. (1996). Researchers nail down leptin receptor. *Science, 271,* 913.

Bouchard, C., & Aderusse, L. (1993). Genetics of obesity. *Ann. Rev. Nutr., 13,* 337–354.

Bouchard, C., Tremblay, A., Despres, J.-P., Nadeau, A., Lupien, P. J., Theriault, G., Dussault, J., Moorjani, S., Pinault, S., & Fournier, G. (1990). The response to long-term overfeeding in identical twins. *New Engl. J. Med., 322,* 1477–1481.

Bouchard, T. (1994). Genes, environment and personality. *Science, 264,* 1700–1701.

Bower, B. (2000). Gene implicated in development of autism. *Sci. News, 158,* 90.

Feldman, M. W., & Lewontin, R. (1975). The heritability hangup. *Science, 190,* 1163–1166.

Harrison, G. A., & Owens, J. J. T. (1964). Studies on the inheritance of human skin color. *Ann. Hum. Genet., 28,* 27–37.

Horgan, J. (1995). Get smart, take a test. *Sci. Am., 273,* 12–14.

International Molecular Genetic Study of Autism Consortium. (1998). A full genome screen for autism with evidence for linkage to a region on chromosome 7q. *Human Mol. Genet., 7,* 571–578.

Rice, T. (2000). Tracking of familial resemblance for resting blood pressure over time in the Quebec family study. *Hum. Biol., 72,* 415–431.

Stern, C. (1970). Model estimates for the number of gene pairs involved in pigmentation variability in Negro-Americans. *Hum. Heredity, 20,* 165–168.

Stunkard, A. J., Foch, T., & Hrubec, Z. (1986). A twin study of human obesity. *JAMA, 256,* 51–54.

Travis, J. (2000). Obesity hormone tackles wound healing. *Sci. News, 157,* 55.

Vogel, S. (1999). Why we get fat. *Discover, 20,* 94–99.

© M. Coleman/Visuals Unlimited

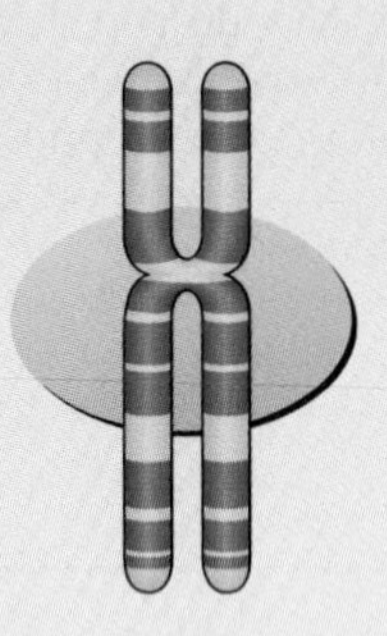

Cytogenetics: Karyotypes and Chromosome Aberrations

Chapter Outline

In 1953 Jerome Lejeune, a French physician, began working on a condition first described in 1866 by John Langdon Down, an English physician. Affected individuals have a distinctive physical appearance and are mentally retarded. Lejeune suspected that there was a genetic link to this phenotype.

He began by studying the fingerprints and palm prints of children who had this condition, known as Down syndrome, and comparing them with the prints of unaffected children. The prints from Down syndrome children showed a high frequency of abnormalities. Because fingerprints and palm prints are laid down very early in development, they serve as a record of events that take place early in embryogenesis. From his studies on print patterns, Lejeune was convinced that the drastic changes he saw in Down syndrome children were not caused by abnormalities in one or two genes. Instead he reasoned that many genes must be involved, perhaps even an entire chromosome.

Lejeune decided to examine the chromosomes of Down syndrome children, but he lacked access to the necessary equipment and techniques. Working with a colleague, he began culturing cells from Down syndrome children in 1957. Lejeune studied chromosomes from the cells of Down syndrome patients using a microscope that had been discarded by the bacteriology laboratory in the hospital where he worked. He repaired the instrument by inserting a foil wrapper from a candy bar into the gears so that the image could be focused. His chromosome counts showed that cells from Down syndrome children contained 47 chromosomes, whereas cells from unaffected children contained 46 chromosomes. He published a short paper in 1959 that reported that Down syndrome is caused by the presence of an extra chromosome. This chromosome was later identified as chromosome 21.

This remarkable discovery was the first human chromosome abnormality to be identified and marked an important turning point in human genetics. The discovery made clear that genetic disorders can be associated with changes in chromosome number, not just mutations of single genes inherited in Mendelian fashion.

THE HUMAN CHROMOSOME SET

Human chromosomes are usually studied using a light microscope and are photographed in metaphase of mitosis. For convenience, the chromosomes are cut out from a photograph and are arranged in pairs according to size and centromere lo-

▶ **FIGURE 6.1** A human karyotype showing replicated chromosomes from a cell in the metaphase of mitosis. This female has 46 chromosomes, including two X chromosomes.

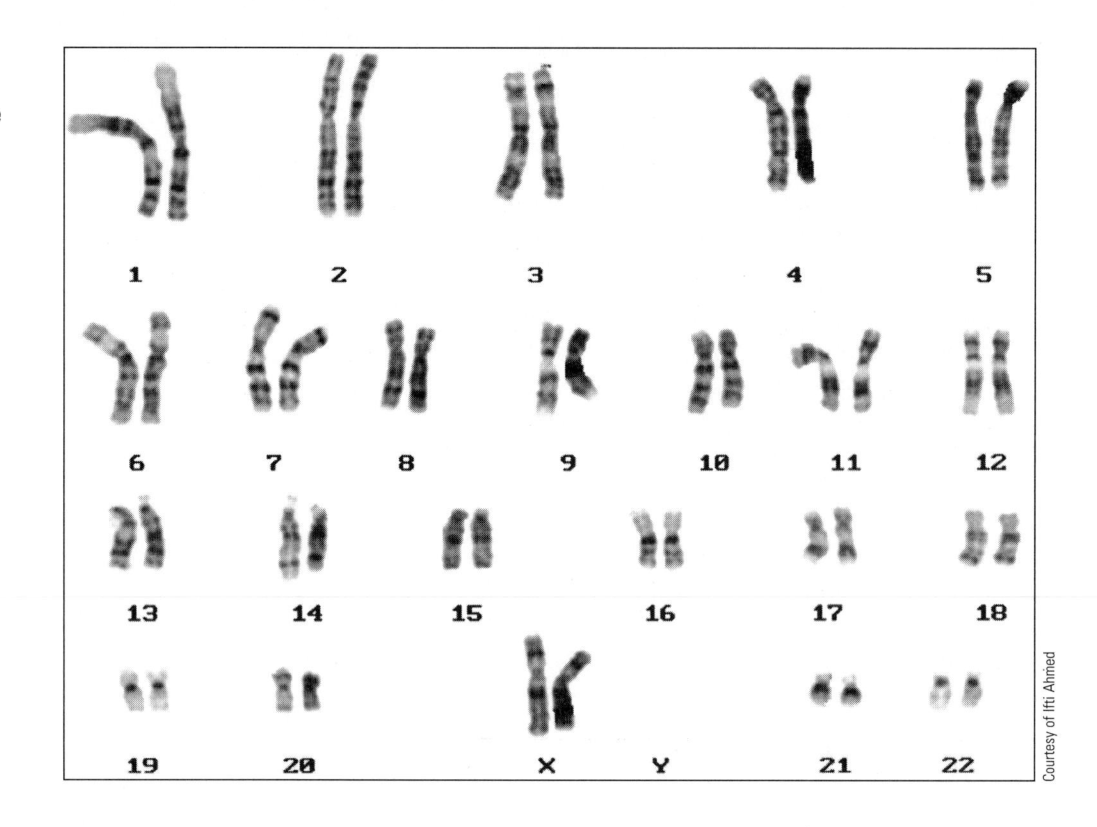

Karyotype The chromosomal complement of a cell line or a person, photographed at metaphase and arranged in a standard sequence.

Sex chromosomes Chromosomes involved in sex determination. In humans, the X and Y chromosomes are the sex chromosomes.

Autosomes Chromosomes other than the sex chromosomes. In humans, chromosomes 1 to 22 are autosomes.

cation to form a **karyotype** (▶ Figure 6.1). In human karyotypes, one pair of chromosomes do not always match. Members of this pair are involved in the process of sex determination and are known as **sex chromosomes.** There are two types of sex chromosomes, X and Y. Females have two homologous X chromosomes, and males have a nonhomologous pair consisting of one X and one Y chromosome. Both males and females have two copies of the remaining 22 pairs of chromosomes, known as **autosomes.**

By convention, the chromosomes in the human karyotype are numbered and arranged into seven groups, A through G. Each group is defined by size and centromere location. Staining produces unique band patterns for each chromosome, allowing individual chromosomes to be clearly identified. The standardized G-banding pattern for the human chromosome set is shown in ▶ Figure 6.2. Chromosome banding patterns are used to identify specific regions on each chromosome (▶ Figure 6.3). For identification of regions, the short arm of each chromosome is designated the p arm, and the long arm the q arm. Each arm is subdivided into numbered regions beginning at the centromere. Within each region, the bands are identified by number. Thus, any region in the human karyotype can be identified by a descriptive address, such as 1q2.4. This address consists of the chromosome number (1), the arm (q), the region (2), and the band (4) (Figure 6.3).

Karyotypic analysis of banded chromosomes is a powerful tool for chromosomal studies and is one of the basic techniques in human genetics (see Concepts and Controversies: Making a Karyotype). Chromosome analysis has many important applications in human genetics, some of which are discussed in later sections. These include prenatal diagnosis, gene mapping, and cancer research.

ANALYZING CHROMOSOMES AND KARYOTYPES

Chromosomal banding methods use stains and dyes to produce a pattern of bands that is different for each chromosome (although homologous chromosomes have the same pattern). One of the most common staining methods is G-banding, in which

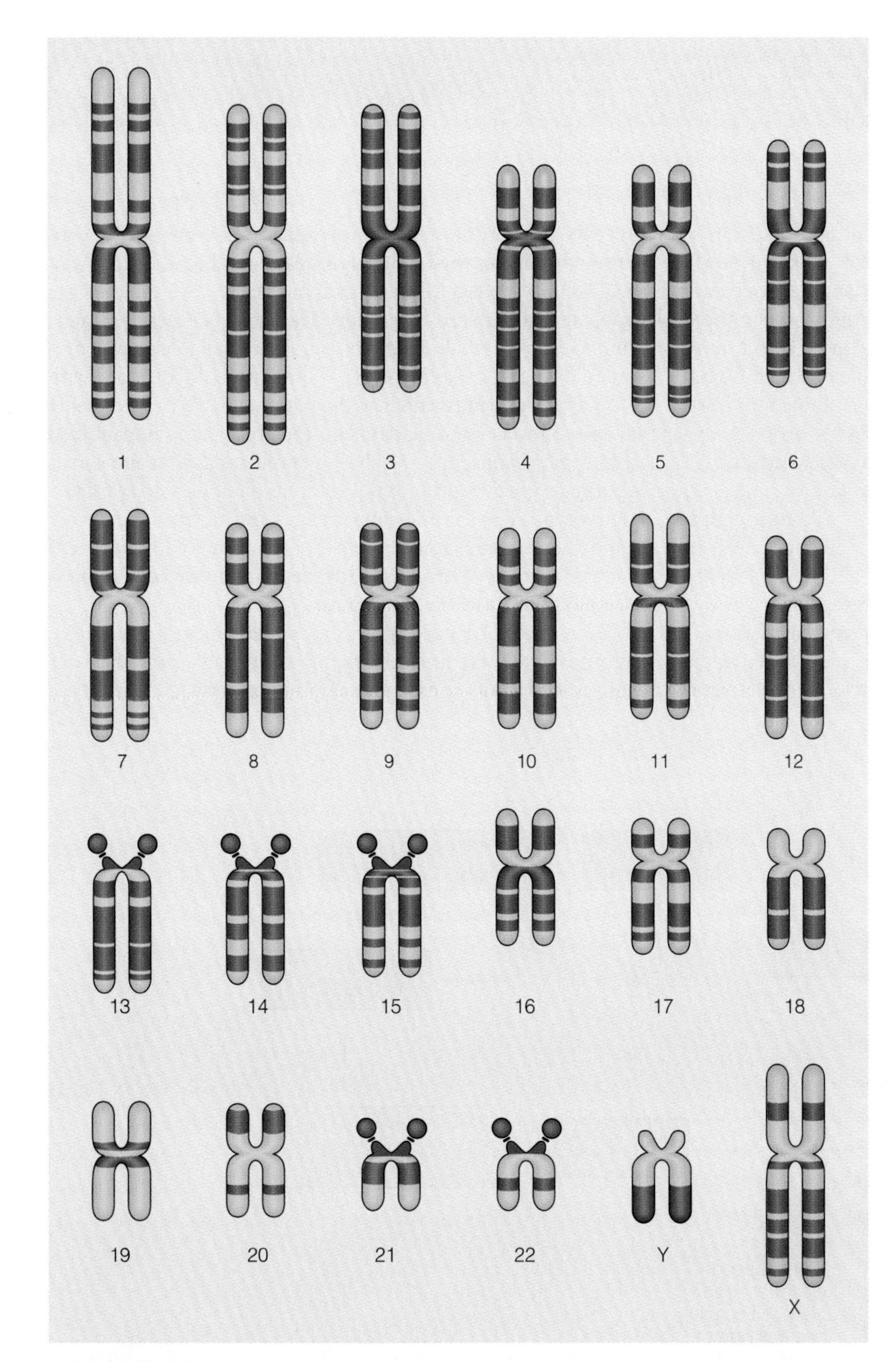

FIGURE 6.2 A karyogram of the human chromosome set, showing the distinctive banding pattern of each chromosome.

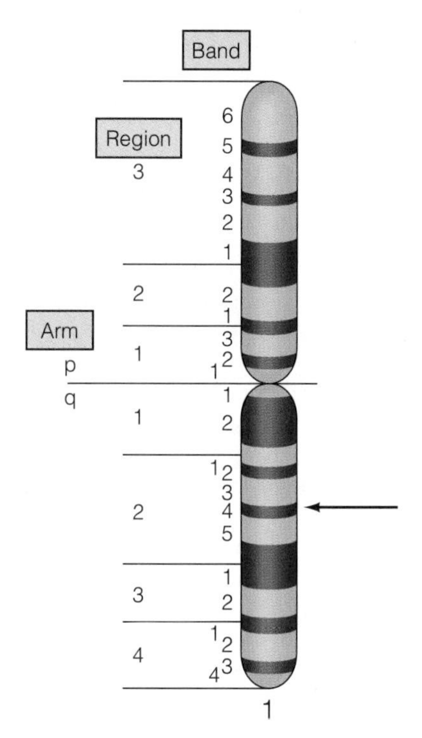

FIGURE 6.3 The system of naming chromosome bands. Each autosome is numbered from 1 to 22. The sex chromosomes are X and Y. Within a chromosome, the short arm is the p arm, and the long arm is the q arm. Each arm is divided into numbered regions. Within each region, numbers designate the bands. The area marked by the arrow is designated as 1q2.4 (chromosome 1, long arm, region 2, band 4).

chromosomes are first treated with an enzyme (trypsin) to partially digest some chromosomal proteins and are then exposed to Giemsa stain (a mixture of dyes). The resulting pattern of bands is visible under the microscope and is used to identify individual chromosomes in cytogenetic analysis. Usually, metaphase chromosomes are stained because this is the stage of maximum chromosomal condensation.

Normally, cytogeneticists use metaphase chromosomes that have about 550 bands. More detailed banding patterns can be produced by staining chromosomes at early

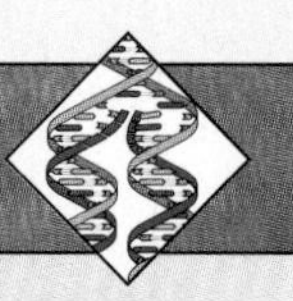

Concepts and Controversies

Making a Karyotype

Chromosomal analysis can be performed using cells from a number of sources, including white blood cells (lymphocytes), skin cells (fibroblasts), amniotic fluid cells (amniocytes), and chorionic villus cells (placental cells). One of the most common methods of preparing cells begins with collecting a blood sample from a vein in the arm. A few drops of the collected blood are added to a flask containing a nutrient growth medium. Because the lymphocytes in the blood sample do not normally divide, a mitosis-inducing chemical, such as phytohemagglutinin, is added to the flask, and the cells are grown for 2 or 3 days at body temperature (37°C) in an incubator. At the end of this time, the drug Colcemid is added to stop dividing cells at metaphase. Over a period of about 2 hours of Colcemid treatment, cells entering mitosis are arrested in metaphase.

Then, the blood cells are concentrated by centrifugation, and a salt solution is added to break open the red blood cells (which are nondividing) and to swell the lymphocytes. This treatment prevents the chromosomes in dividing cells from sticking together and makes the lymphocytes easier to break open. After fixation in a mixture of methanol and acetic acid, the cells are dropped onto a microscope slide. The impact causes the fragile lymphocytes to break open, spreading the metaphase chromosomes. The chromosome preparation is partially digested with the enzyme trypsin to enhance the banding pattern. After being stained to reveal banding patterns, the preparation is examined with a microscope, and the cluster of metaphase chromosomes is photographed. A print of this photograph is used to construct a karyotype. The chromosomal images are cut from the photograph and arranged according to size, centromere location, and banding pattern. A trained cytogeneticist can detect the presence of chromosomal abnormalities.

The process of cutting chromosomal images from photographs by hand and arranging them into karyotypes is tedious and time consuming. Computer-assisted karyotype

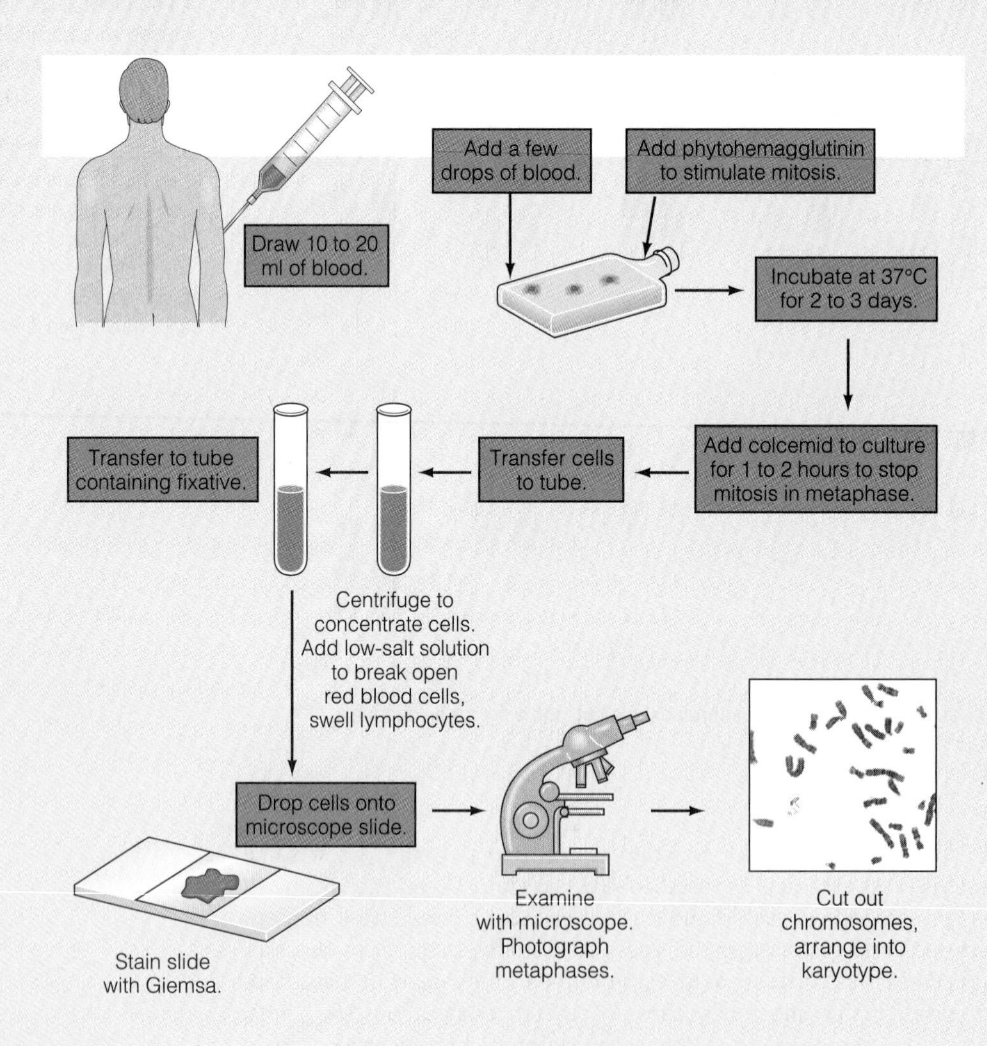

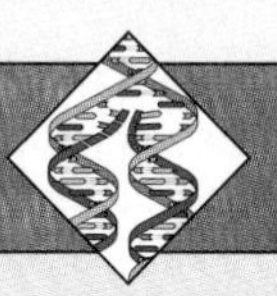

preparation is now used in most cytogenetic laboratories. In this system, a television camera and a computer are linked to a microscope. As metaphase chromosomes are located using the microscope, the image is recorded by the camera, digitized, and transmitted to the computer, where it is processed into a karyotype and printed. In the photographs shown here, a metaphase array of chromosomes has been printed by such a system. At the right is the computer-derived karyotype. Note that the computer has straightened the chromosomes as they were processed and arranged into a karyotype.

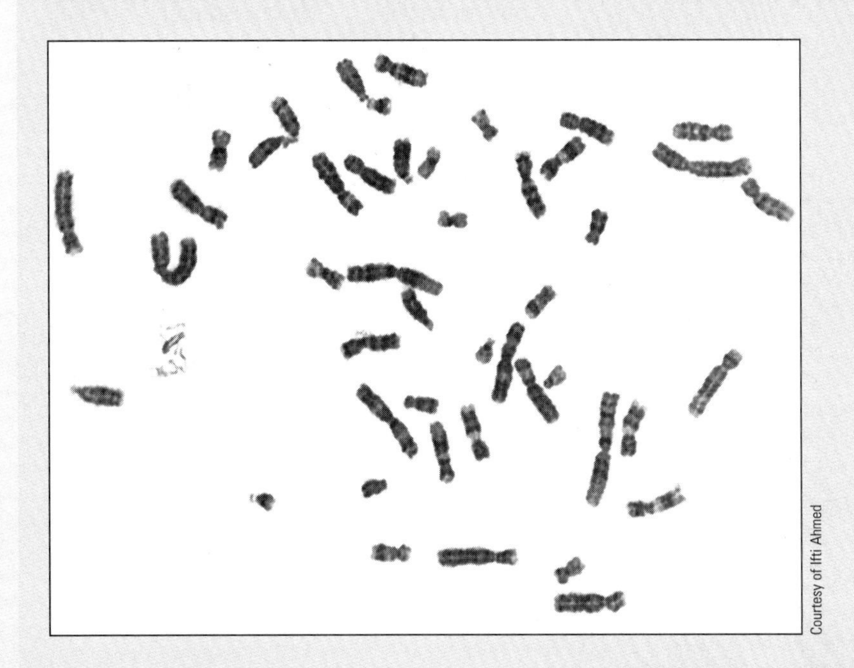

Courtesy of Ifti Ahmed

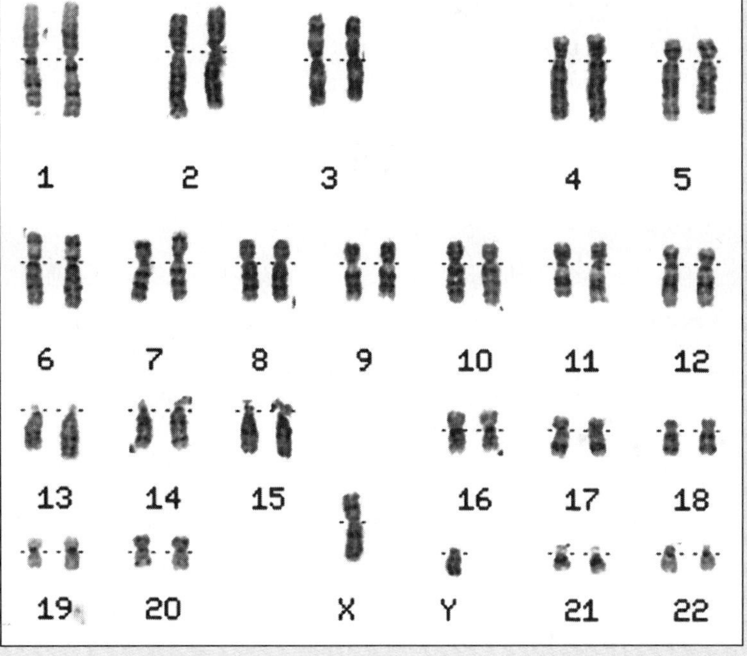

Courtesy of Ifti Ahmed

FIGURE 6.4 Four common staining procedures used in chromosomal analysis. Most karyotypes are prepared using G-banding. Q-banding and R-banding produce a pattern of bands that is the reverse of those in G-banded chromosomes.

Banding technique	Appearance of chromosomes
G-banding — Treat metaphase spreads with enzyme that digests part of chromosomal protein. Stain with Giemsa stain. Observe banding pattern with light microscope.	Darkly stained G bands.
Q-banding — Treat metaphase spreads with the chemical quinacrine mustard. Observe fluorescent banding pattern with a special ultraviolet light microscope.	Bright fluorescent bands upon exposure to ultraviolet light; same as darkly stained G bands.
R-banding — Heat metaphase spreads at high temperatures to achieve partial denaturation of DNA. Stain with Giemsa stain. Observe with light microscope.	Darkly stained R bands correspond to light bands in G-banded chromosomes. Pattern is the reverse of G-banding.
C-banding — Chemically treat metaphase spreads to extract DNA from the arms but not the centromeric regions of chromosomes. Stain with Giemsa stain and observe with light microscope.	Darkly stained C band centromeric region of the chromosome corresponds to region of constitutive heterochromatin.

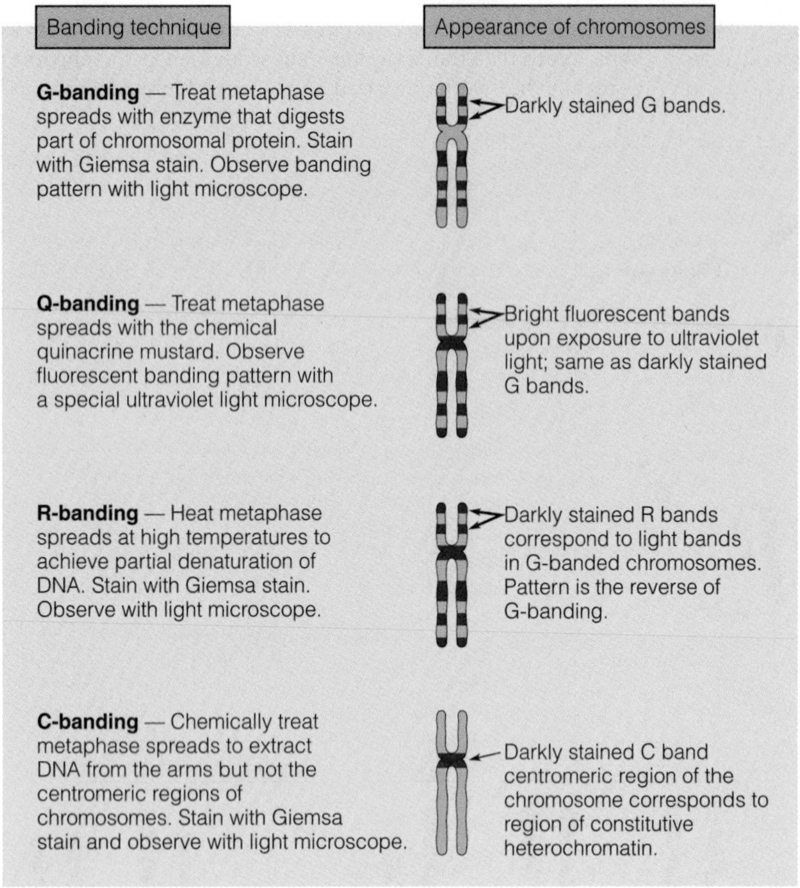

metaphase or late prophase. In these earlier stages, the chromosomes are longer and less condensed, and about 2,000 bands can be identified in the normal human karyotype. Some other banding methods are reviewed in Figure 6.4.

In analyzing karyotypes, a standard system of terms is used. A karyotype is described by (1) the number of chromosomes, (2) the sex chromosome status, (3) the presence or absence of an individual chromosome, and (4) the nature and extent of any structural abnormality. The symbols for structural alterations include **t** for a **translocation, dup** for **duplication,** and **del** for **deletion.** These structural aberrations are discussed later in the chapter. If a male has a deletion in the short arm of chromosome 5, but otherwise is chromosomally normal, this would be represented as 46,XY,del(5p). Table 6.1 lists some chromosomal aberrations using this system of terminology.

Chromosome analysis is a painstaking procedure done by cytogeneticists who study karyotypes stained to reveal banding patterns. To make it easier to spot abnormalities, scientists are now using a technique called chromosome painting. This method uses markers tagged with fluorescent dyes. The markers attach to chromosome-specific DNA sequences. This labeling can use more than one DNA sequence and more than one fluorescent marker to produce a unique pattern for each of the 24 human chromosomes (22 autosomes and the X and Y chromosomes). Using this method, it is easier and faster to find chromosomal abnormalities (Figure 6.5).

Obtaining Cells for Chromosome Studies

Almost any cell with a nucleus (mature red blood cells have no nuclei) can be used to make a karyotype. In adults, white blood

Table 6.1 **Chromosomal Aberrations**

Chromosomal Abnormality	Syndrome	Phenotype
46,del(4p)	Wolf-Hirschhorn syndrome	Mental retardation; midline facial defects consisting of broad nose, wide-set eyes, small lower jaw, and cleft palate; heart, lung, and skeletal abnormalities common; severely reduced survival
46,del(11)(p13)	WAGR syndrome	Tumors of the kidney (Wilms tumor) and of the gonad (gonadoblastoma); aniridia (absence of the iris); ambiguous genitalia; mental retardation
46,t(9;22)(q34a11)	CML (chronic myelogenous leukemia)	Enlargement of liver and spleen; anemia; excessive, unrestrained growth of white cells (granulocytes) in the bone marrow
46,t(8;14)	Burkitt's lymphoma	Malignancy of B lymphocytes that mature into the antibody-producing plasma cells; solid tumors, typically in the bones of the jaw and organs of the abdomen

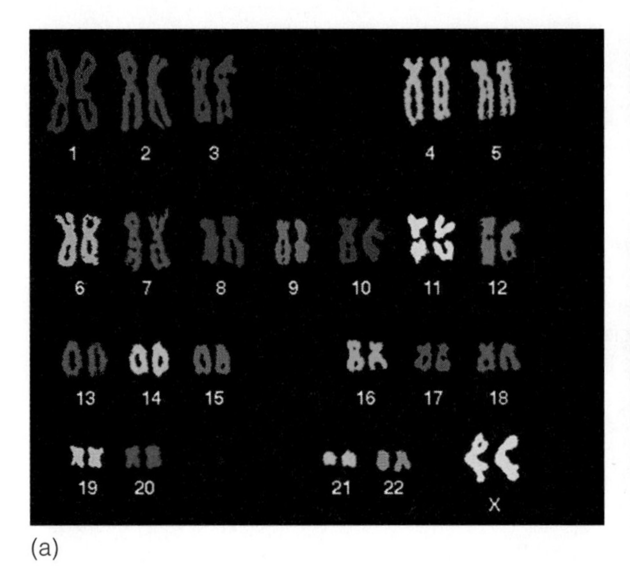

(a)

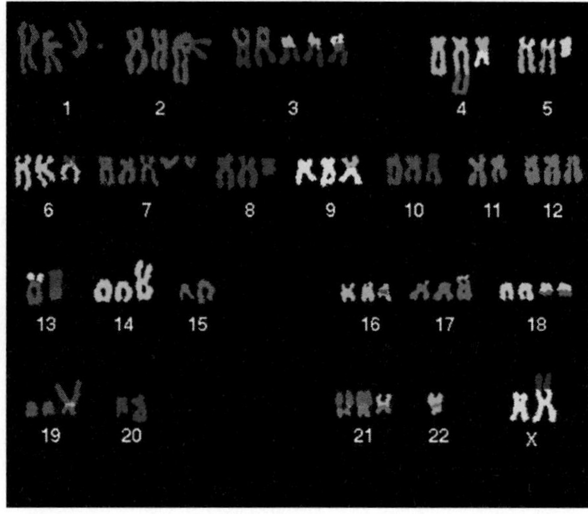

(b)

FIGURE 6.5 Chromosome painting with five different-colored markers in a normal cell (a) produces a pattern that highlights the translocations and deletions found in cancer cells (b).

cells (lymphocytes), skin cells (fibroblasts), and cells from biopsies or surgically removed tumor cells are routinely used for chromosome studies.

Using **amniocentesis** and **chorionic villus sampling** to collect cells from embryos and fetuses, chromosomal abnormalities can be detected before birth. A less invasive method to collect fetal cells is now under development. This technique collects fetal cells that cross into the mother's circulatory system (see Concepts and Controversies: Using Fetal Cells from the Mother's Blood).

Translocation A chromosomal aberration in which a chromosomal segment is transferred to another, nonhomologous chromosome.

Duplication A chromosomal aberration in which a segment of a chromosome is repeated and therefore is present in more than one copy within the chromosome.

Deletion A chromosomal aberration in which a segment of a chromosome is missing.

Amniocentesis A method of sampling the fluid surrounding the developing fetus by inserting a hollow needle and withdrawing suspended fetal cells and fluid; used in diagnosing fetal genetic and developmental disorders; usually performed in the sixteenth week of pregnancy.

Chorionic villus sampling A method of sampling fetal chorionic cells by inserting a catheter through the vagina or abdominal wall into the uterus. Used in diagnosing biochemical and cytogenetic defects in the embryo. Usually performed in the eighth or ninth week of pregnancy.

Amniocentesis Collects Cells from the Fluid Surrounding the Fetus

Amniocentesis is routinely used to collect fetal cells for analysis. To do this, the fetus and the placenta are located by ultrasound, and a needle is inserted through the abdominal and uterine walls (avoiding the placenta and fetus) into the amniotic sac surrounding the fetus (Figure 6.6). Approximately 10 to 30 ml of fluid is withdrawn using a syringe.

The amniotic fluid is composed mostly of fetal urine that contains cells shed from the skin, respiratory tract, and urinary tract of the fetus. The cells are isolated from the fluid by centrifugation (Figure 6.7). The fluid can be tested for abnormal levels of compounds that indicate the presence of a genetic disorder. More than 100 genetic biochemical disorders can be diagnosed by amniocentesis.

Amniotic cells can be analyzed for biochemical defects or grown in the laboratory for chromosome analysis. Once a karyotype has been prepared, it is possible to diagnose the sex of the fetus and search for any chromosomal abnormalities.

Amniocentesis is not usually performed until the sixteenth week of pregnancy. Before this time, there is very little amniotic fluid, and contamination of the sample with maternal cells is often a problem. Several studies in the United States, Britain, and Canada have shown that amniocentesis involves a small but measurable risk to both the fetus and mother. There is a risk of maternal infection and a slight increase (less than 1%) in the probability of a spontaneous abortion. To offset these risks, amniocentesis is normally used only under the following conditions:

- Advanced maternal age. Because the risk for children with chromosome abnormalities increases dramatically after the age of 35, amniocentesis is recommended in pregnant women who are 35 years or older. The majority of all amniocentesis cases are performed because of advanced maternal age.

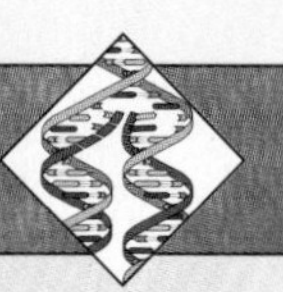

Concepts and Controversies

Using Fetal Cells from the Mother's Blood

More than a century has passed since placental cells from the fetus were first discovered in the circulatory systems of pregnant women. In 1969, cytogeneticists observed cells with a Y chromosome in the blood of women who later gave birth to male infants. Since then, research has been directed at finding ways to recover and use fetal cells from the maternal circulation for prenatal diagnosis. The goal is to carry out genetic analysis on fetal cells recovered from the maternal circulation without using the invasive procedures of amniocentesis and chorionic villus sampling (CVS). Use of fetal blood for prenatal diagnosis would lower the risk of injury to the mother and fetus. Several types of fetal cells enter the maternal circulation, including placental cells; white blood cells; and immature, nucleated red blood cells. These cells probably enter the bloodstream in detectable amounts between the sixth and twelfth week of pregnancy. But because less than 1 in every 100,000 cells in the mother's blood is from the fetus, collecting enough fetal cells from a blood sample is one of the challenges facing those working to develop this technique.

Several methods are being tried to isolate fetal cells from maternal blood, but no single technique has emerged as the best. Even though these methods are still under development, several studies have found it possible to diagnose chromosomal abnormalities in fetal cells collected from maternal blood. In addition, fetal cells have been used to diagnose some genetic disorders, including sickle cell anemia and thalassemia, and to determine fetal blood types.

The separation of fetal cells may be most useful for detecting trisomies of chromosomes 13, 18, 21, X, and Y, which together account for more than 95% of all fetal aneuploidies. The wider use of this technique will depend on further refinements in separating fetal cells, and in growing them in the laboratory. If these barriers can be overcome, fetal blood sampling may gradually replace more invasive procedures for prenatal chromosome analysis.

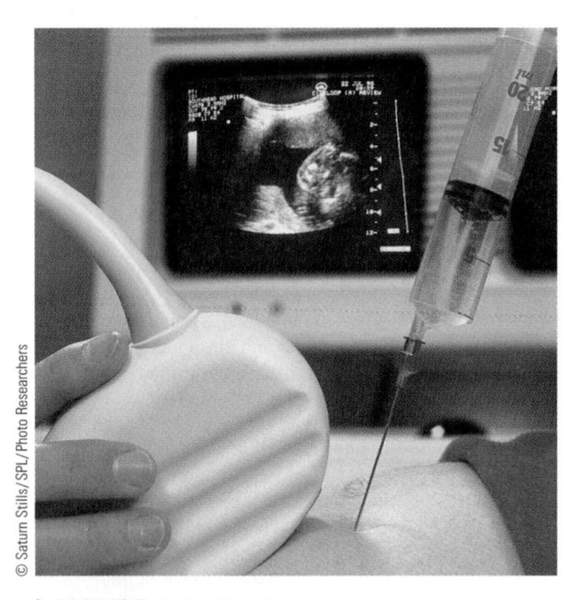

FIGURE 6.6 A patient undergoing amniocentesis.

- A previous child with a chromosomal aberration. If a previous child had a chromosome abnormality, the recurrence risk is about 1% to 2%, and amniocentesis is recommended.
- A parent with a chromosome rearrangement. If either parent carries a chromosomal translocation or other rearrangement that can cause an abnormal karyotype in the child, amniocentesis should be considered.
- X-linked disorder. If the mother is a carrier of an X-linked biochemical disorder that cannot be diagnosed prenatally and if she is willing to abort if the fetus is male, amniocentesis is recommended.

Chorionic Villus Sampling Retrieves Fetal Tissue from the Placenta

Chorionic villus sampling (CVS) is another method of collecting cells for analysis. This technique has several advantages over amniocentesis. CVS can be performed earlier in the pregnancy (8 to 10 weeks, compared to 16 weeks for amniocentesis). Results from chromosome analysis using CVS are available within a few hours or a few days, rather than a week to 10 days as is required when cells are collected by amniocentesis. For CVS, a small flexible catheter is inserted through the vagina or abdomen into the uterus, guided by ultrasound images. A small sample of chorionic villi, a fetal tissue that forms part of the placenta, is obtained by suction (Figure 6.8). This tissue is mitotically active and can be used immediately to prepare a karyotype. As with amniocentesis, enough material is usually obtained by CVS to allow biochemical testing or extraction of DNA for molecular analysis. The use of recombinant DNA techniques for prenatal diagnosis is discussed in Chapter 19.

CVS is more specialized than amniocentesis and at present is used less often. Although early studies indicated that the procedure posed a higher risk to mother and

Web-Integrated Toolbox

Analyzing Karyotypes

Along with the construction and analysis of pedigrees, the construction and analysis of karyotypes is one of the basic methods in human genetics. Karyotypes are important in diagnosing many genetic disorders, including Down syndrome, and for identifying whether there is a risk of having more children with a genetic disorder. Karyotypes are also used in cancer diagnosis and design of treatment for several types of cancer, including leukemia and breast cancer.

Karyotypes are usually constructed using mitotic cells that have been treated with chemicals to stop cell division at metaphase. The chromosomes are dropped onto a microscope slide, treated with an enzyme, and stained to produce chromosome bands. Other methods use recombinant DNA technology to "paint" chromosomes and parts of chromosomes, making it easier to detect some chromosome abnormalities. Often, karyotypes are prepared from several family members to help detect chromosomal aberrations and to determine their origin.

The book's Web site contains a link to exercises that will help you develop the skills needed to analyze karyotypes. These exercises explain how to construct and analyze a karyotype, beginning with a photograph of chromosomes taken at metaphase and stained to produce bands. The exercise takes you through the process in a series of steps. Working with karyotypes will help you understand how cytogenetics is used by itself and used together with other techniques such as pedigree analysis to diagnose genetic disorders, to predict genetic risk to future generations, and to treat cancer.

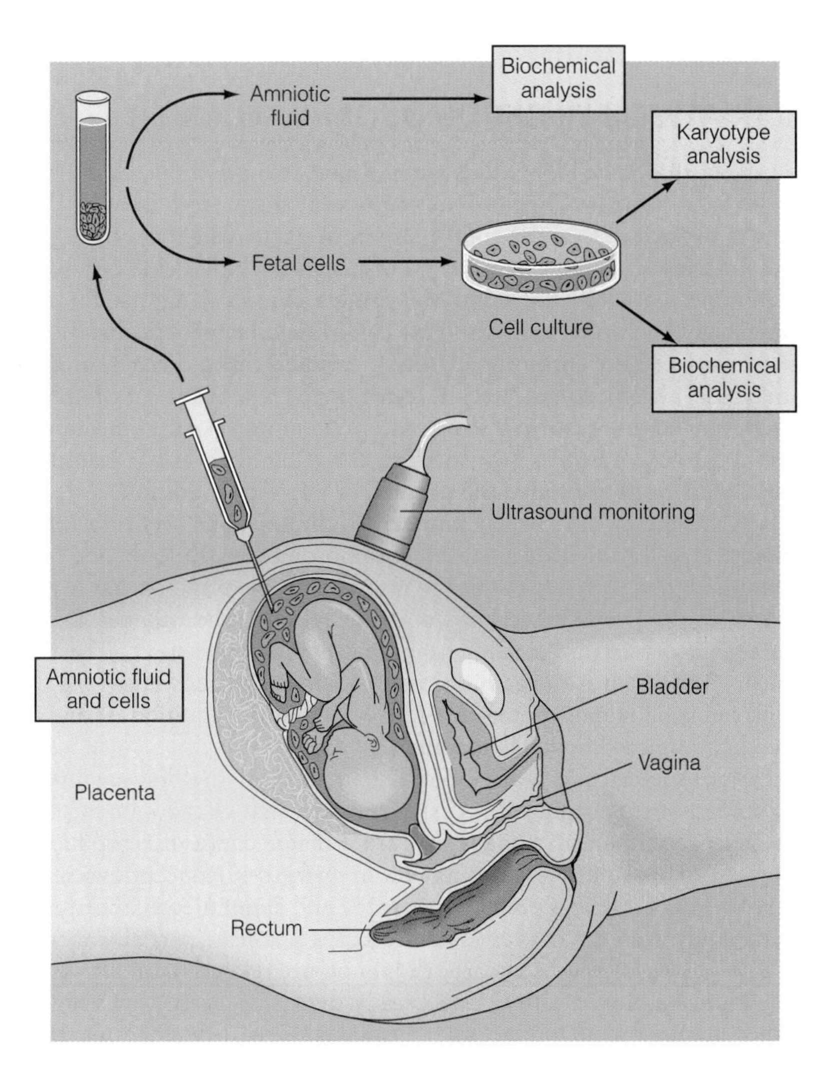

FIGURE 6.7 In amniocentesis, a syringe needle is inserted through the abdominal wall and uterine wall to collect a small sample of the amniotic fluid. The fluid contains fetal cells that can be collected and used for prenatal chromosomal or biochemical analysis.

FIGURE 6.8 The chorionic villus sampling technique. A catheter is inserted into the uterus through the vagina to remove a sample of fetal tissue from the chorion. Cells in the tissue can be used for chromosomal or biochemical analysis.

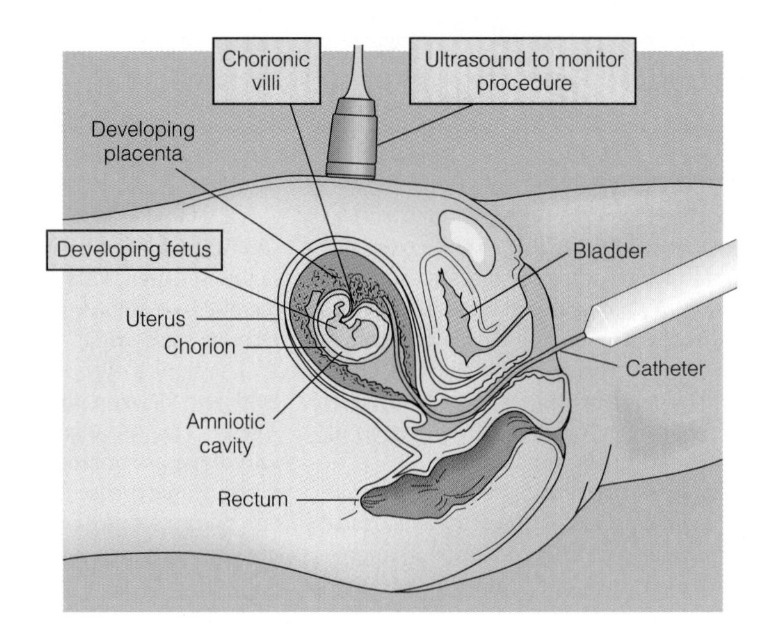

fetus than amniocentesis, improvements in instrumentation and technique have lowered the risk somewhat. CVS offers early diagnosis of genetic diseases, and if termination of pregnancy is elected, maternal risks are lower at 9 to 12 weeks than at 16 weeks.

VARIATIONS IN CHROMOSOMAL NUMBER

At the birth of a child, anxious parents have two questions: Is it a boy or a girl, and is the baby normal? The term *normal* usually means free from all birth defects. The causes of such defects, of course, are both environmental and genetic. Among the genetic causes, we have considered disorders such as sickle cell anemia or Marfan syndrome, caused by the mutation of single genes. Changes in chromosome number or changes in chromosome structure can cause other genetic disorders. Such changes may involve entire chromosome sets, individual chromosomes, or alterations within individual chromosomes. Recall from Chapter 2 that the set of 46 chromosomes in each somatic cell is called the diploid, or $2n$, number of chromosomes. Similarly, the set of 23 chromosomes (constituting the n number) is the haploid set. Together, these conditions are called the normal, or euploid, condition.

A change in the normal number of chromosome sets in a cell is called **polyploidy.** A cell with three sets of chromosomes is triploid, one with four sets is tetraploid, and so forth. A change in chromosome number that involves less than a whole chromosome set is known as **aneuploidy.** The number of chromosomes in aneuploid cells is not a simple multiple of the haploid set. The simplest form of aneuploidy involves the gain or loss of a single chromosome. Loss of a single chromosome is known as **monosomy** ($2n - 1$), and the addition of one chromosome to the diploid set is known as **trisomy** ($2n + 1$).

Polyploidy A chromosomal number that is a multiple of the normal haploid chromosomal set.

Aneuploidy A chromosomal number that is not an exact multiple of the haploid set.

Monosomy A condition in which one member of a chromosomal pair is missing; having one less than the diploid number ($2n = 1$).

Trisomy A condition in which one chromosome is present in three copies, whereas all others are diploid; having one more than the diploid number ($2n + 1$).

Since the discovery of trisomy 21 in 1959 as the first example of aneuploidy in humans, cytogenetic studies have revealed that changes in chromosome number are fairly common in humans and are a major cause of reproductive failure. It is now estimated that as many as one in every two conceptions may be aneuploid and that 70% of early embryonic deaths and spontaneous abortions are caused by aneuploidy. About 1 in every 170 live births is at least partially aneuploid, and from 5% to 7% of all deaths in early childhood are related to aneuploidy.

Humans have a rate of aneuploidy that is up to 10 times higher than that of other mammals, including other primates. This difference may represent a considerable reproductive disadvantage for our species. Understanding the causes of aneuploidy in humans remains one of the great challenges in human genetics.

Polyploidy Changes the Number of Chromosomal Sets

Abnormalities in the number of haploid chromosomal sets can arise in several ways: (1) errors in meiosis during gamete formation, (2) events at fertilization, or (3) errors in mitosis following fertilization. Polyploidy can result when cleavage of the cytoplasm (cytokinesis) fails to occur in the last stage of cell division in either mitosis or meiosis. Recall that mitotic divisions precede meiosis in the ovary and testis. If the chromosomes replicate and sister chromatids divide during mitosis, but there is no cytoplasmic division, the result is a tetraploid cell that contains four copies of each chromosome. If this cell undergoes meiosis, the resulting gametes will contain a diploid set of chromosomes instead of the normal haploid set. Fusion of this polyploid gamete with a normal haploid gamete will produce a triploid zygote (▶ Figure 6.9).

A diploid gamete can also arise through errors in meiosis. An error in meiosis I can result in the failure of homologous chromosomes to separate, producing diploid gametes after meiosis II. Alternatively, if all chromosomes move to the same pole after centromere separation in meiosis II, diploid gametes will also result.

Polyploidy can also be produced at fertilization by **dispermy,** the simultaneous fertilization of a haploid egg by two haploid sperm. The result is a zygote that contains three haploid chromosome sets, or triploidy. Some common polyploid conditions are discussed in the following sections.

Dispermy Fertilization of haploid egg by two haploid sperm, forming a triploid zygote.

Triploidy The most common form of polyploidy in humans is **triploidy,** found in 15% to 18% of spontaneous abortions. Three kinds of triploid chromosome sets are observed: 69,XXY; 69,XXX; and 69,XYY. Approximately 75% of all cases of triploidy have two sets of paternal chromosomes. Accidents in male gamete formation do not occur this often, so most triploid zygotes probably arise as the result of dispermy. Although biochemical changes that accompany fertilization normally prevent such fertilizations, this system is not fail-safe.

Triploidy A chromosomal number that is three times the haploid number, having three copies of all autosomes and three sex chromosomes.

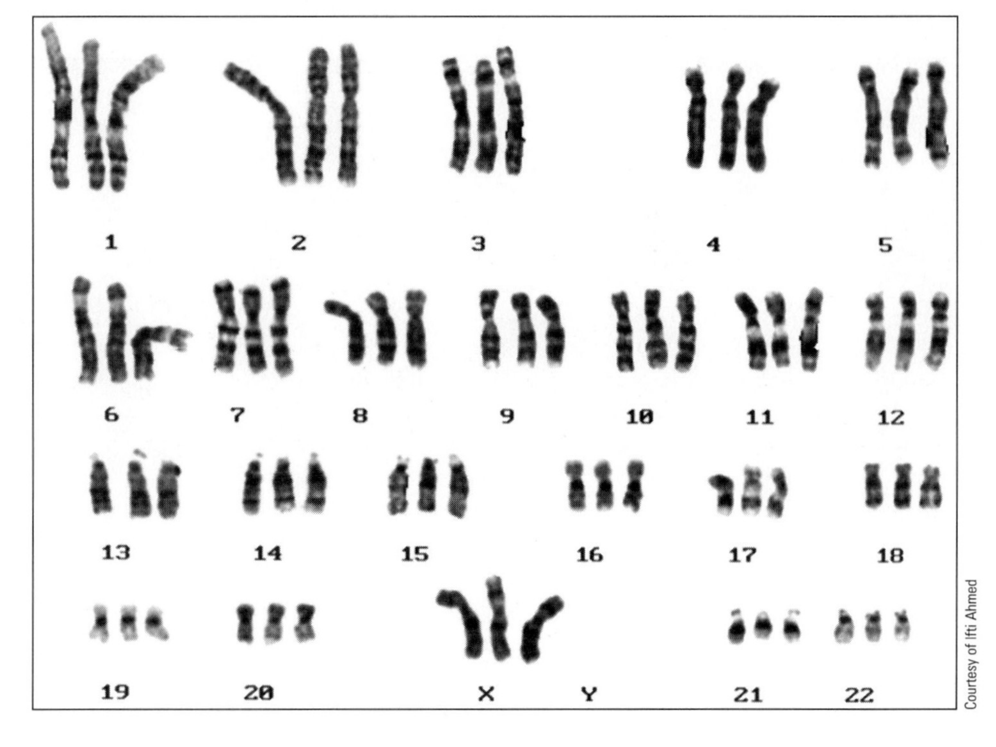

▶ **FIGURE 6.9** The karyotype of a triploid individual contains three copies of each chromosome.

Almost 1% of all conceptions are triploid, but over 99% of these die before birth, and only 1 in 10,000 live births is triploid. Triploid infants have only limited survival, and most die within a month. The phenotype of triploid newborns includes multiple abnormalities: an enlarged head; fusion of fingers and toes (syndactyly); and malformations of the mouth, eyes, and genitals (Figure 6.10). The high rate of embryonic death and failure to survive as a newborn indicates that triploidy is incompatible with life and should be regarded as a lethal condition.

Tetraploidy A chromosomal number that is four times the haploid number, having four copies of all autosomes and four sex chromosomes.

Tetraploidy Tetraploidy is present in about 5% of all spontaneous abortions and is extremely uncommon in live births. The sex chromosome constitution of all tetraploid embryos is either XXXX or XXYY. Tetraploidy can result from a failure of cytoplasmic division (cytokinesis) in the first mitotic division after fertilization. Later rounds of mitosis result in a tetraploid embryo. If tetraploidy arises sometime after the first mitotic division, two different cell lines coexist in the embryo, one a normal diploid line and the other a tetraploid cell line. Such mosaic individuals survive somewhat longer than complete tetraploids, but the condition is still life threatening.

In summary, polyploidy in humans can arise by at least two different mechanisms, but it is inevitably lethal. Polyploidy does not involve the mutation of any genes, but only changes in the number of gene copies. How this quantitative change in gene number is related to lethality in development is unknown.

Aneuploidy Changes the Number of Individual Chromosomes

Nondisjunction The failure of homologous chromosomes to separate properly during meiosis or mitosis.

As defined earlier, aneuploidy is the addition or deletion of individual chromosomes from the normal diploid set of 46. Aneuploidy can be caused in several ways. The most important is **nondisjunction,** a process in which chromosomes fail to separate properly at anaphase. Although this failure can occur in either meiosis or mitosis, nondisjunction in meiosis is the leading cause of aneuploidy in humans. There are two cell divisions in meiosis, and nondisjunction can occur in either the first or second division with different genetic consequences (Figure 6.11).

If nondisjunction occurs in meiosis I, all of the gametes will be abnormal. These gametes will carry both members of a chromosomal pair or neither

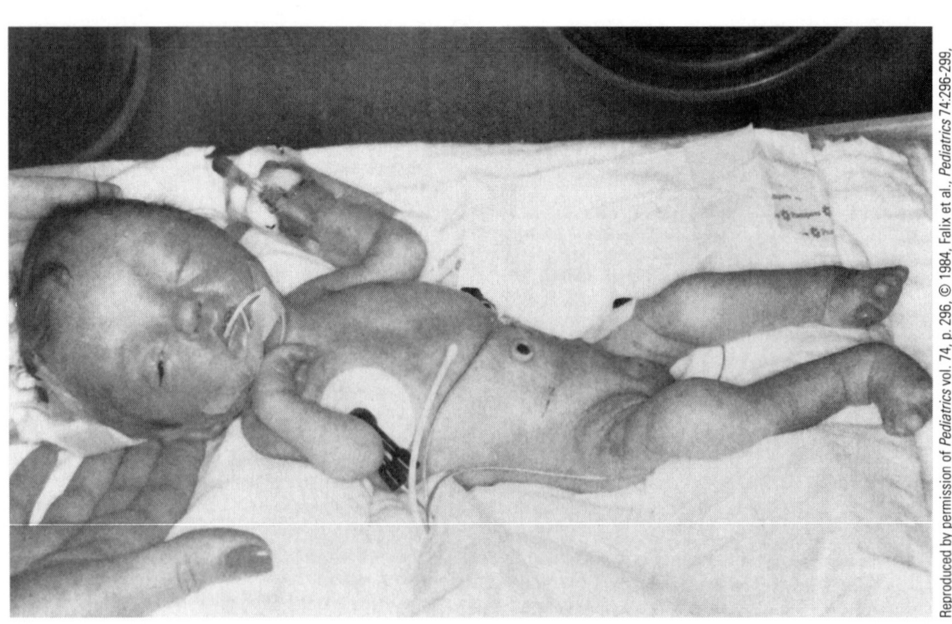

FIGURE 6.10 A triploid infant, showing the characteristic enlarged head.

member of the pair. Nondisjunction in meiosis II results in two normal gametes and two abnormal gametes (Figure 6.11). Gametes that are missing a copy of a given chromosome will produce a monosomic zygote. Those with an extra copy of a chromosome will produce a trisomic zygote.

Aneuploid individuals have distinct and characteristic phenotypic features. Those with a given form of aneuploidy, such as Down syndrome, tend to resemble each other more closely than their own brothers and sisters. The phenotypic effects of aneuploidy range from minor physical symptoms to devastating and lethal deficiencies in major organ systems. Among survivors, phenotypic effects often include behavioral deficits and mental retardation. In the following section, we look at some of the important features of autosomal aneuploid phenotypes. Then we consider the phenotypic effects of sex chromosome aneuploidy.

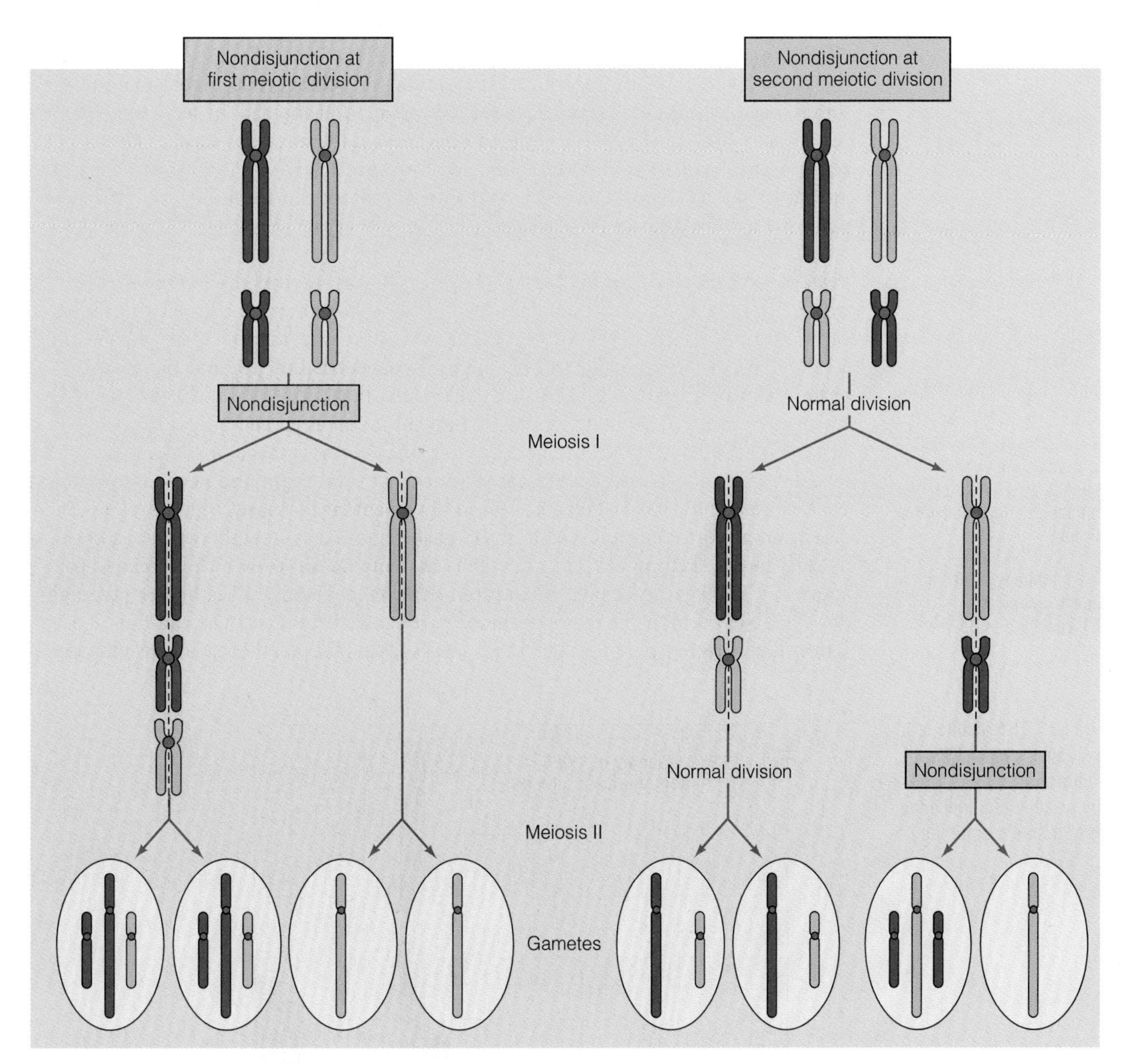

FIGURE 6.11 Nondisjunction in the first meiotic division (left) results in four abnormal gametes. Two gametes carry both members of a chromosome pair, and two are missing one chromosome. Nondisjunction in the second meiotic division (right) produces two normal gametes and two abnormal gametes. One gamete carries both members of a chromosome pair, and one is missing a chromosome.

Autosomal Monosomy Is a Lethal Condition

Aneuploidy during gamete formation should produce equal numbers of monosomic and trisomic gametes and embryos. However, autosomal monosomies are only rarely observed among spontaneous abortions and live births. The likely explanation is that the majority of autosomal monosomic embryos are lost very early, even before pregnancy is recognized.

Autosomal Trisomy Is Relatively Common

Most autosomal trisomies are lethal conditions; but unlike monosomy, the presence of an extra chromosome allows varying degrees of development. Autosomal trisomy is found in about 50% of all cases of chromosomal abnormalities in fetal death. The findings also indicate that autosomes are differentially involved in trisomy (◗ Figure 6.12). Trisomies for chromosomes 1, 3, 12, and 19 are rarely observed in spontaneous abortions, whereas trisomy for chromosome 16 accounts for almost one-third of all cases. As a group, the acrocentric chromosomes (13 to 15, 21, and 22) are represented in 40% of all spontaneous abortions. Reasons for this differential involvement include differences in the rate of nondisjunction, differences in the rate of fetal death before recognition of pregnancy, or a combination of factors. Only a few autosomal trisomies result in live births (trisomy 8, 13, and 18). Trisomy 21 (Down syndrome) is the only autosomal trisomy that allows survival into adulthood. In the following section, we will review some autosomal trisomies.

Trisomy 13 The presence of an extra copy of chromosome 13 that produces a distinct set of congenital abnormalities resulting in Patau syndrome.

Trisomy 18 The presence of an extra copy of chromosome 18 that results in a clinically distinct set of invariably lethal abnormalities known as Edwards syndrome.

Trisomy 21 Aneuploidy involving the presence of an extra copy of chromosome 21, resulting in Down syndrome.

Trisomy 13: Patau Syndrome (47,+13) Trisomy 13 was discovered in 1960 by cytogenetic analysis of a malformed child. The karyotype indicated the presence of 47 chromosomes, and the extra chromosome was identified as chromosome 13 (47,+13). Only 1 in 15,000 live births involves trisomy 13, and the condition is lethal. Half of all affected individuals die in the first month, and the mean survival time is 6 months. The phenotype of trisomy 13 involves facial malformations (◗ Figure 6.13), eye defects, extra fingers or toes, and feet with large protruding heels. Internally, there are usually severe malformations of the brain and nervous system, as well as congenital heart defects. The involvement of so many organ systems indicates that abnormal development in this case begins early in embryogenesis, perhaps as early as the sixth week. Parental age is the only factor known to be related to trisomy 13. The age of parents of children who have trisomy 13 is higher (averaging about 32 years) than the average age of parents who have normal children. The relationship between parental age and aneuploidy is discussed later in this chapter.

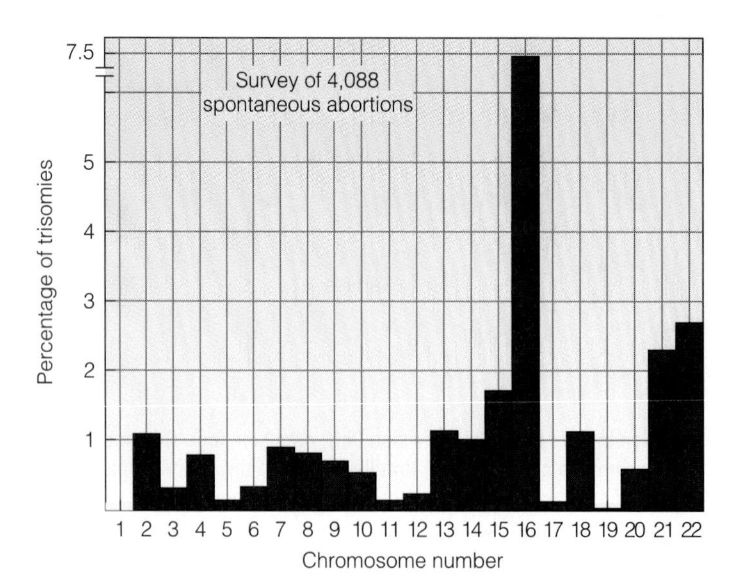

◗ **FIGURE 6.12** The results of a cytogenetic survey of over 4,000 spontaneous abortions show a wide variation in the presence of individual chromosomes in trisomic embryos.

Trisomy 18: Edwards Syndrome (47,+18) In 1960 John Edwards and his colleagues reported the first case of **trisomy 18** (47,+18). Affected infants are small at birth, grow very slowly, and are mentally retarded. For reasons still unknown, 80% of all trisomy 18 births are female. Clenched fists, with the second and fifth finger overlapping the third and fourth fingers, and malformed feet are also characteristic (◗ Figure 6.14). Heart malformations are almost always present, and heart failure or pneumonia usually causes death. Trisomy 18 occurs with a frequency of 1 in 11,000 live births, and the average survival time is 2 to 4 months. As in trisomy 13, advanced maternal age is a factor predisposing to trisomy 18.

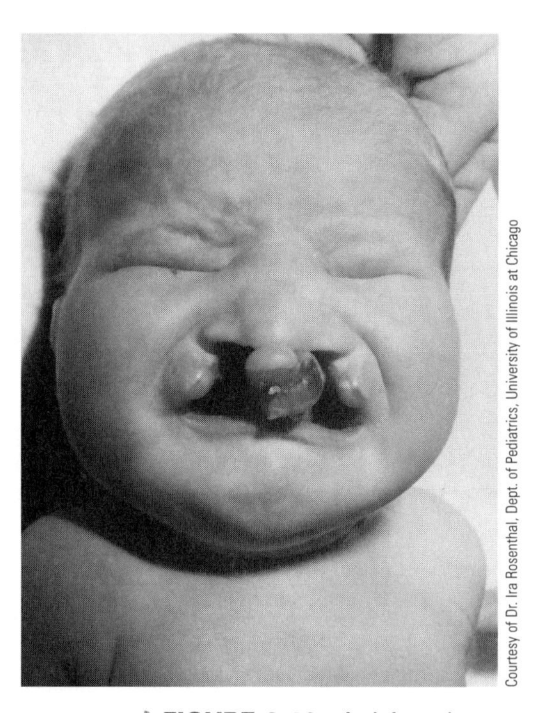
Courtesy of Dr. Ira Rosenthal, Dept. of Pediatrics, University of Illinois at Chicago

◗ **FIGURE 6.13** An infant who has trisomy 13, showing a cleft lip and palate (the roof of the mouth).

Trisomy 21: Down Syndrome (47,+21) The phenotypic features of **trisomy 21** (MIM/OMIM 190685) were first described by John Langdon Down in 1866. He called the condition "mongolism" because of the distinctive skin fold, known as an epicanthic fold, in the corner of the eye (◗ Figure 6.15). To remove the racist implications inherent in the term, Lionel Penrose and others changed the designation to Down syndrome. As described in the chapter opening, trisomy 21 was the first chromosomal abnormality discovered in humans. In 1959, Jerome Lejeune and his colleagues discovered that the presence of an extra copy of chromosome 21 is the underlying cause of Down syndrome. Trisomy 21 has also been observed in other primate species, including the chimpanzee.

Down syndrome, one of the most common chromosomal defects in humans, occurs in about 0.5% of all conceptions and in 1 in 900 live births (Figure 6.15). It is a leading cause of childhood mental retardation and heart defects in the United States. Affected individuals have a wide skull that is flatter than normal at the back. The eyelids have an epicanthic fold, and the iris contains spots, known as Brushfield spots. The tongue is often furrowed and protruding, causing the mouth to remain partially open. Physical growth, behavior, and mental development are retarded, and approximately 40% of all affected individuals have congenital heart defects. In

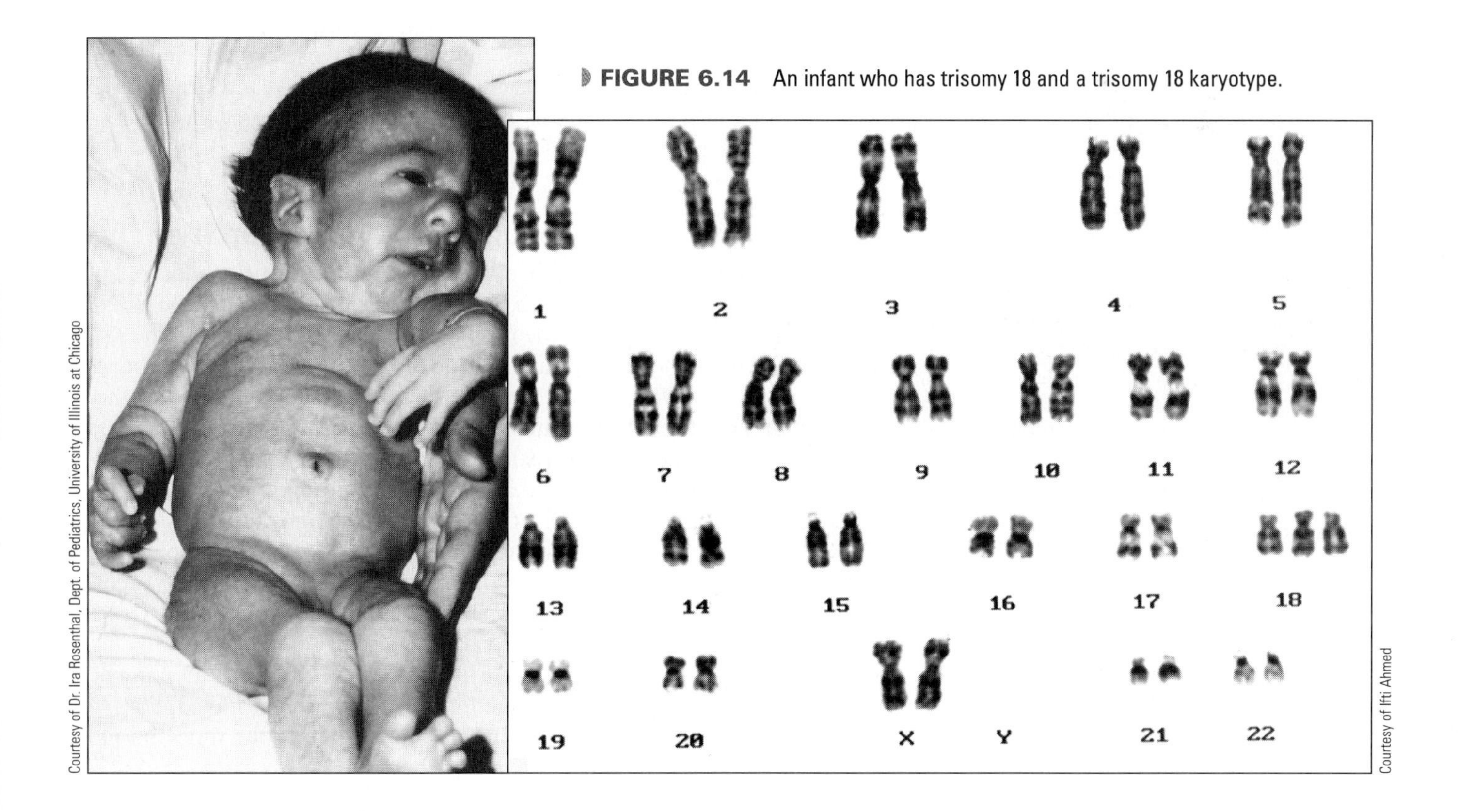

Courtesy of Dr. Ira Rosenthal, Dept. of Pediatrics, University of Illinois at Chicago

Courtesy of Ifti Ahmed

◗ **FIGURE 6.14** An infant who has trisomy 18 and a trisomy 18 karyotype.

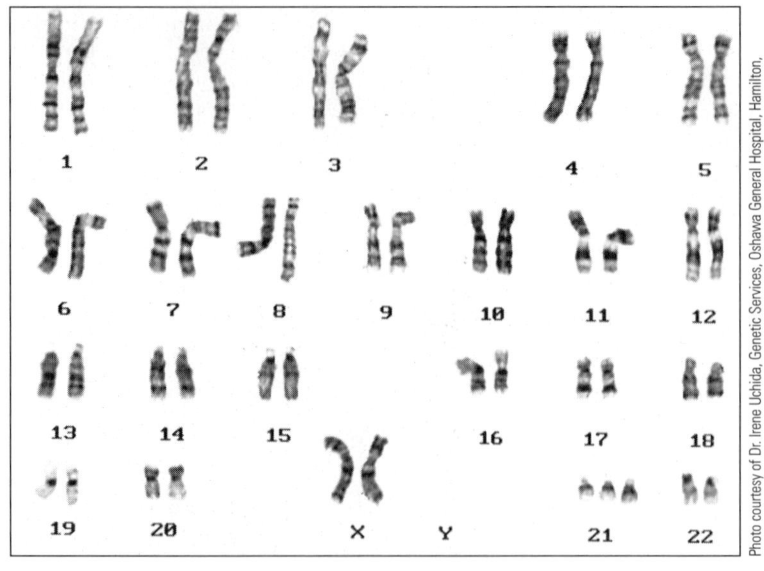

FIGURE 6.15 A child who has trisomy 21. The karyotype shows this child has three copies of chromosome 21.

addition, children with Down syndrome are prone to respiratory infections and contract leukemia at a rate far above the normal population. In the past decade, improvements in medical care have increased survival rates dramatically, so that many affected individuals survive into adulthood, although few reach the age of 50 years. In spite of these handicaps, many individuals who have Down syndrome lead rich, productive lives and can serve as an inspiration to us all.

WHAT ARE THE RISKS FOR AUTOSOMAL TRISOMY?

The causes of autosomal trisomies such as Down syndrome are unknown, but a variety of genetic and environmental factors have been proposed, including radiation, viral infection, hormone levels, and genetic predisposition. To date the only factor clearly related to autosomal aneuploidy is advanced maternal age. In fact, a relationship between maternal age and Down syndrome was well established 25 years before the chromosomal basis for the condition was discovered.

FIGURE 6.16 The relationship between maternal age and the frequency of trisomy 21 (Down syndrome). The risk increases rapidly after 35 years of age.

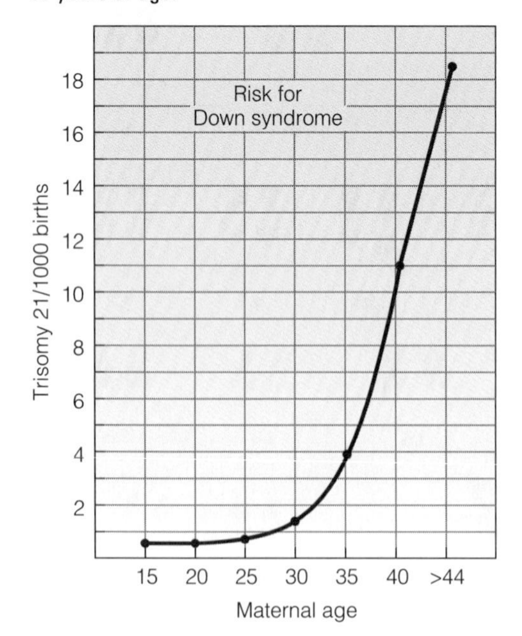

Maternal Age Is the Leading Risk Factor for Trisomy

Young mothers have a low probability of having trisomy 21 children, but the risk increases rapidly after the age of 35 years. At maternal age of 20, the incidence of Down syndrome is 0.05%; by maternal age of 35 the risk has climbed to 0.9%; and at maternal age of 45 years, 3% of all newborns have trisomy 21 (Figure 6.16). The effect of maternal age on other aneuploidies has been documented, and the relationship between advanced maternal age and autosomal trisomy is very striking (Figure 6.17). Paternal age has also been proposed as a factor in autosomal trisomy, but the evidence is weak, and no clear-cut link has been demonstrated.

Advanced maternal age as a risk factor for autosomal aneuploidy is supported by studies on the parental origin of the nondisjunction event, using chromosome banding and recombinant DNA technology. Occasionally, chromosomes have minor variations in banding patterns. By examining banded chromosomes from the trisomic child and the parents, the parental origin of the nondisjunction can often be determined. For trisomy 21, the nondisjunction is maternal about 94% of the time and paternal about 6% of the time.

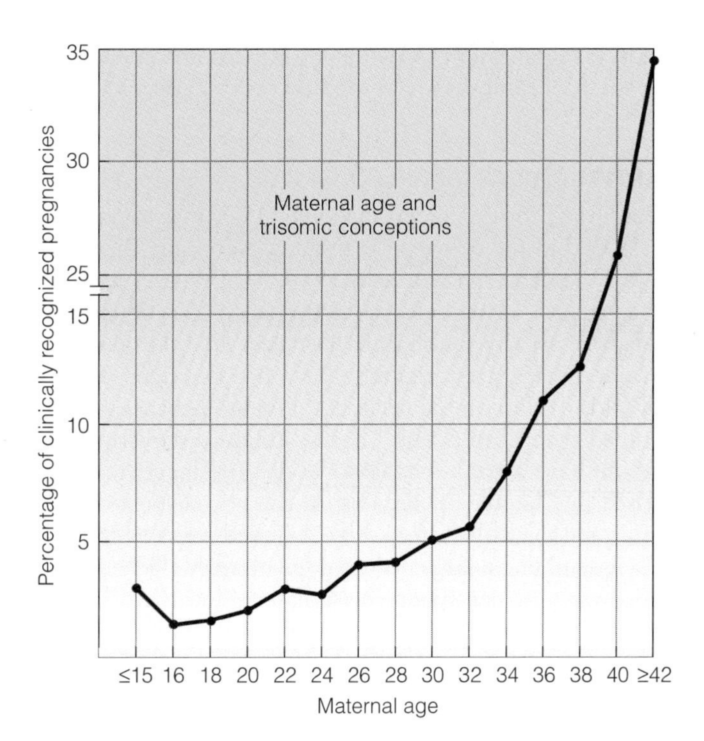

FIGURE 6.17 Maternal age is the major risk factor for autosomal trisomies of all types. By age 42, about one in three identified pregnancies is trisomic.

Many other factors have been suggested to influence aneuploidy, including radiation; the use of contraceptives, fertility drugs, alcohol, or tobacco; or the presence of certain alleles at a number of loci. Aside from advanced maternal age, no other factor has been clearly linked to the incidence of aneuploidy. Overall, the paternal contribution to autosomal trisomy is about 7%. In autosomal trisomies, the great majority of nondisjunction events take place at meiosis I in oocytes.

Why Is Maternal Age a Risk Factor?

One idea about the relationship between maternal age and nondisjunction focuses on the duration of meiosis in females. Recall from Chapter 2 that primary oocytes are formed early in development and enter the first meiotic prophase well before birth. Meiosis I is not completed until ovulation, so that eggs produced at age 40 have been in meiosis I for more than 40 years. During this time, metabolic errors or environmental agents may damage the cell so that aneuploidy results when meiosis resumes. However, it is not yet known whether the age of the egg is directly related to the increased frequency of nondisjunction.

A second idea to explain the increased risk of aneuploid children in older mothers is related to the interaction between the implanting embryo and the uterine environment. According to this idea, the embryo–uterine interaction normally results in the spontaneous abortion of chromosomally abnormal embryos, a process called maternal selection. As women age, maternal selection becomes less effective, allowing more chromosomally abnormal embryos to become implanted and develop. There may well be other factors in addition to age of the egg and maternal selection that play a role in the relationship between maternal age and autosomal aneuploidy, and more research is needed to clarify the underlying mechanisms.

ANEUPLOIDY OF THE SEX CHROMOSOMES

Sex chromosome aneuploidy is more common than autosomal aneuploidy. The overall incidence of sex chromosome anomalies in live births is 1 in 400 for males and 1 in 650 for females. Unlike the situation with autosomes, in which monosomy is always fatal, monosomy for the X chromosome is a viable condition. Monosomy for the Y chromosome (45,Y), however, is always lethal.

Turner Syndrome (45,X)

Monosomy for the X chromosome (45,X) was reported as a chromosomal disorder in 1959, but the cytogenetic findings were only the finishing touch to a larger piece of genetic detective work. In 1954, Paul Polani, working to understand the causes of congenital heart defects, examined three women who suffered from a defect of the aorta usually found only in males. Polani noted that all three women also had phenotypic features originally described by Henry Turner in 1938: short stature, extra folds of skin on the neck, and low levels of sexual development. Examining cells scraped from inside their cheeks, Polani discovered that, like males, these females did not have a Barr body (Barr bodies are inactivated X chromosomes, and are discussed in Chapter 7). Because of this, he suspected that females who have **Turner syndrome** might be affected by X-linked traits as frequently as males.

Turner syndrome A monosomy of the X chromosome (45,X) that results in female sterility.

Assembling a large group of females who lacked ovarian development, including those with Turner syndrome, Polani tested them for color blindness, a sex-linked trait. Four of the 25 females tested were colorblind, a result that was much higher than expected in a population of females and similar to that expected in males. In a paper published in 1956 he suggested that Turner syndrome females might have only one X chromosome and in effect be hemizygous for traits on the X chromosome.

After a careful cytogenetic study, Polani and Charles Ford published a paper in 1959 confirming that Turner females are indeed 45,X. The unraveling of the chromosomal basis of Turner syndrome illustrates a basic property of scientific research: work in one area—in this case, congenital heart disease—often leads to significant findings in another field.

Turner syndrome females are short and wide-chested with underdeveloped breasts and rudimentary ovaries (Figure 6.18). At birth, puffiness of the hands and feet is prominent, but this disappears in infancy. As reported by Polani, such individuals also have narrowing, or coarctation, of the aorta. There

Klinefelter syndrome Aneuploidy of the sex chromosomes involving an XXY chromosomal constitution.

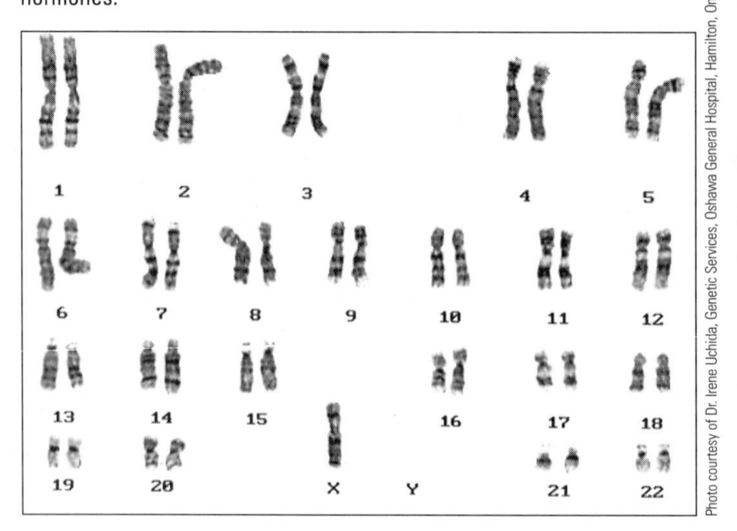

Photo courtesy of Dr. Irene Uchida, Genetic Services, Oshawa General Hospital, Hamilton, Ontario, Canada

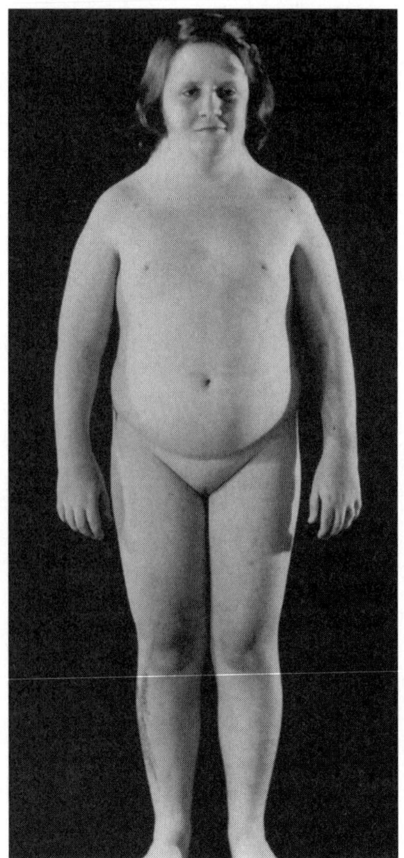

Courtesy of Dr. Ira Rosenthal, Dept. of Pediatrics, Univ. of Illinois at Chicago

FIGURE 6.18 A girl who has Turner syndrome and the associated karyotype, containing a single X chromosome. Turner syndrome is characterized by short stature, a broad chest, and lack of sexual development. Stature and sexual development can be treated using hormones.

is no mental retardation associated with this syndrome, although evidence suggests that Turner syndrome is associated with reduced skills in interpreting spatial relationships. This chromosomal disorder occurs with a frequency of 1 in 10,000 female births, and although affected newborns suffer no life-threatening problems, 95% to 99% of all 45,X embryos die before birth. It is estimated that 1% of all conceptions are 45,X. Another notable feature of this syndrome is that in 75% of all cases, the nondisjunction apparently originates in the father.

The phenotypic impact of the single X chromosome in Turner syndrome is strikingly illustrated in a case of identical twins, one of them 46,XX and the other 45,X. This situation apparently arose by mitotic nondisjunction after fertilization and twinning. These twins (◗ Figure 6.19) were shown to be identical on the basis of blood types and chromosome banding studies. In spite of being identical twins, they have significant differences in height, sexual development, hearing, dental maturity, and performance on tests that measure numerical skills and space perception. Although some environmental factors may contribute to these differences, the major role of the second X chromosome in normal female development is apparent. These and results from other studies on individuals with Turner syndrome indicate that a second X chromosome is necessary for normal development of the ovary, normal growth patterns, and development of the nervous system. Complete absence of an X chromosome in the absence or presence of a Y chromosome is always lethal, emphasizing that the X chromosome is an essential component of the karyotype.

From Weiss, E., et al. (1982). Monozygotic twins discordant for Ulrich-Turner syndrome. *Am. J. Med. Genet.* 13: 389–399.

◗ **FIGURE 6.19** Monozygotic twins, one of which has Turner syndrome. The twin who has Turner syndrome (left) is 45,X; the normal twin (right) is 46,XX.

Klinefelter Syndrome (47,XXY)

The phenotype of **Klinefelter syndrome** was first described in 1942, and Patricia Jacobs and John Strong reported the XXY chromosomal condition in 1959. The frequency of Klinefelter syndrome is approximately 1 in 1,000 male births. The phenotypic features of this syndrome do not develop until puberty. Affected individuals are male but show poor sexual development and have very low fertility. Some degree of breast development occurs in about 50% of the cases (◗ Figure 6.20). A degree of subnormal intelligence appears in some affected individuals.

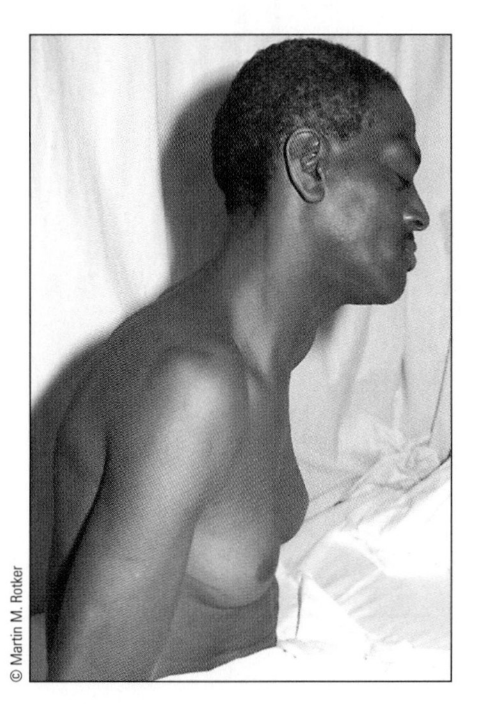

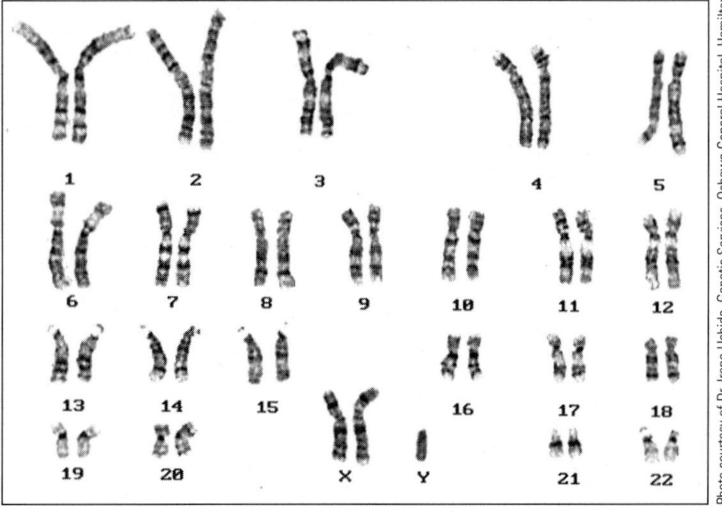

◗ **FIGURE 6.20** A young man who has Klinefelter syndrome and its characteristic karyotype. In some cases, Klinefelter syndrome is associated with breast development.

Photo courtesy of Dr. Irene Uchida, Genetic Services, Oshawa General Hospital, Hamilton, Ontario, Canada

A significant fraction of Klinefelter males are mosaics who have XY and XXY cell lines in the body. In these cases, nondisjunction occurred after fusion of chromosomally normal gametes, during mitosis of embryonic cells.

Overall, about 60% of the cases are the result of maternal nondisjunction, and advanced maternal age is known to increase the risk of affected offspring. Other forms of Klinefelter have XXYY, XXXY, and XXXXY sex chromosome sets. Additional X chromosomes in these karyotypes increases the severity of the phenotypic symptoms and brings on clear-cut mental retardation.

XYY Syndrome (47,XYY)

XYY karyotype Aneuploidy of the sex chromosomes involving XYY chromosomal constitution.

In 1965, a cytogenetic survey of 197 males institutionalized for violent and dangerous antisocial behavior aroused a great deal of interest in the scientific community and the popular press. The findings indicated that nine of these males (about 4.5% of the males in the survey) were **XYY karyotype** (Figure 6.21). These individuals were all above average in height, all suffered personality disorders, and seven of the nine were of subnormal intelligence. Subsequent studies indicated that the frequency of XYY males in the general population is 1 in 1,000 male births (about 0.1% of the males in the general population) and that the frequency of XYY individuals in penal and mental institutions is significantly higher than in the population at large.

Early investigators associated the tendency to violent criminal behavior with the presence of an extra Y chromosome. In effect, this would mean that some forms of violent behavior were brought about by genetic predisposition. In fact, the XYY karyotype has been used on several occasions as a legal defense (unsuccessfully, so far) in criminal trials. The question is this: Is there really a direct link between the XYY condition and criminal behavior? There is no strong evidence to support such a link. In fact, the vast majority of XYY males lead socially normal lives. In the United States, long-term studies of the relationship between antisocial behavior and the 47,XYY karyotype were discontinued. Researchers feared that identifying children with potential behavioral problems might lead parents to treat them differently and result in behavioral problems as a self-fulfilling prophecy.

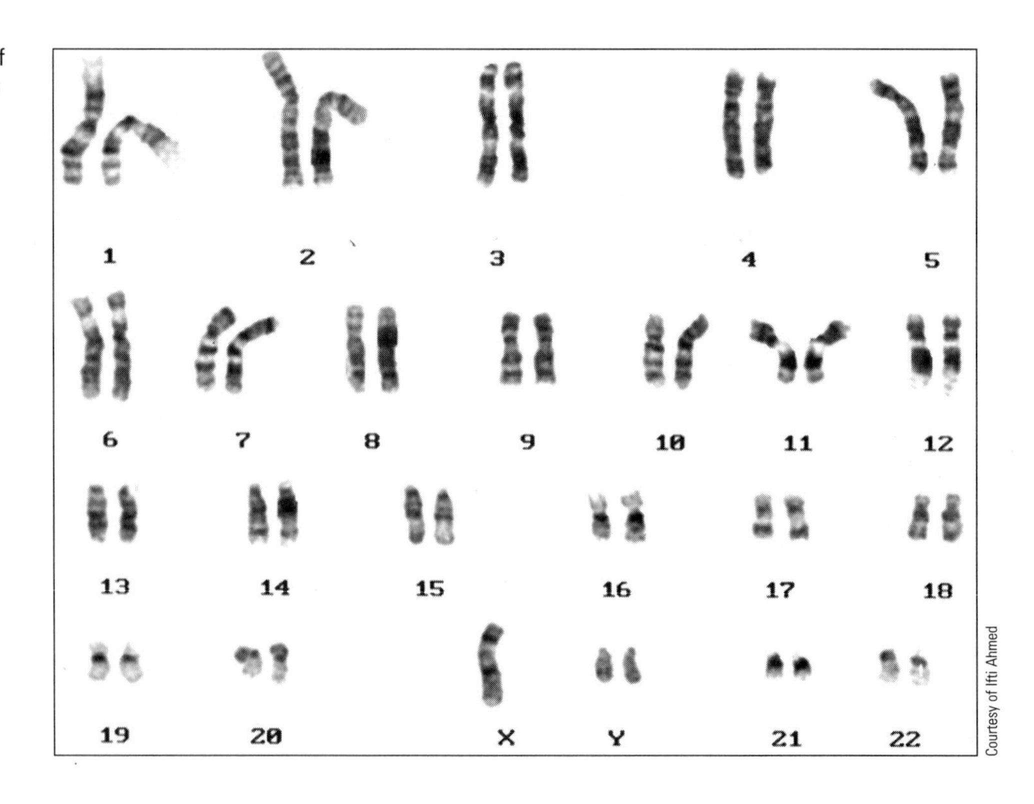

FIGURE 6.21 The karyotype of an XYY male. Affected individuals are usually taller than normal, and some, but not all, suffer from personality disorders.

Aneuploidy of the Sex Chromosomes: Some Conclusions

Several conclusions can be drawn from the study of sex chromosome disorders. First, at least one copy of an X chromosome is essential for survival. Embryos without any X chromosomes (44,-XX and 45,OY) are not observed in studies of spontaneous abortions. They must be eliminated even before pregnancy is recognized, emphasizing the role of the X chromosome in normal development. The second general conclusion is that the addition of extra copies of either sex chromosome interferes with normal development and causes both physical and mental problems. As the number of sex chromosomes in the karyotype increases, the phenotype becomes more severe, indicating that a balance of sex chromosomes is essential to normal development in both males and females.

STRUCTURAL ALTERATIONS WITHIN CHROMOSOMES

Now that we have discussed changes in chromosome number, we will focus on structural changes within and between chromosomes that result in an abnormal phenotype. Changes in structure can involve one, two, or more chromosomes and result from the breakage and reunion of chromosomal parts. In some cases the pieces generated by chromosome breaks are rejoined to restore the original structure. In others, an array of abnormal chromosomes results. Breaks can occur spontaneously through errors in replication or recombination. Environmental agents, such as ultraviolet light, radiation, viruses, and chemicals, can also produce them. Structural alterations that result from breaks include deletions (loss of chromosome segments), duplications (extra copies of a chromosome segment), translocations (moving a segment from one chromosome to another, nonhomologous chromosome), and **inversions** (reversing a chromosome segment). These changes are summarized in Figure 6.22. Rather than considering how such aberrations are produced, we will concentrate on the phenotypic effects of these structural alterations and on how such changes in chromosomal structure can be used to provide information about the location and action of genes.

Inversions Chromosomal aberrations in which a chromosomal segment has been rotated 180° from its usual orientation.

Deletions Involve Loss of Chromosomal Material

Deletion of more than a small amount of chromosomal material has a detrimental effect on the developing embryo, and deletion of an entire autosome is lethal. Consequently, only a few viable conditions are associated with large-scale deletions. Some of these are listed in Table 6.2.

Cri du chat Syndrome An infant with a deletion in the short arm of chromosome 5 was first reported in 1963. This condition occurs in 1 of 100,000 births (Figure 6.23). It is the loss of genes in the deleted chromosome that produces the abnormal phenotype, not the presence of any mutant genes. Affected infants are mentally retarded, with defects in facial development, gastrointestinal malformations, and abnormal throat structures. Affected infants have a cry that sounds like a cat meowing, hence the name **cri du chat syndrome** (MIM/OMIM 123450) (Figure 6.24). This deletion of a chromosome segment affects the motor and mental development of affected individuals but does not seem to be life threatening.

Cri du chat syndrome A deletion of the short arm of chromosome 5 associated with an array of congenital malformations, the most characteristic of which is an infant cry that resembles a meowing cat.

By correlating phenotypes with chromosomal breakpoints in affected individuals, two regions associated with this syndrome have been identified on the short arm of chromosome 5 (Figure 6.23). Loss of chromosome segments in 5p15.3 results in abnormal larynx development; deletions in 5p15.2 are associated with mental retardation and other phenotypic features of this syndrome. This correlation indicates that genes controlling larynx development may be located in 5p15.3, and genes important in the development or function of the nervous system are located in 5p15.2.

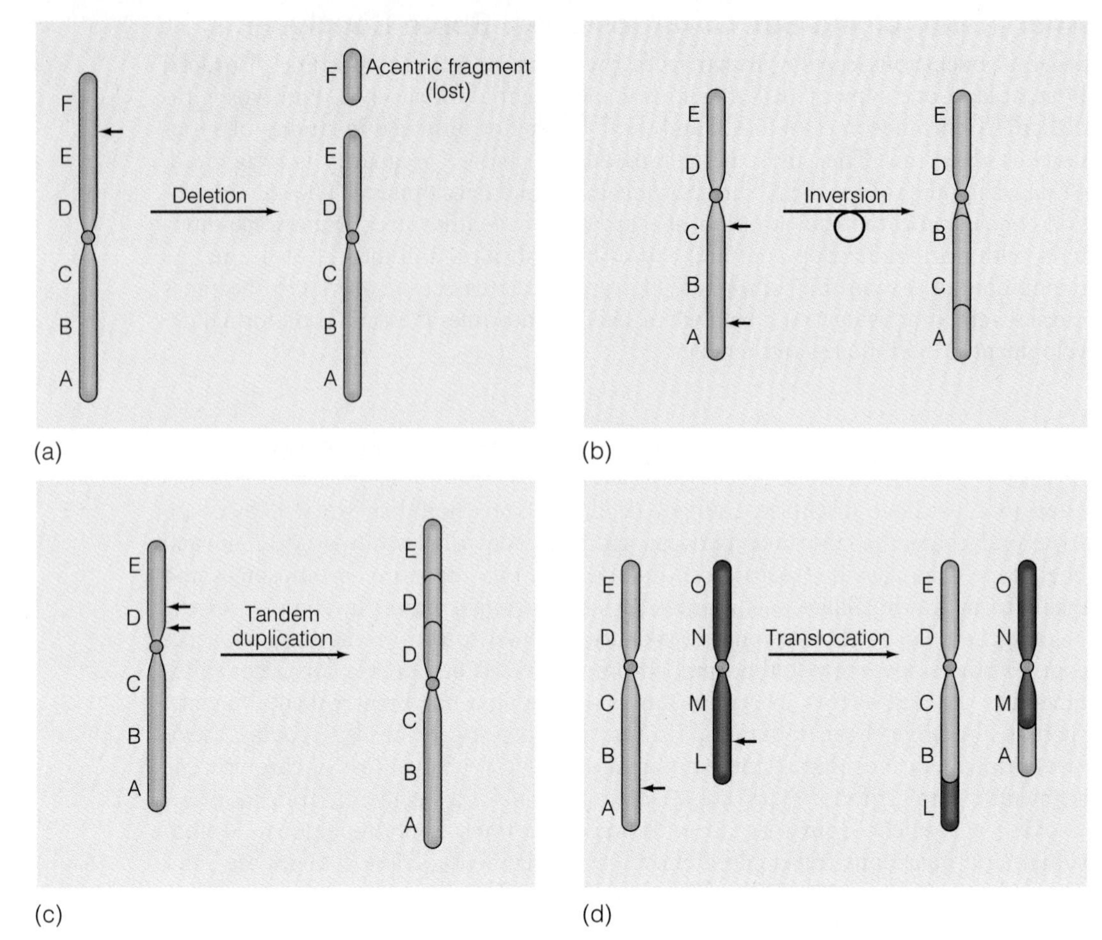

FIGURE 6.22 Some of the common structural abnormalities seen in chromosomes. (a) In a deletion, part of the chromosome is lost. This can occur at the tip of the chromosome as shown, or an internal segment can be lost. (b) In an inversion, the order of part of the chromosome is reversed. This does not change the amount of genetic information carried by the chromosome, only its arrangement. (c) A duplication has a chromosomal segment repeated (in this example, gene *D* and its surrounding region are duplicated). (d) In a translocation, chromosomal parts are exchanged.

Prader-Willi syndrome A disorder associated with chromosome 15 characterized by uncontrolled eating and obesity.

Prader-Willi Syndrome A constellation of physical and mental symptoms known as **Prader-Willi syndrome** (MIM/OMIM 176270) has been correlated with deletions in the long arm of chromosome 15. As infants, affected individuals are weak and do not feed well because of a poor sucking reflex. However, by the age of 5 or 6 years, these children develop an uncontrollable compulsion to eat resulting in obesity and related health problems, such as diabetes. Left untreated, victims literally eat themselves to death. Other symptoms include poor sexual development in males, behavioral problems, and mental retardation.

Prader-Willi patients have deletions in the long arm of chromosome 15 between bands q11 and q13. The size of the deletion is variable but always includes band 15q11.2 (Figure 6.25).

It is estimated that Prader-Willi syndrome affects from 1 in 10,000 to 1 in 25,000 people, and males predominate. The cause of the eating disorder is unknown but may be related to disturbances in endocrine function. Treatment includes behavior modification with constant supervision of access to food.

Table 6.2 Chromosomal Deletions

Deletion	Syndrome	Phenotype
5p-	Cri du chat syndrome	Infants have catlike cry, some facial anomalies, severe mental retardation
11q-	Wilms tumor	Kidney tumors, genital and urinary tract abnormalities
13q-	Retinoblastoma	Cancer of eye, increased risk of other cancers
15q-	Prader-Willi syndrome	Infants: weak, slow growth; children and adults: obesity, compulsive eating

Translocations Involve Exchange of Chromosomal Parts

A translocation results when a chromosome segment is moved to another nonhomologous chromosome. There are two major types of translocations: **reciprocal translocations** and **Robertsonian translocations.** If there is an exchange between two nonhomologous chromosomes, the event is a reciprocal translocation. There is no gain or loss of genetic information in such exchanges, only the rearrangement of gene sequences. In some cases, there are no phenotypic effects, and the translocation is passed through a family for generations. However, cells carrying translocations can produce genetically unbalanced gametes that show duplicated or deleted chromosomal segments. If these gametes participate in fertilization, the result is embryonic death or abnormal offspring.

Reciprocal translocations Chromosomal aberrations involving the exchange of chromosomal material between two nonhomologous chromosomes. This changes the chromosomal location of the exchanged genes, but not the number of genes in the genome.

Robertsonian translocation Breakage in the short arms of acrocentric chromosomes followed by fusion of the long parts into a single chromosome.

About 5% of all cases of Down syndrome involve a Robertsonian translocation, most often between chromosomes 21 and 14. In this type of translocation, centromeres of two chromosomes fuse, and chromosomal material is lost from the short arms (Figure 6.26). The carrier of such a translocation is phenotypically normal, even though the short arms of both chromosomes may be lost. This carrier is actually aneuploid and has only 45 chromosomes. But because the carrier has two copies of the long arm of chromosome 14 and two copies of the long arm of chro-

(a) (b) (c)

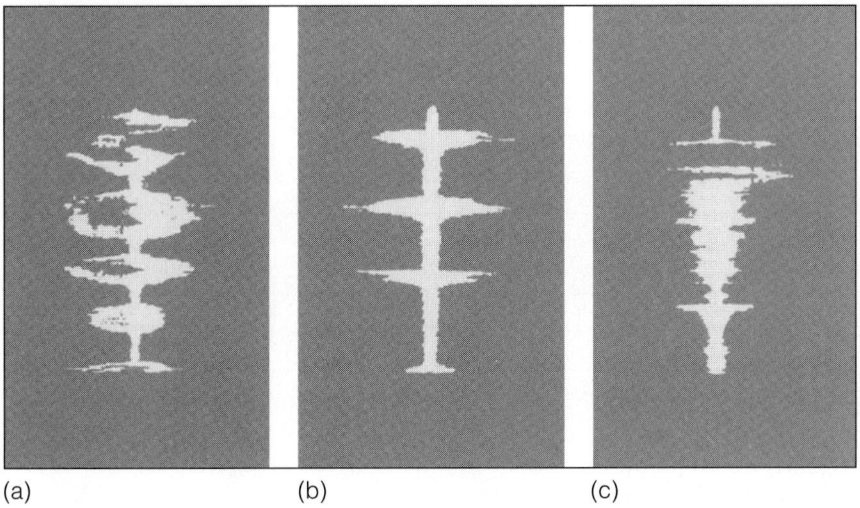

FIGURE 6.24 Sound recordings of (a) a normal infant, (b) an infant who has cri du chat syndrome, and (c) a cat. The cry of the affected infant is much closer in sound pattern to that of the cat than to that of a normal infant, giving rise to the name of the syndrome, cry of the cat.

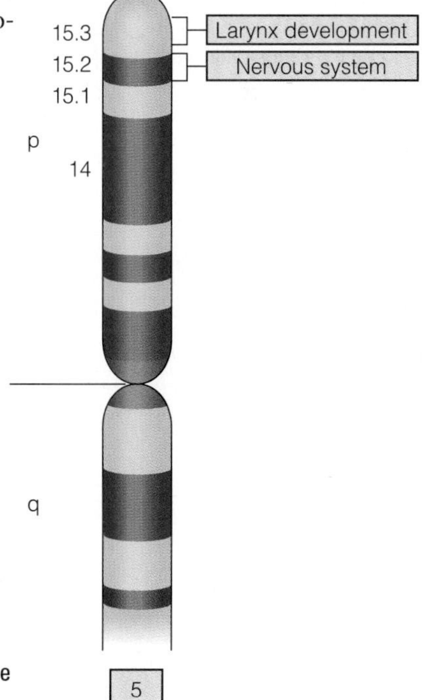

FIGURE 6.23 A deletion of part of chromosome 5 is associated with cri du chat syndrome. By comparing the region deleted with its associated phenotype, investigators have identified regions of the chromosome that carry genes involved in developing the larynx.

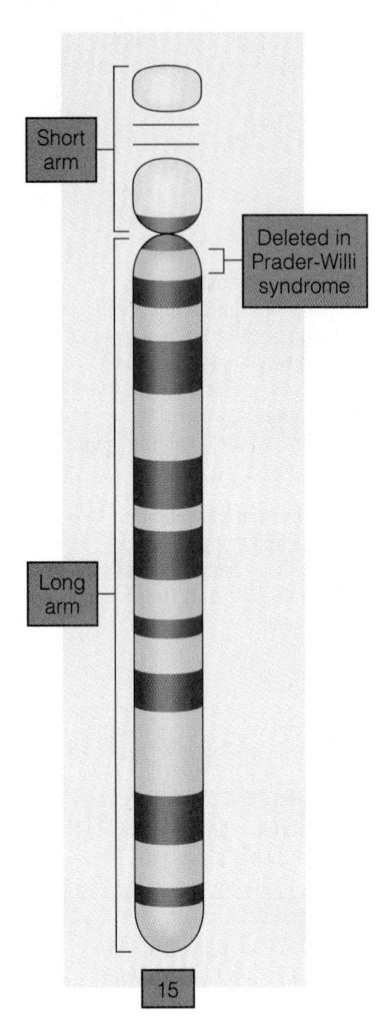

FIGURE 6.25 Diagram of chromosome 15, showing the region deleted in Prader-Willi syndrome.

mosome 21 (a normal 14, a normal 21, and a translocated 14/21), there is no phenotypic effect. At meiosis the carrier produces six types of gametes in equal proportions (Figure 6.26). Three of these produce zygotes that do not develop. Of the remaining three, one will produce a Down syndrome child, one is a translocation heterozygote, and one is chromosomally normal.

Although it might seem that translocation heterozygotes have a one in three, or 33%, risk of having a Down syndrome child, the observed frequency is somewhat lower. It is important to remember that this risk does not increase with maternal age. In addition, there is also a one in three chance of producing a phenotypically normal translocation carrier, who is at risk of producing children with Down syndrome. For this reason it is important to karyotype a Down syndrome child and the parents to determine whether a translocation is involved. This information is essential in counseling parents about future reproductive risks.

Cytogenetic analysis of translocations can be used to identify chromosomal regions that are most critical to the expression of aneuploid phenotypes. The study of a large number of translocations involving chromosome 21 has correlated chromosomal regions on the long arm with the symptoms associated with Down syndrome (Figure 6.27). With this information available, efforts are now focused on identifying and characterizing the genes in this chromosomal segment to understand how a change in gene dosage produces such serious phenotypic effects. This effort has

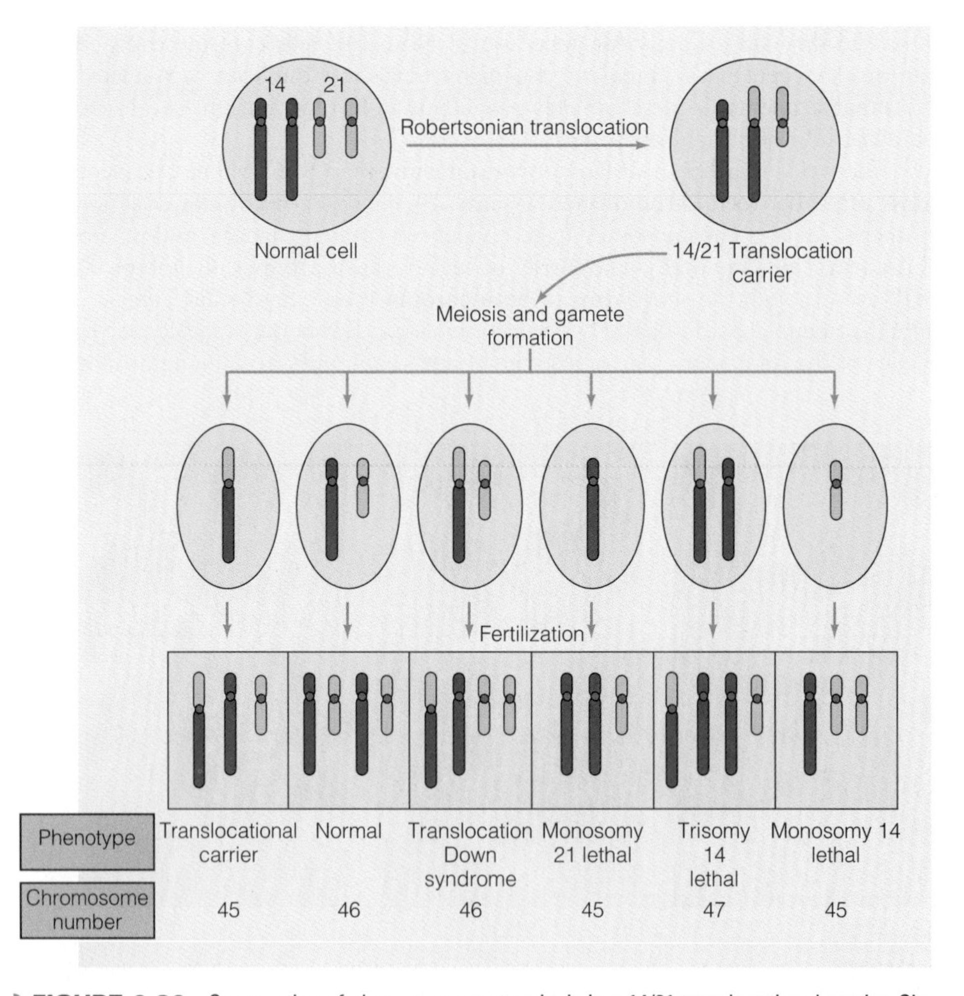

FIGURE 6.26 Segregation of chromosomes at meiosis in a 14/21 translocational carrier. Six types of gametes are produced. When these gametes fuse with those of a normal individual, six types of zygotes are produced. Of these, two (translocational carrier and normal) have a normal phenotype, one is Down syndrome, and three are lethal combinations.

been partly superseded by results from the Human Genome Project, in which all the genes on the long arm of chromosome 21 have been sequenced. However, the cytogenetic analysis will be valuable in helping identify genes on chromosome 21 involved in the phenotype of Down syndrome.

Leukemia
Low-set ears
Mild mental retardation
Severe mental retardation
Skin fold over eye
Protruding tongue
Spots on iris
Abnormal hand, palm prints
Slanted eyes
Heart defects
21

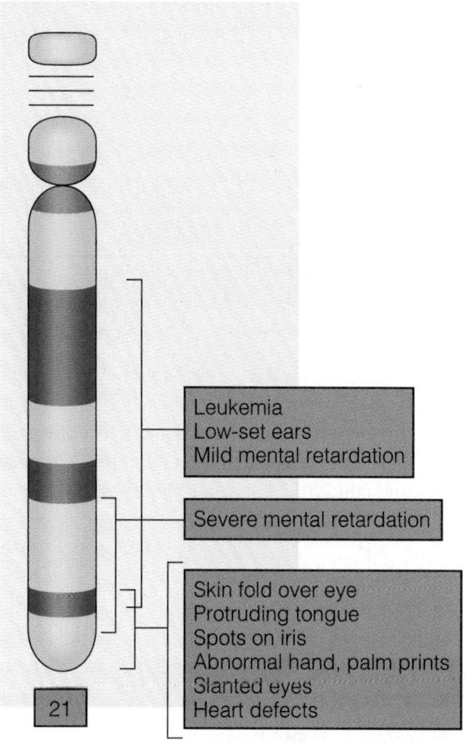

▶ **FIGURE 6.27** Chromosome 21, showing regions associated with various phenotypic features of Down syndrome. These assignments have been made by comparing various deletions in the chromosome with the phenotypes they produce.

WHAT ARE SOME CONSEQUENCES OF ANEUPLOIDY?

Aneuploidy is the most common chromosomal abnormality in humans, and has several important consequences. We will discuss three of these consequences: pregnancy loss, birth defects, and cancer.

As outlined earlier in this chapter, aneuploidy is a major cause of spontaneous abortions (see Figure 6.12). Table 6.3 summarizes some of the major chromosomal abnormalities found in miscarriages. These include triploidy, monosomy for the X chromosome (45,X), and trisomy 16. It is interesting to compare the frequency of chromosomal abnormalities found in spontaneous abortions with those in live births. Triploidy is found in 17 of every 100 spontaneous abortions but in only about 1 in 10,000 live births; 45,X is found in 18% of chromosomally abnormal abortuses but in only 1 in 7,000 to 10,000 live births.

Comparison of the number of chromosomal abnormalities detected by CVS (performed at 10 to 12 weeks of gestation) versus amniocentesis (at 16 weeks of gestation) show that the abnormalities detected by CVS are two to five times more frequent than those detected by amniocentesis, which in turn are about two times higher than those found in newborns. This decrease in the frequency of chromosomal abnormalities over developmental time is evidence that chromosomally abnormal embryos and fetuses are eliminated by spontaneous abortion throughout pregnancy (▶ Figure 6.28).

Birth defects are another consequence of chromosomal abnormalities. The frequencies of chromosomal aberrations in newborns are shown in Table 6.4.Trisomy 16, which is common in spontaneous abortions, is not found among infants, indicating that fetuses with this condition are not viable. Only trisomies for chromosomes 13, 18, and 21 occur with any frequency in live births. Trisomy 21 occurs with a frequency of about 1 in 900 births, but cytogenetic surveys of spontaneous

Table 6.3 **Chromosomal Abnormalities in Spontaneous Abortions**

Abnormality	Frequency (%)
Trisomy 16	15
Trisomies, 13, 81, 21	9
XXX, XXY, XYY	1
Other	27
45,X	18
Triploidy	17
Tetraploidy	6
Other	7

Note: Adapted from T. Hassold. (1986). Trends in Genetics 2, *105–110. Adapted with permission of the publisher.*

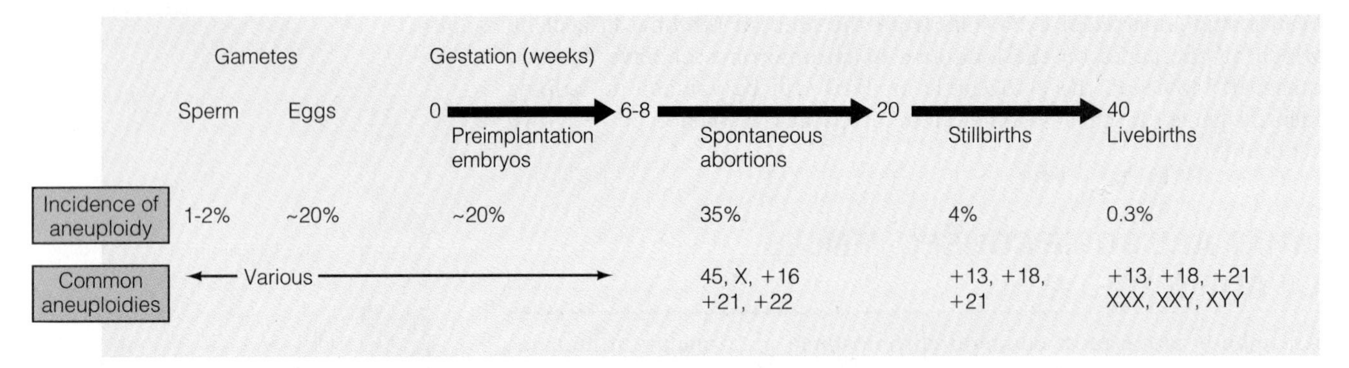

FIGURE 6.28 The frequency of aneuploidy changes dramatically over developmental time. Between 6 to 8 weeks and 20 weeks, about 35% of spontaneous abortions are aneuploid. Around 20 weeks, the frequency falls by an order of magnitude to about 4% in stillbirths. The frequency decreases again by an order of magnitude, with about 0.3% of newborns being aneuploid. [Modified from Hassold, T. & Hunt, P. (2001). To err (meiotically) is human: The genesis of human aneuploidy. *Nature Reviews Genetics 2,* 280–291 (Table 1, p. 283), with permission. Copyright 2001 Macmillan Magazines, Ltd.]

abortions indicate that about two-thirds of such conceptions are lost by miscarriage. Similarly, over 99% of all 45,X conceptions are lost before birth. Overall, although selection against chromosomally abnormal embryos and fetuses is efficient, the high rate of nondisjunction in humans means there is a significant reproductive risk for chromosomal abnormalities. Over 0.5% of all newborns are affected with an abnormal karyotype.

The relationship between cancer and its accompanying chromosomal changes is a third consequence of chromosomal abnormalities. A significant number of cancers, especially leukemias, are associated with specific chromosomal translocations. Solid tumors have a wide range of chromosomal abnormalities, including aneuploidy, translocations, and duplications. Evidence suggests that these abnormalities may arise during a period of genomic instability that precedes or accompanies the transition of a normal cell into a malignant cell. The chromosomal changes that accompany the development of cancer are discussed in Chapter 14.

OTHER FORMS OF CHROMOSOMAL ABNORMALTIES

In some cases, the karyotype and individual chromosomes appear to be normal, but the phenotype is abnormal. However, careful analysis reveals a subtle chromosome abnormality. One of these cases is uniparental disomy, in which both members of a

Table 6.4 Chromosomal Abnormalities in Newborns

Abnormality	Approximate Frequency
45,X	1/7500
XXX	1/1200
XXY	1/1000
XYY	1/1100
Trisomy 13	1/15,000
Trisomy 18	1/11,000
Trisomy 21	1/900
Structural abnormalities	1/400

SIDEBAR

Fragile Sites and Cancer

More than 175,000 cases of lung cancer are diagnosed each year in the United States, and more than 150,000 deaths are related to this form of cancer. It is thought that the short arm of chromosome 3 carries one or more genes that when mutated are involved in the development of lung cancer. A gene called *FHIT,* located at the fragile site FRA3p14.2, has recently been shown to be abnormal in many cases of lung cancer. *FHIT* encodes an enzyme involved in nucleotide metabolism.

In lung cancer cells taken from tumors and in laboratory-grown cell lines, 76% of the samples showed a loss of one or both *FHIT* alleles. Up to 80% of the samples were defective in the expression of the *FHIT* gene, mostly caused by deletions or rearrangements of the *FHIT* gene.

Expression of the fragile site at 3p14.2 is much higher in the circulating blood cells of lung cancer patients, indicating that this site has been damaged. Fragile sites are susceptible to breakage and chromosomal damage by physical, chemical, or biological agents. Tobacco smoke contains compounds that interfere with the replication of DNA and might lead to abnormalities in genes at fragile sites, such as *FHIT.* These abnormalities, along with other mutant gene, are associated with the development of lung tumors.

chromosome pair are derived from one parent, resulting in an abnormal phenotype. Another is the presence of fragile sites, which appear only when cells are grown in the laboratory and certain chemicals are added to the growth medium.

Uniparental Disomy

Normally, meiosis ensures that one member of each chromosomal pair is derived from the mother, and the other member is from the father. On rare occasions, however, a child gets both copies of a chromosome from one parent, a condition known as **uniparental disomy** (UPD). This condition can arise in several ways, all of which involve two chromosomal errors in cell division (▶ Figure 6.29). These errors can occur in meiosis or in mitotic divisions following fertilization.

Uniparental disomy A condition in which both copies of a chromosome are inherited from a single parent.

UPD has been identified in some unusual situations. These include females affected with rare X-linked disorders such as hemophilia; father-to-son transmission of rare, X-linked disorders in which the mother is homozygous normal; and children who are affected with rare autosomal recessive disorders, but in which only one parent is heterozygous. Prader-Willi syndrome and Angelman syndrome (MIM/OMIM 105830) can be caused by deletions in region 15q11.12 or by UPD. If both copies of chromosome 15 are derived from the mother, Prader-Willi syn-

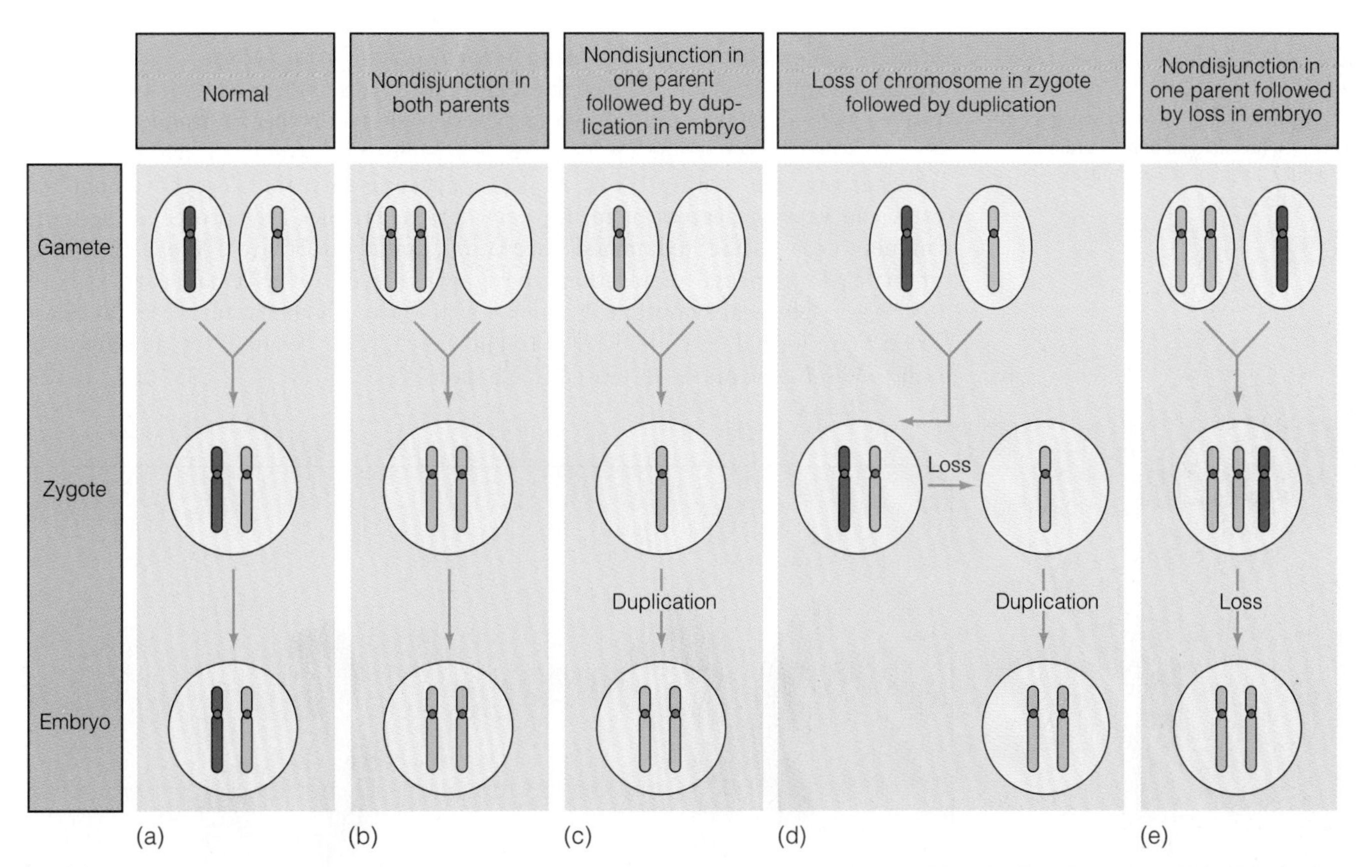

▶ **FIGURE 6.29** Uniparental disomy can be produced by several mechanisms, involving nondisjunction in meiosis or nondisjunction in the zygote or early embryo. (a) Normally, gametes contain one copy of each chromosome, and fertilization produces a zygote carrying two copies of a chromosome, one derived from each parent. (b) Nondisjunction in both parents, in which one gamete carries both copies of a chromosome and the other gamete is missing a copy of that chromosome. Fertilization produces a diploid zygote, but both copies of one chromosome are inherited from a single parent. (c) Nondisjunction in one parent, resulting in the loss of a chromosome. This gamete fuses with a normal gamete to produce a zygote monosomic for a chromosome. An error in the first mitotic division results in duplication of the monosomic chromosome, producing uniparental disomy. (d) Normal gametes can fuse to produce a normal zygote, which loses a chromosome at the first mitotic division. A second nondisjunction restores the lost copy, but both copies are derived from one parent. (e) Nondisjunction in one parent produces a gamete that carries both copies of a chromosome. Fusion with a normal gamete produces a trisomic zygote, which loses one copy of the chromosome in nondisjunction, resulting in uniparental disomy.

drome results. If both copies of chromosome 15 are derived from the father, Angelman syndrome results. The origin of these disorders by UPD is discussed in detail in Chapter 11.

Fragile Sites Appear as Gaps or Breaks in Chromosomes

Fragile sites are structural features of chromosomes that become visible when cells are grown in the laboratory under certain conditions. These sites appear as gaps or breaks at specific sites on a chromosome and are inherited as codominant traits. Fragile sites are classified by their frequency in the population. Rare sites are found in less than 5% of individuals, and common sites are found in almost all individuals. Over 100 fragile sites have been identified in the human genome. Chromosome breaks often occur at fragile sites, producing chromosome fragments, deletions, and other aberrations. The molecular nature of most fragile sites is unknown but is of great interest because they represent regions susceptible to breakage. Almost all studies of fragile sites have been carried out on cells in tissue culture, and it is currently not known whether such sites are expressed in meiotic cells. Several fragile sites are located on the X chromosome (Figure 6.30). Two of these rare sites, FRAX E and FRAX A, are associated with genetic disorders. We will briefly outline the relationship between FRAX A and the fragile-X syndrome.

Fragile-X Syndrome Is Associated with Mental Retardation A rare fragile site near the tip of the long arm of the X chromosome is associated with an X-linked form of mental retardation known as Martin-Bell syndrome, or **fragile-X** (MIM/OMIM 309500) syndrome (Figure 6.31). Males with fragile-X syndrome have long, narrow faces with protruding chins and large ears; they also have enlarged testes and varying degrees of mental retardation. Fragile-X syndrome is the most common form of inherited mental retardation, and about 3% to 5% of males institutionalized for mental retardation have fragile-X syndrome. Female carriers have no clear-cut physical symptoms but as a group have a higher rate of mental retardation than normal individuals. The fragile-X syndrome is caused by an alteration in the *FMR-1* gene and is discussed in Chapter 11.

Fragile X An X chromosome that carries a nonstaining gap, or break, at band q27; associated with mental retardation in males.

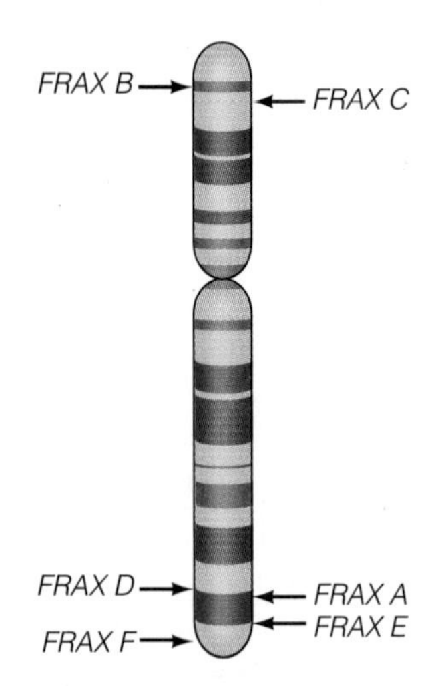

FIGURE 6.30 The fragile sites on the human X chromosome. Sites B, C, and D are common sites and found on almost all copies of the X chromosome. A, E, and F are rare sites; expression of A is associated with fragile-X syndrome.

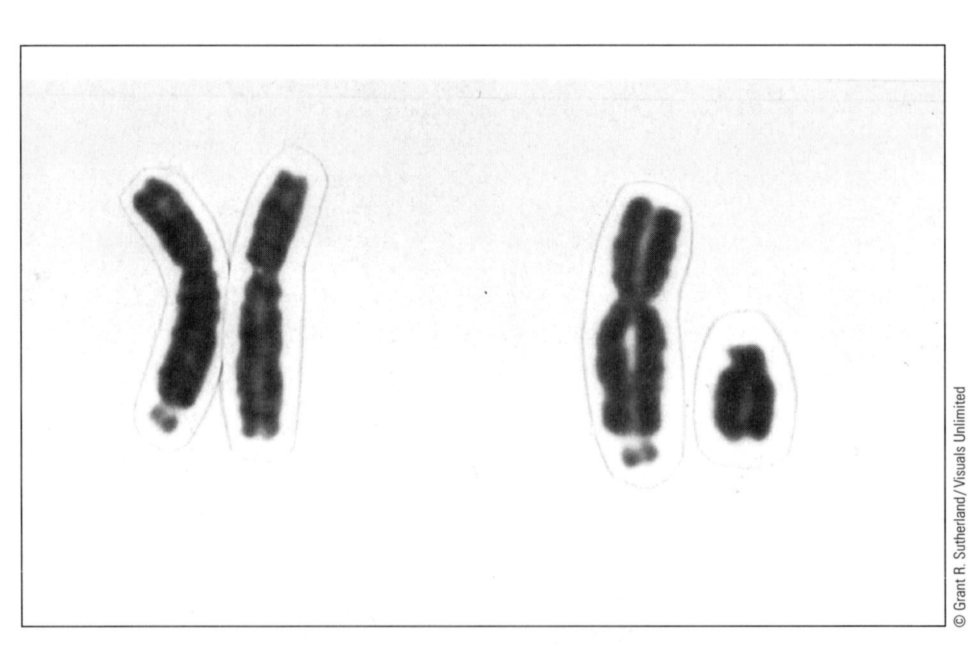

FIGURE 6.31 The fragile-X syndrome in a carrier female (left) and a male (right). This syndrome causes the lower tip of the X chromosome to appear as a fragile piece hanging by a thread. The nature of the mutation that causes this syndrome is discussed in Chapter 11.

Summary

The Human Chromosome Set

1. Human chromosomes are analyzed by the construction of karyotypes. A system of identifying chromosome regions allows any region to be identified by a descriptive address. Chromosome analysis is a powerful and useful technique in human genetics.

Analyzing Chromosomes and Karyotypes

2. The study of variations in chromosomal structure and number began in 1959 with the discovery that Down syndrome is caused by the presence of an extra copy of chromosome 21. Since then the number of genetic diseases related to chromosomal aberrations has steadily increased. The development of chromosome banding and techniques for identifying small changes in chromosomal structure have contributed greatly to the information that is now available.

Variations in Chromosome Number

3. There are two major types of chromosomal changes: a change in chromosomal number and a change in chromosomal arrangement. Polyploidy and aneuploidy are major causes of reproductive failure in humans. Polyploidy is rarely seen in live births, but the rate of aneuploidy in humans is reported to be more than 10-fold higher than in other primates and mammals. The reasons for this difference are unknown, but this represents an area of intense scientific interest.

What Are the Risks for Autosomal Trisomy?

4. The loss of a single chromosome creates a monosomic condition, and the gain of a single chromosome is called a trisomic condition. Autosomal monosomy is eliminated early in development. Autosomal trisomy is selected against less stringently, and cases of partial development and live births of trisomic individuals are observed. Most cases of autosomal trisomy greatly shorten life expectancy, and only individuals who have trisomy 21 survive into adulthood.

Aneuploidy of the Sex Chromosomes

5. Aneuploidy of sex chromosomes involves both the X and Y chromosome. Studies of sex chromosome aneuploidies indicate that at least one copy of the X chromosome is required for development. Increasing the number of copies of the X or Y chromosome above the normal range causes progressively greater disturbances in phenotype and behavior, indicating the need for a balance in gene products for normal development.

Structural Alterations Within Chromosomes

6. Changes in the arrangement of chromosomes include deletions, duplications, inversions, and translocations. Deletions of chromosomal segments are associated with several genetic disorders, including cri du chat and Prader-Willi syndromes. Translocations often produce no overt phenotypic effects but can result in genetically imbalanced and aneuploid gametes. We discussed a translocation resulting in Down syndrome that in effect makes Down syndrome a heritable genetic disease, potentially present in one in three offspring.

What Are Some Consequences of Aneuploidy?

7. Aneuploidy is the leading cause of reproductive failure in humans, resulting in spontaneous abortions and birth defects. In addition, aneuploidy is associated with most cancers.

Other Forms of Chromosome Abnormalities

8. Uniparental disomy (UPD) is a condition in which both copies of a chromosome are inherited from a single parent. UPD is associated with several genetic diseases. Fragile sites appear as gaps, or breaks, in chromosome-specific locations. One of these fragile sites on the X chromosome is associated with a common form of mental retardation that affects a significant number of males.

Case Studies

CASE 1

Michelle was a 42-year-old Caucasian woman who had declined counseling and amniocentesis at 16 weeks of pregnancy, but was referred for genetic counseling following an abnormal ultrasound at 20 weeks gestation. Following the ultrasound, a number of findings suggested a possible chromosome abnormality in the fetus. The ultrasound showed swelling under the skin at the back of the fetus's neck; shortness of the femur, humerus, ear length; and underdevelopment of the middle section of the fifth finger. Michelle's physician performed an amniocentesis, and a referral was made to the genetics program. The couple felt they did not want genetic counseling before receiving results of the cytogenetic analysis.

This was Michelle's third pregnancy; she and her husband Mike had a 6-year-old daughter and a 3-year-old son. At their next session, the counselor informed the couple that the results revealed trisomy 21, explored their understanding of Down syndrome, and elicited their experiences with people with disabilities. She also reviewed the clinical concerns revealed by the ultrasound and associated anomalies (mild to severe mental retardation, cardiac defects, and kidney problems). The options available to the couple were outlined. They were provided with a booklet written for parents making choices following the prenatal diagnosis of Down syndrome. After a week of careful deliberation with their family, friends, and clergy, they elected to terminate the pregnancy.

1. Do you think that this couple had the right to terminate the pregnancy given the prenatal diagnosis? If not, under what circumstances would a couple have this right? What other options were available to the couple?
2. Should physicians discourage a 42-year-old woman from having children because of an increased chance of a chromosomal abnormality?

CASE 2

The genetic counselor was called to the nursery for a consultation on a newborn who was described as "floppy with a weak cry." The counselor noted that the newborn's chart indicated that he was having feeding problems and had not gained weight since his delivery 15 days earlier. The counselor noted several other findings during his evaluation. The infant had almond-shaped eyes, a small mouth with a thin upper lip, downturned corners of the mouth, and a narrow face. The infant was born with undescended testes and a small penis. The counselor suspected that this child had the genetic disorder known as Prader-Willi syndrome.

Prader-Willi syndrome is caused by the absence of a small region on the long arm of chromosome 15. It is always the lack of the paternal copy of this region that causes Prader-Willi syndrome. This absence can occur in three ways: deletion of a segment of the paternal chromosome 15, a mutation on the paternal chromosome 15, or maternal uniparental disomy—in other words, both copies of chromosome 15 are from the mother and none are contributed by the father.

The child and his parents were tested for a deletion in the long arm of chromosome 15 (15q11-q13) by fluorescence in situ hybridization (FISH) and uniparental disomy 15 (polymerase chain reaction [PCR]). In this case, maternal disomy was detected by PCR—which is the cause of Prader-Willi syndrome in about 30% of the cases.

1. Why is a copy of the paternal chromosome 15 needed to prevent Prader-Willi syndrome?
2. Are there any treatments for Prader-Willi syndrome? What steps should the family now take to cope with the diagnosis?
3. Explain to the parents how maternal disomy happens during gamete formation and/or in mitosis following fertilization.

Questions and Problems

Analyzing Chromosomes and Karyotypes

1. Originally, karyotypic analysis relied on size and centromere placement to identify chromosomes. Because many chromosomes are similar in size and centromere placement, the identification of individual chromosomes was difficult, and chromosomes were placed into eight groups, identified by letters A–G. Today, each human chromosome can be readily identified.
   a. What technical advances led to this improvement in chromosome identification?
   b. List two ways this improvement can be implemented.
2. What clinical information does a karyotype provide?

3. Given the following karyotype, is this a male or female? Normal or abnormal? What would the phenotype of this individual be?

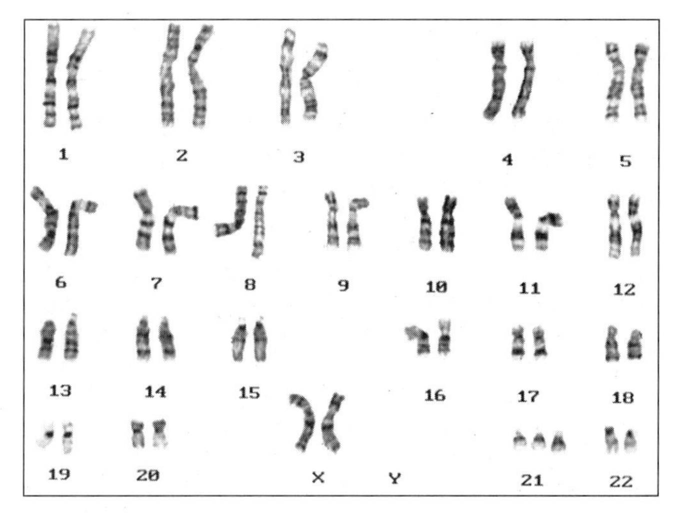

4. A colleague emails you a message that she has identified an interesting chromosome variation at 21q13. In discussing this discovery with a friend who is not a cytogeneticist, explain how you would describe the location, defining each term in the chromosome address 21q13.
5. What are the two prenatal diagnosis techniques used to detect genetic defects in a baby before birth? Which technique can be performed earlier, and why is this an advantage?
6. What are some conditions that warrant prenatal diagnosis?

Variations in Chromosome Number—Polyploidy

7. Discuss the following sets of terms:

   trisomy and triploidy
   aneuploidy and polyploidy

8. What chromosomal abnormality can result from dispermy?
9. Tetraploidy may result from:
   a. lack of cytokinesis in meiosis II.
   b. nondisjunction in meiosis I.
   c. lack of cytokinesis in mitosis.
   d. nondisjunction in mitosis in the early embryo.
   e. none of the above.
10. A cytology student believes he has identified an individual suffering from monoploidy. The instructor views the cells under the microscope and correctly dismisses the claim. Why was the claim dismissed? What type of cells were being viewed?
11. An individual is found to have some tetraploid liver cells but diploid kidney cells. Be specific in explaining how this condition might arise.
12. A spermatogonial cell undergoes mitosis before entering the meiotic cell cycle en route to the production of sperm. However, during mitosis the cytoplasm fails to divide, and only one daughter cell is produced. A resultant sperm eventually fertilizes a normal ovum. What is the chromosomal complement of the embryo?
13. A teratogen is an agent that produces nongenetic abnormalities during embryonic or fetal development. Suppose a teratogen is present at conception. As a result, during the first mitotic division the centromeres fail to divide. The teratogen then loses its potency and has no further effect on the embryo. What is the chromosomal complement of this embryo?
14. As a physician, you deliver a baby with protruding heels and clenched fists with the second and fifth fingers overlapping the third and fourth fingers.
   a. What genetic disorder do you suspect the baby has?
   b. How do you confirm your suspicion?

Variations in Chromosome Number—Aneuploidy

15. Describe the process of nondisjunction and when it takes place during cell division.
16. A woman gives birth to monozygotic twins. One boy has a normal genotype (46,XY), but the other boy has trisomy 13 (47,+13). What events—and in what sequence—led to this situation?
17. Assume that a meiotic nondisjunction event is responsible for an individual who is trisomic for chromosome 8. If two of the three copies of chromosome 8 are absolutely identical, at what point during meiosis did the nondisjunction event take place?
18. Hypothetical human conditions have been found to have a genetic basis. Suppose a hypothetical genetic disorder responsible for condition 1 is similar to Marfan syndrome. The defect responsible for condition 2 resembles Edwards syndrome. One of the two conditions results in more severe defects, and death occurs in infancy. The other condition produces a mild phenotypic abnormality and is not lethal. Which condition is most likely lethal, and why?
19. What is the genetic basis and phenotype for each of the following disorders (use proper genetic notation)?

   Edwards syndrome
   Patau syndrome
   Klinefelter syndrome
   Down syndrome

20. The majority of nondisjunction events leading to Down syndrome are maternal in origin. Based on the duration of meiosis in females, speculate on the possible reasons for females contributing aneuploid gametes more frequently than males.
21. Name and describe the theory that deals with embryo–uterus interaction that explains the relationship between advanced maternal age and the increased frequency of aneuploid offspring.

22. If all the nondisjunction events leading to Turner syndrome were paternal in origin, what trisomic condition might be expected to occur at least as frequently?

Structural Alterations Within Chromosomes:

23. Identify the type of chromosomal aberration described in each of the following cases:

    Loss of a chromosome segment
    Extra copies of a chromosome segment
    Reversal in the order of a chromosome segment
    Movement of a chromosome segment to another, nonhomologous chromosome

24. Describe the chromosomal alterations and phenotype of cri du chat syndrome and Prader-Willi syndrome.
25. A geneticist discovers that a girl with Down syndrome has a Robertsonian translocation involving chromosomes 14 and 21. If she has an older brother who is phenotypically normal, what are the chances that he is a translocation carrier?
26. Albinism is caused by an autosomal recessive allele of a single gene. An albino child is born to phenotypically normal parents. However, the paternal grandfather is albino. Exhaustive analysis suggests that neither the mother nor her ancestors carry the allele for albinism. Suggest a mechanism to explain this situation.

Other Forms of Chromosomal Abnormalities

27. Fragile-X syndrome causes the most common form of inherited mental retardation. What is the chromosomal abnormality associated with this disorder? What is the phenotype of this disorder?

Internet Activities

The following activities use the resources of the World Wide Web to enhance the topics covered in this chapter. To investigate the topics described below, log on to the book's home page (www.brookscole.com/biology_d) and follow the *Student Resources* link to your subject and text.

1. *Identifying Chromosomes.* The University of Arizona's Biology Project provides a chromosome karyotyping activity (www.biology.arizona.edu/human_bio/activities/karotyping/karotyping.html). In this exercise, you have the opportunity to create part of a human karyotype. In the first part of the activity you will be arranging chromosomes onto a karyotyping sheet; once you have completed the karyotype, you will interpret the results of your efforts. Read the introductory material, and then proceed to the "Patient Histories."
   a. What are the karyotypes of the three patients? What diagnoses did you make?
   b. What kinds of chromosomal features did you use in making your karyotyping decisions? Were some chromosomes harder to identify than others? Why? How would you go about resolving the identity of a mystery chromosome if you were working in a karyotyping laboratory? (Hint: refer to the Concepts and Controversies section of this chapter, re-read the section on how karyotypes are prepared, and consider the number of cells/nuclei per slide.) How serious might the consequences of a mistake in karyotyping be? Why?

   Further Exploration: To read more about the latest high-tech methods in karyotyping, go to The Biology Project's "New Methods for Karyotyping" Web site (www.biology.arizona.edu/human_bio/current/new_karotyping/new_karotyping.html).

2. *Exploring a Chromosomal Defect.* The chromosomal abnormality called fragile-X syndrome, discussed in this chapter, is the leading genetic cause of mental retardation. Go to the "Your Genes, Your Health" Web site maintained by the Dolan DNA Learning Center at Cold Spring Harbor Laboratory (http://vector.cshl.org/ygyh/mason.index) and click on the "Fragile-X syndrome" link. (If you want to find out about hemophilia or Marfan syndrome, there are links at this same site.) For this exercise, you should choose the "What causes it?" link.
   a. Where is the site of fragile-X mutation? What is the gene involved called? How does this gene differ in people who have fragile-X syndrome compared to people without fragile-X syndrome? Look at the photograph of fragile-X chromosomes in Figure 6.31 of this text. How might the structure of the mutant form of the gene as described on this Web site relate to the "fragile" appearance of the chromosomes in the photograph?
   b. We'll continue to discuss various aspects of fragile-X syndrome in later chapters of this text. If you would like to investigate some of this information now, go to the fragile-X Internet Activities for chapters 7 and 11.

   Further Exploration: To find out more about general aspects of fragile-X syndrome, from current research to how to get involved with support groups, go to the FRAXA (Fragile X Research Foundation) Web site (http://www. fraxa.org/html/about_testing.htm).

For Further Reading

Borgaonkar, D. S. (1997). *Chromosome variation in man: A catalogue of chromosomal variants and anomalies* (8th ed.). New York: Wiley-Liss.

Dellarco, V., Voytek, P., & Hollaender, A. (1985). *Aneuploidy: Etiology and mechanisms.* New York: Plenum.

Faix, R., Barr, M., Jr., & Waterson, J. (1984). Triploidy: Case report of a live-born male and an ethical dilemma. *Pediatrics, 74,* 296–299.

Friedmann, T. (1971, November). Prenatal diagnosis of genetic disease. *Sci. Am., 225,* 34–42.

Grouchy, J. de. (1984). *Clinical atlas of human chromosomes.* New York: Wiley.

Jacobs, P. A., & Hassold, T. J. (1995). The origin of numerical chromosome abnormalities. *Adv. Genet., 33,* 101–133.

Rowley, J. D. (1998). The critical role of chromosome translocations in human leukemias. *Annu. Rev. Genet., 32,* 495–519.

Weiss, E., Loevy, H., Saunders, A., Pruzansky, S., & Rosenthal, I. (1982). Monozygotic twins discordant for Ullrich-Turner syndrome. *Am J. Med Genet., 13,* 389–399.

Williams, R. (1999). Testing for birth defects. *FDA Consumer, 33,* 22.

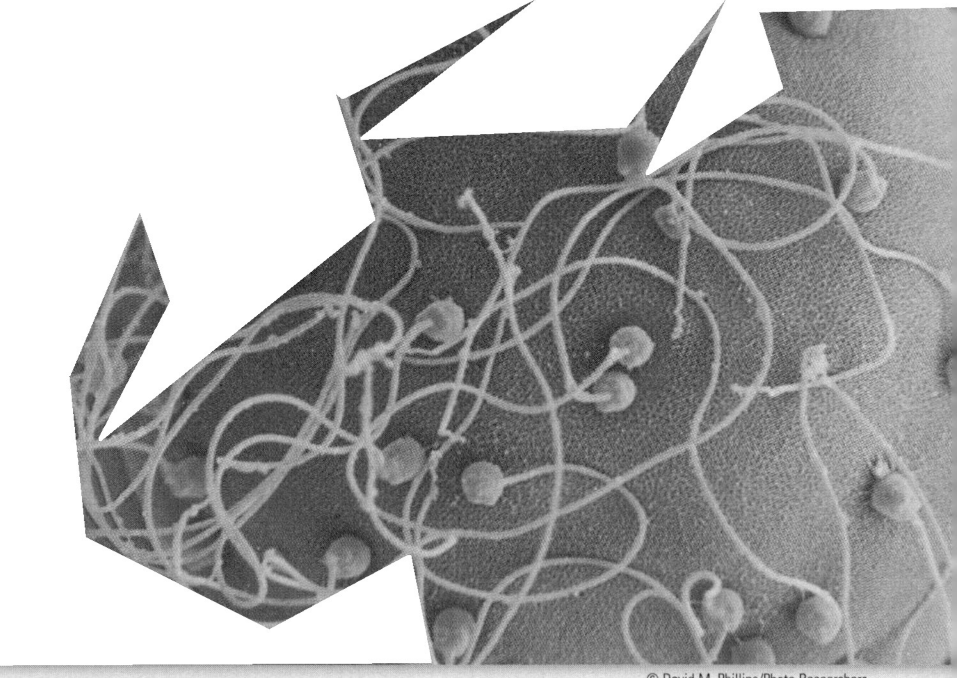

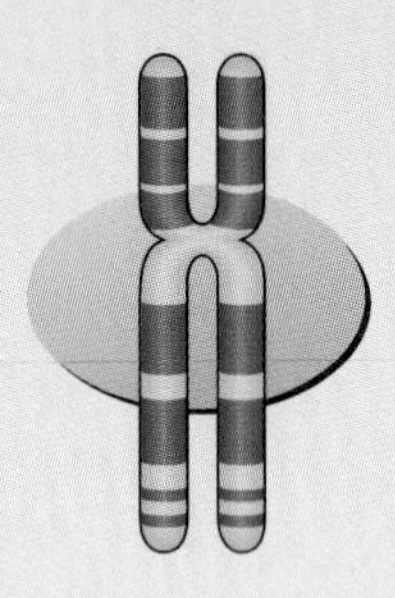

Development and Sex Determination

Chapter Outline

At 11:47 PM on July 25, 1978, a 5 lb, 12 oz. baby girl named Louise Brown was born in Oldham, England. Unlike the millions of other children born that year, her birth was an international media event, and she was called the "baby of the century" and a "miracle baby." Although some praised her birth as a medical miracle, others talked of her birth as the beginning of an era of genetically engineered children and even human–animal hybrids raised in artificial wombs. Why all the fuss? Louise was the first human born after in vitro fertilization (IVF), a procedure in which an egg is fertilized outside the body (in vitro literally means in glass) and the developing embryo is implanted into the uterus for development.

The development of IVF by Patrick Steptoe and Robert Edwards was a long, slow process. After 9 years, and over 80 failed attempts, Steptoe repeated his procedure once more in November 1977 by recovering an egg from Lesley Brown. He first made a small incision (about ½ in. long) in her abdomen. Then he inserted a tube-like laparoscope to examine the ovary, and made another small incision to remove an egg. He gave the egg to Edwards, who put it in a sterile glass dish and mixed it with semen from Louise's father. The dish containing the fertilized egg in a solution of nutrients was placed in an incubator for 2½ days, after which the developing embryo was implanted into Lesley's uterus through a tube inserted into the vagina. For the first time, the procedure was a success, and Louise was born on July 25 of the following year. In the United States, IVF is now a routine procedure available at over 350 clinics, and more than 45,000 babies started life in a glass dish.

This procedure, which now seems so commonplace, was a remarkable achievement for its time and was the first step in the development of methods, collectively called assisted reproductive technology (ART), used to help infertile couples. Some of these methods are outlined in this chapter. The use of ART along with recombinant DNA techniques has created a powerful new technology for sex selection, the diagnosis of genetic disorders, and gene therapy, which we will cover in Chapters 13 and 19.

In this chapter, we focus on three topics: the biology of human development, the genetics and biology of sex determination, and some of the genetic differences between males and females. We will begin the chapter by discussing the major features of human development and ART technology. The central portion of the chapter covers genetic aspects of sex determination and sex differentiation. The chapter concludes with an examination of how patterns of gene expression differ in males and females.

A SURVEY OF HUMAN DEVELOPMENT: FERTILIZATION TO BIRTH

Human reproduction depends on the integrated action of the endocrine system and the reproductive organs. Males and females each possess a pair of gonads and associated accessory glands and ducts. In males, the testes produce spermatozoa and sex hormones. The ovaries of females produce eggs or ova and female sex hormones. Within the gonads, cells produced by meiosis mature into gametes (review this process in Chapter 2).

Fertilization The fusion of two gametes to produce a zygote.

Fertilization, the fusion of male and female gametes, usually occurs in the upper third of the oviduct (▶ Figure 7.1). Sperm deposited in the vagina swim through

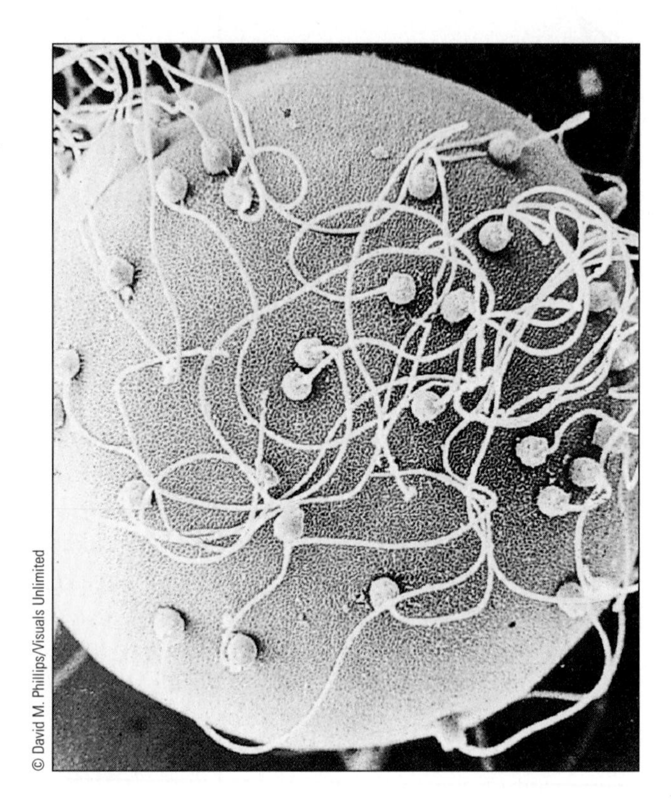

▶ **FIGURE 7.1** Scanning electron micrograph of an egg surrounded by sperm. Usually, only one sperm enters the egg.

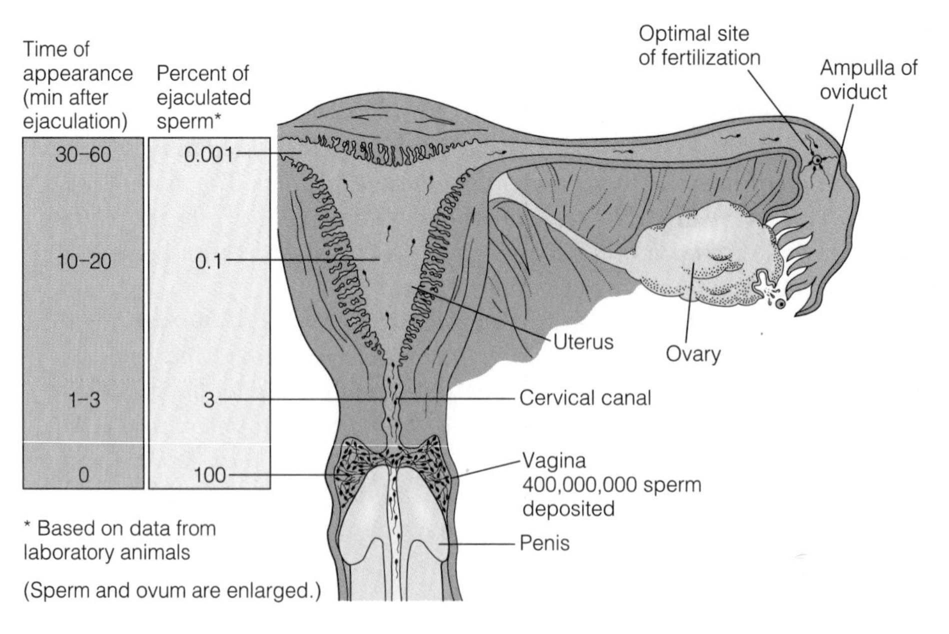

▶ **FIGURE 7.2** The relative amounts of sperm and time of transport from the vagina into the oviduct.

the cervix, up the uterus, and into the oviduct. About 30 minutes after ejaculation, sperm are present in the oviduct (▸ Figure 7.2). Sperm travel this distance by swimming, using whiplike contractions of their tails, and are assisted by muscular contractions of the uterus.

Usually only one sperm fertilizes the egg, but many other sperm assist in this process (▸ Figure 7.3) by helping to trigger chemical changes in the egg. During fertilization, a sperm binds to receptors on the surface of the egg (technically, a secondary oocyte) and fuses with the cell's outer membrane. This fusion triggers a series of chemical changes in the membrane and prevents any other sperm from entering the oocyte. As a sperm enters the cytoplasm, its presence reinitiates meiosis in the egg, and the second meiotic division is completed (recall from Chapter 2 that in females, meiosis is halted after the first division). After meiosis, the haploid sperm nucleus fuses with the haploid oocyte nucleus, forming a diploid cell, called a **zygote.**

Zygote The fertilized egg that develops into a new individual.

The zygote is swept along by cilia lining the walls of the oviduct, and travels down the oviduct to the uterus over the next 3 to 4 days. Development, in the form of cell divisions, begins while the zygote is in the oviduct. Once cell division begins, the zygote becomes an embryo. The embryo, consisting of a small number of cells, descends into the uterus and floats unattached in the uterine interior for several days, drawing nutrients from the uterine fluids. Cell division continues during this time and the embryo enters a new stage of development; it is now called a **blastocyst** (▸ Figure 7.4).

Blastocyst The developmental stage at which the embryo implants into the uterine wall.

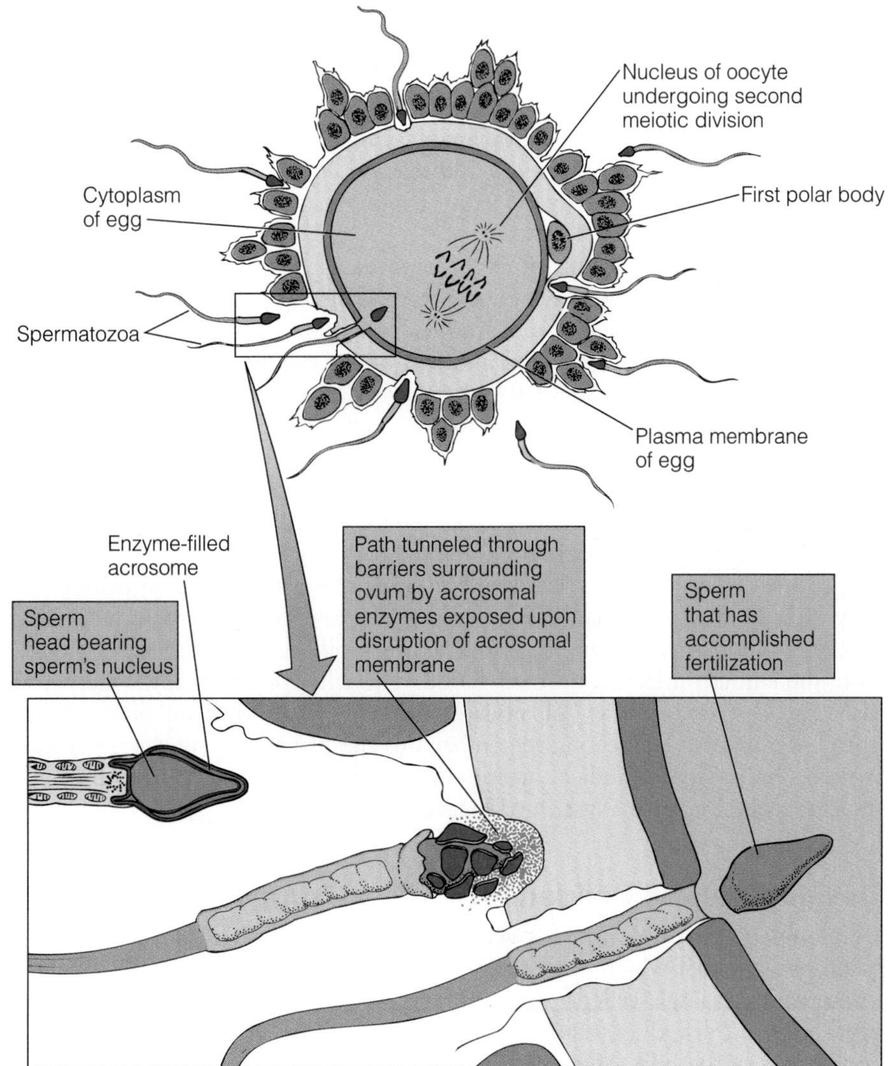

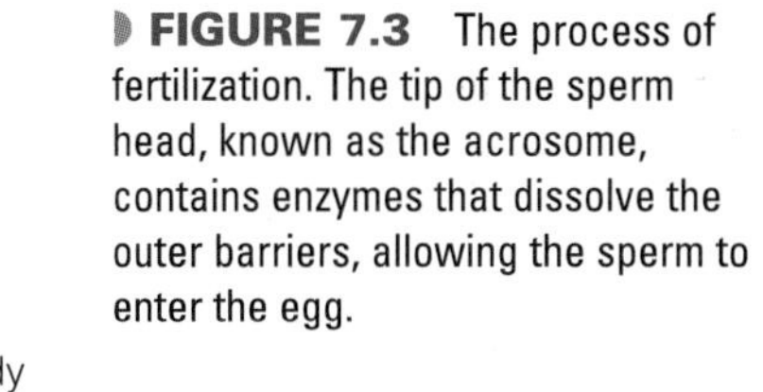

▸ **FIGURE 7.3** The process of fertilization. The tip of the sperm head, known as the acrosome, contains enzymes that dissolve the outer barriers, allowing the sperm to enter the egg.

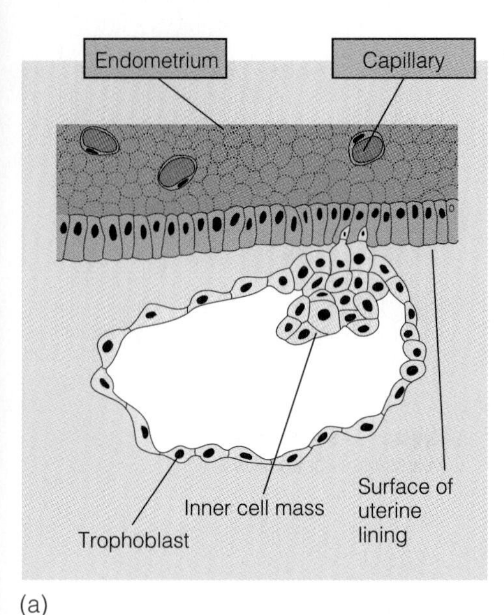

(a)

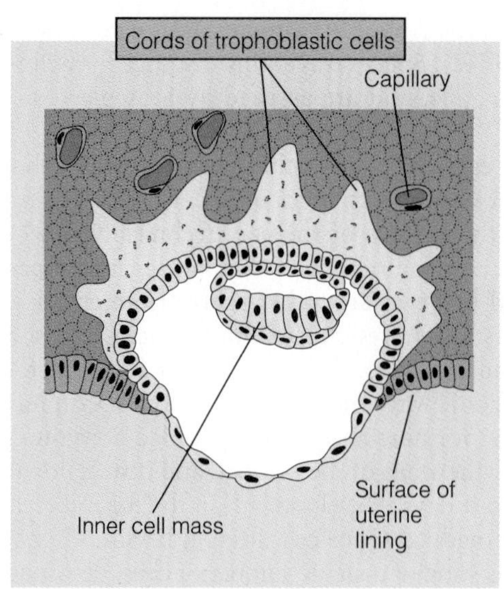

(b)

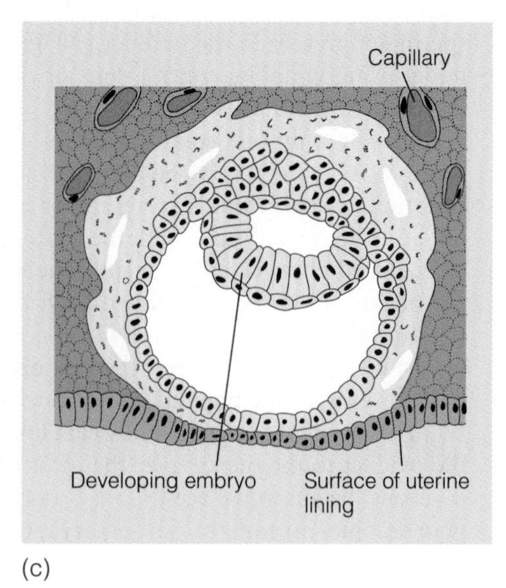

(c)

FIGURE 7.4 The process of implantation. (a) A blastocyst attaches to the endometrial lining of the uterus. (b) As the blastocyst implants, cords of chorionic cells form. (c) When implantation is complete, the blastocyst is buried in the endometrium.

Inner cell mass A cluster of cells in the blastocyst that gives rise to the embryonic body.

Trophoblast The outer layer of cells in the blastocyst that gives rise to the membranes surrounding the embryo.

Chorion A two-layered structure formed from the trophoblast.

A blastocyst, made up of about 100 cells, has several parts: the **inner cell mass,** an internal cavity, and an outer layer of cells (the **trophoblast**). While the embryo is growing to form the blastocyst, the cells lining the uterus (called the endometrium) enlarge and differentiate, preparing for attachment of the embryo to the uterus. During the process of attachment, called implantation, the embryo's trophoblast sticks to the endometrium and releases enzymes that dissolve endometrial cells, allowing finger-like growths from trophoblasts to lock the embryo into place (Figure 7.4).

By 12 days after fertilization, the embryo is firmly embedded, and the trophoblast has formed a two-layered structure, called the **chorion.** Once formed, the chorion makes and releases a hormone called human chorionic gonadotropin (hCG). This hormone prevents breakdown of the uterine lining and stimulates endometrial cells to release hormones that help maintain the pregnancy. Excess hCG is eliminated in the urine. Home pregnancy tests work by detecting the excess hCG levels as early as the first day of a missed menstrual period.

As the chorion grows and expands, it forms a series of fingerlike projections called villi that extend into endometrial cavities filled with maternal blood. Capillaries from the embryo's developing circulatory system extend into the villi. The blood of the embryo and the maternal pools of blood are separated from each other only by a thin layer of cells. Food molecules and oxygen cross easily from the mother's blood into the embryo, and waste molecules and carbon dioxide move from the embryo into the mother's blood. The villi eventually form the placenta, a disc-shaped structure that will nourish the embryo throughout prenatal development. Membranes connecting the embryo to the placenta form the umbilical cord, which contains two umbilical arteries and a single umbilical vein as extensions of the embryo's circulatory system (Figure 7.5).

Development is Divided into Three Trimesters

Development in the period between fertilization and birth is divided into three trimesters, each of which lasts about 12 weeks. During the 36 to 38 weeks of development, the single-celled zygote undergoes 40 to 44 rounds of mitosis, resulting in trillions of cells, all of which become organized into the highly specialized tissues and organs of the fully developed fetus.

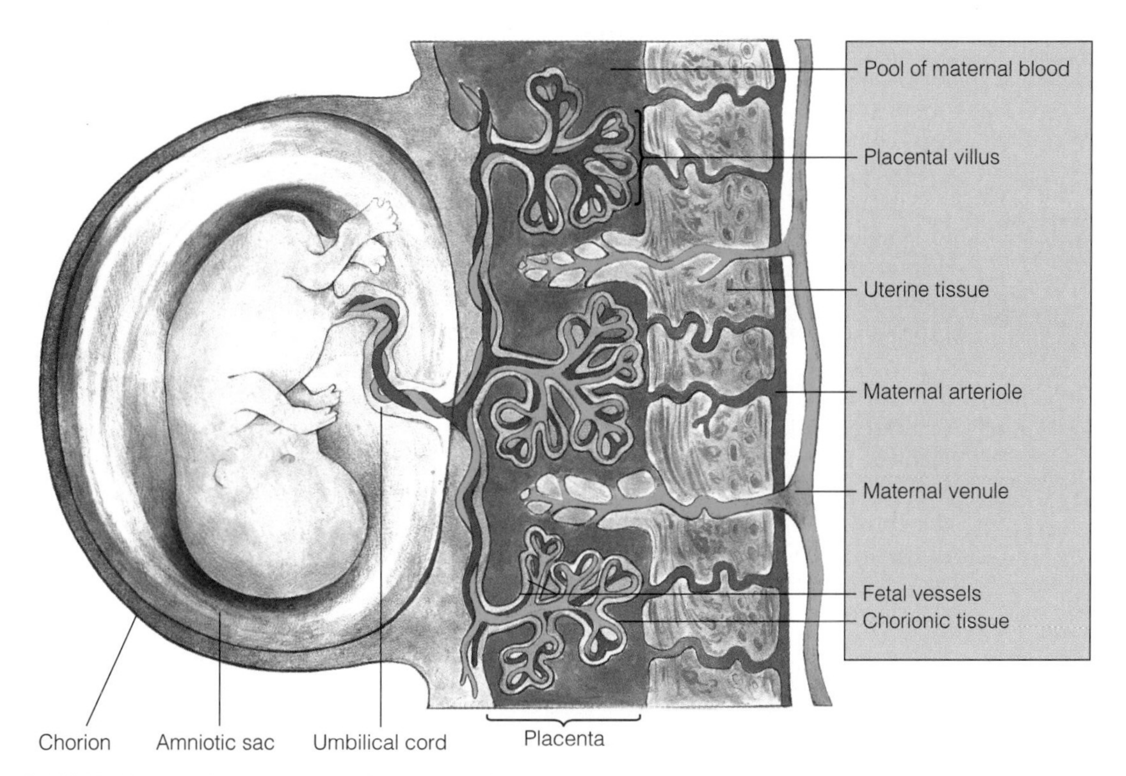

FIGURE 7.5 Maternal and embryonic structures interact to form the placenta.

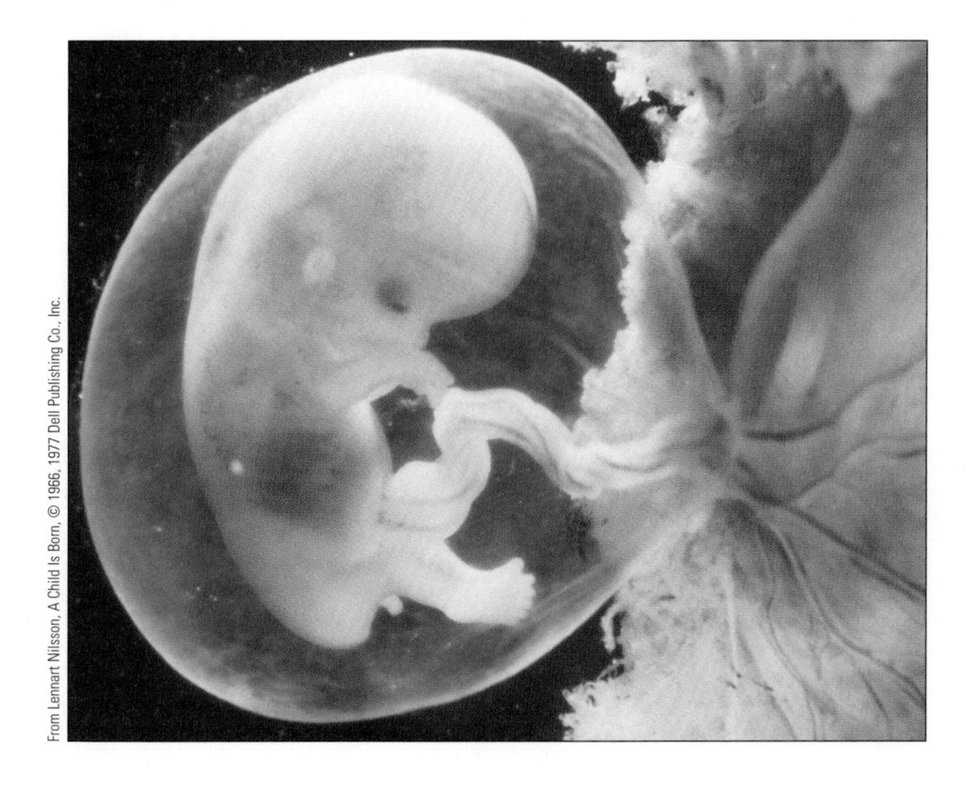

FIGURE 7.6 A human embryo near the end of the first trimester of development.

Organ Formation Occurs in the First Trimester The first 12 weeks of development is a period of radical change in the size, shape, and complexity of the embryo (Figure 7.6). In the week after implantation, basic tissue layers are formed, and by the end of the third week, organ systems are beginning to take shape. By 4 weeks, the embryo is about 5-mm long (about one-fifth of an inch), and much of the body is composed of paired segments.

During the second month, the embryo grows dramatically to a length of about 3 cm (about 1½ in.) and undergoes a 500-fold increase in size. Most of the major organ systems, including the heart, are formed. Limb buds develop into arms and legs, complete with fingers and toes. The head is very large in relation to the rest of the body because of the rapid development of the nervous system.

By about 7 weeks, the embryo is called a fetus. Although chromosomal sex (XX in females and XY in males) is determined at the time of fertilization, the fetus is neither male nor female at the beginning of the third month. During the third month, specific gene sets are activated, and sexual development is initiated. This process is discussed in detail later in this chapter.

The Second Trimester Is a Period of Organ Maturation In the second trimester, major changes include an increase in size and the further development of organ systems (◗ Figure 7.7). Bony parts of the skeleton begin to form, and the heartbeat can be heard with a stethoscope. Fetal movements begin in the third month, and by the fourth month, the mother can feel movements of the fetus's arms and legs. At the end of the second trimester, the fetus weighs

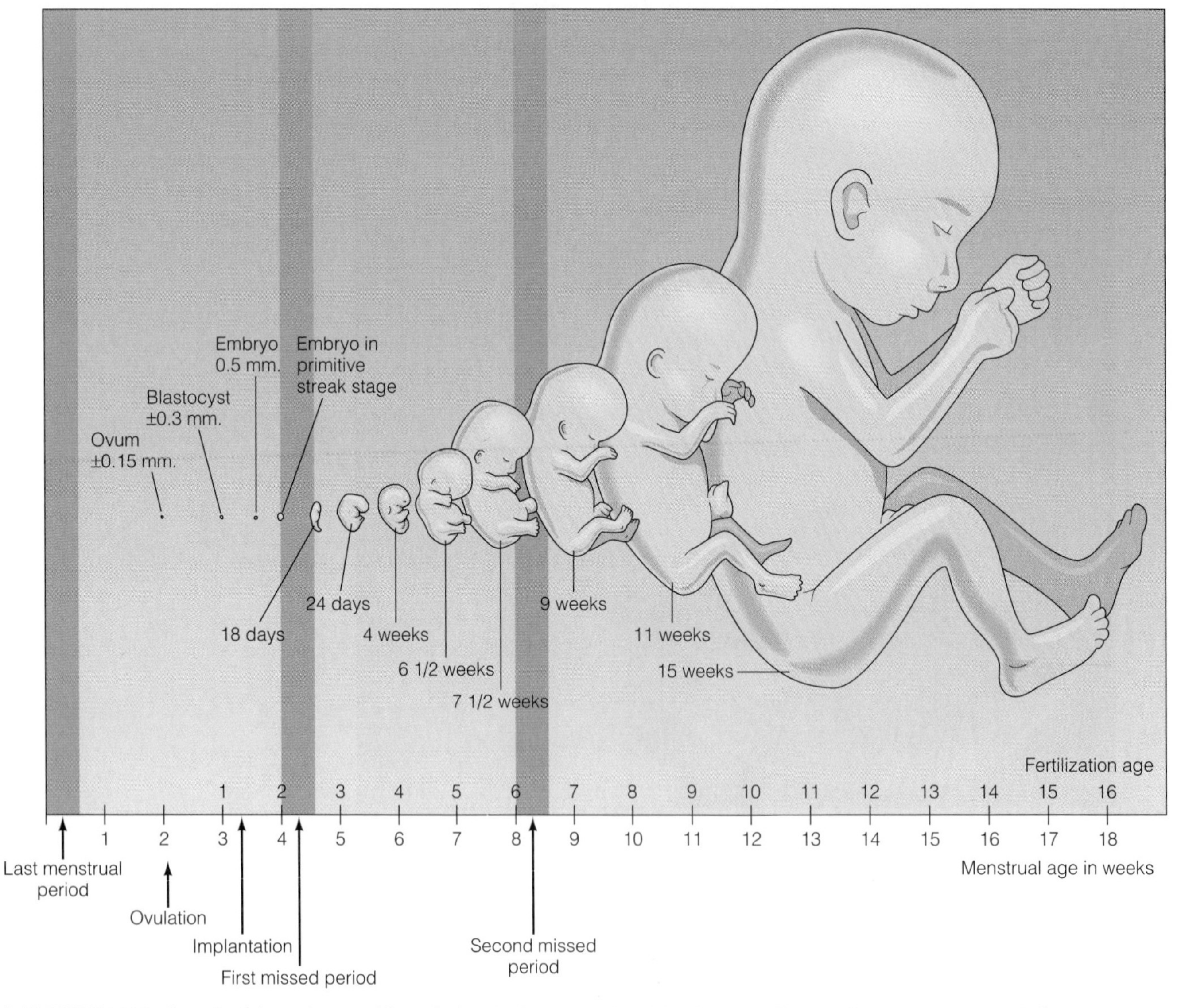

◗ **FIGURE 7.7** Growth of the embryo and fetus during the first 16 weeks of development. The drawings represent actual sizes.

about 700 g (27 oz.) and is 30 to 40 cm (about 13 in.) long. It has a well-formed face, its eyes can open, and it has fingernails and toenails.

Rapid Growth Takes Place in the Third Trimester The fetus grows rapidly in the third trimester, and the circulatory system and the respiratory system mature to prepare for air breathing. During this period of rapid growth, maternal nutrition is important because a large fraction of the protein the mother eats is used for growth and development of the fetal brain and nervous system. Similarly, much of the calcium in the mother's diet is used to develop the fetal skeletal system.

The fetus doubles in size during the last 2 months, and chances for survival outside the uterus increase rapidly during this time. In the last month, antibodies pass from the mother to the fetus, conferring temporary immunity on the fetus. In the first months after birth, the baby's immune system matures, and as it begins to make its own antibodies, the maternal antibodies disappear. At the end of the third trimester, the fetus is about 50 cm (19 in.) long and weighs from 2.5 to 4.8 kg (5.5 to 10.5 lb).

Birth Is Hormonally Induced

Birth is a hormonally induced process. During the last trimester, the cervix softens and the fetus shifts downward, usually with its head pressed against the cervix. Mild uterine contractions start during the third trimester, but at the start of the birth process, they become more frequent and intense. Release of the hormone oxytocin from the pituitary gland helps stimulate uterine contractions. During labor, the cervical opening dilates in stages to allow passage of the fetus, and uterine contractions expel the fetus. The head usually emerges first. If any other body part enters the birth canal first, the result is a breech birth. A short time after delivery, a second round of uterine contractions begins delivery of the placenta. These contractions separate the placenta from the lining of the uterus, and the placenta is expelled through the vagina.

TERATOGENS ARE A RISK TO THE DEVELOPING FETUS

Teratogens are physical and chemical agents that produce embryonic and/or fetal abnormalities. Teratogens produce nongenetic birth defects, not heritable changes. In 1960 only four or five agents were known to be teratogens. The discovery that a tranquilizer, thalidomide, caused limb defects in unborn children helped focus attention on this field. Today, 30 to 40 teratogenic agents are known, and another 10 to 12 chemicals are strongly suspected of being teratogens. Some of these are listed in Table 7.1.

Teratogen Any physical or chemical agent that brings about an increase in congenital malformations.

Radiation, Infectious Agents, and Chemicals Can Be Teratogens

Radiation, especially medical x-rays, can act as a teratogen. Women of childbearing age should not have abdominal x-rays unless they know they are not pregnant. Pregnant women should avoid all unnecessary x-rays, and all females should have abdominal shielding for most x-ray procedures.

Some viruses are teratogens, including the measles virus and the virus that causes genital herpes. These viruses can cause severe brain damage and mental retardation in a developing fetus. The damaging effects of the herpes virus occur only when the mother becomes infected with herpes during pregnancy. There is no damage to the fetus when the mother has been infected before pregnancy and has a recurring attack during pregnancy. Herpes virus I (associated with cold sores) does not appear to be teratogenic. Some infectious organisms, such as *Toxoplasma gondii*, which is transmitted to humans by cats, are teratogenic and can result in a stillborn child or a child with mental retardation or other disorders.

Table 7.1 Human Teratogens

Known	Possible
Radiation	Cigarette smoking
Fallout	High levels of vitamin A
x-rays	Lithium
	Zinc deficiency
Infectious Agents	
Cytomegalovirus	
Herpes virus II	
Rubella virus	
Toxoplasma gondii (spread in cat feces)	
Maternal Metabolic Problems	
Phenylketonuria (PKU)	
Diabetes	
Virilizing tumors	
Drugs and Chemicals	
Alcohol	
Aminopterins	
Chlorobiphenyls	
Coumarin anticoagulants	
Diethylstilbestrol	
Tetracyclines	
Thalidomide	

Many chemicals, including medications, such as the antibiotic tetracycline, are teratogens. Case 1 at the end of this chapter discusses drugs with teratogenic effects.

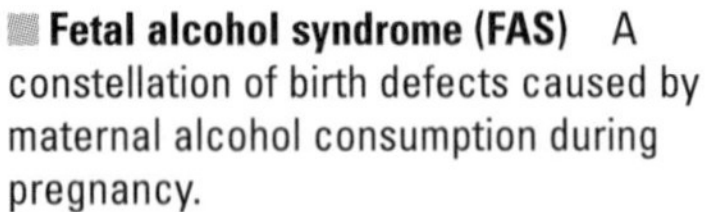

Fetal alcohol syndrome (FAS) A constellation of birth defects caused by maternal alcohol consumption during pregnancy.

Fetal Alcohol Syndrome: A Preventable Tragedy

Exposure of the fetus to alcohol is one of the most serious and the most widespread teratogenic problems and is the leading preventable cause of birth defects. Alcohol consumption during pregnancy can result in spontaneous abortion, growth retardation, facial abnormalities (Figure 7.8), and mental retardation. This collection of defects is known as **fetal alcohol syndrome (FAS).** In milder forms, the condition is known as fetal alcohol effects. The incidence of FAS is about 1.9 affected infants per 1,000 births, and the incidence for fetal alcohol effects is about 3.5 affected infants per 1,000 births.

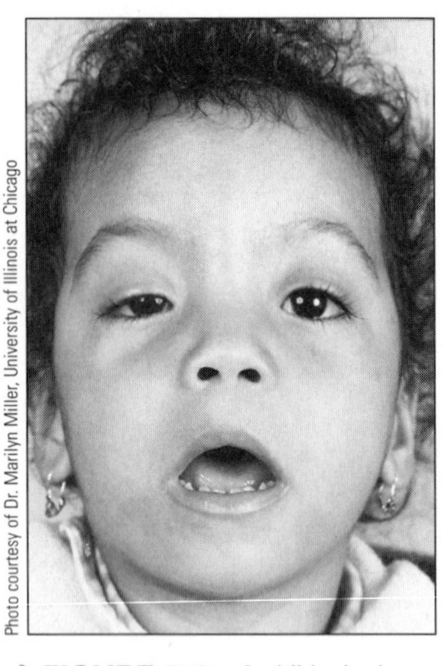

Photo courtesy of Dr. Marilyn Miller, University of Illinois at Chicago

FIGURE 7.8 A child who has fetal alcohol syndrome has misshapen eyes, flat nose, and face characteristic of this condition.

The teratogenic effects of alcohol can occur at any time during pregnancy, but weeks 8 to 12 are particularly sensitive periods. Even in the third trimester, alcohol can seriously impair fetal growth. Most studies show that the consumption of one or more drinks per day is associated with an increased risk of having a child with growth retardation. However, because fetal damage is related to blood alcohol levels, thinking about averages can be misleading. Having six drinks in one day and no drinks the rest of the week may pose a greater risk to the fetus than having one drink each day of the week. To emphasize the risks of alcohol consumption during pregnancy, the Surgeon General of the United States requires that all alcohol containers carry this warning: "Drinking during pregnancy may cause mental retardation and other birth defects. Avoid alcohol during pregnancy." The American Academy of Pediatrics, in a policy statement, states that: "because there is no known safe amount of alcohol consumption during pregnancy, the Academy recommends abstinence from alcohol for women who are pregnant or who are planning a pregnancy."

The economic cost of FAS is enormous. The lifetime cost of caring for a child with FAS exceeds $1.4 million, and annual estimates for the overall costs to society range into billions of dollars. The mental retardation associated with FAS is estimated to account for 11% of the cost of caring for treating all institutionalized, mentally retarded individuals. The emotional costs and social effects are difficult to estimate. Insight into the struggles of a family with an FAS child is recorded by Michael Dorris in his book, *The Broken Cord*.

Although the actions of alcohol as a teratogen are now well known, work is needed to resolve the degree of risk involved with other chemicals and substances that are suspected teratogens and to identify new teratogens among the thousands of chemicals currently used. More importantly, research is needed to investigate the genetic basis for susceptibility to teratogenic agents and to develop tests to identify those who are susceptible to teratogens.

CONTROLLING REPRODUCTION: CONTRACEPTION AND ASSISTED REPRODUCTIVE TECHNOLOGY (ART)

Over the past 50 years, steady advances in genetics, physiology, and molecular biology have allowed us to control many aspects of human reproduction. The techniques developed can be used to enhance the chances of conception (**assisted reproductive technology,** or **ART**) or to reduce and even eliminate the chances of conception (contraception). Reproductive technologies can also correct defective functions of the reproductive system and manipulate the physiology of reproduction.

Assisted reproductive technologies (ART) The collection of techniques used to help infertile couples have children.

Contraception Uncouples Sex from Pregnancy

To uncouple sexual intercourse from fertilization and pregnancy, one of the three stages in reproduction (release and transport of gametes, fertilization, and implantation) must be blocked. Aside from complete abstinence from sexual intercourse (Figure 7.9), nothing is completely successful in preventing pregnancy or sexually transmitted diseases (STDs). Methods that physically prevent the release and transport of gametes (such as **vasectomy** and **tubal ligation**) are the most effective in preventing pregnancy in sexually active individuals. In tubal ligation, the oviduct is cut, and the ends are tied off to prevent sperm from reaching eggs released from the ovaries. In vasectomy, the vas deferens is withdrawn through a small incision in the scrotum and cut. The cut ends are sealed to prevent transport and release of sperm.

Vasectomy A contraceptive procedure in men in which the vas deferens is cut and sealed to prevent the transport of sperm.

Tubal ligation A contraceptive procedure in women in which the oviducts are cut, preventing eggs from reaching the uterus.

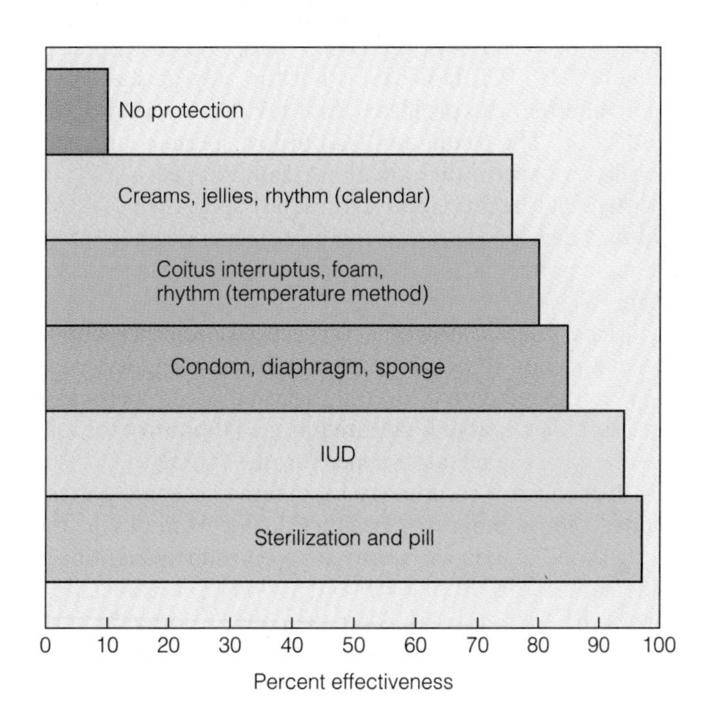

FIGURE 7.9 Effectiveness of various methods of birth control. The percent effectiveness is a measure of how many women in a group of 100 will not become pregnant in a year when using a given method of birth control.

Manipulating the hormonal cycle that controls the maturation and release of the egg is also an effective method of controlling pregnancy. The most common birth control pills contain a combination of hormones that prevent ovulation (the release of an egg from the ovary). These pills must be taken daily for a certain number of days each month. Another method of manipulating the hormonal cycle uses time-release capsules (sold as Norplant) that are implanted under the skin. These implants release hormones slowly and offer long-term suppression of ovulation.

To prevent fertilization, both physical and chemical barriers can be used. Condoms are latex or gut sheaths and pouches worn over the penis or inserted into the vagina before intercourse. These barriers prevent sperm from entering the female reproductive tract. Only latex condoms prevent STDs, including AIDS. Diaphragms are caps that fit over the opening of the cervix and are inserted before intercourse. They are designed to prevent sperm from entering the cervix and swimming up the uterus to the oviducts. Chemical barriers, such as spermicidal jelly or foam placed in the vagina just before intercourse, kill sperm on contact. Contraceptive sponges, filled with spermicides, use a combination of physical and chemical barriers. Condoms treated with a chemical spermicide are more successful at preventing pregnancy than either condoms or chemical barriers alone.

RU-486, a drug developed in Europe, is now available in the United States as a contraceptive. This drug, chemically related to reproductive hormones, interferes with events following fertilization and may inhibit implantation.

Reproductive Technologies Expand Childbearing Options

About one in six couples want to have children but are infertile. Physical and physiological conditions can prevent or interfere with gamete production, fertilization, or implantation and can affect either males or females. Several technologies to reduce or overcome these problems have been developed in the last two decades. In females, blocked oviducts, often the result of untreated STDs, are the leading cause of infertility. In males, low sperm count, low motility, and blocked ducts are leading causes of infertility. Overall, about 40% of all cases of infertility are related to problems in the male reproductive tract, and about 40% are associated with the female tract. In the remaining 20% of cases, the cause of infertility is unknown.

■ ***In vitro* fertilization (IVF)** A procedure in which gametes are fertilized in a dish in the laboratory, and the resulting zygote is implanted in the uterus for development.

The development of ***in vitro* fertilization (IVF)** has greatly increased the chances for successful pregnancies and offers hope for many infertile couples. Many variations of this technology are now widely available; some of them are summarized in ▶ Figure 7.10. IVF, however, is not the only option. For example, if the male partner is infertile, the problem often involves spermatogenesis, the process of sperm production. The results can lead to low sperm count, sperm that move slowly (low motility), or sperm that are abnormally shaped.

Artificial insemination with donor sperm was one of the first methods of ART and was developed to overcome problems of male infertility. In its simplest form, the male partner is infertile, and the female receives sperm collected from a donor (Figure 7.10).

■ **Intracytoplasmic sperm injection (ICSI)** To overcome defects in sperm motility or fertilization, an egg is fertilized by microinjection of individual sperm.

Many couples however, now choose another option, called **intracytoplasmic sperm injection (ICSI)**. In this procedure, an egg is recovered from the female partner and injected with a carefully selected single sperm from the male partner (▶ Figure 7.11). The zygote is cultured in the laboratory for a few days and transferred to the uterus of the female partner for development.

Reproductive technology has altered accepted patterns of reproduction and redefined the meaning of parenthood. For example, in the United States, surrogate motherhood and its variations are a reproductive option. Surrogate parenthood can take many forms. In one version, a woman is artificially inseminated by sperm and carries the child to term. After the child is born, she surrenders the child to the father and his mate. In this case, the surrogate is both the genetic and gestational mother of the child. In another version, a couple provides both the egg and the

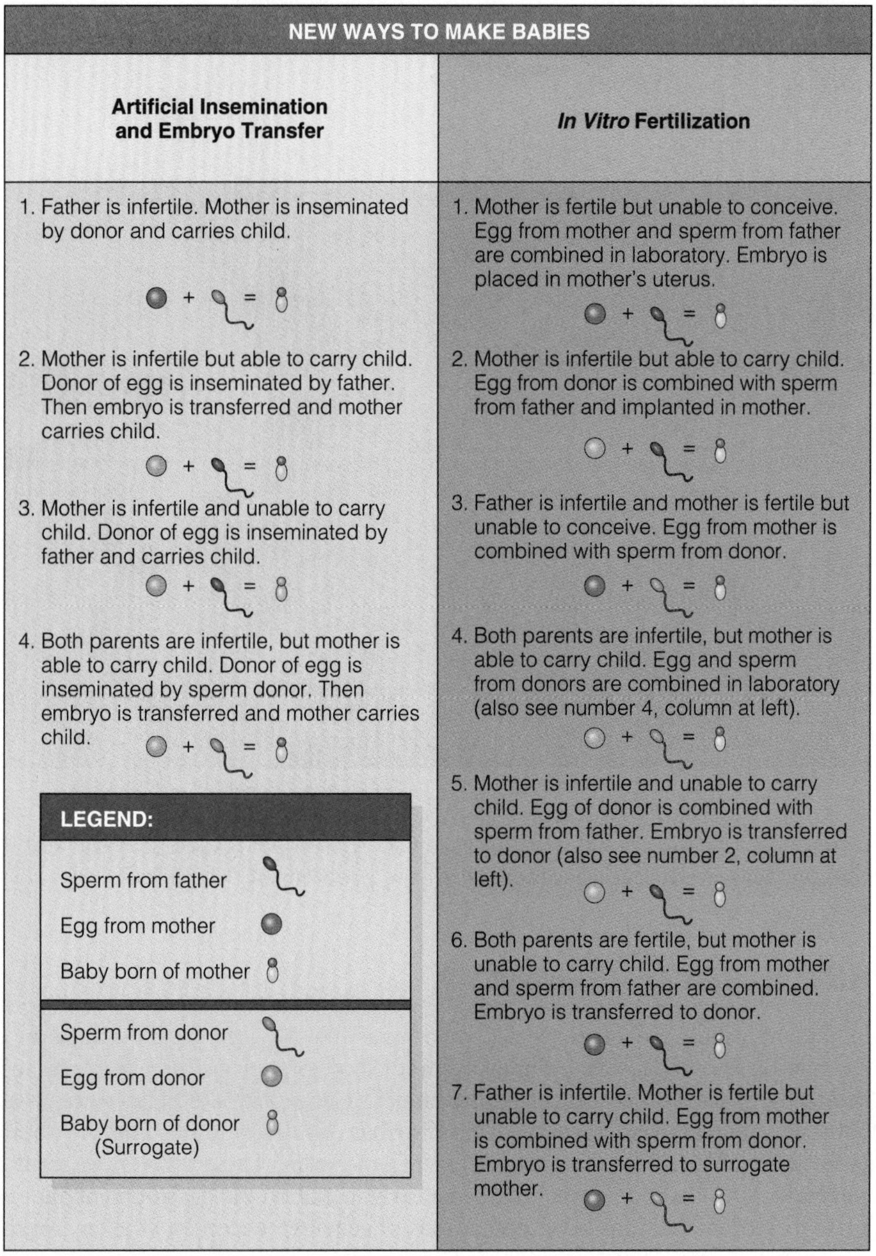

NEW WAYS TO MAKE BABIES

Artificial Insemination and Embryo Transfer	*In Vitro* Fertilization
1. Father is infertile. Mother is inseminated by donor and carries child.	1. Mother is fertile but unable to conceive. Egg from mother and sperm from father are combined in laboratory. Embryo is placed in mother's uterus.
2. Mother is infertile but able to carry child. Donor of egg is inseminated by father. Then embryo is transferred and mother carries child.	2. Mother is infertile but able to carry child. Egg from donor is combined with sperm from father and implanted in mother.
3. Mother is infertile and unable to carry child. Donor of egg is inseminated by father and carries child.	3. Father is infertile and mother is fertile but unable to conceive. Egg from mother is combined with sperm from donor.
4. Both parents are infertile, but mother is able to carry child. Donor of egg is inseminated by sperm donor. Then embryo is transferred and mother carries child.	4. Both parents are infertile, but mother is able to carry child. Egg and sperm from donors are combined in laboratory (also see number 4, column at left).
	5. Mother is infertile and unable to carry child. Egg of donor is combined with sperm from father. Embryo is transferred to donor (also see number 2, column at left).
	6. Both parents are fertile, but mother is unable to carry child. Egg from mother and sperm from father are combined. Embryo is transferred to donor.
	7. Father is infertile. Mother is fertile but unable to carry child. Egg from mother is combined with sperm from donor. Embryo is transferred to surrogate mother.

LEGEND:

Sperm from father
Egg from mother
Baby born of mother

Sperm from donor
Egg from donor
Baby born of donor (Surrogate)

FIGURE 7.10 Some of the ways gametes can be combined to produce babies.

sperm for IVF. A surrogate is implanted with the developing embryo and serves as the gestational mother but is genetically unrelated to the child she bears.

Following the discovery that it is the age of the egg, not the reproductive system, that is responsible for infertility as women age, women are now becoming mothers in their late fifties and early sixties. After hormonal treatment, they are implanted with zygotes produced by IVF of eggs donated by younger women. In addition, fertilized eggs can now be collected and frozen for later use, separating fertilization from development. This allows younger women to collect eggs while they are young, when the risks for chromosome abnormalities in the offspring are low. The eggs can be fertilized by IVF and the resulting embryos, frozen in liquid nitrogen, can be stored for years. The embryos can be thawed and implanted over a period of years, including menopause, allowing women to extend their childbearing years.

These and other unconventional means of generating a pregnancy have developed more rapidly than the social conventions and laws governing their use. In the process, controversy about the moral, ethical, and legal grounds for

▶ **FIGURE 7.11** One method of assisted reproduction is intracytoplasmic sperm injection (ISCI), in which single sperm (in needle) is injected into the egg.

using these techniques has arisen but is not yet resolved. Case 2 at the end of this chapter deals with a couple faced with the problem of infertility.

HOW IS SEX DETERMINED?

In humans, as in many other species, we can see obvious differences between the sexes, a condition known as sexual dimorphism. In some organisms the differences are limited to the gonads; in others, including humans, secondary sex characteristics such as body size, muscle mass, patterns of fat distribution, and amounts and distribution of body hair emphasize the differences between the sexes. These differences are the outcome of a long chain of events that begin early in embryonic development and involve a network of interactions between patterns of gene expression and the environment.

The Environment Can Help Determine Sex

What determines maleness and femaleness? The answer is a complex interaction between genes and the environment. In some organisms, environmental factors play a major role (▶ Figure 7.12). For example, in some reptiles such as turtles or crocodiles, the temperature of the nest in which the eggs develop determines sex. Higher temperatures produce females; lower temperatures produce males. In other reptiles, the opposite is true: Higher incubation temperatures produce males, and lower temperatures result in females.

Chromosomes Can Help Determine Sex

In humans, on the other hand, whether someone is male or female is determined in stages beginning at fertilization, when the sex chromosomes carried by the gametes combine in the zygote. As discussed in Chapter 2, females have two X chromosomes (XX) and males have an X and a Y chromosome.

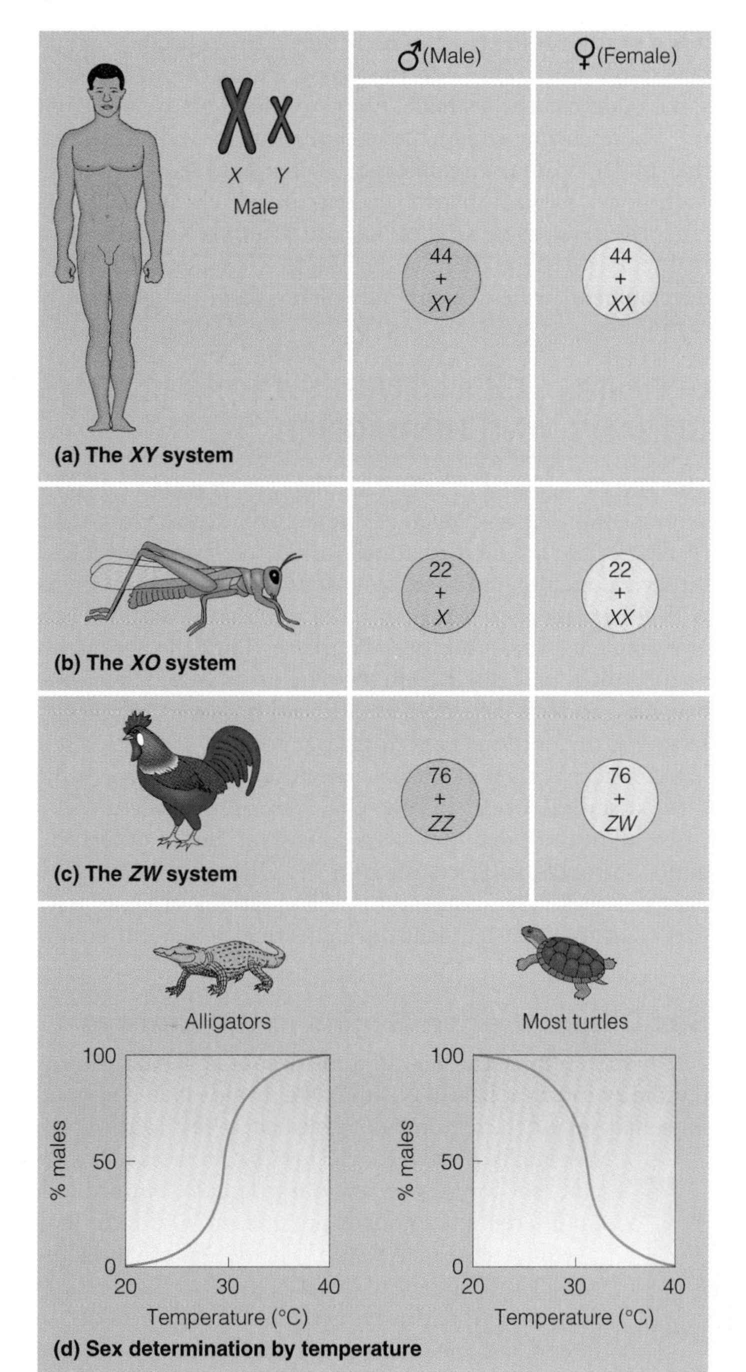

FIGURE 7.12 (a–c) Animals have several mechanisms of sex determination that involve chromosomes. (d) In some reptiles, the temperature at which the egg is incubated determines the sex of the offspring.

Although saying that females are XX and that males are XY seems straightforward, it does not provide all the answers to the question of what determines maleness and femaleness. Is a male a male because he has a Y chromosome or because he does not have two X chromosomes? Can someone be XY and develop as a female? Can someone be XX and develop as a male? These questions are still not completely resolved, but the discovery of answers began about 40 years ago when some people who carry an abnormal number of sex chromosomes were identified. In rare cases, individuals with only 45 chromosomes (45,X) are born. They have only one X chromosome and are female. At about the same time, males who carry two X chromosomes along with a Y chromosome were discovered (47,XXY). From the study of these and other individuals with abnormal numbers of sex chromosomes, it is clear that some females have only one X chromosome and that some males can have more than one X chromosome. Furthermore, anyone who has a Y chromosome is almost always male, no matter how many X chromosomes he may have.

These chromosome studies, and similar studies in mice, led to the conclusion that under normal circumstances being male is associated with the presence of a Y chromosome and that the absence of a Y chromosome results in a female phenotype. However, these studies indicate that two X chromosomes are required for normal female development, and a single X chromosome is required for normal male development (review aneuploidy of sex chromosomes in Chapter 6).

DEFINING SEX IN STAGES: CHROMOSOMES, GONADS, AND HORMONES

The XX-XY method of sex determination provides a genetic framework for the developmental events that guide the embryo toward the male or female phenotypes (◗ Figure 7.13). The formation of male or female reproductive structures depends on several factors, including gene action, interactions within the embryo, interaction with other embryos that may be in the uterus, and interactions with the maternal environment. As a result of these interactions, the chromosomal sex (XX or XY) of an individual may differ from the phenotypic sex. These differences arise during embryonic and fetal development and can produce a phenotype opposite to the chromosomal sex, intermediate to the phenotypes of the two sexes, or a phenotype that has characteristics and genitalia of both sexes. The sex of an individual can be defined at several levels: chromosomal sex, gonadal sex, and phenotypic sex. In most cases, all of these definitions are consistent, but in others they are not (see Concepts and Controversies: Sex Testing in the Olympics—Biology and a Bad Idea). To understand these variations and the interactions of genes with the environment, let's first consider what happens during normal sexual differentiation.

SRY A gene, called the sex-determining region of the Y, located near the end of the short arm of the Y chromosome, plays a major role in causing the undifferentiated gonad to develop into a testis.

Sex Differentiation Begins in the Embryo

The first step in sex differentiation occurs at fertilization with the formation of a diploid zygote that has an XX or XY chromosomal constitution. Although the chromosomal sex of the zygote is established at fertilization, the embryo that develops is sexually neutral for the first 7 or 8 weeks. The external genitalia of early embryos are neither male nor female, but are indifferent. Internally, two undifferentiated gonads are present, and both male and female reproductive duct systems develop. The two internal duct systems are the Wolffian (male) and Müllerian (female) ducts (◗ Figure 7.14). At about 7 weeks, developmental pathways activate different sets of genes and cause the undifferentiated gonads to develop as testes or ovaries, establishing the gonadal sex of the embryo. This process takes place over the next 4 to 6 weeks. Although it is convenient to think of only two pathways, one leading to males and the other to females, there are many alternate pathways that produce intermediate outcomes in gonadal sex and in sexual phenotypes, some of which we consider in the following paragraphs.

If a Y chromosome is present in the embryo, the action of genes on the Y chromosome causes the indifferent gonad to begin development as a testis. Products from these activated genes stimulate the growth and differentiation of a small collection of cells within the indifferent gonad, forming a testis. A gene called the sex-determining region of the Y (*SRY*) (MIM/OMIM 480000), located on the short arm of the Y chromosome, plays a major role in starting the cascade of gene action leading to testis development. Other genes on the Y chromosome and on autosomes also play important roles at this time.

◗ **FIGURE 7.13** A cascade of gene action that begins in the seventh week of gestation results in the development of the male and female sexual phenotype.

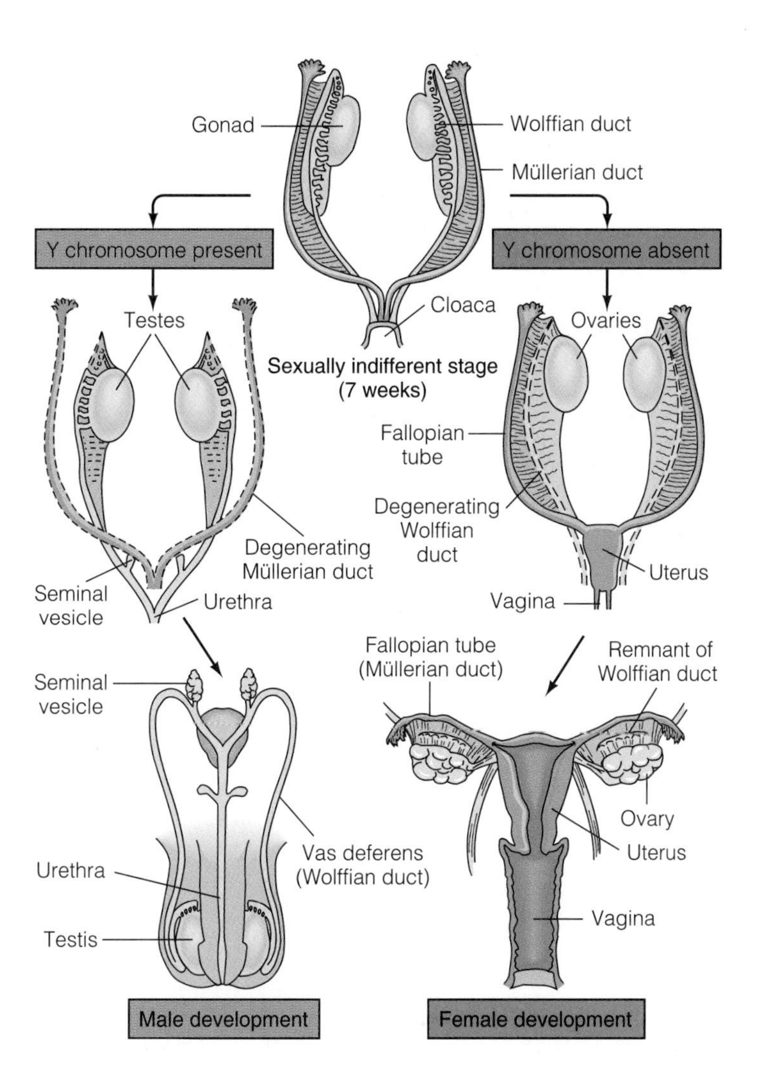

FIGURE 7.14 Two duct systems (Wolffian and Müllerian) are present in the early embryo. They enter different developmental pathways in the presence and absence of a Y chromosome.

Once testis development is initiated, cells in the testis secrete two hormones, **testosterone** and **Müllerian inhibiting hormone (MIH).** Along with patterns of gene expression, these hormones control further sexual differentiation. Testosterone stimulates the Wolffian ducts to form the male internal duct system that will carry sperm. These ducts include the epididymis, seminal vesicles, and vas deferens. The MIH secreted by the developing testis stops further development of female duct structures and causes the Müllerian ducts to degenerate (Figure 7.14).

Testosterone A steroid hormone produced by the testis; the male sex hormone.

Müllerian inhibiting hormone (MIH) A hormone produced by the developing testis that causes the breakdown of the Müllerian ducts in the embryo.

In embryos with two X chromosomes, the absence of the Y chromosome and the presence of the second X chromosome cause the embryonic gonad to develop as an ovary. Development begins as cells along the outer edge of the gonad divide and push into the interior, forming an ovary. Because the ovary does not produce testosterone, the Wolffian duct system fails to develop and degenerates (Figure 7.14). In the absence of MIH, the Müllerian duct system develops to form the fallopian tubes, uterus, and parts of the vagina.

Hormones Help Shape Male and Female Phenotypes

After gonadal sex has been established, the third phase of sexual differentiation, the appearance of sexual phenotype, begins (Figure 7.15). In males, testosterone is metabolized and converted into another hormone, dihydrotestosterone (DHT), which helps directs formation of the external genitalia. Under the influence of DHT and testosterone, the genital folds and genital tubercle develop into the

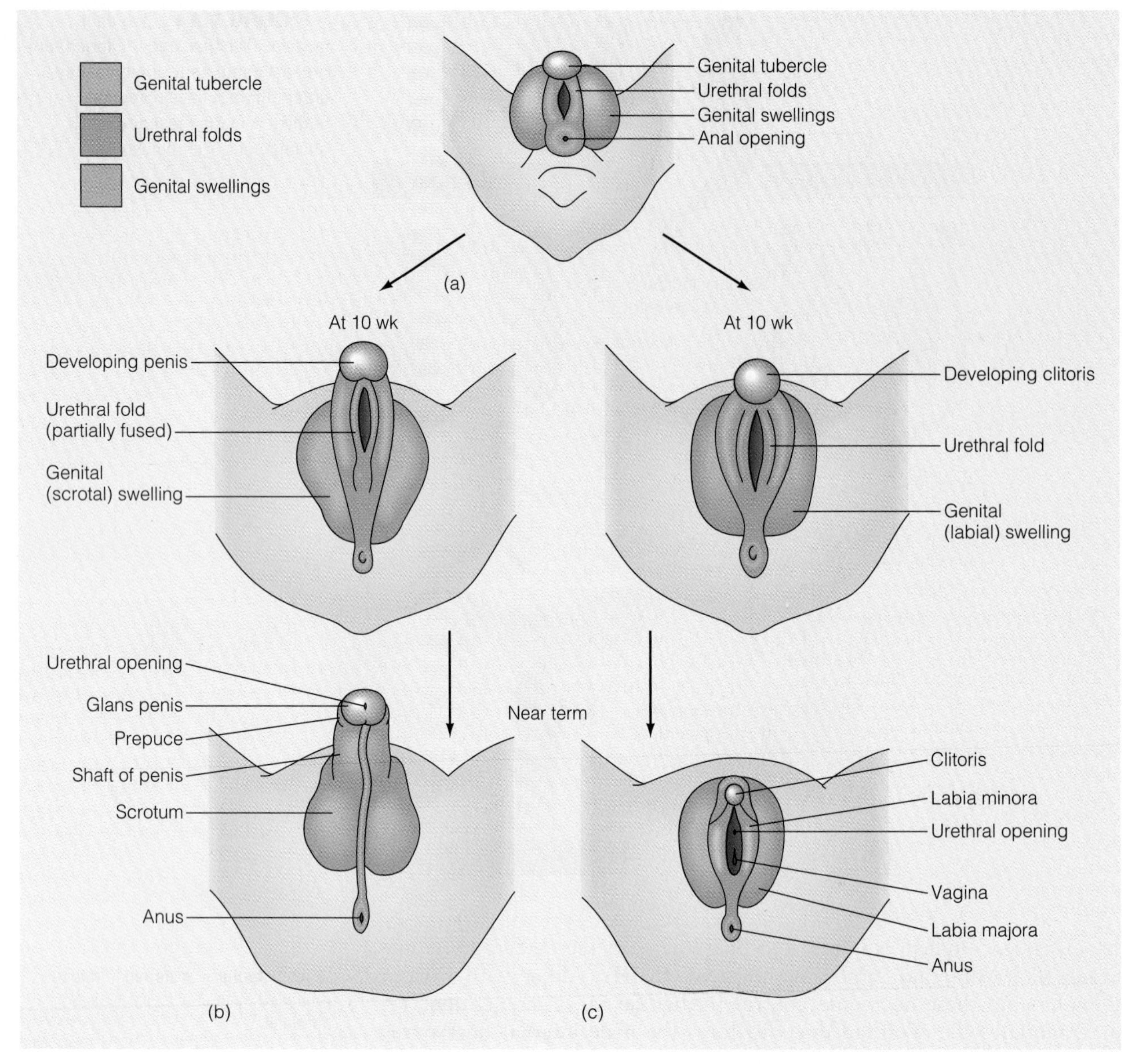

FIGURE 7.15 Steps in the development of phenotypic sex from the undifferentiated stage (a) to the male (b) or female (c) phenotype. The male pathway of development takes place in response to the presence of testosterone and dihydrotestosterone (DHT). Female development takes place in the absence of these hormones.

penis, and the surrounding labioscrotal swelling forms the scrotum. In females, no DHT is present, and so the genital tubercle develops into the clitoris, the genital folds form the labia minora, and the labioscrotal swellings form the labia majora (Figure 7.15).

In terms of gene action, it is important to remember that the development of gonadal sex and the sexual phenotype results from different developmental pathways (Figure 7.16). In males, this pathway involves the action of several genes on the Y chromosome, the presence of at least one X chromosome, and expression of several autosomal genes. In females, this pathway involves the presence of two X chromosomes, the absence of Y chromosome genes, and presumably other autosomal genes. These distinctions indicate that there may be important differences in the way genes in these pathways are activated, and they may provide clues in the search for genes that regulate these pathways.

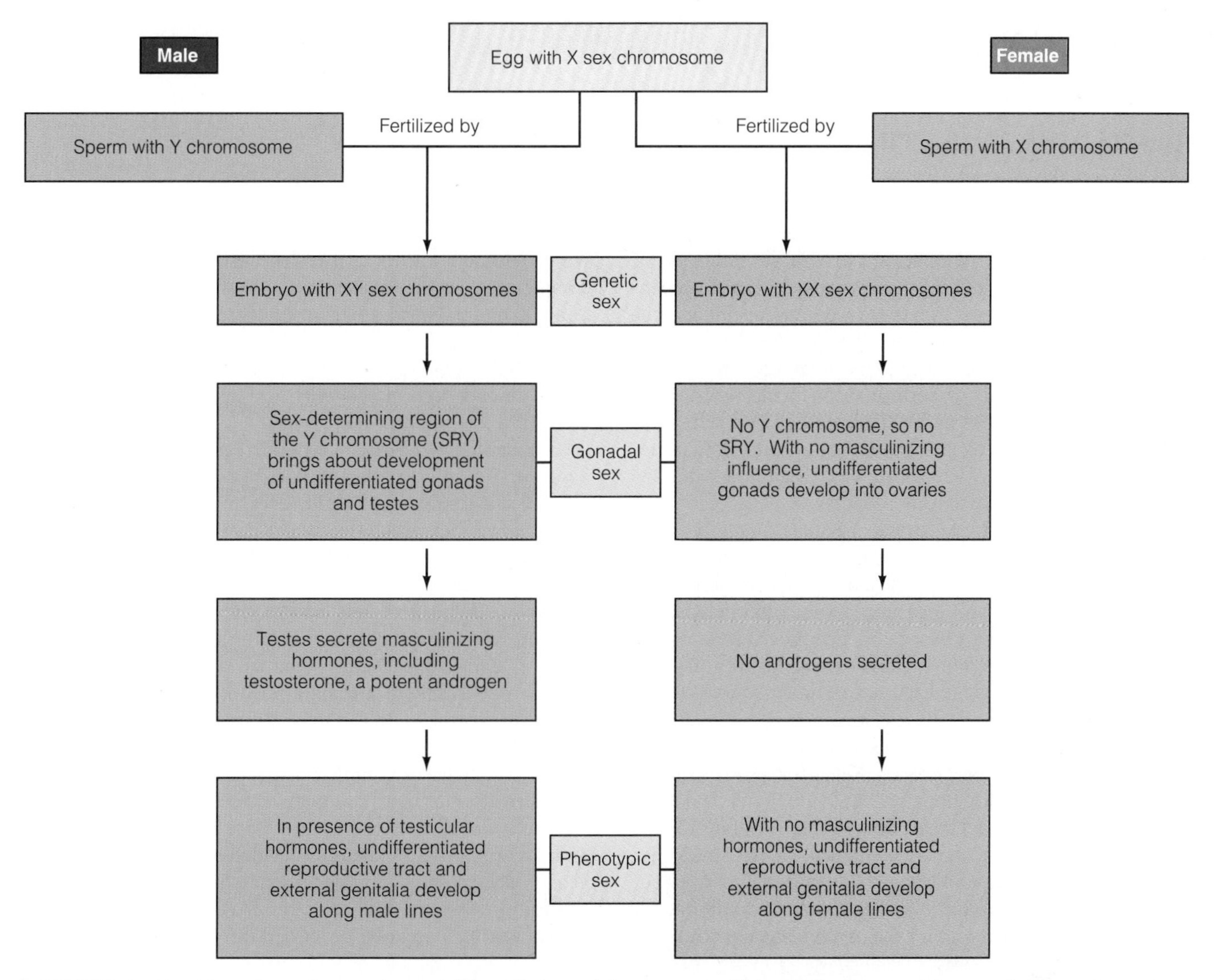

FIGURE 7.16 The major pathways of sexual differentiation and the stages at which genetic sex, gonadal sex, and phenotypic sex are established.

MUTATIONS CAN UNCOUPLE CHROMOSOMAL SEX FROM PHENOTYPIC SEX

Developmental pathways that begin with the indifferent gonad often result in a gonadal and/or sexual phenotype that differs from the chromosomal sex of XX for females and XY for males. These outcomes can result from several causes: chromosomal events that exchange segments of the X and Y chromosomes, mutations that affect the ability of cells to respond to the products of Y chromosome genes, or action of autosomal genes that control events on the X and/or Y chromosome. In addition, interactions between the embryo and maternal hormones in the uterus and the presence of other embryos in the uterus can affect the outcome of both the gonadal sex and the sexual phenotype.

Part of our understanding of sexual development is derived from the study of variations in this process, including single-gene mutations that produce altered sexual phenotypes. We briefly consider three situations in which there is a lack of agreement among chromosomal sex, gonadal sex, and sexual phenotype.

First, let's discuss hermaphrodites. True hermaphrodites are organisms that possess both ovaries and testes and their associated duct systems. In some species, such

Concepts and Controversies

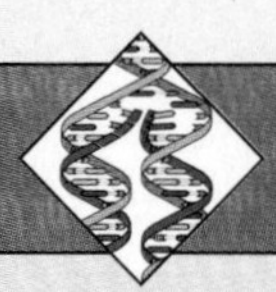

Sex Testing in the Olympics—Biology and a Bad Idea

Success in amateur athletics, including the Olympics, is often a prelude to financial rewards and acclaim as a professional athlete. Because the stakes are so high, several methods are used to guard against cheating in competition. Competitors in many international events are required to submit urine samples (collected while someone watches) for drug testing. In other cases, this is done at random in an attempt to eliminate the use of steroids or performance-enhancing drugs. In the 1960s, rumors about males attempting to compete as females led the International Olympic Committee (IOC) to require sex testing of all female athletes, beginning with the 1968 Olympic Games.

The IOC's test involved analysis of Barr bodies in cells collected by scraping the inside of the mouth. In genetic females (XX), the inactivated X chromosome forms a Barr body, which can be stained and seen under a microscope. Genetic males (XY) do not have a Barr body. The procedure is noninvasive, and females are not required to submit to a physical examination of their genitals. If sexual identity were called into question as a result of the test, a karyotype was required, and if necessary, a gynecological examination followed.

In both theory and practice, the IOC's test was a bad idea for several reasons. Barr body testing is unreliable and leads to both false positive and false negative results. It fails to take into account phenotypic females who are XY with androgen insensitivity and other conditions that result in a discrepancy between chromosomal and phenotypic sex. In addition, the test does not take into account the psychological, social, and cultural factors that enter into one's identity as a male or a female. Ironically, no men attempting to compete as females were identified, but the test unfairly prevented females from competition. Of the more than 6,000 women athletes tested, 1 in 500 had to withdraw from competition as a consequence of failing the sex test. The Spanish hurdler Maria Martinez Patino led a courageous fight against sex testing. She has complete androgen insensitivity, was raised as a female, and competed as a female.

In response to criticism, the IOC and the International Amateur Athletic Federation (IAAF) reconsidered the question of sex testing and instituted a new test, based on recombinant DNA technology, to detect the presence of the male-determining gene *SRY*, which is carried on the Y chromosome. This test was instituted at the 1992 Winter Olympics. A positive test makes the athlete ineligible to compete as a female. However, again the test was flawed because it fails to recognize several chromosomal combinations that result in a female phenotype, even though an *SRY* gene is present. At the 1996 Summer Olympic Games in Atlanta, 8 of 3,387 females were *SRY* positive. Of these, 7 of the 8 had partial or complete androgen insensitivity. Again, no males attempting to compete as females were identified.

Finally, in the face of criticism from medical professionals and athletes, in 1999, the IOC decided to abandon the use of genetic screening of female athletes at the 2000 Olympic Games in Australia. However, the IAAF still retains the option of testing a competitor should the question arise.

as the earthworm, this condition is normal. In humans, a true hermaphrodite has ovarian and testicular tissue in separate gonads or in a single combined gonad. Cytogenetic examination of several hermaphrodites has shown that they are sex chromosome mosaics: Some cells in the body are XX, and others are XY or XXY. In other cases, only XY cells are found. These individuals have mixed masculine and feminine genitals; the degree of maleness or femaleness depends on whether testicular or ovarian tissue predominates in gonads.

Androgen Insensitivity and Sex Phenotype

Androgen insensitivity An X-linked genetic trait that causes XY individuals to develop into phenotypic females.

Pseudohermaphroditism An autosomal genetic condition that causes XY individuals to develop the phenotypic sex of females.

Androgen insensitivity (MIM/OMIM 313700) is an X-linked trait in which XY individuals develop as phenotypic females. In this case, chromosomal sex (XY) is opposite from phenotypic sex (Figure 7.17), a condition that is caused by a mutation in a gene, called the androgen receptor (*AR*), that maps to the long arm of the X chromosome in the Xq11-13 region. The *AR* gene encodes a protein that interacts with male hormones (testosterone and DHT) inside the cell and changes the pattern of gene expression. Some mutations cause complete insensitivity and a female phenotype; some cause partial insensitivity, resulting in some male phenotype; and others produce a wide range of phenotypes, even in siblings.

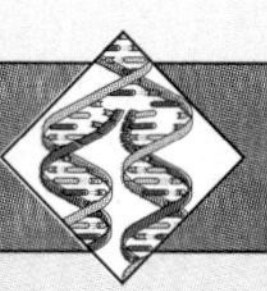

Concepts and Controversies

Joan of Arc—Was It Really John of Arc?

Joan of Arc, the national heroine of France, was born in a village in northeastern France in 1412, during the Hundred Years' War. At the age of 13 or 14, she began to have visions that directed her to help fight the English at Orleans. Following victory, she helped orchestrate the crowning of the new king, Charles VII. During a siege of Paris, the English captured Joan, and in 1431 she was tried for heresy. Although her trial was technically a religious one conducted by the English-controlled Church, it was clearly a political trial. Shortly after being sentenced to life imprisonment, she was declared a relapsed heretic, and on May 30, 1431, she was burned at the stake in the marketplace at Rouen.

In 1455 Pope Callistus formed a commission to investigate the circumstances of her trial, and a Trial of Rehabilitation took place over a period of 7 months in 1456. The second trial took testimony from over 100 individuals who knew Joan personally. Extensive documentation from the original trial and the Trial of Rehabilitation exists. This material has served as the source for the more than 100 plays and countless books written about her life. Although the story of her life is well known, perhaps more remains to be discovered. From an examination of the original evidence, R. B. Greenblatt proposed that Joan had phenotypic characteristics of androgen insensitivity. By all accounts, Joan was a healthy female who had well-developed breasts. Those living with her in close quarters testified that she never menstruated, and physical examinations conducted during her imprisonment revealed a lack of pubic hair. Although such circumstantial evidence is not enough for a diagnosis, it provides more than enough material for speculation. This speculation also provides a new impetus for those medicogenetic detectives who prowl through history, seeking information about the genetic makeup of the famous, infamous, the notorious, and the obscure.

In complete androgen insensitivity, testis formation is induced normally in XY individuals, and testosterone and MIH production begin as expected. MIH causes degeneration of the Müllerian duct system so that no internal female reproductive tract is formed. However, a mutation in the receptor gene blocks the ability of cells to respond to testosterone or DHT. As a result, development proceeds as if there were no testosterone or DHT present. The Wolffian duct system degenerates and the genitalia develop as female structures. Individuals with this condition are chromosomal males but phenotypic females who do not menstruate and have well-developed breasts and very little pubic hair (see Concepts and Controversies: Joan of Arc—Was It Really John of Arc?).

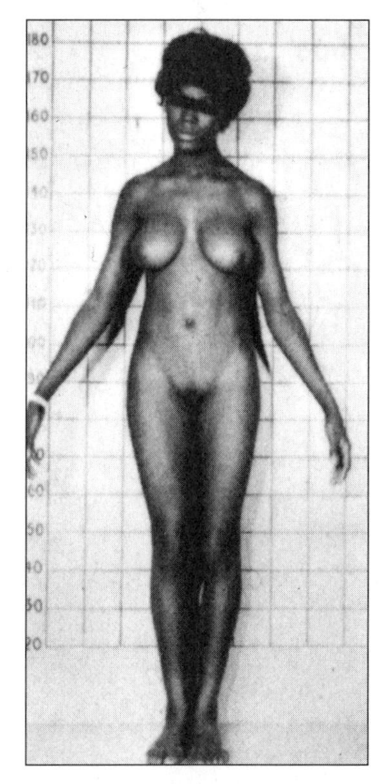

FIGURE 7.17 A phenotypic female who has an XY chromosomal constitution and androgen insensitivity. (From Zourlas, P. et al, 1965, Clinical histologic and cytogenetic findings in male hermaphroditism, *Obstetrics and Gynecology* 25:768–778, Figure 7.)

Changing Sex Phenotypes at Puberty

Pseudohermaphrodites are individuals in whom the phenotypic sex does not match the sex chromosome set, but who later change phenotypic sex. Mutations in an autosomal gene cause one form of **pseudohermaphroditism** (MIM/OMIM 264300). In XY individuals, this mutation prevents the conversion of testosterone to DHT. The Y chromosome initiates development of testes, and the Wolffian ducts form the male duct system. MIH secretion prevents the development of female ducts. However, the failure to produce DHT results in genitalia that are essentially female. The scrotum resembles the labia, a blind vaginal pouch is present, and the penis resembles a clitoris. Although chromosomally male, these individuals are identified and raised as females.

At puberty, however, these females change into males. The testes move down into a developing scrotum, and what resembled a clitoris develops into a functional penis. The voice deepens, a beard grows, and muscle mass increases. In most cases, sperm production is normal. What causes these changes? This phenotype is altered by the increased levels of testosterone secretion that accompany puberty. This condition is rare, but in a group of small villages in the Dominican Republic, more than 30 such cases are known. The high incidence is the result of common ancestry through intermarriage (Figure 7.18). In 12 of the 13 families in these villages, a line of descent can be traced from a single individual (I-3).

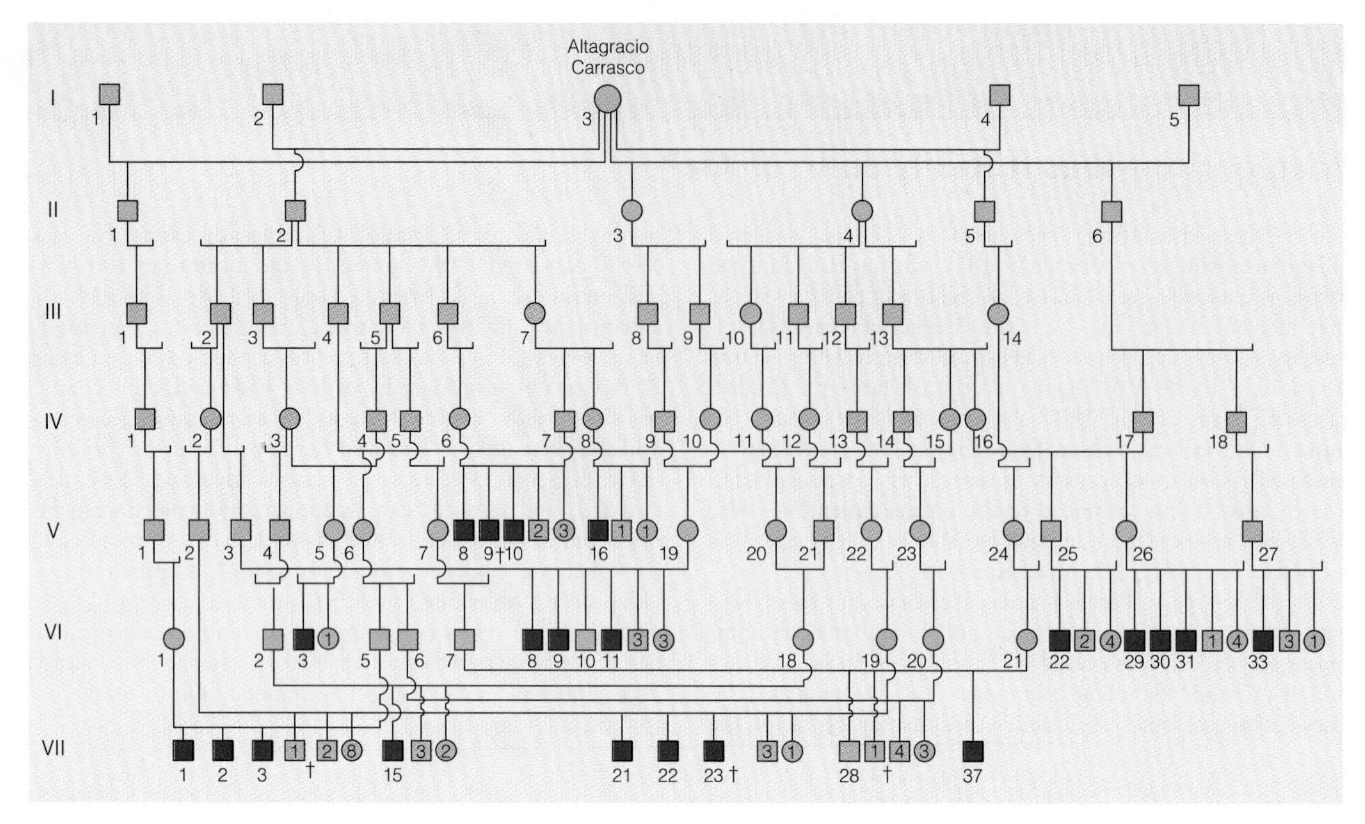

FIGURE 7.18 Pedigree of pseudohermaphroditism in several generations of village residents in the Dominican Republic. Although the condition was first diagnosed in generation V, members of earlier generations were probably affected.

This mutation and the one that produces androgen insensitivity show the importance of a normal pattern of gene expression in development. Not only is the process of sexual development clearly under genetic control, but the ability to respond to the hormonal environment is critical to normal sexual differentiation.

EQUALIZING X CHROMOSOMES IN MALES AND FEMALES

Because females have two copies of all genes on the X chromosome and males have only one copy of these genes, at first glance it would seem that females should have higher levels of all products encoded by these genes. Is this true, or is there a way to equalize the expression of genes on the X chromosome between males and females?

Expression of Genes on the X Chromosome

In Chapter 4, we discussed hemophilia A, an X-linked genetic disorder in which clotting factor VIII is missing. Because normal females have two copies of the clotting factor gene and normal males have only one, we can ask whether blood from females has twice as much of this clotting factor as blood from males. The answer is straightforward: Careful measurements indicate that females have the same amount of this clotting factor as males. In fact, the same is true for all X chromosome genes tested: The amount of gene product is the same in males and females. A process called dosage compensation equalizes the amount of X chromosome gene products in both sexes. How that is accomplished in humans and how it came to be understood is an interesting story.

Mice, Barr Bodies, and X Inactivation

The explanation of how **dosage compensation** works in female mammals is known as the **Lyon hypothesis,** named after Mary Lyon, a British geneticist. She proposed that dosage compensation works by inactivating almost all of the genes on one of the X chromosomes in females. She based this idea on both genetic and cytological evidence.

The genetic evidence came from her studies on coat color in mice. In female mice heterozygous for X-linked coat-color genes, Lyon found that the coat color was unique. It was not the same as either homozygous parent, nor was it a blend of the parent's coat color. Instead the fur had patches of the two parental colors in a random arrangement. Males, hemizygous for either gene, never showed such patches and had coats of uniform color. This genetic evidence suggested to Lyon that in heterozygous females, both alleles were active but not in the same cells.

The cytological observations were made in 1949 by Murray Barr and his colleagues. Barr was studying nerve cells from cats. He saw a small, dense spot on the inside of the nuclear membrane in cells from female cats, but not in cells from male cats (Figure 7.19). Susumo Ohno suggested that this spot, called the **Barr body,** is actually an inactivated, condensed X chromosome. Mary Lyon summarized the genetic and cytological findings in her hypothesis:

- One X chromosome is genetically active in the body cells (not the germ cells) of female mammals. The second X chromosome is inactivated and tightly coiled to form the Barr body.
- The inactivated chromosome can come from the mother or thc father.
- Inactivation takes place early in development. After four to five rounds of cell division following fertilization, each cell of the embryo randomly inactivates one X chromosome.
- This inactivation is permanent (except in germ cells), and all descendants of a given cell will have the same X chromosome inactivated.
- The random inactivation of one X chromosome in females equalizes the activity of X-linked genes in males and females.

Dosage compensation A mechanism that regulates the expression of sex-linked gene products.

Lyon hypothesis The proposal that dosage compensation in mammalian females is accomplished by partially and randomly inactivating one of the two X chromosomes.

Barr body A densely-staining mass in the somatic nuclei of mammalian females. An inactivated X chromosome.

Females Can Be Mosaics for X-Linked Genes

The Lyon hypothesis means that female mammals are actually mosaics, constructed of two different cell types: Some cells express genes from the mother's X chromosome, and some cells express genes from the father's X chromosome. The pattern of coat color that Lyon observed in the heterozygous mice is a result of this inactivation. In females heterozygous for X-linked coat-color genes, patches of one color are inter-

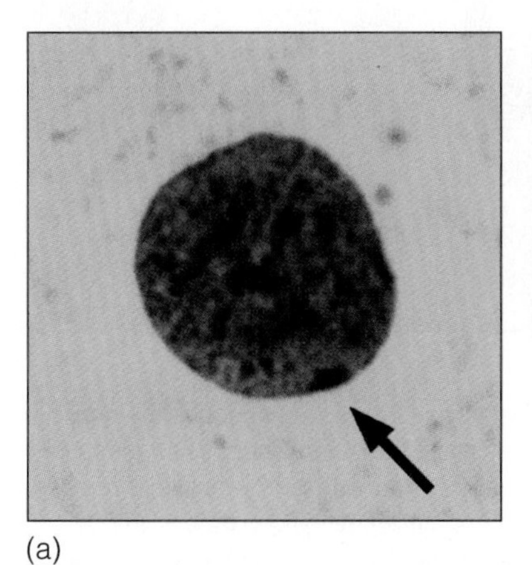

(a)

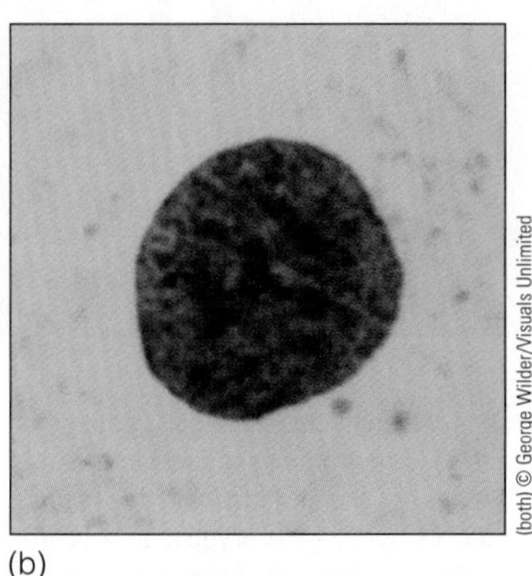

(b)

(both) © George Wilder/Visuals Unlimited

FIGURE 7.19 (a) Nucleus from a female cell showing a Barr body (arrow). (b) Nucleus from a male cell shows no Barr body.

Courtesy of Dr. Martin Feather and Dr. C. Bird

▶ **FIGURE 7.20** The differently colored patches of fur on this tortoiseshell cat result from X-chromosome inactivation.

spersed with patches of another color. According to the Lyon hypothesis, each patch represents a group of cells descended from a single cell in which the inactivation event occurred.

An example of this mosaicism is the tortoiseshell cat (▶ Figure 7.20). In cats, a gene for coat color on the X chromosome has two alleles, a dominant allele (*O*) that produces an orange/yellow color and a recessive allele (*o*) that produces a black color. Heterozygous females (*O*/*o*) have patches of orange/yellow fur mixed with patches of black fur, called a tortoiseshell pattern. Cells expressing either the orange/yellow allele or the black allele cause this pattern. A cat with a tortoiseshell pattern on a white background is called a calico cat (white fur on the chest and abdomen in such cats is controlled by a different, autosomal gene). Therefore, tortoiseshell cats (and calico cats) are invariably female because males have only one X chromosome and would be either be all orange/yellow or all black.

Mosaicism has also been demonstrated in human females. A gene on the X chromosome encodes the information for an enzyme called G6PD (MIM/OMIM 305900) and has two alleles. Each allele produces a distinct form of the enzyme. Skin cells from heterozygous females (identified by pedigree analysis) were isolated and grown individually. Each culture, grown from a single cell, showed only one form or the other of the enzyme (▶ Figure 7.21).

Sex-influenced genes Loci that produce a phenotype conditioned by the sex of the individual.

Pattern baldness A sex-influenced trait that acts like an autosomal dominant trait in males and an autosomal recessive trait in females.

How and When X Chromosomes Are Inactivated

The process of X inactivation has presented researchers with several puzzling questions. How does the cell count the number of X chromosomes in the nucleus? If there are two X chromosomes in the nucleus, how is one X chromosome cho-

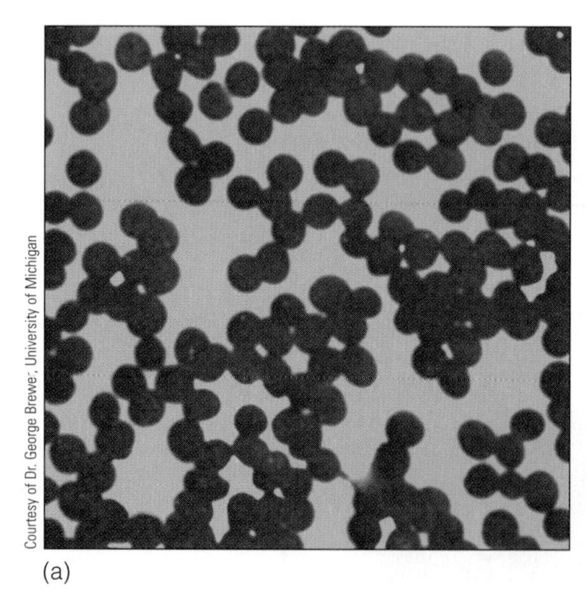
Courtesy of Dr. George Brewe; University of Michigan

(a)

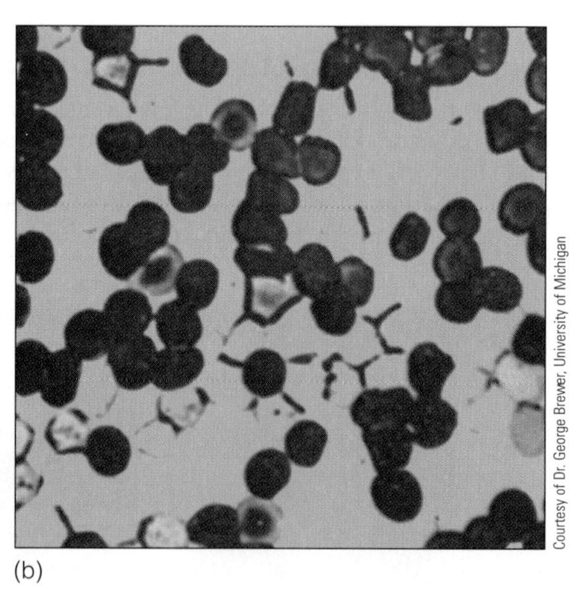
Courtesy of Dr. George Brewer, University of Michigan

(b)

▶ **FIGURE 7.21** Dosage compensation in humans. The X-linked gene that encodes the enzyme G6PD has two alleles. One produces an active form of the enzyme; the other produces an inactive form. (a) Blood cells from an individual who has two active alleles stained to show enzymatic activity. (b) Blood cells from a heterozygous female. The stained cells have an active allele. The unstained cells contain an inactive allele. Heterozygotes are mosaics composed of two cell types, one with G6PD activity and one with no activity.

sen to be turned off, but not the other? Finally, how is the chromosome inactivated? Detailed answers to these questions are not yet available, but we know that inactivation begins from a region on the X chromosome called the X inactivation center (Xic). The Xic contains several genes, one of which is called *XIST*. If the *XIST* gene on an X chromosome is expressed, the chromosome becomes coated with *XIST* RNA (Figure 7.22) and becomes coiled and inactivated. How the *XIST* gene on only one of the two X chromosomes in a female embryo is turned on is still a puzzle.

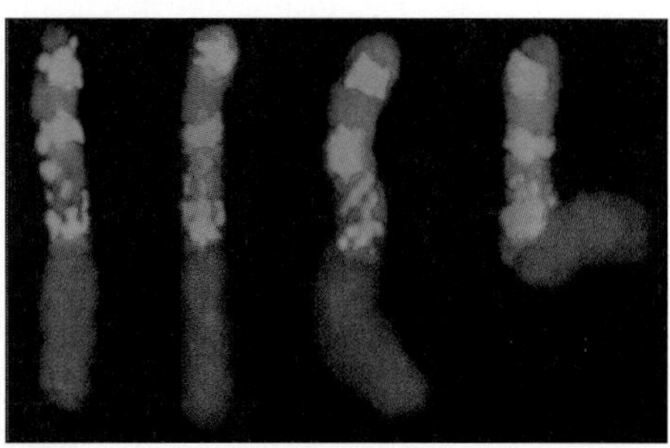

FIGURE 7.22 In female mammals, expression of the *XIST* gene coats one X chromosome with *XIST* RNA (green) in the gene-rich regions between G bands as a first step in X inactivation. This photograph shows several X chromosomes coated with *XIST* RNA; other chromosome regions are stained red. (Courtesy of Dr. Neil Brockdorff, Imperial College School of Medicine, London. From Duthie S. et al, 1999, Xist RNA exhibits a banded localization of the inactive X chromosome and is excluded from autosomal material in cis, *Human Molecular Genetics* 8:195–204.)

In humans, both X chromosomes are active in the zygote and early embryo. Random inactivation of one X chromosome usually occurs in the blastocyst stage, when the embryo has about 32 cells. Because there are only a small number of cells in the embryo at the time of inactivation, is it possible that all or almost all the cells might inactivate the mother's or father's X chromosome? This could cause females to express X-linked traits for which they are heterozygous. In fact, this imbalance has been seen a number of times in female monozygotic twins, one of whom has an X-linked recessive trait, whereas the other does not. In the pedigree shown in Figure 7.23, two female identical twins are heterozygotes for red-green color blindness through their colorblind father. One of the twins has normal color vision, and the other has red-green color blindness. The colorblind twin has three sons, two who have normal vision and one who is colorblind (see pedigree).

Molecular testing of skin cells from the colorblind twin showed that almost all of the active X chromosomes were from her father and carry the allele for colorblindness. In the twin with normal vision, the opposite situation is observed; almost all of the active X chromosomes are maternal in origin.

The colorblindness in one twin and normal color vision in the other twin can be explained by X inactivation associated with twinning. One twin was formed from a small number of cells, most of which had the paternal X inactivated. The other twin originated from a small number of cells, most of which had the maternal X chromosome inactivated.

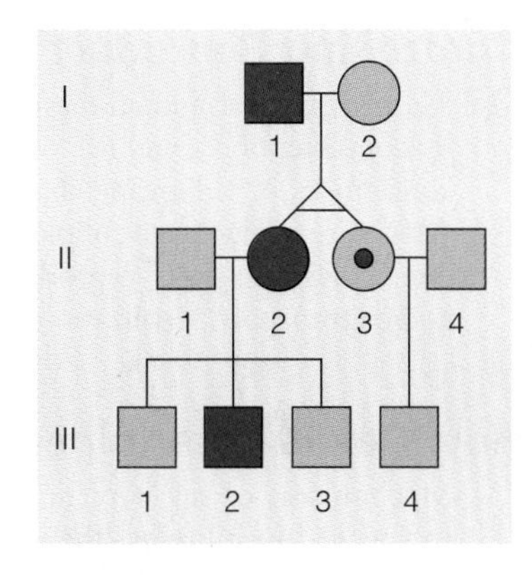

FIGURE 7.23 Pedigree showing monozygotic female twins (II-2 and II-3) discordant for colorblindness. The twins inherited the allele for colorblindness from their father. Almost all of the active X chromosomes in the colorblind twin carry the mutant allele. Almost all of the active X chromosomes in the twin who has normal vision carry the allele for normal vision.

SEX-INFLUENCED AND SEX-LIMITED TRAITS

Sex-influenced genes are expressed in both males and females but have phenotypic frequencies much different than would be predicted by Mendelian ratios. These are usually autosomal genes and illustrate the effect of the hormonal environment on the level of gene expression. **Pattern baldness** (MIM/OMIM 109200) is an example of sex-influenced inheritance (Figure 7.24). This trait is expressed more often in males than in females. The gene acts as an autosomal dominant in males and as an autosomal recessive in females, and the pattern of expression is related to the different levels of testosterone present. In this case, the hormonal environment and the genotype interact in determining phenotypic expression of this gene.

Sex-limited genes are expressed only in one sex, whether they are inherited in an autosomal or sex-linked pattern. An autosomal dominant trait that controls **precocious puberty** (MIM/OMIM 176410) is expressed in heterozygous males but not in heterozygous females. Affected males undergo puberty at 4 years of age or earlier. Heterozygous females are unaffected but pass this trait on to half of their sons, making it hard to distinguish this trait from a sex-linked gene. Genes that deal with traits such as breast development in females and facial hair in males are other examples of sex-limited genes, as are virtually all other genes that deal with secondary sexual characteristics.

Sex-limited genes Loci that produce a phenotype in only one sex.

Precocious puberty An autosomal dominant trait expressed in a sex-limited fashion. Heterozygous males are affected, but heterozygous females are not.

FIGURE 7.24 Pattern baldness as an autosomal dominant trait in males, and an autosomal recessive trait in females. The level of baldness is related to testosterone levels and other environmental influences.

© Leonard Lessin/Peter Arnold, Inc.

Summary

Fertilization to Birth

1. Human development begins with fertilization and the formation of a zygote. Cell divisions in the zygote form an early embryonic stage called the blastocyst. The embryo implants in the uterine wall, and a placenta develops to nourish the embryo.

Controlling Reproduction

2. Many aspects of human reproduction can be controlled by contraception to reduce or eliminate the chances of conception, or enhancing chances of conception by using assisted reproductive technologies. These methods control one or more stages of reproduction: gamete production and/or transport, fertilization and implantation.

How Sex Is Determined

3. Mechanisms of sex determination vary from species to species. In humans, the presence of a Y chromosome is associated with male sexual development, and the absence of a Y chromosome is associated with female development.

Mutations Can Uncouple Chromosomal Sex from Phenotypic Sex

4. Early in development, the Y chromosome signals the indifferent gonad to begin development as a testis. Hormones secreted by the testis control later stages of male sexual differentiation, including the development of phenotypic sex.

Sex-Influenced and Sex-Limited Traits

5. In sex-influenced and sex-limited inheritance, the sex of the individual affects whether and the degree to which the trait is expressed. This is true for autosomal and sex-linked genes. Sex hormone levels modify expression of these genes, giving rise to altered phenotypic ratios.

Case Studies

CASE 1

Melissa was referred to genetic counseling at 16 weeks of her pregnancy because of a history of epileptic seizures. She takes medication (valproic acid) for her seizures and has not had an attack for the last 3 years. Her physician became concerned when he learned that she was still taking this medication, against his advice, during her pregnancy. He wanted her to speak to a counselor about the possible effects of this medication on the developing fetus. The counselor took a detailed family history, which indicated that Melissa was the only family member with seizures and that no other genetic conditions were apparent in the family. The counselor asked Melissa why she continued to take valproic acid during her pregnancy. Melissa stated she was "afraid her child would be like her, if she didn't take her medicine." Melissa went on to say that she was teased as a child when she would have her "fits," and she wanted to prevent that from happening to her children.

Case Studies (continued)

With this in mind, the counselor reviewed the process of fetal development and why it is best that a physician carefully evaluate all medications that a woman takes while she is pregnant. Melissa's medication has been shown to cause spina bifida—affecting almost twice as many children who were exposed to it than children who are not exposed. Using illustrations, the counselor explained that spina bifida is a defect that occurs when the neural tube fails to completely close during embryonic development. The failure to fold exposes part of the spinal area when an infant is born. Valproic acid could also cause problems in the heart and in the genitals.

The counselor explained that prenatal diagnosis using ultrasound, and possibly amniocentesis, could help determine whether the baby's tube has closed properly.

Postscript: Melissa elected to have an ultrasound, which showed that the baby did not have a neural tube defect. However, she was offered an amniocentesis to rule out a possible false negative result of the ultrasound. She declined the amniocentesis, and Melissa delivered a healthy baby boy.

1. As a counselor, you have taken Melissa's family history. How can you address Melissa's fears that her child will develop epilepsy because she did?
2. From a genetics perspective, is Melissa at greater risk for having a child with epilepsy than someone without epilepsy?
3. Women taking valproic acid have a 1% to 2% risk of having a child with a neural tube defect. Does the fact that Melissa had a normal child increase the risk that her next child will be affected? Why or why not?
4. The neural tube forms and closes during the first trimester of pregnancy. What does this suggest about Melissa's medication program in future pregnancies?

CASE 2

Jan, a 32-year-old woman, and her husband, Darryl, have been married for 7 years. They attempted to have a baby on several occasions. Five years ago they had a first-trimester miscarriage followed by an ectopic pregnancy later that same year. Jan continued to see her OB/GYN physician for infertility problems but was very unsatisfied with the response. After four miscarriages, she went to see a fertility specialist who diagnosed her with severe endometriosis and polycystic ovarian disease (detected by hormone studies). The infertility physician explained that these two conditions were hampering her ability to become pregnant and thus making her infertile. She referred Jan to a genetic counselor. At the appointment, the counselor explained to Jan that one form of endometriosis is a genetic disorder, inherited as an autosomal dominant trait, and that polycystic ovarian disease can also be a genetic disorder and is the most common reproductive disorder among women. The counselor recommended that a detailed family history of both Jan and Darryl would help establish whether Jan's problems have a genetic component, and whether any of her daughters would be at risk for one or both of these disorders. In the meantime, Jan is in the process of taking hormones, and she and Darryl are considering alternative modes of reproduction.

1. Using the information in Figure 7.10, explain the reproductive options that are open to Jan and Darryl.
2. Would ISCI be an option? Why or why not?
3. Jan is concerned about using ART. She wants to be the genetic mother and have Darryl be the genetic father of any children they have. What methods of ART would you recommend to this couple?

Questions and Problems

Fertilization to Birth

1. The gestation of a fetus occurs over 9 months and is divided into three trimesters. Describe the major events that occur in each trimester. Is there a point at which the fetus becomes more "human"?
2. FAS is caused by alcohol consumption during pregnancy. It can result in spontaneous abortion, growth retardation, facial abnormalities, and mental retardation. How does FAS affect all of us, not just the unlucky children born with this syndrome? What steps need to be taken to prevent this syndrome?

Controlling Reproduction

3. Explain how the following methods prevent conception.
   a. vasectomy
   b. tubal ligation
   c. birth control pills
4. RU486 is a controversial birth control method. What makes it different from other methods?
5. How does IVF differ from artificial fertilization?
6. What do you think are the legal and ethical issues surrounding the use of IVF? How can these issues be

resolved? What should be done with the extra gametes that are removed from the woman's body but never implanted in her uterus?

7. Researchers are currently learning how to transfer sperm-making cells from fertile male mice into infertile male mice in the hopes of learning more about reproductive abnormalities. These donor spermatogonia cells have developed into mature spermatozoa in 70% of the cases and some recipients have gone on to father pups. This new advance opens the way for a host of experimental genetic manipulations. It also offers enormous potential for correcting human genetic disease. One potentially useful human application of this procedure is treating infertile males who wish to be fathers.
   a. Do you foresee any ethical or legal problems with the implementation of this technique? If so, elaborate on them.
   b. Could this procedure have the potential for misuse? If so, explain how.

How Sex is Determined

8. Describe, from fertilization, the major pathways of normal male sexual development; include the stages in which genetic sex, gonadal sex, and phenotypic sex are determined.
9. Which pathway of sexual differentiation (male or female) is regarded as the default pathway? Why?
10. The absence of a Y chromosome in an early embryo causes:
    a. the embryonic testis to become an ovary.
    b. the Wolffian duct system to develop.
    c. the Müllerian duct system to degenerate.
    d. the indifferent gonad to become an ovary.
    e. the indifferent gonad to become testis.
11. Assume that human-like creatures exist on Mars. As in the human population on Earth, there are two sexes and even sex-linked genes. The gene for eye color is an example of one such gene. It has two alleles. The purple allele is dominant to the yellow allele. A purple-eyed female alien mates with a purple-eyed male. All of the male offspring are purple-eyed, whereas half of the female offspring are purple-eyed and half are yellow-eyed. Which is the heterogametic sex?

Mutations Can Uncouple Chromosomal Sex from Phenotypic Sex

12. Give an example of a situation in which genetic sex, gonadal sex, and phenotypic sex do not coincide. Explain why they do not coincide.
13. How can an individual who is XY be phenotypically female?
14. Discuss whether the following individuals are (1) gonadally male or female, (2) phenotypically male or female (discuss Wolffian/Müllerian ducts and external genitalia), and (3) sterile or fertile.
    a. XY, homozygous for a recessive mutation in the testosterone gene, which renders the gene nonfunctional
    b. XX, heterozygous for a dominant mutation in the testosterone gene, which causes continuous production of testosterone
    c. XY, heterozygous for a recessive mutation in the MIH gene
    d. XY, homozygous for a recessive mutation in the SRY gene
    e. XY, homozygous for a recessive mutation in the MIH gene

Sex-Influenced and Sex-Limited Traits

15. It has been shown that hormones interact with DNA to turn certain genes on and off. Use this fact to explain sex-linked and sex-influenced traits.
16. What method of sex testing did the International Olympic Committee previously use? What method did they subsequently use? Do either of these methods conclusively test for "femaleness"? Explain.
17. Explain why pattern baldness is more common in males than in females yet the gene resides on an autosome.

Equalizing X Chromosomes in Males and Females

18. Calico cats are almost invariably female. Why? (Explain genotype and phenotype of calico females and the theory of why calicos are female.)
19. How many Barr bodies would the following individuals have?
    a. normal male
    b. normal female
    c. Klinefelter male
    d. Turner female
20. Males have only one X chromosome and therefore only one copy of all genes on the X chromosome. Each gene is directly expressed, thus providing the basis of hemizygosity in males. Females have two X chromosomes, but one is always inactivated. Therefore females, like males, have only one functional copy of all the genes on the X chromosome. Again, each gene must be directly expressed. Why, then, are females not considered hemizygous, and why are they not afflicted with sex-linked recessive diseases as often as males?
21. Individuals with an XXY genotype are sterile males. If one X is inactivated early in embryogenesis, the genotype of the individual effectively becomes XY. Why will this individual not develop as a normal male?

Internet Activities

The following activities use the resources of the World Wide Web to enhance the topics covered in this chapter. To investigate the topics described below, log on to the book's home page (www.brookscole.com/biology_d) and follow the *Student Resources* link to your subject and text.

1. *Embryological Development.* The Web site *The Visible Embryo* (http://www.visembryo.com) provides free images and descriptions of human developmental stages from conception to stage 23. (Descriptions only are available for stages beyond 10 weeks.) Follow the stages and read about the development of the embryo.
   a. At what stage does sexual differentiation begin? At which point would ovaries or testes be distinguishable?
   b. The drug thalidomide, once given to help reduce morning sickness, caused birth defects such as poorly developed or missing arms and legs. By what stage are the basics of arm development essentially complete? Would thalidomide be likely to cause arm problems after this point? What is the critical stage for the development of lower limbs?
   c. In general, would you expect teratogens (discussed in this chapter) to have more significant whole-body effects for the embryo if exposure occurs early in pregnancy or later in pregnancy? Why? What are the most critical stages for exposure to alcohol as a teratogen (see text)? Which critical structures and organs are developing during these times?

Further Exploration: To check out the *Morphing Embryos* video at Nova Online's *Odyssey of Life* Web site, go to http://www.pbs.org/wgbh/nova/odyssey/clips/.

2. *Further Exploration of a Chromosomal Defect.* Fragile X syndrome, which you may have researched as part of the Chapter 6 Internet Activities, affects males and females differently. Go to the *Your Genes, Your Health* Web site maintained by the Dolan DNA Learning Center at Cold Spring Harbor Laboratory (http://vector.cshl.org/ygyh/mason/index) and click on the "Fragile X Syndrome" link. At this page, choose the "How is it inherited?" link and explore how males and females inherit and display fragile X syndrome.
   a. How is fragile X syndrome inherited in male children? Which parent(s) can a boy inherit the syndrome from? Do all males that have the mutant gene display fragile X? Why or why not?
   b. How is fragile X syndrome inherited in female children? From which parent(s) can a girl inherit the syndrome? Do all females who have the mutant gene display fragile X? Why or why not? What is X-linked inactivation? What is an inactivated X chromosome in a nucleus called? How is the pattern of X-inactivation determined? What is the factor that determines whether or not a female child who carries the fragile X mutation will display significant mental impairment?

Further Exploration: To explore the complexities of the genetics of coat color in cats, including the genetics of X-linked characteristics such as tortoiseshell coat patterns, you can try "The Cat Color FAQ" (http://www/fanciers.com/other-faqs/color-genetics.html#intro).

For Further Reading

Brenner, C., & Cohen, J. (2000). The genetic revolution in artificial reproduction: A view of the future. *Hum. Reprod., 15*(Suppl. 5), 111–116.

Gottlieb, B., Pinsky, L., Beitel, L, & Trifiro, M. (1999). Androgen insensitivity. *Am. J. Med. Genet., 89,* 210–217.

Greenblatt, R. B. (1981). Case history: Jeanne d'Arc—Syndrome of feminizing testes. *Br. J. Sex Med., 8,* 54.

Heard, E., Clerc, P., & Avner, P. (1997). X-chromosome inactivation in mammals. *Ann. Rev. Genet., 31,* 571–610.

Lyon, M. F. (1962). Sex chromatin and gene action in the mammalian X-chromosome. *Am. J. Hum. Genet., 14,* 135–148.

Mittwoch, U. (2000). Two thousand years of questioning sex determination. *Cytogenet. Cell Genet., 91,* 186–191.

Perlman, S. E., Richmond, D. M., Sabatini, M. M., Krueger, H., & Rudy, S. J. (2001). Contraception: Myths, facts and methods. *J. Reprod. Med., 46*(Suppl. 2): 169–177.

Simpson, J. L., Ljungqvist, A., Ferguson-Smith, M. A., de la Chapelle, A., Elsas, L. J., Ehrhardt, A. A., Genel, M., Ferris, E. A., Carlson, A. (2000). Gender verification in the Olympics. *JAMA, 284,* 1568–1569.

Courtesy of Dmitry Pruss, National Institutes of Health

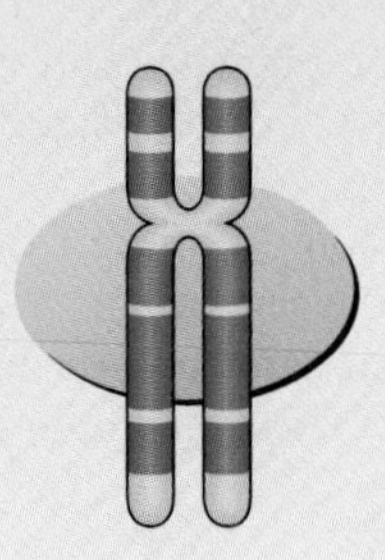

DNA Structure and Chromosomal Organization

Chapter Outline

Early in the 1860s, a German chemist named Frederick Miescher set out to study the chemical and cellular mechanisms associated with life. For his work, Miescher learned that the pus of infected wounds was made up of white blood cells that had scavenged the infecting bacteria. In the days before antibiotics, wound infections were common. Miescher visited local surgical clinics to collect discarded bandages. He scraped the pus from these bandages and developed a method of separating the pus cells from debris and bandage fragments by washing with a salt solution.

In his early experiments Miescher recovered a chemical substance from the nuclei of the pus cells. To analyze this substance further, he decided to purify the nuclei. He broke open the cells by treating them with a protein-digesting substance called pepsin. To obtain pepsin, he prepared extracts of pig stomachs (a good source of pepsin, which functions in digestion). He then treated the pus cells for several hours with the extract of pig stomach. He noted that gray sediment collected at the bottom of the flask. Under the microscope this sediment turned out to be pure nuclei. Miescher was therefore the first to isolate and purify a cellular organelle.

By chemically extracting the purified nuclei, Miescher obtained a substance he called nuclein. Chemical analysis revealed that it contained hydrogen, carbon, nitrogen, oxygen, and phosphorus. Miescher showed that nuclein was found in the nuclei of other cell types, including kidney, liver, sperm, and yeast. He regarded it as an important component of most cells. Many years later it was shown that his nuclein contained deoxyribonucleic acid (DNA).

At about the same time Miescher was carrying out his experiments, Mendel outlined the rules for the inheritance of traits and developed the notion of what we now call genes. In the 1880s, August Weismann and others emphasized the importance of the nucleus in heredity. At the turn of the twentieth century, Walter Sutton and Theodore Boveri noted that the behavior of genes in inheritance paralleled that of nuclear components (the chromosomes) in meiosis. Later workers confirmed that, in fact, genes are part of chromosomes, and it was generally agreed that the genetic material was to be found in the nucleus. Through all of this work, however, the most basic question remained unanswered: What is the nature of the genetic material?

The answer to this question takes us from the work of Mendel on the transmission of traits to the level of nucleic acid molecules as the chemical components of cells involved in storing, expressing, and transmitting genetic information. The path to this answer runs from the experiments of Miescher through the experiments of Avery and his colleagues in the 1940s and beyond. Although several lines of evidence pointed to nucleic acids, especially DNA, as the chemical answer to this question, most theories were originally based on proteins as the molecular carriers of genetic information. The general requirements for the genetic material, however, were clear and unambiguous. Any chemical component of the nucleus that carries genetic information must have a structure that explains the observed properties of genes: replication, information storage, expression of the stored information, and mutation. It was not until the middle of the twentieth century that this issue was resolved.

In this chapter we examine the events that led to the confirmation of DNA as the molecule that carries genetic information, and we consider the work of Watson and Crick on the organization and structure of DNA. We also explore what is known about the way DNA is incorporated into the structure of chromosomes.

DNA CARRIES GENETIC INFORMATION

Research in the first few decades of the twentieth century established that genes exist and are carried on chromosomes. But what is a gene? As is often the case in science, the answer to this question came from an unexpected direction, the study of an infectious disease.

At the beginning of the twentieth century, pneumonia was a serious public health problem and was the leading cause of death in the United States. Medical research of that era studied this infectious disease to develop an effective treatment, perhaps in the form of a vaccine. The unexpected outgrowth of this research was the discovery of the chemical nature of the gene.

DNA Mediates the Transfer of Genetic Traits in Bacterial Strains

Transformation The process of transferring genetic information between cells by DNA molecules.

Transforming factor The molecular agent of transformation; DNA.

By the 1920s, it was known that a bacterial infection could cause pneumonia. One form of pneumonia is caused by the bacterium *Streptococcus pneumoniae.* Two strains of this species were known; in one strain (strain S), the cells were contained in a capsule. Strain S was infective and caused pneumonia (that is, was a virulent strain). The other strain (strain R) did not form a capsule and was not infective. Fredrick Griffith studied these strains, and the results of his experiment are straightforward and easily interpreted. Griffith showed that mice injected with living cells of strain S developed pneumonia and soon died, but mice injected with live cells from strain R did not develop pneumonia and lived. In another set of experiments, Griffith found that if he first killed the strain S cells with heat before injection, the mice survived and did not develop pneumonia (Figure 8.1).

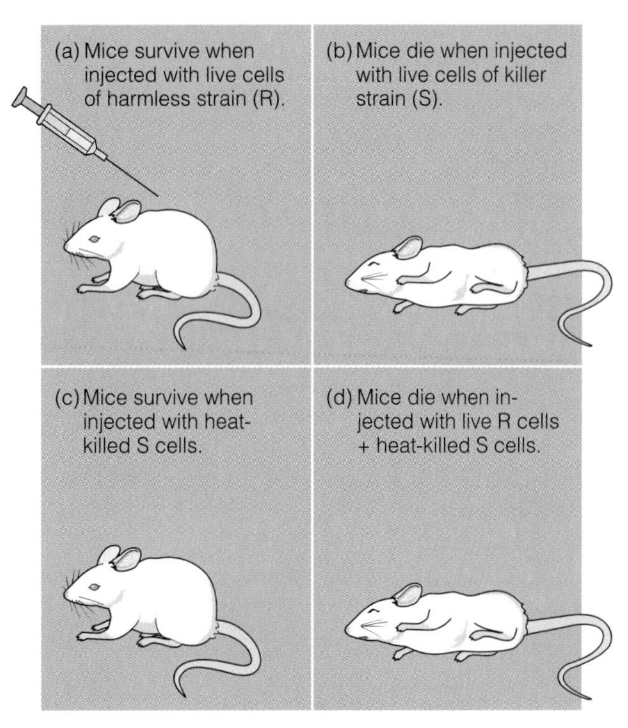

FIGURE 8.1 Griffith discovered that the ability to cause pneumonia is a genetic trait that can be passed from one strain of bacteria to another. (a) Mice injected with strain R do not develop pneumonia. (b) Mice injected with strain S develop pneumonia and die. (c) When the S strain cells are killed by heat treatment before injection, mice do not develop pneumonia. (d) When mice are injected with a mixture of heat-killed S cells and live R cells, they develop pneumonia and die. Griffith concluded that the live R cells acquired the ability to cause pneumonia from the dead S cells.

For Griffith, the most intriguing result was obtained when mice were injected with a mixture of heat-killed strain S cells and live cells from strain R. These mice developed pneumonia and died. Griffith recovered live strain S bacteria from the bodies of the dead mice. When grown in the laboratory, the progeny of these cells were always strain S. After further experiments, Griffith concluded that the living cells from strain R were transformed into strain S cells in the bodies of the mice. He proposed that hereditary information had somehow passed from the dead strain S cells into the strain R cells, allowing them to make a capsule and become virulent. He called this process **transformation** and the unknown material the **transforming factor.**

In 1944, a team at the Rockefeller Institute in New York that included Oswald Avery, Colin MacLeod, and Maclyn McCarty discovered that the transforming factor is DNA. Maclyn McCarty recounts the story of this discovery in a readable memoir: *The Transforming Principle: Discovering That Genes Are Made of DNA.*

In a series of experiments that stretched over 10 years, Avery and his colleagues extended the work of Griffith on transformation. Griffith had shown that the ability to cause pneumonia is associated with the presence of a capsule surrounding the bacterial cell. When heat-killed strain S bacteria are mixed with living cells of strain R, a small fraction of the R cells acquire the ability to form a capsule, become S cells, and cause pneumonia. The bacteria that have acquired the ability to form a capsule transmit this trait to all their offspring, indicating that the trait is a permanent change and is heritable.

Concepts and Controversies

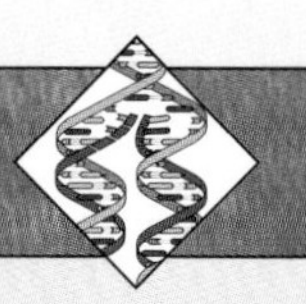

DNA as a Commercial Product

The magazine ad for the perfume reads: "Where does love originate? Is it in the mind? Is it in the heart? Or in our genes?" A perfume named DNA was marketed in a helix-shaped bottle. There is no actual DNA in the fragrance, but the molecule is invoked to sell the idea that love emanates from the genes. Seem strange? Well, how about jewelry that actually contains DNA from your favorite celebrities? In this line of products, a process called the polymerase chain reaction (PCR) amplifies DNA in a single hair or cheek cell. The resulting solution contains millions of copied DNA molecules and is added to small channels drilled into acrylic earrings, pendants, or bracelets. The liquid can be colored to contrast with the acrylic and be more visible. Just as people wear T-shirts with pictures of Elvis or Einstein, they can now wear jewelry containing DNA from their favorite entertainer, poet, composer, scientist, or athlete. For dead heroes, the DNA can come from a lock of hair; in fact, a single hair will do.

Or, how about using DNA as a protection against counterfeit clothes? In the 2000 Summer Olympic Games in Sydney, DNA extracted from cheek cells swabbed from Australian athletes was amplified by PCR, and mixed with the ink used to print souvenir shirts. More than 2,000 different types of items were created, and DNA testing of the labels was used to ensure that everything sold at the Games was genuine. Based on the results at Sydney, officials of the 2002 Winter Olympics in Salt Lake City considered using DNA markers for its souvenirs.

Want music composed from the base sequence of DNA? Composers have translated the four bases of DNA (adenine, guanine, cytosine, and thymine) into musical notes. Long sequences of bases, retrieved from computer databases, are converted into notes, transferred to sheet music, and played by instruments or synthesizers as the music of the genes. Those who have listened to this music say that DNA near chromosomal centromeres sounds much like the music of Bach or other Baroque composers, but that music from other parts of the genome has a contemporary sound.

From a scientific standpoint, this fascination with DNA may be a little difficult to understand, but DNA has clearly captured popular fancy and is being used to sell an ever-increasing array of products. DNA has name recognition. Over the past 40 years, DNA has moved from scientific journals and textbooks to the popular press and even the comic strips. The relationship between genes and DNA is now well known enough to be used in commercials and advertisements. In a few years, this fascination will probably fade and be replaced with another fad, but for now, if you want it to sell, relate it to DNA.

In their experiments, Avery and his colleagues separated the chemical components of S cells into chemical classes including carbohydrates, fats, proteins, and nucleic acids. Each component was mixed with live R cells and injected into mice. Mice only got pneumonia and died when injected with a mixture of S cell DNA and live R cells. Avery concluded that DNA from S cells was responsible for transforming the R cells into S cells. To confirm that DNA was responsible for transformation, they treated the DNA with enzymes that destroy protein and ribonucleic acid (RNA) before injection. This treatment removed any residual protein or RNA from the preparation, but did not affect transformation. As a final test, the DNA preparation was treated with deoxyribonuclease, an enzyme that digests DNA, and the transforming activity was abolished.

The work of Avery and his colleagues produced two important conclusions. In the bacterium they studied

- DNA carries genetic information. Only DNA transfers heritable information from one strain to another strain.
- DNA controls the synthesis of specific products. Transfer of DNA also results in the transferring of the ability to synthesize a specific gene product (in the form of a capsule).

DNA Is a Part of Chromosomes

Transformation cannot be performed on eukaryotic organisms, so it was not possible to duplicate Avery's results with organisms, such as *Drosophila* or mice. Because it was generally accepted that the chromosomes contain genetic information, in eukaryotic organisms indirect evidence was used to strengthen the link between DNA and chromosomes. Staining cells with certain dyes indicates that DNA is found in the nucleus and is part of chromosomes. In addition, DNA is found all along the length of chromosomes, just as genetic maps indicate that genes are distributed all along the chromosome. Most importantly, the concentration of DNA within the cells of a given eukaryotic organism is correlated with the number of chromosomes carried by the cell. Diploid (2*n*) somatic cells contain twice as many chromosomes as the haploid (*n*) gametes. Measurements of DNA concentration indicate that somatic cells have twice as much DNA as gametes. These and other forms of indirect evidence support the idea that DNA is the genetic material of eukaryotic organisms. More recently, direct evidence for the role of DNA as the genetic material has come from the development of **recombinant DNA technology.** DNA segments from organisms such as humans can be spliced into bacterial DNA, and under the proper conditions this hybrid DNA molecule can direct the synthesis of a human gene product. The synthesis of human proteins in bacteria requires the presence of specific human DNA sequences, providing direct evidence for the role of DNA as the genetic material in higher organisms. Case study 1 deals with problems raised by patenting DNA sequences. In fact, as awareness of DNA and its role in genetics has grown, DNA is even being used to sell products (see Concepts and Controversies: DNA as a Commercial Product).

Recombinant DNA technology Technique for joining DNA from two or more different organisms to produce hybrid, or recombined, DNA molecules.

WATSON, CRICK, AND THE STRUCTURE OF DNA

Recognition that DNA is the molecule that carries genetic information helped fuel the efforts to understand the chemical structure of DNA and other nucleic acids. From the mid-1940s through 1953, several laboratories made significant strides in unraveling the structure of DNA, culminating in the Watson-Crick model for the DNA double helix in 1953. James Watson has documented the scientific, intellectual, and personal intrigue that characterized the race to discover the structure of DNA in his book, *The Double Helix*. This book, and others on the same topic, provide a rare glimpse into the ambitions, jealousies, and rivalries that entangled scientists who were involved in the dash to a Nobel prize.

Reviewing Some Basic Chemistry

The structure of DNA in the Watson-Crick model and the structure of proteins shown in a later chapter are described and drawn using chemical terms and symbols. For this reason, we will pause here for a brief review of the terminology and definition of some chemical terms.

All matter is composed of atoms; the different types of atoms are known as elements (of which there are 114). In nature, atoms are combined into molecules, which are units of two or more atoms chemically bonded together. Molecules can be represented by formulas that indicate how many of each type of atom are present. The type of atom is indicated by a symbol for the element it represents: H for hydrogen, N for nitrogen, C for carbon, O for oxygen, and so forth. For example, a water molecule, composed of two hydrogen atoms and one oxygen atom, has its chemical formula represented as H_2O:

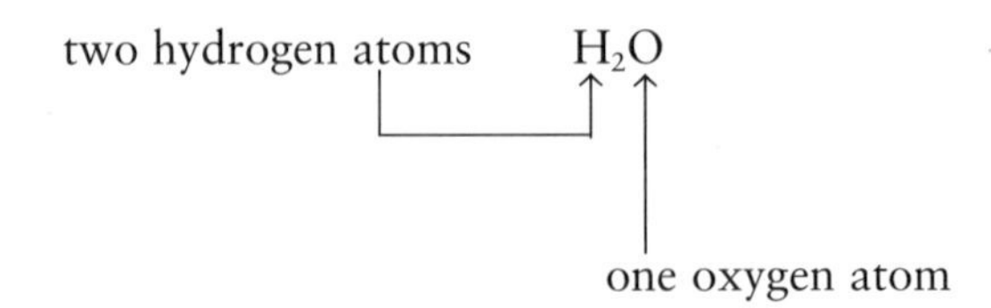

Many molecules in cells are large and have more complex formulas. A molecule of glucose contains 24 atoms and is written as

$$C_6H_{12}O_6$$

The atomic components of molecules are held together by links called **covalent bonds.** In its simplest form, a covalent bond is a pair of electrons shared between two atoms. Sharing two or more electrons can form more complex covalent bonds. ▶ Figure 8.2 shows how such bonds are written in chemical structures. A second type of atomic interaction can also take place. These interactions involve a weak attraction known as a **hydrogen bond.** In living systems, hydrogen bonds make an important contribution to the three-dimensional shape and functional capacity of biological molecules. Hydrogen bonds are weak interactions between two atoms (one of which is always hydrogen), that carry partial but opposite electrical charges. Hydrogen bonds are usually represented in structural formulas as dotted or dashed lines that connect two atoms (Figure 8.2).

Although individual hydrogen bonds are weak and easily broken, they hold molecules together by sheer force of numbers. As we see in a following section, hydrogen bonds hold together the two strands in a DNA molecule, and they are also responsible for the three-dimensional structure of proteins (Chapter 9).

Covalent bonds Chemical bonds that result from electron sharing between atoms. Covalent bonds are formed and broken during chemical reactions.

Hydrogen bond A weak chemical bonding force between hydrogen and another atom.

Nucleotides: The Building Blocks of Nucleic Acids

There are two types of nucleic acids in biological organisms: **DNA** and **RNA.** Both are made up of strings of subunits known as nucleotides. A **nucleotide** contains a **nitrogen-containing base** (either a **purine** or a **pyrimidine**), a **pentose sugar** (either ribose or deoxyribose), and a phosphate group. The phosphate groups are strongly acidic and are why DNA and RNA are called acids. Both purines and pyrimidines have the same six-atom ring, but purines have an additional three-atom ring. The purine bases **adenine** (A) and **guanine** (G) are found in both RNA and DNA (▶ Figure 8.3c). The pyrimidine bases are **thymine** (T), found in DNA; **uracil** (U), found in RNA; and **cytosine** (C), found in both RNA and DNA. RNA has four bases (A, G, U, C), and DNA has four bases (A, G, T, C).

The sugars in nucleic acids contain five carbon atoms (which is why they are called pentoses). The sugar in RNA is known as **ribose,** and the sugar in DNA is **deoxyribose.** The difference is a single oxygen atom that is present in ribose and absent in deoxyribose (Figure 8.3b).

Nucleotides are composed of a base covalently linked to a sugar, which in turn is covalently linked to a phosphate group (Figure 8.3d). Nucleotides are named according to the base and sugar they contain (Table 8.1). Nucleotides can be linked

Deoxyribonucleic acid (DNA) A molecule consisting of antiparallel strands of polynucleotides that is the primary carrier of genetic information.

Ribonucleic acid (RNA) A nucleic acid molecule that contains the pyrimidine uracil and the sugar ribose. The several forms of RNA function in gene expression.

Nucleotide The basic building block of DNA and RNA. Each nucleotide consists of a base, a phosphate, and a sugar.

Nitrogen-containing base A purine or pyrimidine that is a component of nucleotides.

Purine A class of double-ringed organic bases found in nucleic acids.

Pyrimidine A class of single-ringed organic bases found in nucleic acids.

Pentose sugar A five-carbon sugar molecule found in nucleic acids.

Adenine and guanine Nitrogen-containing purine bases found in nucleic acids.

Thymine, uracil, and cytosine Nitrogen-containing pyrimidine bases found in nucleic acids.

Ribose and deoxyribose Pentose sugars found in nucleic acids. Deoxyribose is found in DNA, ribose in RNA.

Covalent bonds

Single covalent bond

Double covalent bond

(a)

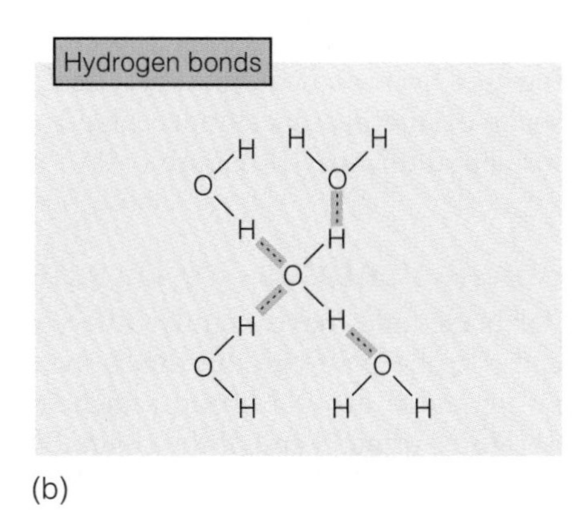

(b)

▶ **FIGURE 8.2** Representation of chemical bonds. (a) Covalent bonds are represented as solid lines that connect atoms. Depending on the degree of electron sharing, there can be one (left) or more (right) covalent bonds between atoms. Once formed, covalent bonds are stable and are broken only in chemical reactions. (b) Hydrogen bonds are usually represented as dotted lines that connect two or more atoms. As shown, water molecules form hydrogen bonds with adjacent water molecules. These are weak interactions that are easily broken by heat and molecular tumbling and can be re-formed with other water molecules.

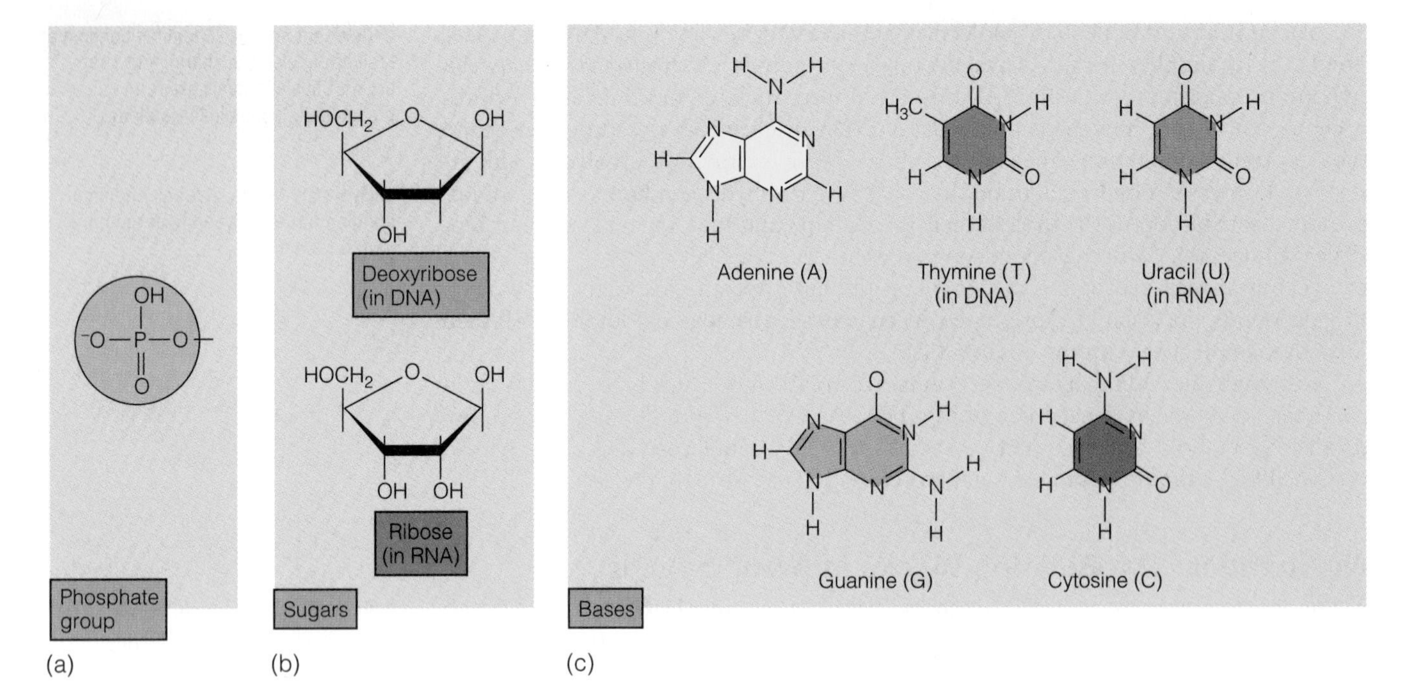

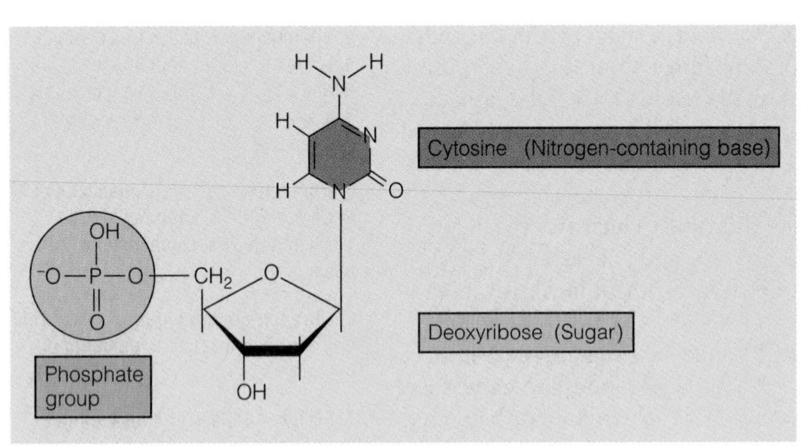

FIGURE 8.3 DNA is made up of subunits called nucleotides. Each nucleotide is composed of phosphorus (a), a sugar (b), and a base (c). (d) A nucleotide.

together by a covalent bond between the phosphate group of one nucleotide and the sugar of another nucleotide. Chains of nucleotides called polynucleotides can be formed in this way (Figure 8.4a).

Polynucleotides Are Directional Molecules

Polynucleotides have slightly different structures at each end of the chain. At one end is a phosphate group; this is the 5′ (pronounced "five prime") end. At the opposite end is an OH group attached to the sugar molecule; this is known as the 3′ ("three prime") end of the chain. By convention, nucleotide chains are written beginning with the 5′ end, such as 5′-CGATATGCGAT-3′, and are usually labeled to indicate which end is which.

DNA IS A DOUBLE HELIX

In the early 1950s James Watson and Francis Crick began to work out the structure of DNA. To build their model, they sifted through and organized the information that was already available about DNA. Their model is based on two types of information

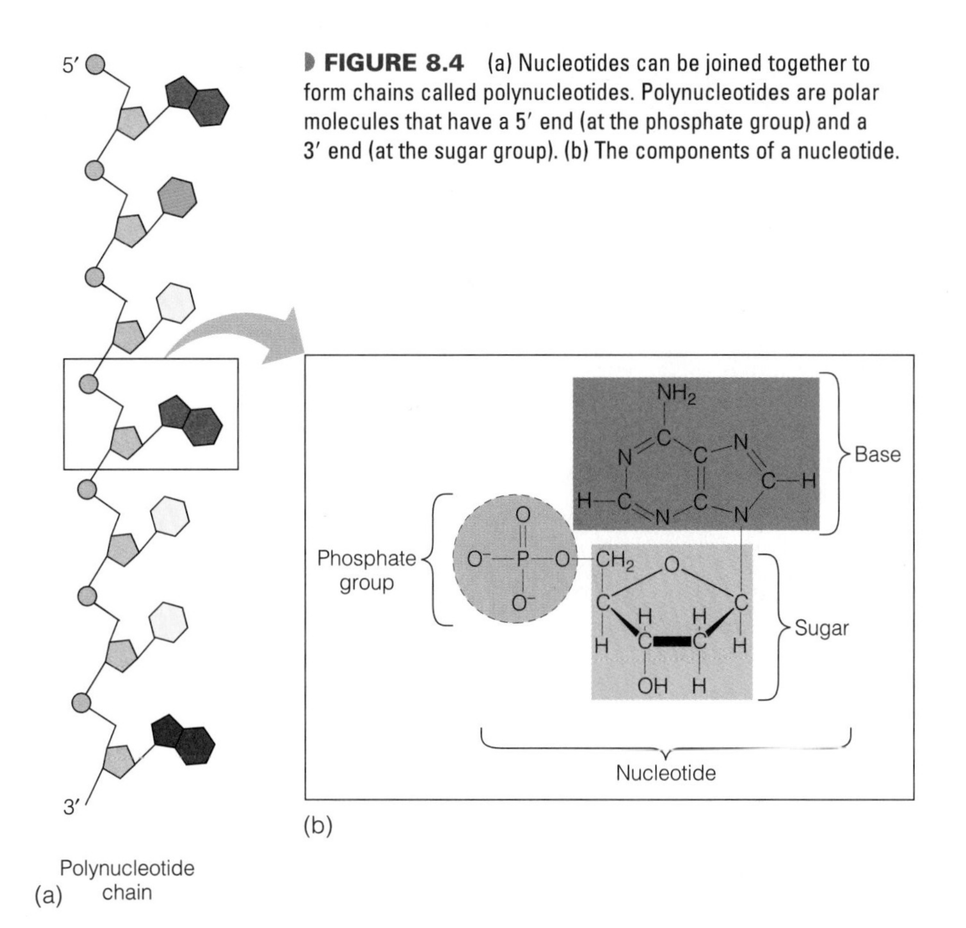

FIGURE 8.4 (a) Nucleotides can be joined together to form chains called polynucleotides. Polynucleotides are polar molecules that have a 5′ end (at the phosphate group) and a 3′ end (at the sugar group). (b) The components of a nucleotide.

Table 8.1

Nucleotides of RNA and DNA

RNA
Ribonucleotides
Adenylic acid
Cytidylic acid
Guanylic acid
Uridylic acid
DNA
Deoxyribonucleotides
Deoxyadenylic acid
Deoxycytidylic acid
Deoxyguanylic acid
Deoxythymidylic acid

about DNA: x-ray crystallography, which provides information about the physical structure of the molecule, and information about the nucleotide composition of DNA.

In x-ray crystallography, molecules are crystallized and placed in an x-ray beam. As the x-rays pass through the crystal, some hit the atoms in the crystal and are deflected at an angle. The pattern of x-rays emerging from the crystal can be recorded on photographic film and analyzed to produce information about the organization and shape of the crystallized molecule.

Maurice Wilkins and Rosalind Franklin obtained x-ray crystallographic pictures from highly purified DNA samples. Their pictures indicated that the DNA molecule has the shape of a helix of constant diameter (Figure 8.5). The x-ray films also provided information about the distances between the stacked bases within the molecule.

Erwin Chargaff and his colleagues extracted DNA from a variety of organisms and analyzed the amounts of the four bases in these samples. The results indicated that DNA from all organisms tested had common properties, which became known as Chargaff's rule:

- There is a 1:1 relationship between the amount of adenine and the amount of thymine and a 1:1 relationship between the amount of guanine and the amount of cytosine.

Using the information from x-ray and chemical studies, Watson and Crick built a series of wire and metal models of DNA structures. Eventually they succeeded in producing a model that incorporated all of the information from the x-ray and chemical studies (Figure 8.6). This model has the following features:

- DNA is composed of two polynucleotide chains running in opposite directions.
- The two polynucleotide chains are coiled around a central axis to form a double helix.

These two features fit the x-ray results of Rosalind Franklin and Maurice Wilkins.

FIGURE 8.5 An x-ray diffraction photograph of a DNA crystal. The central x-shaped pattern is typical of helical structures, and the darker areas at the top and bottom indicate a regular arrangement of subunits in the molecule. Watson and Crick used this and other photographs to construct their model of DNA.

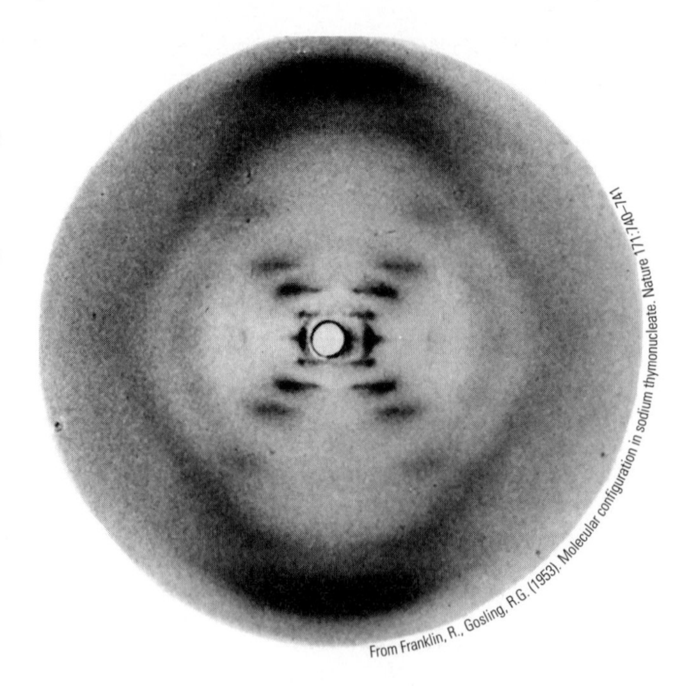

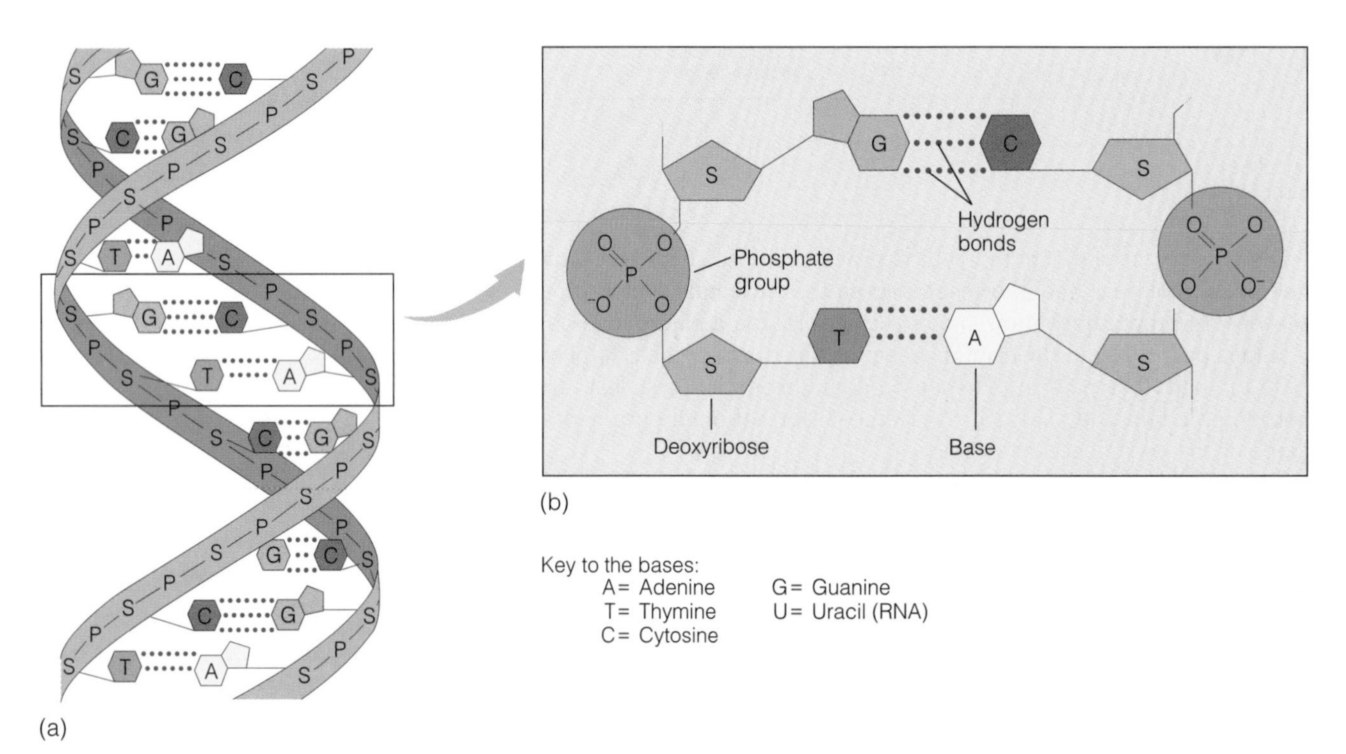

FIGURE 8.6 (a) The Watson-Crick model of DNA. Two polynucleotide strands are coiled around a central axis, forming a helix. (b) Hydrogen bonds between the bases hold the two strands together. In the molecule, A always pairs with T on the opposite strand, and C always pairs with G.

- In each chain, sugar and phosphate groups are linked together to form the backbone of the chain. The bases face inward, where they are paired by hydrogen bonds to bases in the opposite chain.
- Base pairing is highly specific: A in one chain pairs only with T in the opposite chain, and C pairs only with G. Each set of hydrogen-bonded bases is called a base pair. The pairing of A with T and C with G fits the results obtained by Chargaff.
- The base pairing of the model makes the two polynucleotide chains of DNA complementary in base composition (Figure 8.7). If one strand has the se-

Web-Integrated Toolbox

The Structure of a DNA Molecule

Understanding the structure of a DNA molecule is an essential step in learning about molecular genetics. Almost everything a gene is and does, including replication, the storage and transfer of genetic information, and mutation, is closely related to the structure of DNA. Knowing how a DNA molecule is built makes it easier to understand its many functions.

Some knowledge of the language and symbols used in chemistry is required to analyze DNA structure: single letters represent the atoms in DNA (C for carbon, N for nitrogen, etc.) and lines and dots represent chemical bonds. There are two levels of DNA structure. The first level is the subunit structure of DNA. DNA is composed of chemical subunits called nucleotides. At the second level, two chains of nucleotides form a helix, with specific nucleotide pairings between the chains. Once you are familiar with the components and the structure of DNA in some detail, it is a short step to understanding how DNA functions as the genetic material.

The book's Web site contains a link to exercises that will help you learn how a DNA molecule is put together from subunits (nucleotides). These exercises explain the structure of DNA and its subunits, and the relationship between structure and function.

quence 5′-ACGTC-3′, the opposite strand must be 3′-TGCAG-5′, and the double-stranded structure would be written as

5′-ACGTC-3′
3′-TGCAG-5′

Three important properties of this model should be considered:

- In this model, genetic information is stored in the sequence of bases in the DNA. The linear sequence of bases has a high coding capacity. A molecule n bases in length has 4^n combinations. That means that a sequence of 10 nucleotides has 4^{10}, or 1,048,576, possible combinations of nucleotides. The complete set of genetic information carried by an organism (its genome) can be expressed as base pairs of DNA (Table 8.2). Genomic sizes vary from a few thousand nucleotides (in viruses) that encode only a few genes to billions of

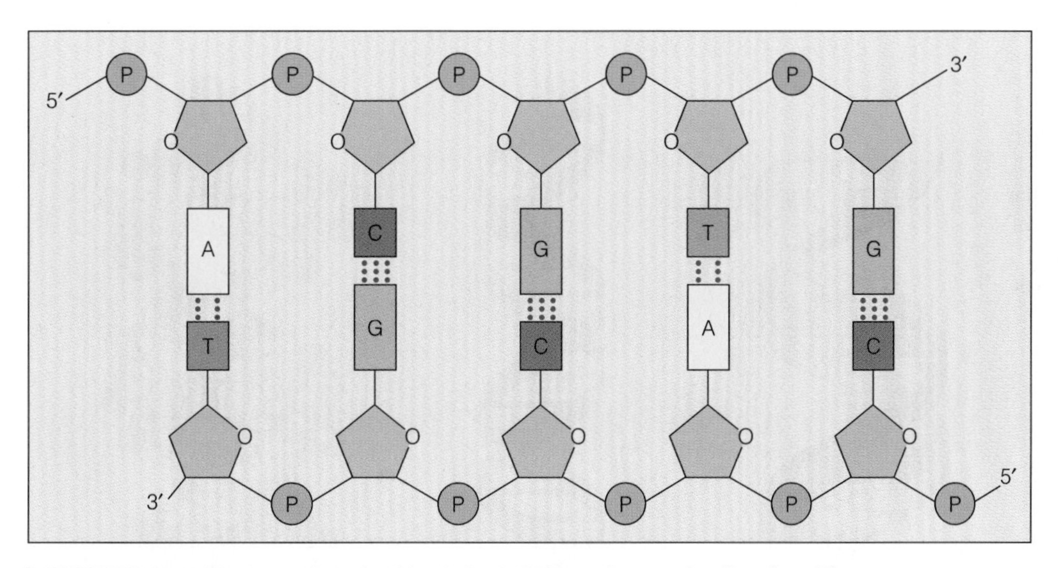

FIGURE 8.7 The two polynucleotide chains in DNA run in opposite directions. The top strand runs 5′ to 3′ and the bottom strand runs 3′ to 5′. The base sequences in each strand are complementary. An A in one strand pairs with a T in the other strand, and a C in one strand is paired with a G in the opposite strand.

SIDEBAR

DNA Organization and Disease

Huntington disease (HD) is an autosomal dominant disease of the nervous system, characterized by involuntary movements, psychiatric and mood disorders, dementia, and death. HD is caused by an increase in the size of a cluster of nucleotide (CAG) repeats in the *HD* gene. The cluster tends to expand further when passed on by the father.

Measuring the number of CAG repeats can identify those at risk for HD. Normal individuals have 10 to 29 CAG repeats, and those with HD have 40 or more repeats. Those with 36 to 39 repeats are at risk for HD, and many of these individuals develop the disease. People with 30 to 35 repeats do not get HD, but males with 30 to 35 repeats may pass on an expanded HD gene to their offspring, who may become affected.

Individuals from families that have HD can have presymptomatic testing that follows recommendations established by the Huntington Disease Society and involves pretest and posttest visits with a neurologist, geneticist, and psychiatrist or psychologist.

Template The single-stranded DNA that serves to specify the nucleotide sequence of a newly synthesized polynucleotide strand.

Table 8.2 Genome Size in Various Organisms

Organism	Species	Genome Size in Nucleotides
Bacterium	*E. coli*	4.6×10^6
Yeast	*S. cerevisiae*	1.2×10^7
Fruit fly	*D. melanogaster*	1.7×10^8
Tobacco plant	*N. tabacum*	4.8×10^9
Mouse	*M. musculus*	2.7×10^9
Human	*H. sapiens*	3.2×10^9

nucleotides that encode 30,000 to 40,000 genes (as in humans). The human genome consists of about 3×10^9, or 3 billion, base pairs of DNA, distributed over 24 chromosomes (22 autosomes and two sex chromosomes).

- The model offers a molecular explanation for mutation. Because genetic information can be stored as a linear sequence of bases in DNA, any change in the order or number of bases in a gene can result in a mutation that produces an altered phenotype. This topic is explored in more detail in Chapter 11.
- As Watson and Crick observed, the complementary strands in DNA can be used to explain the molecular basis of the replication of genetic information that takes place before each cell division. In such a model, each strand can be used as a **template** to reconstruct the base sequence in the opposite strand. This topic is discussed later in this chapter.

The Watson-Crick model was described in a brief paper in *Nature* in 1953. Although their model was based on the results of other workers, Watson and Crick correctly incorporated the physical and chemical data into a model that also could be used to explain the properties expected of the genetic material (▶ Figure 8.8). Present day applications, including genetic engineering, gene mapping, and gene therapy can be traced directly to this paper.

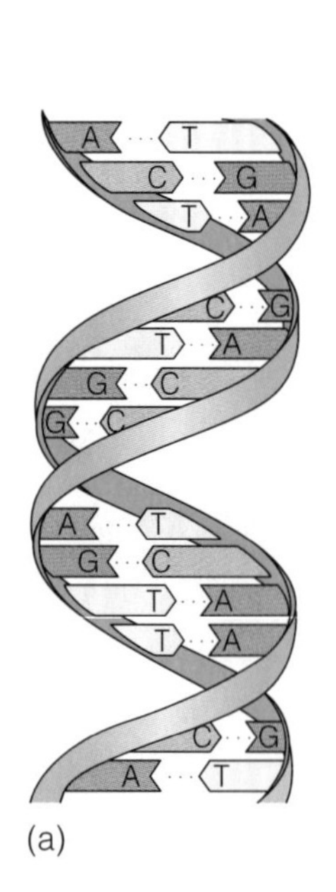

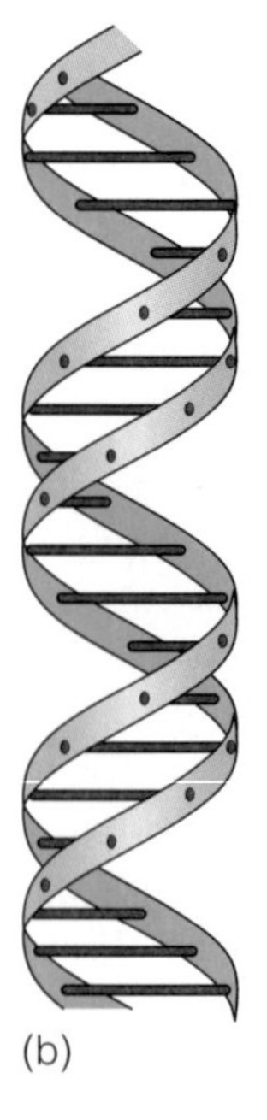

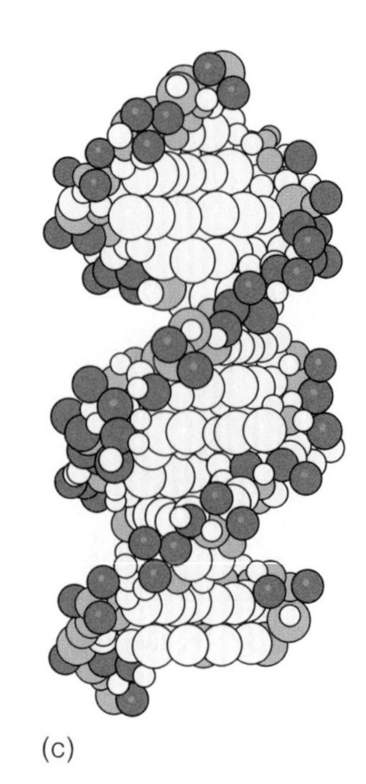

▶ **FIGURE 8.8** Three different ways of representing DNA. (a) A model that shows the complementary base pairing in the interior of the molecule. (b) A model that shows the double helix wound around a central axis. (c) A space-filling model that shows the relative sizes of the atoms in the molecule.

At the time there was no direct evidence to support the Watson-Crick model, but in subsequent years it has been confirmed by experimental work in laboratories worldwide. The 1962 Nobel Prize for Medicine or Physiology was awarded to Watson, Crick, and Wilkins for their work on the structure of DNA. Although Rosalind Franklin provided much of the x-ray data for the Watson-Crick model, she did not receive a share of the prize. There has been some controversy over this, but because only living individuals are eligible, she could not have shared in the prize. Franklin died of cancer in 1958. Her role in the discovery of the structure of DNA is presented in her biography, *Rosalind Franklin and the Discovery of DNA.*

Table 8.3

Differences Between DNA and RNA

	DNA	RNA
Sugar	Deoxyribose	Ribose
Bases	Adenine Cytosine Guanine Thymine	Adenine Cytosine Guanine Uracil

RNA IS A SINGLE-STRANDED NUCLEIC ACID

Another type of nucleic acid, RNA is found in both the nucleus and the cytoplasm. DNA functions as a repository of genetic information. RNA transfers genetic information from the nucleus to the cytoplasm (in a few viruses, RNA also functions to store genetic information). The nucleotides in RNA differ from those in DNA in two respects: The sugar in RNA nucleotides is ribose (deoxyribose in DNA), and the base uracil is used in place of the base thymine (Table 8.3).

Although two complementary strands of RNA can form a double helix similar to the one found in DNA, the RNA present in most cells is single-stranded, and a complementary strand is not made (Figure 8.9). RNA molecules can fold back on themselves, however, and form double-stranded regions. The functions of RNA are considered in more detail in Chapter 9.

FROM DNA MOLECULES TO CHROMOSOMES

Although an understanding of DNA structure represents an important development in genetics, it provides no immediate information about the way a chromosome is organized or about what regulates the cycle of condensation and decondensation of the chromosomes as the cell moves from interphase to mitosis and back. This problem is significant because the spatial arrangement of DNA plays an important role in regulating the expression of genetic information. In addition, putting billions of nucleotides of DNA into the 46 human chromosomes requires packing a little more than 2 m (about 6.5 feet) of DNA into a nucleus that measures about 5 micrometers in diameter. Within this cramped environment, the chromosomes unwind and become dispersed during the interphase. In this condition they undergo replication, gene expression, homologous pairing during meiosis, and contraction and coiling to become visible again during prophase. An understanding of chromosomal organization is necessary to understand these processes. In contrast to what is known about chromosomes in the nucleus, details of the structure and even the nucleotide sequence of the mitochondrial chromosome are well understood, and we will first consider this simpler system.

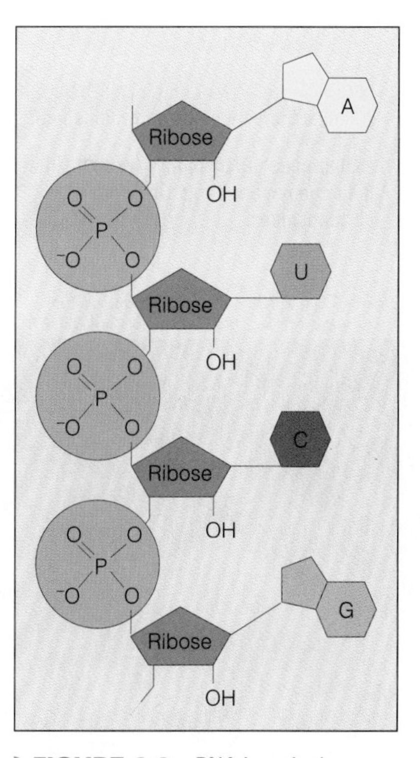

FIGURE 8.9 RNA is a single-stranded polynucleotide chain. RNA molecules contain a ribose sugar instead of a deoxyribose and have uracil (U) in place of thymine (T).

The Mitochondrial Chromosome Is a Circular DNA Molecule

As discussed in Chapter 2, mitochondria are cytoplasmic organelles involved in energy conversion. They contain DNA that encodes genetic information. Because the sperm does not contribute any cytoplasm in fertilization, we inherit our mitochondria from our mothers. A number of genetic disorders resulting from mutations in mitochondrial DNA are known (see Chapter 4), and all offspring of affected mothers are also affected. In case study 2, a family deals with a mitochondrial mutation in their child.

The complete nucleotide sequence of the mitochondrial chromosome is known (Figure 8.10). The chromosome is a circular DNA molecule containing 16,569 base pairs. It encodes 13 proteins, as well as several RNAs used in protein synthesis. Most mitochondria contain 5 to 10 copies of the chromosome, and

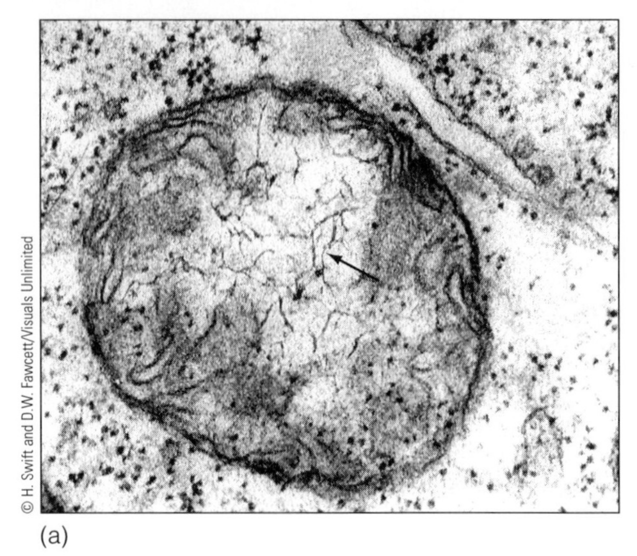

(a)

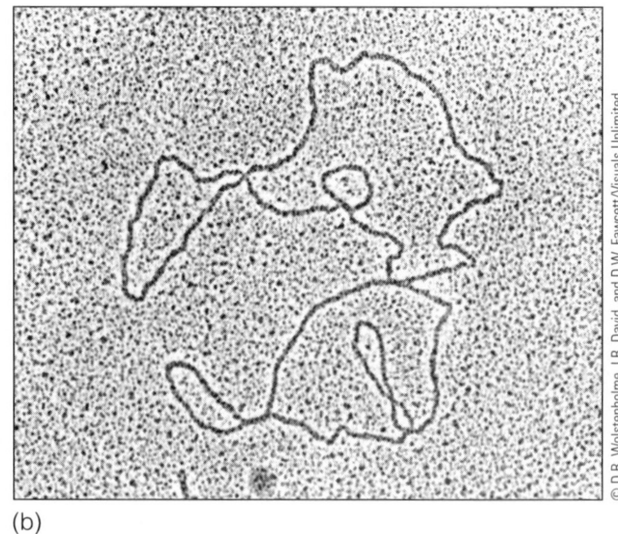

(b)

FIGURE 8.10 The mitochondrial genome. (a) An electron micrograph shows filaments of DNA in the mitochondrial matrix (arrow). (b) Mitochondrial DNA occurs in circles of 5 to 6 μm in contour length.

because each cell has hundreds of mitochondria, thousands of mitochondrial chromosomes are present in each human cell. The DNA of the mitochondrial chromosome is not complexed with proteins and physically resembles the chromosomes of bacteria and other prokaryotes. This similarity reflects the evolutionary history of mitochondria and their transition from free-living prokaryotes to intracellular organelles.

Nuclear Chromosomes Have a Complex Structure

Chromatin The complex of DNA and proteins that makes up a chromosome.

Histones DNA-binding proteins that help compact and fold DNA into chromosomes.

A combination of biochemical, molecular, and microscopic techniques has provided a great deal of information about the organization and structure of human chromosomes. In humans and other eukaryotes, each chromosome contains a single DNA molecule. This DNA is combined with proteins to form **chromatin.** The major class of proteins in chromosomes is **histones.** They play a major role in chromosomal structure and gene regulation. Five types of histones are complexed with

FIGURE 8.11 Nucleosomes. (a) An electron micrograph that shows the association of DNA with histones to form nucleosomes connected by threads of DNA in chromatin from chicken blood cells. (b) A diagram that shows how DNA coils around the outside of a histone cluster in nucleosomes.

(a) (b)

DNA to form small spherical bodies known as **nucleosomes,** connected to each other by thin threads of DNA (▶ Figure 8.11). Nucleosomes consist of DNA wound around a core of eight histone molecules.

Nucleosomes A bead-like structure composed of histones wrapped with DNA.

Winding DNA around the histones shortens the length of the DNA molecule by a factor of six or seven. But because mitotic chromosomes are compacted by a factor of 5,000 to 10,000, there are obviously several levels of organization between the nucleosome and the chromosome, each of which involves additional folding and/or coiling of the DNA molecule. Several models have been proposed to explain how nucleosomes are organized into more complex structures. Most of these models are based on the idea that DNA/protein complexes (chromatin) fold into loops and fibers extending from a central protein scaffold or matrix. Studying chromosomes by electron microscopy has partly clarified the subchromosomal levels of organization. ▶ Figure 8.12 is a scanning electron micrograph of a mitotic chromosome. At high magnification, individual fibers can be observed. These fibers appear to be the result of coiling and folding the chromatin fibers.

A model of chromosomal organization, incorporating these features is shown in ▶ Figure 8.13.

The Nucleus Has a Highly Organized Architecture

The interphase nucleus is not a disorganized bag containing a diploid set of chromosomes and several nucleoli. Instead, the nucleus has an organized internal structure in which each chromosome occupies a distinct region, called a chromosome territory (▶ Figure 8.14). Chromosome territories do not overlap with one another, and are separated by spaces called interchromosomal domains. Nuclear organization is closely linked with function. As a result, these territories are not fixed; chromosomes move around in the nucleus at different times of the cell cycle. Some of these movements may be associated with DNA replication and chromosome duplication that takes place during S phase (review the cell cycle in Chapter 2). It has been proposed that DNA replication takes place at certain sites within the nucleus called "replication factories," and chromosomes move to these sites for replication.

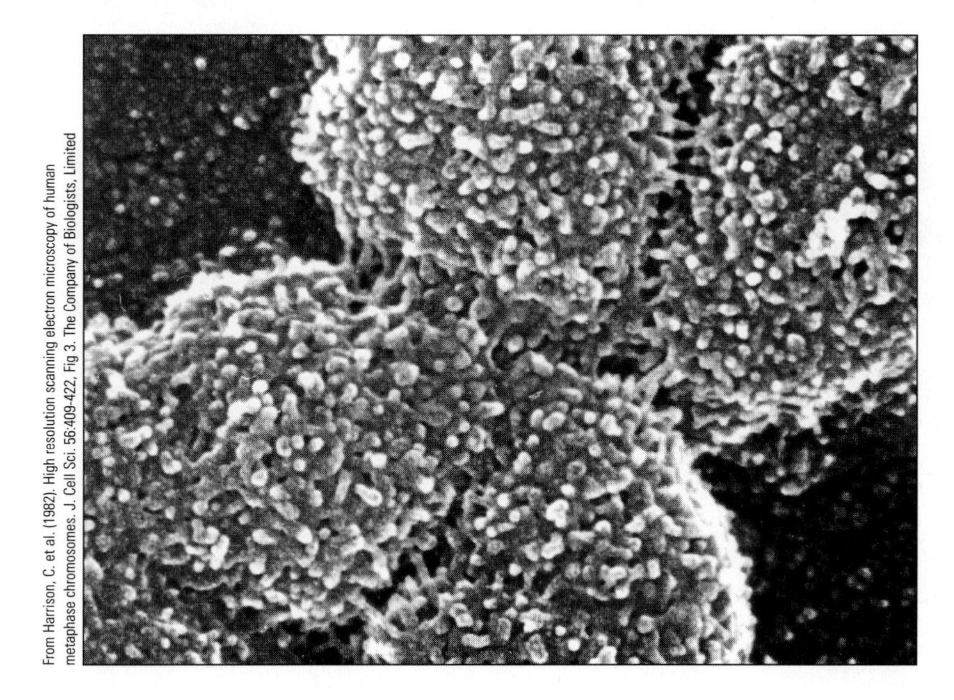

▶ **FIGURE 8.12** A scanning electron micrograph of the centromeric region of a human chromosome. Small fibrils whose diameter is about 30 nanometers result from the supercoiling of chromatin.

FIGURE 8.13 A model of chromosomal structure beginning with a double-stranded DNA molecule. The DNA is first coiled into nucleosomes. Then, the nucleosomes are coiled again and again into fibers that form the body of the chromosome. Chromosomes undergo cycles of coiling and uncoiling in mitosis and interphase, so their structure is dynamic.

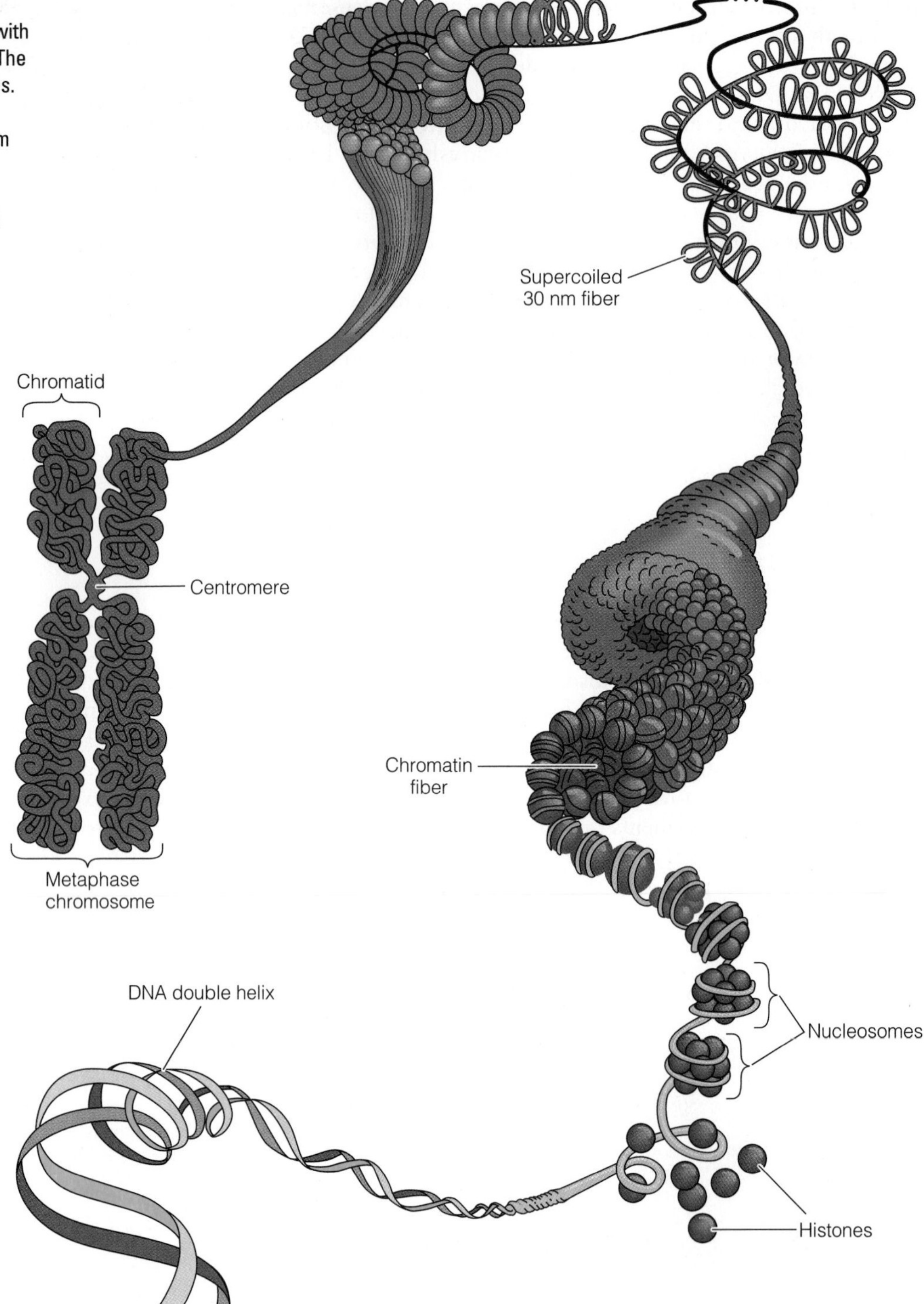

DNA REPLICATION DEPENDS ON COMPLEMENTARY BASE PAIRING

Between mitotic divisions, all cells replicate their DNA during the S phase of the cell cycle, so that each daughter cell will receive a complete set of genetic information. In their paper on the structure of DNA, Watson and Crick note that "It has not escaped our notice that the specific pairing we have postulated immediately suggests a possible copying mechanism for the genetic material." In a subsequent paper, they proposed a mechanism for DNA replication that depends on the complementary base pairing in the polynucleotide chains of DNA. If a DNA helix is unwound, each strand can serve as a template or pattern for synthesizing a new, complementary strand (Figure 8.15). This process is known as **semiconservative replication** because one old strand is conserved in each new molecule, and one new strand is synthesized.

DNA replication in all cells, from bacteria to humans, is a complex process requiring the action of more than a dozen different enzymes. In humans, replication begins at sites called origins of replication that are present at intervals along the length of the chromosome. At these sites, proteins unwind the double helix by breaking the hydrogen bonds between bases in adjacent strands. This opens the molecule to the action of the enzyme **DNA polymerase.** DNA polymerase reads the sequence in the strand being copied and links complementary nucleotides together to form a newly synthesized strand. The

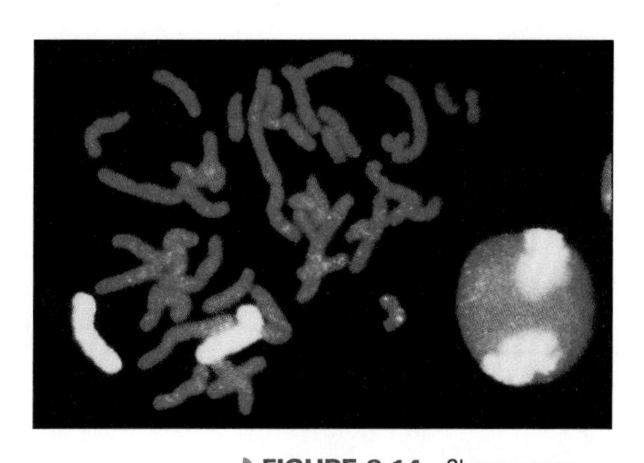

FIGURE 8.14 Chromosome painting highlights both copies of human chromosome 4 in a metaphase spread (left), and shows the chromosome territories occupied by chromosome 4 in the interphase nucleus (right). In the nucleus, each chromosome occupies a distinct territory, separated from other chromosomes by a region called the interchromosome domain, a region that is free of chromosomes. (Photo by Thomas Cremer. Courtesy of William C. Earnshaw from Lamond, A. I., and Earnshaw, W. C. Structure and function in the nucleus. Reprinted with permission from *Science,* 280:547–553 © American Association for the Advancement of Science.)

Semiconservative replication A model of DNA replication that provides each daughter molecule with one old strand and one newly synthesized strand. DNA replicates in this fashion.

DNA polymerase An enzyme that catalyzes the synthesis of DNA using a template DNA strand and nucleotides.

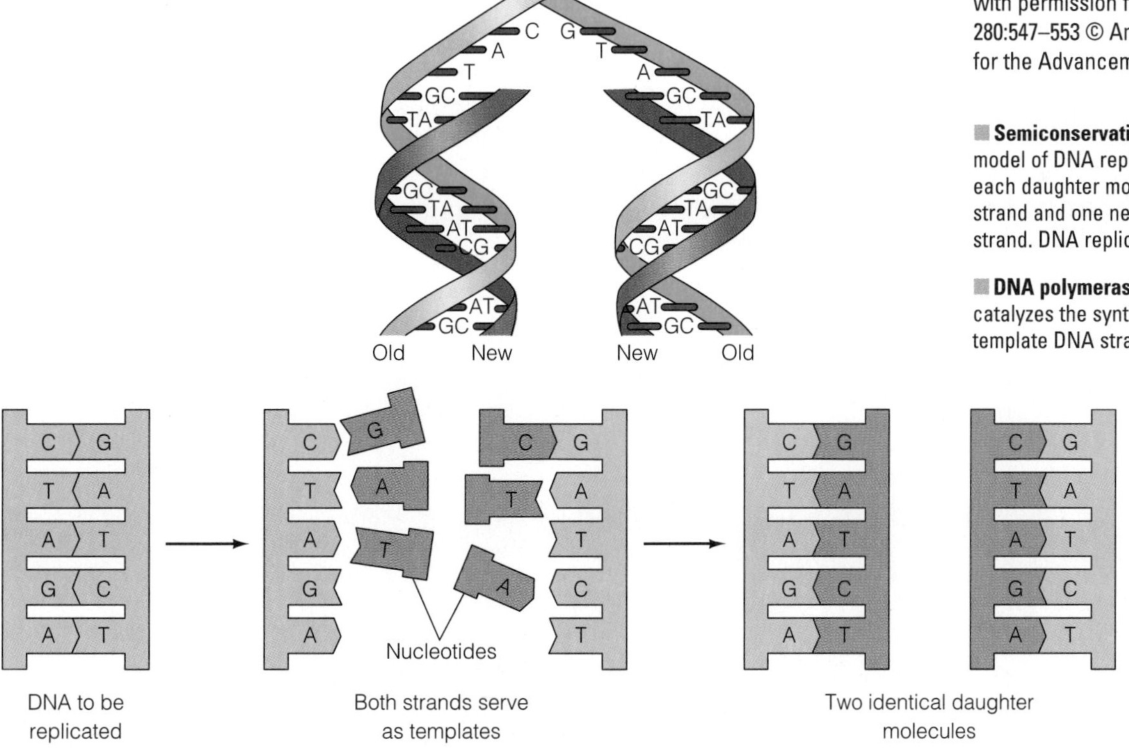

FIGURE 8.15 In DNA replication, the two polynucleotide strands uncoil, and each is a template for synthesizing a new strand. As the strands uncoil, complementary base pairing occurs with the template strand. These are linked together by DNA polymerase, assisted by other enzymes that help uncoil the DNA and seal up gaps in the new strands. A completed DNA molecule contains one new strand and one old strand.

completed DNA molecule contains one old strand (the strand that was copied) and one new strand (complementary to the old strand).

From the chromosomal perspective, recall that each chromosome contains one double-stranded DNA helix running from end to end. When replication is finished, the chromosome consists of two sister chromatids, joined at a common centromere. Each chromatid contains a DNA molecule that consists of one old strand and one new strand. When the centromeres divide at the beginning of anaphase, each chromatid becomes a separate chromosome that contains an accurate copy of the genetic information in the parental chromosome.

Summary

DNA Carries Genetic Information

1. At the turn of the twentieth century, scientists identified chromosomes as the cellular components that carry genes. This discovery focused efforts to identify the molecular nature of the gene on the chromosomes and the nucleus. Biochemical analysis of the nucleus began around 1870 when Frederick Miescher first separated nuclei from cytoplasm and described nuclein, a protein/nucleic acid complex now known as chromatin.

DNA Mediates the Transfer of Genetic Traits

2. Originally, proteins were regarded as the only molecular component of the cell with the complexity to encode the genetic information. This changed in 1944 when Avery and his colleagues demonstrated that DNA is the genetic material in bacteria.

Watson, Crick, and the Structure of DNA

3. In 1953 Watson and Crick constructed a model of DNA structure that incorporated information from the chemical studies of Chargaff and the x-ray crystallographic work of Wilkins and Franklin. They proposed that DNA is composed of two polynucleotide chains oriented in opposite directions and held together by hydrogen bonding to complementary bases in the opposite strand. The two strands are wound around a central axis in a right-handed helix.
4. The mitochondrial chromosome, carrying genes that can cause maternally transmitted disorders, is a circular DNA molecule.

DNA Is a Double Helix

5. Within chromosomes, DNA is coiled around clusters of histones to form structures known as nucleosomes. Supercoiling of nucleosomes may form fibers that extend at right angles to the axis of the chromosome. The structure of chromosomes must be dynamic to allow the uncoiling and recoiling seen in successive phases of the cell cycle, but the details of this transition are still unknown.

From DNA Molecules to Chromosomes

6. In DNA replication, strands are copied to produce semiconservatively replicated daughter strands.

CASE 1

There is considerable debate over the patenting of genes. This controversy is largely fueled by the work of the Human Genome Project and the biotechnology industry. Despite numerous meetings and publications on the subject, U.S. patent laws have not developed a policy that allows maximum innovation from biotech inventions. The first gene patents, issued in the 1970s, were granted for genes whose full nucleotide sequence was known, the protein product was known, and the protein's function was well understood.

Since then, genome projects have produced new ways of finding genes. Short sequences, only 25 to 30 nucleotides in length, called expressed sequence tags (ESTs), can be used to identify genes, but provide no information about the entire gene, the product, the function of the

Case Studies (continued)

product, or its association with any genetic disorder. Using gene-hunting software, researchers can take a short sequence of DNA and use it to search gene databases, turning up theoretical information about the sequence. For example, the sequence may belong to a gene encoding a plasma membrane protein or is similar to one in yeast that is involved in cell-cell signaling. At the present time, there are tens of thousands of ESTs and gene-hunting patent applications filed at the U.S. Patent Office.

The unresolved question, at the moment, is how much do you need to know about a gene and its usefulness to file a patent application. How should utility be defined? The diagnosis of disease certainly meets the definition of utility. Many discoveries have identified disease genes such as for cystic fibrosis, fragile X syndrome, breast cancer, colon cancer, and obesity. Many of these discoveries have patents based on diagnostic utility. An increasing number of patent applications are being filed for discoveries of hereditary disease genes. These discoveries frequently lack immediate use for practical therapy, however, because gene discovery does not always include knowledge of gene function or a plan for developing a disease therapy.

The impact of a decision about gene patents is enormous. Pharmaceutical and biotech companies have invested hundreds of millions of dollars in identifying genes to be used in developing diagnostic tests and drugs. Without patents, it is unlikely that companies will invest in developing these drugs. On the other hand, patenting genes can lead to royalty-based gene testing with exorbitant fees, and licensing arrangements requiring payment to companies who own the patent on a particular gene. As the results of the Human Genome Project redefine health care, these issues are important to everyone.

1. What is a patent?
2. Is patenting a gene different from patenting another product or invention? Should patents be awarded for genes under any circumstances? Explain.
3. If patenting genes were not allowed, do you think it would slow gene research in a significant way?

CASE 2

A 34-year-old woman and her 1-month-old newborn were seen by a genetic counselor in the neonatal intensive care unit in a major medical center. The neonatologist was suspicious that the newborn boy had a genetic condition and requested a genetic evaluation. The newborn was very pale, was failing to thrive, had diarrhea, and had markedly increased serum cerebrospinal fluid lactate levels. In addition, he had severe muscle weakness with chart notes describing him as "floppy," and he had already had two seizures since birth. The neonatologist reported that the infant was currently suffering from liver failure, which would probably result in his death in the next few days. The panel of tests performed on the infant led the neonatologist and genetic counselor to the diagnosis of Pearson syndrome. The combination of marked metabolic acidosis and abnormalities in bone marrow cells is highly suggestive of Pearson syndrome.

Pearson syndrome is associated with a large deletion of the mitochondrial (mt) genome. The way the deleted mtDNA molecules are distributed over individual cells during mitosis is not known. However, it is assumed that during cell division daughter cells randomly receive mitochondria carrying wild type (WT) or mutant mtDNA. Mitochondrial DNA is, theoretically, transmitted only to offspring through the mother via the large cytoplasmic component of the oocyte. Nearly all cases of Pearson syndrome arise from new mutational events. Mitochondria have extremely poor DNA repair mechanisms, and mutations accumulate very rapidly. Most infants with Pearson syndrome die before age 3, often due to infection, or liver failure.

A diagnosis of Pearson syndrome results in an extremely grave prognosis for the patient. Unfortunately, at this point, treatment can be directed only toward symptomatic relief.

1. How would a large deletion in the mitochondrial genome cause a disease?
2. Why doesn't the mother have the disease if she has mutant mitochondrial DNA?
3. How would you react to hearing this diagnosis? How would you counsel a couple through this kind of situation?

Questions and Problems

DNA Carries Genetic Information

1. Until 1944, which cellular component was thought to carry genetic information?
   a. carbohydrate
   b. nucleic acid
   c. protein
   d. chromatin
   e. lipid
2. Why do you think nucleic acids were not originally considered to be carriers of genetic information?

3. The experiments of Avery and his co-workers led to the conclusion that:
   a. bacterial transformation occurs only in the laboratory.
   b. capsule proteins can attach to uncoated cells.
   c. DNA is the transforming agent and is the genetic material.
   d. transformation is an isolated phenomenon in *E. coli.*
   e. DNA must be complexed with protein in bacterial chromosomes.
4. In the experiments of Avery, what was the purpose of treating the transforming extract with enzymes?
5. For this question, read the following experiment and interpret the results to form your conclusion. Experimental data: S bacteria were heat-killed and cell extracts were isolated. The extracts contained cellular components including lipids, proteins, DNA, and RNA. These extracts were mixed with live R bacteria and then injected together into mice along with various enzymes (proteases, RNAases, and DNAases) Proteases degrade proteins, RNAases degrade RNA, and DNAases degrade DNA.

S extract + live R cells	kills mouse
S extract + live R cells +1 protease	kills mouse
S extract +1 live R cells +1 RNAase	kills mouse
S extract + live R cells + DNAase	mouse lives

   Based on these results, what is the transforming principle?
6. Recently, scientists discovered that a rare disorder called polkadotism is caused by a bacterial strain, polkadotiae. Mice injected with this strain (P) develop polka dots on their skin. Heat-killed P bacteria and live D bacteria, a nonvirulent strain, do not produce polka dots when separately injected into mice. However, when a mixture of heat-killed P cells and live D cells were injected together, the mice developed polka dots. What process explains this result? Describe what is happening in the mouse to cause this outcome.

DNA Is a Part of Chromosomes

7. Nucleosomes are complexes of:
   a. nonhistone protein and DNA.
   b. RNA and histone.
   c. histones, nonhistone proteins, and DNA.
   d. DNA, RNA, and protein.
   e. amino acids and DNA.
8. Discuss the levels of chromosomal organization with reference to the following terms:
   a. nucleotide
   b. DNA double helix
   c. histones
   d. nucleosomes
   e. chromatin

DNA Is a Double Helix

9. List the pyrimidine bases, the purine bases, and the base pairing rules for DNA.
10. In analyzing the base composition of a DNA sample, a student loses the information on pyrimidine content. The purine content is A = 27% and G = 23%. Using Chargaff's rule, reconstruct the missing data, and list the base composition of the DNA sample.
11. The basic building blocks of nucleic acids are:
    a. nucleosides.
    b. nucleotides.
    c. ribose sugars.
    d. amino acids.
    e. purine bases.
12. Adenine is a:
    a. nucleoside.
    b. purine.
    c. pyrimidine.
    d. nucleotide.
    e. base.
13. Polynucleotide chains have a 5′ and a 3′ end. Which groups are found at each of these ends?
    a. 5′ sugars, 3′ phosphates
    b. 3′ OH, 5′ phosphates
    c. 3′ base, 5′ phosphate
    d. 5′ base, 3′ OH
    e. 5′ phosphates, 3′ bases
14. DNA contains many hydrogen bonds. Describe a hydrogen bond and how this type of chemical bond holds DNA together.
15. Watson and Crick received the Nobel Prize for:
    a. generating x-ray crystallographic data of DNA structure.
    b. establishing that DNA replication is semiconservative.
    c. solving the structure of DNA.
    d. proving that DNA is the genetic material.
    e. showing that the amount of A equals the amount of T.
16. State the properties of the Watson-Crick model of DNA in the following categories:
    a. number of polynucleotide chains
    b. polarity (running in same or opposite directions)
    c. bases on interior or exterior of molecule
    d. sugar/phosphate on interior or exterior of molecule
    e. which bases pair with which
    f. right- or left-handed helix
17. Using Figure 8.4 as a guide, draw a dinucleotide composed of C and A. Next to this, draw the complementary dinucleotide in an antiparallel fashion. Connect the dinucleotides with the appropriate hydrogen bonds.
18. A beginning genetics student is attempting to complete an assignment to draw a base pair from a DNA molecule. The drawing is incomplete, and the student does not know how to finish. He or she asks for your advice. The assignment sheet shows that the drawing is to contain three hydrogen bonds, a purine, and a pyrimidine. From your knowledge of the pairing rules and the number of hydrogen bonds in A/T and G/C base pairs, what base pair do you help the student draw?

RNA Is a Single-Stranded Nucleic Acid

19. What is the purpose of making an RNA copy of the DNA in gene expression?
20. How does DNA differ from RNA with respect to the following characteristics?
    a. number of chains
    b. bases used
    c. sugar used
    d. function
21. RNA is ribonucleic acid, and DNA is *deoxy*ribonucleic acid. What exactly is deoxygenated about DNA?

DNA Replication

22. What is the function of DNA polymerase?
    a. It degrades DNA in cells.
    b. It adds RNA nucleotides to new strand.
    c. It coils DNA around histones to form chromosomes.
    d. It adds DNA nucleotides to a replicating strand.
    e. none of the above
23. Which of the following statements is *not* true about DNA replication?
    a. It occurs during the M phase of the cell cycle.
    b. It makes a sister chromatid.
    c. It denatures DNA strands.
    d. It occurs semiconservatively.
    e. It follows base pairing rules.
24. Make the complementary strand to the following DNA template, and label both strands as 5′ to 3′, or 3′ to 5′ (P = phosphate in the diagram). Draw an arrow showing the direction of synthesis of the new strand. How many hydrogen bonds are in this double strand of DNA?

    template P - AGGCTCG - OH
    new strand:

25. How does DNA replication occur in a precise manner to ensure that identical genetic information is put into the new chromatid? See Figure 8.15.

Internet Activities

The following activities use the resources of the World Wide Web to enhance the topics covered in this chapter. To investigate the topics described below, log onto the book's home page (www.brookscole.com/biology_d) and follow the *Student Resources* link to your subject and text.

1. *Experimenting with the Structure of DNA.* The Genetic Science Learning Center (a joint project of the University of Utah and the Utah Museum of Natural History) provides general genetics information to students and the community. Go to http://gslc.genetics.utah.edu/basic/howto/index.html to discover "How to Extract DNA from Anything Living." This is an activity that you *can* do at home, but even if you choose not to try the experiment, you can still learn from the experimental design and discussion.
    a. After you've blended the peas (or other biological material), water, and salt, the next step is to add detergent. What is the purpose of adding detergent at this step? Where was the DNA prior to this step? Where will the DNA be after this step?
    b. In the next step, you are asked to add meat tenderizer to the solution. Meat tenderizer contains enzymes. What is the purpose of this step?
    c. In the third step, you add alcohol to the solution. What happens? What *is* that stringy stuff in the photographs and in your test tube?
    d. In the second step, you added meat tenderizer, which contains enzymes that break down proteins. How would you expect the results of this extraction experiment to be different if you added an enzyme that breaks the bonds in DNA?

Further Exploration: The Genetic Science Learning Center home page (http://gslc.genetics.utah.edu/index.html) has a variety of review materials, interesting visuals, and fun activities.

2. *How Do Scientific Advances Occur? Access Excellence* (www.accessexcellence.org) is a Web site for "health and bioscience teachers and learners" run by the National Health Museum. Follow the "About Biotech" link to the *Biotech Chronicles* and then to "Pioneer Profiles." Read the profiles of Rosalind Franklin and James Watson. Then, from the home page, follow the "Activities Exchange" link to the *Classic Collection* and read "A Visit with Dr. Francis Crick."
    a. It's interesting to note that Francis Crick studied mathematics and physics before his curiosity motivated him to move into the study of biology. How might his background in math and physics have helped prepare him to work on the problem of DNA structure?
    b. How objective is the process of science? How much of a role do you think personality and interpersonal dynamics might play in determining how and when discoveries in science are made—and who makes them? In his book *The Double Helix*, James Watson writes of an international race to determine the structure of DNA; he describes Rosalind Franklin as a dowdy underling with little or no interest in the DNA problem. In contrast, Rosalind Franklin's biographer states that Franklin herself was an established and respected researcher who

was known to be pursuing the DNA question, and that her critical x-ray diffraction data were provided to Crick and Watson without her knowledge. In the interview, what perspective do the interviewer and Dr. Crick seem to have? How might the particular interactions and personalities of the different researchers (as portrayed in the interview and in the two profiles) have played a role in this discovery?

Further Exploration: The "On-line Biology Book" has a good overview of DNA and molecular genetics that goes through the process of the discovery of DNA. (http://gened.emc.maricopa.edu/bio/bio181/BIOBK/BioBookDNAMOLGEN.html)

3. *Is The Pursuit of Science Always Objective and Unbiased? Access Excellence* (www.accessexcellence.org) was first developed in 1993 by the pioneering biotechnology company Genentech. In 1999, the Web site was donated to the nonprofit National Health Museum but is still partially funded and intellectually supported by Genentech.
   a. Are there any biases you can detect in the description of *Access Excellence*? What would be some potentially positive and potentially negative aspects of using educational information provided by a company in an emerging (and sometimes controversial) health technology field?

For Further Reading

Felsenfeld, G. (1985, October). DNA. *Sci. Am., 253,* 58–67.

Judson, H. F. (1979). *The eighth day of creation: The makers of the revolution in biology.* New York: Simon & Schuster.

Kornberg, R., & Klug, A. (1981, February). Nucleosomes. *Sci. Am., 244,* 52–79.

McCarty, M. (1986). *The transforming principle: Discovery that genes are made of DNA.* New York: Norton.

Olby, R. (1974). *The path to the double helix.* Seattle: University of Washington Press.

Sayre, A. (1975). *Rosalind Franklin and DNA.* New York: Norton.

Watson, J. D. (1968). *The double helix.* New York: Atheneum.

Watson, J. D., & Crick, F. H. C. (1953). Molecular structure of nucleic acids: A structure for deoxyribose nucleic acid. *Nature, 171,* 737–738.

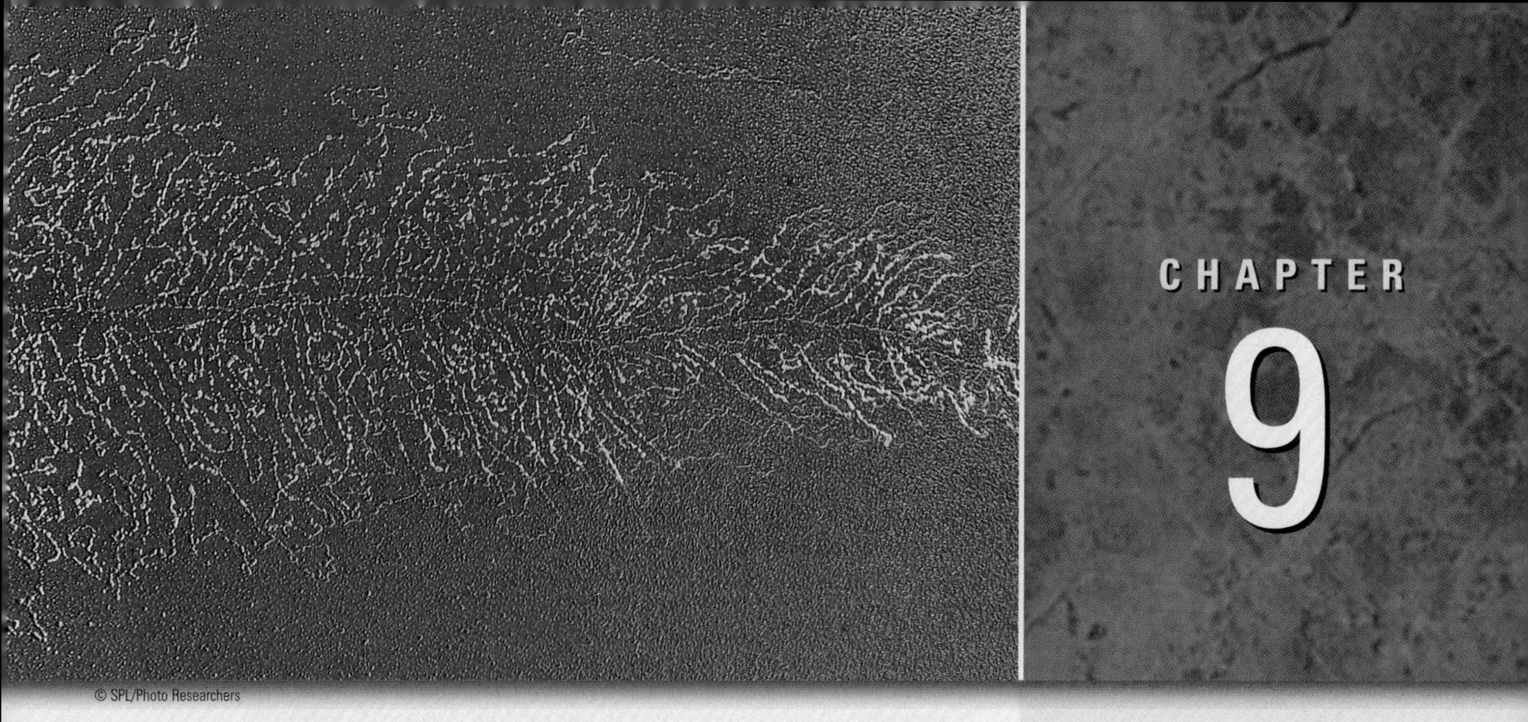

© SPL/Photo Researchers

CHAPTER

9

Gene Expression: How Proteins Are Made

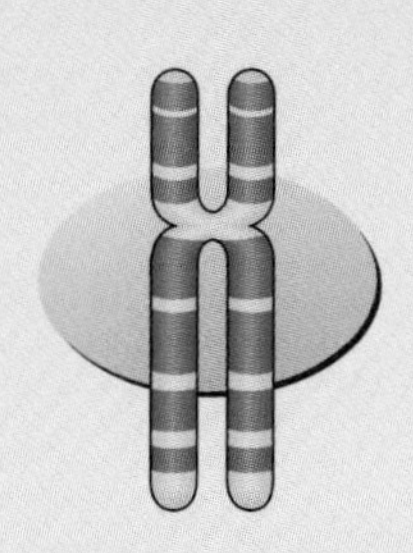

Chapter Outline

Just after World War II, Charles Dent began analyzing the molecules excreted in human urine. To do this, he spotted a sample near the bottom of a piece of filter paper and dried it thoroughly. He placed the paper in a jar containing a small amount of solvent, so that the bottom of the paper just touched the solvent. The top of the jar was sealed, and gradually the solvent moved up the filter paper by capillary action. As the solvent moved upward, the molecules in the urine sample also migrated upward. The molecules in the urine migrated at different rates and were separated from each other. When the solvent reached the top of the paper, Dent removed the filter paper and sprayed it to visualize the amino acids in the sample, which showed up as a series of spots.

Dent and another scientist, Henry Harris, used this technique, called paper chromatography, to examine the amino acids excreted by patients with a genetic disorder called cystinuria. In cystinuria, a defect in kidney function causes large quantities of the amino acid cystine to appear in the urine. In some case, the concentration is so high that cystine crystals or even stones are excreted.

Harris and Dent showed that there are two forms of cystinuria, one characterized by the excretion of cystine, lysine, and three other amino acids, and a second form marked by the excretion of just cystine and lysine. Through a combination of genetic and biochemical studies, Harris established that the first type is caused by a homozygous recessive genotype and the second by a heterozygous condition. This work had three important consequences: (1) it demonstrated that some genetic disorders have biochemical phenotypes, (2) it helped establish human biochemical genetics as a field of genetic research, and (3) it helped forge the link between genes and phenotypes through biochemical processes.

In this chapter, we will examine the relationship between genes and proteins and the role of DNA as a carrier of genetic information. We also discuss the transfer of genetic information from the sequence of nucleotides in DNA to the sequence of amino acids in protein.

THE LINK BETWEEN GENES AND PROTEINS

What are proteins? How and where are they made? How are proteins related to genes and in particular to human genetics? As we see in this chapter, proteins like blood-clotting factors and digestive enzymes are gene products. Proteins are the intermediate between genes and phenotype. The phenotypes of a cell, tissue, and organism are all the result of protein function. When these functions are absent or changed, the result is a mutant phenotype, which we describe as a genetic disorder.

At the turn of the twentieth century, Archibald Garrod recognized the relationship between proteins and phenotype after his studies on a condition called **alkaptonuria** (MIM/OMIM 203500). Infants with alkaptonuria can be identified soon after birth because their urine turns black. Garrod found that affected infants put out large quantities of a compound he called alkapton (now called homogentisic acid) in their urine. He reasoned that in unaffected individuals, homogentisic acid is converted into other products and does not build up in the urine. In alkaptonuria, however, the conversion is blocked, causing the buildup of homogentisic acid, which is excreted in the urine. He called this condition an "inborn error of metabolism." Garrod also discovered that alkaptonuria is a genetic disorder and is inherited as an autosomal recessive trait.

Alkaptonuria An autosomal recessive trait with altered metabolism of homogentisic acid. Affected individuals do not produce the enzyme needed to metabolize this acid.

Genes and Metabolism

Biochemical reactions, like the conversion of homogentisic acid, are organized into chains of chemical reactions known as metabolic pathways. In a pathway, an enzyme converts one compound into another. This second compound is converted into a third, and so forth (◗ Figure 9.1). Each reaction is carried out by a different enzyme. Enzymes are protein molecules that serve as biological catalysts to carry out specific biochemical reactions. Loss of activity in a single enzyme can disrupt an entire pathway, causing an alteration in the cell's biochemistry and producing an altered phenotype. Case 1 at the end of the chapter deals with a disorder caused by the loss of a single enzyme by mutation.

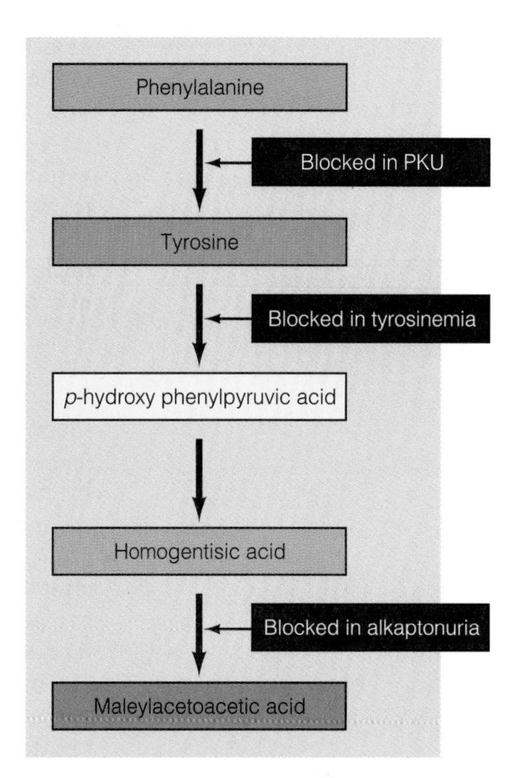

◗ **FIGURE 9.1** A metabolic pathway beginning with the amino acid phenylalanine. The enzyme-catalyzed steps are shown as arrows. Mutations cause blocks in this pathway, leading to genetic disorders, including phenylketonuria (PKU), tyrosinemia, and alkaptonuria.

The Relationship Between Genes and Enzymes

George Beadle and Edward Tatum established the connection between enzymes and the phenotype in the late 1930s and early 1940s through their work on *Neurospora,* a common bread mold (◗ Figure 9.2). Using *Neurospora,* they demonstrated that mutation of a single gene caused loss of activity in a single enzyme, resulting in a mutant phenotype. This connection between a mutant gene, a mutant enzyme, and a mutant phenotype was a key step in understanding that genes produce phenotypes through the action of proteins. Beadle and Tatum received the Nobel Prize in 1958 for their work.

GENETIC INFORMATION IS STORED IN DNA

Because proteins are the products of genes, and genes are made up of DNA, information encoded in DNA must control the kinds and amounts of proteins present in the cell. But just how is genetic information carried in DNA? The Watson-Crick

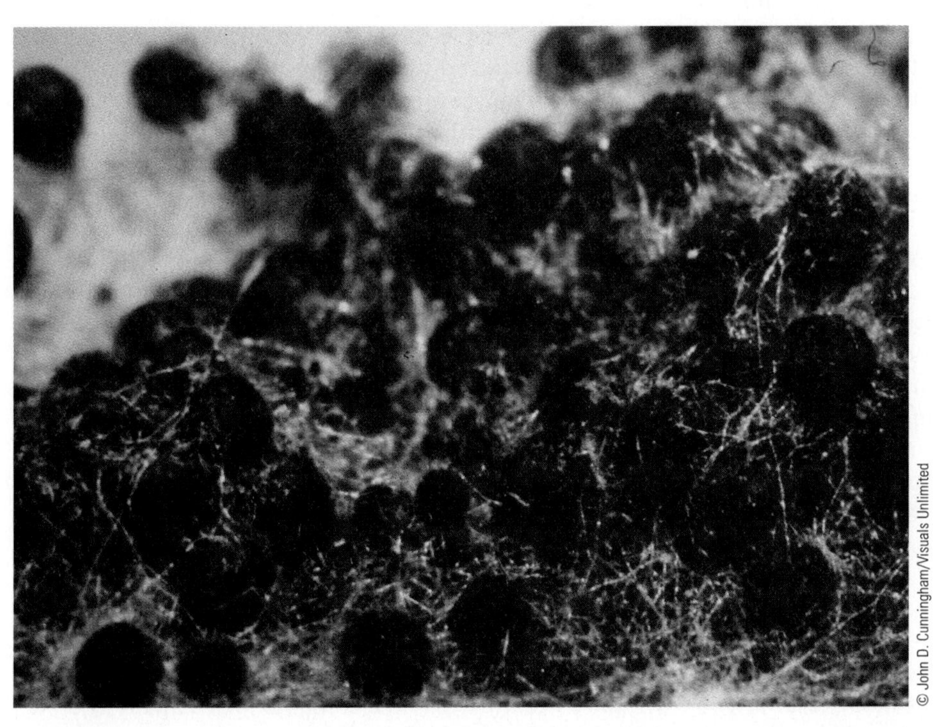

◗ **FIGURE 9.2** A culture of the mold *Neurospora crassa,* an organism used by George Beadle and Edward Tatum in their work on biochemical genetics.

model of DNA provides an explanation for the way genetic information is encoded in the structure of the molecule. Watson and Crick proposed that genetic information is encoded as the sequence of nucleotides in DNA. The amount of information stored in any cell is related to the number of nucleotide pairs in the DNA carried within the cell. This number ranges from a few thousand base pairs in some viruses to more than 3 billion base pairs in humans and more than twice that amount in some amphibians and plants. A gene typically consists of hundreds or thousands of nucleotides. Each gene has a beginning and end, marked by specific nucleotide sequences, and a molecule of DNA can contain thousands of genes.

How do genes (in the form of DNA) control the production of proteins? Proteins are linear molecules assembled from subunits called amino acids. Twenty different types of amino acids are used to assemble proteins. The diversity of proteins found in nature results from the number of possible combinations of these 20 different amino acids. Because each amino acid position in a protein can be occupied by any of 20 amino acids, the number of different combinations is 20^n, where n is the number of amino acids in the protein. In a protein composed of only 5 amino acids, 20^5, or 3,200,000, combinations are possible, each of which has a different amino acid sequence and a potentially different function. Most proteins are actually composed of several hundred amino acids, so literally billions and billions of combinations are possible.

If DNA has only four different nucleotides (A,T, C, and G) and there are 20 different amino acids in proteins, how can only four nucleotides encode the information for all 20 amino acids?

THE GENETIC CODE: THE KEY TO LIFE

Information that spells out the number and order of the amino acids that make up a specific protein is encoded in the nucleotide sequence of a gene. Because DNA is composed of only four different nucleotides, at first glance it may seem difficult to envision how the information for literally billions of different combinations of 20 different amino acids can be carried in DNA. How exactly does DNA encode genetic information? Let's start with the simplest case. If each nucleotide encoded the information for one amino acid, only four different amino acids could be inserted into proteins (four nucleotides, taken one at a time, or 4^1). If a sequence of two nucleotides encoded the information for one amino acid, only 16 combinations would be possible (four nucleotides, taken two at a time, or 4^2). On the other hand, a sequence of three nucleotides allows 64 combinations (four nucleotides, taken three at a time, or 4^3), which doesn't seem right, because there are 44 more combinations than the 20 needed to encode amino acids.

The problem of how many nucleotides are needed to encode one amino acid was answered in a series of experiments using mutants of a bacteriophage called T4, conducted by Francis Crick, Sidney Brenner, and their colleagues. By analyzing mutations in T4 genes, they discovered that a sequence of three nucleotides encodes the information for one amino acid. They also found that some amino acids could be specified by more than one combination of three nucleotides. This built-in redundancy in the code uses up most of the other 44 combinations. This work established that in DNA a genetic code consists of a linear series of nucleotides, read three at a time, and that each triplet specifies an amino acid.

After Crick and Brenner's work, the question of which sets of three nucleotides encodes which amino acids was quickly worked out, and the coding information contained in all 64 triplet combinations was established (Table 9.1). By convention, the genetic code is written in the form of mRNA, and each group of three nucleotides is called a **codon.** Table 9.1 shows that 61 of these combinations actually code for amino acids, whereas three (UAA, UAG, and UGA) do not encode amino acids, but are called stop codons. One codon, AUG, has two functions. It encodes the infor-

Codon Triplets of nucleotides in mRNA that encode the information for a specific amino acid in a protein.

Table 9.1 Codons on Messenger RNA and Their Corresponding Amino Acids

Codon	Amino Acid	Codon	Amino Acid	Codon	Amino Acid	Codon	Amino Acid
AAU	Asparagine	CAU	Histidine	GAU	Aspartic acid	UAU	Tyrosine
AAC		CAC		GAC		UAC	
AAA	Lysine	CAA	Glutamine	GAA	Glutamic acid	UAA	Stop*
AAG		CAG		GAG		UAG	
ACU	Threonine	CCU	Proline	GCU	Alanine	UCU	Serine
ACC		CCC		GCC		UCC	
ACA		CCA		GCA		UCA	
ACG		CCG		GCG		UCG	
AGU	Serine	CGU	Arginine	GGU	Glycine	UGU	Cysteine
AGC		CGC		GGC		UGC	
AGA	Arginine	CGA		GGA		UGA	Stop*
AGG		CGG		GGG		UGG	Tryptophan
AUU	Isoleucine	CUU	Leucine	GUU	Valine	UUU	Phenylalanine
AUC		CUC		GUC		UUC	
AUA		CUA		GUA		UUA	Leucine
AUG	Methionine**	CUG		GUG		UUG	

**Stop codons signal the end of the formation of a polypeptide chain.*
***Codon has two functions: specifies the amino acid methionine and serves as the start codon, marking the beginning of a polypeptide chain.*

mation for the amino acid methionine, and it also serves as the start codon, the first codon in a gene, marking the beginning of a coding sequence for a specific protein.

An important feature of the genetic code is that the same codons are used for the same amino acids in viruses and organisms, including bacteria, algae, fungi, ciliates, and multicellular plants and animals. The universal nature of the genetic code means that the genetic code was established very early during the evolution of life on this planet. The existence of a universal code is strong evidence that all living things are closely related and may have evolved from a common ancestor.

There are, however, some rare exceptions to the universal nature of the genetic code. These exceptions involve stop codons. In some species, such as the ciliates *Tetrahymena* and *Paramecium,* UAG and UAA are not stop codons. Instead, they encode glutamine. These and some other species of ciliates have only one stop codon (UGA). Other more substantial alterations in the genetic code have taken place in mitochondria and point to a divergent evolutionary pathway for this organelle.

Now that we know how the information for proteins is encoded in DNA, let's turn our attention to another question: How is the linear sequence of nucleotides in a gene converted into the linear sequence of amino acids in a protein? In humans and other eukaryotes, almost all the cell's DNA is found in the nucleus (some is in the mitochondria), and almost all proteins are found in the cytoplasm. This means that the process of information transfer from gene to gene product must be indirect.

TRACING THE FLOW OF GENETIC INFORMATION FROM NUCLEUS TO CYTOPLASM

The transfer of genetic information from nucleotides in a DNA molecule into the amino acids of a protein has two major steps. First, the information encoded in a gene is copied into a molecule of RNA known as **messenger RNA (mRNA).** This step is called **transcription** (Figure 9.3). In humans and other eukaryotes, the mRNA moves to the cytoplasm, where the information encoded in the nucleotide sequence of the mRNA is converted into the amino acid sequence of a protein. This step is called **translation** (Figure 9.3).

mRNA A single-stranded complementary copy of the nucleotide sequence in a gene.

Transcription Transfer of genetic information from the base sequence of DNA to the base sequence of RNA, mediated by RNA synthesis.

Translation Conversion of information encoded in the nucleotide sequence of an mRNA molecule into the linear sequence of amino acids in a protein.

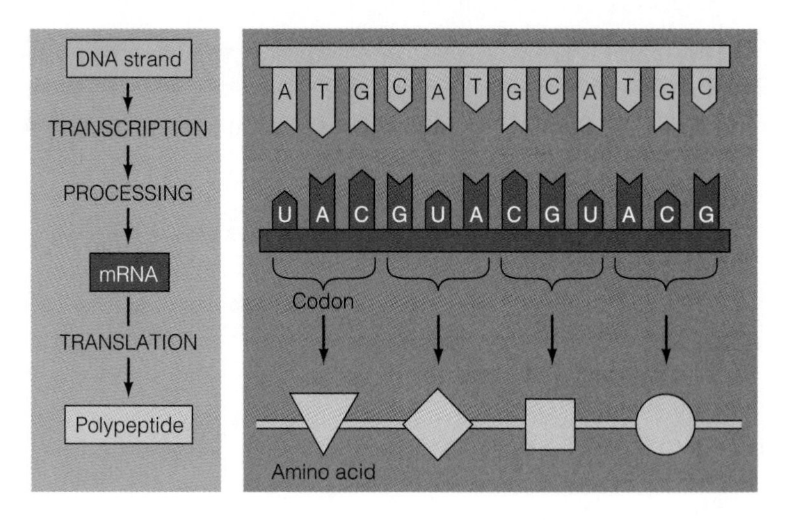

FIGURE 9.3 The flow of genetic information. One strand of DNA is transcribed into a strand of mRNA. The mRNA is processed and moves to the cytoplasm, where it is converted into the amino acid sequence of a protein.

The amino acid sequence, in turn, determines the structural and functional characteristics of the protein and its role in phenotypic expression. In the next sections, we will examine the steps of protein synthesis in more detail.

Transcription Produces Genetic Messages

In the nucleus, transcription begins when a section of a DNA double helix in a chromosome unwinds, and one strand is used as a template to make an mRNA molecule. Transcription has three stages: initiation, elongation, and termination. In the first step, an enzyme called RNA polymerase binds to a specific nucleotide sequence in the DNA, just outside a gene. This sequence is called a **promoter region.** After the polymerase is bound, the two strands of DNA in the gene unwind, exposing the DNA strand that will be a template for RNA synthesis.

Promoter region A region of a DNA molecule to which RNA polymerase binds and initiates transcription.

In elongation, RNA polymerase links RNA nucleotides to form an RNA molecule (Figure 9.4). The rules of base pairing in transcription are the same as in DNA replication, with one exception: an A on the DNA template ends up as a U in the RNA transcript (recall from Chapter 8 that there is no T in RNA, so there is no A:T pairing in RNA). In humans, 30 to 50 nucleotides per second are added to an mRNA molecule. In bacteria, mRNA is synthesized at rates up to 500 nucleotides per second, about 10 times faster than in humans.

The end of the gene is marked by a nucleotide sequence called a **terminator region.** When the RNA polymerase reaches this point, it detaches from the DNA template strand, the mRNA molecule is released, the DNA strands re-form a double helix, and transcription is terminated (Figure 9.4). The length of the mRNA transcript depends on the size of the gene. Most transcripts in humans are about 5,000 nucleotides long, although lengths up to 200,000 nucleotides have been reported.

Terminator region The nucleotide sequence at the end of a gene that signals the end of transcription.

Most Human Genes Have a Complex Internal Organization

Many, if not most, human genes contain nucleotide sequences that are transcribed but not translated into the amino acid sequence of a protein. These sequences, called **introns,** can vary in number from zero to 75 or more. Introns also vary in size, ranging from about 100 nucleotides up to more than 100,000. Most research has shown that introns have no function.

Introns DNA sequences present in some genes that are transcribed but are removed during processing and therefore are not present in mature mRNA.

The groups of nucleotides in a gene that are transcribed and translated into the amino acid sequence of a protein are called **exons.** The internal organization of a typical human gene is shown in Figure 9.5. The combination of exons and introns determines the length of a gene, and often the exons are only a small fraction

Exons DNA sequences that are transcribed, joined to other exons during mRNA processing, and translated into the amino acid sequence of a protein.

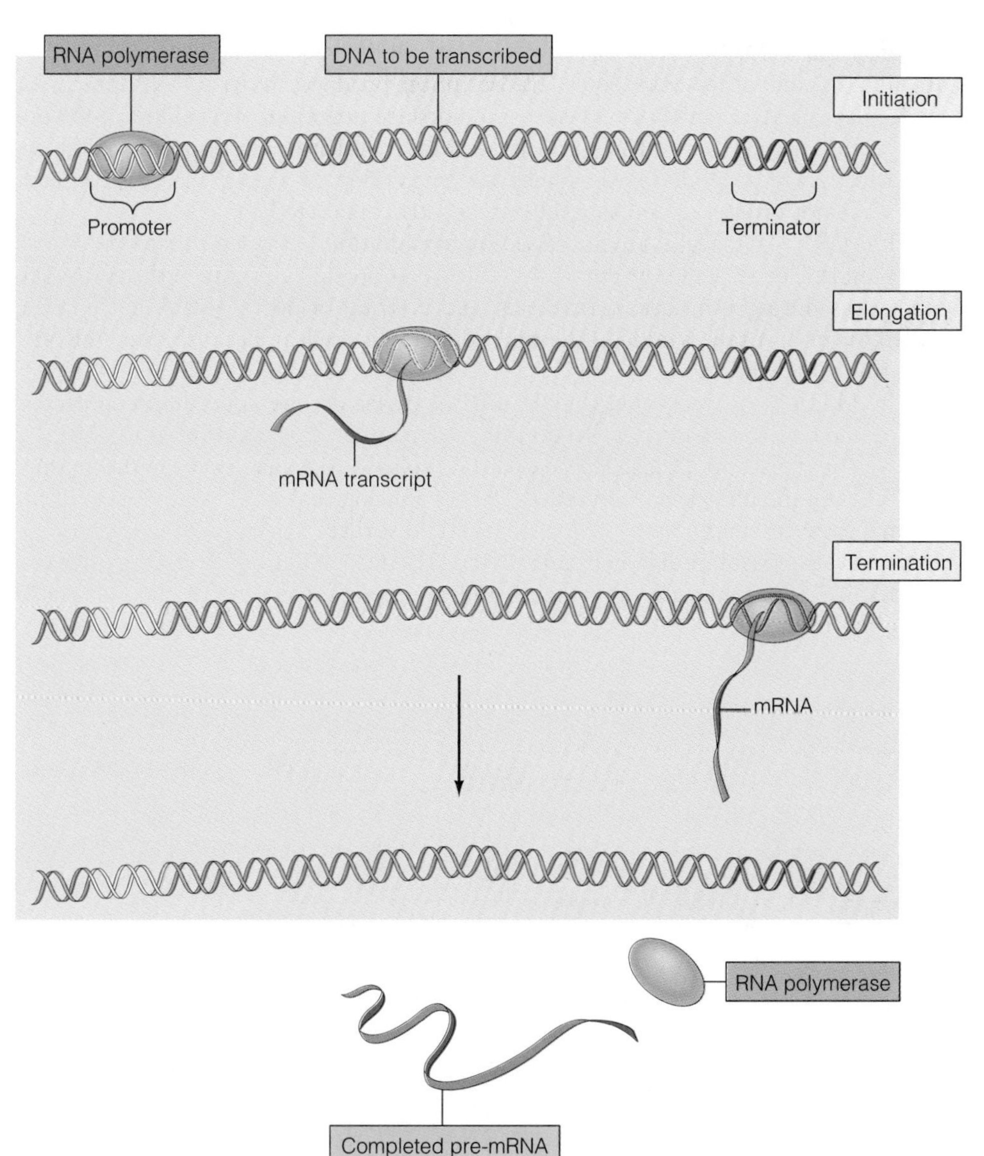

FIGURE 9.4 Transcription begins when the enzyme RNA polymerase attaches to a promoter sequence that marks the beginning of a gene. One strand of DNA (the template strand) is transcribed into a complementary RNA molecule. Transcription ends when the RNA polymerase reaches a terminator sequence that marks the end of the gene.

of the total nucleotides in a gene. For example, the dystrophin gene (see Chapter 4 for a discussion of muscular dystrophy and dystrophin) is more than 2 million nucleotides in length and contains 79 introns, which make up more than 99% of the gene. Most genes are not as long as the dystrophin and do not have as many introns. The alpha globin gene (discussed in Chapter 10) is 800 nucleotides in length and has two introns that make up 30% of the gene.

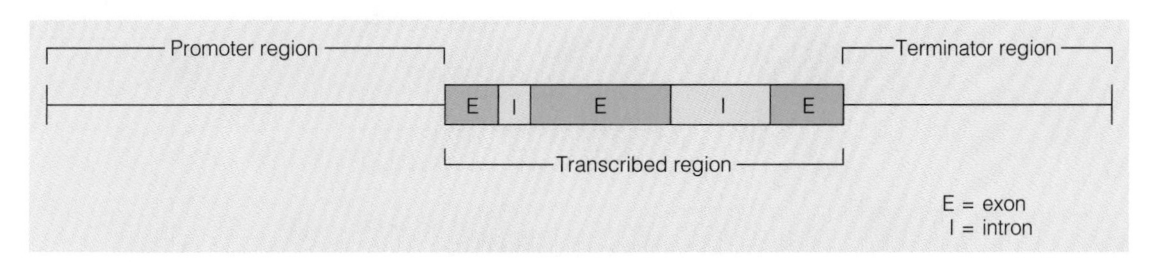

FIGURE 9.5 Organization of a typical eukaryotic gene. RNA polymerase binds to the promoter region for transcription. The part of the gene that is transcribed consists of all the introns and exons. Transcription ends in the terminator region. Only the sequences in the exons appear in the mature mRNA and are translated into the amino acid sequence of a protein.

Messenger RNA Is Processed and Spliced

In humans and other eukaryotes, transcription produces large mRNA precursor molecules called pre-mRNAs. These precursors are processed and spliced in the nucleus to produce mature mRNA molecules. Mature mRNA is transported to the cytoplasm, where it binds to ribosomes for translation (◗ Figure 9.6). In bacteria, mRNA is not processed and spliced, but is translated directly.

Cap A modified base (guanine nucleotide) attached to the 5′ end of eukaryotic mRNA molecules.

Poly-A tail A series of A nucleotides added to the 3′ end of mRNA molecules.

Pre-mRNA molecules are processed by the addition of nucleotides to the 5′ and the 3′ ends. The sequence at the 5′ end, known as a **cap,** helps attach the mRNA to ribosomes during translation. At the 3′ end a string of 30 to 100 A nucleotides, called the **poly-A tail,** is added. The role of the tail is unclear, because some mRNAs lack this modification.

In addition to processing, the pre-mRNA molecules are spliced and shortened during the removal of introns. To do this, enzymes cut the transcript at the borders between introns and exons. The exons are spliced together to form the mature mRNA, and the introns are discarded. Proper splicing of pre-mRNA is essential for normal gene function. Several human genetic disorders are caused by splicing defects. In a hemoglobin disorder called β-thalassemia, mutations at the intron/exon border lowers the efficiency of splicing and results in a deficiency in the amount of β-globin produced.

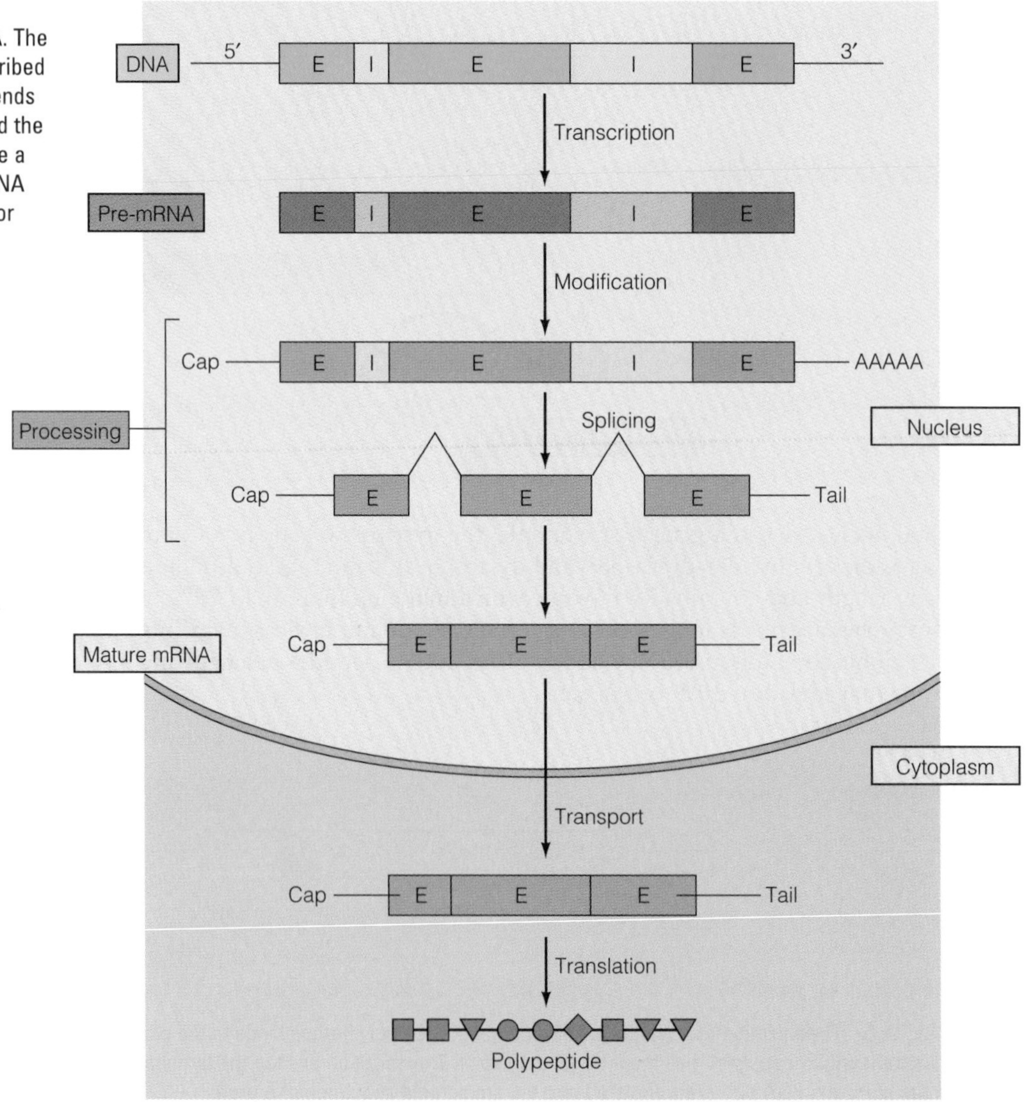

◗ **FIGURE 9.6** Steps in the processing and splicing of mRNA. The template strand of DNA is transcribed into a pre-mRNA molecule. The ends of this molecule are modified, and the introns are spliced out to produce a mature mRNA molecule. The mRNA is then moved to the cytoplasm for translation.

After processing and splicing, the mRNA is carried from the nucleus to the cytoplasm, where the encoded information is translated into the linear series of amino acids in a protein.

TRANSLATION REQUIRES THE INTERACTION OF SEVERAL COMPONENTS

Converting the nucleotide sequence in mRNA into the amino acid sequence of a protein requires interaction among many cytoplasmic components, each of which has a separate, specialized job. Before we examine the details of translation, let's look at the components.

Proteins are assembled from amino acids. Twenty different amino acids are used in proteins. Each amino acid has an **amino group** (NH_2), a **carboxyl group** (COOH), and an **R group** (Figure 9.7). R groups are side chains that are different for each amino acid. Some R groups are positively charged, some carry a negative charge, and others are electrically neutral. The 20 amino acids found in proteins and their abbreviations are listed in Table 9.2.

Amino acids are linked together by covalent **peptide bonds** formed between the amino group of one amino acid and the carboxyl group of another amino acid (Figure 9.8). Two linked amino acids form a dipeptide, three form a tripeptide, and 10 or more make a **polypeptide.** Each polypeptide (and protein)

Amino group A chemical group (NH_2) found in amino acids and at one end of a polypeptide chain.

Carboxyl group A chemical group (COOH) found in amino acids and at one end of a polypeptide chain.

R group A term used to indicate the position of an unspecified group in a chemical structure.

Peptide bond A covalent chemical link between the carboxyl group of one amino acid and the amino group of another amino acid.

Polypeptide A molecule made of amino acids joined together by peptide bonds.

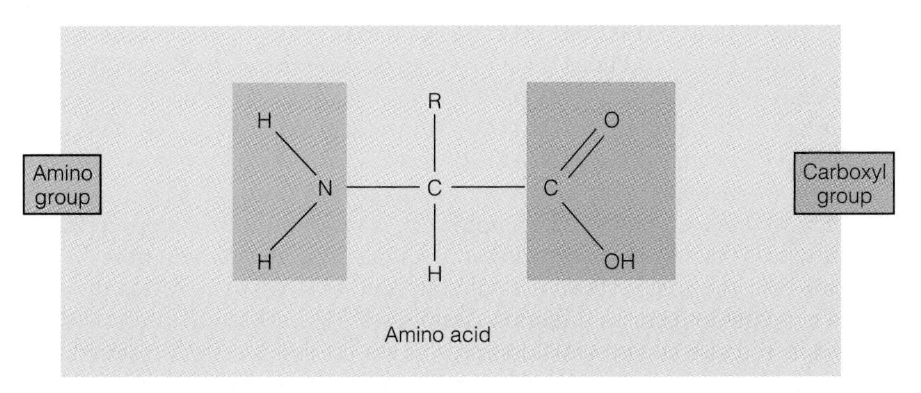

FIGURE 9.7 An amino acid, showing the amino group, the carboxyl group, and the chemical side chain known as an R group. The R groups differ in each of the 20 amino acids used in protein synthesis.

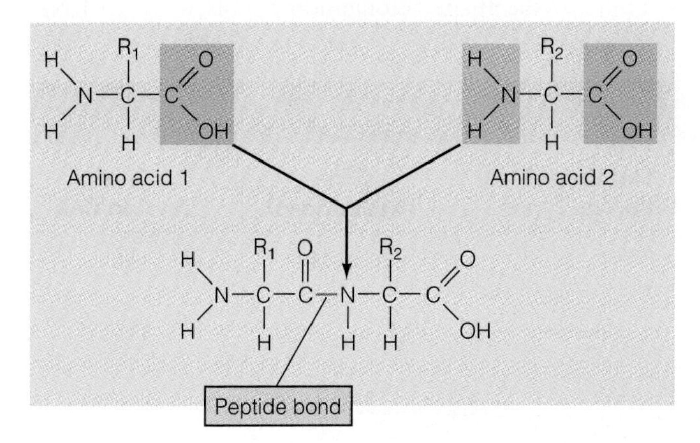

FIGURE 9.8 Formation of a peptide bond between two amino acids. An equivalent of a water molecule (H_2O) is split off during covalent bond formation. An enzyme catalyzes this reaction.

Table 9.2

Amino Acids Commonly Found in Proteins

Amino Acid	Abbreviation
Alanine	ala
Arginine	arg
Asparagine	asn
Aspartic acid	asp
Cysteine	cys
Glutamic acid	glu
Glutamine	gln
Glycine	gly
Histidine	his
Isoleucine	ile
Leucine	leu
Lysine	lys
Methionine	met
Phenylalanine	phe
Proline	pro
Serine	ser
Threonine	thr
Tryptophan	trp
Tyrosine	tyr
Valine	val

N-terminus The end of a polypeptide or protein that has a free amino group.

C-terminus The end of a polypeptide or protein that has a free carboxyl group.

Transfer RNA (tRNA) A small RNA molecule that contains a binding site for a specific type of amino acid and a three-base segment known as an anticodon that recognizes a specific base sequence in messenger RNA.

Ribosomal RNA (rRNA) RNA molecules that form part of the ribosome.

Anticodon A group of three nucleotides in a tRNA molecule that pairs with a complementary sequence (known as a codon) in an mRNA molecule.

Start codon A codon present in mRNA that signals the location for translation to begin. The codon AUG functions as a start codon.

has a free amino group at one end, known as the **N-terminus**, and a free carboxyl group, called the **C-terminus** at the other.

The nucleotide sequence of the mRNA codons are converted into the amino acid sequence of a protein with the help of two other components of the cytoplasm: ribosomes and **transfer RNAs (tRNAs)** (Table 9.3).

Ribosomes are cellular organelles with two subunits. Each subunit contains a type of RNA called **ribosomal RNA (rRNA)** combined with proteins. Ribosomes can exist floating in the cytoplasm or attached to the outer membrane of the endoplasmic reticulum (ER) (review organelles in Chapter 2). At either location, ribosomes are the site of protein synthesis.

Transfer RNA (tRNA) molecules are adapters that translate mRNA codons into the amino acid sequence of a polypeptide. A tRNA molecule is a small (about 80 nucleotides) single-stranded molecule folded back on itself to form several looped regions (Figure 9.9). tRNAs match the codons in mRNA with the proper amino acids for incorporation into a protein. As adapters, tRNA molecules have two tasks: (1) bind to the appropriate amino acid; and (2) recognize the proper codon in mRNA. The folded structure of tRNA molecules allows them to perform both tasks. A loop at one end of the molecule contains three nucleotides called an **anticodon.** The anticodon recognizes and pairs with a specific codon in an mRNA molecule. The other end of the tRNA contains a site that binds the amino acid specified by the codon (Figure 9.9).

There are 20 different amino acids and at least 20 different types of tRNA. Each tRNA carries an anticodon that has a specific nucleotide sequence. For example, if the anticodon is CCC, the amino acid glycine should bind at the other end of the molecule. However, tRNA molecules cannot recognize and bind amino acids by themselves. This task is carried out by an enzyme that recognizes a specific amino acid and its proper tRNA and links them together. Because there are 20 amino acids, there is a family of 20 such enzymes to bind amino acids to the proper tRNAs. In turn, because of base-pairing rules, tRNAs with the anticodon CCC can bind only to the codon GGG on an mRNA molecule. In this way, the three nucleotides in each mRNA codon are matched with the proper amino acid to form the new protein.

Now that we have examined the components, we will outline the steps in translation. Translation has three steps: initiation, elongation, and termination. In the first step, mRNA, the small ribosomal subunit, and a tRNA that carries the first amino acid combine to form an initiation complex (Figure 9.10). Because AUG is the **start codon** and also encodes methionine, this amino acid is usually inserted first in all proteins. The small ribosomal subunit binds at the start codon (AUG), and the anticodon (UAC) of a tRNA that carries methionine binds to the start codon. Initiation is completed when a large ribosomal subunit binds to the small subunit.

Elongation begins when amino acids are added to the growing protein. Ribosomes have two tRNA binding sites, the P site and the A site. During initiation, a tRNA carrying methionine binds to the P site. Elongation begins when a tRNA

Table 9.3 **RNA Classes**

Class	Function	Number of Different Types	Size (Nucleotides)	% of RNA in Cell
Ribosomal RNA (rRNA)	Structural, functional component of ribosomes	3	120 to 4800	90
Messenger RNA (mRNA)	Carries genetic information from DNA to ribosomes	Many thousands	300 to 10,000	3 to 5
Transfer RNA (tRNA)	Adapter recognizes nucleotide triplets and amino acids. Transports amino acids to ribosomes	50 to 60	75 to 90	5 to 7

molecule that carries the second amino acid pairs with the mRNA codon in the A site (▸ Figure 9.11). When the second amino acid is in position, an enzyme joins the two amino acids together with a peptide bond. After this bond is formed, the tRNA in the P site is released and moves out of the ribosome. At this point, the A site contains a tRNA with its attached amino acids. The ribosome then moves down the mRNA, and the tRNA is moved to the P site (Figure 9.11). This movement places the third mRNA codon into the A site, where it is recognized by the anticodon of a tRNA that carries the third amino acid. A peptide bond is formed between the second and third amino acid, and the process repeats itself, adding

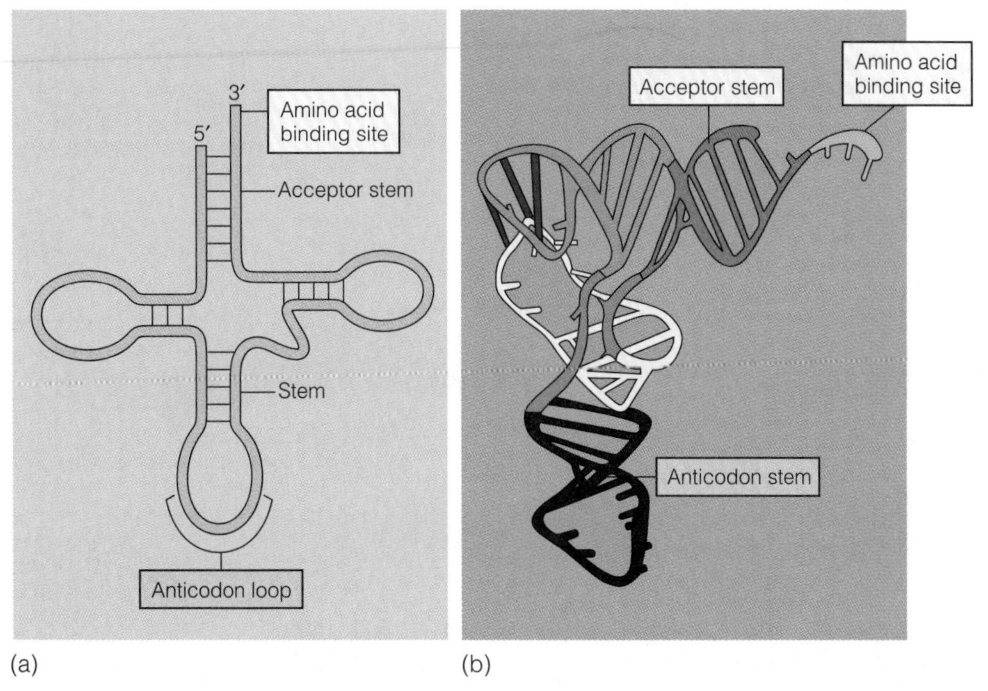

▸ **FIGURE 9.9** (a) Transfer RNA acts as a molecular adapter. It recognizes mRNA codons (at the tRNA anticodon loop) and can bind the appropriate amino acid (at the amino acid binding site). (b) The three-dimensional structure of a tRNA molecule.

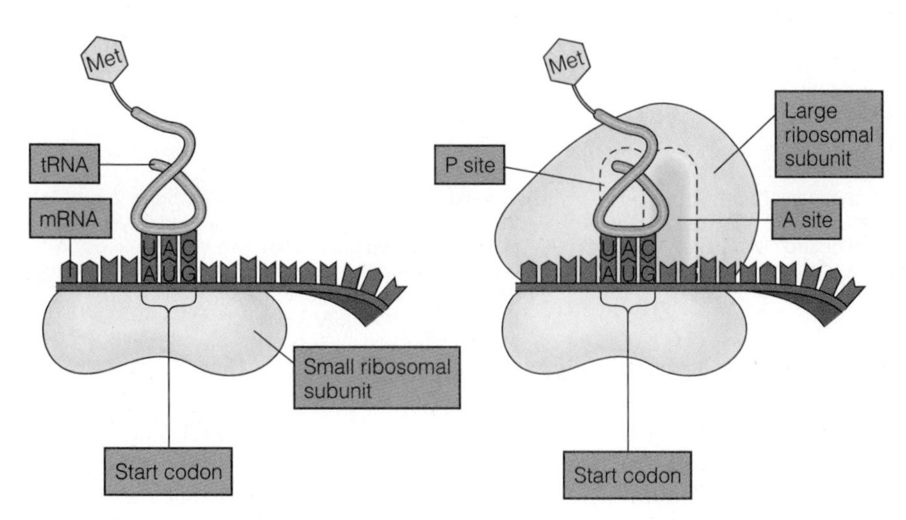

▸ **FIGURE 9.10** Initiation of translation. An mRNA molecule in the cytoplasm binds to a small ribosomal subunit. A tRNA that carries an amino acid (usually methionine) binds to the first codon (the start codon) in the mRNA. In a second step, a large ribosomal subunit binds to the small subunit, forming the initiation complex.

FIGURE 9.11 Elongation during translation. (a) After initiation, the anticodon of a second tRNA binds to the second mRNA codon, which occupies the A site of the ribosome. (b) During peptide bond formation, the two amino acids are linked together by a covalent chemical bond. When the peptide bond is formed, the tRNA in the P site is released, leaving this site empty. (c) In translocation, the ribosome shifts down the mRNA by one codon, moving the tRNA that carries the growing polypeptide chain to the P site and moving the next mRNA codon into the A site. (d) A tRNA that carries an amino acid binds to the codon in the A site, and another peptide bond is formed. This process continues until a stop codon in the mRNA (UAA) is reached. When the stop codon occupies the A site, the translation complex comes apart, and the ribosome, mRNA, and completed polypeptide are separated.

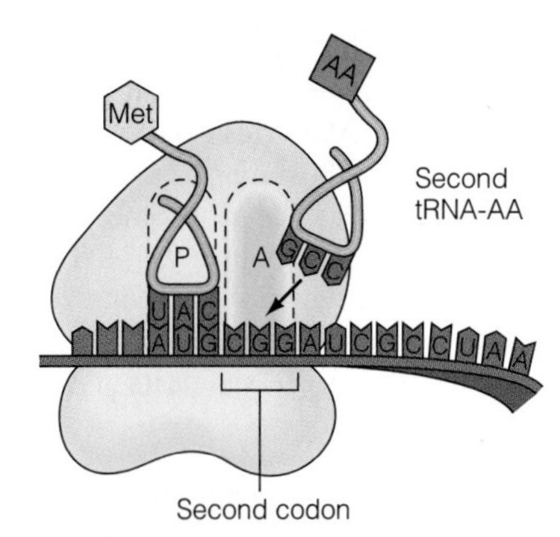

(a) As the first step in elongation, a tRNA-AA complex binds to the codon in the A site.

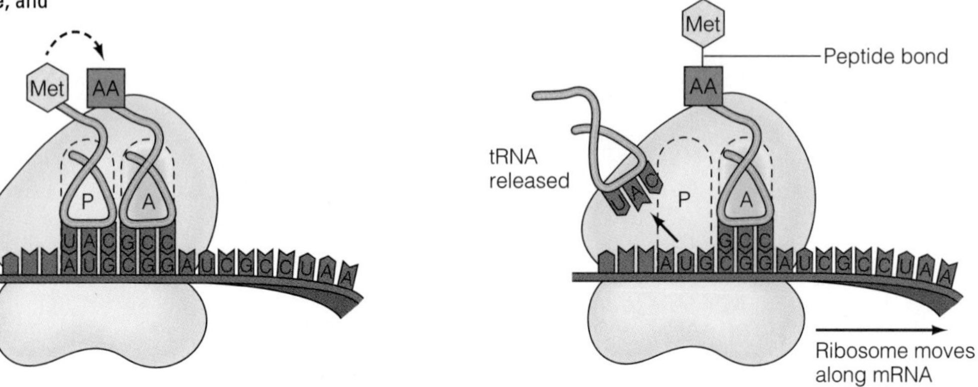

(b) An enzyme catalyzes the formation of a peptide bond between the two amino acids. The dipeptide that forms is attached to the second tRNA. This frees up the first tRNA, which vacates the P site.

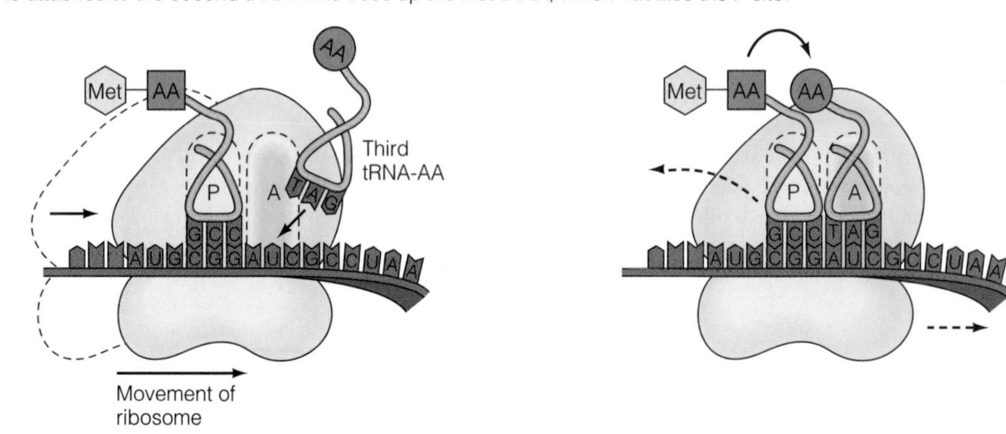

(c) The ribosome moves down the mRNA. The tRNA-dipeptide now occupies the P site and another tRNA-AA complex can occupy the A site.

(d) The dipeptide is linked by a peptide bond to the third amino acid. This frees the second tRNA. The ribosome moves down one more codon, exposing the A site and freeing it up for the addition of another tRNA-AA. This process repeats itself until the terminator codon (UAA) is reached.

Stop codons Codons present in mRNA that signal the end of a growing polypeptide chain. The codons UAA, UGA, and UAG function as stop codons.

amino acids to the growing chain of amino acids (Figure 9.12). Elongation continues until the ribosome reaches a **stop codon.** Stop codons (UAG, UGA, and UAA) do not code for amino acids, and there are no tRNA molecules with anticodons for stop codons. At this point, polypeptide synthesis is ended, and the polypeptide, mRNA, and tRNA are released from the ribosome. The role of antibiotics in interfering with protein synthesis is discussed in Concepts and Controversies: Antibiotics and Protein Synthesis.

Concepts and Controversies

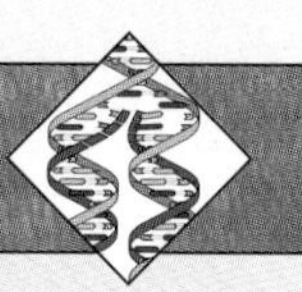

Antibiotics and Protein Synthesis

Antibiotics are chemicals produced by microorganisms as defense mechanisms. The most effective antibiotics work by interfering with essential biochemical or reproductive processes. Many antibiotics block or disrupt one or more stages in protein synthesis. Some of these are listed here.

Tetracyclines are a family of chemically related compounds used to treat several types of bacterial infections. Tetracyclines interfere with the initiation of translation. The tetracycline molecule binds to the small ribosomal subunit and prevents binding of the tRNA anticodon in the first step in initiation. Both eukaryotic and prokaryotic ribosomes are sensitive to the action of tetracycline, but this antibiotic cannot pass through the plasma membrane of eukaryotic cells. Because it can enter bacterial cells to inhibit protein synthesis, it will stop bacterial growth, helping the immune system fight the infection.

Streptomycin is used in hospitals to treat serious bacterial infections. It binds to the small ribosomal subunit but does not prevent initiation or elongation; however, it does affect the efficiency of protein synthesis. When streptomycin binds to a ribosome, it changes the way codons in the mRNA interact with the tRNA anticodons. As a result, incorrect amino acids are incorporated into the growing polypeptide chain. In addition, streptomycin causes the ribosome to fall off the mRNA at random, preventing the synthesis of complete proteins.

Puromycin is not used clinically, but has played an important role in studying the mechanism of protein synthesis in the research laboratory. The puromycin molecule has the same size and shape as a tRNA-amino acid complex. As a result, it enters the ribosome and is incorporated into a growing polypeptide chain. Once puromycin is added to the polypeptide, further synthesis is terminated because no peptide bond can be formed with an amino acid, and the shortened polypeptide falls off the ribosome.

Chloramphenicol was one of the first broad-spectrum antibiotics introduced. Eukaryotic cells are resistant to its actions, and it was widely used to treat bacterial infections. However, its use is now limited to external applications and serious infections. Chloramphenicol destroys cells in the bone marrow, the source of all blood cells. This antibiotic binds to the large ribosomal subunit in bacteria and inhibits the formation of peptide bonds. Another antibiotic, erythromycin, also binds to the large ribosomal subunit and inhibits the movement of ribosomes along the mRNA.

Almost every step of protein synthesis can be inhibited by one antibiotic or another. Work on designing new, synthetic antibiotics to fight infections is based on our knowledge of how the nucleotide sequence of mRNA is converted into the amino acid sequence of a protein.

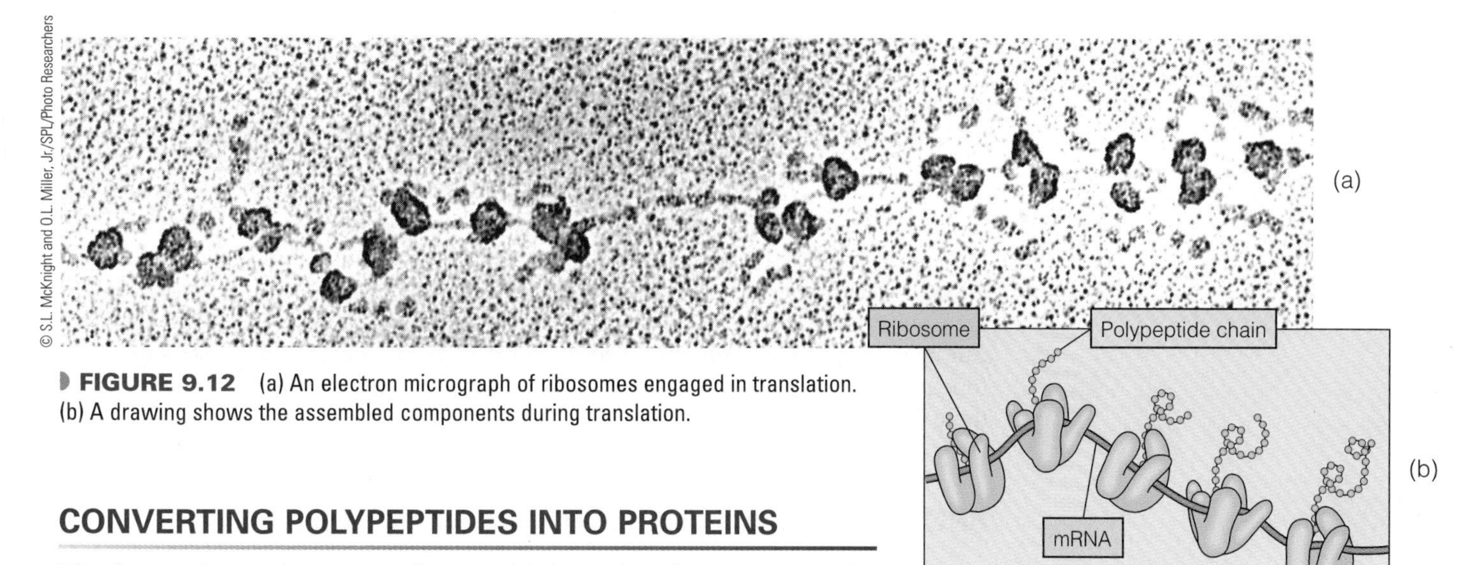

FIGURE 9.12 (a) An electron micrograph of ribosomes engaged in translation. (b) A drawing shows the assembled components during translation.

CONVERTING POLYPEPTIDES INTO PROTEINS

The linear chain of amino acids assembled on the ribosome is a polypeptide. Once formed, the polypeptide can have several different fates. Polypeptides made on the outer surface of the ER move inside the ER, where they are folded, chemically modified, and transported to the Golgi complex for packaging and secretion from the cell. Other polypeptides, made on cytoplasmic ribosomes, are folded, remain in the cell, and function in the cytoplasm or the nucleus (Figure 9.13).

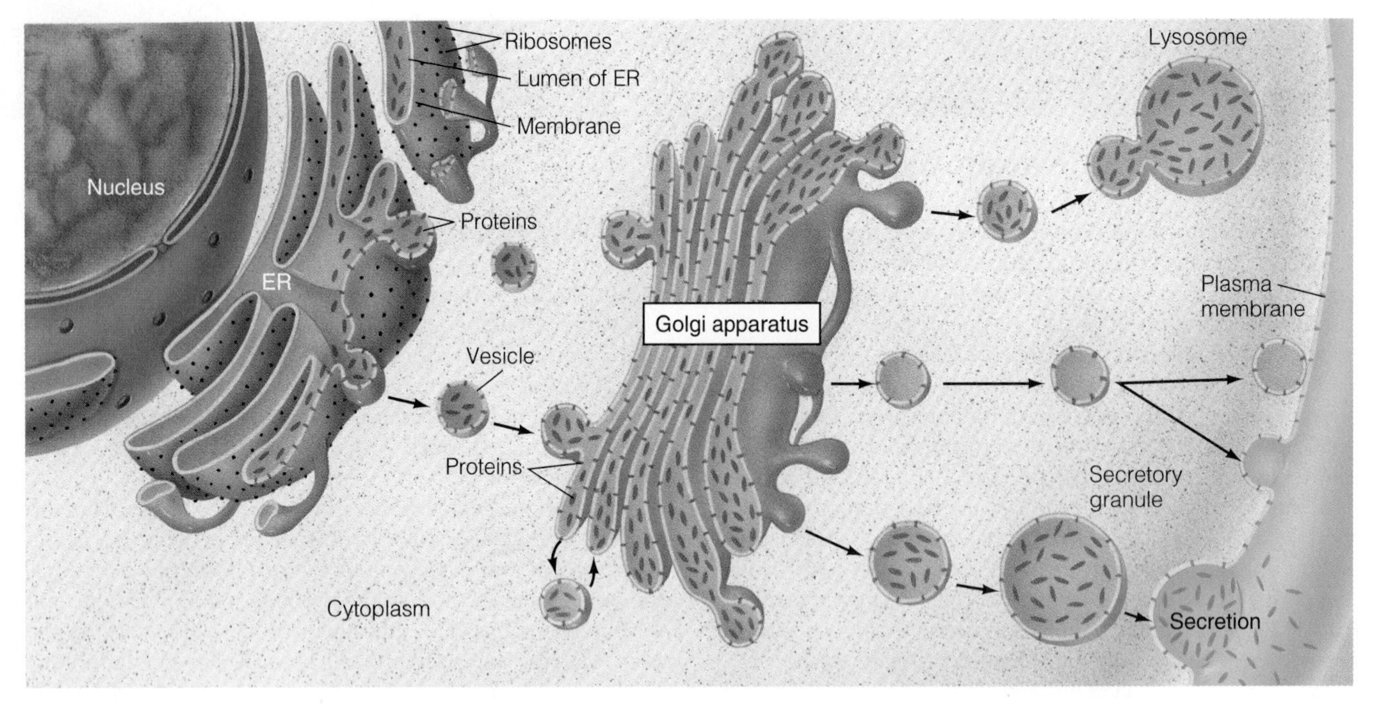

FIGURE 9.13 Processing, sorting, and transport of proteins synthesized in a eukaryotic cell. Proteins synthesized on ribosomes attached to endoplasmic reticulum (ER) are transferred into the ER, where they are folded and chemically modified. Many of these proteins are transferred to the Golgi apparatus in vesicles. In the Golgi the proteins are further modified, sorted, and packaged into vesicles for delivery to other parts of the cell and are incorporated into organelles such as lysosomes or are transported to the surface for insertion into the plasma membrane. Proteins can also be packaged into vesicles for secretion.

After the polypeptide is synthesized, it folds into a three-dimensional shape, determined by its amino acid sequence. Polypeptide folding is guided by proteins called molecular chaperones. Mutations in genes can alter folding and lead to genetic disorders, as discussed later. Polypeptides can be chemically modified after they are synthesized; this process is called post-translational modification. Over 200 different ways of modification have been identified. Some of these include attaching lipids or sugars to the polypeptide, chemically changing some of the amino acids in the polypeptide, or even removing some amino acids. Once a polypeptide is folded, is modified, and becomes functional, it is called a protein.

The many ways that polypeptides are altered after they are translated helps explain one of the surprises from the Human Genome Project (this project is discussed in Chapter 13). Scientists predicted that the human genome would contain about 100,000 genes, but only about 35,000 genes have been identified. In other words, we have only about 15,000 genes more than corn plants. However, because polypeptides can be post-translationally modified in so many ways to change their structure and function, human cells can make several hundred thousand different proteins from these 35,000 genes.

PROTEIN STRUCTURE AND FUNCTION ARE RELATED

Primary structure The amino acid sequence in a polypeptide chain.

Secondary structure The pleated or helical structure in a protein molecule that is brought about by the formation of bonds between amino acids.

The amino acid sequence of a protein determines its three-dimensional shape and its function. There are four levels of protein structure. The sequence of amino acids is the **primary structure** (Figure 9.14). The next two levels of structure are determined to a great extent by the R groups of each amino acid. The R groups in different parts of the protein interact with each other via hydrogen bonding to form a pleated or coiled **secondary structure.** Most proteins have both types of

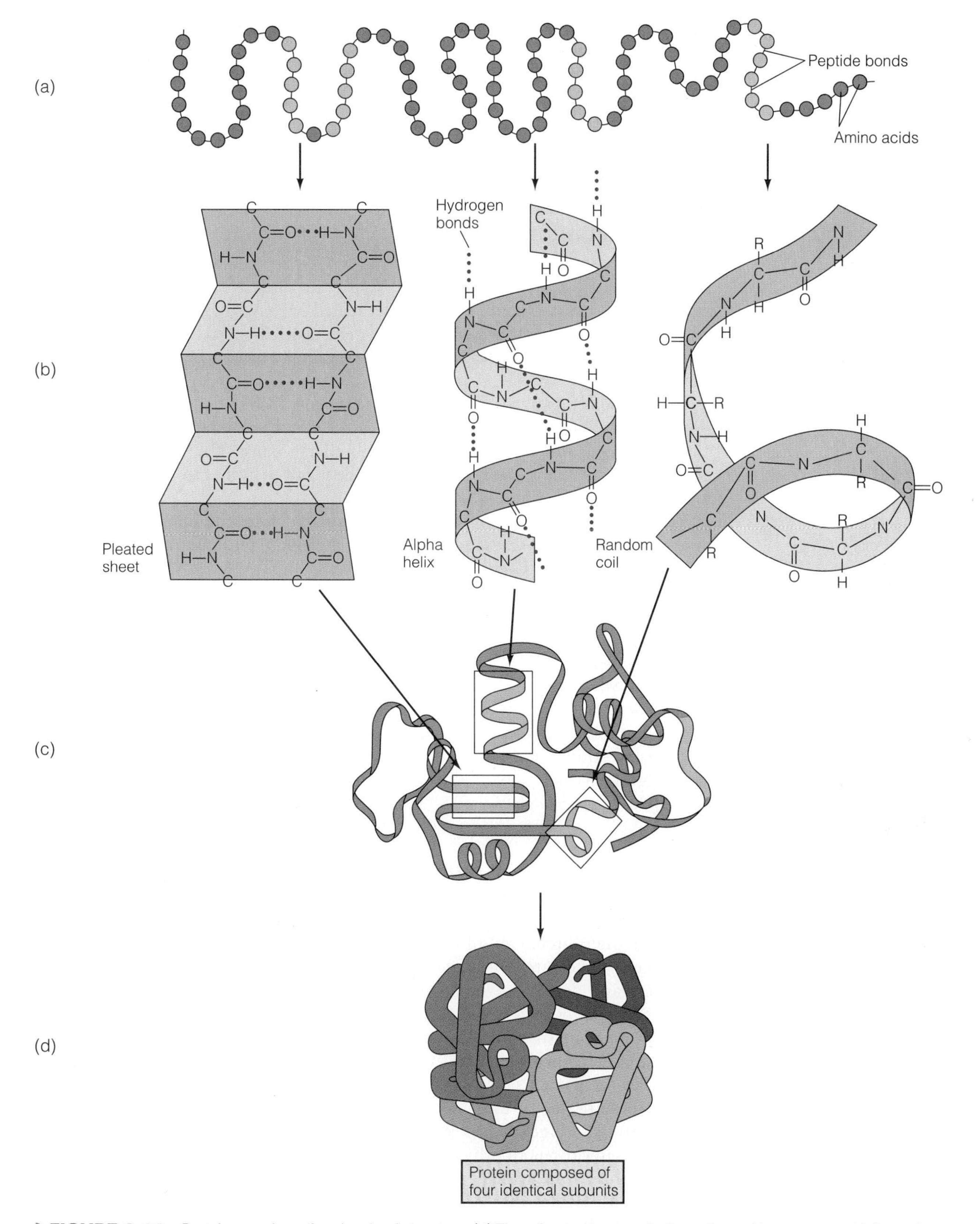

FIGURE 9.14 Proteins can have four levels of structure. (a) The primary structure is the amino acid sequence, which to a large extent determines the other levels of structure. (b) The secondary level of structure can be a pleated sheet, alpha helix, or a random coil. (c) Folding of the secondary structures into a functional three-dimensional shape is the tertiary level of structure. (d) Some functional proteins are made up of more than one polypeptide chain. These interactions are the quaternary level of protein structure.

Tertiary structure The three-dimensional structure of a protein molecule brought about by folding on itself.

Quaternary structure The structure formed by the interaction of two or more polypeptide chains in a protein.

secondary structure. Folding the helical or pleated regions back onto themselves creates **tertiary structure.** Some functional proteins are composed of more than one polypeptide chain, and this level of interaction is known as the **quaternary structure.** When a polypeptide forms a functional, three-dimensional structure, it is called a protein.

It is this three-dimensional conformation, ultimately determined by its DNA-controlled primary structure, that determines a protein's function.

Protein Folding and Disease

Some mutations alter the folding pattern of polypeptides and result in a genetic disorder. Several disorders, including Alzheimer disease (MIM/OMIM 104300 and other numbers), a metabolic disorder called MPS VI (MIM/OMIM 253200), and cystic fibrosis, are associated with defects in the final stages of folding. This defect prevents the formation of a functional protein, producing a mutant phenotype.

The cystic fibrosis gene encodes a protein (called CFTR) of 1,440 amino acids (review cystic fibrosis in Chapter 4). The most common mutation is a deletion of phenylalanine as amino acid 508. This single change causes the polypeptide to fold improperly. As a result, the CFTR polypeptide is not delivered to the plasma membrane and is degraded in the ER.

Prion A protein folded into an infectious conformation that is the cause of several disorders, including Creutzfeldt-Jakob disease and mad-cow disease.

Mad-cow disease A prion disease of cattle.

Under certain conditions, some proteins can refold and change their three-dimensional shape. Diseases associated with protein refolding are called **prion** diseases. In humans, Creutzfeldt-Jakob disease (CJD) (MIM/OMIM 123400), Gerstmann-Straussler disease (MIM/OMIM 137440), and fatal familial insomnia (MIM/OMIM 600072) are prion diseases. In cattle, bovine spongiform encephalopathy, also known as **mad-cow disease,** is a prion disease. Prion diseases are all degenerative diseases of the nervous system, leading to an early death. They begin when one or a small number of proteins refold into a disease-causing shape. These refolded proteins gradually cause other proteins of the same type to refold, producing the disease phenotype. Prion diseases like mad-cow disease are infectious, and the disease is transmitted when refolded proteins are transferred from one individual to another. Case 2 at the end of the chapter deals with CJD.

Proteins Have Many Functions

Proteins are the most abundant type of molecules found in any cell. They participate in a wide range of functions (summarized in Table 9.4). Myosin and actin are contractile proteins found in muscle cells. Hemoglobin is a transport protein that shuttles oxygen to cells. The immune system depends on protein antibodies to identify and destroy invading organisms such as bacteria. Connective tissue and hair are rich in structural proteins such as keratin and collagen. Histones are structural proteins that form a complex with DNA to help form chromosomes. Many hormones, such as insulin, are proteins. The receptors on the surface of cell membranes are proteins.

Enzymes are one of the most important groups of proteins in the cell. They act as catalysts in biochemical reactions (Figure 9.15a). Enzymes accelerate the rate of a chemical reaction by reducing the energy needed to carry out the reaction. The three-dimensional shape of the enzyme creates a region called the active site. Molecules that can fit into the active site are known as substrates. When they bind to the active site, they undergo a chemical change. Enzymes are usually named for their substrate, with the suffix *ase* added. The enzyme that

Table 9.4 Some Biological Functions of Proteins

Protein Function	Examples	Occurrence or Role
Catalysis	Lactate dehydrogenase	Oxidizes lactic acid
	Cytochrome C	Transfers electrons
	DNA polymerase	Replicates and repairs DNA
Structural	Viral-coat proteins	Sheath around nucleic acid of viruses
	Glycoproteins	Cell coats and walls
	α-Keratin	Skin, hair, feathers, nails, and hoofs
	β-Keratin	Silk of cocoons and spider webs
	Collagen	Fibrous connective tissue
	Elastin	Elastic connective tissue
Storage	Ovalbumin	Egg-white protein
	Casein	A milk protein
	Ferritin	Stores iron in the spleen
	Gliadin	Stores amino acids in wheat
	Zein	Stores amino acids in corn
Protection	Antibodies	Form complexes with foreign proteins
	Complement	Complexes with some antigen-antibody systems
	Fibrinogen	Involved in blood clotting
	Thrombin	Involved in blood clotting
Regulatory	Insulin	Regulates glucose metabolism
	Growth hormone	Stimulates growth of bone
Nerve impulse transmission	Rhodopsin	Involved in vision
	Acetylcholine receptor protein	Impulse transmission in nerve cells
Motion	Myosin	Thick filaments in muscle fiber
	Actin	Thin filaments in muscle fiber
	Dynein	Movement of cilia and flagella
Transport	Hemoglobin	Transports O_2 in blood
	Myoglobin	Transports O_2 in muscle cells
	Serum albumin	Transports fatty acids in blood
	Transferrin	Transports iron in blood
	Ceruloplasmin	Transports copper in blood

catalyzes the breakdown of the sugar lactose is called lactase, and the enzyme that catalyzes the conversion of the amino acid phenylalanine to tyrosine is called phenylalanine hydroxylase. The relationship between enzymes and genetic disorders is explored in Chapter 10.

The function of all proteins depends ultimately on the amino acid sequence of the polypeptide chain. The nucleotide sequence of DNA determines the amino acid sequence of proteins. If protein function is to be maintained from cell to cell and from generation to generation, the nucleotide sequence of a gene must be maintained. Changes in the nucleotide sequence of DNA (a mutation) produce mutant genes that in turn produce mutant proteins with altered or impaired functions (Figure 9.15b). These alterations result in an altered phenotype. Alkaptonuria, the condition described at the beginning of this chapter, is caused by a mutation that alters the function of an enzyme. The phenotypic consequences of mutational changes in DNA are discussed in Chapter 11.

FIGURE 9.15 The folded protein has an active site. Substrates bind to enzymes at active sites. (a) The enzyme acts as a catalyst to carry out a chemical reaction, converting the substrate into a product. (b) Mutation changes the folding pattern, preventing the binding of the substrate, and no product is made.

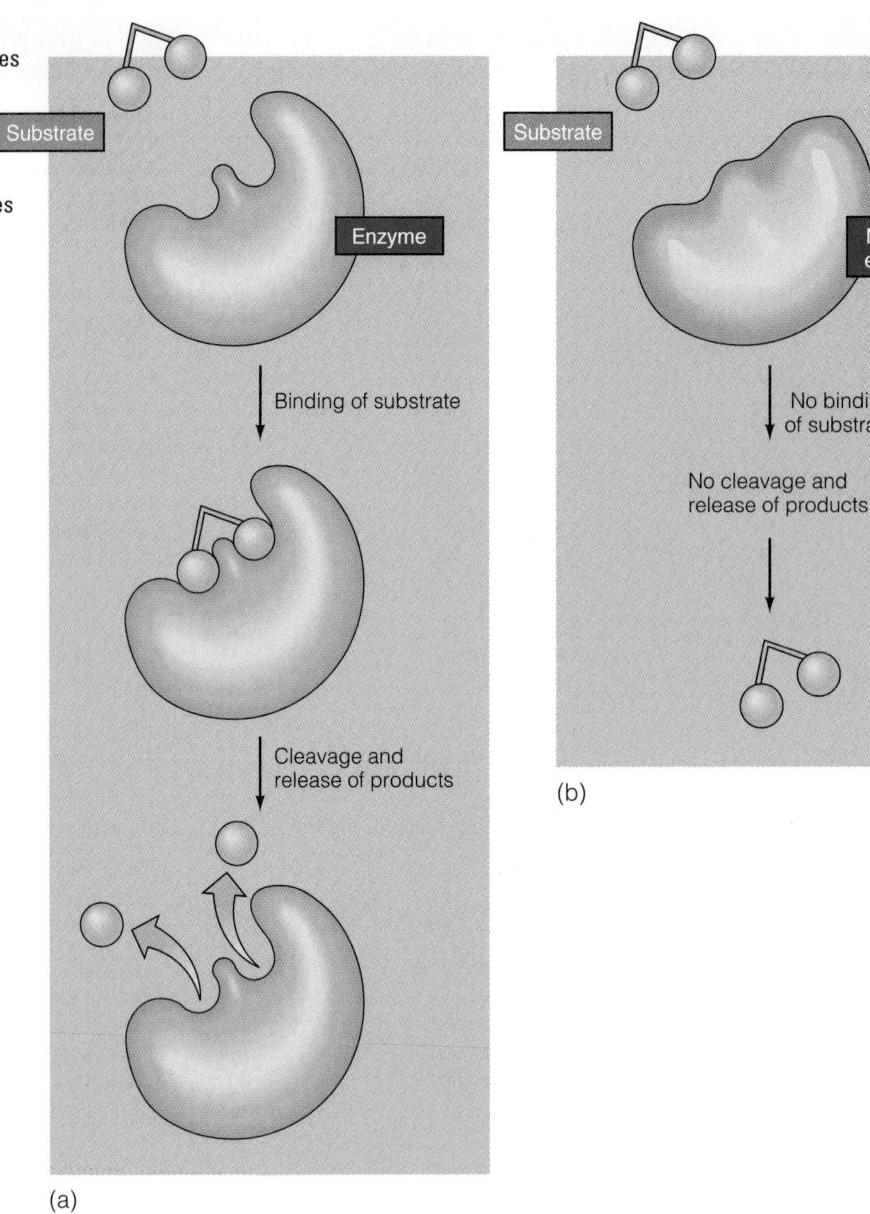

Summary

The Link Between Genes and Proteins

1. At the beginning of the last century, Garrod proposed that genetic disorders result from biochemical alterations.

Genes and Metabolism

2. Using *Neurospora*, Beadle and Tatum showed that mutations can produce a loss of enzyme activity and a mutant phenotype. Beadle proposed that genes control the synthesis of proteins and that protein function is responsible for producing the phenotype.

Genetic Information Is Stored in DNA

3. In proposing their model, Watson and Crick maintained that DNA stores genetic information in its nucleotide sequence.

Tracing the Flow of Genetic Information from Nucleus to Cytoplasm

4. The processes of transcription and translation require the interaction of many components, including ribosomes, mRNA, tRNA, amino acids, enzymes, and energy sources. Ribosomes are the workbenches on which protein synthesis occurs. tRNA molecules are

adapters that recognize amino acids and the nucleotide sequence in mRNA, the gene transcript.

Transcription Produces Genetic Messages

5. In transcription, one of the DNA strands is used as a template for making a complementary strand of RNA, called mRNA. The information transferred to RNA is encoded in sets of three nucleotides, called codons. Of the 64 possible codons, 61 code for amino acids, and three are stop codons.

Translation Requires the Interaction of Several Components

6. Translation requires the interaction of tRNA molecules, amino acids, ribosomes, mRNA, and energy sources. Within the ribosome, tRNA anticodons bind to complementary codons in the mRNA. The ribosome moves along the mRNA, linking amino acids and producing a growing polypeptide chain. At termination, this polypeptide is released from the ribosome and undergoes a conformational change to produce a functional protein.

Converting Polypeptides into Proteins

7. After synthesis, polypeptides fold into a three-dimensional shape, often assisted by other proteins, called chaperones. Mutations in chaperones can cause genetic disorders. Polypeptides can be chemically modified in many different ways, producing functionally different proteins from one polypeptide.

Protein Structure and Function Are Related

8. Four levels of protein structure are recognized, three of which are results of the primary sequence of amino acids in the backbone of the protein chain.

Proteins Have Many Functions

9. Although proteins perform a wide range of tasks, enzyme activity is one of the primary tasks. Enzymes function by lowering the energy of activation required in biochemical reactions. The products of these biochemical reactions are inevitably involved in producing phenotypes.

Case Studies

CASE 1

A genetic counselor was called to the pediatric ward to examine an infant who was diagnosed with a genetic disorder of sugar metabolism, galactosemia. All newborns in the infant's state are screened for this and other genetic disorders. The infant was 3 weeks old and was admitted to the hospital because of failure to thrive and severe jaundice (yellowing of the skin resulting from liver problems). Upon examination, the physician determined that the infant had an enlarged liver (hepatomegaly), cataracts, and constant diarrhea and vomiting when fed milk. *E. coli* infection is a common cause of death in infants who have galactosemia, and cultures were drawn on this infant. Laboratory results confirmed that the infant had a deficiency of the enzyme galactose-1-phosphate uridyltransferase and was infected with *E. coli*.

The counselor took a detailed family history and explained this condition to the parents. She indicated that this condition is due to the inheritance of a mutant gene from each parent (the trait is autosomal recessive) and that there is a 25%, or one in four, chance that with each pregnancy they have together that they will bear a child with this condition. The counselor explained that there is wide variability in phenotype, ranging from very mild to severe. A blood test could determine which variant of this disease they carry.

1. What exists in blood that can be tested for a variant of a disease-causing gene?
2. What are possible treatments for this disease?

CASE 2

Recurring cases of mad-cow disease have plagued the United Kingdom during the past 4 years. What started out as a topic of interest to a few cell biologists has blossomed into a huge public interest story. Mad-cow disease is caused by a prion, an infectious particle that consists only of protein. The media reported that cows were dying all over England from a mysterious disease. Initially, however, there was little interest in finding whether humans could be affected. For 10 years, the British government maintained that this unusual disease could not be transmitted to humans. However, in March 1996, the government did an about-face and announced that bovine spongiform en-

(continued)

cephalopathy (BSE), commonly known as mad-cow disease, can be transmitted to humans. BSE and a similar condition known as Creutzfeldt-Jakob disease (CJD) eat away at the nervous system, destroying the brain and essentially turning it into a sponge. Victims suffer dementia; confusion; loss of speech, sight, and hearing; convulsions; coma; and death. Prion diseases are always fatal, and there is no treatment. Precautionary measures taken in Britain to prevent this disease in humans may have begun too late; many of the victims today may have contracted it a decade ago when the BSE epidemic began, and the incubation period is long (CJD has an incubation period of 10 to 40 years).

1. How can a prion replicate itself without genetic material?
2. What measures have been taken to stop BSE?
3. If you were traveling in Europe, would you eat beef? Give sound reasons why or why not.

Questions and Problems

The Link Between Genes and Proteins

1. The genetic material has to store information and be able to express it. What is the relationship between DNA, RNA, proteins, and phenotype?
2. Define replication, transcription, and translation. In what part of the cell does each process occur?

The Genetic Code: The Key to Life

3. If the genetic code used four bases at a time, how many amino acids could be encoded?
4. If the genetic code uses triplets, how many different amino acids can be coded by a repeating RNA polymer composed of UA and UC (UAUCUAUCUAUC. . . . .)
   a. one
   b. two
   c. three
   d. four
   e. five
5. What is the start codon? What are the stop codons? Do any of these code for amino acids?

Transcription Produces Genetic Messages

6. Is an entire chromosome made into an mRNA during transcription?
7. The 5′ promoter and the 3′ terminator regions of genes are important in:
   a. coding for amino acids.
   b. gene regulation.
   c. structural support for the gene.
   d. intron removal.
   e. anticodon recognition.
8. What are the three modifications made to pre-mRNA molecules before they become mature mRNAs ready to be used in protein synthesis? What is the function of each modification?
9. The pre-mRNA transcript and protein made by several mutant genes were examined. The results are given below. Determine where in the gene a likely mutation lies: the 5′ flanking region, exon, intron, cap on mRNA, or ribosome binding site.
   a. normal length transcript, normal length nonfunctional protein
   b. normal length transcript, no protein made
   c. normal length transcript, normal length mRNA, short nonfunctional protein
   d. normal length transcript, longer mRNA, longer nonfunctional protein
   e. transcript never made
10. The following segment of DNA codes for a protein. The uppercase letters represent exons. The lowercase letters represent introns. The lower strand is the template strand. Draw the primary transcript and the mRNA resulting from this DNA.

    GCTAAATGGCAaaattgccggatgacGCACATTGACTCGGaatcgaGGTCAGATGC
    CGATTTACCGTtttaacggcctactgCGTGTAACTGAGCCttagctCCAGTCTACG

Translation Requires the Interaction of Several Components

11. Briefly describe the function of the following in protein synthesis.
  a. rRNA
  b. tRNA
  c. mRNA

12. What is the difference between codons and anticodons?

13. Determine the percent of the gene below that will code for the protein product. Gene length is measured in kilobases (kb) of DNA. Each kilobase is 1,000 bases long.

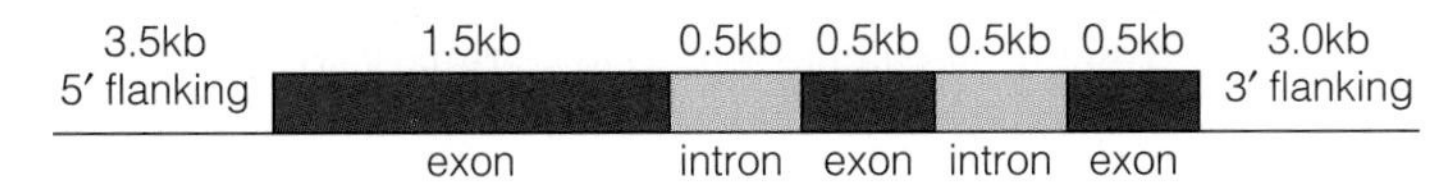

14. How many kilobases of the following DNA strand will code for the protein product?

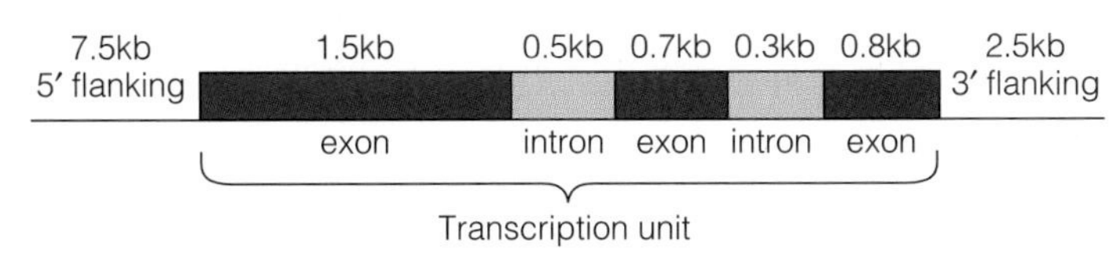

15. Write the anticodon(s) for the following amino acids:
  a. met
  b. trp
  c. ser
  d. leu

16. Given the following tRNA anticodon sequence, derive the mRNA and the DNA template strand. Also, make the protein that is encoded by this message.
tRNA: UAC UCU CGA GGC
mRNA:
DNA:
protein:
How many hydrogen bonds would be present in the DNA segment?

17. Given the following mRNA, write the double-stranded DNA segment that served as the template. Indicate both the 5′ and the 3′ ends of both DNA strands. Also make the tRNA anticodons and the protein that is encoded by the mRNA message.
DNA:
mRNA: 5′ - CCGCAUGUUCAGUGGGCGUAAACACUGA - 3′
protein:
tRNA:

18. The following is a portion of a protein:
met-trp-tyr-arg-gly-pro-thr-
Various mutant forms of this protein have been recovered. Using the normal and mutant sequences, determine the DNA and mRNA sequences that code for this portion of the protein, and explain each of the mutations.
  a. met-trp-
  b. met-cys-ile-val-val-leu-gln-
  c. met-trp-tyr-arg-ser-pro-thr-
  d. met-trp-tyr-arg-gly-ala-val-ile-ser-pro-thr-

19. The following is the structure of glycine. Draw a tripeptide composed exclusively of glycine. Label the N-terminus and C-terminus. Draw a box around the peptide bonds.

H
H_2N—C—C(=O)OH
H

20. Each of the following functions either in transcription or in translation. Indicate in which category each of the following functions: RNA polymerase, ribosomes, nucleotides, tRNA, pre-mRNA, DNA, A site, TATA box, anticodon, amino acids.

Protein Structure and Function Are Related

21. Proteins have many critical functions in the human body. Some of these functions include:
    a. transporting oxygen.
    b. hormonal signalling.
    c. carrying out enzymatic reactions.
    d. destroying invading bacteria.
    e. all of the above.
22. Enzyme X normally interacts with substrate A and water to produce compound B.
    a. What would happen to this reaction in the presence of another substance that resembles substrate A and was able to interact with enzyme X?
    b. What if a mutation in enzyme X changed the shape of the active site?
23. Do mutations in DNA alter proteins all the time?
24. Can a mutation change a protein's tertiary structure without changing its primary structure? Can a mutation change a protein's primary structure without affecting its secondary structure?

Internet Activities

The following activities use the resources of the World Wide Web to enhance the topics covered in this chapter. To investigate the topics described below, log on to the book's home page (http://www.brookscole.com/biology_d), and follow the *Student Resources* link to your subject and text.

1. *Review of Gene Expression.* At the "Cell Biology Topics 1: Ribosome" site (http://www.cytochemistry.net/Cell-biology/ribosome.htm), review the basics of translation after the mRNA leaves the nucleus.
2. *Quiz Yourself.* At the University of Arizona's "The Biology Project: Molecular Biology" (http://www.biology.arizona.edu/molecular_bio/molecular_bio.html), click on the "Nucleic Acids" link to access quizzes on DNA replication, transcription, and translation. Correct answers are rewarded with brief overviews; if you answer incorrectly, you will be linked to a short tutorial that will help you solve the problem.
3. *Control of Gene Expression.* At the "On-line Biology: Control of Gene Expression" page (http://gened.emc.maricopa.edu/Bio/BIO181/BIOBK/BioBookGENCTRL.html), read about the control of gene expression in bacteria, viruses, and eukaryotes.
    a. How many different proteins and protein factors are involved in the various steps of gene expression? What would be the possible effects of a mutation that changed one of these proteins? Consequently, would you expect to see greater similarity or less similarity in the DNA sequences that code for these proteins in different organisms?
    b. In some cases the expression of multiple genes is controlled by a single protein factor, as in the operon model of transcriptional regulation proposed by Jacob and Monod. What might be the benefits of such a comparatively streamlined mechanism for the control of gene expression?
    c. Compare the genome sizes for various eukaryotes. What percentage of the average eukaryotic genome actually codes for protein? What percentage of the human genome codes for protein? What function, if any, does the noncoding portion of the genome serve?

For Further Reading

Al-Karadaghi, S., Kristensen, O., & Liljas, A. (2000). A decade of progress in understanding the structural basis of protein synthesis. *Prog. Biophys. Mol. Biol., 73,* 167–193.

Beadle, G. W., & Tatum, E. L. (1941). Genetic control of biochemical reactions. *Neurospora. Proc. Natl. Acad. Sci. U S A, 27,* 499–506.

Chambon, P. (1981, May). Split genes. *Sci. Am., 244,* 60–71.

Crick, F. H. C. (1962, October). The genetic code. *Sci. Am., 207,* 66–77.

Darnell, J. (1985). RNA. *Sci. Am., 253,* 68–87.

Dobson, C. M. (2001). The structural basis of protein folding and its links with human disease. *Philos. Trans. R. Soc. Lond. B Biol. Sci., 356,* 133–145.

Doolittle, R. F. (1985, October). Proteins. *Sci. Am., 253,* 88–99.

Garrod, A. (1902). The incidence of alkaptonuria: A study in chemical individuality. *Lancet, 2,* 1666–1670.

Lee, T. I., & Young, R. A. (2000). Transcription of eukaryotic protein-coding genes. *Annu. Rev. Genet., 34,* 77–137.

Schimmel, P., & Alexander, R. (1998). All you need is RNA. *Science, 281,* 658–659.

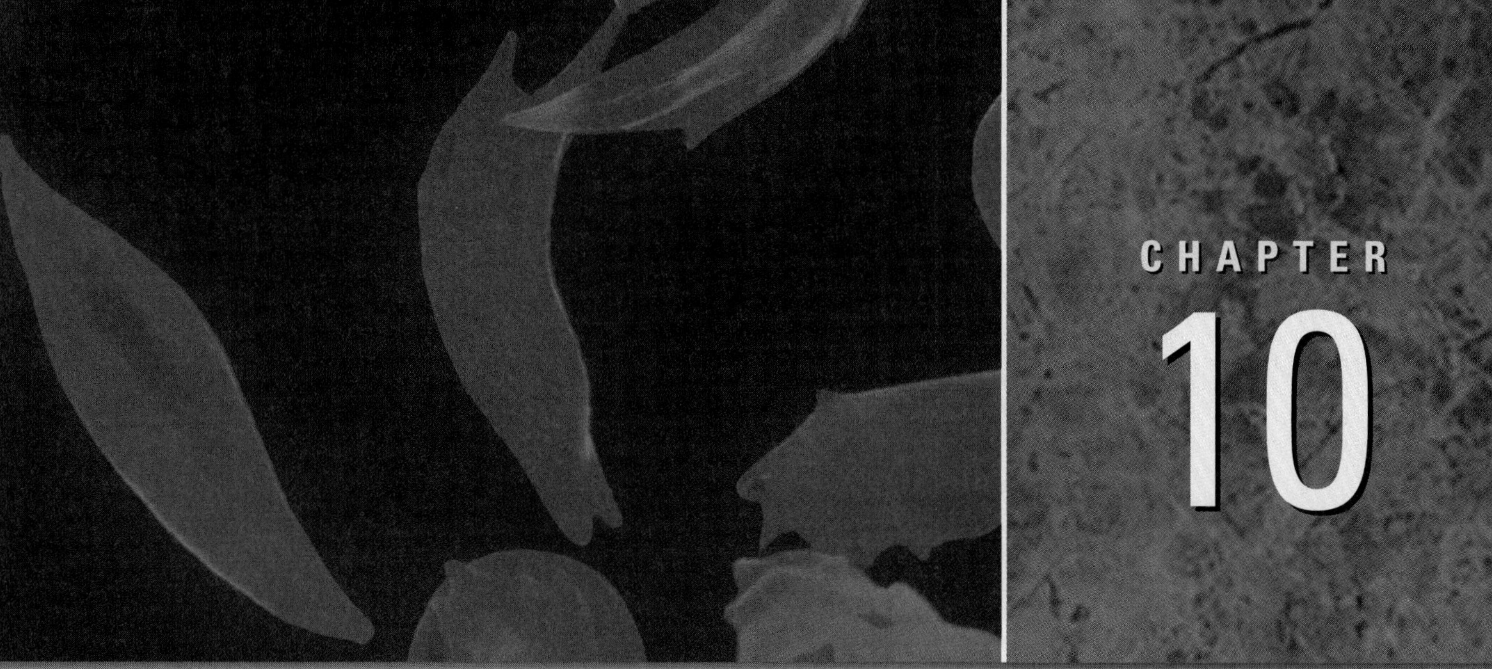

From Proteins to Phenotypes

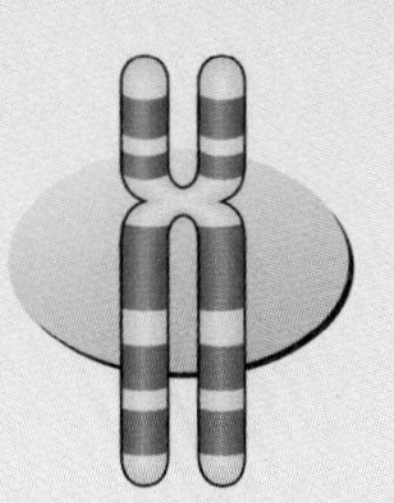

Chapter Outline

The field of human biochemical genetics was founded in part by a young Norwegian mother with two mentally retarded children. The first child, a girl, did not walk until nearly 2 years of age and spoke only a few words. She also had a musty odor that could not be washed away. Her younger brother was also slow to develop and never learned to walk or talk. He had the same musty odor as his sister. To learn why both her children were retarded and had a musty odor, the mother went from doctor to doctor but to no avail. Finally, in the spring of 1934, the persistent woman took the two children, then aged 4 and 7 years, to Dr. Asbjorn Fölling, a biochemist and physician.

Because the urine from these children had a musty odor, Fölling first tested the urine for signs of infection, but there was none. He discovered that the urine reacted with ferric chloride to produce a green color, indicting the presence of an unknown chemical. Beginning with 20 L of urine collected from the children, he worked to isolate and identify the unknown substance. Over the next 3 months, he managed to purify the compound and worked out its chemical structure. The chemical in the children's urine was a compound called phenylpyruvic acid. To confirm his finding, Fölling synthesized and purified phenylpyruvic acid from laboratory chemicals and showed that the compound from the urine and his synthetic phenylpyruvic acid had the same physical and chemical properties.

Fölling proposed that the phenylpyruvic acid in the urine was produced by a metabolic disorder that affected the breakdown of the amino acid phenylalanine. He further proposed that the accumulation of phenylpyruvic acid in the bodies of the children caused their mental retardation. To confirm this, he examined the urine of several hundred retarded patients and normal individuals. He found phenylpyruvic acid in the urine of eight retarded individuals, but never in the urine of normal individuals. Less than 6 months after he began working on the problem, Fölling submitted a manuscript for publication that described this metabolic disorder called phenylketonuria (PKU). His work helped establish the relationship between a gene product and a phenotype, and PKU is now regarded as a prototype for metabolic genetic disorders.

As we discussed in the last chapter, the discovery of biochemical mutations in fungi by George Beadle and Edward Tatum in 1941 was an important step in understanding how gene expression results in a phenotype. We also considered how DNA encodes information for the chemical structure of proteins. In this chapter we will show how protein function is related to the phenotype, and how mutations that change or eliminate protein function produce an abnormal phenotype.

PROTEINS ARE THE LINK BETWEEN GENES AND THE PHENOTYPE

As outlined in Chapter 9, proteins are among the most important molecules in the cell. They are essential parts of all structures and biological processes carried out in every cell type. For example, proteins are part of membrane systems and the internal skeleton of cells. They are the glue that holds cells and tissues together. Proteins carry out biochemical reactions; destroy invading microorganisms; and act as hormones (Figure 10.1), receptors, and transport molecules. Even the replication of DNA and the expression of genes depend on the action of proteins.

The many different functions of proteins are matched by their enormous variation in amino acid composition. Because any one of 20 amino acids can occupy each position in a protein (amino acid #1, amino acid #2, and so forth), an enormous number of different proteins can be made. For example, in a protein containing 100 amino acids, 20^{100} different molecules, each with a unique sequence, can be made using the 20 different amino acids found in proteins.

As we will see in this chapter, there is a direct link between a person's genotype, the proteins the person makes, and his or her phenotype. Mutations that alter the amino acid sequence of a protein can produce phenotypes that range from insignificant to lethal. We will examine this link using examples from the role of proteins as enzymes and as transport molecules. In addition, we will explore how variations in the proteins we make affect our reactions to drugs and to environmental chemicals.

FIGURE 10.1 Portrait of a dwarf by Goya. Some genetic forms of dwarfism are caused by mutations in genes that encode proteins that act as growth hormones, receptors, and growth factors.

ENZYMES AND METABOLIC PATHWAYS

Enzymes are proteins that carry out biochemical reactions. They convert molecules known as **substrates** into **products** by catalyzing chemical reactions (Figure 10.2). In the cell, enzymatic reactions do not occur randomly; they are connected to form chains of reactions called *biochemical pathways* (Figure 10.3a). The sum of all the biochemical reactions going on in a cell is called **metabolism,** and chains of biochemical reactions are often called metabolic pathways.

In a metabolic pathway, the product of one reaction serves as the starting point (substrate) for the next reaction. If a mutation shuts down one step in a pathway, all the reactions beyond that point are shut down, because there is no

Substrate The specific chemical compound that is acted upon by an enzyme.

Product The specific chemical compound that is the result of enzymatic action. In biochemical pathways, a compound can serve as the product of one reaction and the substrate for the next reaction.

Metabolism The sum of all biochemical reactions by which cells convert and utilize energy.

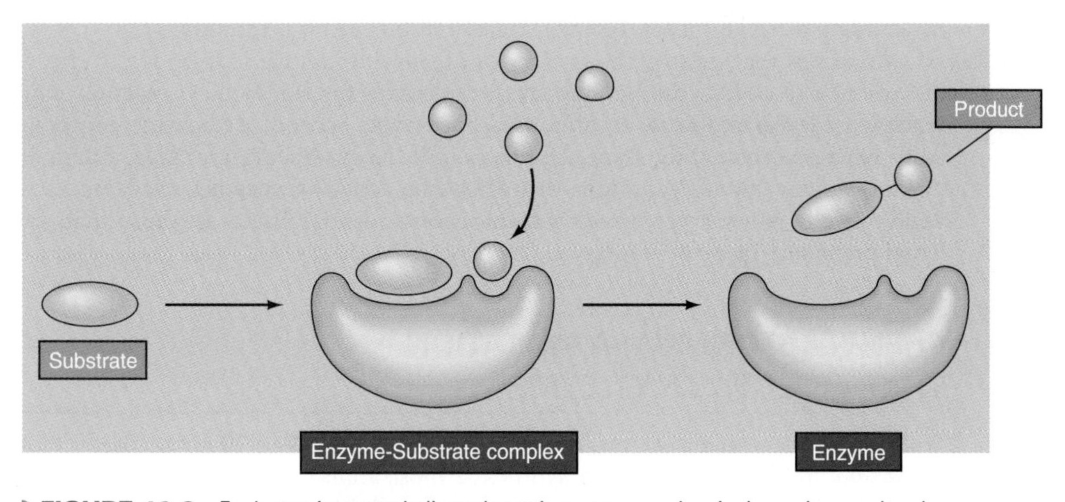

FIGURE 10.2 Each step in a metabolic pathway is a separate chemical reaction catalyzed by an enzyme, in which a substrate is converted to a product.

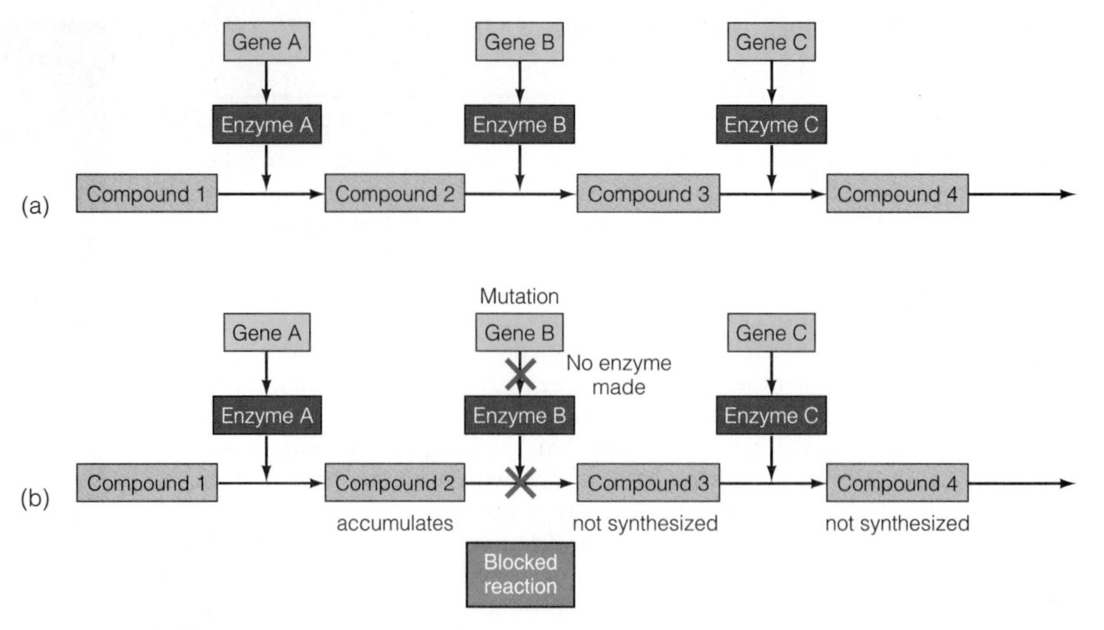

FIGURE 10.3 (a) The sequence of reactions in a metabolic pathway. In this pathway, compound 1 is present in the diet and is converted in the body into compound 2, which is then converted into compound 3. Finally, compound 3 is converted into compound 4. A specific enzyme catalyzes each of these reactions. Each enzyme is the product of a gene. (b) In this pathway, a mutation in gene B leads to the production of a defective protein that cannot function as an enzyme. As a result, compound 2 cannot be converted into compound 3. Because no compound 3 is made, compound 4 will not be produced, even though enzyme C is present. Compound 1 is supplied by the diet and is converted into compound 2, which accumulates because it cannot be metabolized.

substrate for reactions beyond the blocked reaction (Figure 10.3b). If one reaction is shut down by mutation, it also results in the accumulation of substrates and products in the part of the pathway leading up to the block.

Sir Archibald Garrod was the first to propose that human genetic disorders and metabolism are related (see Concepts and Controversies: Garrod and Metabolic Disease). Garrod studied a disorder called **alkaptonuria** (MIM/OMIM 203500), in which large amounts of a compound called homogentisic acid are excreted in urine. Garrod called this heritable condition an **inborn error of metabolism.** His work represented a pioneering study in applying Mendelian genetics to humans and in understanding the relationship between genes and biochemical reactions.

Alkaptonuria A relatively benign autosomal recessive genetic disorder associated with the excretion of high levels of homogentisic acid.

Inborn error of metabolism The concept advanced by Archibald Garrod that many genetic traits result from alterations in biochemical pathways.

Mutations that destroy the activity of an enzyme can cause phenotypic effects in several ways. First, the buildup of the substrate for the blocked reaction may reach toxic levels and cause an abnormal phenotype. Second, the enzyme may control a reaction essential for some cell function. If this product is not made, a mutant phenotype may result. Mutations that affect the action of enzymes can produce a wide range of phenotypes, ranging from inconsequential effects to those that are lethal prenatally or early in infancy.

PHENYLKETONURIA: A MUTATION THAT AFFECTS AN ENZYME

To make all the proteins we need to maintain life, we need all 20 amino acids. Humans, like most eukaryotes, can synthesize most amino acids using a number of metabolic pathways. However, we cannot make all 20 amino acids; some must be included in our diet. The amino acids we cannot synthesize are called **essential amino acids.** In humans, there are nine essential amino acids: histidine, isoleucine,

Essential amino acids Amino acids that cannot be synthesized in the body and must be supplied in the diet.

Concepts and Controversies

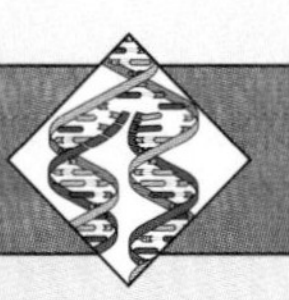

Garrod and Metabolic Disease

Sir Archibald Garrod was a distinguished physician, Oxford professor, and physician to the royal family. Garrod was widely trained; he was skilled in biology, biochemistry, and clinical medicine. In studying alkaptonuria, he showed that the condition causes the excretion of abnormal amounts of a compound called homogentisic acid, which causes the urine to turn black upon exposure to air.

Garrod proposed that the presence of homogentisic acid in the urine of affected individuals was caused by a metabolic block, which prevented metabolism of the compound. He also speculated that the metabolic block was caused by an enzyme deficiency. Garrod observed that 60% of the affected individuals were the result of first-cousin marriages and that the parents were unaffected. This led him to propose that alkaptonuria is inherited as a recessive Mendelian trait. Remember, Garrod did his work just a few years after Mendel's work was rediscovered and widely publicized. Garrod studied other metabolic disorders and published his findings in a book, *Inborn Errors of Metabolism,* that is widely regarded as a milestone in human genetics.

Why was Garrod's work, like that of Mendel, not immediately appreciated? From our perspective, it may seem strange, but Garrod worked in three areas that at the time were almost totally isolated from one another: genetics, medicine, and chemistry. Chemists had no interest in studying heredity, physicians regarded the conditions he studied as too rare to be important, and geneticists cared little about either metabolism or medicine.

leucine, lysine, methionine, phenylalanine, threonine, tryptophan, and valine. In other words, our diet has to be varied enough to provide 9 of the 20 amino acids.

Metabolism of Phenylalanine and PKU

Phenylalanine is an essential amino acid and is the starting point for a network of metabolic pathways. Here we will focus on what happens when the first step in the phenylalanine metabolic pathway is blocked by a mutation that prevents the conversion of phenylalanine to another amino acid, tyrosine. Failure to convert phenylalanine to tyrosine results in a genetic disorder called **phenylketonuria** (MIM/OMIM 261600), or PKU. This is the disorder described at the beginning of the chapter. About one in every 12,000 newborns has PKU. In almost all cases, this is caused by a mutation in a gene for the enzyme phenylalanine hydroxylase (PAH), which converts phenylalanine to tyrosine.

Phenylketonuria (PKU) An autosomal recessive disorder of amino acid metabolism that results in mental retardation if untreated.

In individuals with PKU, phenylalanine that enters the body in protein-containing food cannot be converted to tyrosine and builds up to high levels (▶ Figure 10.4). If untreated, newborns with high levels of phenylalanine become severely mentally retarded, have enhanced reflexes that cause their arms and legs to move in a jerky fashion, develop epileptic seizures, and never learn to talk. Because the skin pigment melanin is also a product of the blocked metabolic pathway (Figure 10.4), most people with PKU usually have lighter hair and skin color than their siblings or other family members. Even though there are two pathways leading from phenylalanine, blockage of the pathway to tyrosine overloads the alternate pathway, leading to high levels of phenylacetic acid (responsible for the musty odor of affected individuals) and other compounds that contribute to the clinical phenotype.

How does the failure to convert phenylalanine to tyrosine produce mental retardation and the other aspects of the phenotype? These effects are not caused by the failure to convert phenylalanine to tyrosine because adequate amounts of tyrosine are available in the diet. The problem is caused by the accumulation of high levels of phenylalanine and its metabolic byproducts (Figure 10.4) while the nervous system is maturing during infancy.

The human brain and nervous system continue to develop after birth and require a constant supply of amino acids for protein synthesis. Special proteins (trans-

FIGURE 10.4 The metabolic pathway that leads from the essential amino acid phenylalanine. Normally, phenylalanine is converted to tyrosine and from there to many other compounds. A metabolic block, caused by a mutation in the gene encoding the enzyme phenylalanine hydroxylase, prevents the conversion of phenylalanine to tyrosine and, in homozygotes, produces the phenotype of phenylketonuria (PKU). The diagram also shows other metabolic diseases produced by mutations in genes encoding enzymes in this pathway.

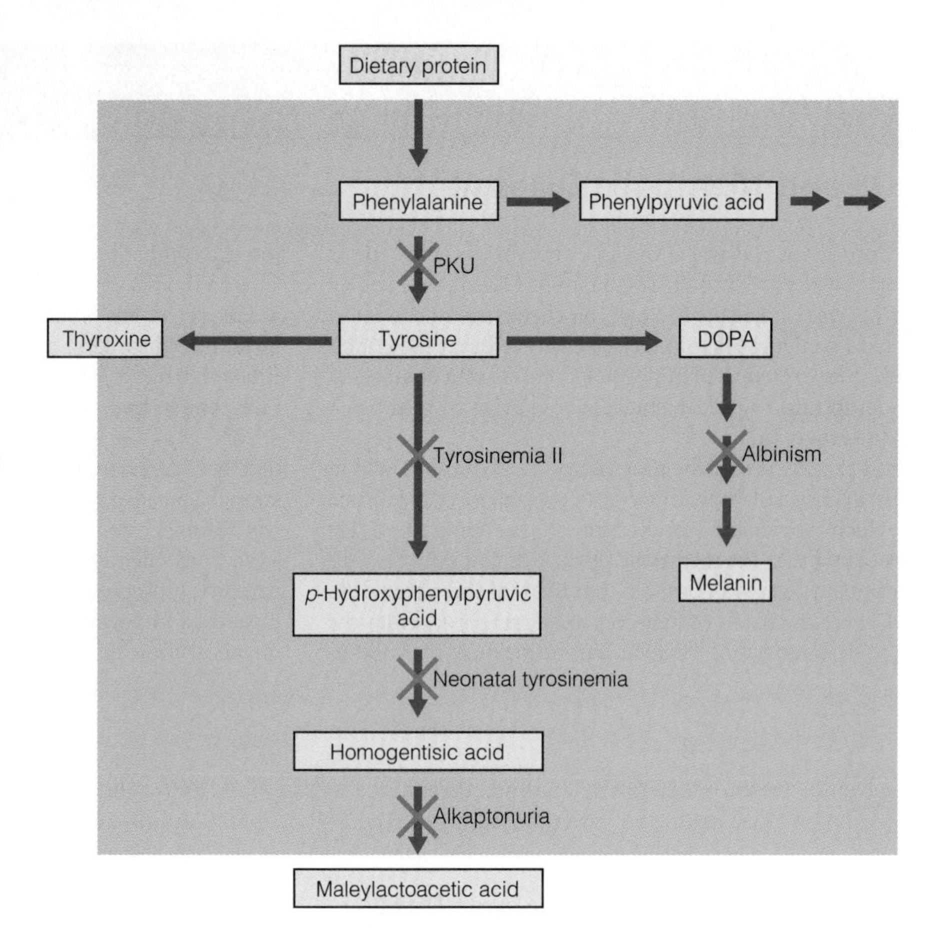

porters) embedded in the plasma membrane of cells move amino acids into the cell. Phenylalanine and seven other amino acids (called the neutral amino acids) are transported by one of these systems. As phenylalanine accumulates in the fluid surrounding cells of the maturing nervous system, phenylalanine molecules outnumber those of the seven other amino acids, and the transport system takes in too much phenylalanine. It is not clear whether the damage to the nervous system is the result of transporting too much phenylalanine, not enough of other amino acids, or the accumulation of secondary metabolites like phenylpyruvic acid in cells of the nervous system. The result, however, is brain damage, mental retardation, and the other neurological symptoms that result in the phenotype of PKU.

Treating PKU with a Diet Low in Phenylalanine

Most people with PKU have heterozygous mothers and were unaffected during prenatal development because the mother has enough PAH in her body to metabolize any excess phenylalanine that accumulates during fetal development. PKU homozygotes suffer neurological damage and become retarded only if fed on a normal diet after birth.

PKU is a genetic disorder, but it is also an environmental disease. If phenylalanine is not present in the environment (diet), there is no abnormal phenotype. In the early 1950s, PKU was treated using a diet with very low levels of phenylalanine (see Concepts and Controversies: Dietary Management and Metabolic Disorders). This same treatment is widely used today and has been successful in reducing the effects of this disease. However, managing PKU by controlling dietary intake is both difficult and expensive.

One major problem is that phenylalanine is present in many protein sources, and it is impossible to eliminate all protein from the diet. The protein restriction means

Concepts and Controversies

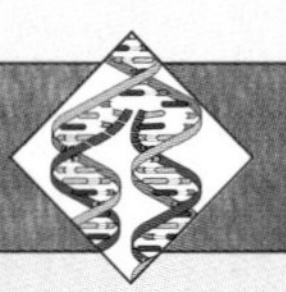

Dietary Management and Metabolic Disorders

In several metabolic diseases, a diet is used to prevent full expression of the mutant phenotype. These diets can be manipulated to replace metabolites that are not produced or to prevent the buildup of toxic compounds. This method is really one form of gene therapy. Dietary modification is used with varying degrees of success in several metabolic conditions, including phenylketonuria (PKU), galactosemia, tyrosinemia, homocystinuria, and maple syrup urine disease.

The diet for each disorder usually is available in two versions, one for infants with low levels of the restricted component and another version for older children and adults that usually contains higher levels of the restricted compound and other nutrients. For PKU, a formula is prepared from enzymatically digested proteins or synthetic mixtures of amino acids. In addition, the formula contains fats, usually in the form of corn oils, and carbohydrates from sugar, cornstarch, or corn syrup. Vitamin and mineral supplements are also added. In one popular formula for PKU, casein (a protein extracted from milk) is enzymatically digested into individual amino acids. The mixture of amino acids is treated to remove phenylalanine. This process also removes two other amino acids, tyrosine and tryptophan. These two amino acids are added back to the mixture along with sources of fat, carbohydrates, vitamins, and minerals. Affected individuals use the powder at each meal as a source of amino acids and have no protein in their diet. This means they cannot eat any meat (hamburgers, chicken, and so forth) or any dairy products (milk, ice cream, and so forth). A typical menu for a school-aged child is shown in the next column.

Until the early 1980s, this protein-restricted diet was followed for 6 to 9 years. The rationale was that development of the nervous system is completed by this age and that elevated levels of phenylalanine that accompany a normal diet would have no impact on intellectual development or behavior. This decision was also partly economic because the diet can cost more than $5,000 a year. Standard practice now is to continue the diet through adolescence, and some clinicians recommend continuing the diet for life. This decision is based on research indicating that withdrawal of the diet can be deleterious and leads to a decline in intellectual ability and abnormal changes in electroencephalographic patterns.

Breakfast
2 to 3 cups dry rice cereal
1 to 2 bananas
6 oz. formula

Lunch
1 to 2 cans vegetable soup
3 crackers
1 cup fruit cocktail
4 oz. formula

Dinner
2 cups low-protein noodles
1 to 2 cups meatless spaghetti sauce
1 cup of salad (lettuce)
French dressing
4 oz. formula

Snack
1 to 2 cups popcorn
1 tablespoon margarine

that meat, fish, milk, cheese, bread, cake, and nuts cannot be eaten. The diet is hard to follow, because it means that affected children cannot eat hamburgers, chicken nuggets, pizza, or ice cream. Instead, they have to eat a dietary supplement containing a synthetic mixture of amino acids (with very low levels of phenylalanine) along with vitamins and minerals. The supplement is foul-smelling and bad-tasting; in most cases, it must be continued for life.

The challenge on a PKU diet is maintaining a blood level of phenylalanine that is high enough to permit normal development of the nervous system and yet low enough to prevent mental retardation. To avoid the phenotypic consequences of PKU, dietary treatment must be started in the first month after birth. After that time the brain is damaged, and treatment is less effective.

The first sign of PKU is abnormal levels of phenylalanine in the blood and urine of newborns. Since the 1960s, newborns in the United States have been routinely tested for PKU by analyzing blood or urine for phenylalanine levels. By the mid-1970s, many countries of the world were testing newborns for PKU.

Over 100 million infants have been screened, and over 10,000 cases of PKU have been detected and treated with a low phenylalanine diet. In the United States, all states require screening of newborns for PKU, so the number of untreated cases is very low. Screening and treatment with a low phenylalanine diet allows PKU homozygotes to lead essentially normal lives.

There is some controversy about how long the low-phenylalanine diet must be continued. Some studies suggest that PKU homozygotes can begin to eat a normal diet at about 10 to 14 years of age without any effects on intellect or behavior. Some treatment centers disagree with this recommendation and recommend that the treatment be continued for life. Recent findings indicate that parts of the brain continue to develop into adulthood. If confirmed, these results will probably require that the PKU diet be extended well into adulthood.

Women with PKU: Having Children

As PKU children treated with diet therapy have matured and reached reproductive age, the question arises: Can a woman homozygous for PKU have an unaffected child? The answer to this seems straightforward. If she has a child with a homozygous normal man, the child would be heterozygous and unaffected. If she has a child with a heterozygote, the chances are 50% that the child will be unaffected. Only if she has a child with a homozygous PKU man will the child have a 100% chance of being affected.

The real answer is that all the children born to homozygous PKU women who eat a regular diet during pregnancy will be mentally retarded, no matter what their genotype. A pregnant PKU female on a normal diet will accumulate high levels of phenylalanine in her blood. The excess phenylalanine will cross the placenta and damage the nervous system of the developing fetus in a way that is independent of the child's genotype.

To get around this problem, it is recommended that PKU females stay on a low-phenylalanine diet all through life or return to the diet before becoming pregnant and maintain the diet throughout pregnancy. In addition, PKU females have other reproductive options, including *in vitro* fertilization and the use of surrogate mothers (see Chapter 7).

OTHER METABOLIC DISORDERS IN THE PHENYLALANINE PATHWAY

The mutation in the pathway from phenylalanine to tyrosine is not a unique occurrence. Several other genetic disorders are associated with mutations that cause blocks in pathways leading from phenylalanine. For example, the phenylalanine pathway leads to the production of the thyroid hormones thyroxine and triiodothyronine. A mutation that causes a block in this pathway leads to the autosomal recessively inherited disorder called genetic goitrous cretinism (Figure 10.4). Newborn homozygotes are unaffected because during prenatal development, maternal thyroid hormones cross the placenta and promote normal growth. In the weeks following birth, physical development is slow, mental retardation begins, and the thyroid gland greatly enlarges. This condition is caused by the failure to synthesize an essential product, thyroid hormone, not by the accumulation of a metabolic intermediate as in PKU. If diagnosed early, affected infants can be treated with thyroid hormone.

In this same network of metabolic pathways, a mutation in a gene encoding an enzyme leads to the buildup of homogentisic acid and causes alkaptonuria, an autosomal recessive condition. This is the disorder first investigated by Garrod at the beginning of the twentieth century. Mutations in other genes controlling enzymes in this network (Figure 10.4) result in neonatal tyrosinemia (MIM/OMIM 276700), tyrosinemia II (MIM/OMIM 276600), albinism (MIM/OMIM 203100), and other disorders.

GENES AND ENZYMES OF CARBOHYDRATE METABOLISM

Mutations in genes that control enzymes are not limited to amino acid metabolic pathways. Other pathways, including those of lipid metabolism, nucleic acid metabolism, and carbohydrate metabolism are also affected. We will briefly illustrate this by considering some mutations in carbohydrate metabolism.

Carbohydrates are organic molecules that include sugars, starches, glycogens, and celluloses. The simplest carbohydrates are sugars called monosaccharides (Figure 10.5a). Glucose, galactose, and fructose are all monosaccharides and are important as energy sources for the cell. The chemical joining of two monosaccharides produces a disaccharide (Figure 10.5b). Some common disaccharides are maltose (two glucose units, used in brewing beer), sucrose (a glucose and fructose unit, the sugar you buy at the store), and lactose (a glucose and a galactose unit, found in milk). Larger combinations of sugars form long chains known as polysaccharides; these include glycogen, starch, and cellulose.

Many different enzymes catalyze the reactions that generate and metabolize glucose to produce energy in our cells. Mutations that cause metabolic blocks in any of these reactions can have serious phenotypic consequences. Some genetic disorders associated with the metabolism of glycogen are listed in Table 10.1. We will look at two examples of mutations that affect enzymes of carbohydrate metabolism.

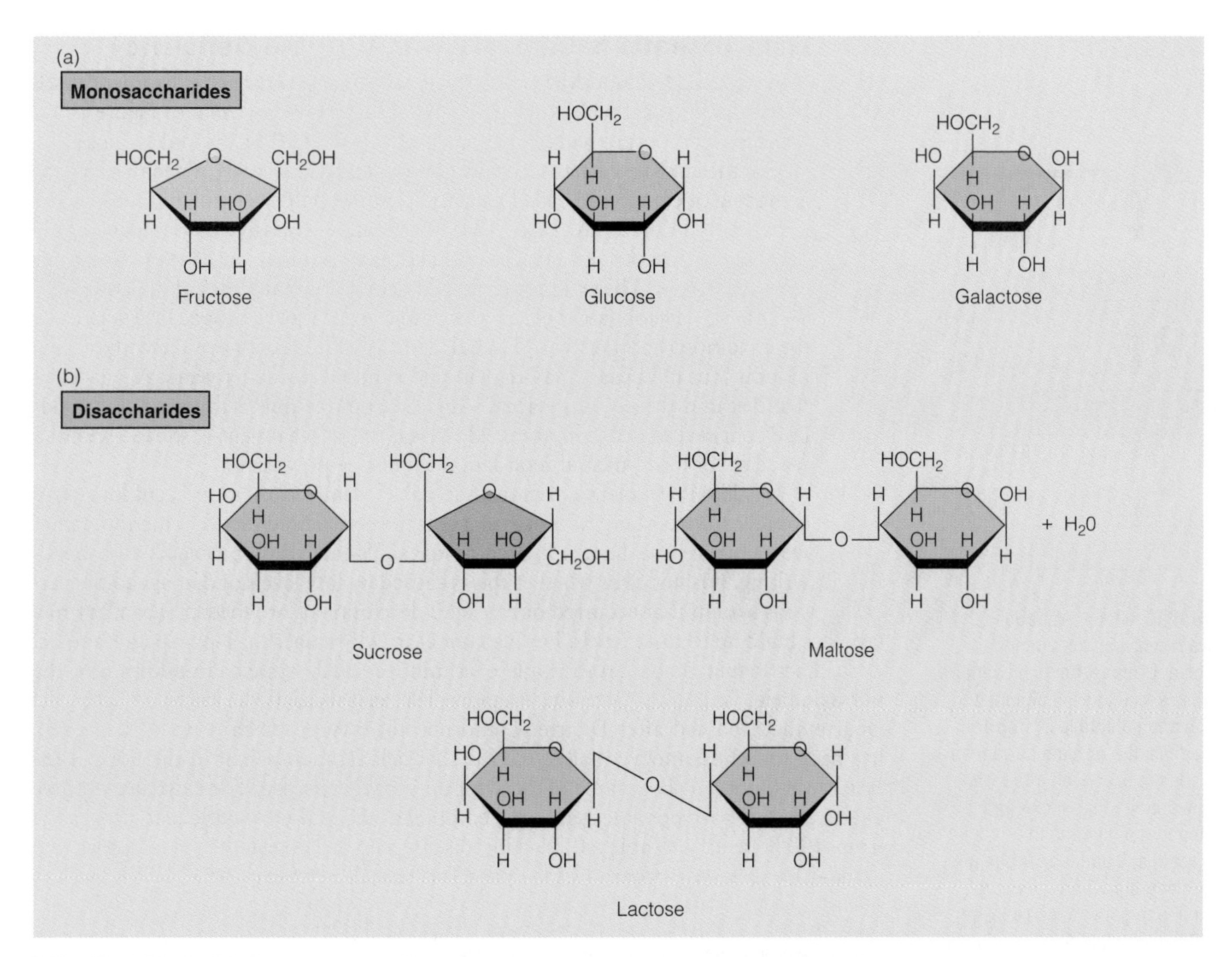

FIGURE 10.5 (a) The structures for three common monosaccharides. (b) Structures for three disaccharides. Mutations in genes controlling metabolism of sugars are associated with genetic disorders.

Table 10.1 Some Inherited Diseases of Glycogen Metabolism

Type	Disease	Metabolic Defect	Inheritance	Phenotype	MIM/OMIM Number
I	Glycogen storage disease—Von-Gierke disease	Glucose-6-phosphatase deficiency	Autosomal recessive	Severe enlargement of liver, often recognized in second or third decade of life; may cause death due to renal disease	232200
II	Pompe disease	Lysosomal glucosidase deficiency	Autosomal recessive	Accumulation of membrane-bound glycogen deposits. First lyosomal disease known. Childhood form leads to early death	232300
III	Forbes disease, Cori disease	Amylo 1, 6 glucosidase deficiency	Autosomal recessive	Accumulation of glycogen in muscle, liver. Mild enlargement of liver, some kidney problems	232400
IV	Amylopectinosis, Andersen disease	Amylo 1, 4 transglucosidase deficiency	Autosomal recessive	Cirrhosis of liver, eventual liver failure, death	232500

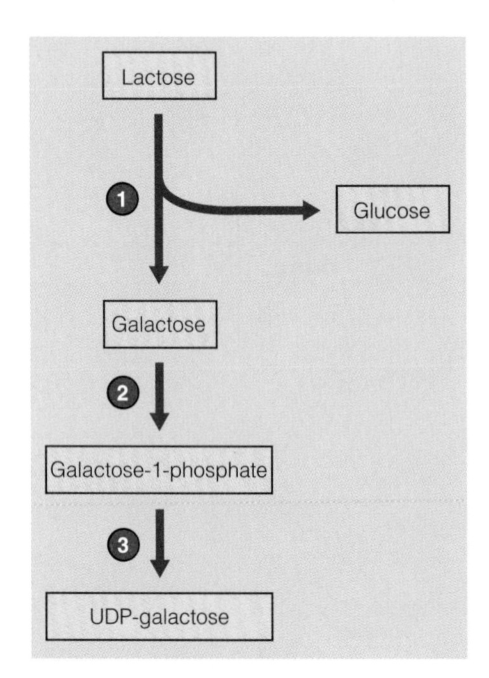

FIGURE 10.6 Metabolic pathway involving lactose and galactose. Lactose, the main sugar in milk, is enzymatically digested to glucose and galactose in step 1. In many adults, the enzyme for digesting lactose is not produced. Step 2 is the conversion of galactose into galactose-1-phosphate. In galactosemia, a mutation in the gene that controls step 3 prevents the conversion of galactose-1-phosphate into UDP-galactose. As a result, the concentration of galactose-1-phosphate rises in the blood, causing mental retardation and blindness.

Galactosemia Is Caused by an Enzyme Deficiency

Galactosemia (MIM/OMIM 230400) is an autosomal recessive disorder caused by the inability to break down galactose, a sugar found in milk (Figure 10.6). Galactosemia occurs with a frequency of 1 in 57,000 births and is caused by lack of the enzyme galactose-1-phosphate uridyl transferase. When this enzyme is missing, a compound called galactose-1-phosphate accumulates and reaches toxic levels in the body. Like PKU, homozygous recessive individuals usually have a heterozygous mother and are unaffected at birth but develop symptoms a few days later. The symptoms include dehydration and loss of appetite; later the infants develop jaundice, cataracts, and mental retardation. In severe cases the condition is progressive and fatal. Seriously affected infants die within a few months, but mild cases may remain undiagnosed for many years. A galactose-free diet and the use of galactose- and lactose-free milk substitutes and foods lead to a reversal of symptoms. However, unless treatment is started within a few days of birth, mental retardation cannot be prevented.

Unlike PKU, dietary treatment in galactosemia does not prevent long-term complications. Many of those on a galactose-restricted diet develop difficulties with balance and have impaired motor skills, including problems with handwriting. It is not clear whether this is caused by low levels of damage to the nervous system that occurred during fetal development or whether secondary metabolites build up to toxic levels. It is also not clear why galactose-1-phosphate is toxic.

Galactosemia is also an example of a multiple-allele system. In addition to the normal allele, *G*, and the recessive mutant allele, *g*, a third allele, known as *GD* (the Duarte allele, named after Duarte, California, the city in which it was discovered), has been found. Homozygous *GD/GD* individuals have only half of the normal enzymatic activity but show none of the symptoms of the disease. The existence of three alleles produces six possible genotypic combinations, and enzymatic activities range from 100% to 0% (Table 10.2). This disease can be detected in newborns, and mandatory screening programs in some states test all newborns for galactosemia.

Lactose Intolerance Is a Genetic Variation

Human milk is about 7% lactose, which serves as a major energy source for the nursing infant. The first step in metabolizing lactose is controlled by the enzyme lac-

Table 10.2 Multiple Alleles of Galactosemia

Genotype	Enzyme Activity (%)	Phenotype
G^+/G^+	100	Normal
G^+/G^D	75	Normal
G^D/G^D	50	Normal
G^+/g	50	Normal
G^D/g	25	Borderline
g/g	0	Galactosemia

tase, which splits lactose into two sugars, glucose and galactose (Figure 10.6). In many parts of the world, lactase levels decline during mid- to late childhood to less than 10% of the levels found in infants. The decline in adult lactase levels is inherited as an autosomal recessive trait.

Adults with low lactase levels are unable to fully digest the lactose found in milk and other dairy products. If these adults eat lactose-containing foods, the result is a series of intestinal symptoms that include bloating, cramps, gas, and diarrhea. This condition is called lactose intolerance, and most lactase-deficient adults learn to avoid dairy products. Lactose intolerance is not considered a genetic disorder but only a variation in gene expression. Most human populations have low adult lactase levels, but the frequency of lactose intolerance varies from 0 to 100%. In Chapter 18, we will explore the role of natural selection in controlling the frequency of lactose intolerance in human populations.

Galactosemia A heritable trait associated with the inability to metabolize the sugar galactose. If it is left untreated, high levels of galactose-1-phosphate accumulate, causing cataracts and mental retardation.

MUTATIONS IN RECEPTOR PROTEINS

Although many, and perhaps most, proteins function as enzymes, proteins have many other roles, including signal receptors and transducers. These proteins are usually embedded in the plasma membrane of the cell, and mutations in receptor function can have drastic consequences. For example, in androgen insensitivity (discussed in Chapter 7), a mutation in a gene encoding a receptor makes cells unable to respond to the presence of the hormone testosterone, causing a genotypic male to develop into a phenotypic female. Other genetic disorders associated with receptors, including familial hypercholesterolemia, are listed in Table 10.3.

MUTATIONS THAT AFFECT TRANSPORT PROTEINS

Hemoglobin, an iron-containing protein in red blood cells, transports oxygen from the lungs to the cells of the body. The hemoglobin molecule occupies a central position in human genetics. The study of hemoglobin variants led to an understanding of the molecular relationship between genes, proteins, and human disease. (1) The discovery that there are variations in the amino acid composition of hemoglobin was the first example of inherited variations in protein structure. (2) The study of hemoglobin in sickle cell anemia was the first direct proof that mutations result in a change in the amino acid sequence of proteins. (3) The nature of the mutation in sickle cell anemia provided evidence that a change in a single nucleotide is enough to cause a genetic disorder. (4) Understanding the organization of the globin gene clusters has helped scientists understand how genes evolve and how gene expression is regulated.

Table 10.3 Some Heritable Traits Associated with Defective Receptors

Disease	Defective/Absent Receptor	Inheritance	Phenotype	MIM/OMIN Number
Familial hypercholesterolemia	Low-density lipoprotein (LDL)	Autosomal dominant	Elevated levels of cholesterol in blood, atherosclerosis, heart attacks; early death	144010
Pseudohypoparathyroidism	Parathormone (PTH)	X-linked dominant	Short stature, obesity, round face, mental retardation	300800
Diabetes insipidus	Vasopressin receptor defect	X-linked recessive	Failure to concentrate urine; high flow rate of dilute urine, severe thirst, dehydration; can produce mental retardation in infants unless diagnosed early	304800
Androgen insensitivity	Testosterone/DHT receptor	X-linked recessive	Transformation of genotypic male into phenotypic female; malignancies often develop in intraabdominal testes	313700

Heritable defects in globin structure or synthesis are well understood at the molecular level and are truly "molecular diseases," as Linus Pauling called them. In this section we consider the structure of the hemoglobin molecule, the organization of the globin genes, and some genetic disorders related to globin structure and synthesis.

Hemoglobin is composed of four protein (globin) molecules. A heme group is cradled within each globin molecule; heme is an organic molecule containing an iron atom (Figure 10.7). In the lungs, oxygen enters the red blood cell and binds to the iron for transport to cells of the body. Although there are several different kinds of globin molecules (and hemoglobins), the heme group is the same in all cases.

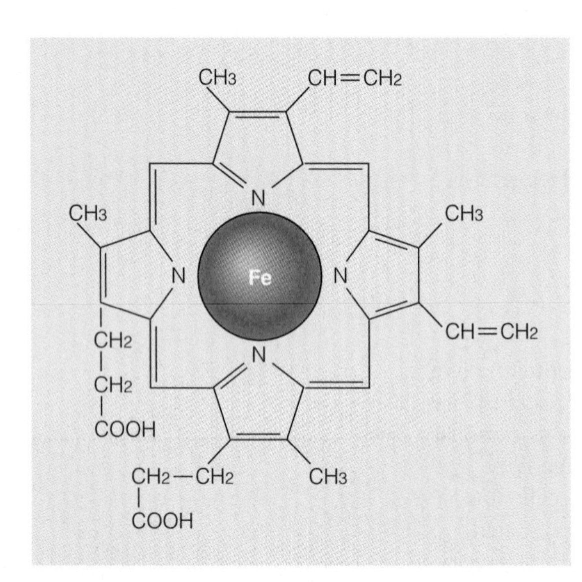

FIGURE 10.7 A heme group is a flat molecule that inserts into the folds of a globin polypeptide. Each heme group carries an iron atom, which binds oxygen in the lungs for transport to the cells and tissues of the body.

Adult hemoglobin (called Hb A) is made up of four globin molecules: two alpha globins and two beta globins (Figure 10.8). Alpha globin is encoded in a gene cluster on chromosome 16 (Figure 10.9); beta globin is encoded in a gene cluster on chromosome 11 (Figure 10.10). Each red blood cell contains about 280 million molecules of hemoglobin, and there are between 4 and 6×10^{12} red blood cells in each liter of blood. Each red blood cell is replaced every 120 days, so hemoglobin production is a major metabolic event in the body, with millions of new hemoglobin molecules produced each second of each day.

Unlike most genes we carry, there are two copies of the alpha-globin gene (designated $alpha_1$ and $alpha_2$) in the alpha-gene cluster on each chromosome 16 (Figure 10.9), along with three related genes: the zeta gene, pseudozeta, and pseudoalpha-1 genes. **Pseudogenes** are nonfunctional copies of genes whose nucleotide sequence is similar to a functional gene, but with mutations that prevent their expression.

Pseudogenes Nonfunctional genes that are closely related (by DNA sequence) to functional genes present elsewhere in the genome.

The alpha- and beta-globin genes are descended from a common ancestral gene. As a result, they have similar organizations, and their gene products, the alpha and beta globins, are similar in size and amino acid composition. Alpha globin contains 141 amino acids and the beta-globin molecule has 146 amino acids, and the amino acid sequence of the two polypeptides is very similar. Because of these similarities, the alpha and beta genes fold into similar configurations, and each cradles the heme group inside the folds of the polypeptide chain.

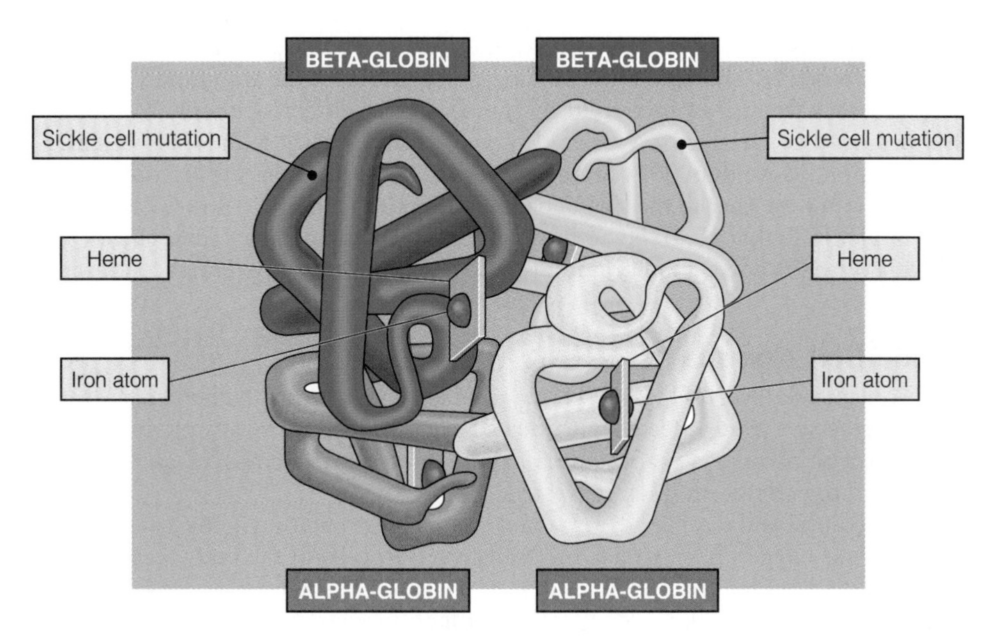

FIGURE 10.8 A functional hemoglobin molecule is composed of two alpha-globin polypeptides and two beta-globin polypeptides. Each globin molecule carries a heme group within its folds. The location of the mutation in beta globin that is responsible for sickle cell anemia is shown near the start of each beta chain.

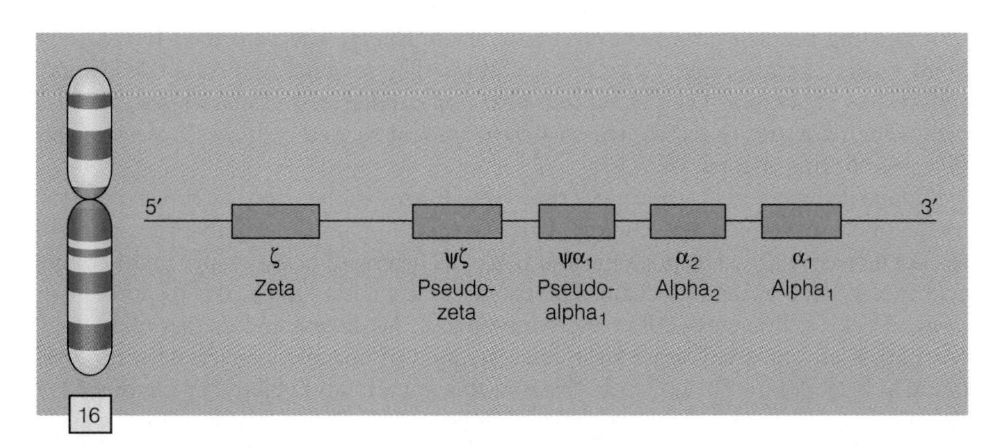

FIGURE 10.9 The chromosomal location and organization of the alpha-globin cluster. Each copy of chromosome 16 contains two copies of the alpha-globin gene (alpha$_1$ and alpha$_2$), two nonfunctional versions (called pseudogenes), and a zeta gene, which is active only during early embryonic development.

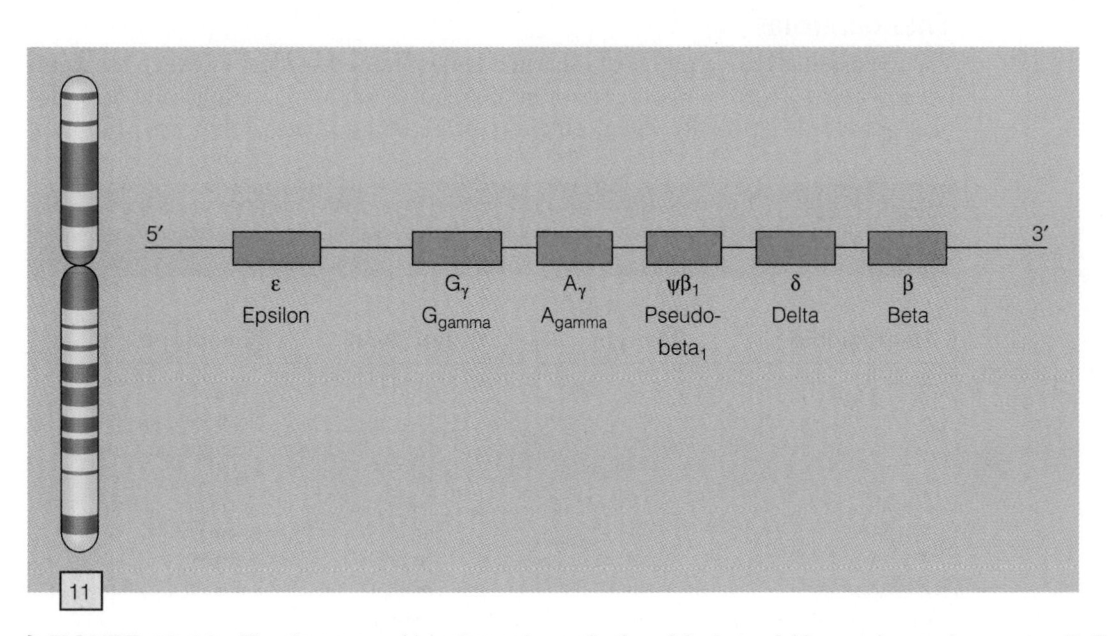

FIGURE 10.10 The chromosomal location and organization of the beta-globin complex on chromosome 11. Each copy of chromosome 11 has an epsilon gene, active during embryonic development; two gamma genes (G_{gamma} and A_{gamma}), active in fetal development; and a delta gene and a beta gene, which are transcribed after birth.

Hemoglobin variants Alpha and beta globins with variant amino acid sequences.

Thalassemias Disorders associated with an imbalance in the production of alpha or beta globin.

Genetic disorders of hemoglobin fall into two categories: the **hemoglobin variants,** which involve changes in the amino acid sequence of the globin polypeptides, and the **thalassemias,** characterized by imbalances in globin synthesis. More than 400 hemoglobin variants have been identified, each of which is caused by a different mutation. More than 90% of all variants are caused by the substitution of one amino acid for another in the globin chain, and more than 60% of these are found in beta globin (Table 10.4). Some hemoglobin variants have no visible phenotype, whereas others produce mild symptoms, and still others result in lethal conditions.

Sickle Cell Anemia

Sickle cell anemia (MIM/OMIM 141900) is a hemoglobin variant inherited as an autosomal recessive trait. Affected individuals have a wide range of symptoms, including weakness, abdominal pain, kidney failure, and heart failure (Figure 10.11), which lead to early death if left untreated.

This painful and disabling condition is caused by a mutation in the gene for beta globin. After oxygen is unloaded and transferred to cells in the body, hemoglobin molecules with mutant beta-globin subunits come out of solution. The insoluble hemoglobin molecules stick together and form long tubular structures inside the cell (Figure 10.12). These tubes distort and harden the membrane of the red blood cell, twisting the cell into a characteristic sickle shape. The deformed blood cells break easily and are removed from circulation. The lowered number of red blood cells results in anemia. The sickled cells also clog capillaries and small blood vessels, producing pain and tissue damage. Micrographs of sickled cells are shown on the first page of this chapter.

Linus Pauling and his colleagues showed that hemoglobin from sickle cell patients has an abnormal amino acid content. Later, Vernon Ingram showed that the only difference between normal hemoglobin and sickle cell hemoglobin is a change in the amino acid at position 6 in the beta chain. This change in a single amino acid is the molecular basis of sickle cell anemia. All of the symptoms of the disease and its inevitably fatal outcome if left untreated derive from this alteration of one amino acid out of the 146 found in beta globin. The molecular basis of this mutation is explored in Chapter 11.

Thalassemias

The thalassemias are a group of inherited hemoglobin disorders in which a mutant phenotype results from an alteration in the relative amounts of alpha and beta globins produced. Normally, equal amounts of alpha and beta globin are produced,

Table 10.4 Beta-Globin Chain Variants with Single Amino Acid Substitutions

Hemoglobin	Amino Acid Position	Amino Acid	Phenotype
A_1	6	glu	Normal
S	6	val	Sickle cell anemia
C	6	lys	Hemoglobin C disease
A_1	7	glu	Normal
Siriraj	7	lys	Normal
San Jose	7	gly	Normal
A_1	58	tyr	Normal
HbM Boston	58	his	Reduced O_2 affinity
A_1	145	cys	Normal
Bethesda	145	his	Increased O_2 affinity
Fort Gordon	145	asp	Increased O_2 affinity

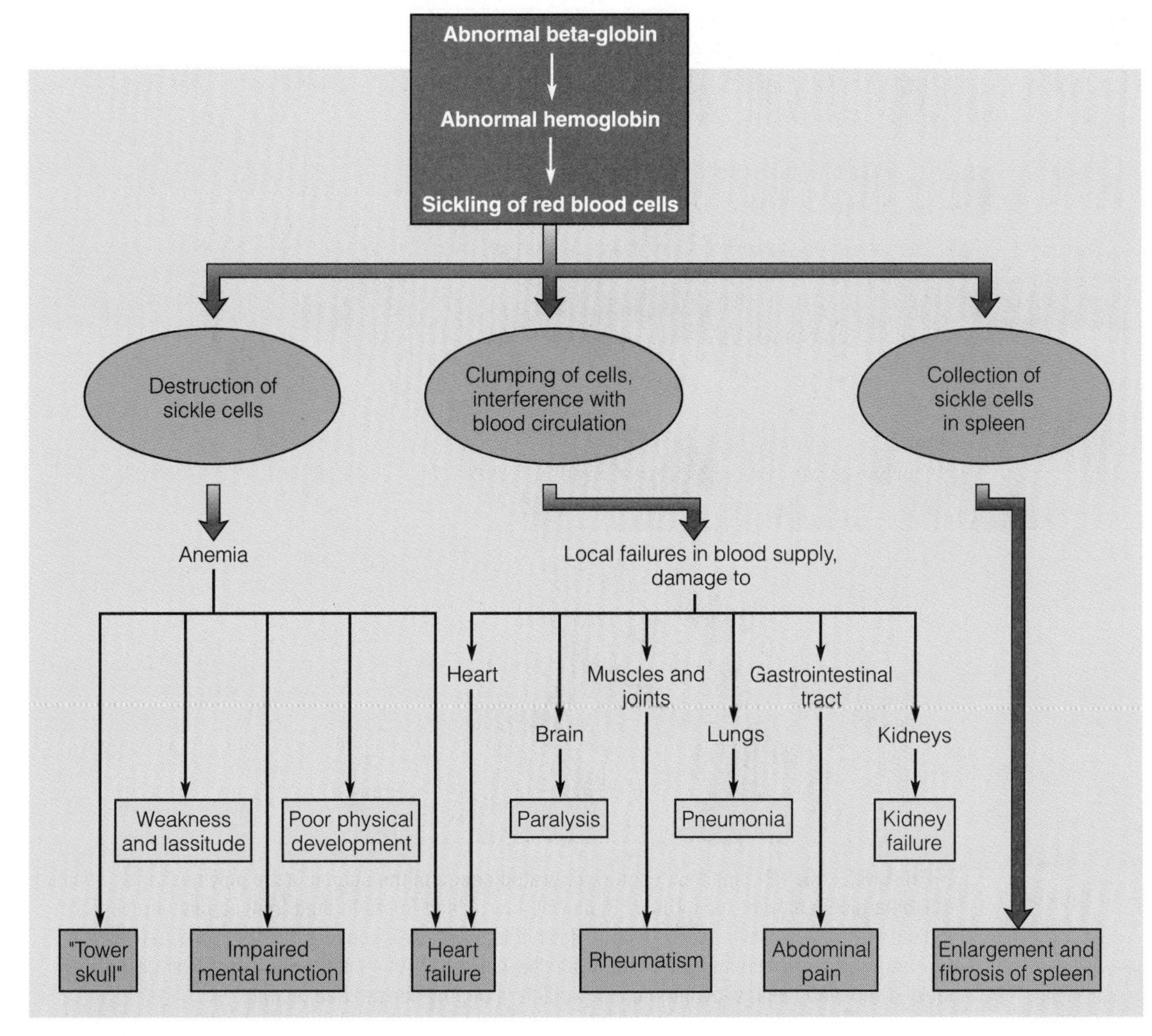

FIGURE 10.11 The cascade of phenotypic effects resulting from the mutation that causes sickle cell anemia. Affected homozygotes have effects at the molecular, cellular, and organ levels, all resulting from the substitution of a single amino acid in the beta-globin polypeptide chain.

and normal hemoglobin molecules contain two molecules of each type of globin. In thalassemia, the synthesis of alpha or beta globin is reduced or absent, causing the formation of hemoglobin molecules with an abnormal number of alpha- or beta-globin components. These hemoglobin molecules do not bind oxygen efficiently and can have serious and even fatal effects.

Thalassemias are common in several parts of the world, especially the Mediterranean region and Southeast Asia, where up to 20% or 30% of the population can be affected. The name "thalassemia" is derived from the Greek word *thalassa,* for "sea," emphasizing that this condition was first described in people living around the Mediterranean Sea.

There are two types of thalassemia: **alpha thalassemia** (MIM/OMIM 141800), in which the synthesis of alpha globin is reduced or absent, and **beta thalassemia** (MIM/OMIM 141900), which affects the synthesis of beta chains (Table 10.5). Both conditions have more than one cause, and although inherited as autosomal recessive traits, both alpha and beta thalassemia have phenotypic effects in the heterozygous condition.

Alpha thalassemia is caused by the deletion of one or more alpha-globin genes. Six genotypes are possible, five of which have symptoms ranging from mild to lethal (Figure 10.13). There are several forms of beta thalassemia, but most do not

Alpha thalassemia Genetic disorder associated with an imbalance in the ratio of alpha and beta globin caused by reduced or absent synthesis of alpha globin.

Beta thalassemia Genetic disorder associated with an imbalance in the ratio of alpha and beta globin caused by reduced or absent synthesis of beta globin.

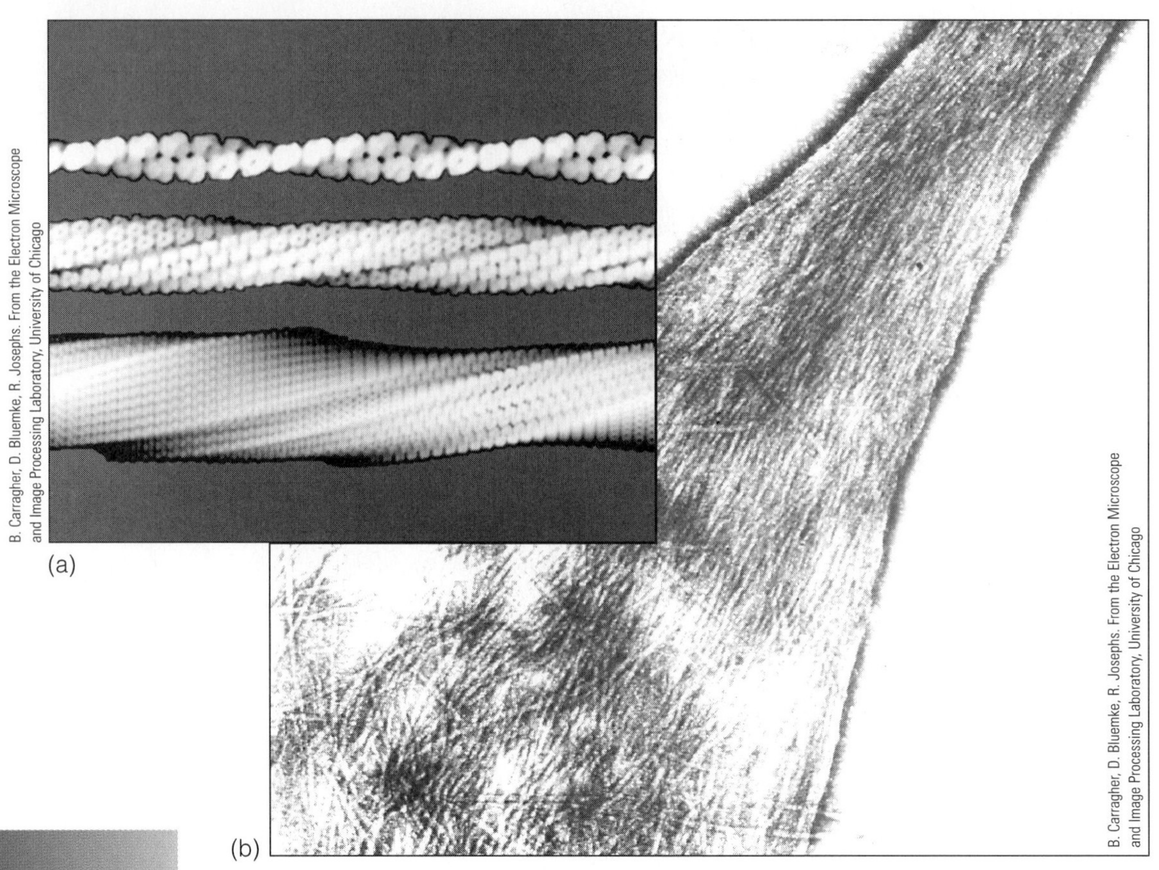

FIGURE 10.12 (a) A computer-generated image of the stages in the polymerization of sickle cell beta globin to form rods. Upper: A pair of intertwined fibers formed from stacked hemoglobin molecules. Middle: Seven pairs of fibers form the polymer responsible for distorting red blood cells. Lower: A large fiber composed of many smaller fibers. (b) An electron micrograph of a ruptured sickled red blood cell, showing the internal fibers of polymerized hemoglobin.

involve deletions of the gene. In some forms of beta thalassemia a mutation lowers the efficiency of processing the beta-globin pre-mRNA into a mature mRNA. In β^0-thalassemia, a mutation at the junction between an intron and an exon interferes with normal mRNA splicing, resulting in very low levels of functional mRNA and, in turn, low levels of beta globin. Low levels of beta globin result in the formation of hemoglobin molecules with three or four alpha globins.

Treatment of Hemoglobin Disorders by Gene Switching

If untreated, sickle cell anemia is a fatal disease, and most affected individuals die by the age of 2 years. Even with an understanding of the molecular basis of the disease, treatments are only partially successful in relieving the symptoms. Recently, a discovery that certain anticancer drugs change patterns of gene expression led to a new and effective treatment for sickle cell anemia. The drug hydroxyurea shuts off cell division and is used to treat cancer patients. As a side effect, patients develop elevated levels of fetal hemoglobin. This combination of two alpha globins and two gamma globins is usually seen only in developing fetuses. Gamma globins are members of the beta cluster and are switched off at birth, when the beta gene is activated (Figure 10.14).

Treatment with hydroxyurea reactivates these genes and makes fetal hemoglobin reappear in the red blood cells. Because sickle cell anemia is caused by a defect in beta globin, switching on another member of the beta cluster (gamma globin) would produce fetal hemoglobin and reduce the amount of hemoglobin with mu-

SIDEBAR

Population Genetics of Sickle Cell Genes

Sickle cell anemia is a genetic disorder caused by an alteration in the gene for beta globin, a component of hemoglobin, changing normal hemoglobin (HbA) to a mutant form (HbS). Individuals homozygous for sickle cell anemia who receive treatment often die prematurely (median age approximately 45.6 years).

The high frequency of heterozygotes in West Africa indicates that they have a competitive edge over homozygotes in certain environments. West Africa is an area where malaria is widespread, and malaria has been a powerful force in changing genotype frequencies, because resistance to malaria is about 25% greater in heterozygotes than in those with the homozygous normal genotype.

In the United States, the gene for hemoglobin S is decreasing due to early screening and testing of those at risk and because malaria is not present to enhance the survival of heterozygotes. However, if the Earth's atmosphere warms, malaria may reemerge in the United States and become a force in changing genotype frequencies.

Table 10.5 Summary of Thalassemias

Type of Thalassemia	Nature of Defect
α-Thalassemia-1	Deletion of two alpha-globin genes/haploid genome
α-Thalassemia-2	Deletion of one alpha-globin gene/haploid genome
β-Thalassemia	Deletion of beta and delta genes/haploid genome
Nondeletion α-thalassemia	Absent, reduced, or inactive alpha-globin mRNA
β^0-Thalassemia	Absent, reduced, or inactive beta-globin mRNA. No beta-globin produced
β^+-Thalassemia	Absent, reduced, or inactive beta-globin mRNA. Reduced beta-globin production

tant beta globin. This in turn would reduce the number of sickled red blood cells. When individuals with sickle cell anemia are treated with hydroxyurea, fetal hemoglobin is produced and replaces some of the defective hemoglobin in red blood cells, relieving many of the disorder's symptoms. Other drugs, including sodium butyrate, also switch on the synthesis of fetal hemoglobin. In some patients treated with sodium butyrate, up to 25% to 30% of the hemoglobin in the blood is fetal hemo-

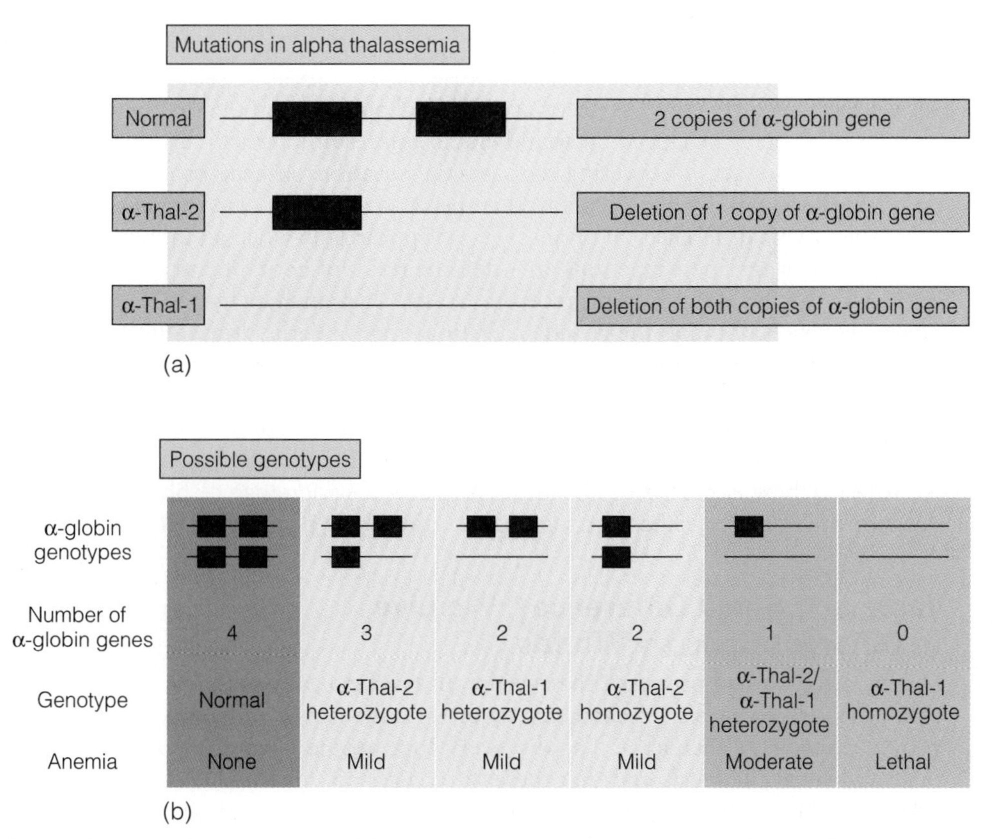

FIGURE 10.13 Deletions of alpha-globin genes in alpha thalassemia. (a) Normally, each copy of chromosome 16 carries two copies of the alpha-globin gene (normal). One copy is deleted in the *alpha-thal-2* allele, and both copies are deleted in the *alpha-thal-1* allele. (b) These three alleles can be combined to form six genotypic combinations that have zero to four copies of the alpha-globin gene. Genotypes that have one copy deleted have moderate anemia and other symptoms, and genotypes that have no copies of the gene are lethal.

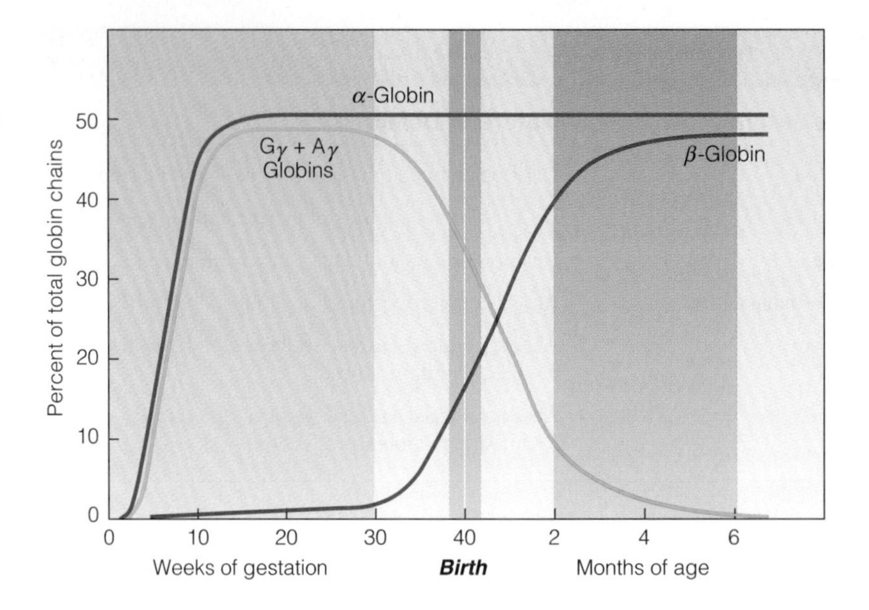

FIGURE 10.14 Patterns of globin gene expression during development. The alpha genes are switched on early in development and continue throughout life. The G_{gamma} and A_{gamma} genes, members of the beta family, are active during fetal development and switch off just before birth. The beta-globin gene is switched on at birth and is active throughout life. Sickle cell anemia and beta thalassemia are caused by mutations that affect beta globin. Research aimed at treating these conditions is directed at switching on the gamma genes, producing fetal hemoglobin to correct the conditions.

globin. Because sodium butyrate and related chemicals are less toxic than hydroxyurea, they are being used in clinical trials to treat both sickle cell anemia and beta thalassemia by switching on genes that are normally turned off at birth.

PHARMACOGENETICS

As we have seen in previous sections of this chapter, variations in the type and amount of proteins produced by an individual can result in genetic disorders of metabolism. We are also discovering that genetic variations in gene products can and do affect the way individuals react to therapeutic drugs. Like some metabolic disorders, phenotypic differences in drug reactions appear only when an individual is exposed to the drug. These reactions are often the result of heritable variations in proteins and can be dominant or recessive traits. A branch of genetics known as **pharmacogenetics** studies the genetic variations that underlie drug responses.

Pharmacogenetics A branch of genetics concerned with the inheritance of differences in the response to drugs.

Differences in drug responses can produce a range of phenotypic responses: drug resistance, toxic sensitivity to low doses, development of cancer after prolonged exposure, or an unexpected reaction to a combination of drugs. Some of these variations are harmless, whereas others can be life threatening. In this section, we consider how exposure to chemicals produces a wide range of phenotypes and describe the role of specific proteins in generating these phenotypes (if known).

Taste and Smell Differences: We Live in Different Sensory Worlds

Shortly after Garrod proposed that we are all biochemically unique individuals because of our genotypes, researchers began to demonstrate differences in the way people respond to chemicals. The discovery that people have different abilities to taste and smell chemicals and that these differences are inherited was the first indication that there might be important genetic differences in responses to drugs used to treat diseases. Some of these differences are discussed below.

The first pharmacogenetic trait was discovered as a byproduct of work in the 1930s on artificial sweeteners. In searching for sugar substitutes, workers at DuPont discovered that some people cannot taste the chemical phenylthiocarbamide (PTC), whereas others find it to be very bitter or taste like sand. Shortly thereafter, it was found that the ability to taste PTC depends on a single pair of alleles and that genotypes *TT* and *Tt* represent tasters, whereas those who have genotype *tt* are nontasters. The ability to taste PTC varies from population to population. In the United States, about 30% of adult whites are nontasters, whereas only about 3% of U.S. blacks are nontasters. Later work showed that the ability to taste PTC is more complex than originally thought. When PTC solutions at various dilutions are used, a wide bimodal range of tasters can be detected. It appears that modifying genes affect the threshold of taste sensitivity. The origin and significance of this variation is unknown, but may be important in determining food preferences. Some food plants contain compounds similar to PTC and a related compound, PROP. These plants, including kale, cabbage, broccoli, and Brussels sprouts are bitter tasting to some people. So, if you don't like broccoli or Brussels sprouts, you may be able to blame it on your genotype.

Other evidence indicates that PTC/PROP tasters may live in a different taste world than nontasters. For example, capsaicin, the compound that gives hot peppers their hot taste, has a more intense taste to PTC/PROP tasters; sucrose (table sugar) and artificial sweeteners are more intensely sweet to tasters. In addition, tasters have more food dislikes than nontasters and usually do not like foods such as black coffee, dark beer, anchovies, and strong cheeses.

Are there relationships between our genotypes, taste preferences, and our overall diets? For example, do tasters choose fruits and vegetables lower in cancer-*fighting* compounds, or do they choose foods that are lower in cancer-*causing* compounds? Is there a relationship between genotype, diet preference, and obesity? More research is needed to answer these and other questions related to taste preferences.

The ability to smell is mediated by a family of somewhere between 100 and 1,000 different membrane proteins that are distributed over the surface of cells in the nose and sinuses. There are many combinations of alleles for these proteins, so that each of us lives in a slightly different sensory world of smell. In fact, our sensory worlds can be so different that some people cannot smell the odor released by skunks (MIM/OMIM 270350).

The garden flower *Verbena* comes in a variety of colors, including red and pink (◗ Figure 10.15). Blakeslee discovered that people differ in their ability to smell the fragrances produced by these varieties. About two-thirds of the people he tested could smell a fragrance in the pink flowers but not the red ones. The remaining one-third could detect a smell in the red flowers but not the pink ones.

Although the genetics of taste and smell demonstrate that different genotypes may be responsible for our food preferences and the ability to smell flowers, the importance of pharmacogenetics lies in determining the genetic foundations for the wide range of reactions to therapeutic drugs.

◗ **FIGURE 10.15** Pink and red verbena flowers. Many people can smell the fragrance from the pink flowers but not the red ones. Others can smell the fragrance from the red flowers but not the pink ones.

Drug Sensitivities Are Genetic Traits

During the past 50 years, tens of thousands of new drugs have been developed. As these were tested on human volunteers and put into general use, distinctive patterns of response to these chemicals were identified. Subsequent work has shown that many of the differences in drug response are genetically controlled. We will examine two examples of drug sensitivities.

Succinylcholine Sensitivity Succinylcholine is used as a muscle relaxant and as a short-acting anesthetic (called suxamethonium). Soon after its introduction about 50 years ago, it became apparent that some people took hours rather than minutes to recover from a small dose of the drug. Normally the drug is broken down to an inactive form by the enzyme serum cholinesterase. Those who take a long time to recover from the drug have a form of serum cholinesterase that breaks down the drug very slowly, prolonging the effect of the anesthetic (MIM/OMIM 177400). Pedigree analysis indicates that this trait is inherited in an autosomal recessive manner. In a study of a Canadian population, the frequency of heterozygotes was 3% to 4%, and the sensitive recessive homozygote frequency was about 1 in 2,000. Use of succinylcholine on sensitive individuals can lead to paralysis of the respiratory muscles and can cause death.

Primaquine Sensitivity and Favism During World War II, several new drugs were developed to protect military personnel from malaria. Two of these drugs were primaquine and pamaquine. When these drugs were given to members of certain ethnic groups, including U.S. blacks, African blacks, and those from the European shores of the Mediterranean, such as Greeks, the result was a massive destruction of red blood cells (hemolytic anemia). A similar response to eating fava beans (*Vicia fava*) has been known since ancient times in Greece, Italy, and nearby countries.

Both responses are caused by a deficiency in an X-linked enzyme, glucose-6-phosphate dehydrogenase (G6PD). The link between antimalarial drugs and fava beans is that both produce peroxides in the blood. The enzyme G6PD, in conjunction with glutathione, another compound in serum, break down and inactivate the peroxides. If G6PD activity is reduced or absent, the high concentration of peroxides causes red blood cells to break down, producing a life-threatening anemia.

There are many other drugs and chemicals that cause lysis of red blood cells in G6PD-deficient individuals, including naphthalene, the active ingredient in mothballs. More than 400 million people worldwide are affected by G6PD deficiency. The reasons for the high frequency of this genetic variant are discussed in Chapter 18.

ECOGENETICS

Ecogenetics A branch of genetics that studies genetic traits related to the response to environmental substances.

The scope of pharmacogenetics has expanded to study genetic differences in reactions to chemicals in food, occupational exposure, and industrial pollution, leading to the development of the branch of genetics known as **ecogenetics.** It is well known that the health risks of environmental chemicals involve the properties of the chemical itself, as well as the dose and the length of exposure. It is now clear that the overall risks to environmental chemicals also depends upon genetically determined variations in the protein systems involved in transport, metabolism, and excretion of these chemicals and their metabolic byproducts.

What Is Ecogenetics?

Ecogenetics is the study of genetic variation that affects responses to environmental chemicals. Although there are more than 500,000 different chemicals used in manufacturing and agriculture, few have been tested for their toxicity or ability to cause cancer. The recognition that some members of a population may be sensitive or resistant to environmental chemicals has important consequences for research, medicine, and public policy. In this section, we will focus on the ecogenetics of pesticides.

Pesticide Metabolism

Insects, weeds, fungi, and other pathogens destroy about 35% of the world's crops. After harvesting, another 10% to 20% is destroyed in storage. Chemical agents, including herbicides, insecticides, and fungicides, are used to control these pests. In the United States, around 65% of the insecticides used each year are applied to two crops, cotton and corn.

Agricultural insecticides include a group of chemicals called organophosphates, which includes parathion, an insecticide used for more than 50 years. Exposure to parathion and other organophosphates can occur on the job (agricultural workers and forestry workers) or from eating contaminated food. In the human body, parathion is chemically inert, but is enzymatically converted to a compound called paraoxon. Paraoxon is a toxic chemical that disrupts the transmission of signals in the nervous system. Paraoxon is broken down by paraoxonase, an enzyme found in blood serum.

The gene for paraoxonase (*PON1*) has two alleles (*A* and *B*). The *A* allele encodes a protein that has high levels of enzymatic activity, and the *B* allele encodes an enzyme that has low levels of activity. The two proteins differ in a single amino acid at position 192. It has been suggested that people homozygous for high activity (*A/A*) are resistant to the effects of parathion. Conversely, those homozygous for low activity (*B/B*) are highly sensitive to parathion poisoning (MIM/OMIM 168820).

Paraoxonase is also involved in metabolizing toxic nerve gas agents, such as somain and sarin (sarin is the nerve gas released in the Tokyo subway in 1995). However, in this case, the allele effects are reversed. The *B/B* homozygote has high activity, and the *A/A* genotype has low activity. The wide range in the frequency of the *A* and *B* alleles in different populations suggests that ethnic groups may differ in their sensitivity to poisoning by insecticides and nerve gases, a rather grim reminder of the role of proteins in producing a phenotype.

The constellation of genes present within each person is the result of the random combination of parental genes and the sum of changes brought about by recombination and mutation. This genetic combination confers a distinctive phenotype upon each of us. Garrod referred to this metabolic uniqueness as chemical individuality. Understanding the molecular basis for this individuality remains one of the great challenges of human biochemical genetics.

Summary

Proteins Are the Link Between Genes and the Phenotype

1. In the early part of the twentieth century, Sir Archibald Garrod's studies on the metabolic diseases cystinuria, albinism, and alkaptonuria provided the first hints that gene products control the phenotype. These diseases and many others result from mutations that block biochemical reactions. In alkaptonuria, Garrod argued, the normal reaction is blocked by the lack of a needed enzyme. He speculated that the inability to carry out this reaction results from a gene inherited as a recessive trait.

Enzymes and Metabolic Reactions

2. Biochemical reactions in the cell are linked together to form metabolic pathways. Mutations that block one reaction in a pathway can produce a mutant phenotype in several ways.

Phenylketonuria: A Mutation That Affects an Enzyme

3. Phenylalanine is an essential amino acid and the starting point for a network of metabolic reactions. A mutation in a gene encoding the enzyme that controls

the first step in this network causes phenylketonuria (PKU). The phenotype is caused by the buildup of phenylalanine and the products of secondary reactions.

Other Metabolic Diseases in the Phenylalanine Pathway

4. The mutation causing PKU is only one of several genetic disorders caused by the mutation of genes in the phenylalanine pathway. Others include defects of thyroid hormone, albinism, and alkaptonuria, the disease investigated by Garrod.

Genes and Enzymes of Carbohydrate Metabolism

5. Mutations in genes encoding enzymes can affect metabolic pathways of other biological molecules, including carbohydrates. Galactosemia is a genetic disorder caused by lack of an enzyme in sugar metabolism. Lactose intolerance is not a genetic disorder, but a genetic variation that affects millions of adults worldwide.

Mutations in Receptor Proteins

6. Defects in receptor proteins, transport proteins, structural proteins, and other nonenzymatic proteins can cause phenotypic effects in the heterozygous state, and many show an incompletely dominant or dominant pattern of inheritance. Mutations in receptor proteins cause familial hypercholesterolemia.

Mutations That Affect Transport Proteins

7. In 1949, James Neel identified sickle cell anemia as a recessively inherited disease. This disorder is caused by a mutation in a gene encoding beta globin, a protein that transports oxygen from the lungs to cells and tissues of the body. Other mutations cause thalassemia, an imbalance in the production of globins, affecting the transport of oxygen within the body.

Pharmacogenetics

8. Differences in the reactions to therapeutic drugs represents a "hidden" set of phenotypes that are not revealed until exposure occurs. Understanding the genetic basis for these differences is the concern of pharmacogenetics and may lead to customized drug treatment for infections and other diseases.

Ecogenetics

9. Ecogenetics is the study of genetic variation that affects responses to environmental chemicals. The fact that some members of a population may be sensitive or resistant to environmental chemicals including pesticides has important consequences for research, medicine, and public policy.

CASE 1

A couple was referred for genetic counseling because they wanted to know the chances of having a child with dwarfism. Both the man and woman had achondroplasia, the most common form of short-limbed dwarfism. The couple knew that this condition is inherited as an autosomal dominant trait, but they were unsure what kind of physical manifestations a child would have if it inherited both genes for the condition. They were each heterozygous for the *FGFR3* gene that causes achondroplasia, but wanted information on the chances of having a child homozygous for the *FGFR3* gene. The counselor briefly reviewed the phenotypic features of individuals who have achondroplasia. These include the facial features (large head with prominent forehead; small, flat nasal bridge; and prominent jaw), very short stature, and shortening of the arms and legs. Physical examination and skeletal x-ray films are used to diagnosis this condition. Final adult height is approximately 4 feet.

Because achondroplasia is an autosomal dominant condition, a person who has this condition has a 1 in 2, or 50%, chance of having children with this condition. However, approximately 75% of individuals who have achondroplasia are born to parents of average size. In these cases, achondroplasia is due to a new mutation or genetic change. This couple is at risk for having a child with two copies of the changed gene or double homozygosity. Infants who have homozygous achondroplasia are either stillborn or die shortly after birth. The counselor recommended prenatal diagnosis via serial ultrasounds. In addition, a DNA test is available to detect the homozygous condition. Achondroplasia occurs in 1 in every 14,000 births.

1. What is the chance this couple will have a child with 2 copies of the dominant mutant gene? What is the chance that the child will have normal height?

Case Studies (continued)

2. Should the parents be concerned about the heterozygous condition as well as the homozygous mutant condition?
3. Why would the achrondroplasia gene be more susceptible to mutation than other genes?

CASE 2

Tina is 12 years old. Although symptomatic since infancy, she was not diagnosed with acid maltase deficiency (AMD) until she was 10 years old. The progression of her disease has been slow and insidious. She has great difficulty walking and breathing due to severe muscle weakness. She relies on a respirator to assist her breathing. Tina has severe scoliosis (curvature of the spine), which further restricts her breathing and causes even greater difficulty in walking. She is extremely tired and suffers from constant muscle pain. Although she is very bright and thinks like a normal teenager, her body won't let her function like one. She can no longer attend school. The future is bleak for Tina and other children like her. The cause of death in the childhood form of AMD is frequently due to complications from respiratory infections, which are a constant threat. Life expectancy in this form of AMD is only to the second or third decade of life.

AMD, also called glycogen storage disease type II (or Pompe disease), is an autosomal recessive condition that is genetically transmitted from carrier parents to their child. When both parents are carriers (that is, they are heterozygous), there is a 25% chance during each pregnancy that the child will have two abnormal genes and be affected.

Normally, glycogen is synthesized from sugars and is stored in the muscle cells for future use. The acid maltase enzyme breaks down the glycogen in the muscle cells. Someone with AMD lacks this enzyme, and glycogen is not broken down but gradually builds up in the muscle tissues, leading to progressive muscle weakness and degeneration. There is no treatment or cure for AMD. Enzyme replacement and gene therapy are tools that may be useful in the future but have been unsuccessful in the past. However, a new treatment for enzyme replacement is being tested and offers new hope for children like Tina.

1. Should researchers continue with gene therapy even if it has not worked in the past? Who should fund this work?
2. Tina is 12 years old and her life expectancy is 20 to 30 years. What accommodations are needed to help her live as fulfilling and comfortable a life as possible?

Questions and Problems

Enzymes and Metabolic Pathways

1. Many individuals with metabolic diseases are normal at birth but show symptoms shortly thereafter. Why?
2. List the ways in which a metabolic block can have phenotypic effects.
3. Enzymes have all of the following characteristics, except:
   a. they act as biological catalysts.
   b. they are proteins.
   c. they carry out random chemical reactions.
   d. they convert substrates into products.
   e. they can cause genetic disease.

For questions 4 through 6 refer to the following hypothetical pathway in which substance A is converted to substance C by enzymes 1 and 2. Substance B is the intermediate produced in this pathway:

```
      enzyme       enzyme
        1            2
A  -------->  B  -------->  C
```

4. a. If an individual is homozygous for a null mutation in the gene that codes for enzyme 1, what will be the result?
   b. If an individual is homozygous for a null mutation in enzyme 2, what will be the result?
   c. What if an individual is heterozygous for a dominant mutation where enzyme 1 is overactive?
   d. What if an individual is heterozygous for a mutation that abolishes the activity of enzyme 2 (a null mutation)?
5. a. If the first individual in question 4 married the second individual, would their children be able to convert substance A into substance C?
   b. Suppose each of the aforementioned individuals were heterozygous for an autosomal dominant mutation. List the phenotypes of their children with respect to compounds A, B, and C. (Would the compound be in excess, not present, etc.?)
6. An individual is heterozygous for a recessive mutation in enzyme 1 and heterozygous for a recessive mutation in enzyme 2. This individual marries an individual of

the same genotype. List the possible genotypes of their children. For every genotype, determine the activity of enzyme 1 and 2, assuming that the mutant alleles have 0% activity and the normal alleles have 50% activity. For every genotype, determine if compound C will be made. If compound C is not made, list the compound that will be in excess.

Questions 7 to 11 refer to a hypothetical metabolic disease in which protein E is not produced. Lack of protein E causes mental retardation in humans. Protein E's function is not known, but it is found in all cells of the body. Skin cells from eight individuals who cannot produce protein E were taken and were grown in culture. The defect in each of the individuals is the result of a single recessive mutation. Each individual is homozygous for her or his mutation. The cells from one individual were grown with the cells from another individual in all possible combinations of two. After a few weeks of growth the mixed cultures were assayed for the presence of protein E. The results are given in the following table. A plus sign means that the two cell types produced protein E when grown together (but not separately); a minus sign means that the two cell types still could not produce protein E.

	1	2	3	4	5	6	7	8
1	−	+	+	+	+	−	+	+
2	−	+	+	+	+	−	+	
3	−	+	+	+	+	−		
4	−	+	+	+	+			
5	−	+	+	+				
6	−	+	+					
7	−	+						
8	−							

7. a. Which individuals seem to have the same defect in protein E production?
   b. If individual 2 married individual 3, would their children be able to make protein E?
   c. If individual 1 married individual 6, would their children be able to make protein E?
8. a. Assuming that these individuals represent all possible mutants in the synthesis of protein E, how many steps are there in the pathway to protein E production?
   b. Compounds A, B, C, and D are known to be intermediates in the pathway for production of protein E. To determine where the block in protein E production occurred in each individual, the various intermediates were given to each individual's cells in culture. After a few weeks of growth with the intermediate, the cells were assayed for the production of protein E. The results for each individual's cells are given in the following table. A plus sign means that protein E was produced after the cells were given the intermediate listed at the top of the column. A minus sign means that the cells still could not produce protein E even after being exposed to the intermediate at the top of the column.

	Compounds				
Cells	A	B	C	D	E
1	−	−	+	+	+
2	−	+	+	+	+
3	−	−	−	+	+
4	−	−	−	−	+
5	+	+	+	+	+
6	−	−	+	+	+
7	−	+	+	+	+
8	−	−	−	+	+

9. Draw the pathway leading to the production of protein E.
10. Denote the point in the pathway in which each individual is blocked.
11. a. If an individual who is homozygous for the mutation found in individual 2 and heterozygous for the mutation found in individual 4 marries an individual who is homozygous for the mutation found in individual 4 and heterozygous for the mutation found in individual 2, what will be the phenotype of their children?
    b. List the intermediate that would build up in each of the types of children who could not produce protein E.

Phenylketonuria: A Mutation That Affects an Enzyme

12. Essential amino acids are:
    a. amino acids the human body can synthesize.
    b. amino acids humans need in their diet.
    c. amino acids in a box of Frosted Flakes.
    d. amino acids that include arginine and glutamic acid.
    e. amino acids that cannot harm the body if not metabolized properly.
13. Suppose that in the formation of phenylalanine hydroxylase mRNA, the exons of the pre-mRNA fail to splice together properly, and the resulting enzyme is nonfunctional. This produces an accumulation of high levels of phenylalanine and other compounds, which causes neurological damage. What phenotype and disease would be produced in the affected individual?
14. PKU is an autosomal recessive disorder that causes mental retardation. In individuals afflicted with PKU, high levels of the essential amino acid phenylalanine are present because of a deficiency in the enzyme phenylalanine hydroxylase. If phenylalanine was not an essential amino acid, would diet therapy (the elimination of phenylalanine from the diet) work?
15. Phenylketonuria and alkaptonuria are both autosomal recessive diseases. If a person with PKU marries a person with AKU, what will be the phenotype of their children?

Genes and Enzymes of Carbohydrate Metabolism

16. The normal enzyme required for converting sugars into glucose is present in cells, but the conversion never takes place, and no glucose is produced. What could have occurred to cause this defect in a metabolic pathway?

17. Knowing that individuals who are homozygous for the GD allele show no symptoms of galactosemia, is it surprising that galactosemia is a recessive disease? Why?

Mutations in Receptor Proteins

18. Familial hypercholesterolemia is caused by an autosomal dominant mutation in the gene that produces the LDL receptor. The LDL receptor is present in the plasma membrane of cells and binds to cholesterol and helps remove it from the circulatory system for metabolism in the liver. What is the phenotype of the following individuals?

    HH
    Hh
    hh

19. Suppose the gene for the LDL receptor has been isolated by recombinant DNA techniques. Could you treat this disease by producing LDL receptor and injecting it into the bloodstream of affected individuals? Why or why not?
20. If a chromosomal male has a defect in the cellular receptor that binds the hormone testosterone, what condition results? What is the genotype and phenotype of this individual?

Mutations That Affect Transport Proteins

21. Describe the quaternary structure of the blood protein hemoglobin.
22. A person was found to have very low levels of functional beta globin mRNA and therefore very low levels of the beta globin protein. Name this person's disease and explain what mutation may have occurred in the conversion of pre-mRNA into mRNA.
23. If an extra nucleotide is present in the first exon of the beta globin gene, what effect would it have on the amino acid sequence of the globin polypeptides? Would the globin most likely be fully functional, partly functional, or nonfunctional? Why?
24. Transcriptional regulators are proteins that bind to promoters (the 5′-flanking regions of genes) to regulate their transcription. Assume a particular transcription regulator normally promotes transcription of gene X, a transport protein. If a mutation makes this regulator gene nonfunctional, would the resulting phenotype be similar to a mutation in gene X itself? Why?
25. Mutations in the alpha thalassemia genes can result in a variety of abnormal phenotypes. If a heterozygous alpha thalassemia-1 man marries a heterozygous alpha thalassemia-2 woman, what will be the phenotypes of their offspring? (Refer to Figure 10.13.)

Pharmacogenetics

26. Explain why there are variant responses to drugs and why these responses act as heritable traits.

Ecogenetics

27. Ecogenetics is a branch of genetics that deals with genetic variation that underlies reactive differences to drugs, chemicals in food, occupational exposure, industrial pollution, and other substances. Cases have arisen in which workers claim that exposure to a certain agent has made them feel ill (whereas other workers are unaffected). Although claims like these are not always justified, what are some concrete examples that prove that variation to certain substances exist in the human population?

Internet Activities

The following activities use the resources of the World Wide Web to enhance the topics covered in this chapter. To investigate the topics described below, log on to the book's home page (www.brookscole.com/biology_d) and follow the *Student Resources* link to your subject and text.

1. *Sickle Cell Anemia.* At the "Sickle Cell Case Study" site (http://chroma.mbt.washington.edu/outreach/genetics/sickle/sickle-back.html) read about the genetics of sickle cell disease and the relationship between sickle cell disease and malaria. Read, too, about current research in sickle cell anemia. Why would you not expect a fetus that *has the sickle cell genotype* to show symptoms of sickle cell disease before birth?
2. *Enzyme Replacement Therapy and Pompe Disease.* At Applied Biosystems' "Biobeat" site (http://www.appliedbiosystems.com/biobeat/pompe/) access and read the article on enzyme replacement therapy in the treatment of Pompe disease.
   a. What are lysosomal storage diseases? Why are there so many different kinds?
   b. Why do mutations in noncoding intron DNA sequences contribute to Pompe disease?
   c. Within the article, click on the "alpha-glucosidase" link and then scroll down to take a look at the register of mutations in this enzyme. How many are there? Note that they vary from mild to very severe in their effects. Why might this be?

For Further Reading

Garrod, A. E. (1902). The incidence of alkaptonuria: A study in chemical individuality. *Lancet, 2,* 1616–1620.

Neel, J. V. (1949). The inheritance of sickle cell anemia. *Science, 110,* 64–66.

Pauling, L., Itoh, H., Singer, S. J., & Wells, I. C. (1949). Sickle cell anemia: A molecular disease. *Science, 110,* 543–548.

Scriver, C. R., Beaudet, A. L. Sly, W. S., & Valle, D. (2001). *The Metabolic Basis of Inherited Disease,* 8th ed. New York: McGraw-Hill.

Vella, F. (1980). Human hemoglobins and molecular disease. *Biochem. Educ., 8,* 41–53.

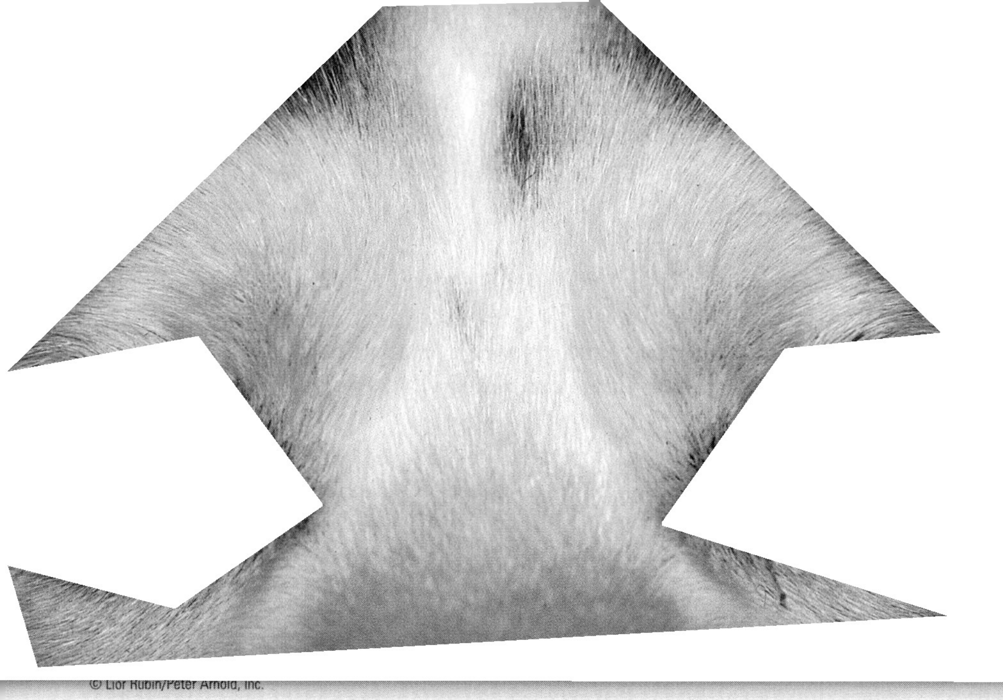

Mutation: The Source of Genetic Variation

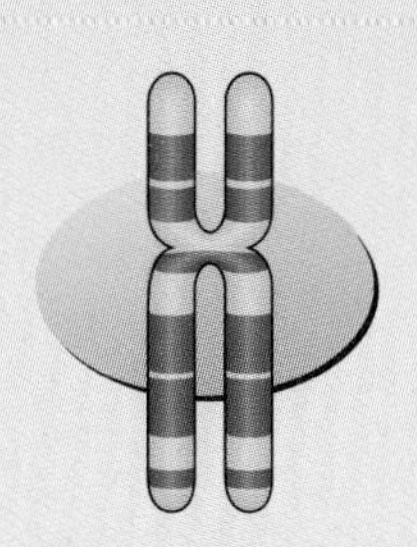

Chapter Outline

MUTATION, GENOTYPES, AND PHENOTYPES
MUTATIONS CAN CAUSE A RANGE OF PHENOTYPES
GENOMIC IMPRINTING IS A REVERSIBLE ALTERATION OF THE GENOME

Linus Pauling, a two-time Nobel Prize winner, once recalled that when he first heard a description of how red blood cells change shape in sickle cell anemia, he had the idea that sickle cell anemia is really a molecular disease. He thought this disorder must involve an abnormality of the hemoglobin molecule caused by a mutated gene.

Early in 1949, Linus Pauling and his student Harvey Itano began a series of experiments to determine whether there is a difference between normal hemoglobin and sickle cell hemoglobin. They obtained blood samples from people who had sickle cell anemia and from unaffected individuals. They prepared hemoglobin from these blood samples, placed it in a tube with an electrode at each end, and passed an electrical current through the tube. Hemoglobin from individuals with sickle cell anemia migrated toward the cathode, indicating that it had a positive electrical charge. Samples of normal hemoglobin migrated in the opposite direction (toward the anode), indicating that it had a net negative electrical charge. In the same year, James Neel, working with sickle cell patients in the Detroit area, demonstrated that sickle cell anemia is an autosomal recessive trait.

Pauling and his colleagues published a paper on their results and incorporated Neel's findings into their discussion. They concluded that a mutant gene involved in the synthesis of hemoglobin caused sickle cell anemia (and the heterozygous condition known as sickle cell trait). The idea that a genetic disorder could be caused by a defect in a single molecule was revolutionary. Pauling's idea about a molecular disease helped start the field of human biochemical genetics and played a key role in understanding the molecular nature of mutations.

After Watson and Crick worked out the structure of DNA, Crick was anxious to prove that mutant genes produce mutant proteins whose amino acid sequences differ from those of the normal protein. He persuaded Vernon Ingram to look for such differences. Ingram settled on hemoglobin as the protein he would analyze because of Pauling's work. Ingram cut hemoglobin into pieces using the enzyme trypsin and separated the 30 resulting fragments. He noticed that normal hemoglobin and sickle cell hemoglobin differed in only one fragment, a peptide about 10 amino acids long. Ingram then worked out the amino acid sequence in this fragment. In 1956, he reported that there is a difference of only a single amino acid (glutamine in normal hemoglobin and valine in sickle cell hemoglobin) between the two proteins. This finding confirmed the relationship between a mutant gene and a mutant gene product, but it raised a more basic question: What is the nature of mutation? In this chapter, we will examine the nature of mutation, how mutations are counted, and the relationship between mutation and DNA.

MUTATIONS ARE HERITABLE CHANGES

Mutation is the source of all genetic variation in humans and other organisms. The results of mutations can be classified in a number of ways. From a genetic perspective, mutations that produce dominant alleles are expressed in the heterozygous condition; mutations to recessive alleles are expressed only when homozygous. Mutations can also be ranked in categories, such as the severity of the phenotype or the age of onset. For our purposes, two general categories of mutations are the most useful: mutations that affect chromosomes and mutations that change the nucleotide sequence of a gene. Chromosomal aberrations were discussed in Chapter 6. In this chapter we focus on mutational changes in single genes, that is, changes in the sequence or number of nucleotides in DNA. First, we consider how mutations are detected and then investigate at what rate these mutations take place. Finally, we examine how mutation works at the molecular level.

MUTATIONS CAN BE DETECTED IN SEVERAL WAYS

How do we know that a mutation has taken place? In humans, the sudden appearance of a dominant mutation in a family can be observed in a single generation. But mutation of a dominant allele to a recessive allele can be detected only in the homozygous condition, posing a challenge for human geneticists, because its phenotype may appear only after many generations.

If an affected individual appears in an otherwise unaffected family, the first question is whether the trait is caused by genetic or nongenetic factors. For example, if a mother is exposed to the rubella virus (which causes a form of measles) early in pregnancy, the fetus may have a phenotype similar to those seen in some biochemical genetic disorders. The phenotype of rubella infection is produced not by mutation but by the effect of the virus on the developing fetus. To determine whether the phenotype is caused by a genetic disorder, geneticists depend on pedigree analysis and the study of births over several generations (a family history).

If a mutant allele is dominant, is fully penetrant (expressed in all who carry the mutant allele) and appears in a family that has no previous history of this condition in previous generations, we can presume that a mutation has taken place. In the pedigree shown in Figure 11.1, severe blistering of the feet appeared in one out of six children, although the parents were unaffected. The trait was transmitted by the affected female to six of her eight children and was passed to the succeeding generation as an autosomal dominant condition. A reasonable explanation for this pedigree is that a mutation to a dominant allele occurred and that individual II-5 was heterozygous for this dominant allele. On the other hand, a number of uncertainties can affect this conclusion.

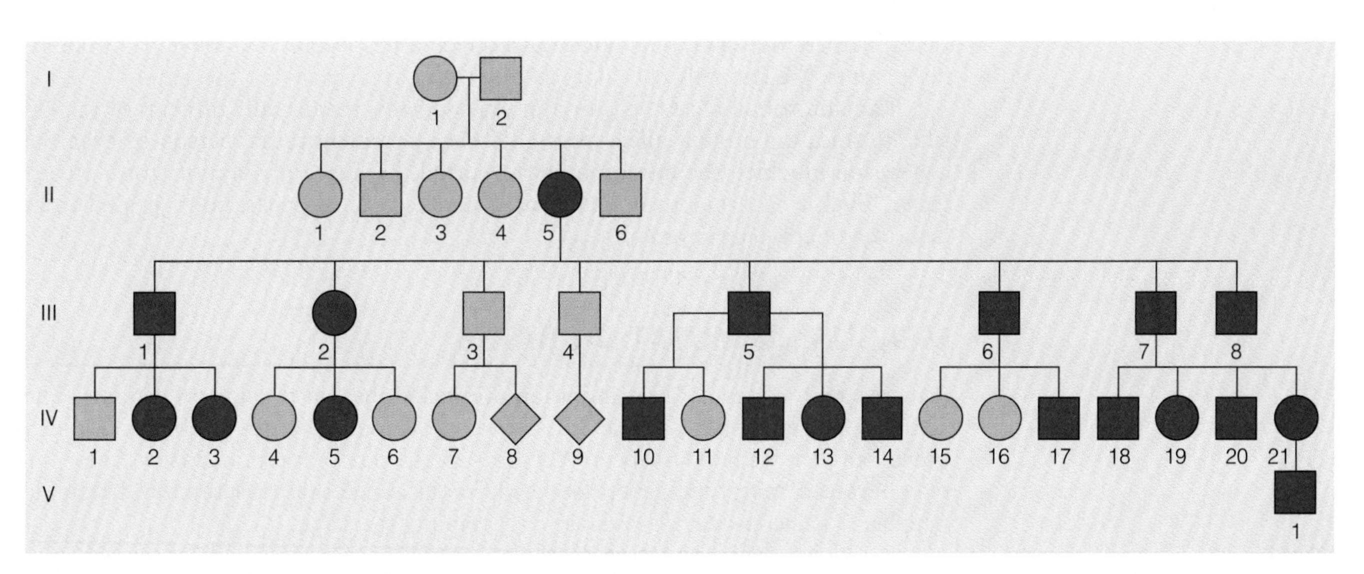

FIGURE 11.1 A dominant trait, foot blistering, appeared (II-5) in a family that had no previous history of this condition. The trait is transmitted through subsequent generations in an autosomal dominant fashion.

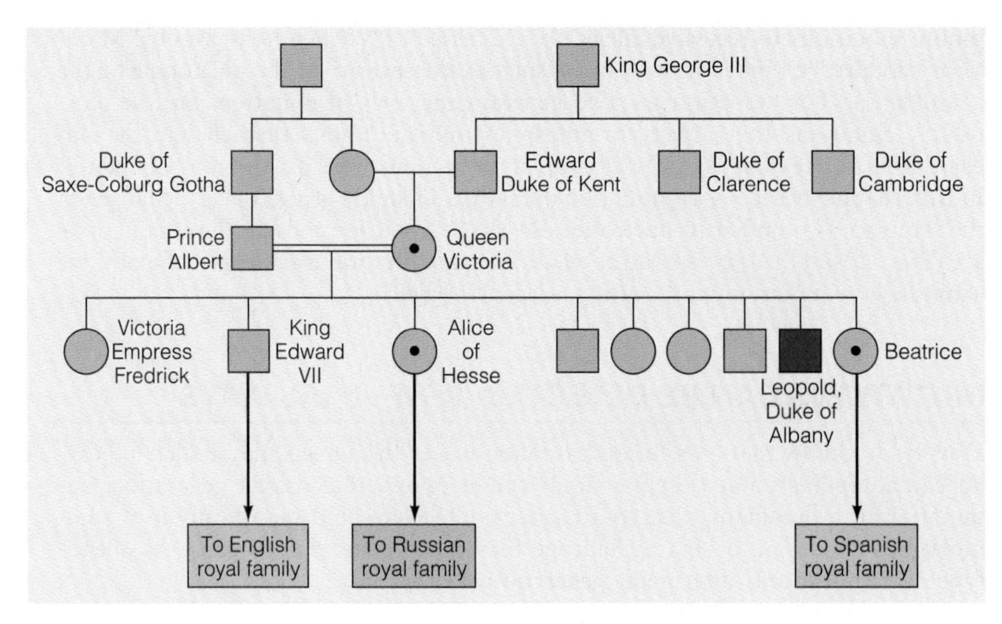

FIGURE 11.2 Pedigree of Queen Victoria of Britain, showing her immediate ancestors and children. Because she passed the mutant allele for hemophilia on to three of her children, she was probably a heterozygote rather than the source of the mutation.

For example, if the child's father is not the husband in the pedigree but is an affected male, then it would only seem that a mutational event had taken place.

If mutation results in an allele that is recessive and sex linked, it often can be detected by examining males in the family line. But it can be difficult to determine whether a heterozygous female who transmits a trait to her son is the source of the mutation or is only passing on a mutation that arose in an ancestor. The X-linked form of hemophilia that spread through the royal families of Western Europe and Russia in the nineteenth and twentieth centuries probably originated with Queen Victoria (Figure 11.2; see Concepts and Controversies: Hemophilia and History in Chapter 4). None of the males in previous generations had hemophilia, but one of Victoria's sons was affected, and at least two of her daughters were heterozygous carriers. Because Victoria transmitted the mutant allele to a number of her children, it is reasonable to assume that she was a heterozygous carrier. Her father was not affected, and there is nothing in her mother's pedigree to indicate that she was a carrier. It is therefore likely that Victoria received a newly mutated allele from one of her parents. We can only speculate as to which parent gave her the gene.

If an autosomal recessive trait appears suddenly, it is usually difficult or impossible to trace the mutant allele through a family to identify the person or even the generation in which the mutation first occurred, because only homozygotes are affected. Such a new mutation can remain undetected for generations, passed from heterozygote to heterozygote.

MEASURING MUTATION RATES

Pedigree analysis reveals that mutation does take place in the human genome. The available evidence suggests that it is a rare event, but given the problems outlined previously, can we hope to accurately measure the rate of mutation? If we knew the overall rate of mutation, geneticists could monitor the rate over time to determine whether it is increasing, decreasing, or remaining the same.

Mutation rate The number of events that produce mutated alleles per locus per generation.

Geneticists define **mutation rates** as the number of mutated alleles per gene per generation. Suppose that for a certain gene, 4 out of 100,000 births show a mutation from a recessive to a dominant allele. Because each of these 100,000 individuals carries two copies of the gene, we have sampled 200,000 copies of the gene. The four births represent four mutated genes (we are assuming that the newborns are heterozygotes for a dominant mutation and carry only one mutant allele). In this case, the mutation rate is 4/200,000, or 2/100,000. In scientific notation this would be written as 2×10^{-5} per allele per generation.

If the locus were X-linked and if 100,000 male births were examined and four mutants were discovered, this would represent a sampling of 100,000 copies of the gene (because the males have only one copy of the X chromosome). Excluding contributions from female carriers, the mutation rate in this case would be 4/100,000, or 4×10^{-5} per allele per generation.

Mutation Rates for Specific Genes

Is there a way to measure directly the rate of mutation for a gene? For certain dominant alleles, the answer is yes, under certain conditions. To ensure the measurement is accurate, the mutant phenotype must

- never be produced by recessive alleles
- always be fully expressed and completely penetrant so that mutant individuals can be identified
- have clearly established paternity
- never be produced by nongenetic agents such as drugs or infection
- be produced by dominant mutation of only one gene

One dominant mutation, achondroplasia, fulfills these requirements. Achondroplasia (MIM/OMIM 100800) is a dominant form of dwarfism that produces short arms, short legs, and an enlarged skull. A diagnosis by x-ray examination can be done shortly after birth. Several population surveys have used mutations in this gene to estimate the overall mutation rate in humans. One survey found 7 achondroplastic children of unaffected parents in a total of 242,257 births. From these data, the mutation rate for achondroplasia has been calculated at 1.4×10^{-5}.

Although the mutation rate for achondroplasia can be measured directly, it is not clear whether the mutation rate in this gene is typical for all human genes. Maybe this gene has an inherently high rate of mutation, or an unusually low rate of mutation. To get an accurate picture of the mutation rate in humans, it is important to measure the mutation rate in a number of different genes before making any general statements. As it turns out, two other dominant mutations have widely different rates of mutations.

Neurofibromatosis (MIM/OMIM 162200), an autosomal dominant condition, is characterized by pigmentation spots and tumors of the skin and nervous system (described in Chapter 4). About 1 in 3,000 births are affected. Many of these births (about 50%) occur in families that have no previous history of neurofibromatosis, indicating that this locus has a high mutation rate. In fact, the calculated mutation rate in this disease is as high as 1 in 10,000 (1×10^{-4}), one of the highest rates so far discovered in humans. For Huntington disease (MIM/OMIM 143100), the mutation rate has been calculated as 1×10^{-6}, a rate 100-fold lower than that of neurofibromatosis and 10-fold lower than that of achondroplasia.

Measurements of the mutation rate in several human genes are listed in Table 11.1. The average of the rates is approximately 1×10^{-5}. All the genes listed in the table are inherited as autosomal dominant or X-linked traits. It is almost impossible to measure directly the mutation rates in autosomal recessive alleles by pedigree analysis, but population surveys using recombinant DNA methods are now providing estimates of the rate and type of mutations found in many human genes, including those with autosomal recessive patterns of inheritance. Still, many geneticists feel that to reduce any potential bias, a more conservative estimate of the mutation rate in humans should be used, and by convention 1×10^{-6} is used as the average mutation rate for human genes.

Table 11.1 Mutation Rates for Selected Genes

Trait	Mutants/Million Gametes	Mutation Rate	MIM/OMIM Number
Achondroplasia	10	1×10^{-5}	100800
Aniridia	2.6	2.6×10^{-6}	106200
Retinoblastoma	6	6×10^{-6}	180200
Osteogenesis imperfecta	10	1×10^{-5}	166200
Neurofibromatosis	50–100	$0.5–1 \times 10^{-4}$	162200
Polycystic kidney disease	60–120	$6–12 \times 10^{-4}$	173900
Marfan syndrome	4–6	$4–6 \times 10^{-6}$	154700
Von Hippel-Landau syndrome	<1	1.8×10^{-7}	193300
Duchenne muscular dystrophy	50–100	$0.5–1 \times 10^{-4}$	310200

Why Do Genes Have Different Mutation Rates?

The factors that influence the mutation rate and contribute to its observed variation include the following:

- *Size of the gene.* Larger genes are bigger targets for mutation. The gene for neurofibromatosis (*NF-1*) is an extremely large gene and has a high mutation rate. The *NF-1* gene extends over 300,000 base pairs of DNA. The gene for Duchenne and Becker muscular dystrophy, the largest gene identified to date in humans, contains more than 2 million base pairs. Both these genes have high mutation rates.
- *Nucleotide sequence.* In some genes, short nucleotide repeats are present in the DNA. In the gene for fragile-X syndrome, the sequence CGG is repeated some 6 to 50 times in normal individuals. Those who have more than 230 copies experience symptoms of the disorder, and these become more severe as the number of CGG repeats increases. The presence of nucleotide repeats may predispose a gene to mutate at a higher rate.
- *Spontaneous chemical changes.* Among the bases in DNA, cytosine is especially susceptible to chemicals that can change the nucleotide sequence in DNA. These and other chemical changes are discussed in a later section of this chapter. Genes rich in G/C base pairs are more likely to undergo chemical changes than those rich in A/T pairs.

ENVIRONMENTAL FACTORS INFLUENCE MUTATION RATES

In general, mutations can occur as the result of mistakes that occur during normal cellular functions such as DNA replication or by the action of agents that attack DNA or cellular functions from outside the cell. These agents include chemicals and radiation.

Radiation Is One Source of Mutations

Radiation The process by which electromagnetic energy travels through space or a medium such as air.

Ionizing radiation Radiation that produces ions during interaction with other matter, including molecules in cells.

Radiation is a process by which energy travels through space. For example, the heat from a fire travels through space and warms a room. There are two main types of radiation: electromagnetic and corpuscular. Electromagnetic radiation comprises waves of electrical or magnetic energy, whereas corpuscular radiation is composed of atomic and subatomic particles that move through space at high speeds. These particles cause damage when they collide with molecules inside the cell. Both forms of radiation are known as **ionizing radiation** because they form chemically reactive ions when they collide with molecules in cells. However, some forms of radiation can cause mutations without producing ions. For example, ultraviolet (UV) light is a powerful mutagen that causes damage without producing ions. The energy in UV light is absorbed directly by DNA and results in mutations.

Remember that exposure to radiation is unavoidable. Everything in the physical world contains sources of radiation. This includes our bodies, the air we breathe, the food we eat, and the bricks in our houses. Some of this radiation is left over from the birth of the universe, and some has been created by interaction of atoms on Earth with cosmic radiation. These natural sources of radiation are called **background radiation.** We are also exposed to radiation that results from human activity, including medical testing, nuclear testing, nuclear power, and consumer goods.

Background radiation Radiation in the environment that contributes to radiation exposure.

Radiation can cause damage at several levels. As radiation strikes the molecules in cells, electrons are removed, creating ions and charged atoms. Such ionized molecules are highly reactive and can cause mutations in DNA. For example, if cosmic radiation passes through the cell and ionizes a water molecule near a DNA molecule, these ions might react with the DNA to produce a mutation. Often, the cell can repair these mutations. However, if too many mutations accumulate in a cell, the repair system can be overwhelmed, and cell death or cancer can result. In germ cells, mutations that are not repaired are transmitted from generation to generation as newly mutated alleles.

How Much Radiation Are We Exposed To?

Radiation is measured using several physical and biological units. A dose of radiation can be measured in several ways: the amount a person is exposed to, the amount absorbed by the body, or the amount of damage caused. Most often, the dose is expressed as a **rem** (radiation equivalent man), the amount of radiation that causes the same damage as a standard amount of x-rays. Because people are usually exposed to very small amounts of radiation, the dose is usually expressed in **millirems** (1,000 mrem equals 1 rem). At doses in the millirem range, cells can repair most, if not all, of the radiation damage. At doses of approximately 100 rem (100,000 mrem), cells begin to die, and radiation sickness results. At a dosage of 400 rem, about 50% of people will die within 60 days if they are not treated. Case 1 at the end of this chapter discusses the impact of the Chernobyl reactor explosion on the surrounding populations.

Rem The unit of radiation exposure used to measure radiation damage in humans. It is the amount of ionizing radiation that has the same effect as a standard amount of x-rays.

Millirem Each rem is equal to 1,000 millirems.

In the United States, the average person is exposed to 360 mrem/year, 82% of which is from nature (Figure 11.3). Because most people are exposed to much less

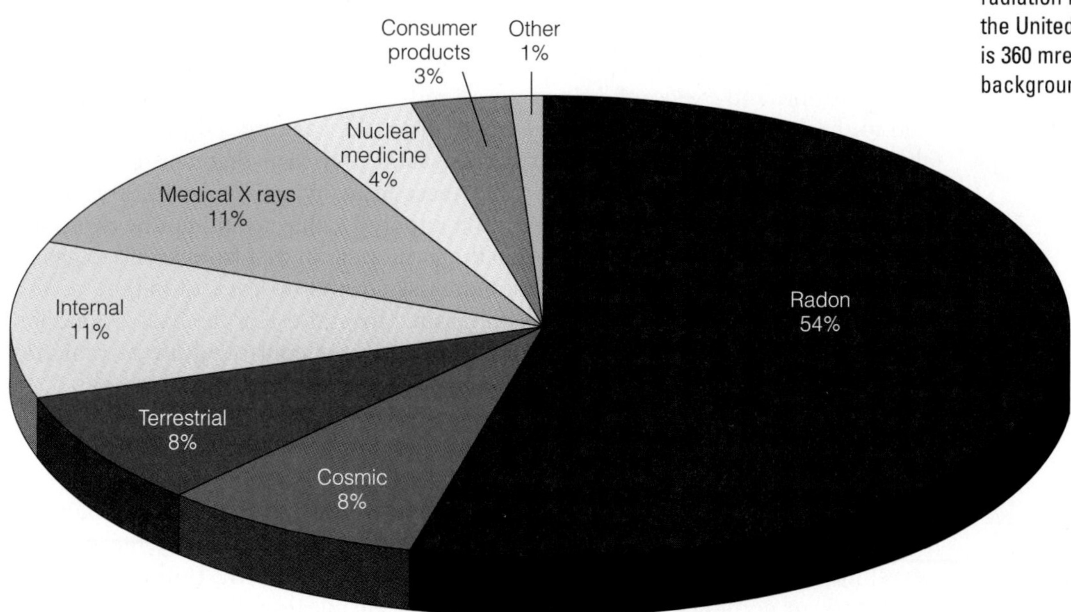

FIGURE 11.3 The sources of radiation received by individuals in the United States. The average dose is 360 mrem, 82% of which is from background radiation.

Concepts and Controversies

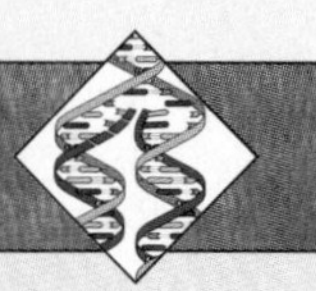

Irradiated Food

During the past 40 years, research has demonstrated that radiation treatments can help preserve food and kill contaminating microorganisms. Irradiation prevents sprouting of root crops, such as potatoes; extends the shelf life of many fruits and vegetables; destroys bacteria and fungi in meat and fish; and kills insects and other pests in spices.

For irradiation, food is placed on a conveyor and moved to a sealed, heavily shielded chamber where it is exposed to a radioactive source. An operator who views the process on video camera delivers the dose. The food itself does not come in contact with the radioactive source, and the food is not made radioactive. Relatively low doses are used to inhibit sprouting of potatoes and to kill parasites in pork. Intermediate doses are used to retard spoilage in meat, poultry, and fish, and high doses can be used to sterilize foods, including meats. The amount of food irradiated varies from country to country, ranging from a few tons of spices to hundreds of thousands of tons of grain.

Irradiated food has been used routinely by NASA to feed astronauts in space, and irradiated foods are sold in more than 20 countries, including the United States. The Food and Drug Administration (FDA) approved the first application for food irradiation in 1964, and approval has been granted for the irradiation of spices, herbs, fruits and vegetables, pork, chicken, and eggs. All irradiated food sold in the United States must be labeled with an identifying logo (shown here).

Public concern about radiation has prevented the widespread sale of irradiated food in this country. Advocates point out that irradiation can eliminate the use of many chemical preservatives, lower food costs by preventing spoilage, and reduce the incidence of food-borne illnesses transmitted by *Salmonella* and *Escherichia coli.* Those opposed to food irradiation argue that irradiation produces chemical changes in food and that the safety of these new chemicals has not been proven, although it should be noted that these same changes occur in foods preserved by other methods. Opponents also point out that treatment may select for radiation-resistant microorganisms.

than 1 rem/year and will not get radiation sickness, what are the major risks from radiation exposure? At levels below 5 rem (5,000 mrem), the major risk is mutations in somatic cells that increase susceptibility to cancer.

To understand how exposure to radiation increases your risk of getting cancer, let's explore some background information. In the United States, the death rate from cancer is about 20%. That is, if 10,000 people are chosen at random from the population, about 20% of them, or 2,000, will die from cancer. If we a take a group of 10,000 people and expose them to a single radiation dose of 1,000 mrem (in addition to the 360 mrem they receive each year from the background), there would be about eight additional cancer deaths in this group over their lifetimes. However, as we examine the overall death rate from cancer in many groups of 10,000, the increase from 2,000 to 2,008 in the radiation-exposed group would not be detectable because the number of cancer deaths fluctuates more than that from group to group. In other words, it is very difficult to estimate risks from low levels of radiation, and, in most cases, the effects cannot be distinguished from those of the background levels of radiation. Overall, risk analysis suggests that the controlled use of radiation is a small risk compared to others in our daily lives.

Chemicals Can Be Mutagens

There are over 6 million chemical compounds now known, and almost 500,000 of these are used in manufacturing processes and are part of everyday life in many ways: in packaging, in food, in building materials, and so forth. Unfortunately, we know little or nothing about whether most of the chemicals cause mutations. Chemicals cause mutations in several ways, and they are often classified by the type of damage they cause to DNA. Some chemicals cause nucleotide substitutions or change the number of nucleotides in a gene, whereas others structurally change the bases in DNA, causing a base pair change after replication. Some of the ways chemicals act as mutagens are discussed here.

Base Analog Mutagenic chemicals that structurally resemble nucleotides and are incorporated by the cell into DNA or RNA are called **base analogs.** The structure of the base analog 5′-bromouracil is similar to that of thymine (Figure 11.4). If 5′-bromouracil is present in cells, converted to a nucleotide and inserted into a DNA molecule in place of thymine, it generates a 5′-Br/A base pair (Figure 11.4). When the DNA replicates the next time, the 5′-bromouridine can undergo a structural rearrangement and serve as a template for the insertions of guanine in the new strand, creating a 5′-Br/G base pair. After another round of replication, the net result is that an A/T base pair is changed into a G/C base pair. Although most people will not come in contact with 5′-bromouridine, other common chemicals are base analogs. For example, the caffeine in your morning coffee is a base analog (intensive investigation has shown that caffeine is rapidly broken down in the body and does not pose much of a threat as a mutagen).

Base analogs A purine or pyrimidine that differs in chemical structure from those normally found in DNA or RNA.

Chemical Modification of Bases Some mutagens attack the bases in a DNA molecule, generating a chemical change in the base sequence of the DNA. This can produce a base pair change. Treatment of DNA with nitrous acid removes a chemical group from cytosine, converting it to uracil (Figure 11.5). What was a G/C base pair is converted into a G/U pair. Uracil has the base pairing properties of thymine (T). In the next round of DNA replication, the U will act as a template for

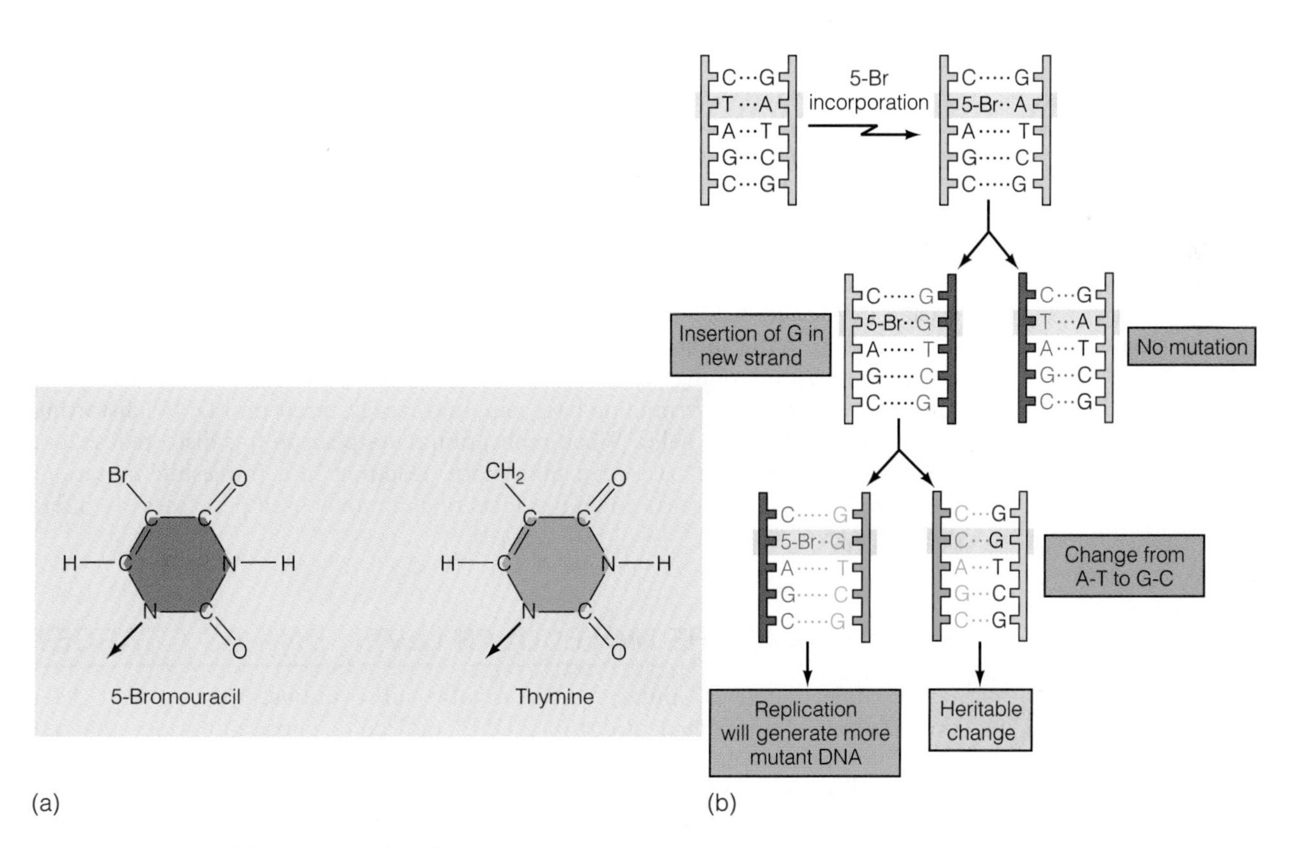

FIGURE 11.4 (a) The structural similarity between thymine and 5-bromouracil. The bromine (Br) group is about the same size as the CH_3 group in the thymine. (b) Rounds of DNA replication showing the incorporation of 5-Br and the base pair change it causes through several rounds of DNA replication. Boxes mark the beginning (T/A) base pair and the resulting (G/C) base pair.

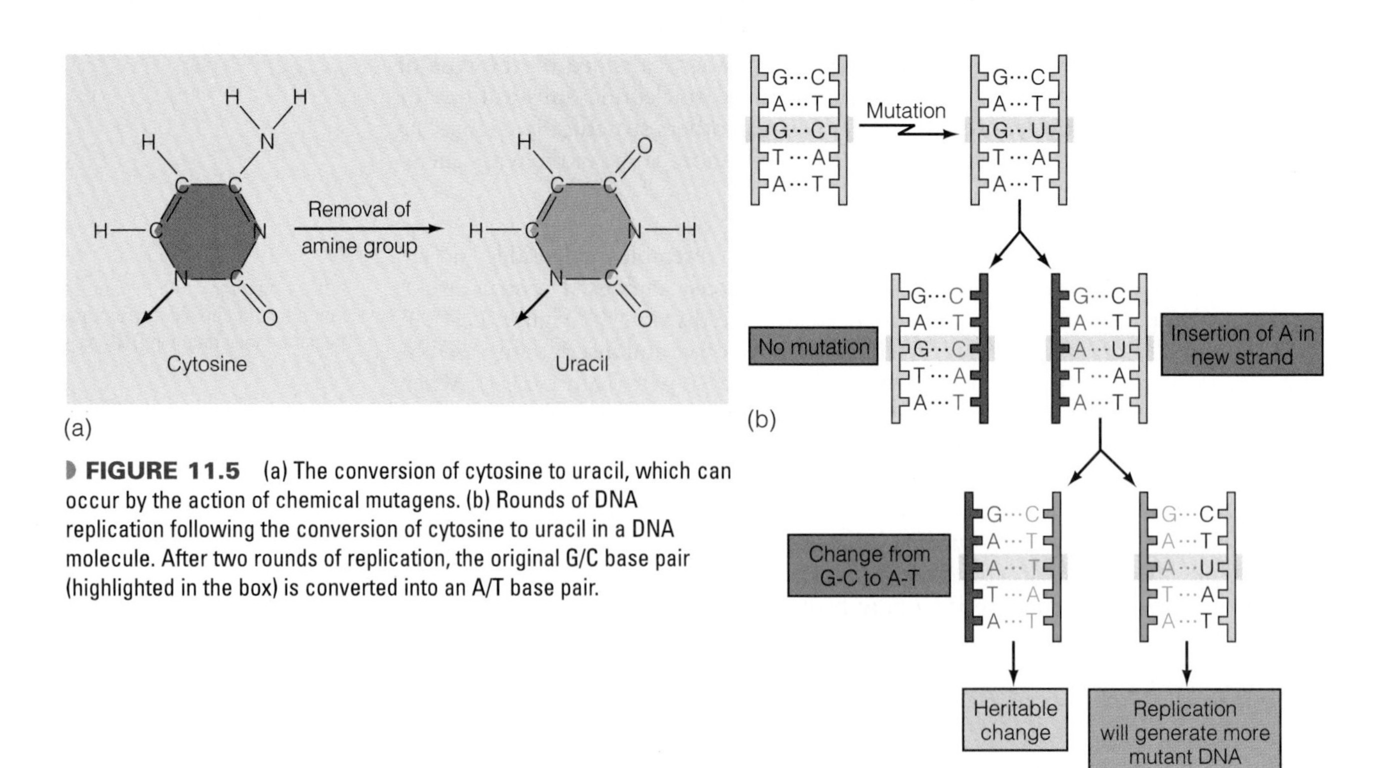

FIGURE 11.5 (a) The conversion of cytosine to uracil, which can occur by the action of chemical mutagens. (b) Rounds of DNA replication following the conversion of cytosine to uracil in a DNA molecule. After two rounds of replication, the original G/C base pair (highlighted in the box) is converted into an A/T base pair.

FIGURE 11.6 The molecular structure of acridine orange, an intercalating agent that inserts into the helical structure of DNA, distorting its shape. Replication in the distorted region can lead to the insertion or deletion of base pairs, producing a mutation.

the insertion of A in the newly synthesized strand, and the G/U base pair is converted to an A/U pair. When A is used as a template for DNA replication in the following round of replication, an A/T pair will be created, replacing the original G/C pair. Nitrates and nitrites used in the preservation of meats, fish, and cheese are converted into nitrous acid in the body. Although studied extensively, how much mutation is caused by these dietary chemicals has been difficult to assess.

Chemicals That Bind to DNA Chemicals that bind directly to DNA generally result in mutations called frameshift mutations (described in a later section). These chemicals, called *intercalating agents,* generally insert themselves into the DNA, distorting the double helix. The distortion can cause a mistake during DNA replication, resulting in the addition or deletion of a base pair. The structure of one of these chemicals, acridine orange is shown in Figure 11.6. This molecule is about the same size as a purine–pyrimidine base pair, and wedges itself into DNA, distorting the shape of the double helix. When replication takes place in this distorted region, deletion or insertion of bases can take place, resulting in a frameshift mutation. Some components and breakdown products of commonly used pesticides are intercalating agents.

MUTATION AT THE MOLECULAR LEVEL: DNA AS A TARGET

Nucleotide substitutions Mutations that involve substitutions of one or more nucleotides in a DNA molecule.

Frameshift mutations Mutational events in which a number of bases (other than multiples of three) are added to or removed from DNA, causing a shift in the codon reading frame.

At the molecular level, mutations can involve substitutions, insertions, or deletions of one or more nucleotides in a DNA molecule. Those mutations that involve an alteration in the sequence but not the number of nucleotides in a gene are called **nucleotide substitutions.** Generally, these involve one or a small number of nucleotides. A second type of mutation, called **frameshift mutations,** causes the *insertion* or *deletion* of one or more bases. Because codons are composed of three bases, changing the number of bases in one codon changes the nucleotide sequence of all following codons and results in large-scale changes in the amino acid sequence of the protein product. We will begin by examining the substitution of one nucleotide for another and then consider frameshift mutations.

Mutations Caused by Nucleotide Substitutions

Hemoglobin is one of the most studied molecules in the human body. Scientists have found several hundred variants of the alpha and beta globins that carry single amino acid substitutions. These variants provide many examples of how changing a single nucleotide in a gene can change the structure and function of a protein. Nucleotide substitutions in coding regions can have a number of outcomes, some of which are described in the following section. In this discussion, keep in mind that the term *codon* refers to the sequence of three nucleotides in mRNA that codes for an amino acid.

Missense mutations are single nucleotide changes that substitute one amino acid for another in a protein. These substitutions do not always affect protein function and often do not have any phenotypic consequences. To illustrate, let's consider amino acid number 6 in beta globin, one of the components of hemoglobin (Figure 11.7). Normal hemoglobin (Hb A) mRNA has the codon GAG at position 6, encoding the amino acid glutamine. A single nucleotide substitution in codon 6 from GAG (glu) to GUG (val) results in Hb S (sickle cell anemia), a condition with a potentially lethal phenotype. A condition known as Hb C (hemoglobin C disease) also has a nucleotide substitution at position 6. In this case, GAG is changed into AAG, and the beta globin has lysine as the sixth amino acid instead of glycine. This change causes a mild set of clinical conditions that are less serious than sickle cell anemia. A change in codon 6 from GAG (glu) to GCG (ala) produces a hemoglobin variant known as Hb Makassar, a substitution that causes no clinical symptoms and is regarded as harmless.

Missense mutations Mutations that cause the substitution of one amino acid for another in a protein.

Sense mutations Mutations that change a termination codon into one that codes for an amino acid. Such mutations produce elongated proteins.

Nonsense mutations Mutations that change an amino acid specifying a codon to one of the three termination codons.

In all three cases, the proteins differ only in the amino acid at position 6: Hb A has glutamic acid (glu), Hb S has valine (val), Hb C has lysine (lys), and Hb Makassar has alanine (ala). The sequence of the other 145 amino acids in the polypeptide is unchanged. In these three examples, single nucleotide changes in the sixth codon of the beta-globin gene result in phenotypes that range from harmless (Hb Makassar), to the mild clinical symptoms of Hb C, to the serious and potentially life-threatening consequences of Hb S (sickle cell anemia).

Other nucleotide substitutions produce proteins that are longer or shorter than normal. **Sense mutations** produce longer-than-normal proteins by changing a termination codon (UAA, UAG, or UGA), into one that codes for amino acids. Several hemoglobin variants with longer-than-normal globin molecules are shown in Table 11.2. In each case, the extended polypeptide chain can be explained by a single nucleotide substitution in the normal termination codon. In hemoglobin Constant Springs-1, the mRNA codon 142 is changed from UAA to CAA, and a glycine codon replaces a stop codon. Thirty more amino acids are inserted before another stop codon is reached.

Nonsense mutations change codons for amino acids into one of the three termination codons: UAA, UAG, or UGA (see Table 9.1). This shortens the protein product. In the McKees Rock variant of beta globin, a change in codon 144 from UAU (tyr) to UAA (termination) results in a beta chain that is 143 amino acids instead of 145. This change has little or no effect on the function of the beta-globin

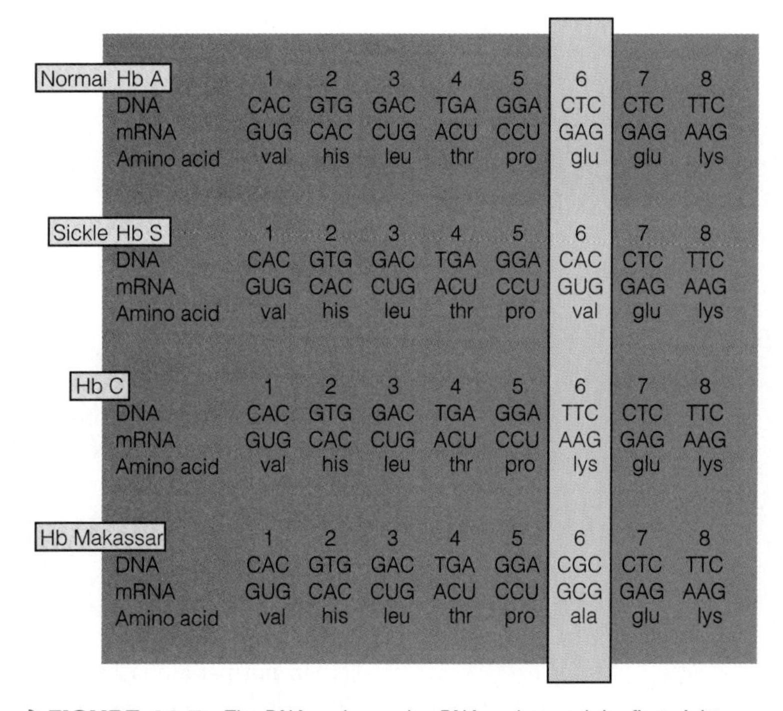

FIGURE 11.7 The DNA code word, mRNA codon, and the first eight amino acids of normal adult hemoglobin (Hb A), sickle cell hemoglobin (Hb S), hemoglobin C (Hb C), and hemoglobin Makassar (Hb Mk). A single nucleotide substitution in codon 6 is responsible for the changes in the two variant forms of hemoglobin.

Table 11.2 Alpha Globins with Extended Chains Produced by Nucleotide Substitutions	
Alpha Globins	**Abnormal Chains**
Constant Springs-1	gln (142) + 30 amino acids
Icaria	lys (142) + 30 amino acids
Seal Rock	glu (142) + 30 amino acids
Koya Dora	ser (142) + 30 amino acids

molecule as a carrier of oxygen. However, other nucleotide substitutions can produce more drastic changes in polypeptide length and have more serious phenotypic effects.

Mutations Caused by Nucleotide Deletions and Insertions

Nucleotide deletions and insertions within a gene can range from the deletion or duplication of one nucleotide to that of an entire gene. As more genes are sequenced in genome projects, deletions and insertions are emerging as a major cause of genetic disorders and in humans account for 5% to 10% of all known mutations. As defined earlier, the insertion or deletion of nucleotides within the coding sequence of a gene causes frameshift mutations. Because codons consist of groups of three bases, adding or subtracting a base from one codon changes the coding sense of all following codons in the gene. A frameshift mutation changes the amino acid sequence of the protein from the site of the mutation to the end of the protein. Suppose that a codon series reads as the following sentence:

THE FAT CAT ATE HIS HAT

A nucleotide (in this case, an A) inserted in the second codon destroys the sense of the remaining message:

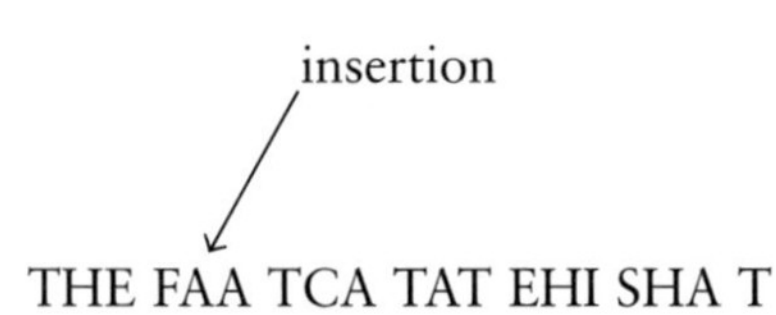

Similarly, a deletion in the second codon can also generate an altered message:

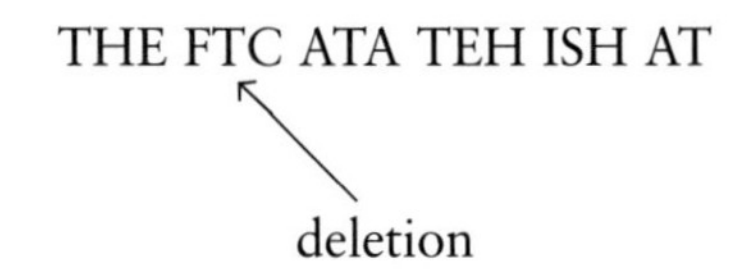

In nucleotide substitutions, the number of nucleotides in the gene remains the same, and usually only one amino acid in the protein is altered. In frameshift mutations, addition or deletion of a single nucleotide changes the number of nucleotides in the gene and usually causes large-scale changes in the amino acid sequence of the protein. Usually, these changes result in a nonfunctional gene product. A hemoglobin variant with an extended chain will serve as an example of an altered protein produced by a frameshift mutation. In this case, the frameshift occurs near the end of the gene and has a minimal impact on the function of the gene product.

In normal alpha hemoglobin, the mRNA codons for the last few amino acids are as follows:

Position number	138	139	140	141	TER
mRNA codon	UCC	AAA	UAC	CGU	UAA
Amino acid	ser	lys	tyr	arg	

Table 11.3 Some Mutations with Expanded Trinucleotide Repeats

Gene	Triplet Repeat	Normal Copy	Copy in Disease	MIM/OMIN Number
Spinal and bulbar muscular atrophy	CAG	12–34	40–62	313200
Spinocerebellar ataxia type 1	CAG	6–39	41–81	164400
Huntington disease	CAG	6–37	35–121	143100
Haw-River syndrome	CAG	7–34	54–70	140340
Machado-Joseph disease	CAG	13–36	68–79	109150
Fragile-X syndrome	CGG	5–52	230–72,000	309550
Myotomic dystrophy	CTG	5–37	50–72,000	160900
Friedreich ataxia	GAA	10–21	200–900	229300

In a variant of alpha hemoglobin called Hb Wayne, the last nucleotide (A) in codon 139 is deleted, producing a frameshift:

Position number	138	139	140	141	142	143	144	145	146	TER
mRNA Codon	UCC	AAU	ACC	GUU	AAG	CUG	GAG	CCU	CGG	UAG
Amino acid	ser	asn	thr	val	lys	leu	gln	pro	arg	

In this case, deletion of a single nucleotide causes a shift in the codon reading frame so that the termination sequence UAA that follows codon 141 is split into two codons, causing new amino acids to be added until another stop codon (generated by the deletion) is reached. The result is an alpha chain variant that has 146 amino acid residues instead of 141.

Trinucleotide repeats A form of mutation associated with the expansion in copy number of a nucleotide triplet in or near a gene.

Allelic expansion Increase in gene size caused by an increase in the number of trinucleotide sequences.

Trinucleotide Repeats, Gene Expansion, and Mutation

Trinucleotide repeats are a recently discovered class of mutations associated with a number of genetic disorders. Trinucleotide repeats are a sequence of three nucleotides repeated several times in consecutive order within a gene (Table 11.3). Mutations that increase the number of repeats are responsible for several genetic disorders. This expansion involves only one of the two alleles of a gene, and the phenomenon is called **allelic expansion.** The potential for expansion is a characteristic of a specific allele and occurs only within that allele. The study of allelic expansion in fragile-X syndrome has explained some aspects of the way this condition is inherited (see Chapter 6 for a review of fragile sites).

In fragile-X syndrome, the phenotype includes mental retardation. About 1% of all males institutionalized for mental retardation have this syndrome. Mothers of affected males are heterozygous carriers and pass the fragile-X chromosome to 50% of their offspring. These carrier mothers are phenotypically normal. In 20% to 50% of all cases, the mutant allele has a low degree of penetrance in males, and these males, with a normal phenotype, are called transmitter males. As discussed below, daughters of transmitter males have a high risk of bearing affected children, producing the type of pedigree shown in Figure 11.8.

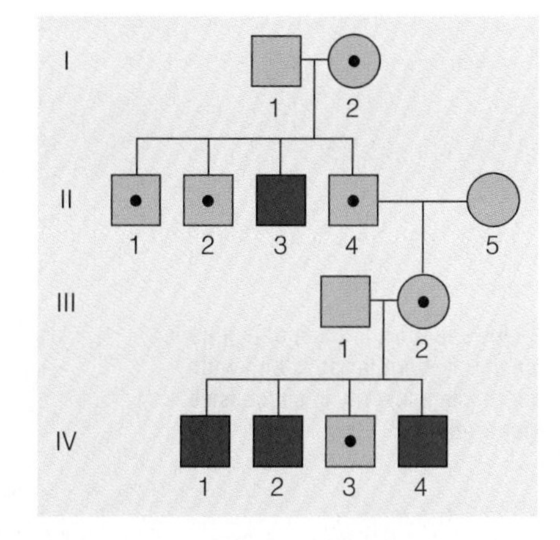

FIGURE 11.8 A pedigree that illustrates the inheritance of fragile-X syndrome. Mothers (I-2) of phenotypically normal but transmitting males (II-4) are phenotypically normal but have some offspring with fragile-X syndrome (II-3). Daughters (III-2) of transmitting males are at high risk of having affected children. Allelic expansion of premutation alleles is more likely when inherited from a female. III-2 has inherited such an allele, which is likely to undergo expansion and affect her children.

The *FMR-1* gene at the fragile-X site always contains CGG repeats. There are three types of *FMR-1* alleles. Normal alleles contain 6 to 52 copies of the CGG repeat. These normal alleles are stable and do not undergo expansion. Premutation alleles have an expanded number of repeats (ranging from 60 to 200). Premutation alleles do not affect expression of the *FMR-1* gene. Males with premutation alleles are phenotypically normal, but are carriers. Children and grandchildren of transmitter males and carrier females are at high risk of

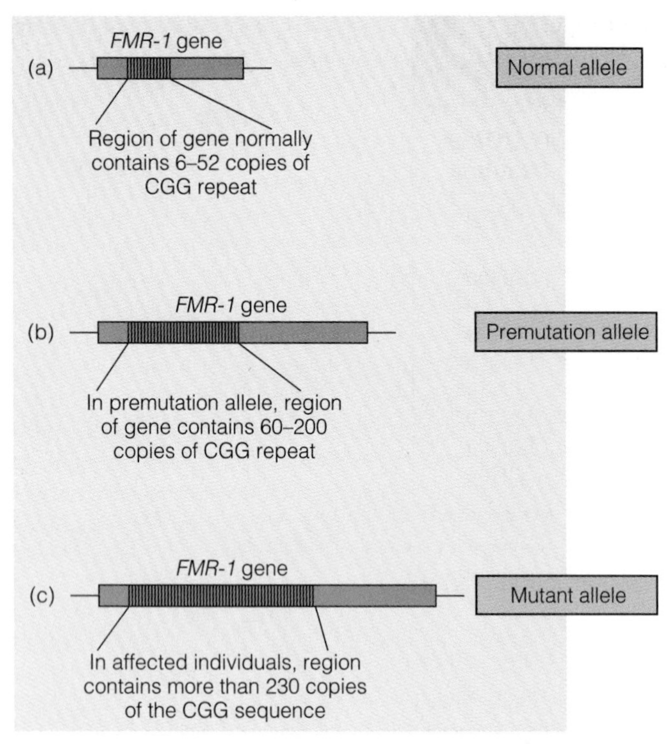

FIGURE 11.9 Allelic expansion in the *FMR-1* gene at the fragile-X locus. (a) The gene normally contains 6 to 52 copies of a CGG trinucleotide repeat. (b) In premutation alleles, this region expands to include 60 to 200 copies of this repeat. (c) Affected individuals have more than 230 copies of the repeat.

being affected. Fully mutant *FMR-1* alleles have more than 230 CGG repeats that prevent expression of the *FMR-1* gene, producing mental retardation in all males and in 60% of females with this allele.

A surprising factor influencing the expansion of the fragile-X gene is the sex of the parent transmitting the mutant allele. When males transmit premutation alleles, the number of CGG repeats is likely to remain constant or even decrease. The change from a premutation allele to a fully mutant allele occurs only when premutation alleles are transmitted by females, (Figure 11.9), resulting in fragile-X syndrome. This transition occurs only after the premutation allele passes through oogenesis or an expansion that takes place during very early embryonic development. Recent evidence indicates that the expansion takes place during oogenesis, but the mechanism has not been fully explained.

More than a dozen genetic disorders are now known to be associated with expansion of trinucleotide repeats. Some of these are listed in Table 11.3. In myotonic dystrophy (DM), the mutation is caused by expansion of a CTG repeat. As with fragile-X syndrome, there is a progressive earlier onset of the disorder in succeeding generations and a correlation between the size of the expanded repeat, the age of onset, and the severity of symptoms. In contrast to the fragile-X syndrome, expansion of trinucleotide repeats in DM and increased probability of clinical symptoms are more likely for male transmission than for female transmission.

Gene Expansion and Anticipation

Five disorders (the first five in Table 11.3) associated with expansion of trinucleotide repeats have some common characteristics. All are progressive, degenerative disorders of the nervous system that are inherited as autosomal dominant traits. All have expanded CAG repeats, and all show a correlation between the increasing size of the repeat and earlier age of onset. In these disorders, mildly affected parents have more seriously affected offspring, who develop symptoms at an earlier age than their parents.

Anticipation Onset of a genetic disorder at earlier ages and with increasing severity in successive generations.

The appearance of more severe symptoms at earlier ages in succeeding generations, called **anticipation**, was first noted for myotonic dystrophy. Although carefully documented by clinicians, geneticists discounted the phenomenon of anticipation because genes were regarded as highly stable entities that have only occasional mutations. This finding means that the concept of mutation must now recognize the existence of unstable genomic regions that undergo these dynamic changes. In other words, when regions of the genome that contain trinucleotide repeats undergo an expansion, this event enhances the chance of further expansions and the development of a disease phenotype. Forms of genome instability similar to trinucleotide expansion might explain genetic disorders that do not show simple Mendelian inheritance (for example, those showing incomplete penetrance or variable expression).

MUTATIONS AND DNA DAMAGE CAN BE REPAIRED

Not every mutation is a permanent genomic change. All cells have enzyme systems that repair damage to DNA. Table 11.4 estimates the rate of spontaneous DNA damage in a typical mammalian cell at 37°C (body temperature). If not repaired, the

Table 11.4 **Rates of DNA Damage in a Mammalian Cell**

Damage	Events/HR
Depurination	580
Depyrimidation	29
Deamination of cytosine	8
Single-stranded breaks	2,300
Single-stranded breaks after depurination	580
Methylation of guanine	130
Pyrimidine (thymine) dimers in skin (noon Texas sun)	5×10^4
Single-stranded breaks from background ionizing radiation	10^{-4}

Table 11.5 **Maximum DNA Repair Rates in a Human Cell**

Damage	Repairs/HR
Single-stranded breaks	2×10^5
Pyrimidine dimers	5×10^4
Guanine methylation	10^4–10^5

accumulated damage would destroy much of the DNA in the cell. Fortunately, humans (and other organisms) have a number of highly efficient DNA repair systems (Table 11.5). However, because the rate of background damage is so high, it is easy to overload the repair systems. One type of damage, the formation of **thymine dimers,** is a major cause of cell death, mutation, and cancer.

Thymine dimer A molecular lesion in which chemical bonds form between a pair of adjacent thymine bases in a DNA molecule.

Cells Have Several DNA Repair Systems

One repair system corrects errors made during DNA replication. In humans, replication proceeds rapidly, with 10 to 20 nucleotides added each second to a DNA strand at each replication site. About 3 billion nucleotides are copied in each round of cell division, and it is little wonder that mistakes occur. Sometimes, the wrong nucleotide is inserted into the newly synthesized strand, resulting in a potential spontaneous mutation. However, DNA polymerase, the enzyme involved in replication, corrects many of these mistakes. In addition to directing DNA replication, the enzyme has a proofreading function. If an incorrect nucleotide is inserted by mistake, the enzyme can detect the mistake and move backward, removing nucleotides until the incorrect nucleotide is eliminated. Then the enzyme inserts the correct nucleotide and moves forward, resuming replication. Those few mistakes that elude the proofreading function of DNA polymerase remain as true spontaneous mutations.

Other repair systems recognize and repair damage to DNA in other phases of the cell cycle. These systems fall into several categories, each controlled by several genes. For example, exposure of DNA to UV light (from sunlight, tanning lamps, and so forth) causes adjacent thymine molecules in the same DNA strand to pair with each other, forming thymine dimers (Figure 11.10). Thymine dimers distort the DNA molecule and can interfere with normal replication. These dimers are corrected by several different DNA repair mechanisms.

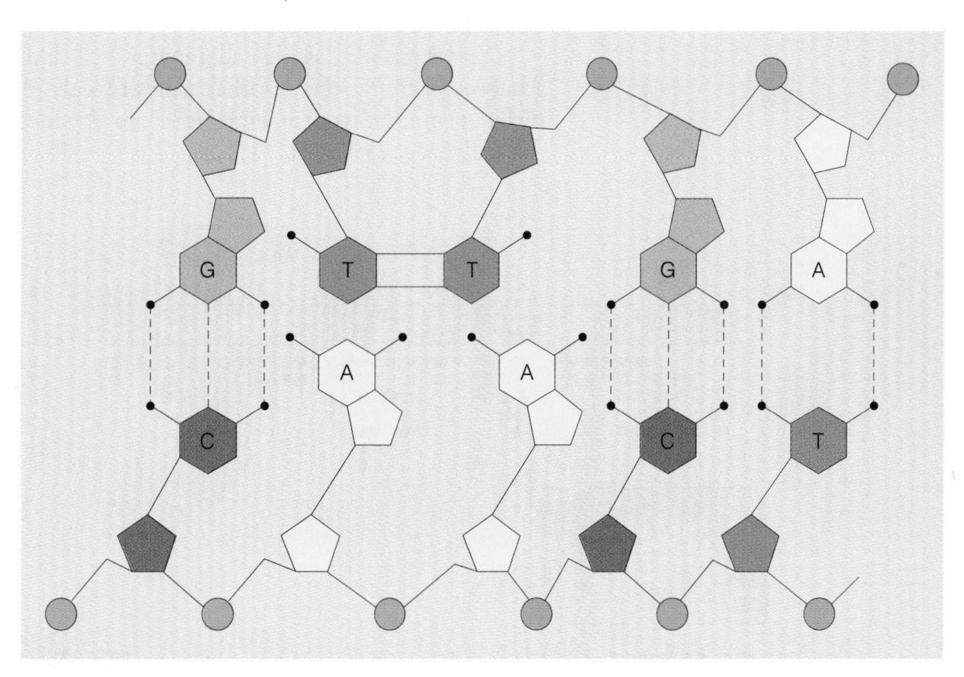

FIGURE 11.10 Thymine dimers are produced when ultraviolet light cross-links two adjacent thymine bases in the same strand of DNA. This structure causes a distortion in the DNA, and errors in replication are likely to occur unless corrected.

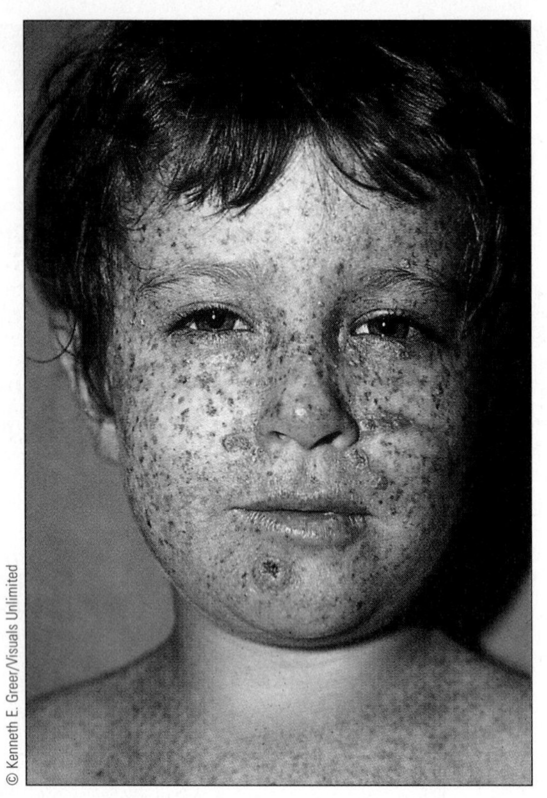

FIGURE 11.11 Child affected with xeroderma pigmentosum. Affected individuals cannot repair damage to DNA caused by ultraviolet light from the sun and other sources.

Genetic Disorders and DNA Repair Systems

Because DNA repair is under genetic control, it, too, can undergo mutation. Several genetic disorders, including xeroderma pigmentosum (XP) (MIM/OMIM 278700), are caused by mutations in genes that repair DNA. XP is an autosomal recessive disorder with a frequency of 1 in 250,000. Affected individuals are extremely sensitive to sunlight (which contains UV light). Even short exposure to the sun causes dry, flaking skin and pigmented spots that can develop into skin cancer (Figure 11.11). Skin cancers are about 1,000 times more common in XP individuals. Early death from cancer is the usual fate of XP individuals who do not take extraordinary measures to protect themselves from UV light.

Mutations in at least eight different genes can cause XP, and all are defective in repairing DNA damaged by UV light. This disease illustrates a functional correlation between mutation and cancer, a topic we consider in detail in Chapter 14.

Several other genetic disorders are characterized by unusual sensitivity to sunlight and/or to other forms of radiation resulting from defects in DNA repair. These include Fanconi anemia (MIM/OMIM 227650), ataxia telangiectasia (MIM/OMIM 208900), and Bloom syndrome (MIM/OMIM 210900). The range of phenotypes seen in these disorders indicates that DNA repair is a complex process that involves many different genes.

MUTATION, GENOTYPES, AND PHENOTYPES

Sickle cell anemia was the first genetic disorder to be analyzed at the molecular level. In this disorder, a nucleotide substitution in codon 6 changes the amino acid in the beta-globin polypeptide at position 6 and produces a distinctive set of clinical symptoms. All affected individuals and all heterozygotes have the same nucleotide substitution.

As it turns out, sickle cell anemia is probably an exception rather than the rule. Analysis of mutations in other genes reveals that more often than not a number of different mutations in a single gene can produce the phenotype associated with a genetic disorder (Figure 11.12).

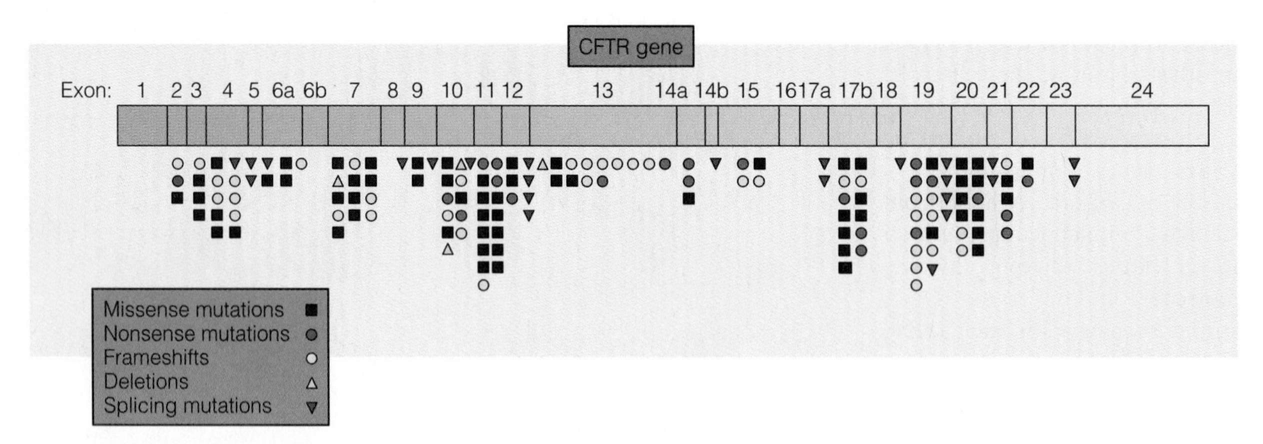

FIGURE 11.12 Distribution of mutations in the cystic fibrosis gene, *CFTR*. More than 500 different mutations have now been discovered. The mutations shown here include nucleotide substitutions, deletions, and frameshift mutations. Any of these mutations in the homozygous condition or in combination with each other (that is, a compound heterozygote) results in the phenotype of cystic fibrosis.

MUTATIONS CAN CAUSE A RANGE OF PHENOTYPES

Cystic fibrosis provides a clear example of the types and numbers of mutations that can occur in a single gene, all of which result in a disease phenotype. More than 500 different mutations have been identified in the cystic fibrosis gene. These include single nucleotide substitutions, nucleotide deletions, and larger deletions that involve one or more regions of the gene. In addition, there are frameshift mutations and splice-site mutations. Mutations are distributed in all regions of the *CFTR* gene (Figure 11.13), strengthening the idea that any mutational event that interferes with expression of a gene produces an abnormal phenotype.

People with cystic fibrosis have a wide range of clinical symptoms. The relationship between the type and location of the *CFTR* mutation and the clinical phenotype has been investigated for a number of mutant alleles. In some mutations, such as the Δ508 deletion (present in 70% of all cases of CF), the CFTR protein is produced but is not inserted into the plasma membrane. As a result, the protein is not functional, chloride ion transport is absent, and clinical symptoms are severe. In other mutations, such as R117, R334, and R347 (Figure 11.13), the CFTR protein is produced and inserted into the membrane but is only partially functional. These mutations are associated with a milder form of cystic fibrosis. With over 500 different mutations known in the CF gene, it is possible for someone with CF to carry two different mutant alleles. This genetic variability further contributes to the phenotypic variability seen in this disease.

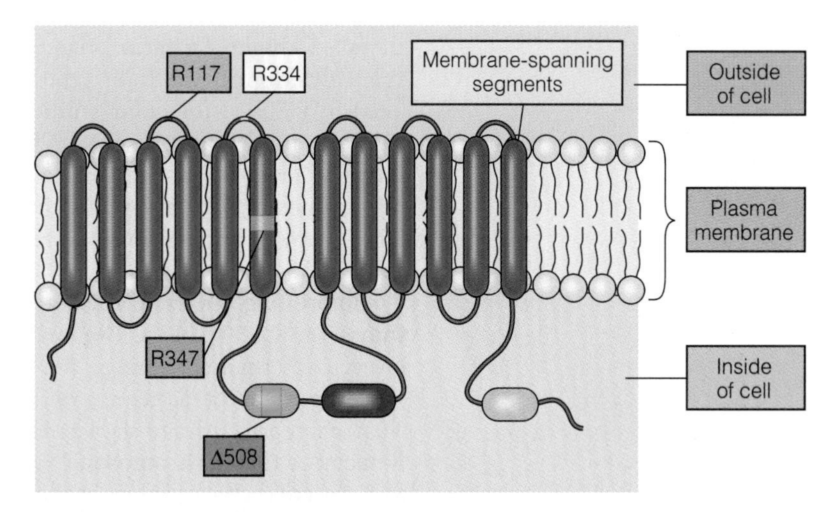

FIGURE 11.13 Mutations in the *CFTR* gene differ in their phenotypic effects. When homozygous, mutations R117, R334, or R347 allow between 5% and 30% of normal activity for the gene product and produce only mild symptoms. These nucleotide substitution mutations are not common: Together they account for about 2% of all cases of cystic fibrosis. The most common mutation in European populations, Δ508, causes an amino acid deletion in a cytoplasmic region of the protein and accounts for 70% of all mutations in the *CF* gene. This mutation inactivates the CFTR protein and is associated with severe symptoms.

GENOMIC IMPRINTING IS A REVERSIBLE ALTERATION OF THE GENOME

We all carry two copies of each gene, one from mom and one from dad. Normally, there is no difference in the expression of the two copies. For certain genes, however, expression depends on whether they are inherited from the mother or the father. This differential expression is called **genomic imprinting.**

Genomic imprinting Phenomenon in which the expression of a gene depends on whether it is inherited from the mother or the father. Also known as genetic or parental imprinting.

The first evidence for imprinting in mammals appeared in mice in which haploid nuclei were transplanted into eggs to produce zygotes carrying an all female or all male genome rather than the usual combination of one male and one female genome. Experimental embryos with only a male genome develop abnormal embryonic structures but have normal placentas. Embryos with only a female genome develop normal embryonic structures but have abnormal placentas. Both conditions are lethal, and we can conclude that both a maternal and a paternal genome are required for normal development.

Genomic imprinting plays a role in several genetic disorders, including Prader-Willi syndrome (PWS) and Angelman syndrome (AS). Most cases of PWS, an autosomal recessive disorder characterized by obesity, uncontrolled appetite, and mental retardation, are associated with a small deletion in the long arm of chromosome 15 (see Figure 6.25). In about 40% of cases, however, no deletion can be detected. Using molecular markers, these nondeletion cases are found

Uniparental disomy A condition in which both copies of a chromosome are inherited from one parent.

to have inherited both copies of chromosome 15 from the mother. This condition, in which both copies of a given chromosome are inherited from a single parent, is called **uniparental disomy.** In such cases, PWS is caused by having no paternal copy of chromosome 15 and therefore having two maternal copies of the chromosome.

Severe mental retardation, uncontrollable puppet-like movements, and seizures of laughter characterize AS. About 50% of affected individuals have a small deletion in the long arm of chromosome 15 (the same region that is deleted in PWS). Molecular studies indicate that AS is a genetic mirror image of PWS. In nondeletion cases of AS, both copies of chromosome 15 come from the father. These two disorders reinforce the idea that paternal and maternal copies of chromosome 15 are required for normal development. The absence of a paternal copy of chromosome 15 results in PWS; the absence of a maternal copy results in AS.

Imprinting does not affect all genes. Only those in certain regions of chromosomes 4p, 8q, 17p, 18p, 18q, and 22q are imprinted. Imprinting is not a mutation or even a permanent change in a gene or a chromosome region. What is affected is the expression of a gene, not the gene itself. Imprinting does not violate the Mendelian principles of segregation or independent assortment, nor is it permanent. Remember that a chromosome received by a female from her father is transmitted as a maternal chromosome in the next generation (▶ Figure 11.14). In each generation, the previous imprinting is erased, and a new pattern of imprinting defines the chromosome as either paternal or maternal. Imprinting events are thought to take place during gamete formation, and the effects are transmitted to all tissues of the offspring. Although imprinting is not strictly a mutational event, it does involve chemical modification of DNA.

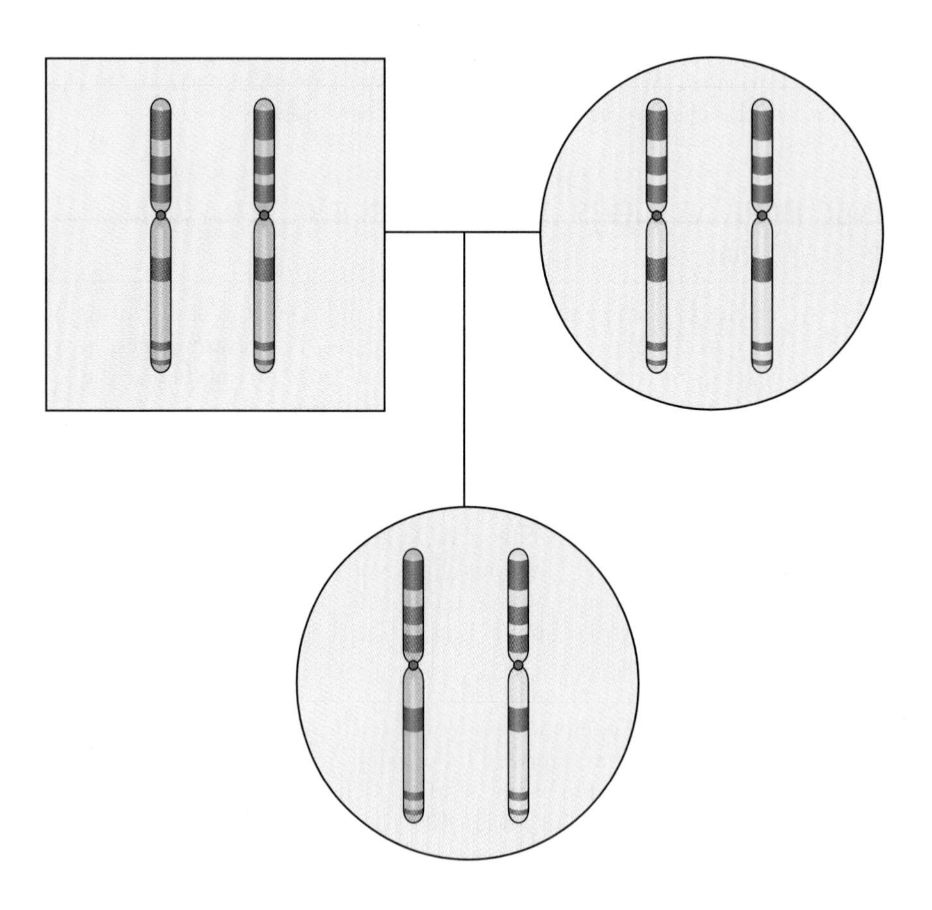

▶ **FIGURE 11.14** Females receive one copy of each chromosome from their fathers. These paternal chromosomes, imprinted as male chromosomes, must be reimprinted as female chromosomes before being transmitted to the next generation.

Summary

Mutations Are Heritable Changes

1. The study of mutations is an essential part of genetics. Without mutations, it would be difficult to determine whether a trait is under genetic control and impossible to determine its mode of inheritance.

Mutations Can Be Detected in Several Ways

2. Mutations can be classified in a variety of ways using criteria such as pattern of inheritance, phenotype, biochemistry, and degrees of lethality.
3. Dominant mutations are the easiest to detect because they are expressed in the heterozygous condition. Accurate pedigree information can often identify the individual in whom the mutation arose. It is more difficult to determine the origin of sex-linked recessive mutations, but an examination of the male progeny is often informative. If the mutation in question is an autosomal recessive, it is almost impossible to identify the original mutant individual.

Measuring Mutation Rates

4. Studies of mutation rates in a variety of dominant and sex-linked recessive traits indicate that mutations in the human genome are rare events, occurring about once in every 1 million copies of a gene. The impact of mutation is diminished by several factors, including the redundant nature of the genetic code, the recessive nature of most mutations, and the lowered reproductive rate or early death associated with many genetic diseases.

Mutation at the Molecular Level: DNA as a Target

5. Molecular analysis of mutations has shown a direct link between gene, protein, and phenotype. Mutations arise spontaneously as the result of errors in DNA replication or as the result of structural shifts in nucleotide bases. Environmental agents, including chemicals and radiation, also cause mutations. Frameshift mutations cause a change in the reading frame of codons, often producing dramatic alterations in the structure and function of proteins.

Mutations and DNA Damage Can Be Repaired

6. Not all mutations cause genetic damage. Cells have a number of DNA repair systems that correct errors in replication and repair damage caused by environmental agents such as ultraviolet light, radiation, or chemicals.

Mutation, Genotypes, and Phenotypes

7. In most genes associated with a genetic disorder, many different types of mutations can cause a mutant phenotype. In the cystic fibrosis gene, more than 500 different mutations have been identified, including deletions, nucleotide substitutions, and frameshift mutations.

Mutation Can Cause a Range of Phenotypes

8. The wide range of mutations found in genetic disorders leads to wide variation in clinical symptoms. Depending on the mutation, symptoms can range from very mild to very severe.

Genomic Imprinting Is a Reversible Alteration of the Genome

9. Genomic imprinting alters expression of normal genes, depending on whether they are inherited maternally or paternally. Imprinting has been implicated in a number of disorders, including Prader-Willi and Angelman syndromes. Not all regions of the genome are affected, and only segments of chromosomes 4, 8, 17, 18, and 22 are imprinted. Genes are not permanently altered by imprinting, but are reimprinted in each generation.

Case Studies

CASE 1

On April 26, 1986, one of the four reactors at the Chernobyl generating station in the Soviet Union melted down. It has been reported that the plant was running with disconnected safety measures. The result was fire, chaos, fear, a cloud of radioactive isotopes spreading across vast reaches of Eastern Europe, and the radioactive contamination of thousands of people.

Unfortunately, this human tragedy is not being investigated as it should be, according to scientists who are trying to learn from it. Cancer prevention specialists claim

Case Studies (continued)

that there is a lack of resources available to look at cancer cases in the population exposed to radiation released from Chernobyl.

There are many obstacles facing scientists who want to study the aftermath of Chernobyl: (1) the dissolution of the Soviet Union split administrative, record-keeping, and medical responsibilities among Belarus, Ukraine, and Russia; (2) a general decline in living standards has reduced the level of medical care in the area; (3) individual doses cannot be reconstructed accurately; estimates on doses received by individuals have been complicated by the fact that some of the dose was external through exposure to radioactive dust and part of it was internal, through eating contaminated food; and (4) little money is available for the kind of large studies that are needed for extracting the best data.

Here are some of the facts—as best we know them—about the Chernobyl meltdown. The accident released 1.85×10^{18} (1,850,000,000,000,000,000) international units of radioactive material. The releases contaminated an estimated 17 million people to some degree. The exact amount of exposure depended on location, wind direction, length of exposure, eating habits, and whether the person was a "liquidator."

These unfortunate heroes were pressed into service in a crude cleanup effort after the accident. One hundred and thirty-four people showed signs of acute radiation sickness immediately after the accident. Many of the 28 people who died from acute radiation sickness had skin lesions covering 50% or more of their bodies. After the fire, 135,000 people evacuated the area around the reactor, and 800,000 liquidators moved in to try to decontaminate the area. Approximately 17% to 45% of the liquidators received doses between 10 and 25 rads. (For comparison, in the United States, the annual dose permitted to the general public is 0.1 rads; nuclear workers are permitted 5 rads.)

Despite the early confusion, some medical information is now available in Chernobyl's aftermath. The most compelling data involves radiation exposure and thyroid cancer, particularly in children. According to the International Chernobyl Conference (April, 1996), radiation exposure caused "a substantial increase in reported cases of thyroid cancer in Belarus, Ukraine, and some parts of Russia, especially in young children." This is thought to be due to exposure to radioactive iodine during the early phases of the accident in 1986. By the end of 1995, approximately 800 cases of thyroid cancer had been reported in children who were under age 15 at the time of diagnosis. To date, three of these thyroid cancer victims have died, and several thousand more cases of thyroid cancer are expected. Ironically, most of the thyroid cancers could have been prevented if people in the contaminated areas had taken iodine tablets immediately after the accident. Most iodine in the body goes to the thyroid gland; if enough normal iodine is available, only a small amount of radioactive isotope irradiates the little gland, whose hormones help regulate growth. More information on the long-term effects of Chernobyl's explosion will be learned in the coming decades, as scientists and health care workers assess the full impact of this disaster.

1. How do you think radiation causes cancer?
2. The liquidators were unfortunately exposed to large amounts of radiation. Are their families at risk even though they were not part of the cleanup effort?
3. What kind of compensation, if any, should liquidators receive from the government for the exposure to radiation?
4. An increase in thyroid cancer was reported after Chernobyl. Are these people at risk of passing a mutant cancer gene to their offspring?

Questions and Problems

Mutation Rates for Specific Genes

1. Define mutation rate.
2. Achondroplasia is an autosomal dominant form of dwarfism caused by a single gene mutation. Calculate the mutation rate of this gene given the following data: 10 achondroplastic births to unaffected parents in 245,000 births.
3. Why is it almost impossible to measure directly the mutation rates in autosomal recessive alleles?
4. What are the factors that influence mutation rates of human genes?
5. Achondroplasia is a rare dominant autosomal defect resulting in dwarfism. The unaffected brother of an individual with achondroplasia is seeking counsel on the likelihood of his being a carrier of the mutant allele. What is the probability that the unaffected client is carrying the achondroplasia allele?

Sources of Mutations

6. Although it is well known that x-rays cause mutations, they are routinely used to diagnose medical problems

including potential tumors, broken bones, or dental cavities. Why is this done? What precautions need to be taken?

7. You are an expert witness called by the defense in a case in which a former employee is suing an industrial company because his son was born with muscular dystrophy, an X-linked recessive disorder. The employee claims that he was exposed to mutagenic chemicals in the work place that caused his son's illness. His attorney argues that neither the employee, his wife, nor their parents have this genetic disorder, and, therefore, the disease in the employee's son represents a new mutation. How would you analyze this case? What would you say to the jury to support or refute this man's case?
8. Bruce Ames and his colleagues have pointed out that, although detailed toxicological analysis has been conducted on synthetic chemicals, almost no information is available about the mutagenic or carcinogenic effects of the toxins produced by plants as a natural defense against fungi, insects, and animal predators. Tens of thousands of such compounds have been discovered, and he estimates that in the United States adults eat about 1.5 g of these compounds each day, levels that are approximately 10,000 times higher than those levels of synthetic pesticides present in the diet. For example, cabbage contains 49 natural pesticides and metabolites, and only a few of these have been tested for their carcinogenic and mutagenic effects.
   a. With the introduction of new foods into the U.S. diet over the last 200 years (mangoes, kiwi fruit, tomatoes, and so forth), has there been enough time for humans to evolve resistance to the mutagenic effects of toxins present in these foods?
   b. The natural pesticides present in plants constitute more than 99% of the toxins we eat. Should diet planning, especially for vegetarians, take into account the doses of toxins present in the diet?

Mutation at the Molecular Level

9. Define and compare the following types of nucleotide substitutions. Which is likely to cause the most dramatic mutant effect?
   a. missense mutation
   b. nonsense mutation
   c. sense mutation
10. If the coding region of a gene (the exons) contains 2,100 base pairs of DNA, would a missense mutation cause a protein to be shorter, longer, or the same length as the normal 700 amino acid protein? What would be the effect of a nonsense mutation? A sense mutation?
11. Two types of mutations discussed in this chapter are (1) nucleotide changes and (2) unstable genome regions that undergo dynamic changes. Describe each type of mutation.
12. What is a frameshift mutation?
13. A frameshift mutation is caused by:
   a. a nucleotide substitution
   b. a three-base insertion
   c. a premature-stop codon
   d. a one-base insertion
   e. a two- base deletion
14. In the gene coding sequence shown, which of the following events will produce a frame shift after the last mutational site?

normal mRNA:	UCC	AAA	UAC	CGU	CGU	UAA
	ser	lys	tyr	arg	arg	stop

   a. insertion of an A after the first codon
   b. deletion of the second codon (AAA)
   c. insertion of TA after the second codon and deletion of CG in the fourth codon
   d. deletion of AC in the third codon.
15. Trinucleotide repeats cause serious neurodegenerative disorders such as Huntington disease, fragile-X syndrome, and myotonic dystrophy (DM). The process of anticipation causes the appearance of symptoms at earlier ages in succeeding generations. Describe the current theory on how anticipation works.
16. Familial retinoblastoma, a rare autosomal dominant defect, arose in a large family that had no prior history of the disease. Consider the following pedigree (the darkly colored symbols represent affected individuals):

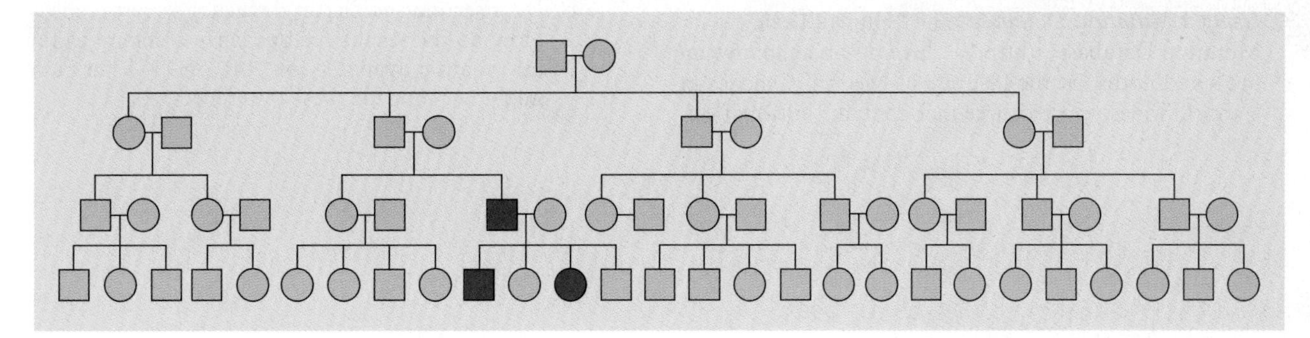

   a. Circle the individual(s) in which the mutation most likely occurred.
   b. Is this individual affected by the mutation? Justify your answer.
   c. Assuming that the mutant allele is fully penetrant, what is the chance that an affected individual will have an affected child?

17. Tay-Sachs disease is an autosomal recessive disease. Affected individuals do not often survive to reproductive age. Why has Tay-Sachs persisted in humans?

Mutations and DNA Damage Can Be Repaired

18. Replication involves a period of time during which DNA is particularly susceptible to the introduction of mutations. If nucleotides can be incorporated into DNA at a rate of 20 nucleotides/second and the human genome contains 3 billion nucleotides, how long would replication take? How is this time reduced so that replication can take place in a few hours?
19. Our bodies are not defenseless against mutagens that alter our genomic DNA sequences. What mechanisms are used to repair DNA?

Mutations Can Cause a Range of Phenotypes

20. The cystic fibrosis gene encodes a chloride channel protein necessary for normal cellular functions. Let us assume that if at least 10% normal channels are present, the affected individual has mild symptoms of cystic fibrosis. Less than 10% normal channels produces severe symptoms. At least 50% of the channels must be expressed for the individual to be phenotypically normal. This gene has various mutant alleles:

allele	molecular defect	% functional channels	symptoms
CF100	deletion in exon	0%	severe
CF1	missense mutation in 5′ flanking region	25%	mild
CF2	nonsense mutation in exon	0%	severe
CF3	missense mutation in exon	5%	mild

Predict the percent of functional channels and severity of symptoms for the following genotypes:
a. heterozygous for CF100
b. homozygous for CF100
c. heterozygous, with one copy of CF100 and one of CF3
d. heterozygous, with one copy of CF1 and one copy of CF3

Internet Activities

The following activities use the resources of the World Wide Web to enhance the topics covered in this chapter. To investigate the topics described below, log on to the book's home page (www.brookscole.com/biology_d) and follow the *Student Resources* link to your subject and text.

1. *Mutant Sequences.* Review the information on Pompe disease from the Internet activities section in Chapter 10. How many mutations have been found in humans for this one enzyme?
2. *Mutation Review.* Work through the Gene Action/Mutation worksheet at the "Access Excellence Activities Exchange" (http://www.accessexcellence.org/AE/AEPC/WWC/1994/gene_action.html) to reinforce the concepts related to the various types of mutations and their consequences.
3. *Using a Mutation Database.* The "Human Gene Mutation Database" (http://archive.uwcm.ac.uk/uwcm/mg/hgmd0.html) at the Institute of Medical Genetics at Cardiff, Wales, is a resource that contains information about mutations identified in human genes. This information includes nucleotide substitutions, missense and nonsense mutations, splicing mutations, and small insertions and deletions. The data provided includes the name and symbol for the gene, its chromosomal location, the mutant sequence codon number, and a reference to the paper that first identified the mutation.
   a. At the Web site, search on "breast cancer." From the list of breast cancer genes, select "BRCA1" and scroll through the list of mutations, noting both the type of mutation and its location. Are the mutations reported here scattered throughout the gene, or are they clustered in certain locations? Because this gene was identified only a few years ago, is the list likely to be comprehensive?
   b. Do a second search, on the gene for cystic fibrosis. How do the results for this entry compare to the information available for "BRCA1"? What factors might account for these differences?

For Further Reading

Ames, B., Profet, M., & Gold, L. S. (1990). Dietary pesticides (99.00% all natural). *Proc. Nat. Acad. Sci. U S A, 87,* 7777–7781.

Ames, B., Profet, M., & Gold, L. S. (1990). Nature's chemicals and synthetic chemicals: Comparative toxicology. *Proc. Nat. Acad. Sci. U S A, 87,* 7782–7786.

Ashley, C. T., & Waren, S. T. (1995) Trinucleotide repeat expansion and human disease. *Annu. Rev. Genet., 29,* 703–728.

Deering, R. A. (1962, December). Ultraviolet radiation and nucleic acids. *Sci. Am., 207,* 135–144.

Moss, T., & Sills, D. (1981). The Three Mile Island nuclear accident: Lessons and implications. *Ann. N.Y. Acad. Sci., 365*(x).

Nachman, M. W., & Crowell, S. (2000). Estimate of the mutation rate per nucleotide in humans. *Genetics, 156,* 297–304.

Sapienza, C. (1995). Genome imprinting: An overview. *Dev. Genet., 17,* 185–187.

Schwartz, E., & Surrey, S. (1986). Molecular biologic diagnosis of the hemoglobinopathies. *Hosp. Pract., 9,* 163–178.

Steele, J. H. (2001). Food irradiation: a public health challenge for the 21st century. *Clin. Infect. Dis., 33,* 376–377.

Tritsch, G. L. (2000). Food irradiation. *Nutrition, 16,* 698–781.

Tsui, L-C. (1992). The spectrum of cystic fibrosis mutations. *Trends Genet., 8,* 392–398.

Warren, S. T. (1996). The expanding world of trinucleotide repeats. *Science, 271,* 1423–1427.

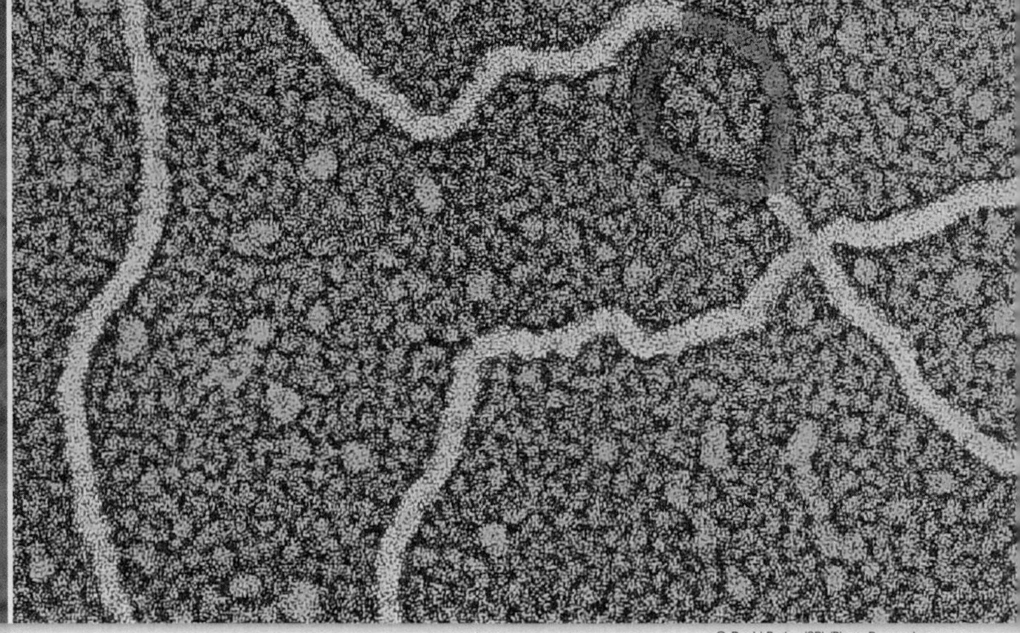

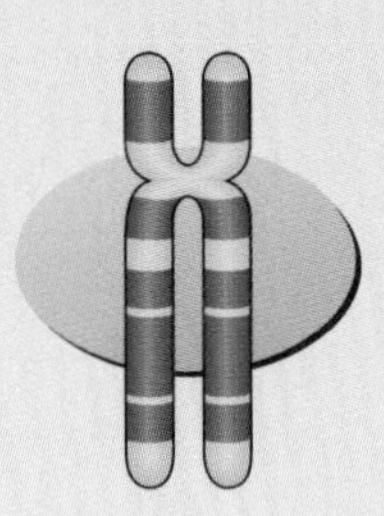

An Introduction to Cloning and Recombinant DNA

Chapter Outline

The story begins simply enough. A boy of Ghanian parents, born in England and therefore a British citizen, moved to Ghana to live with his father. He left behind his mother, two sisters, and a brother. Later, he returned to Britain to live with his mother and siblings; at this point, the story gets complicated. Immigration authorities suspected that the boy was an impostor and thought he was either an unrelated child or a nephew of the boy's mother. Based on their suspicions, the boy was denied residence in Britain. The boy's family fought to help establish his identity so he could live in the country of his birth. Medical tests were done, using the usual genetic markers such as blood types, as well as tests with markers used to match organ donors and recipients. The test results confirmed that the boy was closely related to the woman he claimed was his mother, but the tests could not be used to tell whether she was his mother or an aunt.

The family then turned to Alec Jeffreys, a scientist at the University of Leicester for help. They asked if DNA fingerprinting, a technique developed in Jeffreys's research laboratory, could establish the boy's identity. To complicate the situation even more, neither the mother's sisters nor the boy's father was available for testing, and the mother was not sure about the boy's paternity. Despite these problems, Jeffreys agreed to take on the case. He took blood samples from the boy, the children he claimed were his brother and sisters, and the woman who claimed to be his mother. DNA was extracted from the white blood cells in each sample and treated with enzymes that cut DNA at specific nucleotide sequences. The resulting fragments were separated by size. The pattern of fragments, known as a DNA fingerprint, was analyzed to determine the boy's identity. The results showed two things. The boy has the same father as his brother and his sisters because they all share DNA fragments contributed by the father. The most important question was whether the boy and his "mother" were related.

Jeffreys found that 25 fragments in the woman's DNA fingerprint matched those in the boy's fingerprint, indicating that she was in fact the boy's mother. The chance that they were unrelated was calculated as 2×10^{-15}, or about one in a quadrillion. Faced with this evidence, immigration authorities reversed their position and allowed the boy to remain in Britain and live with his family.

DNA fingerprinting is only one example of the many new techniques developed as part of the ongoing revolution in genetic technology. This revolution began in the 1970s with the discovery that DNA from different organisms could be combined in specific ways to create recombinant DNA molecules. In this and the next chapter, we examine how this revolution began, how recombinant DNA molecules are constructed, and how this technology is being used in areas of human genetics, biology, medicine, agriculture, and the Human Genome Project.

WHAT ARE CLONES?

When a fertilized egg divides to form two cells, these cells usually stay attached to one another and become parts of one embryo. In a small number of cases, the cells separate from each other and form separate embryos that become identical twins. Because they are derived from a common ancestor (in this case a fertilized egg), identical twins are clones. **Clones** are defined as molecules, cells, or individuals derived from a single ancestor. Methods for producing clones are not new; cloned fruit trees were made in ancient times and are still produced today. On the other hand, cloned DNA molecules, cells, and animals have appeared only in recent decades.

Clones Genetically identical organisms, cells, or molecules all derived from a common ancestor.

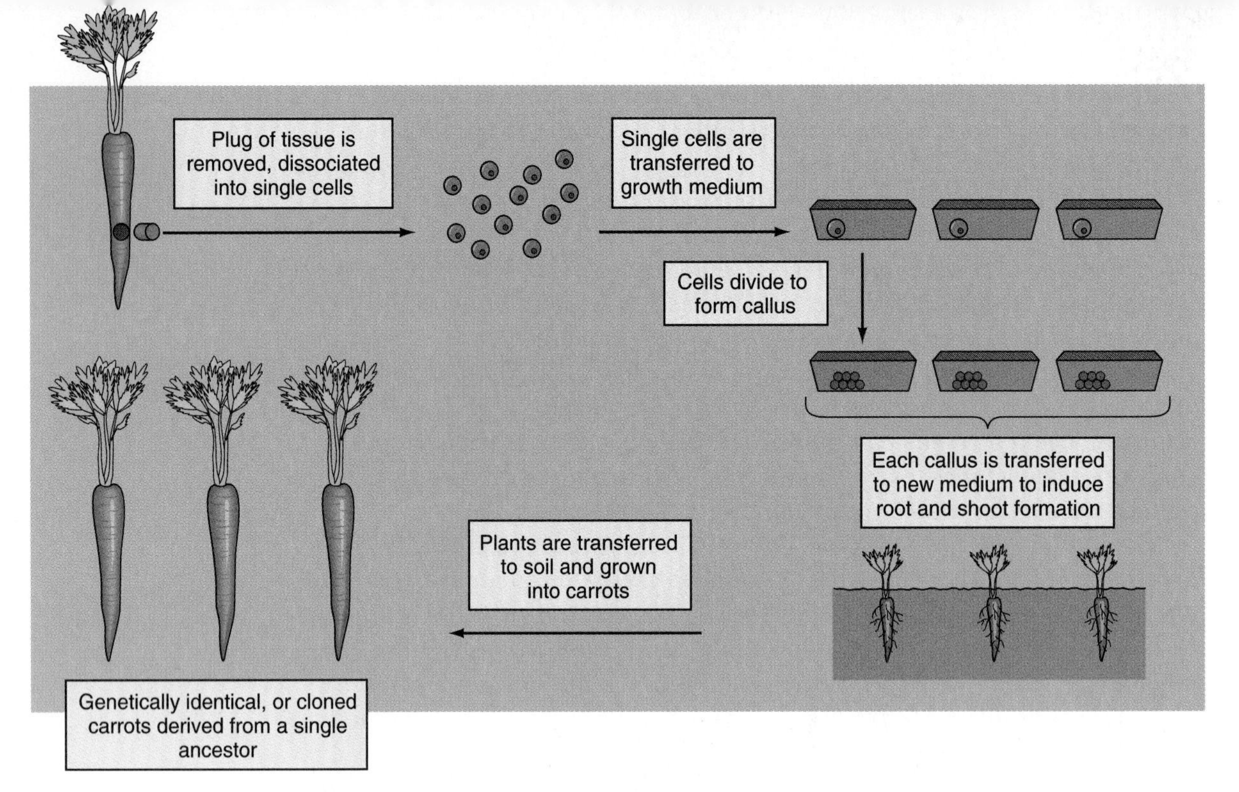

FIGURE 12.1 Cloning carrots. A plug of tissue is removed and separated into single cells. Each cell is placed in a separate dish with growth medium. Each cell divides to form a mass of cells called a callus. Calluses are transferred to a new medium to induce the formation of roots and shoots. The developing plants are then transferred to soil, where they grow into genetically identical copies, or clones, of the original carrot.

The ability to produce large amounts of identical DNA molecules has had a great impact on genetic research. This impact has spread beyond the research lab and is used in many areas, including the criminal justice system, child support cases, archeology, dog breeding, and environmental conservation. In this chapter, we will review the methods used to clone DNA molecules, identify them, and analyze them. In the following chapter, we will examine how cloning is used in research and everyday life. Before we consider the basic techniques of DNA cloning, let's look briefly at how plants and animals can be cloned.

FIGURE 12.2 This cloned plant was grown from a single cell removed from a parental plant.

Plants Can Be Cloned from Single Cells

We have genetically manipulated plants and animals for thousands of years by selective breeding. Organisms with desirable characteristics were chosen for breeding, and the offspring with the best combination of characteristics were used for breeding the next generation. Although this method seems slow and unreliable, Charles Darwin observed that after only a few generations, breeders produced many varieties of pigeons and minks.

In the 1950s, Charles Steward grew individual carrot cells in the laboratory using special nutrients (Figure 12.1). These single cells grew and divided to form a ball of cells known as a callus (Figure 12.2). When the calluses were transferred to a different nutrient medium, they grew into full-sized carrots. Because they were all derived from a single ancestor, the carrots grown from the calluses are all clones. Variations of Steward's method have been used to clone plants of several different species. Cloning is used in the paper and lumber industry to produce trees of uniform size, growth rate, and disease resistance. For example, scientists selected a loblolly pine tree that was resistant to disease, grew rapidly, and had high wood content. A core sample was taken from this tree and separated into single cells. The cells were grown until they formed a callus and then were converted into seedlings. Thousands of these seedlings were planted and grew into a forest of genetically identical trees that mature at the same time. This allows pulpwood companies to plant and harvest trees on a predetermined schedule and is now widely used in timber farming.

Animals Can Be Cloned by Several Methods

Cloning of farm animals such as cattle and sheep moved from the research laboratory to the business world more than 20 years ago. Two of the most widely used methods of cloning animals are embryo splitting and nuclear transfer. To clone an animal by embryo splitting, an egg is collected and fertilized by *in vitro* fertilization (IVF). The embryo develops in the dish to form an embryo containing 8 to 16 cells. A technician then separates the cells from one another using a microscope and very fine needles. The individual cells are grown in the laboratory to form genetically identical embryos. The embryos are then transplanted into surrogate mothers for development. This method extends nature's way of producing identical twins or triplets and can be used to clone any mammalian embryo, including human embryos.

The second method of animal cloning, nuclear transfer, is more difficult but can result in a larger number of cloned offspring. The first successful cloning of mammals by nuclear transfer was done in 1986. In this experiment, sheep eggs were collected, and the nucleus was removed from each egg by dissection under a microscope. One by one, the enucleated eggs were fused with cells from a 16- to 32-cell embryo. The fused cells were grown into embryos and transplanted into surrogate mothers for development. Ideally, if the original embryo contains 16 or 32 cells, then 16 or 32 genetically identical offspring or clones can be produced (Figure 12.3). Case 1 at the end of the chapter deals with cloning.

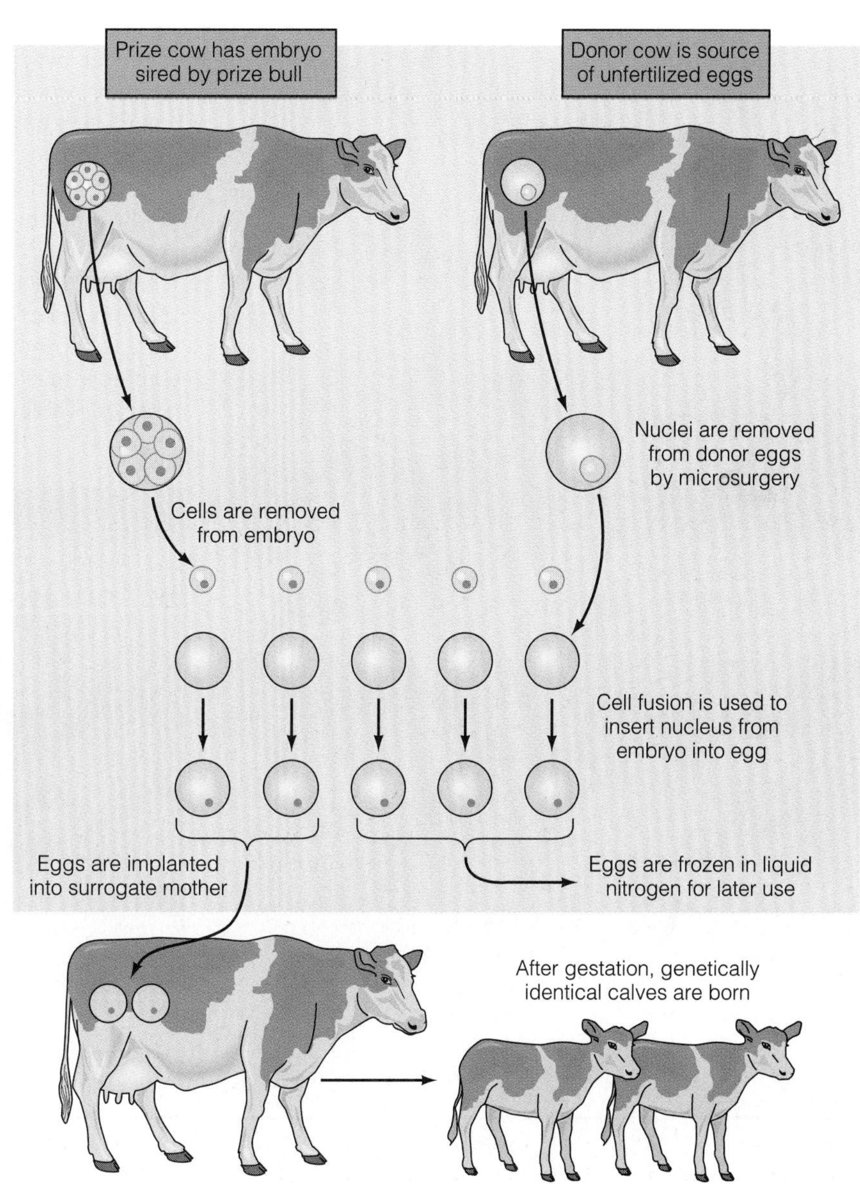

FIGURE 12.3 Mammals, including cows, have been cloned by cell fusion. This process has two stages. First, unfertilized eggs are collected from a donor cow, and the nucleus is removed from each egg by microsurgery. Second, the embryo to be cloned is recovered and separated into single cells. Embryo cells are fused with the donor eggs. Each egg then contains a genetically identical nucleus. These eggs can be frozen in liquid nitrogen for future use or implanted into the uterus of a surrogate mother to develop. All offspring of these eggs would be genetically identical copies, or clones, of the original embryo.

© Le Corre-Ribeiro/Liaison Agency

FIGURE 12.4 Dolly the sheep, cloned by nuclear fusion of an adult cell, with her surrogate mother.

With improvements in the techniques of cell fusion and embryo growth it became possible to perform nuclear transfers with cells from older and older embryos. The cloning of Dolly the sheep, reported in 1997, was the first time that a nucleus from an adult cell (in this case from the udder) was successfully used to produce a cloned animal (Figure 12.4). Cloning Dolly was a significant event because it showed that even nuclei from highly specialized adult cells can direct all stages of development when placed into eggs. Earlier experiments led scientists to the conclusion that adult cells could not be used in cloning experiments. In 1998, a research team successfully cloned more than 2 dozen mice. Instead of using cell fusion, they directly injected nuclei from adult cells into enucleated eggs (Figure 12.5).

Animal cloning has had a great impact on farming. Sheep, cattle, goats, and pigs have all been cloned using either cell fusion or microinjection. It is now possible to produce herds of genetically identical farm animals, all of which have desirable traits, such as superior wool, milk, or meat production. However, cloning animals is not a routine procedure. It is an inefficient process, and only a small percentage of the transferred embryos survive. Some cloned animals are oversized, and others have respiratory and circulatory system problems. Case 2 at the end of the chapter deals with human cloning.

Although the cloning of plants and animals has changed agriculture, the cloning of DNA molecules has revolutionized everything from laboratory research to health care to the food we eat. Using methods of DNA cloning, we can now find genes, map them, and transfer them between species. These methods are used to find carriers of genetic

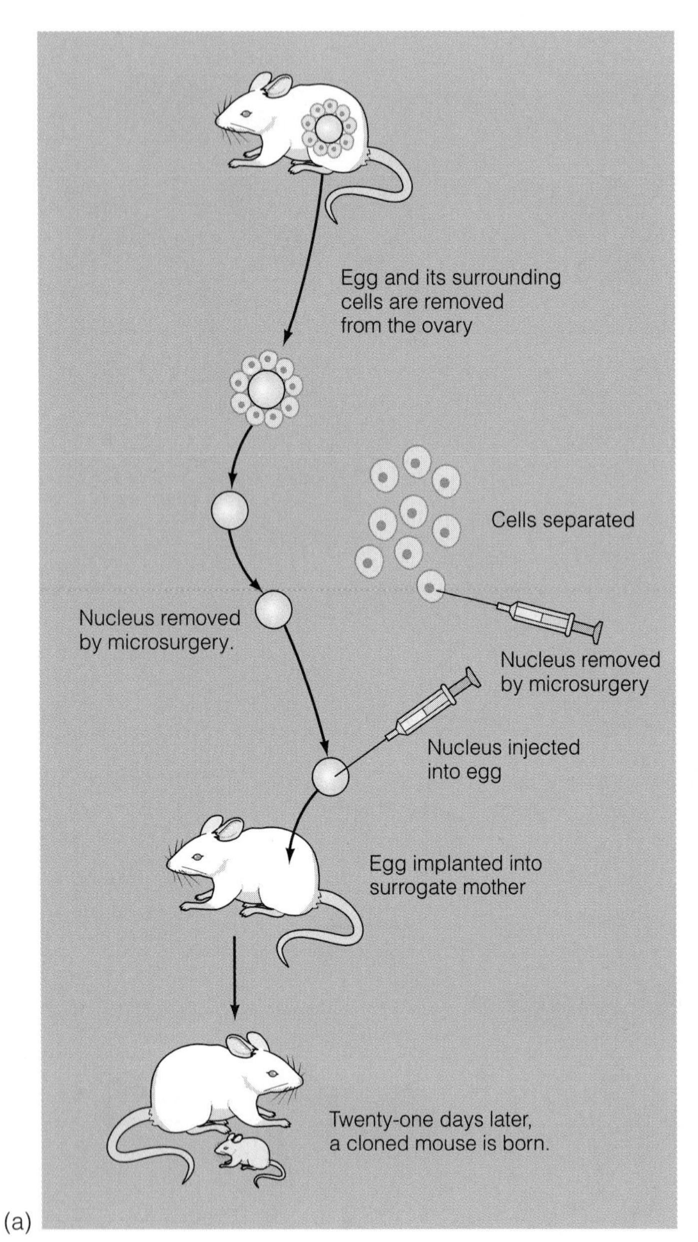

FIGURE 12.5 (a) The strategy for cloning mice by injection of nuclei from adult cells. (b) A cloned mouse (on the bar) and its parent (lower left).

disorders, perform gene therapy, and create disease-resistant food plants. In the rest of this chapter, we will describe how DNA molecules are cloned and analyzed. In the following chapter we will look at some of the ways this technology is being used in genetics, medicine, the criminal justice system, and the biotechnology industry.

CLONING GENES IS A MULTISTEP PROCESS

Cloning DNA (genes) produces a large number of identical molecules, all of which are copies of a single DNA molecule. The methods for cloning DNA molecules are built on discoveries in genetics and biochemistry made in the late 1960s and early 1970s. As a group, these methods are called **recombinant DNA technology.** With recombinant DNA technology, it is possible to

Recombinant DNA technology A series of techniques in which DNA fragments are linked to self-replicating vectors to create recombinant DNA molecules, which are replicated in a host cell.

1. Produce DNA fragments using enzymes that cut DNA at specific base sequences
2. Link these DNA fragments to carrier molecules of DNA, called vectors, to create recombinant DNA molecules
3. Transfer a recombinant DNA molecule to a host cell, where it is copied to form hundreds or thousands of exact copies (clones) of the recombinant molecule
4. Retrieve the cloned DNA fragment from the vector in large quantities
5. Turn on genes in cloned DNA and purify large amounts of the encoded gene product for use

Restriction Enzymes Cut DNA at Specific Sites

Restriction enzymes Enzymes that recognize a specific base sequence in a DNA molecule and cleave or nick the DNA at that site.

Palindrome A word, phrase, or sentence that reads identically in both directions. In DNA, a sequence of nucleotides that reads identically on both strands when read in the 5′ to 3′ direction.

The story of the way recombinant DNA technology was discovered is an example of how basic research in one field often has an unexpected impact on other areas. It might seem odd that mapping the gene for cystic fibrosis and the commercial production of human insulin by bacteria were made possible by research into the way some bacteria can resist infection by viruses, but this is often how science progresses. The discovery of bacterial enzymes that inactivate invading viruses by cutting the virus's DNA into pieces was the beginning of recombinant DNA technology.

In the mid-1970s, Hamilton Smith and Daniel Nathans discovered a number of bacterial enzymes that attach to DNA molecules and cut both strands of DNA at specific nucleotide sequences. In bacteria, the enzymes cut up the DNA of invading viruses, protecting the bacteria against lethal viral infections. These enzymes, called **restriction enzymes,** are a key component in recombinant DNA technology. A restriction enzyme attaches to DNA and moves along the molecule until it finds a specific nucleotide sequence called a recognition site. Once at the recognition site, the enzyme cuts through both strands of the DNA. Each type of restriction enzyme (from different bacteria) has its own recognition and cutting site. The recognition and cutting site for an enzyme isolated from *Escherichia coli,* a bacterium that lives in the human intestine, is shown in Figure 12.6. This recognition and cutting sequence is called a **palindrome,** because it reads the same on either DNA strand (when read in the 5′ to 3′ direction). Words like mom, pop, and radar are palindromes, as are phrases such as "Todd erases a red dot" or "live not on evil." Some restriction enzymes, like *Eco*RI, don't cut straight across the

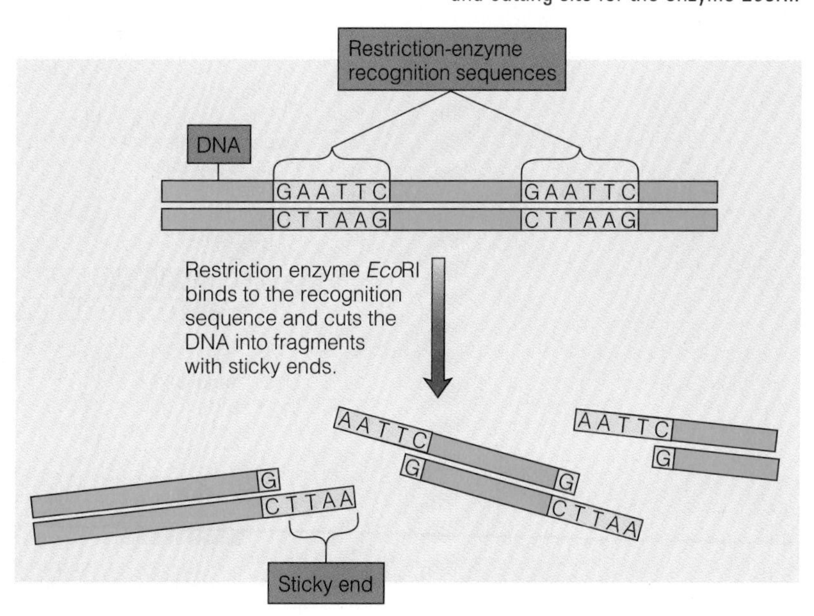

FIGURE 12.6 The recognition and cutting site for the enzyme *Eco*RI.

DNA. Instead they cut the strands at different places, creating single-stranded tails. These single-stranded tails are "sticky" and can reassociate with other DNA molecules with complementary (if necessary, review complementary base pairing in Chapter 8) tails to produce recombinant DNA molecules (Figure 12.7). Restriction enzymes cut DNA molecules into fragments. If DNA fragments from different organisms are mixed together, they can form new combinations, called recombinant DNA molecules. The recognition and cutting sites for several restriction enzymes are shown in Figure 12.8.

Vectors Serve as Carriers of DNA

Vectors Self-replicating DNA molecules that are used to transfer foreign DNA segments between host cells.

Vectors carry DNA fragments into cells where they are copied (cloned). Many of the vectors used in recombinant DNA work are derived from self-replicating, circular DNA molecules called plasmids, which are found in the cytoplasm of bacterial cells

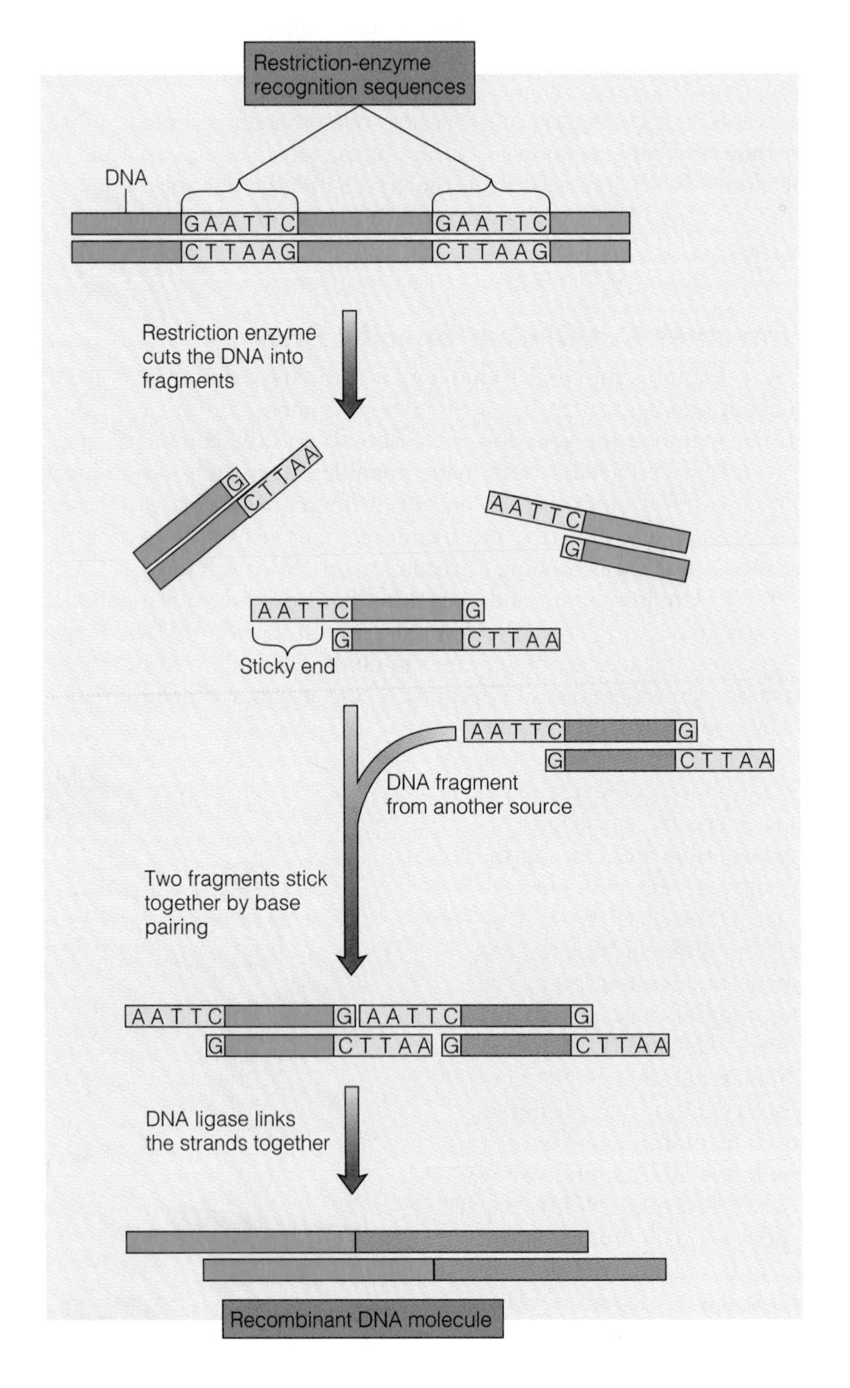

FIGURE 12.7 Recombinant DNA molecules can be created using DNA from two different sources, a restriction enzyme, and DNA ligase. DNA from each source is cut with the restriction enzyme, and the resulting fragments are mixed. In the mixture, the complementary ends of the two types of DNA will associate by forming hydrogen bonds. These linked fragments can be covalently joined by treatment with DNA ligase, creating recombinant DNA molecules.

(▸ Figure 12.9). A map of one such plasmid, pBR322, is shown in ▸ Figure 12.10. The middle section of the diagram shows the position of sites recognized and cut by restriction enzymes that can be used to insert DNA molecules to be cloned.

Steps in the Process of Cloning DNA

Cloning a DNA molecule involves several steps. Let's look at an example. If DNA molecules from a plasmid vector and from human cells are cut with *Eco*RI and then placed together in solution, their cut ends can reassociate, or anneal, to form re-

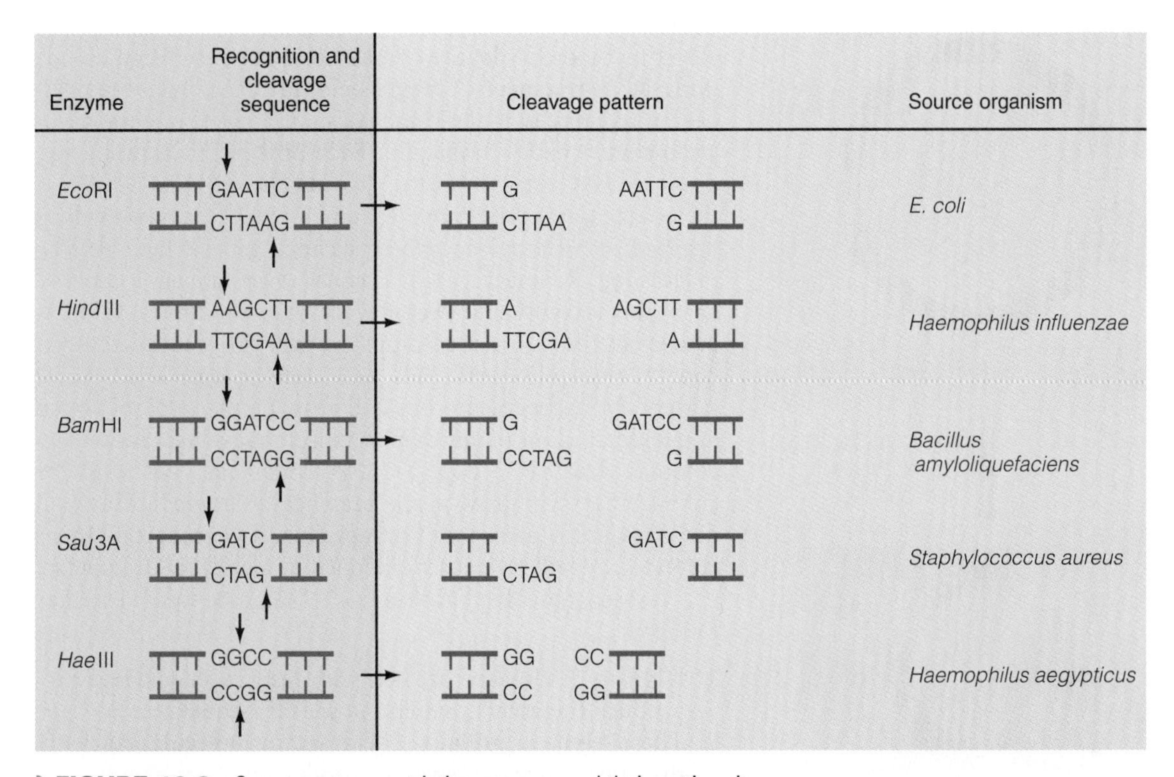

▸ **FIGURE 12.8** Some common restriction enzymes and their cutting sites.

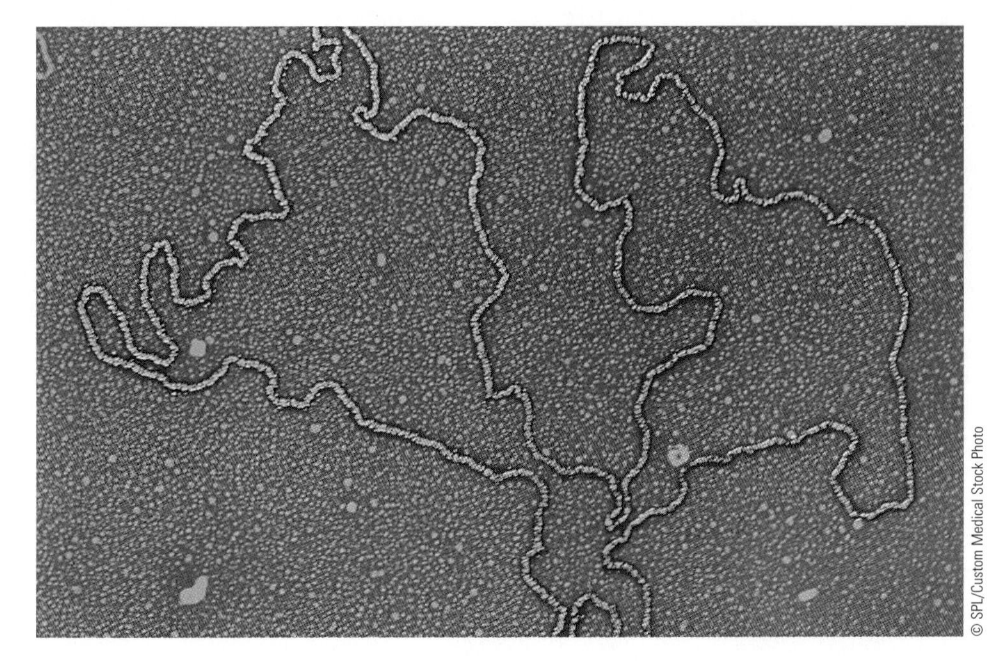

▸ **FIGURE 12.9** A plasmid isolated from a bacterial cell. Plasmids are used as vectors for cloning DNA.

combinant DNA molecules (Figure 12.11). These molecules are only held together by hydrogen bonding. To link the molecules together, an enzyme called DNA ligase is added, creating recombinant DNA molecules composed of human DNA and vector DNA. To make many copies of a recombinant molecule, it is transferred into a bacterial host cell. The host cell is grown on a nutrient plate, where it divides to form a colony (Figure 12.12). Because the cells in each colony are derived from a single ancestral cell, all cells in the colony and the plasmids they contain are clones.

How do we know that a colony actually contains a plasmid, and how do we know if that plasmid carries a piece of the human DNA we want to clone? Scientists use several clever methods to identify colonies that contain plasmids carrying DNA fragments. Our plasmid, pBR322, has been engineered to carry two antibiotic-resistance genes, one for tetracycline and one for ampicillin. Each of these resistance genes also carries a restriction site. If we cut the pBR322 DNA and the human DNA with *Bam*HI, human DNA fragments will be inserted into the tetracycline-resistance gene, which will be inactivated (Figure 12.13). Colonies formed from a cell that carry this recombinant vector will not grow if tetracycline is present in the nutrient medium but will grow on plates that contain ampicillin. If the host cells take up a pBR322 vector with no DNA fragment, they will grow on plates that contain ampicillin and on plates that contain tetracycline. So, to find colonies that carry human DNA fragments, we look for colonies that will grow in the presence of ampicillin but will not grow in the presence of tetracycline. If we collect all those colonies, we will have a set of cloned human DNA fragments.

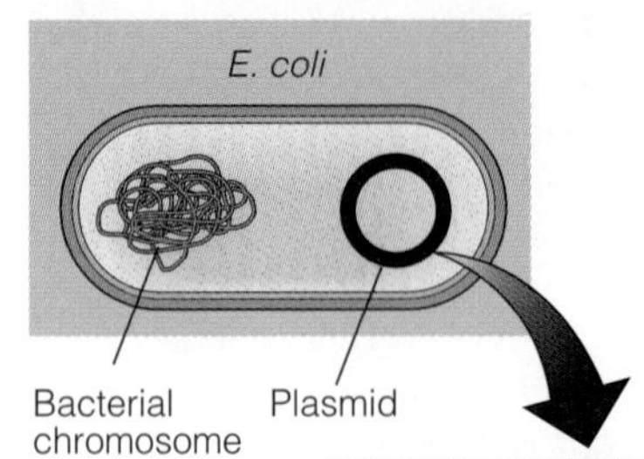

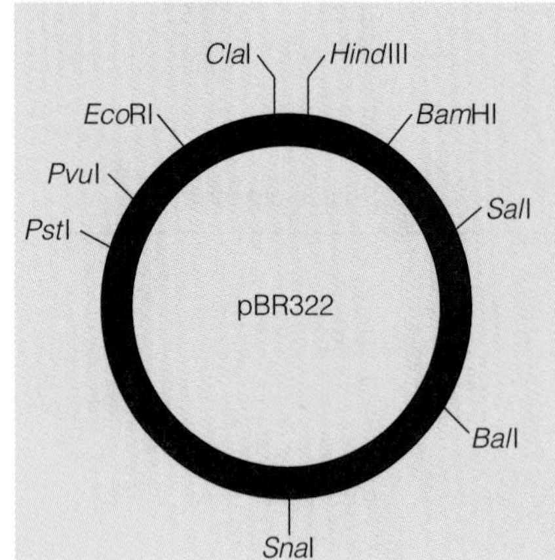

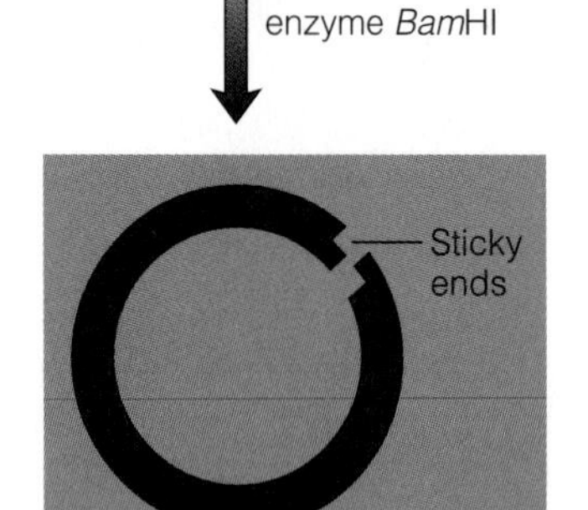

FIGURE 12.10 The plasmid pBR322 contains restriction sites used to carry DNA fragments for cloning.

The steps involved in cloning DNA can be summarized as follows:

1. DNA to be cloned is treated with restriction enzymes to produce fragments that end in specific sequences.
2. These segments are mixed with vector molecules cut with the same restriction enzyme, producing recombinant DNA molecules that are linked together by the enzyme DNA ligase.
3. Plasmids carrying DNA fragments are transferred into bacterial cells, where the recombinant plasmids replicate and produce many copies, or clones, of the recombinant DNA molecule.
4. Colonies carrying recombinant DNA molecules are identified and grown, the host cells are broken open, and the recombinant plasmids are extracted.

In our example, we can isolate the cloned human DNA by treating the recombinant plasmid with the same restriction enzyme used in cloning. The cloned human DNA can then be used in further experiments or transferred to new vectors and host cells to synthesize the protein encoded by the cloned human DNA.

Scientists who work in the field of recombinant DNA technology were among the first to realize that there might be unrecognized dangers in using and releasing recombinant organisms. Accordingly, as DNA cloning began, they called for a moratorium on all such work until safety issues were discussed and resolved (see Concepts and Controversies: Asilomar: Scientists Get Involved). After years of discussion and experimentation, there is general agreement that such work poses little risk, but the episode demonstrates that scientists are concerned about the risks and the benefits of their work.

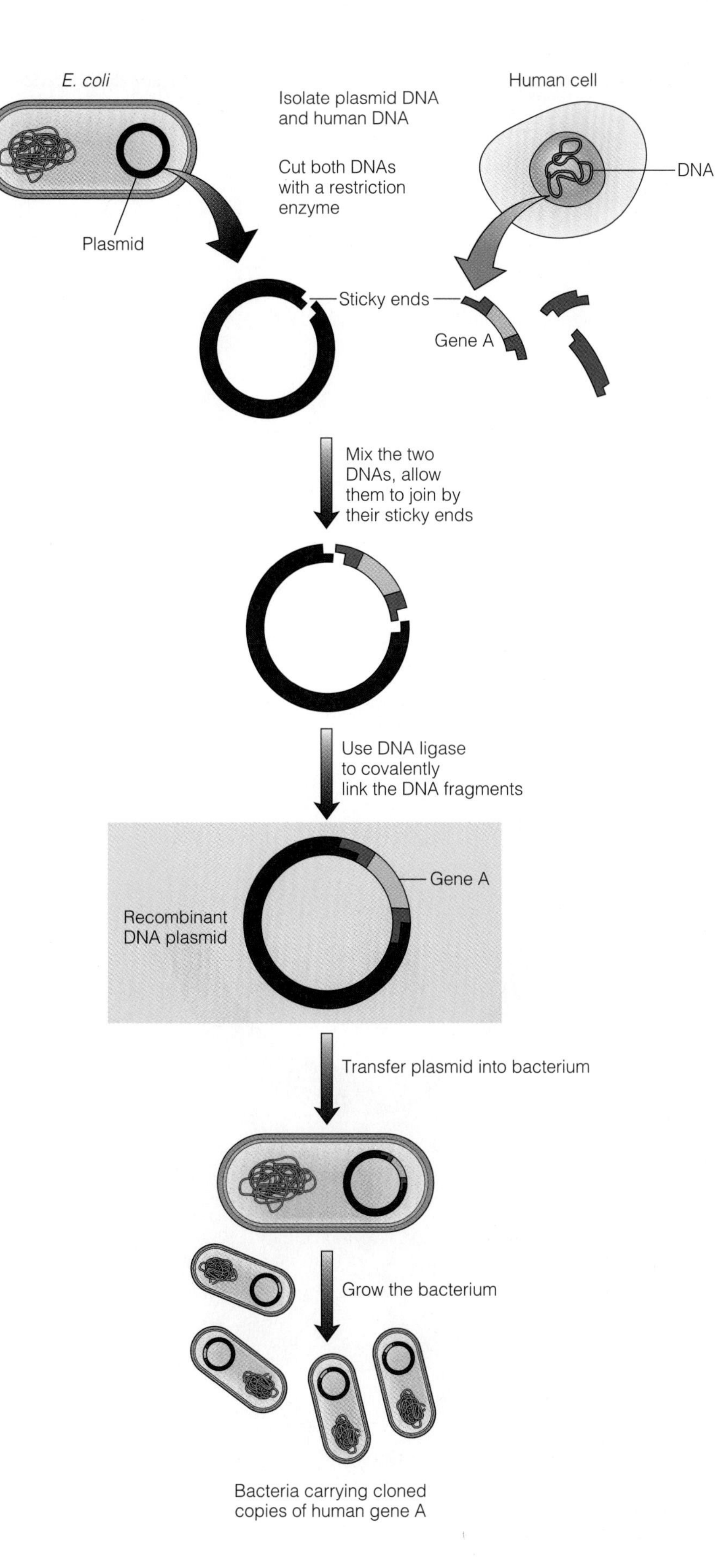

FIGURE 12.11 A summary of DNA cloning. Two types of DNA are isolated: vector DNA in the form of a plasmid and human DNA (containing gene A). Both DNAs are cut with the same restriction enzyme. The cut DNAs are mixed and, after pairing of the sticky ends, are linked together by an enzyme (DNA ligase). The recombinant plasmid DNA is inserted into a bacterial host cell. At each bacterial cell division, the plasmid is replicated, producing many copies, or clones, of the DNA insert carrying the A gene.

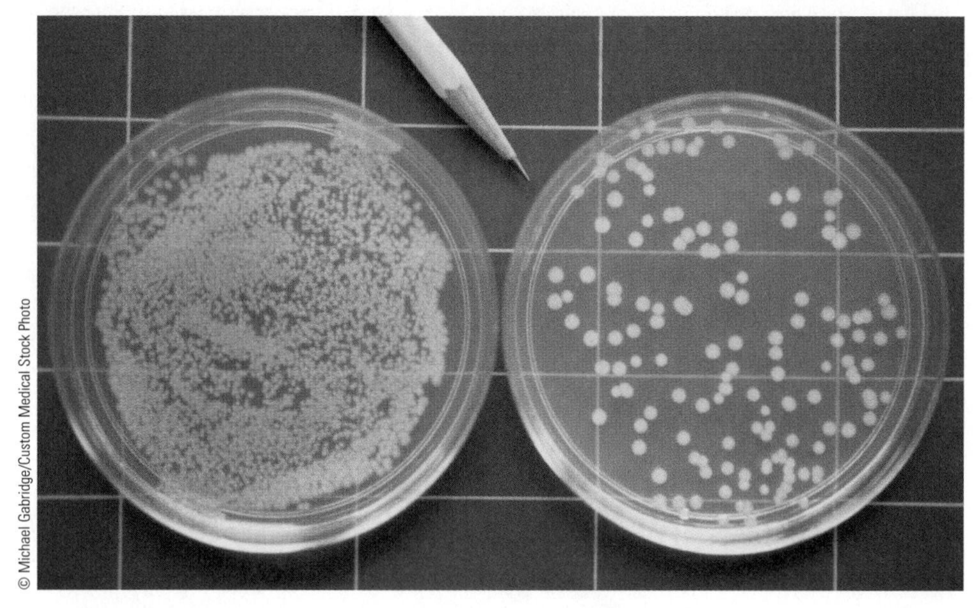

FIGURE 12.12 Colonies of bacteria on Petri plates. Each colony is descended from a single cell. Therefore, each colony is a clone.

FIGURE 12.13 The plasmid pBR322 carries two antibiotic-resistance genes, one for tetracycline and one for ampicillin. If DNA is inserted into the tetracycline gene, bacterial host cells that take up this plasmid will be resistant to ampicillin, but not tetracycline. Cells that have not taken up a plasmid will be killed by either antibiotic.

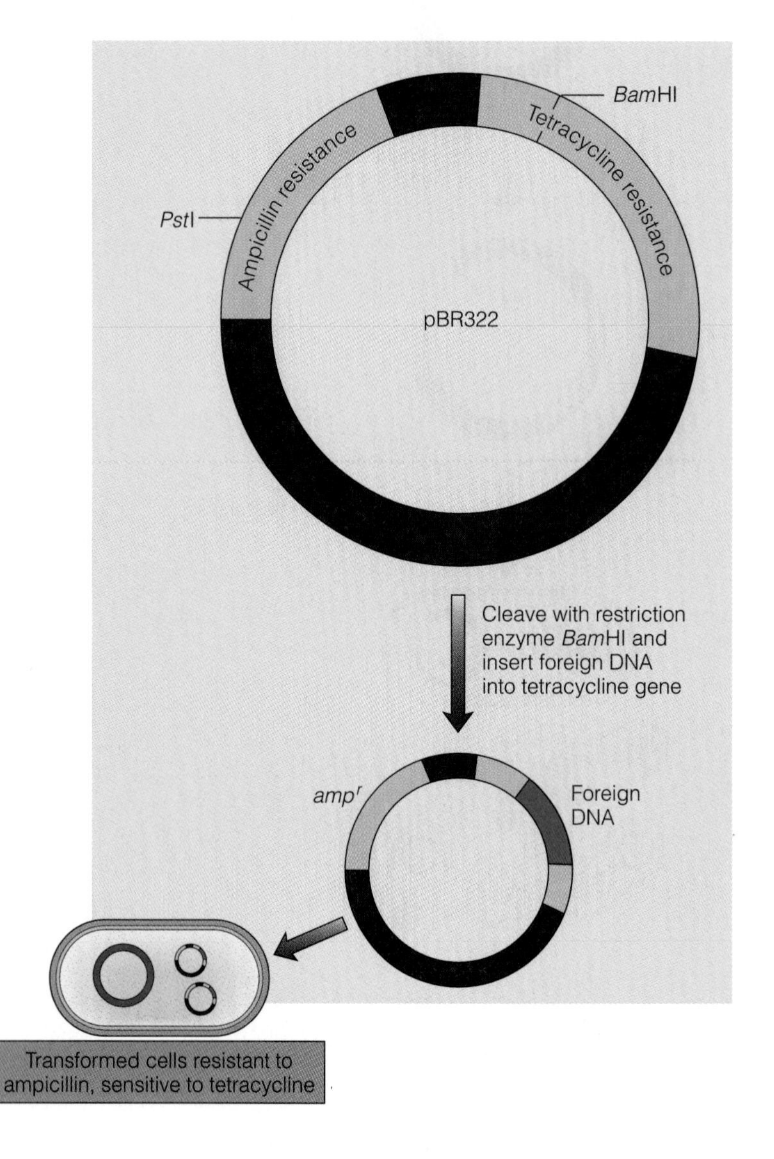

Concepts and Controversies

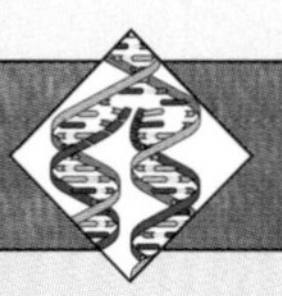

Asilomar: Scientists Get Involved

The first steps in creating recombinant DNA molecules were taken in 1973 and 1974. Scientists immediately realized that modifying the genetic information in *Escherichia coli,* a bacterium that lives in the human gut, could be potentially dangerous. A group of scientists asked the National Academy of Sciences to appoint a panel to assess the risks and the need to control recombinant DNA research. A second group published a letter in *Science* and in *Nature,* two leading scientific journals, calling for a moratorium on certain kinds of experiments until the potential hazards could be assessed. Shortly afterward, an international conference was held at Asilomar, California, to consider whether recombinant DNA technology poses any dangers and whether this form of research should be regulated. In 1975, a set of guidelines resulting from this conference was published under the direction of the National Institutes of Health (NIH), a government agency that sponsors biomedical research in the United States. In 1976, NIH published a new set of guidelines that prohibit certain kinds of experiments and dictated that other types of experiments were to be conducted only with appropriate containment to prevent the release of bacterial cells that carry recombinant DNA molecules. NIH also called for research to accurately assess the risks, if any, associated with recombinant DNA techniques.

In the meantime, legislation was proposed in Congress and at the state and local levels to regulate or prohibit the use of recombinant DNA technology. The federal legislation was withdrawn after exhaustive sessions of testimony and reports. By 1978, research had demonstrated that the common K12 laboratory strain of *E. coli* was much safer for use as a host cell than originally thought. Several projects showed that K12 could not survive in the human gut or outside a laboratory setting. Other work showed that recombinant DNA is produced in nature, and there are no detectable serious effects. In 1982, NIH issued a new set of guidelines that eliminated most of the constraints on recombinant DNA research. No experiments are currently prohibited. The most important lesson from these events is that the scientists who developed the methods were the first to call attention to the possible dangers of recombinant DNA research, and they did so based only on its potential for harm. There were no known cases of the release of recombinant DNA–carrying host cells into the environment. Scientists voluntarily shut down their research work until the situation could be properly and objectively assessed. Only when they reached a consensus that there was no danger did their work resume. Contrary to their portrayal in the popular media, scientists do care about the consequences of their work and do become involved in socially important issues.

CLONED LIBRARIES

Because each cloned fragment of human DNA is relatively small compared with the size of the genome, many clones are needed to hold all the genes carried by humans. A collection of clones that contains all of the DNA sequences (and therefore, all the genes) carried by an individual or a subset of those genes is called a **genetic library.** Libraries can carry all the genes of an individual (a genomic library), the genes from a single chromosome, or only those genes that are expressed in a certain cell type. In a genomic library, the number of clones needed to carry all the genes depends, of course, on the size of the genome and the size and number of fragments carried by the vectors in the library.

Genetic library In recombinant DNA terminology, a collection of clones that contains all of the genetic information in an individual. Also known as a gene bank.

If we made a human genomic library with cloned fragments averaging 1,700 nucleotide pairs in length (1.7 kilobases), we would need about 8.1 million plasmids to make sure our library included all the genetic information from one human cell. Other vectors, like **yeast artificial chromosomes (YACs)**, can carry much larger DNA fragments, and our human genomic library could be carried in just over 3,000 YACs. As we will see in the following sections, it is easier to search through 3,000 clones rather than 8 million clones when looking for a certain gene.

Yeast artificial chromosome (YAC) A cloning vector that has telomeres and a centromere that can accommodate large DNA inserts and that uses the eukaryote yeast as a host cell.

Using recombinant DNA technology, it is now possible to recover any and all genes carried by an organism. These genes can be used in research, screening for genetic disorders, gene therapy, or commercial applications in pharmaceutical or agricultural industries. Genomic libraries from many organisms are now available, including bacteria, yeasts, crop plants, many endangered species, and humans. Genomic libraries are the basic resource used in genome projects, a topic we will discuss in the following chapter.

FINDING A SPECIFIC CLONE IN A LIBRARY

Probe A labeled nucleic acid used to identify a complementary region in a clone or genome.

A genomic library can contain thousands or millions of different clones. Once a library is available, the difficulty is finding the clone that contains a gene of interest. Most often, a specific clone is identified using a labeled nucleic acid molecule called a **probe.** Probes are DNA or RNA molecules with a nucleotide sequence that is complementary to all or part of the gene of interest. Probes can be labeled with radioactivity or chemical groups that generate color or light to indicate the location of a specific clone. The ability of a probe to identify a gene of interest depends on the fact that, under the right conditions, two single-stranded nucleic acid molecules with a complementary base sequence will form a double-stranded hybrid molecule (▶ Figure 12.14). Any bacterial colony containing a cloned DNA fragment that hybridizes to the probe can be identified by the probe's label.

Probes can come from many different sources, including genes isolated from other species. For example, once a clone that carries part of the rat insulin gene was identified, it was used as a probe to search through a cloned library of human genes to find the human insulin gene, because the two genes have very similar nucleotide sequences. If a gene is expressed at high levels in certain cells, the mRNA from those cells can be used as a probe. For example, hemoglobin is almost the only protein made in red blood cells. To make a probe for the globin genes, mRNA was isolated and purified from these cells and used to prepare a probe. These RNA-derived probes were used to recover the human globin genes from a cloned library.

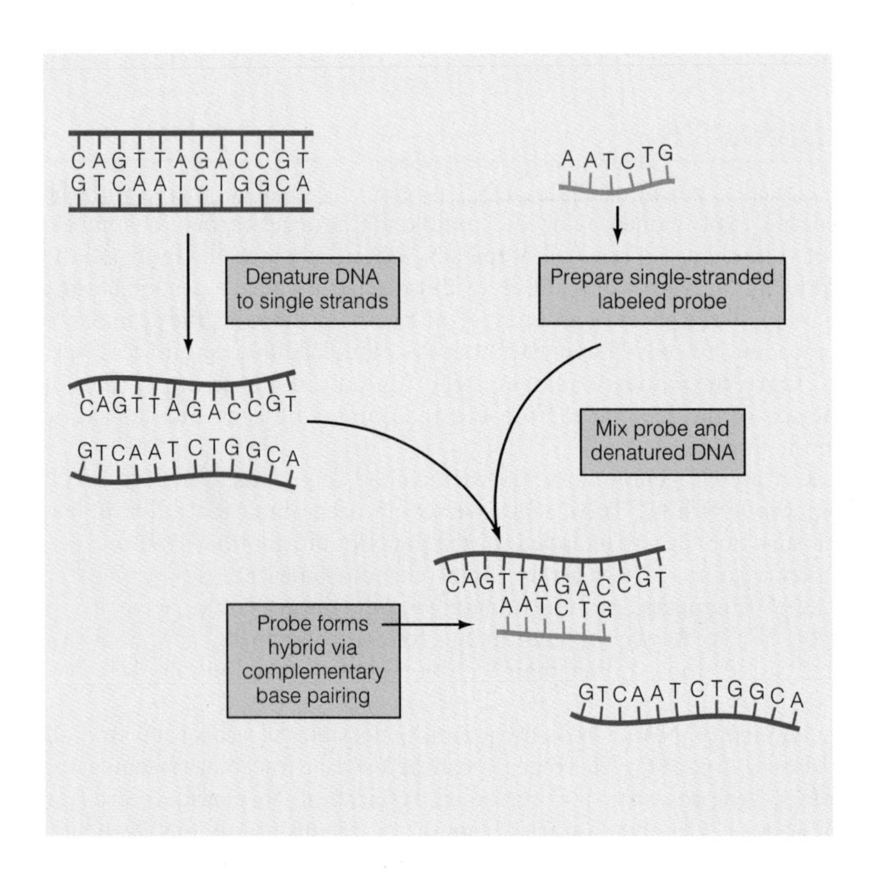

▶ **FIGURE 12.14** A DNA probe is a single-stranded molecule labeled for identification in some way. Both chemical and radioactive labels are used. The single-stranded probe forms a double-stranded hybrid molecule that has complementary regions of the DNA being studied.

A REVOLUTION IN CLONING: THE POLYMERASE CHAIN REACTION

Cloning using vectors and host cells is labor intensive and time consuming. Fortunately, it is not the only way to clone DNA molecules. A technique called the **polymerase chain reaction (PCR)**, invented in 1986, has revolutionized, and in some cases replaced, host cell cloning.

Polymerase chain reaction (PCR) A method for amplifying DNA segments that uses cycles of denaturation, annealing to primers, and DNA-polymerase directed DNA synthesis.

PCR uses single-stranded DNA as a template to make a complementary strand using the enzyme DNA polymerase. This is similar to the way DNA replication works in the cell nucleus. (See Chapter 8 for a discussion of DNA replication.) Once made, the double-stranded molecule is separated into single strands, and each is used to copy a new strand. After every round of replication, the double-stranded products are separated and used as templates. In this way, a DNA fragment can be amplified into thousands or millions of copies. There are several steps in PCR:

1. The DNA to be amplified is heated to break the hydrogen bonds between the two polynucleotide strands, producing two single strands (Figure 12.15).
2. Short nucleotide sequences that act as primers for DNA replication are added, and these primers bind to complementary regions on the single-stranded DNA fragments. Primers can be synthesized in the laboratory and are usually 20 to 30 nucleotides long.
3. The enzyme DNA polymerase is added. It begins at the primers and synthesizes a DNA strand complementary to the region between the primers, a process called primer extension.

This set of three steps is known as a PCR cycle. The cycle is repeated by heating the mixture and separating the double-stranded DNA fragments into single

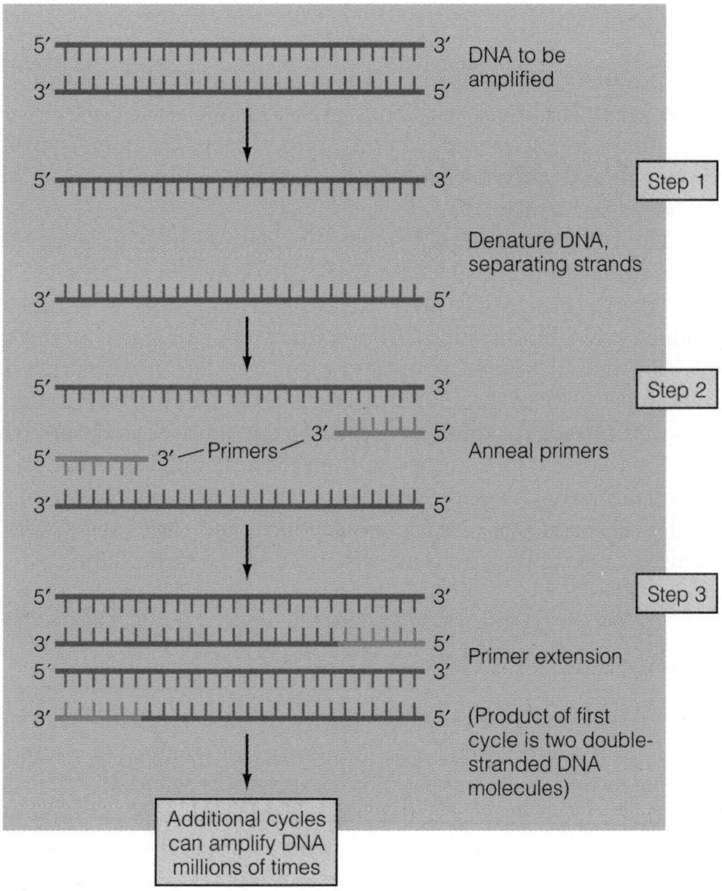

FIGURE 12.15 The polymerase chain reaction (PCR). In step 1, the DNA to be amplified is converted into single strands. In step 2, short single-stranded primers pair with the nucleotides adjacent to the region to be amplified. These primers are made synthetically and are complementary to the nucleotide sequence that flanks the region to be amplified. In step 3, enzymes and nucleotides for DNA synthesis are added, and the primers are extended, forming a double-stranded DNA molecule. The primers are incorporated into the newly synthesized strand. This series of three steps is called a cycle. In the second cycle, the double-stranded molecules produced in the first round of PCR are denatured to form single strands, which are templates for primer binding and another round of DNA synthesis. Repeated cycles can amplify the original DNA sequence by more than a million times.

Table 12.1

DNA Sequence Amplification by PCR

Cycle	Number of Copies
0	1
1	2
5	32
10	1,024
15	32,768
20	1,048,576
25	33,544,432
30	1,073,741,820

FIGURE 12.16 Insects embedded in amber are the oldest organisms from which DNA has been extracted and cloned.

strands, each of which can serve as a template. The end result is that the amount of DNA present doubles with each cycle. After n cycles, there is a 2^n increase in the amount of double-stranded DNA (Table 12.1). The power of PCR is that DNA can be cloned in a test tube instead of a host cell, and millions of copies can be made in hours rather than weeks.

The DNA to be amplified by PCR does not have to be purified and can be present in minute amounts; even a single DNA molecule can serve as a template. DNA from many sources has been used as starting material for the PCR technique, including dried blood, hides from extinct animals such as the quagga (a zebralike African animal hunted to extinction in the late nineteenth century), single hairs, mummified remains, and fossils. To date, the oldest DNA used in the PCR technique has been extracted from insects preserved in amber for about 30 million years (Figure 12.16). The DNA amplified from these samples is being used to study how specific genes have changed over long stretches of evolutionary time. PCR is also used in clinical diagnosis, forensic applications, and other areas, including conservation. Some of these applications are discussed in the next chapter.

ANALYZING CLONED SEQUENCES

Once a cloned sequence containing all or part of a gene has been identified and selected from a library, it can be used for many things. For example, it can be used as a probe to find and study regulatory sequences on adjacent chromosome regions, to investigate the internal organization of the gene, and to study its expression in cells and tissues. For these studies, geneticists routinely use several methods, some of which are outlined in the following sections.

Southern Blotting

Edward Southern discovered a way of using cloned DNA fragments that have been separated by size (using a technique called electrophoresis). Once separated, the fragments are transferred to filters and screened with probes. This procedure, known as a **Southern blot,** has many applications. It is used to find differences in the organization of normal and mutant alleles, to identify related genes in other organisms, and to study the evolution of genes. To make a Southern blot, DNA is extracted and cut into fragments with one or more restriction enzymes. A solution containing the fragments is placed in a gel made of agarose. The DNA fragments are then separated by gel electrophoresis. In this technique, an electric current is passed through the gel. Because the DNA fragments have a negative charge, they migrate through the gel toward the positive pole. The smaller fragments migrate faster than the larger fragments, and the DNA is separated by size. After the fragments are separated, the gel is stained to show the fragments and photographed (Figure 12.17). The gel is chemically treated to convert the DNA into single-stranded fragments. The single-stranded fragments are transferred to a sheet of DNA-binding membrane (made from nitrocellulose or a nylon derivative).

Southern blot A method for transferring DNA fragments from a gel to a membrane filter, developed by Edward Southern for use in hybridization experiments.

To transfer the fragments, the membrane is placed on top of the gel, which is in contact with a buffer solution. Paper towels and a weight are placed on top of the membrane, and buffer flows up through the gel and the nylon membrane by capillary action. As the buffer moves from the gel to the membrane, the DNA fragments are transferred from the gel to the membrane. The DNA fragments stick to the membrane, while the buffer passes through to the paper towels. A thick sponge under the gel acts as a wick to allow buffer to flow through the gel (Figure 12.18).

Once the transfer is completed, the membrane is placed in a food storage bag with the labeled, single-stranded probe that corresponds to the gene of interest. The bag is heat-sealed, and the hybridization reaction occurs. Only those single-stranded DNA fragments embedded in the membrane that are complementary to the base sequence of the probe will form hybrids. The excess and unbound probe is washed away, and the hybridized fragments are visualized. For radioactive probes, a piece of x-ray film is placed next to the filter. The radioactivity in the probe exposes parts of the film. The film is developed, and a pattern of one or more bands is seen (Figure 12.19). These patterns are analyzed and compared with patterns from other experiments.

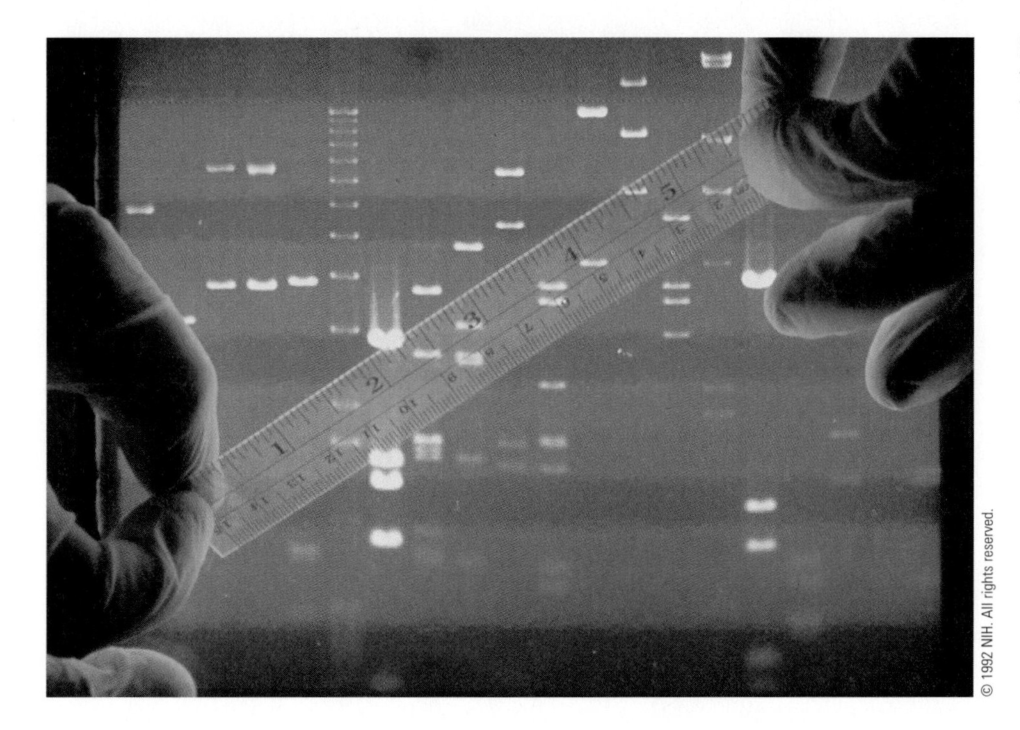

FIGURE 12.17 A gel stained to show the separation of restriction fragments by electrophoresis.

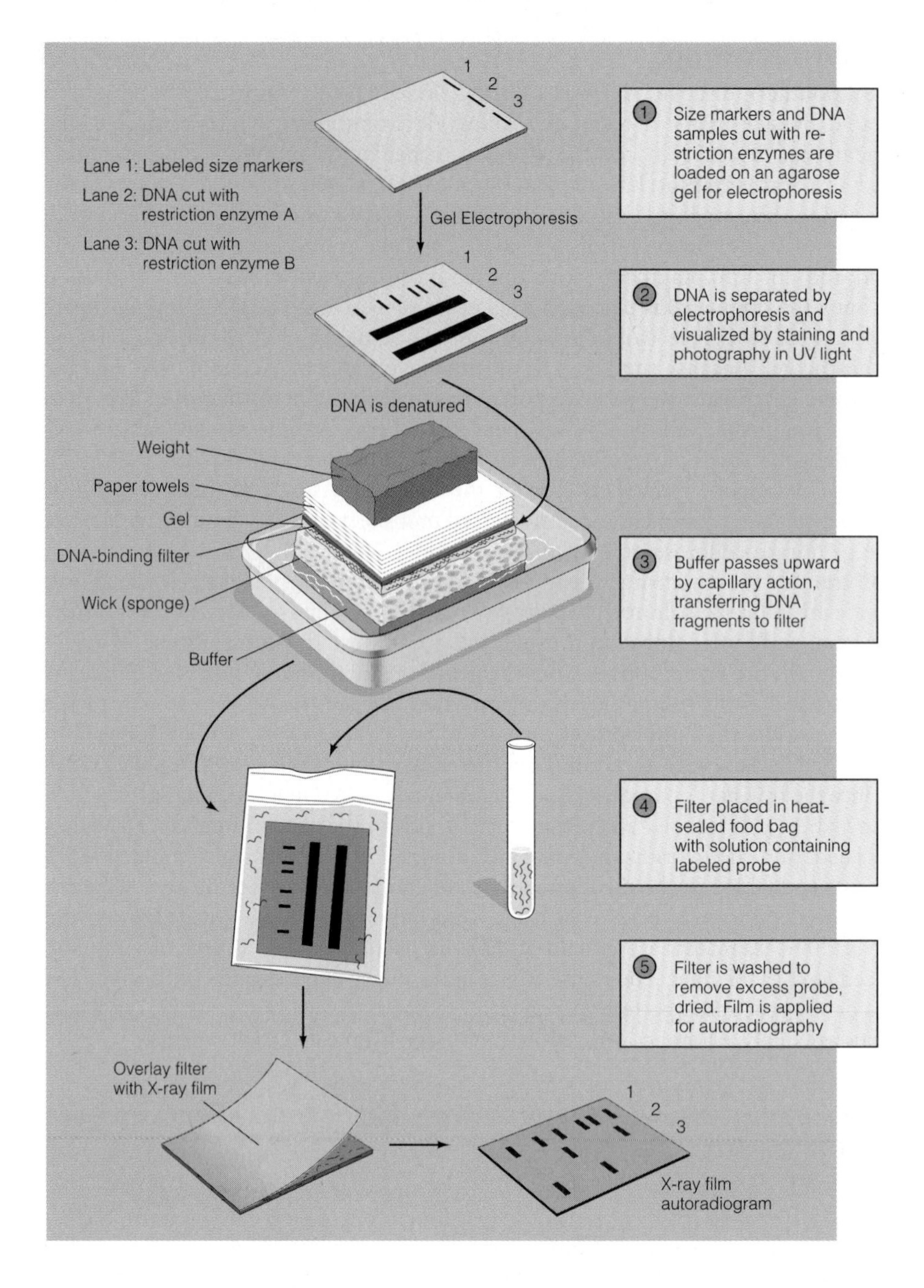

FIGURE 12.18 The Southern blotting technique. DNA is cut with a restriction enzyme, and the fragments are separated by gel electrophoresis. The DNA in the gel is denatured to single strands, and the gel is placed on a sponge that is partially immersed in buffer. The gel is covered with a DNA-binding membrane, layers of paper towels, and a weight. Capillary action draws the buffer up through the sponge, the gel, the DNA-binding membrane, and the paper towels. This movement of buffer transfers the pattern of DNA fragments from the gel to the membrane. The membrane is placed in a heat-sealed food bag with the probe and a small amount of buffer. After hybrids have been allowed to form, the excess probe is washed away, and the regions of hybrid formation are visualized by overlaying the membrane with a piece of x-ray film. After development, regions of hybrid formation appear as bands on the film.

DNA Sequencing

The ability to determine the nucleotide sequence of DNA has greatly enhanced our understanding of gene organization, the regulation of gene expression, and the evolutionary history of genes. It is also the basic method in genome projects, which seek to determine the nucleotide sequence of an entire genome. The success of the Human Genome Project and hundreds of other genome projects is derived from advances in the technology of **DNA sequencing.** There are several ways of sequencing DNA; the most widely used method has been automated and uses a different-colored fluorescent probe for each of the four bases in DNA. After a sequencing reaction is completed, a laser scanner then reads the results, displaying the sequence as a series of colored peaks (Figure 12.20). Banks of sequencing machines are used in large-scale sequencing projects such as the Human Genome Project. Each machine can sequence several hundred thousand nucleotides per day.

DNA sequencing A technique for determining the nucleotide sequence of a fragment of DNA.

Guest Essay: Research and Applications

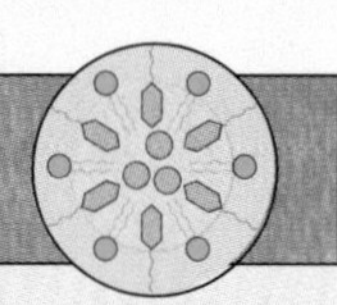

The Human Genome Project: Reading Our Own Genetic Blueprint

Francis Sellers Collins

I developed an early love for mathematics and science and, after learning about the elegant features of the structure of the atom and the chemical bond in a high school chemistry class, I became convinced that I wanted to be a chemist. Accordingly, I majored in chemistry at the University of Virginia, and immediately after graduation went into a Ph.D. program in physical chemistry at Yale, never glancing around to notice what other exciting areas of science I might be missing. Almost by accident, however, I was introduced to molecular biology. To my surprise, I learned that all of my biases about biology being chaotic, intellectually unsatisfying, and devoid of the principles that I prized so much in the physical sciences were really quite unjustified. In fact, it was clear that biology was poised on the threshold of a remarkable revolution, as it became possible to analyze the information content of DNA and the nature of the genetic code.

After a somewhat prolonged and tortuous route, including medical school, I resolved to pursue a career that combined clinical practice and research in medical genetics. Eventually, I got a "real job" as a junior faculty member at the University of Michigan, splitting my time between organizing a research program in human genetics and seeing patients who had genetic diseases. Often frustrated by the lack of information available about the fundamental basis of many genetic diseases, I was particularly drawn to those disorders where the action of a single mutant gene caused a great deal of sickness and distress, but where the cell biology had not been worked out sufficiently to allow direct identification of the responsible gene. Cystic fibrosis (CF) was a particularly puzzling and heart-breaking example of this situation. In collaboration with another researcher in Toronto, my research group began an intense effort to try to identify the CF gene by mapping it to the proper chromosome and then narrowing down the region where the gene must be located until finally the correct candidate was identified. This process, known as positional cloning, had never really succeeded for a problem of this complexity, so it was both gratifying and sobering to participate in such an effort. Many times we thought the problem was just too difficult and were tempted to give up. But finally the strategy worked. The successful cloning of the cystic fibrosis gene in 1989 made better diagnosis available and opened new doors to the design of therapies, including gene therapy. After this, similar efforts in which my research group participated yielded the genes for other puzzling human genetic disorders, including neurofibromatosis (often erroneously referred to as the Elephant Man disease) and Huntington disease.

At about the same time, an international effort to map and sequence all of the human DNA, known as the Human Genome Project, was getting under way. This would make it possible to find all of the genes responsible for human disease, and so I was happy to have the chance to participate by setting up a Human Genome Center at the University of Michigan. When the first U.S. director of the Human Genome Project, Dr. James Watson (the same Watson who with Francis Crick discovered the structure of DNA in 1953) resigned, I was asked to take on this role. Since 1993 I have been director of the National Center for Human Genome Research of the National Institutes of Health, which is the lead agency in the United States responsible for this ambitious and historic effort to map and sequence all of human DNA by the year 2005. So far the project is doing very well and is actually running ahead of schedule and under budget.

Although my own career path was far from focused, I can now look back on all the experiences that I had in science and medicine and see how they helped prepare me for overseeing this remarkable project.

Francis Sellers Collins has been the director of the National Center for Human Genome Research of the National Institutes of Health since 1993. In addition to numerous honors and awards, certifications, and society memberships, he is an associate editor of a number of scientific journals in the area of genetics. He received a Ph.D. from Yale University and an M.D. with honors from the University of North Carolina School of Medicine, Chapel Hill.

DNA sequencing provides information about the size and organization of genes and the nature and function of the gene products they encode. Analysis of mutant genes provides information about how genes are altered by mutational events. Sequencing is also used to study the evolutionary history of genes and species-to-species variation.

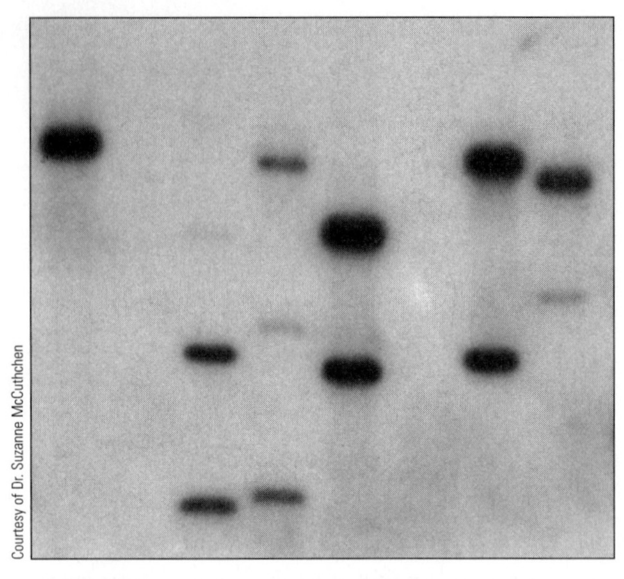

Courtesy of Dr. Suzanne McCuthchen

FIGURE 12.19 The pattern of bands after hybridization on a Southern blot. All bands contain sequences complementary to the nucleotide sequence of the probe.

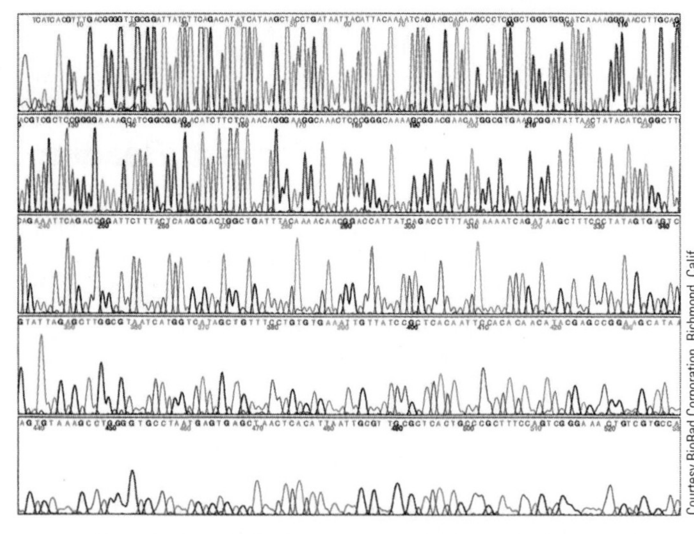

Courtesy BioRad Corporation, Richmond, Calif.

FIGURE 12.20 Using fluorescent dyes and robotic technology, DNA sequencing has been automated. The results are displayed in color, one for each of the four nucleotides in DNA. The results are read from left to right.

Summary

What Are Clones?

1. Cloning is the production of identical copies of molecules, cells, or organisms from a single ancestor. The development and refinement of methods for cloning higher plants and animals represents a significant advance in genetic technology that will speed up the process of improving crops and the production of domestic animals.

Cloning Genes Is a Multistep Process

2. These developments have been paralleled by the discovery of methods to clone segments of DNA molecules. This technology is founded on the discovery that a series of enzymes, known as restriction endonucleases, recognize and cut DNA at specific nucleotide sequences. Linking DNA segments produced by restriction-enzyme treatment with vectors, such as plasmids or engineered viral chromosomes, produces recombinant DNA molecules.
3. Recombinant DNA molecules are transferred into host cells, and cloned copies are produced as the host cells grow and divide. A variety of host cells can be used, but the most common is the bacterium *E. coli*. The cloned DNA molecules can be recovered from the host cells and purified for further use.

Cloned Libraries

4. A collection of cloned DNA sequences from one source is a library. The clones in the library are a resource for work on specific genes.

Finding a Specific Clone in a Library

5. Clones for specific genes can be recovered from a library by using probes to screen the library.

A Revolution in Cloning: The Polymerase Chain Reaction

6. PCR is used to make many copies of a DNA molecule without using restriction enzymes, vectors, or host cells. It is faster and easier than conventional cloning.

Analyzing Cloned Sequences

7. Cloned sequences are characterized in several ways, including Southern blotting and DNA sequencing.

Case Studies

CASE 1

Cloning an animal by nuclear transfer was first done in 1952 using frogs. This and later work showed that the first few cell divisions after fertilization produce cells with nuclei that can form complete embryos when transferred into unfertilized eggs. As an embryo develops further, the cells lose this property, and the success of nuclear transfer rapidly declines. Nuclear transfer in mammals proved to be more difficult. In the 1980s several groups developed transgenic livestock and genetically modified pigs were considered as sources of organs for transplantation to human patients.

If animals can be derived from cells in culture, it should be possible to make genetic modifications, including the removal or addition of specific genes. This has been achieved in mice using embryonic stem (ES) cells, but to date no one has been successful in obtaining ES cells from cattle, sheep, or pigs. Dr. Ian Wilmut at the Roslin Institute in Edinburgh, Scotland, thought that nuclear transfer might provide an alternative to ES cells. A major breakthrough came in 1995 when Wilmut and colleagues produced lambs by nuclear transfer from cells of early embryos.

So, how was Dolly produced? Dolly was the first mammal cloned using a nucleus from an adult animal. Cells were collected from the udder of a 6-year-old ewe and cultured for several weeks in the laboratory. Individual cells were then fused with unfertilized eggs from which the nuclei had been removed. Two hundred seventy-seven of these reconstructed eggs—each now with a diploid nucleus from the adult animal—were cultured for 6 days in temporary recipients. Twenty-nine of the eggs appeared to have developed to the blastocyst stage and were implanted into surrogate Scottish Blackface ewes. One gave rise to a live lamb, Dolly, some 148 days later. Dolly was born on July 5, 1996.

1. Why did the birth of Dolly stun the scientific world, considering that cloning of animals was already taking place?
2. What animal cloning has taken place since Dolly? Have any advancements been made in the cloning technique?
3. Because adult cloning can be done in sheep, does that mean it can be done in humans?
4. What policies or laws need to be in place to regulate animal cloning?

CASE 2

Success in cloning animals has generated controversy over whether or not to clone humans, in the form of embryos or adults. Researchers at the Andrology Institute of America plan to clone humans as a way of helping infertile couples have children. Another organization, Clonaid, is working to clone a human using the nucleus of a cell from a baby that died after having surgery. Others have proposed research programs in what is being called therapeutic cloning. The goal is to create stem-like cells that can be directed to differentiate into functional tissues that can be used in transplantation. Reprogramming a differentiated adult nucleus will be accomplished by transferring it to an unfertilized egg from which the nucleus has been removed. The nuclear-transplanted eggs are grown to the blastocyst stage. However, instead of transferring the embryos for development, the embryo's stem cells are collected. The stem cells will be grown in the lab and, in the presence of growth factors, will differentiate into specialized cells, such as liver cells, kidney cells, or insulin-producing cells of the pancreas. Adding to the ongoing debate about human cloning and embryonic stem cell research, therapeutic cloning will not use embryos left over from IVF procedures, but requires the creation of an embryo that will be destroyed.

1. What are the ethical concerns in cloning human beings? In therapeutic cloning?
2. What diseases could be treated with therapeutic cloning?
3. If therapeutic cloning is allowed for the treatment of disease, will a "slippery slope" occur, leading us to cloning human beings?
4. What laws must be put in place to regulate human cloning? Who should decide what laws are adopted?

Questions and Problems

What Are Clones?

1. Cloning is a general term that is used for whole organisms and single genes. Define what we mean when we say we have a clone.
2. Nuclear transfer to clone cattle is done by which of the following techniques?
   a. An eight-cell embryo is divided into two four-cell embryos and implanted into a surrogate mother.
   b. A 16-cell embryo is divided into 16 separate cells and these cells are allowed to form new 16-cell embryos that are directly implanted into surrogate mothers.
   c. A two-cell embryo is divided into two separate cells and implanted into a surrogate mother.
   d. A 16-cell embryo is divided into 16 separate cells and fused with enucleated eggs. The fused eggs are then implanted into surrogate mothers.
   e. None of the above.
3. Dolly made headlines in 1996 when she was born. Why was her cloning so important, considering that cattle and sheep were already being cloned through embryo splitting?

Cloning Genes Is a Multistep Process

4. What is meant by the term recombinant DNA?
   a. DNA from bacteria and viruses
   b. DNA from different sources that are not normally found together
   c. DNA from restriction enzyme digestions
   d. DNA that can make RNA and proteins
   e. none of the above
5. Restriction enzymes:
   a. recognize specific nucleotide sequences in DNA.
   b. cut both strands of DNA.
   c. often produce single-stranded tails.
   d. do all of the above.
   e. do none of the above.
6. Restriction enzymes are derived from bacteria. Why don't bacterial chromosomes get cut with the restriction enzymes present in the cell? Why do bacteria have these enzymes?
7. The DNA sequence below contains a six-base palindromic sequence that acts as a recognition and cutting site for a restriction enzyme. What is this sequence? Which enzyme will cut this sequence?
   5′ CCGAGTAAGCTTAC 3′
   3′ GGCTCATTCGAATG 5′
8. Why is it an advantage that many restriction enzyme recognition sites are palindromes? Remember that restriction enzymes bind DNA and move along the molecule, looking for a recognition site.
9. Assume restriction enzyme sites as follows: E = *Eco*RI, H = *Hin*dIII, and P = *Pst*I. What size bands would be present when the DNA shown below is cut with
   a. *Eco*RI
   b. *Hin*dIII and *Pst*I
   c. all three enzymes

   2 kb   5 kb   3 kb 3.5 kb   4 kb   6 kb
   ——E——H——P——E——H——
10. Which enzyme is responsible for covalently linking DNA strands together?
    a. DNA polymerase
    b. DNA ligase
    c. *Eco*RI
    d. restriction enzymes
    e. RNA polymerase
11. When cloning human DNA, why is it necessary to insert the DNA into a vector such as a bacterial plasmid?
12. Briefly describe how to clone a segment of DNA. Start with cutting the DNA of interest with a restriction enzyme.

Cloned Libraries

13. A cloned library of an entire genome contains
    a. the expressed genes in an organism.
    b. all the genes of an organism.
    c. only a representative selection of genes.
    d. a large number of alleles of each gene.
    e. none of the above.
14. You are given the task of preparing a cloned library from a human tissue culture cell line. What type of vector would you select for this library and why?

A Revolution in Cloning: The Polymerase Chain Reaction

15. The steps in the polymerase chain reaction (PCR) are:
    a. breaking hydrogen bonds, annealing primers, and synthesizing using DNA polymerase.
    b. restriction cutting, annealing primers, synthesizing using DNA polymerase.
    c. ligating, restriction cutting, and transforming into bacteria.
    d. DNA sequencing and restriction cutting.
    e. transcription and translation.
16. You are running a PCR to generate copies of a fragment of the cystic fibrosis (CF) gene. Beginning with two copies at the start, how much of an amplification of this fragment will be present after six cycles in the PCR machine?
17. Why is PCR so revolutionary? Describe two applications of PCR.

Analyzing Cloned Sequences

18. Name three of the many applications for Southern blotting.
19. A new gene has been isolated in mice that causes an inherited form of retinal degeneration. This disease leads to blindness in affected individuals. You are a researcher in a human vision lab and want to see if a similar gene

exists in humans (genes with similar sequences usually mean similar functions). With your knowledge of the Southern blot procedure, how would you go about doing this? Start with the isolation of human DNA.

20. A base change (A to T) is the mutational event that creates the mutant sickle cell anemia allele of beta globin. This mutation destroys an *Mst*II restriction site that is normally present in the beta globin gene. This difference between the normal allele and the mutant allele can be detected by Southern blotting. Using a labeled beta globin gene as a probe, what differences would you expect to see for a Southern blot of the normal beta globin gene and the mutant sickle cell gene?
21. What exactly is a DNA sequence of a gene?
22. What kind of information can a DNA sequence provide to a researcher studying a disease-causing gene?
23. The cloning methods outlined in this chapter allow researchers to generate many copies of a gene they wish to study through the use of restriction enzymes, vectors, bacterial host cells, the creation of genetic libraries, and PCR. Once useful quantities of a disease-causing gene are available by cloning, what kind of questions do you think should be asked about the gene?

Internet Activities

The following activities use the resources of the World Wide Web to enhance the topics covered in this chapter. To investigate the topics described below, log on to the book's home page (www.brookscole.com/biology_d) and follow the *Student Resources* link to your subject and text.

1. *Biotechnology Review.* At Cold Spring Harbor's *DNA Learning Center* (http://vector.cshl.org/), interactive animations of PCR, Southern blotting and cycle sequencing are provided. Review these techniques by selecting "Resources" and then "Biology Animation Library."
2. *Recombinant DNA problem set.* At the University of Arizona's *Biology Project: Molecular Biology* (http://www.biology.arizona.edu/molecular_bio/problem_sets/Recombinant_DNA_ Technology/recombinant_dna.html) you can work through an interactive problem set on recombinant DNA technology. Correct answers will be summarized; if you answer incorrectly, you will be provided with a tutorial to increase your understanding.
3. *Cloning in the News: A Critical Analysis.* At the *Access Excellence About Biotech* Web site (http://www.accessexcellence.org/AB/BA/casestudy/casestudy.html), Tom Zinnen presents an analysis of the accuracies and inaccuracies in news reporting about Dolly, the first cloned sheep. How did the news reports compare to the statements actually made in the research paper? What kinds of errors were made? What are some of the uncertainties that were not resolved in the paper itself?
4. *Conceiving a Clone.* At the student-created *Conceiving a Clone* Web site (http://library.think.quest.org/24355/) you can access details on cloning and biotechnology, compare different cloning techniques and view animation of cloning processes. Follow the "Interactions" link to create a clone on the Web, test your cloning knowledge, or participate in polls about cloning.

For Further Reading

Drlica, K. (1992). *Understanding DNA and gene cloning: A guide for the curious,* 2nd ed. New York: Wiley.

Jaenisch, R., & Wilmut, I. (2001). Developmental biology. Don't clone humans. *Science, 291,* 2552.

Jeffreys, A. J., Brookfield, J. F., & Semeonoff, R. (1985). Positive identification of an immigration test-case using human DNA fingerprints. *Nature, 317,* 818–819.

Mullis, K. B. (1990). The unusual origin of the polymerase chain reaction. *Sci. Am., 262,* 56–65.

Robertson, J. A. (2001). Human embryonic stem cell research: Ethical and legal issues. *Nat. Rev. Genet., 2,* 74–78.

Southern, E. M. (1975). Detection of specific sequences among DNA fragments separated by gel electrophoresis. *J. Mol. Biol., 98,* 503–517.

Watson, J. D., Gilman, M., Witkowski, J., & Zoller, M. (1992). *Recombinant DNA,* 2nd ed. New York: Scientific American Books.

Wilmut, I. (1998). Cloning for medicine. *Sci. Am., 279,* 58–63.

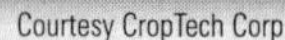
Courtesy CropTech Corp

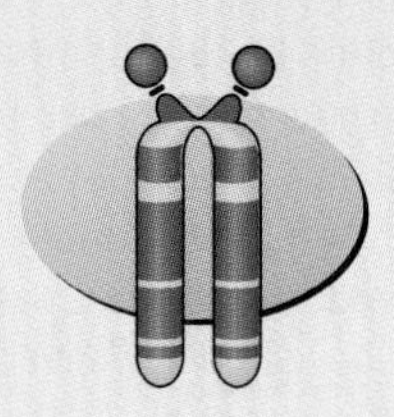

Biotechnology and Genomics

Chapter Outline

Approximately 17 to 20 million years ago, a magnolia leaf fluttered into the cold water of a lake in what is now Idaho. Magnolias and other plants found in warm, humid, temperate climates flourished along the shores of the lake, which was formed when volcanic lava flowed into the streambed of a valley. The leaf sank into the cold, still water down to the oxygen-poor bottom and quickly became covered with mud. Because of the lack of oxygen, the leaf did not deteriorate; instead, over a period of millions of years, it formed a compression fossil as the mud turned to shale.

On a day millions of years later, researchers in the twentieth century split the shale layers. The exposed fossil was still green, although its color faded rapidly in the oxygen-containing atmosphere. The researchers scraped the fossilized leaf into a mortar and ground it with dry ice to form a fine powder. From this powder, the DNA of the fossil leaf was extracted and purified. Using a recombinant DNA technique, the researchers isolated a chloroplast gene from the fossil leaf. This gene encodes the information for an enzyme that controls the production of a sugar that is an intermediate in photosynthesis.

The researchers determined the nucleotide sequence of this fossil gene and compared it with that of similar genes from present-day species of magnolias. This analysis indicated that although there are some differences in the sequence between the fossil gene and the contemporary gene, there has been little overall change as this gene has traveled through the intervening generations. The differences in nucleotide sequence between the fossil magnolia and present-day species were used to establish its relationship to the species we know today. The results indicate that the fossil species is closely related to one of the species of magnolia that grows in the eastern United States.

The ability to isolate and characterize a specific gene from a plant that lived approximately 17 to 20 million years ago and compare it to the gene from species alive today demonstrates the power of recombinant DNA technology. The data gathered in this experiment allow a direct measurement of the rate of mutation in an individual gene over evolutionary time and assist in establishing the evolutionary relationship between present species and their fossil ancestors.

Recombinant DNA technology is having a dramatic impact on many other areas, including molecular biology, pharmaceutical manufacturing, agriculture, diagnosis and treatment of genetic disorders, and criminal investigations. This revolution is just beginning, and the impact of recombinant DNA technology and the accompanying biotechnology industry will continue to spread and profoundly change the lives of many individuals. In this chapter, we discuss some of the changes this technology has brought about in medicine, genetics, and agriculture.

RECOMBINANT DNA TECHNIQUES HAVE REVOLUTIONIZED HUMAN GENE MAPPPING

If we wish to know how many genetic disorders humans have and how to develop treatments for these diseases, we need to map all of the genes in the human genome and find out what these genes do. Genetic mapping is therefore one of the basic activities of human genetics, as it is in other areas of genetics. There are several ways of mapping genes. The oldest method, discussed in Chapter 4, involves detecting linkage between genes (genetic evidence that two or more genes are on the same chromosome) and assigning these linked genes to individual chromosomes. Although genes can be assigned directly to the X chromosome by their unique pattern of inheritance, it is more difficult to map genes to individual autosomes. Only a small number of human genes have been mapped using this approach.

Several new methods of mapping human genes based on recombinant DNA techniques are now available. These methods have revolutionized mapping genes in humans and in organisms used in genetic research. Some of these methods can be used even if there is no information about the nature of the mutant gene or the gene product. We will discuss three of these methods.

RFLP mapping uses markers identified by recombinant DNA techniques and mapped to specific chromosomes. By searching for linkage between markers and a trait, a gene can be assigned to a chromosome. Using markers that cover smaller and smaller regions of a chromosome, a gene can be mapped with great accuracy.

A second method of mapping uses genes that have been successfully identified and cloned. The cloned gene is labeled with a fluorescent dye and directly mapped to a spot on a chromosome by a technique called **fluorescent *in situ* hybridization (FISH).**

Fluorescent *in situ* hybridization (FISH) A method of mapping DNA sequences to metaphase chromosomes by using probes labeled with fluorescent dyes.

The third method is genome sequencing. The genome of the organism is cloned, the clones are sequenced, and the sequence is assembled. Genes are identified by searching the sequence for coding regions. This is the method used in the Human Genome Project and other genome projects.

Finding Restriction Fragment Length Polymorphisms and Using Them as Genetic Markers

One method of gene mapping using recombinant DNA technology begins by identifying variations in the presence or absence of restriction-enzyme cutting sites. We all carry slight variations in the nucleotide sequence of our genomes, mostly in stretches of DNA between genes or in the noncoding introns within genes. These variations create differences in cutting sites for restriction enzymes by creating or destroying restriction enzyme sites. These variations are heritable, can be used as genetic markers, and can be assigned to specific chromosomes and chromosome regions. In turn, this creates a genetic map of variable sites covering the entire human chromosome set. Genetic disorders can then be mapped by showing linkage between these chromosome-specific markers and the disorder.

Recall from Chapter 12 that restriction enzymes recognize a specific nucleotide sequence in DNA and cut both strands of DNA at this sequence. Human DNA contains variations in the form of single nucleotide substitutions. This variation is not associated with any deleterious phenotype and, as mentioned previously, occurs mostly in regions between genes. In some cases this variation creates or destroys a cutting site for a restriction enzyme, altering the pattern of cuts made in the DNA. This creates a heritable difference in the length and number of DNA fragments generated by a restriction enzyme.

This difference is known as a **restriction fragment length polymorphism (RFLP).** RFLPs are heritable and are passed from generation to generation as codominant traits (review codominant inheritance in Chapter 3). In other words, RFLPs can serve as genetic markers. The phenotype of an RFLP is not externally visible, as in albinism, or biochemically detectable, as with a blood type. Rather, the phenotype

Restriction fragment length polymorphism (RFLP) Variations in the length of DNA fragments generated by a restriction endonuclease. Inherited codominantly, RFLPs are used as markers for specific chromosomes or genes.

is present as differences in the number and/or size of DNA fragments. Thousands of RFLPs have been identified in humans, and many (up to several hundred) have been assigned to each chromosome.

For RFLP analysis, a small blood sample (10 to 20 ml) is collected, and the white blood cells (leukocytes) are separated from the red blood cells (erythrocytes). DNA is extracted from the white cells and treated with a restriction enzyme. As a result, the DNA is cut into thousands of **DNA restriction fragments.** These fragments are placed on a gel and separated by size using electrophoresis. The separated fragments are Southern blotted with a probe (◗ Figure 13.1) that identifies a small number of specific fragments (those fragments that contain the sequence complementary to the probe).

DNA restriction fragment A segment of a longer DNA molecule produced by the action of a restriction endonuclease.

If variation has created or destroyed a restriction site, it will show up as a change in the length of a DNA fragment, making it larger or smaller (see Figure 13.1). This

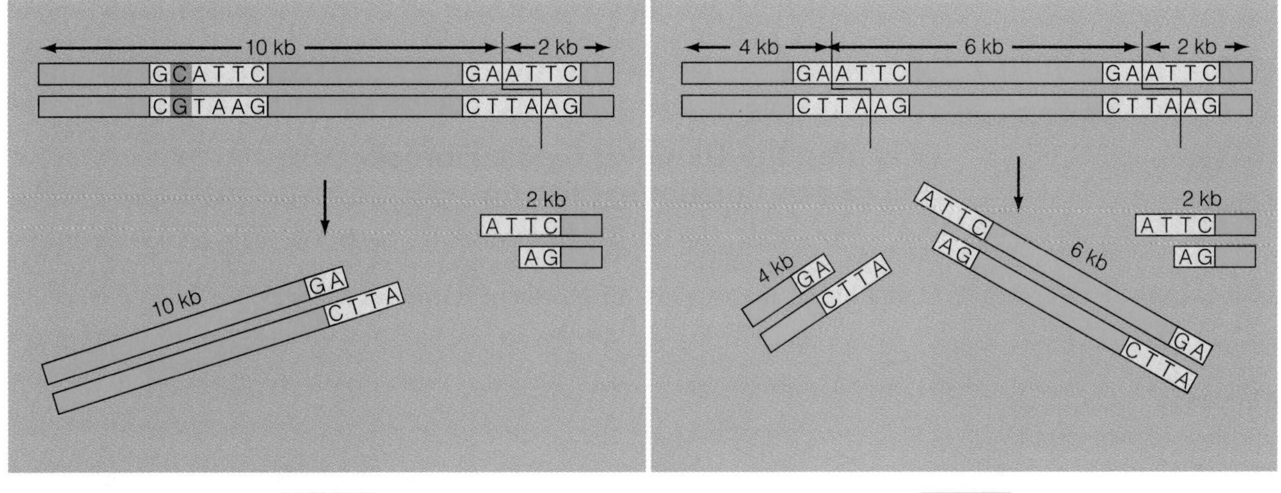

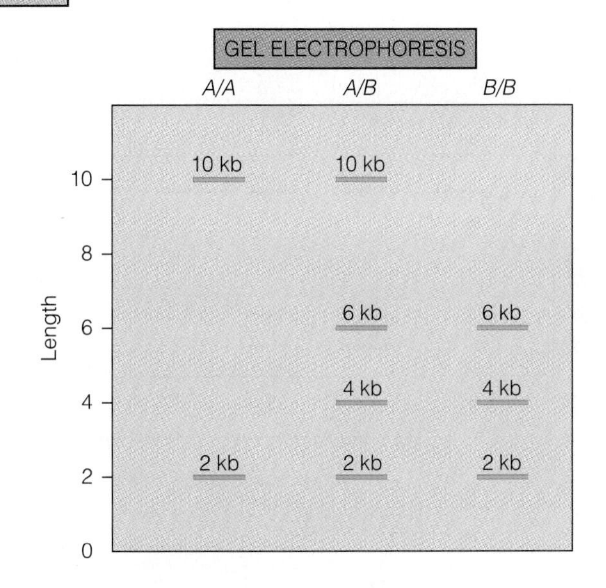

GENOTYPES		*RFLP* FRAGMENT SIZES
Homozygous	*A/A*	10 kb, 2 kb
Heterozygous	*A/B*	10 kb, 6 kb, 4 kb, 2 kb
Homozygous	*B/B*	6 kb, 4 kb, 2 kb

◗ **FIGURE 13.1** Restriction fragment length polymorphisms (RFLPs). The *A* and *B* alleles represent segments of DNA from homologous chromosomes. Arrows indicate recognition and cutting sites for restriction enzymes. Because of variation in nucleotide sequence (highlighted) in one region, a cutting site in *B* is missing in *A*. This variation produces differences in the number and length of DNA fragments in *A* and *B* when they are cut with a restriction enzyme. In *A*, two fragments, one of 10 kb (10,000 base pairs) and one of 2 kb are produced. In *B*, three fragments are produced: one of 6 kb, one of 4 kb, and one of 2 kb. Because these variations in cutting sites are inherited codominantly, there are three possible genotypes: *AA*, *AB*, and *BB*. The allele combination carried by any individual can be determined by restriction digestion of genomic DNA, followed by separation of the fragments by gel electrophoresis. The band patterns for the three combinations are shown as they would appear on a Southern blot.

variation in the cutting pattern is known as an RFLP. Using other methods, RFLPs can be mapped to specific chromosomes and to chromosome regions. Dozens to hundreds of RFLP markers are available for each human chromosome.

Mapping Genes Using RFLPs

We need two things to assign a gene to a particular chromosome using RFLPs: a large, multigenerational family in which a genetic disorder is inherited and a collection of cloned sequences that detects RFLPs (at least one for each human chromosome). First, a family pedigree is constructed and analyzed to determine how the disorder is inherited and to identify affected individuals (▶ Figure 13.2). Then DNA from family members is tested to determine the pattern of inheritance for chromosome-specific RFLP markers, using markers for each chromosome, one at a time. If the genetic disorder and a chromosome-specific RFLP marker are coinherited through several generations, then both the genetic disorder and the RFLP must be near each other on the same chromosome. When this happens, we know that the genetic disorder is on the same chromosome as the RFLP marker.

Mapping the Gene for Cystic Fibrosis: RFLPs and Positional Cloning

Let's illustrate the use of RFLPs in mapping using the gene for cystic fibrosis (*CFTR*). Cystic fibrosis (CF) is an autosomal recessive condition affecting 1 in 2,000 births among those of Northern European ancestry. CF homozygotes de-

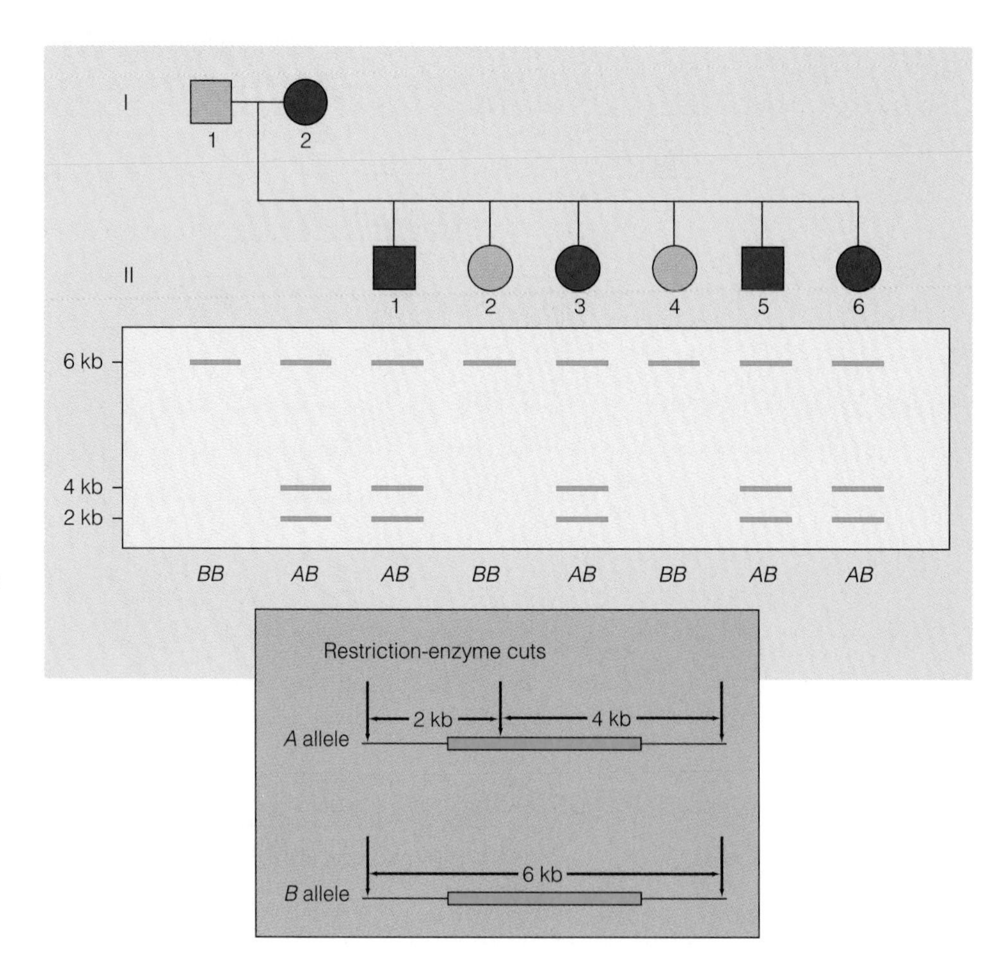

▶ **FIGURE 13.2** Inheritance of an RFLP allele and a dominant trait. A family pedigree is positioned above a gel that shows the distribution of alleles *A* and *B* in members of the family. Members affected by a genetic disorder are shown by filled symbols. The unaffected father (I-1) is homozygous for allele *B*, and the affected mother (I-2) is heterozygous *AB*. Examination of the pedigree and the Southern blot indicates that affected children have inherited a maternal chromosome that carries allele *A* (II-1, II-3, II-5, and II-6), whereas unaffected children (II-2 and II-4) have inherited maternal and paternal chromosomes that carry allele *B*. If this pattern of inheritance is confirmed in large, multigenerational families, the mutant allele for the genetic disorder would be placed on the same chromosome as the *A* allele for the RFLP.

velop obstructive lung disease and infections that lead to premature death. The mapping of the CF gene using RFLPs was accomplished in several steps.

Large families with a history of CF were examined to find linkage between CF and RFLP markers from specific chromosomes. The initial results produced a list of chromosomes in which the *CFTR* gene was *not* linked to any RFLP markers. This list, called an exclusion map, indicates where the gene is not located, allowing researchers to focus their attention on other chromosomes. Finally, linkage was shown between RFLP markers from chromosome 7 and CF, placing the *CFTR* gene somewhere on chromosome 7.

Using RFLPs from different regions of chromosome 7, researchers found that the *CFTR* gene is located near the end of the long arm (▶ Figure 13.3). In the last stage of work, researchers cloned and sequenced over 500,000 base pairs of DNA from the region. They identified the *CFTR* gene by finding a nucleotide sequence that could encode the amino acid sequence for a protein. To confirm that this was the *CFTR* gene, this same sequence was cloned from normal individuals and from CF patients, and the nucleotide sequences were compared. The sequence from CF patients had a three-nucleotide deletion within the gene that was not present in normal individuals. In addition, the normal and mutant proteins were isolated and compared. The protein, called CFTCR, is inserted in the cell membrane and functions in transporting ions (review the inheritance of CF in Chapter 4). In CF, the protein is not present in the membrane or does not function properly.

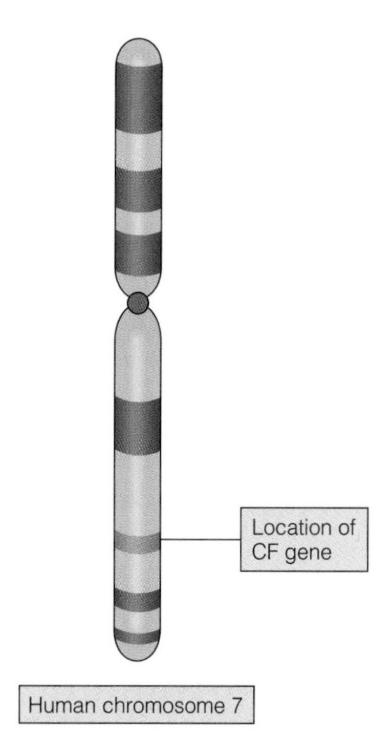

▶ **FIGURE 13.3** The gene for cystic fibrosis was mapped to the long arm of chromosome 7 by RFLP analysis.

This mapping method that begins with RFLP analysis is known as **positional cloning.** The successful mapping and cloning of the *CFTR* gene has several consequences. Studies of the structure and function of the CFTCR protein have the following benefits:

- They can lead to new and effective therapies for CF.
- Heterozygotes can be detected in the population (4% of the Caucasian population), and those couples at risk for bearing a child who has CF can be identified and counseled.
- They confirm the value of using recombinant DNA methods in gene mapping.

Positional cloning Identification and cloning of a gene responsible for a genetic disorder that begins with no information about the gene or the function of the gene product.

Over 100 other genes for heritable disorders have been mapped by RFLP analysis and positional cloning. Some of these are listed in Table 13.1.

Table 13.1 Some of the Genes Identified by Positional Cloning

	MIM/OMIM#		MIM/OMIM#
Chromosome 4		Chromosome 17	
Huntington disease	143100	Breast cancer (*BRCA1*)	113705
Chromosome 5		Neurofibromatosis (*NF1*)	162200
Familial polyposis (APC)	175100	Chromosome 19	
Chromosome 7		Myotonic dystrophy	160900
Cystic fibrosis	219700	Chromosome 21	
Chromosome 11		Amyotrophic lateral sclerosis	105400
Wilms tumor	194070	X Chromosome	
Ataxia-telangiectasia	208900	Duchenne muscular dystrophy	310200
Chromosome 13		Fragile-X syndrome	309550
Retinoblastoma	180200	Adrenoleukodystrophy	300100
Chromosome 16			
Polycystic kidney disease	173900		

Directly Mapping Genes to Chromosomes

RFLP analysis and positional cloning are not the only new methods of gene mapping created by recombinant DNA technology. If a gene has been cloned, it can be mapped directly to its chromosomal location. Recall from Chapter 6 that each member of the human chromosome set can be stained to produce a distinctive set of bands. For direct mapping, two things are needed. First, slides with mitotic chromosomes in metaphase are prepared (see Concepts and Controversies: Making a Karyotype in Chapter 6). Next, the cloned gene is labeled and used as a probe. The most commonly used labels are fluorescent dyes. In fluorescent *in situ* hybridization (FISH), the probe is labeled with a fluorescent dye and hybridized to metaphase chromosomes that have been treated to partially unwind the chromosomes. The probe forms a hybrid with its complementary DNA sequence on the chromosome (this is something like the hybrid that forms in a Southern blot). The location of the hybridized probe is visualized using a microscope. A pair of bright fluorescent spots on homologous chromosomes marks the location of the gene (◗ Figure 13.4).

For high-resolution mapping, the chromosome can be unwound to individual fibers, and the order and distance between two or more genes can be directly observed (◗ Figure 13.5).

The third method of mapping genes using recombinant DNA is by genome sequencing. We will discuss this method in a later section. First, we will discuss one other use for methods that detect variations in DNA sequences, DNA fingerprinting.

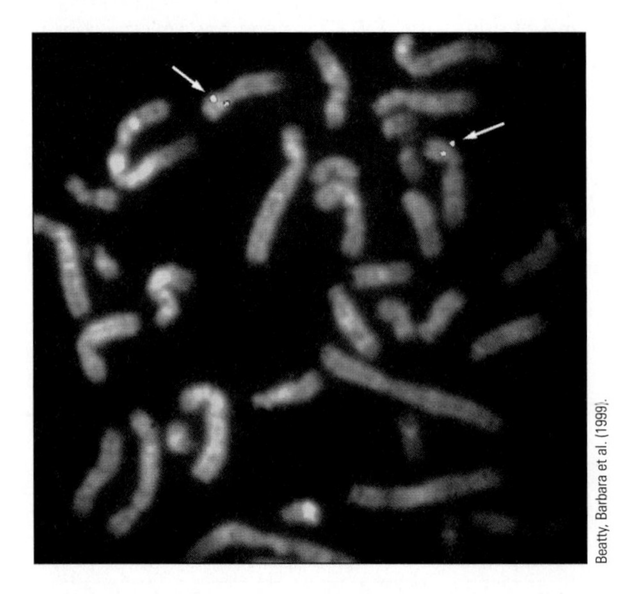

Beatty, Barbara et al. (1999).

◗ **FIGURE 13.4** A gene mapped directly to a homologous pair of human chromosomes by fluorescent *in situ* hybridization (FISH). This gene maps to the short arm of chromosome 9. (Beatty, Barbara, et al. (1999). Chromosomal localization of a phospholipase A2 activating protein, an Ets2 target gene, to 9p21. *Genomics 62*, 529–532.)

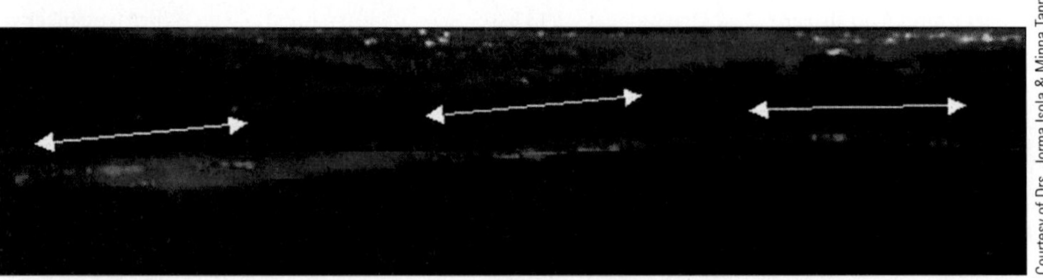

Courtesy of Drs. Jorma Isola & Minna Tanner

◗ **FIGURE 13.5** The arrangement of two different DNA sequences on a chromosome fiber. In this method, the DNA-protein fibers that make up a chromosome are spread onto a microscope slide and hybridized with one sequence marked with a red fluorescent dye and a second sequence tagged with a green fluorescent dye. (Courtesy of Drs. Jorma Isola & Minna Tanner, Univ. of Tampere, Finland, from Järvinen TAH, Tanner M, Bärlund M, Borg Å, Isola J. (1999). Characterization of topoisomerase II gene amplification and deletion in breast cancer. *Genes, Chromosomes and Cancer 26*, 142–150.)

DNA FINGERPRINTING

In addition to gene mapping, RFLP analysis is used in many other applications, including forensics, ecology, archaeology, preservation of endangered species, and even dog breeding. Other nucleotide variations in the human genome are used as genetic markers. In the 1980s, Alec Jeffreys and his colleagues at the University of Leicester discovered a nucleotide variation that depends on changes in the length of repetitive DNA clusters called minisatellites, located at specific chromosomal sites. Other nucleotide variations, including single nucleotide polymorphisms (SNPs), have been discovered in recent years. RFLPs, minisatellites, SNPs, and other nucleotide variations are used in many ways, including court cases, studies of human evolution, and tracing migratory patterns of humans across the world. We will consider only one of these uses, DNA fingerprints.

Minisatellites and Short Tandem Repeats

DNA fingerprinting detects changes in the length of minisatellites. Minisatellites range in size from 14 to 100 nucleotides, and are organized into clusters. For example, the nucleotide sequence

CCTTCCCTTCCCTTCCCTTCCCTTCCCTTC

is made up of six tandemly repeated copies of the five-nucleotide sequence, CCTTC. Clusters of these kinds of repeats are scattered at many sites on all chromosomes. There are many different minisatellites, and the number of repeats at each site is variable, ranging from 2 to more than 100. Each variant is an allele, and they are codominant markers. These minisatellites are called **short tandem repeats (STRs).** Because there are many different STR loci and because each locus can have dozens of alleles, heterozygosity is common.

Short tandem repeats (STRs) Short nucleotide sequences, repeated in tandem, that are clustered at many sites in the genome. The number of repeats at homologous loci is variable.

When restriction enzymes are used to cut STRs and the results are visualized on a Southern blot, a pattern of bands is produced. The term **DNA fingerprint** is used to describe this pattern (Figure 13.6). As with a traditional fingerprint, the pattern is always the same for a given individual. There is so much variation from one per-

DNA fingerprint A pattern of restriction fragments that is unique to an individual.

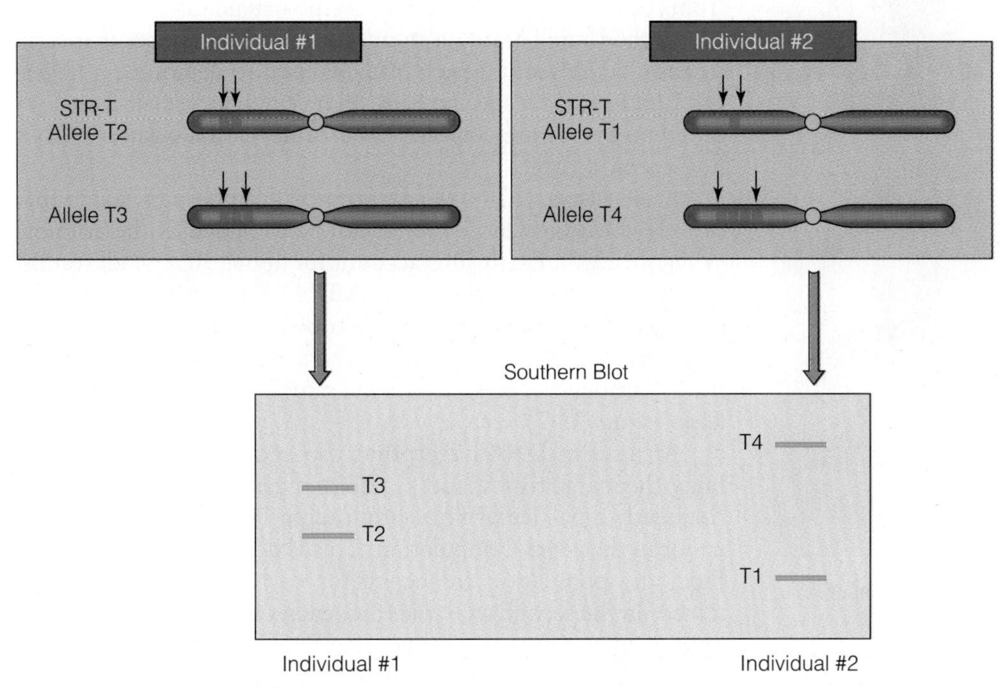

FIGURE 13.6 Short tandem repeats (STRs) and DNA fingerprints.

son to another that if a number of different STRs are analyzed, each person's profile is unique (except, of course, for identical twins).

When the STRs are amplified by polymerase chain reaction (PCR) before blotting, DNA fingerprint analysis can be done using very small samples of DNA, obtained from a variety of sources, including single hairs, envelope flaps, cigarette butts, and the dried saliva on the back of a used postage stamp. DNA fingerprints can also be obtained from very old samples of DNA, increasing its usefulness in legal cases. In fact, DNA fingerprinting has been carried out on mummies that are more than 2,400 years old.

DNA Fingerprinting and the Law

Forensics is the use of scientific knowledge in civil and criminal law. In the United States, forensic DNA fingerprinting was first used in 1987 in a criminal trial. Since then, it has been used in more than 50,000 criminal investigations and trials. DNA fingerprinting is now used in about 10,000 criminal cases every year, 75% of which involve criminal sexual assault. It is also used in about 200,000 civil cases each year for paternity testing. DNA analysis is performed in state and local police crime labs, private labs, and the Federal Bureau of Investigation (FBI) lab in Washington, D.C.

In criminal cases, DNA can be extracted from biological material left at a crime scene. This can include blood, tissue, hair, skin fragments, and semen. DNA fingerprints are prepared and compared with the DNA fingerprints of the victim and any suspects in the case. DNA fingerprints were first used in a criminal case in 1986 to 1987 to solve the rape and murder of two teenage girls in England. Over 4,000 men were DNA fingerprinted during the investigation. The results freed an innocent man who had been jailed for the crime and led to a confession from the killer. This case was described in the book *The Blooding,* by Joseph Wambaugh. In the United States, DNA fingerprints first received widespread attention in the O. J. Simpson criminal trial.

The use of DNA fingerprints in legal cases requires care in collecting and handling samples, accurate identification of bands in the gel, precise methods for determining when bands match, and reliable calculations for the probability of matching bands from the evidence with those from the defendant.

In the United States, a standard set of STR probes is used in forensic DNA testing. Each probe is used to determine a fingerprint pattern for a specific STR. In a criminal case, at least four different probes are typically used to develop a suspect's DNA fingerprint profile.

If the suspect's DNA fingerprint profile does not match that of the evidence, then he or she can be excluded as the criminal (Figure 13.7). Exclusions account for about 30% of all results. On the other hand, if the suspect's DNA fingerprint matches the evidence fingerprint, there are two possibilities: The DNA fingerprints came from the same person, or the fingerprints came from someone else who carries the same combination of STR alleles (Figure 13.7).

Analysis of DNA fingerprints uses a combination of probability theory, statistics, and population genetics to estimate how frequently a particular fingerprint might be found in an individual or in the general population. This is done in a series of steps. First, the population frequency for each STR allele is determined. In the second step, the frequencies for each STR allele are multiplied together to produce the final estimate of frequency. The frequencies are multiplied because STRs on different chromosomes are inherited by independent assortment.

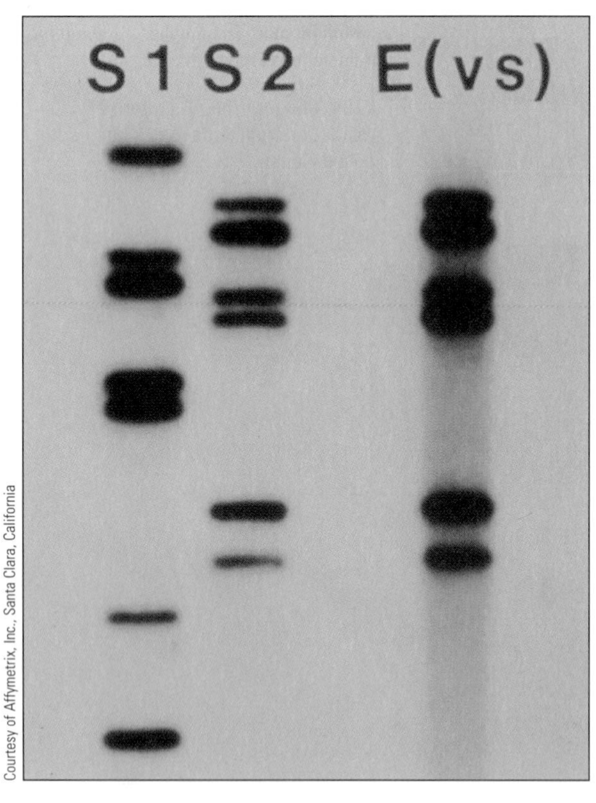

FIGURE 13.7 DNA fingerprinting in a criminal case. DNA fingerprints for two suspects (S1 and S2) and a fingerprint from evidence (E[vs]) are shown. The profile of suspect 1 is different from that of the evidence, but the profile of suspect 2 matches that of the evidence band for band.

Web-Integrated Toolbox

DNA Fingerprints

Recombinant DNA technology has generated many new applications. One of these is DNA fingerprinting. Although the structure of DNA is the same in all of us, there are differences in the type and order of the DNA subunits (nucleotides). Some of these differences are in repeated sequences called short tandem repeats (STRs), which contain up to one hundred nucleotides of DNA. Everyone has STRs, and we inherit them from both parents. As a result, the total STR pattern we each carry is more or less unique. Although analysis of one STR may not reveal this uniqueness, the more STRs we analyze, the more unique the pattern becomes.

DNA fingerprints have many uses, some of which raise important issues of privacy and ethics. STR patterns are used to establish paternity and maternity in cases of child custody, child support, and in situations where adopted children are seeking their biological parents. DNA fingerprints are also used in criminal cases and in gathering evidence at crime scenes. Many states now require that all convicted felons be DNA fingerprinted. DNA fingerprints are also used for personal identification. Some parents have their children's DNA fingerprints recorded with law enforcement agencies in case of kidnapping. There are several companies that will perform DNA fingerprinting of individuals and then post these fingerprints in on-line archives as an aid in genealogy research.

DNA fingerprinting uses recombinant DNA techniques including digestion of DNA with restriction enzymes and Southern blot hybridization to produce a STR pattern. Typically, a DNA fingerprint is assembled from the patterns produced by four or more STRs.

The book's home page contains exercises and links to guide you through the steps in making a DNA fingerprint. These exercises will outline the techniques of making and interpreting DNA fingerprints and their applications in crimes and paternity disputes.

The overall frequency of the combined STR alleles is low, even when the individual frequencies are fairly common. Table 13.2 shows how STR alleles that individually are fairly common in the population can be combined to provide a profile that has a high degree of reliability. The frequency of the allele at locus 1 (1 in 25) combined with the frequency of the allele at locus 2 (1 in 100) means that this combination would be found in 1 in every 2,500 individuals. When the frequency of the first two loci are combined with loci 3 and 4, the frequency of the overall combination is 1 in about 60 million. This means that only one person in a population of 60 million would be expected to have this particular combination of STR alleles. The frequency estimate does not prove the suspect guilty but is a factor that should be considered along with other facts in the case.

Other Applications of DNA Fingerprinting

DNA fingerprints are used in many other applications, including a dispute over the bloodlines of a purebred dog and establishing the fate of Czar Nicholas II and his family (see Concepts and Controversies: Death of a Czar). It has also been used on

Table 13.2 Calculating a DNA Fingerprint Profile Frequency

Allele at Locus	Frequency in Population	Combined Frequency
1	1 in 25 (0.040)	—
2	1 in 100 (0.010)	1 in 2500
3	1 in 320 (0.0031)	1 in 806,000
4	1 in 75 (0.0133)	1 in 60,600,000

Concepts and Controversies

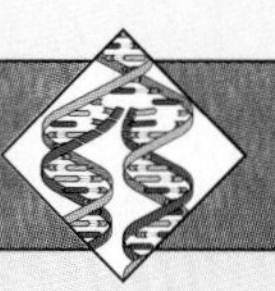

Death of a Czar

Czar Nicholas Romanov II of Russia was overthrown in the Bolshevik Revolution that began in 1917. He and Empress Alexandra (granddaughter of Queen Victoria); their daughters Olga, Tatiana, Marie, and Anastasia; and their son Alexei (who had hemophilia) were taken prisoners. In July 1918, it was announced that the Czar had been executed, but for many years the fate of his family was unknown. In the 1920s, a Russian investigator, Nikolai Solokof, reported that the Czar, his wife and children, and four others were executed at Ekaterinburg, Russia, on July 16, 1918, and their bodies were buried in a grave in the woods near the city. Other accounts indicated that at least one family member, Anastasia, escaped to live in Western Europe or the United States. Over the years, the mystery surrounding the family generated several books and movies.

In the late 1970s, two Russian amateur historians began investigating Solokof's accounts, and, after a painstaking search, nine skeletons were dug from a shallow grave at a site 20 miles from Ekaterinburg in July 1991. All of the skeletons bore marks and bullet wounds indicating violent death. Forensic experts examined the remains and using computer-assisted facial reconstructions and other evidence concluded that the remains included those of the Czar, the Czarina, and three of their five children. The remains of two children were missing: the son, Alexei, and one daughter. DNA analysis was used to confirm the findings.

The investigators used a threefold strategy in the DNA analysis. DNA was extracted from bone fragments and used for sex testing, for DNA typing to establish family relationships, and for mitochondrial DNA testing to trace maternal relationships. The sex testing used a six base pair difference in a gene carried on both the X and Y chromosome. The results indicated that the skeletons were of four males and five females, confirming the results of physical analysis. Probes for five short tandem repeat (STR) sequences were used to test the family relationship. The results showed that skeletons 3 to 7 were a family group; 4 and 7 were parents and 3, 5, and 6 were children. If the remains are those of the Romanovs, the combination of the sex tests and the STR tests establish that the remains of one of the princesses and the Tsarevitch, Alexei, are missing.

To determine whether the remains belonged to the Romanovs, mitochondrial DNA (mtDNA) testing was conducted. Because mtDNA is maternally inherited, living relatives of the Czarina, including Prince Philip, the husband of Queen Elizabeth of England, were included in the tests. This analysis shows an exact match between the remains of the Czarina, the three children, and living relatives. mtDNA from the Czar matched that of two living maternal relatives, confirming that the remains are those of the Czar, his wife, and three of his children. The fate of Alexei and one daughter remains unknown, but historical accounts suggest that their bodies were burned or buried separately. This study overcame several technical challenges but clearly established the identity of the skeletons as those of the Romanovs. The results show the power that genetic technology can bring to solving a mystery of world history. The intrigues, mysteries, and the science surrounding the search for the Romanovs are told by Robert Massie in his book *The Romanovs: The Final Chapter.*

the frozen remains of a wooly mammoth recovered in Siberia to determine the evolutionary relationships between mammoths and modern-day elephants.

By far, the most common use of DNA fingerprinting in the United States is in establishing paternity. Over 30% of the births in the United States are to single mothers, and over 200,000 paternity tests are performed each year, many using DNA fingerprints. These tests are performed to establish legal responsibility for financial support, and efforts by local, state, and federal governments result in support payments of more than $5 billion per year.

THE HUMAN GENOME PROJECT IS AN INTERNATIONAL EFFORT

The development of recombinant DNA technology and its use in gene mapping made it possible to consider mapping all of the genes in the human genome (recall that a genome is the set of genes carried by an individual). After several years of discussion, the Human Genome Project in the United States began in 1991 as a program coordinated by two agencies: the National Institutes of Health and the Department of Energy.

The goal of this project is to map the location of all genes in the human genome and to analyze the nucleotide sequences of these genes. The project has grown into an international effort coordinated by the Human Genome Organization (HUGO). Although not reflected in the name, the Human Genome Project also includes projects to sequence the genomes of organisms used in genetic studies (Table 13.3). These include bacteria, yeast, a roundworm, the fruit fly, and the mouse. The history and timelines for these projects are shown in ▶ Figure 13.8. In addition, genome sequencing projects are underway for literally hundreds of other species (in addition to those in the Human Genome Project), many of which cause disease. To date, the genomes for over 75 species have been completely sequenced, and hundreds more are being sequenced.

Genome Projects Have Created New Scientific Fields

Collectively, the methods used to clone and sequence the human genome or any genome are called **genomics.** Once the sequence is obtained, it must be stored in a database so it can be analyzed to identify genes and their functions. In the case of the human genome, this is not a trivial task; the sequence for the human genome fills more than 200 standard diskettes. Workers in a new field called **bioinformatics** develop and use computer hardware and software for the storage, analysis, and visualization of genomic information. The use of genome databases is now an essential tool for geneticists and other biologists who are mapping genes.

Genomics The collection of techniques used to clone and sequence a genome.

Bioinformatics The use of computers and software to acquire, store, analyze, and visualize the information from genomics.

Table 13.3 Organisms Included in the Human Genome Project

Organism	Genome Size Million base pairs (Mb)	Estimated No. of Genes
Escherichia coli (bacteria)	4.64	4,300
Saccharomyces cerevisiae (yeast)	12	6,500
Caenorhabditis elegans (roundworm)	97	20,000
Arabidopsis thaliana (plant)	120	20,000
Drosophila melanogaster (fruit fly)	170	16,000
Homo sapiens (human)	3,300	35,000

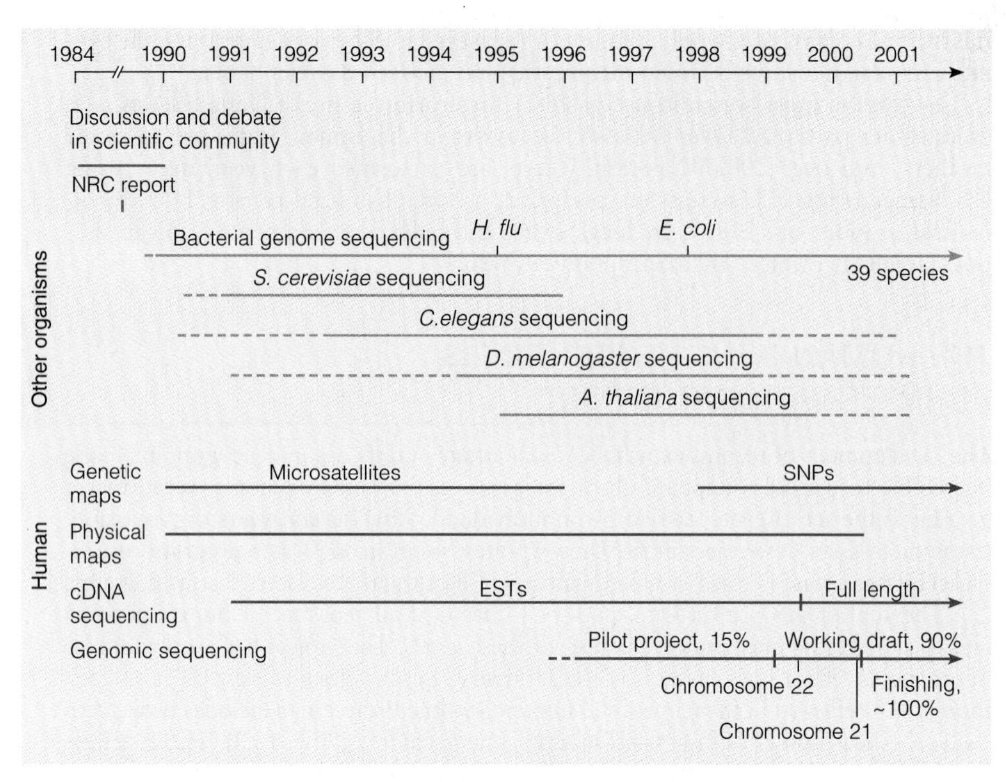

FIGURE 13.8 The history and timelines for the Human Genome Project and some organisms used in genetic research. (Source: From International Human Genome Sequencing Consortium. 2001. Initial sequencing and analysis of the human genome. *Nature* 409:860–921, Fig. 1, p. 862. Adapted with permission from *Nature.* Copyright 2001, Macmillan Magazines Limited.)

In the following sections, we will see how genomics and bioinformatics are used in collecting and analyzing information about the human genome.

Genomics: Sequencing, Identifying, and Mapping an Organism's Genes

Clone-by-clone method A method of genome sequencing that begins with genetic and physical maps and sequences clones after they have been placed in order.

Shotgun cloning A method of genome sequencing that selects clones at random from a genomic library and assembles the sequence using software.

Genomics begins with cloning and sequencing an organism's genome. Geneticists use two strategies for this task. The first, called the **clone-by-clone method,** is used by the publicly funded Human Genome Project (Figure 13.9). In this method, a collection of clones covering the entire genome (the collection is called a genomic library) is established. Second, the clones are pieced together into genetic and physical maps that cover each chromosome. In the third step, each clone is sequenced, and the sequence from each clone is assembled into the genomic sequence.

The second strategy for genome sequencing is called **shotgun cloning.** In this method, a genomic library is prepared, and clones are selected at random from the library and sequenced. Software, called assembler programs, organizes the sequence information into a genome sequence. A privately funded human genome project, coordinated by Celera Corporation, used shotgun cloning to sequence the human genome. Their effort began in September 1999.

In June 2000, the public and private human genome projects announced they had completed a draft of the human genome sequence, and, in February 2001, they each published their results. The sequence is not finished, and much work remains to be completed. There are gaps in the sequence that must be closed, and neither project has started sequencing the 15% of the genome found in heterochromatic regions of chromosomes. These regions surround the centromeres and are also found near the telomeres of all chromosomes.

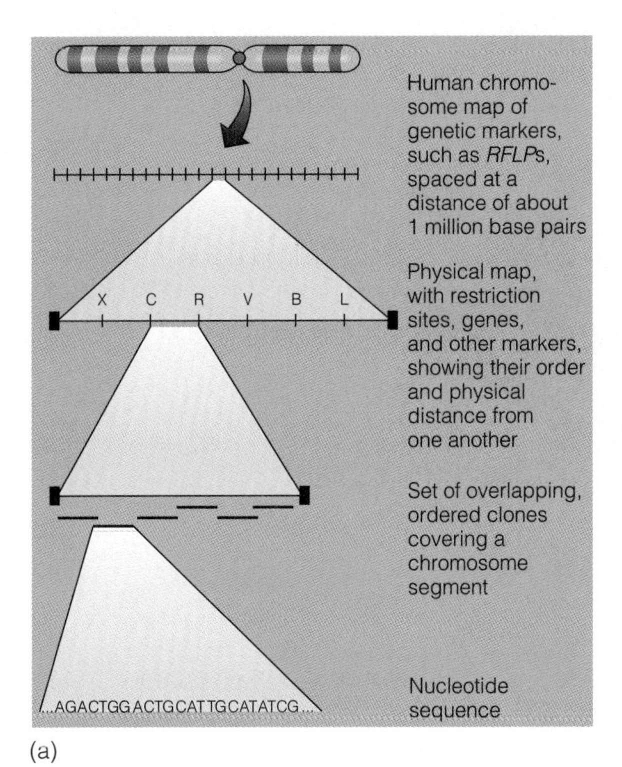

(a)

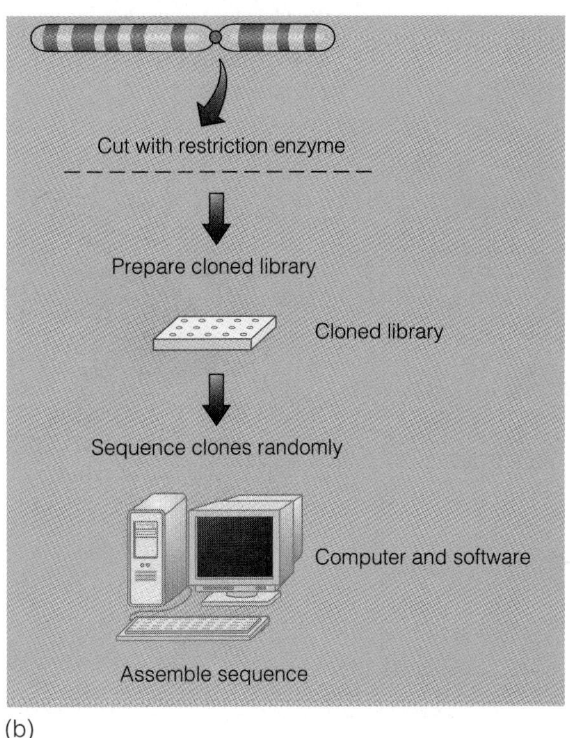

(b)

FIGURE 13.9 Cloning approaches. (a) The clone-by-clone method of genome sequencing. First, a genomic library is used to make genetic maps and physical maps of each chromosome. Clones are organized into overlapping sets, and each clone is sequenced. The genome sequence is assembled as the clones are sequenced. (b) In the shotgun method, a genomic library is prepared, and clones are chosen randomly for sequencing. The sequence is assembled by software programs. The Human Genome Project sponsored by the National Institutes of Health and the Department of Energy used the clone-by-clone method of sequencing. The genome project organized by Celera Corporation used the shotgun method.

Analyzing Genomic Information with Bioinformatics

In addition to the traditional laboratory notebooks, geneticists are now using large-scale databases to store, analyze, visualize, and share the results of their work on genomes. One of the main tasks in bioinformatics is managing the large amount of information generated by genome projects and storing this information in a way that it can be easily used.

One of the first steps in managing genomic information is compiling the sequence and checking it for accuracy. To ensure that a genome sequence is error free, it is sequenced more than once. Using bioinformatics these sequences are compared to each other, and a consensus sequence is derived. The publicly funded genome project sequenced the human genome 12 times, and the privately funded project sequenced the genome approximately 35 times. From this vast amount of information, they each generated a consensus sequence. These consensus sequences are now being compared to derive a single genome sequence.

Once the sequence has been compiled, the next task is to find all the genes. This process is called annotation. Protein-coding genes begin and end with certain nucleotide sequences, and computer software is used to scan a sequence to identify genes. Annotation of the human genome is an ongoing process; when it is completed, we will have a final count of how many genes are in our genome.

After genes are identified by annotation, they are classified into functional groups. Functions have been assigned to about 60% of the genes identified in the human genome (Figure 13.10). Geneticists are working to identify functions for the rest of the genes.

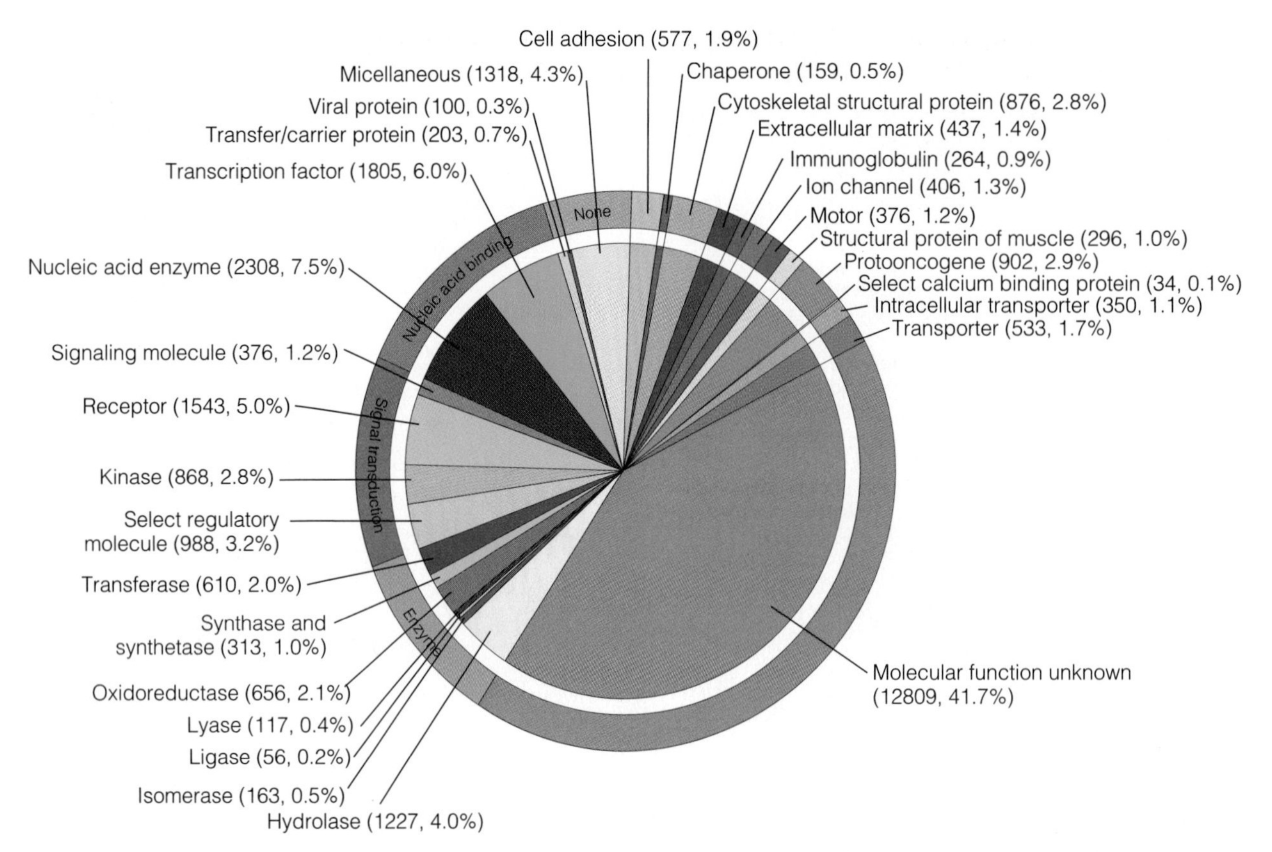

FIGURE 13.10 A preliminary functional assignment for over 26,000 genes in the human genome. Just over 12,000 of these genes (41%) have no known function, and represent the largest class of genes identified to date, emphasizing the work that needs to be completed before we fully understand our genome. The other classes of genes were assigned because they were previously known from other experiments, or are similar to proteins of known function from other organisms. Among the most common genes are those involved in nucleic acid (DNA and RNA) metabolism (7.5% of all the identified genes). (From Venter C et al. (2001). The sequence of the human genome. *Science 291,* 1304–1351, Fig. 15, p. 1335. Copyright 2001 American Association for the Advancement of Science. Reprinted with permission.)

What Have We Learned So Far About the Human Genome?

Several general features of the human genome have been identified in analyzing the sequence information. First, although the genome is large and contains over 3 billion nucleotides of DNA, only approximately 5% of this DNA encodes genetic information. The rest does not code for anything and is spacer DNA. The gene-coding regions are distributed on all 24 chromosomes in gene-rich clusters, separated from each other by long stretches of gene-poor regions. The gene-poor regions correspond to the banded regions of chromosomes (see Chapter 6 for a discussion of banding and karyotypes). Not all the genes in the human genome have been identified, but it appears that humans have 35,000 to 50,000 genes, far fewer than the predicted number of 80,000 to 100,000.

Even in the early stages, findings from the genome projects are changing our concepts of several genetic processes, including mutation and gene function. A new mechanism of mutation, trinucleotide expansion (discussed in Chapter 11), has been identified. This mechanism is important in disorders of the nervous system, and understanding why this is so will play an important role in treating these disorders. Another unexpected finding is that mutations in a single gene can give rise to different genetic disorders, depending on how the gene is affected. For example, the gene *RET* (MIM/OMIM 164761) encodes a protein that is a cell-surface receptor engaged in transferring signals across the plasma membrane. Depending on the type and location of mutations within this gene, four distinct genetic disorders can result: two types of multiple endocrine neoplasia (MIM/OMIM 171400 and MIM/OMIM 162300), familial medullary thyroid carcinoma (MIM/OMIM 155240), and Hirschsprung disease (MIM/OMIM 142623). Another significant discovery is that some mutations that affect DNA repair can destabilize distant regions of the genome and make them susceptible to more mutations, often resulting in cancer (these mutations are discussed in Chapter 14). These and similar discoveries have already had an impact on the diagnosis, treatment, and genetic counseling for several groups of genetic disorders.

Using Genomics and Bioinformatics to Study a Human Genetic Disorder

In studying a genetic disorder, several questions are important:

- Where is the gene located?
- What is the normal function of the protein encoded by this gene?
- How does the mutant protein produce the disease phenotype?

In the case of the CF gene we discussed earlier in this chapter, the function of the gene product was identified in two steps. First, the nucleotide sequence was converted into the predicted amino acid sequence of the protein. Second, the amino acid sequence of the CF protein was compared to protein sequences already in databases. The CF protein has an amino acid sequence similar to that of proteins that are found in membranes and that control the flow of ions into and out of the cell. Because CF patients have problems with chloride ion flow, the protein function in normal and CF individuals could be pinpointed.

What happens if a gene for a genetic disorder is cloned and sequenced, but there is no similar protein in any database to help understand what the protein does, and how a mutant form of the protein causes disease? Let's look at how this problem was solved in the case of Friedreich ataxia (FRDA) (MIM/OMIM 229300), a neurodegenerative disorder inherited as an autosomal recessive trait. FRDA occurs in approximately 1 in 50,000 individuals and is characterized by loss of muscle coordination (ataxia), skeletal deformities, and enlargement of the heart. It is progressive and fatal. Using positional cloning, the gene was mapped to chromosome 9, isolated, cloned, and sequenced. Database scans reached a dead end. The only matches to the FRDA protein, frataxin, were with proteins of unknown function in yeast and a nematode.

To attempt to match frataxin with protein sequences already in databases, researchers searched databases using parts of the frataxin sequence that might have functional significance. These regions, called domains, matched with a protein from a few bacterial species, all related to purple bacteria. Because purple bacteria are the closest living relatives to the ancient bacterial species that evolved into mitochondria, it seemed possible that frataxin might work in the cell's mitochondria.

Using these clues, another research group found that frataxin is located in mitochondria, worked out the three-dimensional structure of the protein (Figure 13.11), and discov-

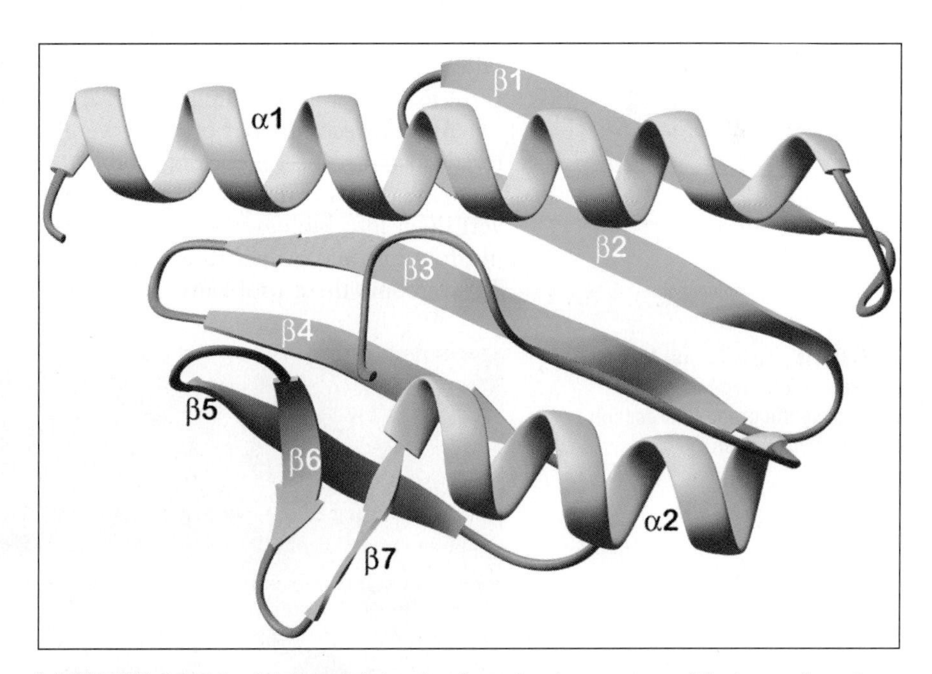

FIGURE 13.11 A 3-D model showing the molecular structure of the human frataxin protein. This protein is necessary for normal mitochondrial function. A decreased production or function of this protein causes the autosomal recessive disorder, Friedreich ataxia. This disorder is one of the first whose molecular basis was uncovered using analysis of genomic information. (Courtesy of Sirano Dhe-Paganon and Steven Shoelson.)

ered that frataxin regulates energy conversion in the mitochondria. With this information in hand, efforts are now focused on developing therapies to treat FRDA. This example illustrates the power of genomics and bioinformatics to help unravel the mysteries of human genetic disorders.

GENE TRANSFER TECHNOLOGY HAS MANY APPLICATIONS

Recombinant DNA technology was originally used to transfer foreign genes into bacterial cells as a way of cloning large amounts of a specific gene for further study. Later, methods were developed for producing eukaryotic proteins in the host bacterial cells. Gene transfer is not limited to using bacteria as host cells; it is also possible to transfer genes between higher organisms (◗ Figure 13.12). In this section, we review some of the current applications of gene transfer technology.

Biopharming: Making Medical Molecules in Plants and Animals

One of the first commercial applications of recombinant DNA technology was the production of proteins to be used in treating human disease. Human insulin, produced by transfer of the human insulin gene into bacteria, was first marketed in 1992 and is widely used to control diabetes. Before this, these proteins were collected from animals, pooled blood samples, or even human cadavers. In some cases, proteins from these sources pose serious and potentially fatal risks. For example, those with hemophilia, an X-linked recessive disorder, cannot manufacture a clotting factor and have episodes of uncontrollable bleeding. In the past, donated blood was used to extract the clotting factor, which was used to treat hemophilia. However, this treatment was used before the time that donated blood was tested for human immunodeficiency virus (HIV) associated with the development of AIDS. As a result, up to 90% of all hemophiliacs treated with clotting factor prepared in this way were infected with HIV. Recombinant DNA technology is now used to make clotting factors, insulin, and many other proteins used for treating disease (Table 13.4). Proteins made by recombinant DNA techniques are not limited to treating human disease and have other applications.

At first, recombinant proteins were produced in bacterial hosts. Although bacterial hosts can be genetically modified to produce human (and other eukaryotic) proteins, the bacteria cannot process and chemically modify these proteins to convert them into biologically active forms. In addition, some eukaryotic proteins do not fold into the proper three-dimensional form in bacterial cells and are inactive. To overcome these problems, a new generation of eukaryotic hosts is now in use.

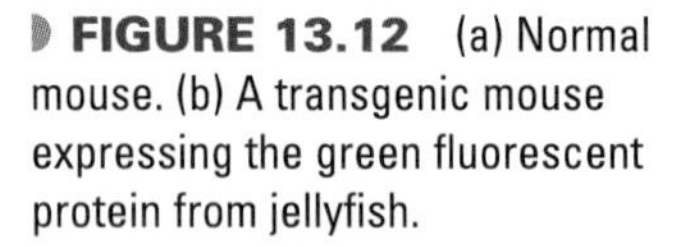

◗ **FIGURE 13.12** (a) Normal mouse. (b) A transgenic mouse expressing the green fluorescent protein from jellyfish.

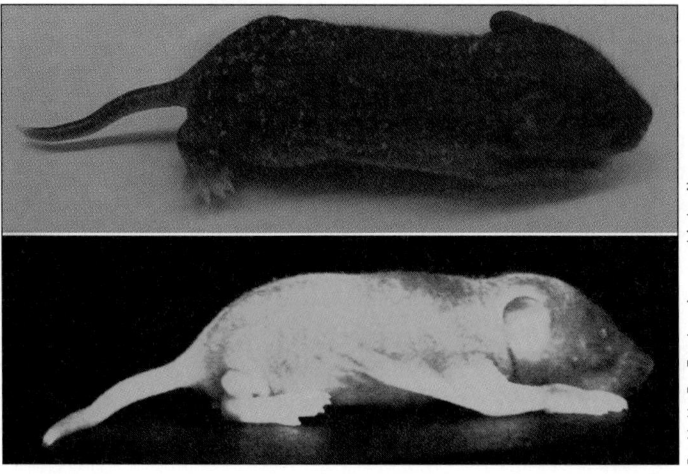

Provided by Drs. Tomokasu Amano and Andras Nagy

Table 13.4 Some Products Made by Recombinant DNA Technology

Product	Use
Atrial natriuretic factor	Treatment for hypertension, heart failure
Bovine growth hormone	Improve milk production in dairy cows
Cellulase	Breakdown cellulose in animal feed
Colony stimulating factor	Treatment for leukemia
Epidermal growth factor	Treatment of burns, improve survival of skin grafts
Erythropoietin	Treatment for anemia
Hepatitis B vaccine	Prevent infection by hepatitis B virus
Human insulin	Treatment for diabetes
Human growth hormone	Treatment for some forms of dwarfism, other growth defects
Interferons (alpha, gamma)	Treatment for cancer, viral infections
Interleukin-2	Treatment for cancer
Superoxide dismutase	Improve survival of tissue transplants
Tissue plasminogen activator	Treatment of heart attacks

This new generation of hosts includes farm animals and plants that are used to make human proteins of pharmaceutical importance.

The heritable form of emphysema (MIM/OMIM 107400) is a progressive and fatal respiratory disorder. It is caused by a mutation in a gene called alpha-1-antitrypsin. To make the gene product, the gene was cloned into a vector to sit next to a DNA sequence that switches on the production of milk proteins. These recombinant vectors were microinjected into fertilized sheep eggs, and the eggs were implanted into female sheep. The resulting transgenic sheep developed normally, and the females produce milk that contains up to one-third of a pound of alpha-1-antitrypsin per gallon of milk (Figure 13.13). This recombinant protein is now in clinical testing, and soon a small herd of sheep will supply the world's need for this protein.

FIGURE 13.13 Transgenic sheep that carry the human gene for alpha-1-antitrypsin secrete the gene product into their milk. The protein can be extracted, purified, and used for treating emphysema.

Human growth hormone, somatotrophin, was one of the first medically important molecules produced in bacterial hosts. However, bacterial hosts do not process the protein, and it requires chemical modification before use. In transgenic tobacco plants carrying the human somatotrophin gene, the hormone is made in its processed form, and once extracted from the plants is ready for use in treating dwarfism and other disorders.

Making Edible Vaccines

Vaccination programs in developing countries are often ineffective because many vaccines require refrigeration for transport and storage, and this is often unavailable in rural areas. In addition, injection-related infections cause medical problems. The Children's Vaccine Initiative, started in 1992, has the goal of developing cheap, edible vaccines made in transgenic food plants for use in developing countries.

The idea is to use edible plant parts, including fruit, as a source of an oral vaccine. Such vaccines would be inexpensive and would not require investment in industrial capacity for production, purification, and distribution. In addition, they would not have to be injected under sterile conditions by trained medical staffs. The first vaccines are targeted against infant diarrhea, which kills 3 million children each year.

Using tobacco plants as a model system, vaccines have been produced against the hepatitis B virus. In clinical trials with a vaccine against diarrhea, volunteers were successfully vaccinated by eating uncooked transgenic potatoes (◗ Figure 13.14). If further tests are successful, this vaccine and others against hepatitis B and diphtheria will be transferred to bananas and other plants that are eaten raw and are grown in many parts of the world. Sometime in the near future, genetically engineered plants will be used to vaccinate infants, children, and adults against many different infectious diseases.

Gene Transfer in Food Crops

The transfer of genes to agriculturally important plants using recombinant DNA methods creates transgenic crop plants. The transferred gene can originate from another plant species or even bacteria. The transferred gene confers new traits on the plant, such as resistance to insects or herbicides. In the media, such plants are often called genetically modified organisms (GMOs) or genetically modified (GM) plants. These terms ignore the fact that all crop plants have been genetically modified over thousands of years by selection and crossbreeding.

Transgenic crop plants resistant to herbicides used in weed control and plants resistant to insects are now available. Farmers use over 100 different herbicides at an annual cost of more than $10 billion. In spite of this effort, weeds destroy more than 10% of all crops worldwide. Herbicide-resistant crops carry genes for resistance to broad-spectrum herbicides such as glyphosate and glufosinate. These herbicides kill all plants in the field except the crop plant and break down quickly in the soil, reducing run-off and environmental impact. Other crops, especially corn, cotton, and potato have been genetically engineered to be resistant to insect pests (◗ Figure 13.15). These crops (known as Bt crops) carry a bacterial gene that produces a toxin released in the insect gut, causing death.

Several new transgenic crops are being developed and will be available in the near future. One of these is a salt-tolerant tomato plant that grows in areas where irrigation water or soils have a high salt level, opening new land for growing this crop. Golden rice, a genetically engineered strain that synthesizes a precursor to

Courtesy Dwayne Kirk, Boyce Thompson Institute for Plant Research

◗ **FIGURE 13.14** A volunteer eats pieces of a potato that has been genetically engineered to produce a vaccine. The use of such modified fruits and vegetables may help reduce disease by making vaccines widely available in some parts of the world, and lower the cost of vaccine production and distribution.

▶ **FIGURE 13.15** Transgenic plants are created by inserting genes from other organisms. Such plants are sometimes called genetically modified or GM organisms. The photograph shows the results of insect infestation on normal cotton bolls (right) and transgenic cotton (left) modified to resist attack by insects.

vitamin A, is now being crossed into other strains and should become available for planting in a year or two. Transgenic canola plants with enhanced vitamin E content are also being developed.

The first transgenic crop plants were grown in 1996. The percentage of acreage devoted to three transgenic crops in the United States over a 3-year period is shown in ▶ Figure 13.16. Worldwide, the percentage of acreage planted in transgenic crops increased 11% in 2000.

There are concerns about the development and use of transgenic crops. Many questions about transgenic crops have been raised: Will Bt-resistant insects develop? Will disease-causing bacteria acquire the antibiotic-resistance genes used as markers in transgenic crops? Is it safe to eat food carrying part of a virus gene used in switching on transgenes? There clearly is a need to consider health and environmental risks as transgenic crops become more common and new ones are developed.

Transgenic Animals as Models for Human Diseases

Gene transfer has been used to create animal models of human diseases, including genetic disorders and infectious diseases. As an example, let's look at the construction of a mouse model of Huntington disease (HD) (MIM/OMIM 143100), a neu-

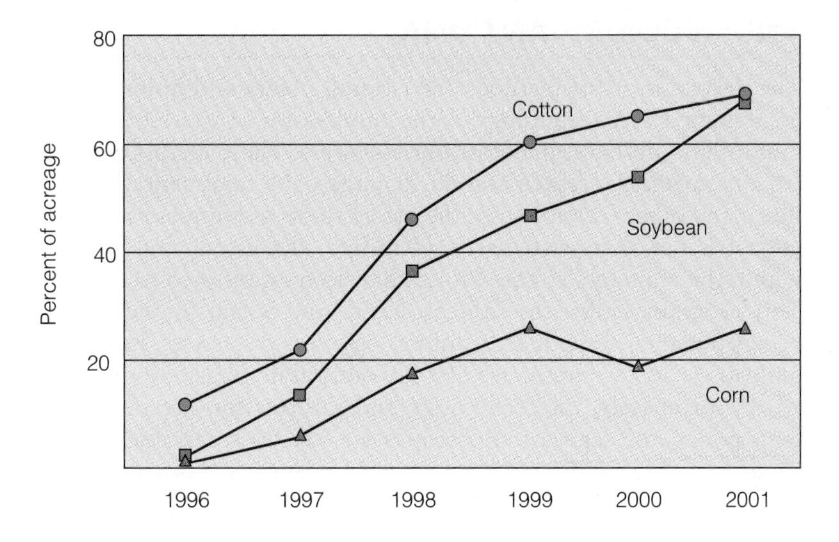

▶ **FIGURE 13.16** U.S. adoption of transgenic crops from 1996 to 2001. (Reprinted with permission of the Department of Soil & Crop Sciences, Colorado State University. http://www.colostate.edu/programs/lifesciences/TransgenicCrops/current.html)

Table 13.5

Some Animal Models of Human Disease

Alcoholism
Amyotrophic lateral sclerosis
Atherosclerosis
Cardiac hypertrophy
Colon cancer
Familial Alzheimer disease
Huntington disease
Lung cancer
Prostate cancer

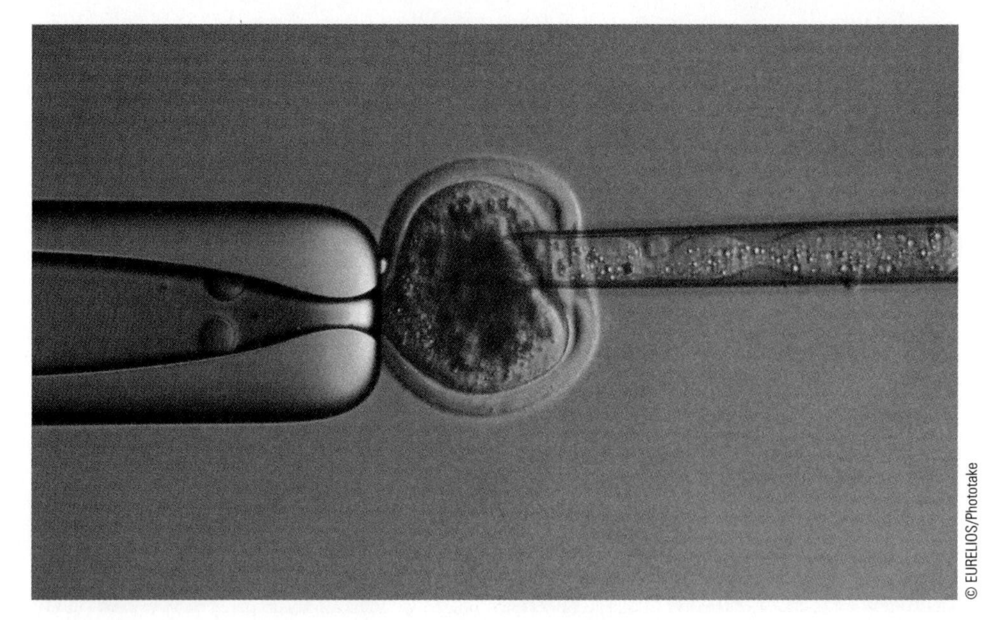

FIGURE 13.17 Microinjection into a mammalian zygote.

rodegenerative disorder inherited as an autosomal dominant trait. First a copy of the mutant HD allele was cloned into a vector next to a promoter sequence that controls expression of the gene. Genes placed next to this promoter will be expressed in the nervous system and other organs. Next, the vector carrying the HD allele was microinjected into the nucleus of a mouse zygote (the zygote was produced by *in vitro* fertilization) (Figure 13.17). Injected zygotes were implanted into foster mothers. Finally, DNA was isolated from the tail tips of newborn mice and genotyped for the presence of the mutant HD allele by Southern blot analysis or by PCR. Mice carrying an active HD allele can be used to study the relationship between the behavioral phenotype and changes in brain structure, to study the molecular events associated with this disorder, and to test drugs that can be used in treatment of this disorder (see Chapter 16). Similar methods have been used to construct models of many human genetic disorders and infectious diseases (Table 13.5).

ETHICAL QUESTIONS ABOUT CLONING AND RECOMBINANT DNA

The development of methods for cloning plants and animals and the use of recombinant DNA technology is the foundation of the biotechnology industry, a multibillion-dollar segment of our economy. These methods are now revolutionizing biomedical research and the diagnosis and treatment of human diseases. Unlike many other technologies, the use of these techniques raises serious social and ethical issues that need to be identified, discussed, and resolved. A program within the Human Genome Project has been established to deal with these issues. This program, called the Ethical, Legal, and Social Implications (ELSI) project, brings together individuals from biology, social sciences, history, law, ethics, and philosophy to explore issues and lay out public policy alternatives.

Unfortunately, the technology and information are developing much faster than policy and legislation. Because we *can* do some things does not mean we *should* do them without a consensus from society. For example, should we permit human cloning as part of programs in assisted reproduction? Should we allow gene transfer to germ cells, altering the genetic makeup of future generation without their consent? Should gene therapy be used as a form of "improving" individuals rather than curing a genetic disease? Should very short children receive recombinant-DNA–produced growth hormone to make them average in height? If it is acceptable to treat abnormally short children, is it ethical to treat children of normal height to enhance their chances of becoming professional basketball or volleyball players?

The current guidelines of the U.S. Food and Drug Administration do not require identifying labels on food produced by recombinant DNA technology. Should such food be labeled and if so, why? It can be argued that food products have been genetically manipulated for thousands of years and that gene transfer is simply an extension of past practices. It can also be claimed that consumers have a need and a right to know that gene transfer has altered food products.

The legal, moral, and ethical implications of cloning and recombinant DNA technology need to be considered carefully. As outlined earlier, some of these issues are being addressed by the Human Genome Project; in other areas, however, discussion, education, and policy formation lag far behind the technology. As citizens, it is our responsibility to become informed about these issues and to participate in formulating policy about the use of this technology.

Summary

Recombinant DNA Techniques Have Revolutionized Human Gene Mapping

1. Gene mapping has been greatly enhanced by using recombinant DNA techniques. Genes can be mapped by several techniques, including positional cloning, direct mapping to chromosomes, and genome sequencing.

DNA Fingerprinting

2. DNA fingerprinting uses variations in the length of repetitive DNA clusters to establish a molecular profile called a DNA fingerprint. DNA fingerprinting is used in cases of criminal and civil law, as well as paternity testing, evolution, conservation, and a wide range of other applications.

The Human Genome Project Is an International Effort

3. The Human Genome Project uses recombinant DNA technology to map all of the estimated 35,000 to 50,000 genes carried in human cells. In the United States, a government-sponsored project and a privately-funded project completed a draft of the human genome sequence in 2000. The sequence is being completed and is being analyzed to determine how many genes we carry, what they do, and where they are located. We have learned much about our genome already and will learn much more when the sequence is completed and analyzed.

Gene Transfer Technology Has Many Applications

4. Genetic engineering is used to manufacture proteins used in treating human diseases. This provides a constant supply that is uncontaminated by disease-causing agents. Gene transfer into crop plants has conferred resistance to herbicides, insect pests, and plant diseases. Gene transfer techniques are being used to create animal models of human diseases as a way of studying the molecular mechanisms associated with the disease process and to develop drugs for treatment of these disorders.

Ethical Questions About Cloning and Recombinant DNA

5. Recombinant DNA technology and its many applications have developed faster than societal consensus, public policy, and laws governing its use. In addition, efforts to inform legislators, members of the legal and medical profession, and the public at large about the applications of this technology have often lagged behind its commercial use. An effort to educate and debate the risks and benefits of how we use genetic engineering will ensure a balanced approach to this technology.

CASE 1

Todd and Shelly Z. were referred for genetic counseling because of advanced maternal age (>40 years) in their current pregnancy. While obtaining the family history, the counselor learned that during their first pregnancy, in 1993, they had elected to have an amniocentesis for prenatal diagnosis of cytogenetic abnormalities because Shelly was 36 years old at the time. During that pregnancy, Shelly and Todd reported that they were also concerned that Todd's family history of cystic fibrosis (CF) increased their risk for having an affected child. Todd's only sister had CF, and she had severe respiratory complications.

The genetic counseling and testing was performed at an outside institution, and the couple had not brought copies of the report with them. They did state that they had completed studies to determine their CF carrier status and that Todd was found to be a CF carrier, but Shelly's results were negative. The couple was no longer concerned about their risk of having a child with CF based on these results. To support their belief, they had a healthy 5-year-old son who had a negative sweat test at the age of 4 months. The counselor explained the need to review the records and scheduled a follow-up appointment.

Review of the test report from 1993 indicated that the only CF mutation tested at that time was the delta F508 mutation. The reports confirmed that Shelly is not a carrier of this mutation and that Todd is. Shelly's report from 1993 indicated that her CF risk status had been reduced from 1 in 25 to about 1 in 300. More information has been learned about the different mutations in the CF gene since the last time they received genetic counseling. The counselor conveyed the information about recent advances in CF testing to the couple, and Shelly decided to have her blood drawn for CF mutational analysis with an expanded panel of mutations. Her results showed that she is a carrier for the W1282X CF mutation. The family was given a 25% risk for CF for each of their pregnancies based on their combined molecular test results. They proceeded with the amniocentesis because of the risks associated with advanced maternal age and requested fetal DNA analysis for CF mutations. The fetal was positive for both parental mutations, indicating that the fetus had a greater than 99% chance of being affected with CF.

1. How is it that the fetus has a greater than 99% chance of being affected with CF if each parent carries a different mutation? Is the fetus homozygous or heterozygous for these mutations?
2. If this child has CF, what are the chances that any future child will have this disease? Does the fact they have a normal five-year-old son affect the chances of having future children with CF?

CASE 2

Can DNA fingerprinting uniquely identify the source of a sample? Because any two human genomes differ at about 3 million sites, no two persons (except identical twins) have the same DNA sequence. Unique identification with DNA typing is therefore possible if enough sites of variation are examined. However, the DNA typing systems used today examine only a few sites of variation and have only limited resolution for measuring the variability at each site. There is a chance that two persons might have DNA patterns (i.e., genetic types) that match at the small number of sites examined. Nonetheless, even with today's technology, which uses 3 to 5 loci, a match between two DNA patterns can be considered strong evidence that the two samples came from the same source. How is DNA fingerprinting currently being used?

- Paternity and maternity testing: Because a person inherits his or her STRs from his or her parents, STR patterns can be used to establish paternity and maternity. The patterns are so specific that a parental STR pattern can be reconstructed even if only the children's STR patterns are known (the more children produced, the more reliable the reconstruction). Parent-child STR pattern analysis has been used to solve standard father-identification cases and more complicated cases of confirming legal nationality and, in instances of adoption, biological parenthood.
- Criminal identification and forensics: DNA isolated from blood, hair, skin cells, or other genetic evidence left at the scene of a crime can be compared, through STR patterns, with the DNA of a criminal suspect to determine guilt or innocence. STR patterns are also useful in establishing the identity of a homicide victim, either from DNA found as evidence or from the body itself. The O. J. Simpson trial used DNA evidence in its investigation.
- Personal identification: The notion of using DNA fingerprints as a sort of genetic bar code to identify individuals has been discussed, but this is not likely to happen any time in the foreseeable future. The technology required to isolate, keep on file, and analyze millions of very specific STR patterns is both expensive and impractical. Social security numbers,

Case Studies (continued)

picture IDs, and other more mundane methods are much more likely to remain the ways to establish personal identification. However, the FBI and other police agencies and the military are enthusiastic proponents of DNA databanks. The practice has civil-liberties implications because police can go on "fishing expeditions" with DNA readouts.

1. A police chief in a large U.S. city has proposed that everyone who is arrested be DNA fingerprinted; this information would be stored in a national database, even if those arrested are later found to be innocent. The chief is actively seeking funding from his state legislature to begin such a program. Do you support such a program? Why or why not? Could there be abuses of this program?
2. Several companies advertise DNA fingerprinting services on the Internet for genealogy research. Given that most ancestors are dead and not available for such testing, how can this service be of value in genealogy? Are there privacy concerns about what the companies do with the results of such testing?

Questions and Problems

Recombinant Techniques Have Revolutionized Human Gene Mapping

1. What does the acronym RFLP stand for? Explain this term word by word.
2. RFLP sites are useful as genetic markers for linkage and mapping studies, as well as for disease diagnosis. Keeping in mind that RFLPs behave as genes, what factors are important in selecting RFLPs for use in such studies?
3. In the examples given in this chapter (Figure 13.2), the RFLP is always inherited with (linked to) the genetic disorder. In practice, however, it is sometimes observed that in some progeny the RFLP is not linked to the genetic disorder, reducing its effectiveness in detecting carrier heterozygotes. How do you explain this observation, and what circumstances are likely to affect the frequency with which this nonlinkage occurs?
4. FISH is a technique that uses a fluorescently labeled probe to bind to metaphase chromosomes. The results are then viewed using a microscope. What kind of probe would you suggest to detect Down syndrome?

DNA Fingerprinting

5. STRs are:
   a. used for DNA fingerprinting.
   b. repeated sequences present in the human genome.
   c. highly variable in copy number.
   d. all of the above.
   e. none of the above.
6. A crime is committed and the only piece of evidence the police are able to gather is a small bloodstain. The forensic scientist at the crime lab is able to:
   a. extract DNA from the blood.
   b. cut the DNA with a restriction enzyme.
   c. separate the fragments by electrophoresis.
   d. transfer the DNA from a gel to a membrane and probe with radioactive DNA.

   Probe 1 is used to visualize the pattern of bands. The forensic scientist compares the band pattern in the evidence (E) with the patterns from the suspects (S1, S2). The first probe is removed, the membrane is hybridized using another probe (Probe 2), and the band patterns are compared. This process is repeated for Probe 3 and Probe 4.

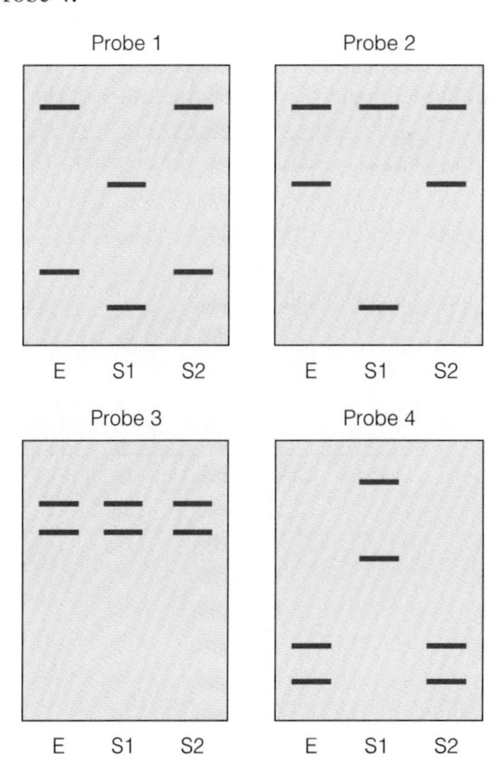

a. Based on the results of this testing, can either of the suspects be excluded as the one who committed the crime?
b. If so, which one? Why?
c. Is the pattern from the evidence consistent with the band pattern of one of the suspects? Which one?

7. You are serving on a jury in a murder case in a large city. The prosecutor has just stated that DNA fingerprinting shows that the suspect must have committed the crime. He says that for cost reasons, two STR probes were used instead of four, but that the results are just as accurate. The first probe detected a locus that has a population frequency of 1 in 100 (1%), and the second probe has a population frequency of 1 in 500 (0.2%).
   a. What is the combined frequency of these alleles in the population?
   b. Does this point to the suspect as the perpetrator of the crime?
   c. Is there any other evidence you want from the DNA lab?
   d. Do you think that most people on juries understand basic probabilities, or do you think that DNA evidence can be used to trick jurors into reaching false conclusions?
8. DNA fingerprinting has been used in criminal identification, establishing blood lines of pure-bred dogs, identifying endangered species, and studying extinct animals. Can you think of other uses for DNA fingerprinting?
9. A paternity test is conducted using PCR to analyze an RFLP that consistently produces a unique DNA fragment pattern from a single chromosome. Examining the results of the Southern blot below, which male(s) can be excluded as the father of the child? Which male(s) could be the father of the child?

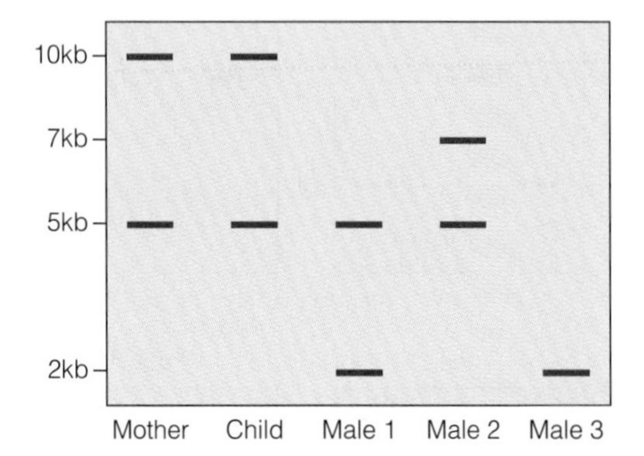

The Human Genome Project Is an International Effort

10. Which of the following is not a goal of the Human Genome Project?
    a. construction of high-resolution genetic maps
    b. sequencing the entire genome
    c. curing all genetic diseases
    d. generating a physical map of each chromosome
    e. obtaining a set of overlapping clones covering the genome
11. Define *genomics* and *bioinformatics*.
12. What are the benefits and drawbacks of the "clone-by-clone" versus the "shotgun" methods of cloning?
13. How will researchers know the final count of genes in the human genome?
14. Genome Data
    a. How many base pairs are in the Human Genome?
    b. What percent of the genome is thought to encode genes?
    c. What is the estimated number of genes in the human genome?
15. For the disease cystic fibrosis, answer the following questions:
    a. Which chromosome does this gene map to?
    b. What is the normal function of protein encoded by this gene?
    c. How does the mutant protein produce the disease phenotype?

Gene Transfer Technology Has Many Applications

16. What organisms are being used to make human proteins? Are some better than others?

Ethical Questions About Cloning and Recombinant DNA

17. What are some of the potential dangers of recombinant DNA technology? How should these potential problems be investigated to determine if they are in fact dangers?
18. Answer the following questions with a "yes" answer and a "no" answer, and support each view with a coherent reason.
    a. Should very short children receive recombinant human growth hormone (HGH) to reach average height?
    b. Should genetically modified foods be labeled in grocery stores?
    c. Is it acceptable for physicians to treat children of normal height with HGH to make them taller if they and their parents want this?

Internet Activities

The following activities use the resources of the World Wide Web to enhance the topics covered in this chapter. To investigate the topics described below, log on to the book's home page (www.brookscole.com/biology_d) and follow the *Student Resources* link to your subject and text.

1. *RFLP Analysis.* This technique can be used to identify and compare human DNA samples on the basis of the DNA "fingerprint." Conclusions from these comparisons can be used in paternity cases or criminal trials. At the University of Arizona's *Biology Project: Human Biology* Web page (http://www.biology.arizona.edu/human_bio/human_bio.html), there are several problem sets and activities to enhance your understanding of RFLP analysis. The two "DNA Forensics" problem sets allow you to try your hand at answering questions about RFLP data, and provide tutorials to explain difficult concepts. The two "Blackett Family DNA Activities" provide opportunities to work with RFLP profiles and the newer STR (short tandem repeat) technique.
2. *Discussing Biotechnology and Genetically Engineered Products.* The *Union of Concerned Scientists* Web site (http://www.ucsusa.org/index.html) provides information and a forum to discuss scientific aspects of biotechnology, as well as the marketing of genetically engineered products. At the Web site, click on the "Genetically Engineered Food" link to find the answers to frequently asked questions about biotechnology, as well as fact sheets on current biotechnology issues. Read the position statement on the "Genetically Engineered Food" Web page. What are the positions—and potential biases—of the scientists in this organization? Do you share them, or do you disagree? And don't forget to check out the "Transgenic Cafe"!
3. *The Gene Therapy Debate.* The safety, efficacy, and ethical concerns surrounding gene therapy have been debated within and outside the biomedical community since the first gene therapy protocol for the genetic disease ADA was carried out. On the *Biotechnology: Ethics and Industry* page maintained by Industry Canada (http://strategis.ic.gc.ca/SSG/bb00010e.html#AP1) you can link to and review a series of Discussion Scenarios designed to highlight different points of view on ethical issues associated with gene therapy use and research. Choose one or two scenarios to focus on or debate with other students. What kinds of risks and what kinds of testing do you feel are acceptable in the development of theses kinds (or any kind) of medical treatment? At the top of this page, clicking on the *Whose Values? Who Decides?* icon will take you to a discussion of ethical decision making. Clicking on highlighted words within the *Basic Background* section or on the *Biotechnology Science Centre* link within the Discussion Scenarios will take you to review information on genetics and biological terminology.

For Further Reading

Anderson, W. F. (2000). The best of times, the worst of times. *Science, 288,* 627–628.

Beaudet, A. L. (1999). Making genomic medicine a reality. *Am. J. Hum. Genet. 64,* 1–13.

Caskey, C. (1993). Presymptomatic diagnosis: A first step toward genetic health care. *Science, 262,* 48–49.

Crystal, R. G. (1995). Transfer of genes to humans: Early lessons and obstacles to success. *Science, 270,* 404–410.

Daniell, H., Streatfield, S. J., & Wycoff, K. (2001). Medical molecular farming: Production of antibodies, biopharmaceuticals, and edible vaccines in plants. *Trends Plant Sci., 6,* 219–226.

deWachter, M. (1993). Ethical aspects of human germ-line therapy. *Bioethics, 7,* 166–177.

Estruch, J. J. (1997). Transgenic plants: An emerging approach to pest control. *Nature Biotech., 15,* 137–141.

Friedman, T. (1989). Progress toward human gene therapy. *Science, 244,* 1275–1281.

Gilbert, W., & Villa-Komaroff, L. (1980, April). Useful proteins from recombinant bacteria. *Sci. Am., 242,* 74–97.

Hubbard, R., & Wald, E. (1993). *Exploding the gene myth.* Boston: Beacon Press.

Kevles, D., & Hood, L. (1992). *The code of codes: Scientific and social issues in the Human Genome Project.* Cambridge, MA: Harvard University Press.

Lapham, E. V., Kozma, C., & Weiss, J. O. (1996). Genetic discrimination: Perspectives of consumers. *Science, 274,* 621–624.

Neufeld, P. J., & Colman, N. (1990, May). When science takes the witness stand. *Sci. Am., 262,* 46–53.

Tacket, C. O., Mason, H. S., Losonsky, G., Clements, J. D., Levine, M. M., & Arntzen, C. J. (1998). Immunogenicity in humans of a recombinant bacterial antigen delivered in a transgenic potato. *Nature Medicine, 4,* 607–609.

Travis, J. (1998). Another human genome project. *Science News, 153,* 334–335.

Velander, W. H., Lubon, H., & Drohan, W. N. (1997). Transgenic livestock as drug factories. *Sci. Amer., 276,* 70–75.

White, R., & Lalouel, J. M. (1988, February). Chromosome mapping with DNA markers. *Sci. Am., 258,* 40–47.

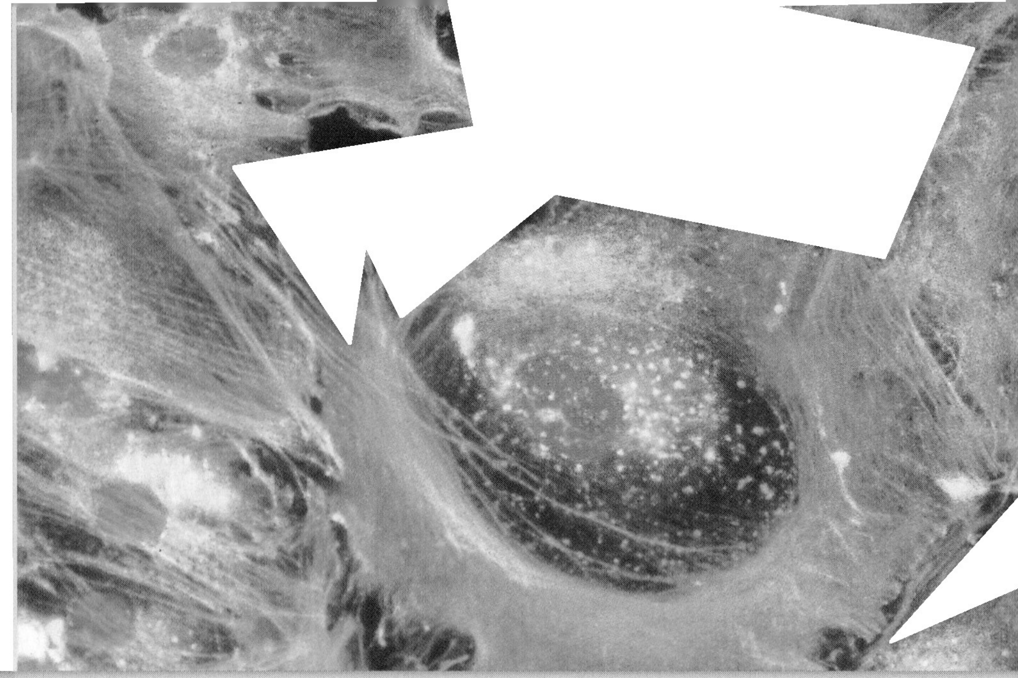

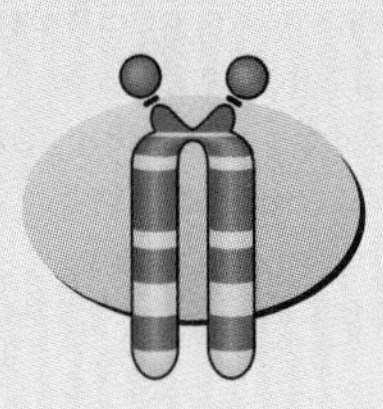

Genes and Cancer

Chapter Outline

Epidemiology and Cancer Links to Environmental Factors
Occupational Hazards and Cancer Risk
GUEST ESSAY A JOURNEY THROUGH SCIENCE
Environmental Factors and Cancer

As you sit in a classroom, a crowded stadium, or a concert hall, look around. According to the American Cancer Society, about one in three of the people you see will develop cancer at some point in their life, and about one in four will die from cancer. Each year about 500,000 people die of cancer, a rate of approximately one death per minute, and almost 1.5 million new cases of cancer are diagnosed annually in the United States (Table 14.1). Currently more than 10 million individuals are receiving medical treatment for cancer in U.S. hospitals and medical centers.

Cancer is a complex disease that affects many different cells and tissues in the body. It is characterized by uncontrolled cell division and by the ability of these cells to spread, or metastasize, to other sites within the body. Unchecked, growth and metastasis result in death, making cancer a devastating and feared disease. Improvements in medical care have reduced deaths from infectious disease and have led to increases in life span, but these benefits have also helped make cancer a major cause of illness and death in our society. The risk of many cancers is age-related, and because more Americans are living longer, they are at greater risk of developing cancer.

The link between cancer and genetic mutations was forged early in the last century by Theodore Boveri, who proposed that normal cells mutate into malignant cells because of changes in chromosome constitution. Four lines of evidence support the idea that cancer has a genetic origin:

1. *More than 50 forms of cancer have some degree of inherited predisposition.*
2. *It has been shown that most carcinogens are also mutagens.*
3. *Some viruses carry mutant genes, known as oncogenes, that promote and maintain the growth of a tumor.*
4. *As Boveri proposed, specific chromosomal changes are found in some cancers, especially leukemia.*

Mutation is a common feature of all cancers. In most cases, these mutations take place in somatic cells, and the mutant alleles are not passed on

Table 14.1 Estimated New Cases of Cancer in the United States, 2002

Site	Number of New Cases
Skin	>800,000
Lung	169,400
Colon-rectum	148,300
Breast (female)	203,500
Prostate	189,000
Urinary System	90,700
Uterus	52,300
Pancreas	30,300
Ovary	23,300

to offspring. In other cases, mutations occur in germ cells and are passed on to succeeding generations as a predisposition to cancer. Cancer, then, is a genetic disorder that acts at the cellular level.

Because mutation is the ultimate cause of cancer and because there is a constant background of spontaneous mutations, there will always be a baseline rate of cancer. The environment (ultraviolet light, chemicals, and viruses) and behavior (diet and smoking) can also play a significant role in cancer risks by increasing the rate of mutation.

In this chapter we examine the relationship between genes and cancer and describe the role of tumor suppressor genes and oncogenes in causing and supporting malignant transformation in cells. We discuss the relationship between leukemia and chromosomal aberrations, and, finally, we analyze the interaction between cancer and the environment to explain how a multitude of factors may initiate the multistep process required for cancer to develop.

CANCERS ARE MALIGNANT TUMORS

An important distinction exists between tumors and cancer. Tumors are abnormal growths of tissue. Benign tumors are self-contained, noncancerous growths that do not spread to other tissues and are not invasive. Benign tumors, including cysts, usually cause problems by increasing in size until they interfere with the function of neighboring organs.

Cancers are malignant tumors with several characteristics. They are clonal in origin; cancers arise from a single cell (usually in somatic tissue). Cancers develop through a series of genetic alterations that result in more aggressive growth with each mutation. Third, cancers are invasive and metastatic. Cells can detach from the primary tumor and move to other sites in the body to form new malignant tumors. The ability to metastasize results from additional mutations in cancer cells. In the following sections we consider the genetic changes that take place within single cells that lead to the formation of malignant tumors.

MUTATIONS IN SPECIFIC GENES CAN PREDISPOSE TO CANCER

Families with high rates of cancer have been known for hundreds of years, but in most cases no clear-cut pattern of inheritance can be identified. How is it, then, that some families have a rate of cancer that is much higher than average? Many explanations have been offered, including multiple gene inheritance, environmental agents, or chance alone (Table 14.2).

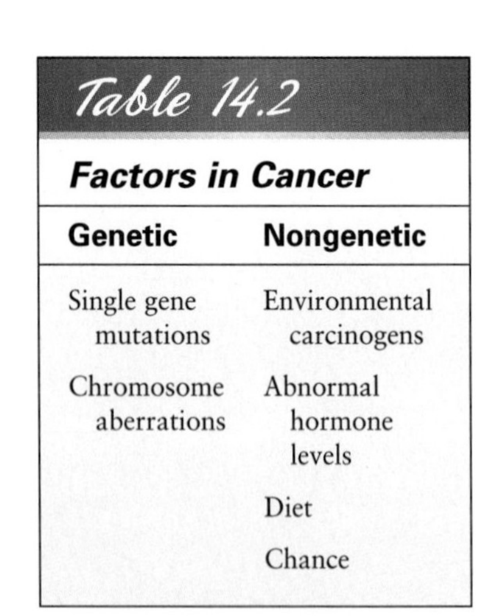
Table 14.2

Factors in Cancer

Genetic	Nongenetic
Single gene mutations	Environmental carcinogens
Chromosome aberrations	Abnormal hormone levels
	Diet
	Chance

Cancer has been linked to mutations in specific genes and is viewed as a disease caused by a small number of mutations in a single cell that take place over time. Although the exact number of mutations varies for different cancers, two mutations may be the minimum number needed to cause cancer (Figure 14.1). In most cases, these mutations accumulate over a period of years, explaining why age is a leading risk factor for many cancers.

Not all cancers are the result of random mutations; some individuals inherit a predisposition to cancer. As a result, these individuals have one mutant gene present in germ cells and all other cells of the body (Table 14.3). One or more additional mutations accumulate spontaneously or by exposure to environmental agents that cause genetic damage, resulting in cancer. In some cases, those who carry a mutant allele predisposing them to cancer have a 100,000-fold increased risk of cancer. If no other mutations take place, however, no malignant tumors develop.

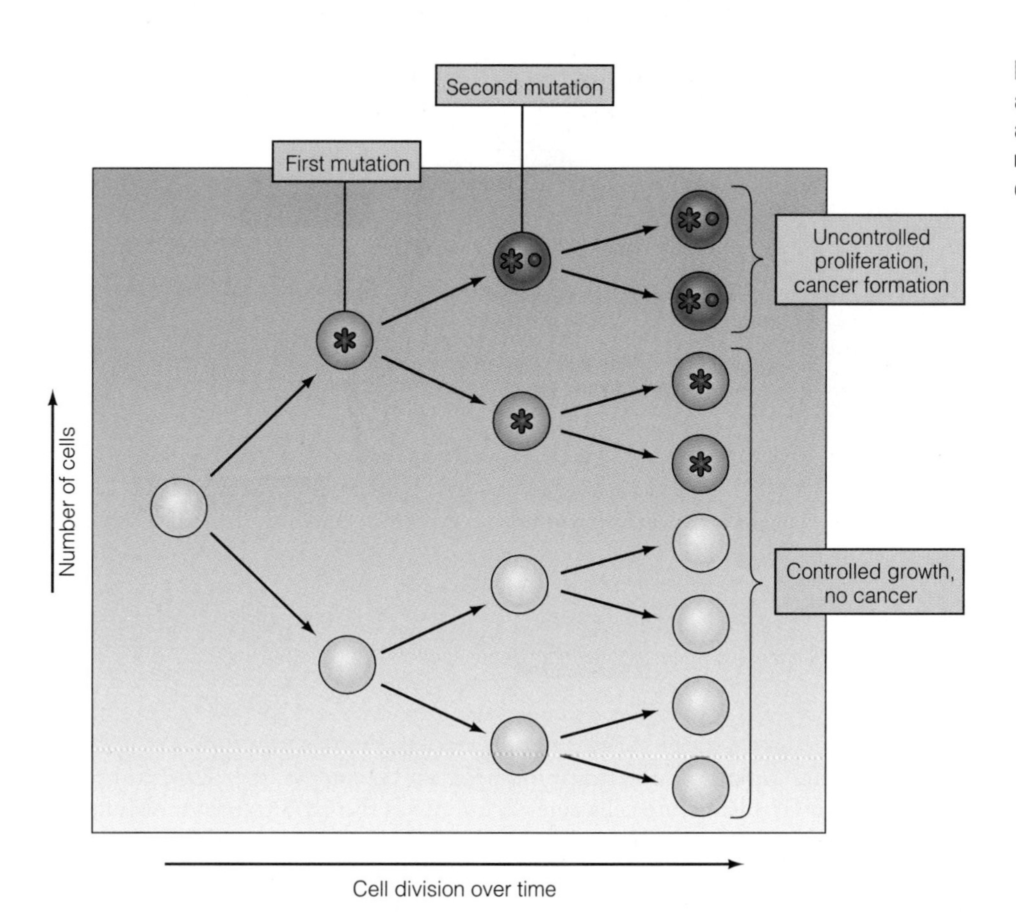

FIGURE 14.1 Over time, cells acquire mutations. Two independently acquired mutations in the same cell may be enough to cause uncontrolled cell growth and cancer.

Table 14.3 **Heritable Predispositions to Cancer**

Disorder	Chromosome	MIM/OMIM Number
Early-onset familial breast cancer	17q	113705
Familial adenomatous polyposis	5q	175100
Hereditary nonpolyposis colorectal cancer	2p	120435
Li-Fraumeni syndrome	17p	151623
Multiple endocrine neoplasia type 1	11q	131100
Multiple endocrine neoplasia type 2	10	171400
Neurofibromatous type 1	17q	162200
Neurofibromatous type 2	22q	101000
Retinoblastoma	13	180200
Von Hippel-Lindau disease	3p	193300
Wilms tumor	11p	194070

TUMOR SUPPRESSOR GENES, ONCOGENES, AND THE CELL CYCLE

Cancer cells are characterized by uncontrolled cell division. These cells bypass checkpoints in the cell cycle that regulate cell division. As a result, studies of the cell cycle are now an important part of cancer research. The 2001 Nobel Prize for Physiology or Medicine was awarded to three scientists who helped establish the link between the cell cycle and cancer.

Recall from Chapter 2 that events in interphase and mitosis define the cell cycle (Figure 14.2). The cycle is regulated at several points. For our discussion, two

FIGURE 14.2 The cell cycle consists of two parts: the interphase and mitosis. The interphase is divided into three stages, G1, S, and G2, which make up the major part of the cycle. Mitosis involves partitioning cytoplasm and replicated chromosomes to daughter cells.

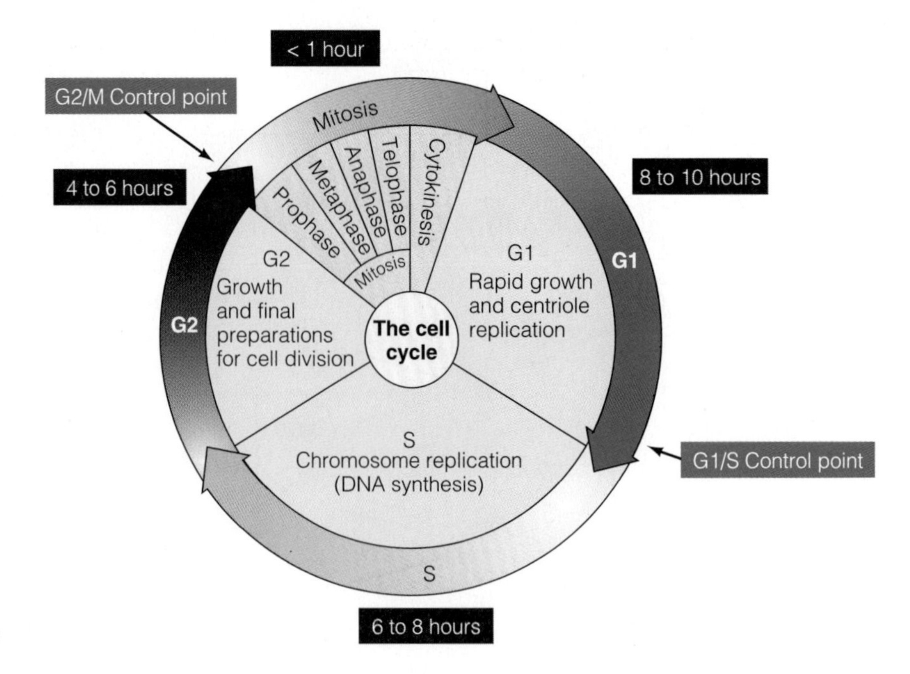

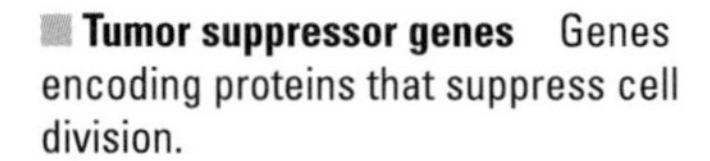

Tumor suppressor genes Genes encoding proteins that suppress cell division.

Proto-oncogenes Genes that initiate or maintain cell division and that may become cancer genes (oncogenes) by mutation.

Oncogenes Genes that induce or continue uncontrolled cell proliferation.

Retinoblastoma A malignant tumor of the eye arising in retinoblasts (embryonic retinal cells that disappear at about 2 years of age). Mature retinal cells do not transform into tumors, so this is a tumor that usually occurs in children. It is associated with a deletion on the long arm of chromosome 13.

checkpoints are important: the transition between G2 and M (the G2/M transition) and a point in G1 just before cells enter S, known as the G1/S transition. Mutations in genes that regulate these control points lead to the formation of cancer. In general, two types of genes regulate the cell cycle: (1) genes that suppress cell division and (2) genes that stimulate cell division. The first type is known as **tumor suppressor genes.** These genes act at either the G1/S or the G2/M control points to inhibit cell division. If these genes are lost or inactivated by mutation, cell division cannot be switched off, and cells divide in an uncontrolled manner. The second class of regulatory genes, called **proto-oncogenes,** act to turn on or maintain cell division. When these genes are active, cells undergo division. If these genes become permanently activated or overproduce their products, uncontrolled cell division results. Mutant forms of proto-oncogenes are called **oncogenes.**

The following examples describe how mutations in tumor suppressor genes control the development of cancer. After we discuss oncogenes and their involvement in cancer, a genetic model of colon cancer will integrate what is known about tumor suppressor genes, oncogenes, and control of the cell cycle.

FIGURE 14.3 A child with retinoblastoma in one eye.

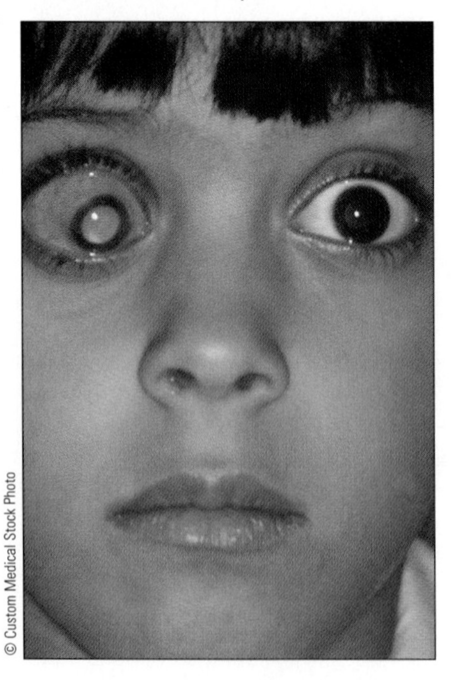

Retinoblastoma Is Caused by Mutation in a Tumor Suppressor Gene

Retinoblastoma (*RB1*) (MIM/OMIM 180200) is a cancer of the eye that affects the retina. It occurs in 1 in 14,000 to 1 in 20,000 births. Although it may be present in infancy, it is most often diagnosed between the ages of 1 and 3 years. There are two forms of retinoblastoma. Hereditary retinoblastoma (accounting for 40% of all cases) is an autosomal dominant trait in which a predisposition to retinoblastoma is inherited. In families with this trait, children have a 50% chance of inheriting the mutant *RB* allele, and 90% of these individuals will develop retinoblastoma, usually in both eyes. In addition, those carrying the mutant allele are at high risk for other cancers, especially osteosarcoma and fibrosarcoma. The second form (60% of all cases) is sporadic retinoblastoma. Affected individuals have not inherited any mutant retinoblastoma alleles. Instead, sometime early in childhood, mutations arise that lead to retinoblastoma. In sporadic cases, tumors usually develop in only one eye and patients are not at high risk for other cancers (Figure 14.3).

Alfred Knudson and his colleagues proposed a two-step model that explains both forms of the disease. According to their model, retinoblastoma develops when two mutant copies of the *RB* gene are present in a single retinal cell. The model predicts that retinoblastoma can arise in two ways. If a child inherits a mutant *RB* allele, all cells of the body, including those in the retina, carry this mutation. If the normal copy of the *RB* gene in a retinal cell becomes mutated, the child will develop retinoblastoma (Figure 14.4). Because all retinal cells already carry one mutant gene, this form of the disease is likely to involve both eyes (bilateral cases) and occur at an earlier age than the sporadic form.

If a child begins life with two normal alleles of the *RB* gene, then both copies of the *RB* gene must become mutated in a single retinal cell for a tumor to develop. Because the chance that two *RB* mutations will occur in the same cell is low, sporadic cases are more likely to involve only one eye (unilateral cases). Because more time is required for two mutations to occur in a single cell, the sporadic form of retinoblastoma arises later in childhood.

Surveys of patients who have retinoblastoma have confirmed the predictions made by the model. Pedigree analysis indicates that most bilateral cases involve an inherited mutant allele, but the vast majority of cases of retinoblastoma in one eye are sporadic and result from acquired mutations. In addition, most cases of retinoblastoma in one eye occur at a later age (although still in childhood) than bilateral cases.

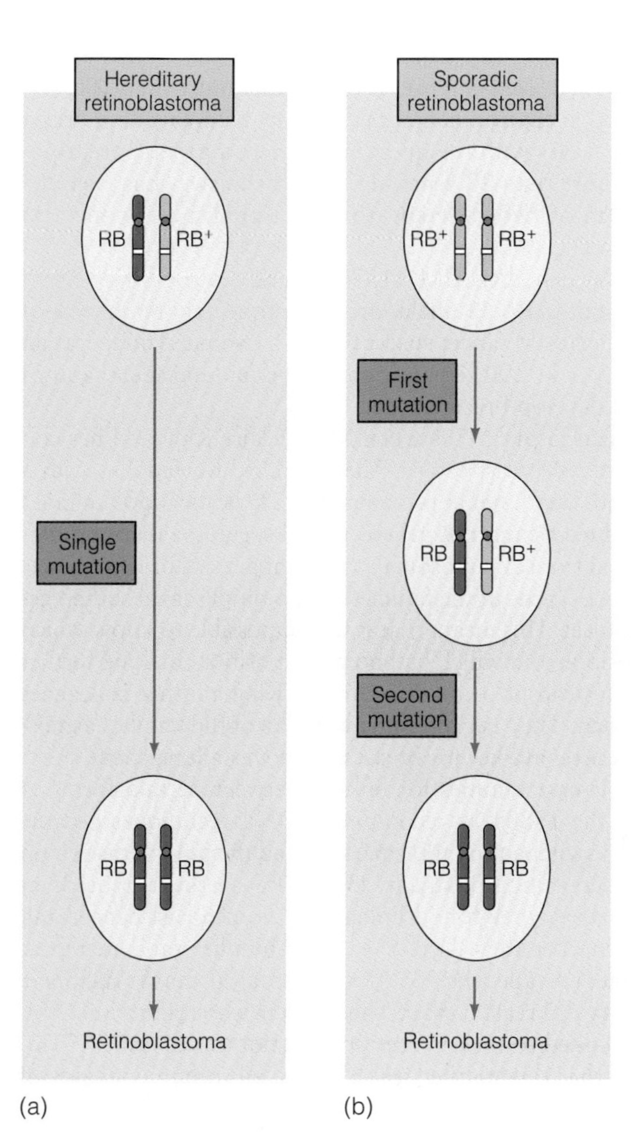

FIGURE 14.4 A model of retinoblastoma. In hereditary cases, one mutation is inherited. (a) Inheriting a single mutation in the normal allele of the *RB* gene causes a predisposition to retinoblastoma. If the normal *RB* allele mutates, retinoblastoma results. (b) Sporadic retinoblastoma requires two independent mutations in the *RB* gene in a single cell.

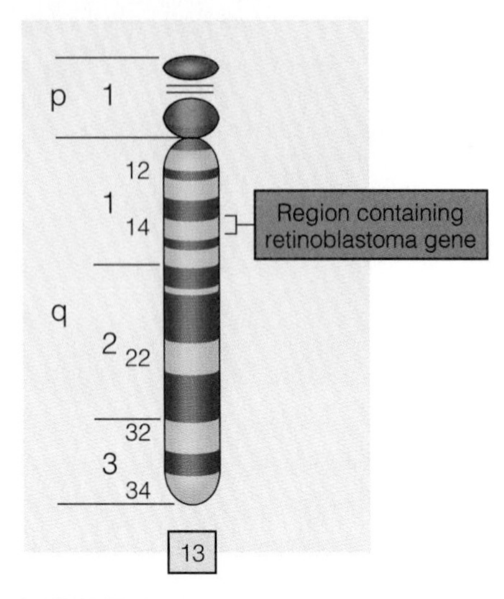

▶ **FIGURE 14.5** A diagram of chromosome 13 showing the retinoblastoma locus.

The *RB* gene is located on chromosome 13 at 13q14 (▶ Figure 14.5) and encodes a protein (pRB) found only in the nucleus. pRB is present in retinal cells and almost all other cell types and tissues of the body. pRB is in the nucleus at all stages of the cell cycle, but its *activity* is regulated synchronously with the cell cycle. The protein is a molecular switch that controls progression through the cell cycle. If pRB is in its active form during G1, the cell will not move into S and will not go on to divide. On the other hand, if pRB protein is inactive in G1, the cell moves to S phase, through G2, and on to mitosis. If both copies of the *RB* gene are deleted or become mutated in a retinal cell, there is no pRB available to regulate cell division, and the cell begins to divide in an uncontrolled manner, forming a tumor.

What if both copies of the *RB* gene become mutated in a cell type other than retinal cells? In bone cells, the result is a cancer known as osteosarcoma. In fact, osteosarcoma cells grown in the laboratory are used to study the role of the *RB* gene in cancer. Osteosarcoma cells carry two mutant alleles of the *RB* gene and do not produce pRB. If a cloned copy of a normal *RB* gene is transferred into the osteosarcoma cells, pRB is produced, and cell division stops. This finding reinforces the idea that the retinoblastoma protein plays a central role in regulating the cell cycle.

The Search for Breast Cancer Genes

Breast cancer is the most common form of cancer in women in the United States. Each year, more than 40,000 women die from breast cancer, and more than 180,000 new cases are diagnosed. Although environmental factors may be involved in breast cancer, geneticists struggled for years with the question: Is there a genetic predisposition to breast cancer? After more than 20 years of work, the answer has been determined: yes. The mutant form of two different genes can predispose women to breast cancer and ovarian cancer.

One of these genes, *BRCA1* (MIM/OMIM 113705), maps to the long arm of chromosome 17. Although mutant *BRCA1* alleles are present in only about 5% of cases, this gene is responsible for breast cancer susceptibility in women younger than 40 years of age. Approximately 1 in 200 females inherit the mutant allele, and of these approximately 90% will develop breast cancer.

The search for *BRCA1* began in the 1970s when Mary Claire King and her colleagues analyzed familial patterns of breast cancer. They searched for families with a clear history of breast cancer and found that approximately 15% of the 1,500 families surveyed had multiple cases of breast cancer. A genetic model suggested that approximately 5% of these cases resulted from an autosomal dominant pattern of inheritance. The model also predicted that two-thirds of the families with multiple cases had no genetic predisposition to breast cancer. This meant that it was impossible to know which families had a genetic predisposition and which had no predisposition, making further research on the gene difficult. Instead of being discouraged, King and her colleagues decided on a brute force approach. They began testing as many of the families as possible, knowing that finding a genetic marker for breast cancer was a long shot.

They began testing multiple-case families for linkage between certain forms of proteins and breast cancer. In the 1980s, as recombinant DNA techniques became widely available, the King team switched to DNA markers and the polymerase chain reaction (PCR) technique for screening. Finally, in 1990, after testing hundreds of families using hundreds of markers, they found linkage between breast cancer and a genetic marker. The 183rd marker they used, a variable-number tandem repeat (VNTR) locus on chromosome 17 called D17S74, was tested on family members from 23 pedigrees with a history of breast cancer. This marker was coinherited with breast cancer and was clearly linked to the disease. Other laboratories quickly confirmed their results and found that this marker was also linked to familial cases of ovarian cancer.

The investigators formed an international consortium and began studying members of 214 families with multiple cases of breast cancer or ovarian cancer. They used restriction fragment length polymorphism (RFLP) markers to narrow the search for the gene to a small region on the long arm of chromosome 17 (Figure 14.6). The *BRCA1* gene was finally identified and cloned in 1994, some 20 years after King began the search. The mutant allele of *BRCA1* is associated with an autosomal dominantly inherited predisposition to breast cancer. Approximately 85% of women inheriting one mutant *BRCA1* allele will develop a mutation in the other *BRCA1* allele and develop breast cancer. Mutations in this gene are responsible for about half of all cases resulting from hereditary predisposition to breast cancer. Women with a *BRCA1* mutation are also at higher risk for ovarian cancer.

A second breast cancer predisposition gene, *BRCA2* (MIM/OMIM 600185), was discovered in 1995. Mutant alleles of this gene cause predisposition to breast cancer that is inherited as an autosomal dominant trait. *BRCA2* maps to q12-13 of chromosome 13 and may be responsible for the majority of inherited predispositions not caused by *BRCA1*. Although mutations in *BRCA1* and *BRCA2* account for most cases associated with a genetic predisposition (approximately 10% to 15% of all cases of breast cancer), mutations in these genes are not associated with the vast majority of breast cancers. There are other genes associated with breast cancer that remain to be discovered.

BRCA1 and *BRCA2* Are Tumor Suppressor Genes

Questions about the functions of both genes remain unanswered, as do questions about how mutations in these genes lead to breast cancer. Each gene encodes a large protein found only in the nucleus. In rapidly dividing cells, expression of *BRCA1* and *BRCA2* is highest at the G1/S boundary and into S phase. The BRCA1 protein is activated when DNA is damaged and has multiple functions; it stops DNA replication and participates in DNA repair. The BRCA2 protein has similar functions, and both proteins bind to RAD51, a protein involved in the repair of double-stranded breaks in DNA molecules.

Because the predisposition to breast cancer is inherited as an autosomal dominant trait and both alleles must be mutant for cancer to develop, these genes are regarded as tumor suppressor genes.

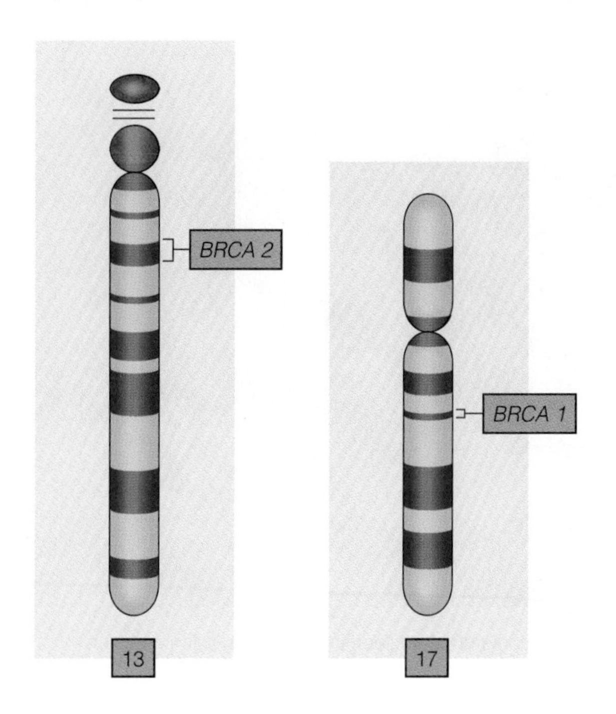

FIGURE 14.6 The chromosome locations for *BRCA1* and *BRCA2*. Together these genes account for the majority of cases of breast cancer associated with a genetic predisposition.

ONCOGENES AND CANCER

Oncogenes are mutant alleles of genes that normally turn on or maintain cell division and are turned off when division stops. In mutant form, the genes are permanently switched on, causing the uncontrolled cell division associated with cancer. The existence of oncogenes was first discovered by work on a virus that causes cancer in chickens.

Rous Sarcoma Virus and the Discovery of Oncogenes

Sarcomas Cancers of connective tissue. One type of sarcoma in chickens is caused by the Rous sarcoma retrovirus.

The understanding of the role of viruses in cancer began with Peyton Rous, who studied malignant tumors, known as **sarcomas,** found in chickens. In 1911 Rous discovered that cell-free extracts from sarcomas would cause tumor formation in healthy chickens. Others claimed that his extract was not cell-free and that he was simply transferring cancer cells into healthy chickens, causing sarcomas to develop. As a result of these criticisms, Rous gradually abandoned the project. Decades later, it was shown that the extract contains a virus, known as the Rous sarcoma virus (RSV) (Figure 14.7). RSV is now one of the most widely studied animal tumor viruses. In belated recognition of his pioneering work on viral tumors, Rous was awarded the Nobel Prize in 1966 (at the age of 85).

Retroviruses Viruses with an RNA genome. During the viral life cycle, the RNA is reverse-transcribed into DNA. The name retrovirus symbolizes this backward order of transcription.

Viruses can be grouped into two classes—DNA viruses or RNA viruses—depending on the nature of the genetic material they carry. Cancer-causing viruses are found in both groups (Table 14.4). DNA viruses that cause cancer include SV40 (simian virus 40), first identified in monkey tumors, and polyoma, a virus that produces tumors in several species of animals, including mice, hamsters, and rats. Some RNA viruses are called **retroviruses** because their single-stranded RNA genome must be converted into a double-stranded DNA molecule before it can replicate. Tumor-causing retroviruses have been found in the chicken, mouse, rat, hamster, and other species. RSV is an example of a tumor-producing retrovirus. The results of work with RSV and similar viruses have been important in understanding the origins of human cancer.

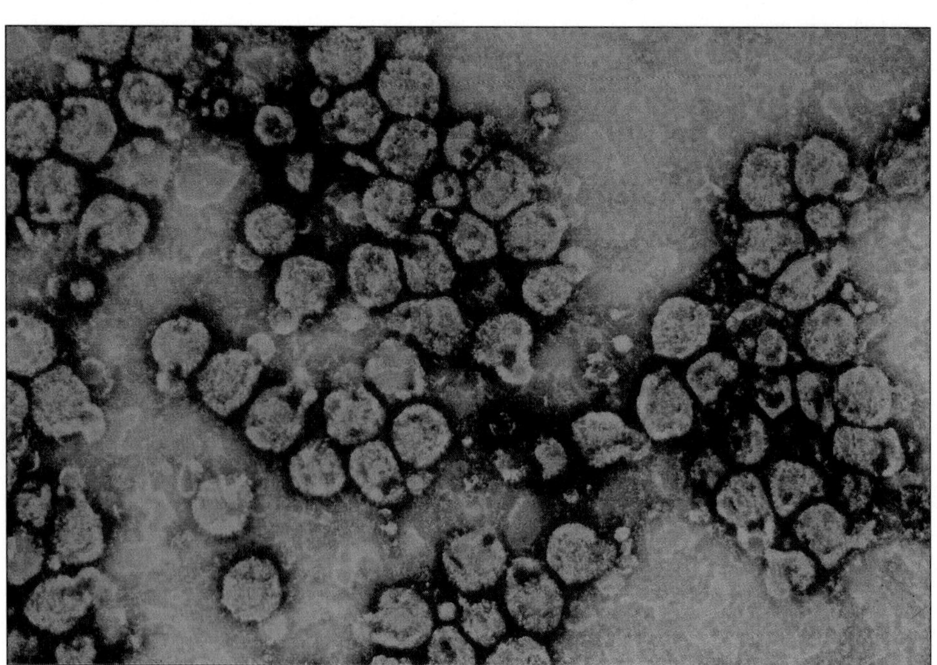

FIGURE 14.7 Particles of the Rous sarcoma virus as seen in the transmission electron microscope.

Table 14.4

Viruses and Cancer

DNA Cancer Viruses	RNA Cancer Viruses
SV40	Rous sarcoma virus
Polyoma	Mouse mammary tumor virus
Adenovirus	

The ability of RSV to cause sarcomas in chickens results from the presence of a single gene (Figure 14.8). Because this gene controls the formation of a cancer, it is called an oncogene. The discovery of the RSV oncogene was an important event in cancer research because it showed that cancer can be caused by mutation in a small number of genes. Many forms of animal cancers, including mouse leukemia, cat leukemia, and mouse breast cancer are caused by retroviruses. More than 20 different oncogenes have been identified in retroviruses, and altogether more than 50 oncogenes have been identified (Table 14.5). Oncogenes are named for the virus in which they were discovered. In RSV, the gene is known as *v-src,* the oncogene in avian erythroblastosis virus is known as *v-erb,* and so forth.

Researchers have shown that oncogenes carried by retroviruses are acquired from animal cells during viral infection. The normal, cellular versions of these genes are called proto-oncogenes, or cellular oncogenes (c-*onc*). Proto-oncogenes are normal genes present in all cells. These genes have the potential to cause cancer if they are mutated or if their usual pattern of expression is altered.

What is the usual function of these genes? Proto-oncogenes directly or indirectly control cell growth, cell division, and cell differentiation, and their gene products are found at many locations within the cell (Table 14.6). For example, the normal allele of the *sis* oncogene is a proto-oncogene that encodes a growth factor called PDGF. The majority of proto-oncogenes regulate cell growth and division, and mutations in these genes lead to uncontrolled proliferation and cancer.

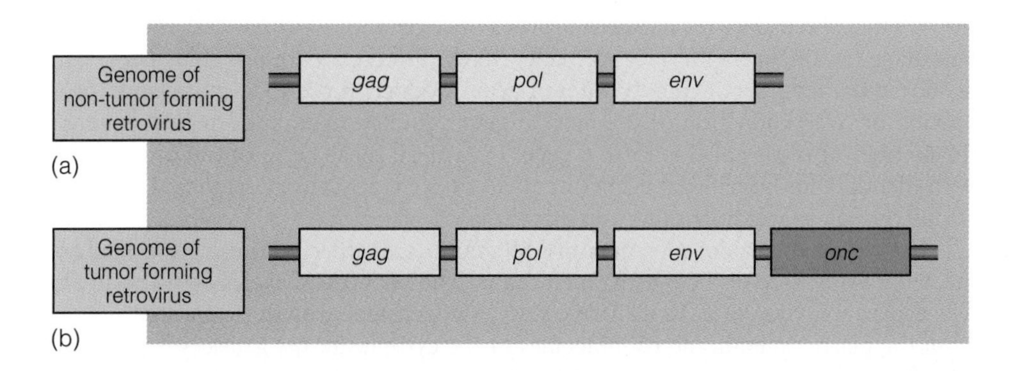

FIGURE 14.8 Organization of the genome in a nontransforming strain of Rous sarcoma virus. (a) The three genes carried by the virus enable the virus to infect cells and replicate but not cause tumors. (b) In a tumor-producing strain, the genome contains an extra gene, an oncogene. This gene is derived from the host cell in a previous infection and gives the virus the ability to produce uncontrolled growth and tumor formation in another host cell.

Table 14.5 Retroviral Oncogenes and Human Proto-Oncogenes

Viral Oncogene	Associated Tumor	Human Proto-Oncogene	Human Chromosome
v-*src*	Sarcoma	c-*src*	20
v-*fps*/v-*fes*	Sarcoma	c-*fps*/c-*fes*	15
v-*myc*	Carcinoma, sarcoma, myelocytoma	c-*myc*	8
v-*erb*-B	Erythroleukemia, sarcoma	c-*erb*-B	7
v-*myb*	Myeloblastic leukemia	c-*myb*	6
v-*mos*	Sarcoma	c-*mos*	8
v-*abl*	B-cell lymphoma	c-*abl*	9
v-*fos*	Sarcoma	c-*fos*	14
v-*Haras*/v-*bas*	Sarcoma, erythroleukemia	c-Ha-*ras*-1	11
v-Ki-*ras*	Sarcoma, erythroleukemia	c-Ki-*ras*-1	6
v-*fms*	Sarcoma	c-*fms*	5

Table 14.6 Cellular Localization of Oncogene (c-onc and v-onc) Gene Products

Gene	Location of c-*onc* Proteins	Location of v-*onc* Proteins
abl	Nucleus	Cytoplasm
erb-B	Plasma membrane	Plasma membrane and Golgi
fos	Cytoplasm	Cytoplasm and membranes
myc	Nucleus	Nucleus
ras	Membranes	Membranes
src	Membranes	Membranes

How Proto-Oncogenes Become Oncogenes

Although oncogenes were discovered in viruses, only a few rare human forms of cancer are caused by virally transmitted oncogenes. In most cases, the conversion from proto-oncogene to oncogene takes place in the nucleus of a somatic cell without the action of a virus.

What is the difference between a proto-oncogene in a normal cell and its mutant allele (an oncogene) in a cancer cell? Many differences are possible. These can include mutations producing an altered gene product, those that cause underproduction or overproduction of the normal gene product, or mutations that increase the number of copies of the normal gene. In fact, all these types of mutations have been identified in human oncogenes or their adjacent regulatory regions. Let's look at one oncogene and how mutation affects it.

The *ras* gene encodes a protein of 189 amino acids that becomes embedded on the cytoplasmic side of the plasma membrane. The *ras* protein cycles between an active and an inactive state. In its active state, *ras* transfers growth-promoting signals from the plasma membrane to molecules in the cytoplasm. *ras* genes isolated from 12 different human tumors show that in each case, a single nucleotide change is the only difference between the mutant oncogene in tumor cells and the proto-oncogene in normal cells. In all 12 cases, a nucleotide substitution causes an amino acid change in the *ras* protein. The mutant gene and the mutant oncoprotein are found only in tumor cells and not in the normal tissue of the patient.

The amino acid change in all 12 mutant genes occurred at one of two places: amino acid 12 or amino acid 61 (◗ Figure 14.9). Changing glycine at position 12 disrupts the structure of the protein and prevents it from folding into its normal shape. This prevents the mutant ras protein from switching to the inactive state. Because the mutant protein is locked into the active state, it is constantly signaling for cell growth. As a result, the cell escapes from growth control and becomes cancerous.

The molecular organization of oncogenes and their products is being studied to develop new methods for the diagnosis and treatment of cancer. For example, if mutant proteins are found in the blood, antibodies that bind to these proteins can be used to detect cancer at a very early stage. Tests are already available for a mutant protein released into the bloodstream by breast cancer cells, and others are being tested in clinical trials.

New strategies for treatment may also be derived from knowledge about oncogenes. In some cases, many copies of an oncogene are present in lung cancer, breast cancer, and cervical cancer. The number of copies of the mutant gene can be measured with recombinant DNA techniques, and these tumors receive more aggressive treatment in order to counteract the increased amount of the mutant protein in these tumors. Other strategies focus on how to turn off oncogene expression. In the lab-

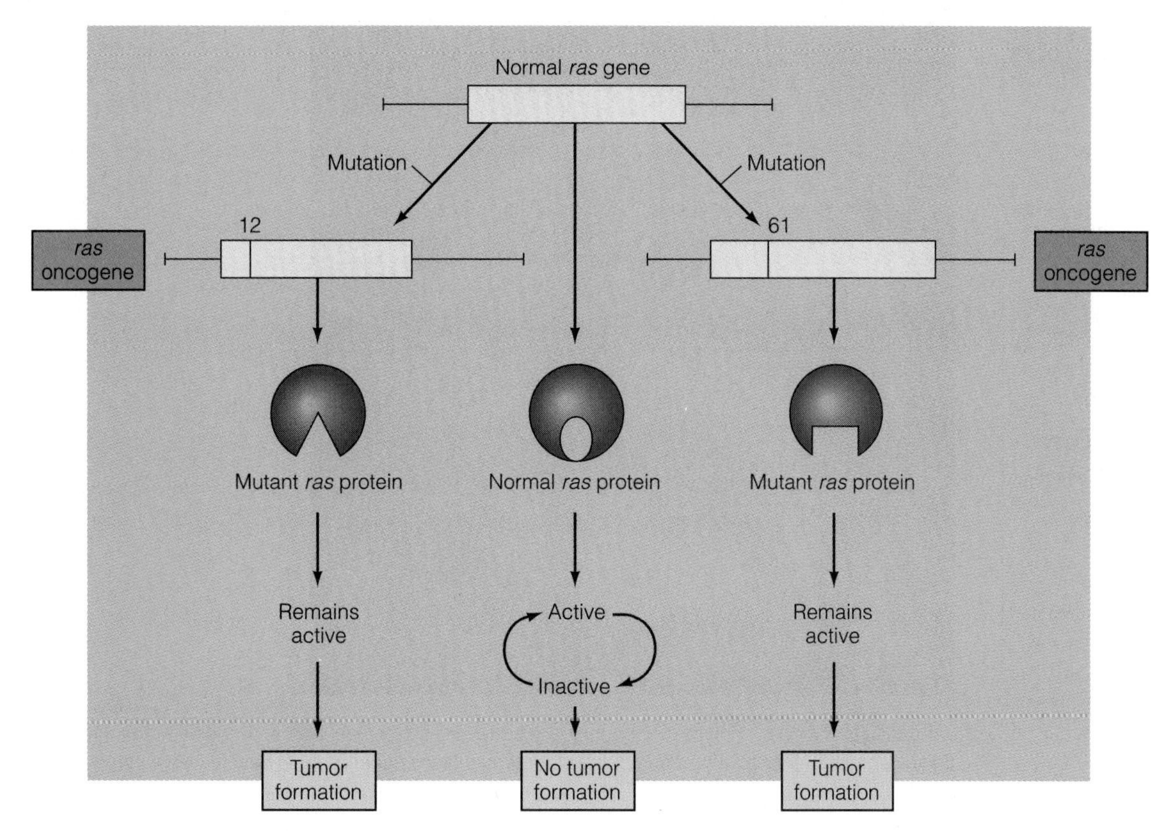

FIGURE 14.9 The *ras* proto-oncogene encodes a protein that receives and transfers signals needed for cell division. Mutations at amino acid positions 12 or 61 cause the formation of an oncoprotein that permanently signals for cell division.

oratory, studies of cancer cells show switching off oncogenes converts cells into noncancer cells. Learning more about how to turn off genes may result in new anticancer drugs that act on the regulatory regions of specific genes.

A GENETIC MODEL FOR CANCER: COLON CANCER

As discussed earlier, cancer requires mutations in a number of specific genes. In retinoblastoma, only two mutational steps are required to convert a normal cell into a cancerous one. In other cases, a half dozen or more mutations are required to initiate formation of a cancer cell. Colon cancer is one of those latter types. The study of colon cancer is a useful model to determine which genes mutate and in what order they mutate as a cell changes from a normal cell to a cancer cell. This form of cancer was selected for several reasons. First, colon/rectal cancer is one of the most common forms of cancer in the United States (Table 14.7). Second, malignant tumors develop from benign tumors, and tumors at all stages of development are available for study. Third, in addition to spontaneous cases, two forms of genetic predisposition to colon cancer are known: an autosomal dominant trait, **familial adenomatous polyposis (FAP)** (MIM/OMIM 175100), and **hereditary nonpolyposis colon cancer (HNPCC)** (MIM/OMIM 120435 and 120436). FAP accounts for only about 1% of all cases of colon cancer but has been useful in deriving the main features of a genetic model for colon cancer that is described later. HNPCC accounts for approximately 15% of all cases and is associated with a form of genomic instability.

To clarify the role of inheritance in colorectal cancer, Randall Burt and his colleagues studied a large pedigree with more than 5,000 members covering six generations. This family has clusters of siblings and relatives with colon cancer. Like

Familial adenomatous polyposis (FAP) An autosomal dominant trait resulting in the development of polyps and benign growths in the colon. Polyps often develop into malignant growths and cause cancer of the colon and/or rectum.

Hereditary nonpolyposis colon cancer (HNPCC) A form of colon cancer associated with genomic instability of microsatellite DNA sequences.

Table 14.7

Colon and Rectal Cancer in the United States

Estimated new cases, 2001	
Colon	98,200
Rectum	37,200
Total	135,400
Mortality (estimated deaths, 1999)	
Colon	48,100
Rectum	8,600
Total	56,700
5-year survival rate (early detection)	
Colon	90%
Rectum	85%

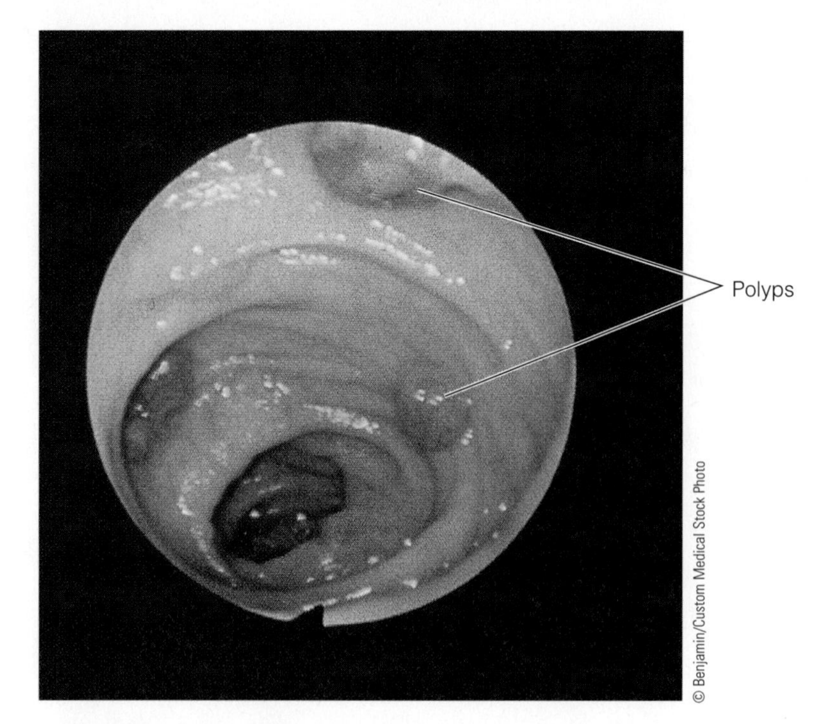

FIGURE 14.10 Polyps in the colon are a precursor to colon cancer. If a cell in one of these polyps acquires enough mutations, it will transform into a cancer cell, leading to colon cancer.

Polyps Growths attached to the substrate by small stalks. Commonly found in the nose, rectum, and uterus.

many other families, this one shows no definite pattern of inheritance for the cancer. However, as part of this study, about 200 family members were examined for intestinal growths (Figure 14.10). These benign tumors, known as **polyps,** usually precede or accompany colon cancer and are regarded as the first step in colon cancer. When polyps and colon cancer are considered a single phenotype, an autosomal dominant pattern of inheritance is clear. This trait is FAP. The results also show that the dominant mutant allele for polyps and cancer (FAP) has a high frequency in the general population (3/1,000).

FAP and Colon Cancer

By analyzing the mutations present at various stages of tumor development, from benign to cancerous, researchers have defined the number and order of steps that change a normal intestinal cell into a malignant cell. The genetic model for colon cancer, shown in Figure 14.11, has two important features: (1) Development of

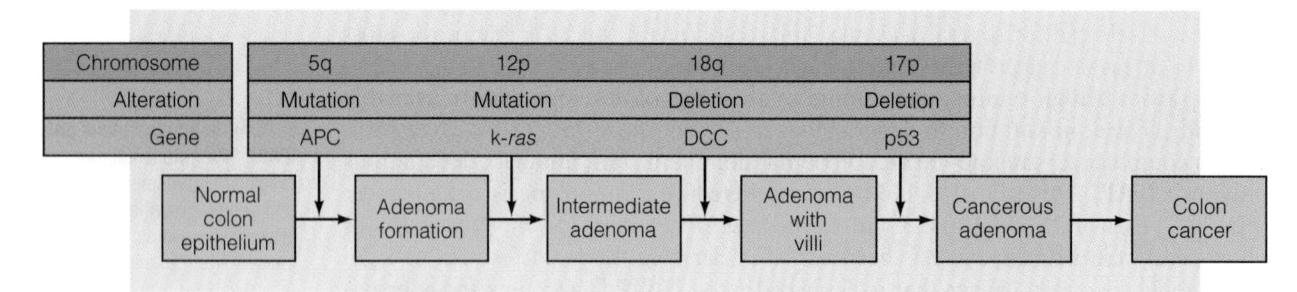

FIGURE 14.11 A model for colon cancer. In this multiple-step model, the first mutation occurs in the *APC* gene, leading to the formation of polyps. Subsequent mutations in genes on chromosomes 12, 17, and 18 cause the transformation of the polyp into a tumor. In this model, the sum of the changes is more important than the order in which the changes take place.

colon/rectal cancer requires five to seven mutations. If fewer mutations are present, benign growths or intermediate stages of malignant tumor formation result. (2) The order of mutations usually follows that shown in the figure, indicating that *both* the number and order of mutations are important in tumor formation.

In cases of FAP-associated colon cancer, a mutation in the *APC* gene on chromosome 5 is the first step. In a normal homozygote, no growth of polyps occurs. In spontaneous cases, mutation of a single copy of *APC* takes place in a single intestinal cell. The cell carrying this mutation partially escapes cell cycle control and divides to form a polyp (a benign tumor). In familial cases, heterozygotes carry a mutant copy of *APC* and develop hundreds or thousands of polyps in the colon and rectum. In either case, the polyps are benign tumors made up of clones of cells, each of which carries a mutant *APC* gene. Because mutation of one copy of *APC* causes polyp formation, the *APC* gene is regarded as a tumor suppressor gene. A single mutation in *APC* is not enough to cause cancer, which develops only after mutation of other genes causes the transition from polyp to colon cancer.

Stages between polyp formation and colon cancer carry an intermediate number of mutations (Figure 14.11). Mutation of one copy of the K-*ras* gene in a polyp cell causes the polyp to form an adenoma, an intermediate stage tumor with finger-like projections. To progress further, *both* alleles of the downstream genes shown in the figure must be mutated in a polyp cell. The 18q region contains a number of genes involved in colon cancer (all of them are tumor suppressor genes), including *DCC, DPC4,* and *JV18-1*. Mutation in both alleles of any of these genes forms late-stage adenomas. In the last stage, mutations in both alleles of the *p53* gene on chromosome 17 cause the late-stage adenoma to become cancerous. Mutations in the *p53* gene are pivotal in the formation of other cancers, including lung, breast, and brain cancer. The *p53* gene is a tumor suppressor gene that is active in regulating the passage of cells from the G1 to the S phase of the cell cycle.

In sum, the model for FAP-associated colon cancer requires a series of mutations that accumulate over time in a single cell. Each mutation confers a slight growth advantage on the cell, first allowing it to proliferate and form a benign tumor, then enlarging the tumor through a series of stages. Eventually, one cell accumulates enough mutations to escape from the cell cycle to form a malignant tumor. Later, additional mutations allow tumor cells to become metastatic and break away to form tumors at remote sites.

Recombinant DNA techniques have been used to identify other cases in which cancer involves a number of mutations at specific chromosomal sites, often on different chromosomes (Table 14.8).

HNPCC and Colon Cancer

Most cancers are caused by mutations in two or more genes that accumulate over time. If one of these mutations is inherited, fewer mutations are required to cause

Table 14.8 Number of Mutations Associated with Specific Forms of Cancer

Cancer	Chromosomal Sites of Mutations	Minimal Number of Mutations Required
Retinoblastoma	13q14	2
Wilms tumor	11p13	2
Colon cancer	5q, 12p, 17p, 18q	5 to 7
Small-cell lung cancer	3p, 11p, 13q, 17p	10 to 15

cancer, resulting in a genetic predisposition to cancer. But if the mutation rate is low (as we have seen in Chapter 11), how do the multiple mutations needed for cancer formation accumulate in a single cell? Work on a second form of colon cancer has provided a partial answer to this question.

There are two forms of HNPCC (*HPNCC1* and *HNPCC2*). *HNPCC1* is on chromosome 2 (2p16), and *HNPCC2* is on chromosome 3 (3p23-21.3). HNPCC-associated colon cancer may be one of the most common genetic disorders, affecting approximately 1 in 200 individuals. Mutations of either of these genes destabilizes the genome, generating a cascade of mutations in microsatellite sequences (review microsatellite sequences in Chapter 13) located on many chromosomes. It has been estimated that cells from *HNPCC* tumors can carry more than 100,000 mutations in microsatellite sequences scattered throughout the genome.

Because of these findings, it has been proposed that the *HNPCC1* and *HNPCC2* genes are *neither* tumor suppressor genes *nor* oncogenes but represent a new class of genes, which when mutated lead to genomic instability and cancer. Proteins encoded by *HNPCC1* and *HNPCC2* repair errors made during DNA replication. When these genes are inactivated by mutation, DNA repair is defective, and microsatellite mutation rates increase by at least 100-fold. This genomic instability promotes mutations in other genes, including the APC gene, eventually leading to colon cancer, as well as other forms of cancer.

Gatekeeper Genes and Caretaker Genes: Insights from Colon Cancer

There are at least two pathways to colon cancer; one begins with a mutation in the *APC* gene, and the other (*HNPCC*) begins with a mutation in a DNA repair gene. Together, these two mechanisms provide insight into which genes can cause normal cells to become cancerous. Mutations in *APC* cause the formation of hundreds or thousands of benign tumors. These benign growths progress slowly to cancer by accumulating mutations in other genes. Because there are thousands of polyps, there is a good chance that at least one benign tumor will progress to colon cancer. In *HNPCC*, polyps accumulate slowly, forming only a small number of benign tumors. However, mutations in these polyps accumulate at a rate two to three times faster than in normal cells, making it almost certain that at least one benign growth will progress to colon cancer.

The different pathways to colon cancer show that two different types of genes are involved in cancer: **gatekeeper genes** and **caretaker genes.** Gatekeeper genes inhibit cell growth and control the cell cycle. Mutation in these genes opens the gate to uncontrolled cell division. Tumor suppressor genes are one type of gatekeeper gene. Many types of oncogenes are also gatekeeper genes.

Gatekeeper genes Genes that regulate cell growth and passage through the cell cycle, for example, tumor suppressor genes.

Caretaker genes Genes that help maintain the integrity of the genome, for example, DNA repair genes.

Caretaker genes, on the other hand, encode proteins that maintain the integrity of the genome, for example, DNA repair genes. Some genetic disorders caused by mutations in DNA repair genes are listed in Table 14.9. The encoded proteins repair

Table 14.9 Human Genetic Disorders Associated with Chromosome Instability and Cancer Susceptibility

Disorder	Inheritance	Chromosome Damage	Cancer Susceptibility	Hypersensitivity
Ataxia telangiectasia	Autosomal recessive	Translocations on 7, 14	Lymphoid, others	X rays
Bloom syndrome	Autosomal recessive	Breaks, translocations	Lymphoid, others	Sunlight
Fanconi anemia	Autosomal recessive	Breaks, translocations	Leukemia	X rays
Xeroderma pigmentosum	Autosomal recessive	Breaks	Skin	Sunlight

damage to DNA caused by mistakes in DNA replication or damage caused by environmental agents (ultraviolet light, for example). Mutation of a caretaker gene does not directly lead to tumor formation but creates a state of genomic instability that increases the mutation rate of all genes, including gatekeeper genes. This insight about gene types and genomic instability may also explain why many forms of cancer become associated with chromosomal instability and aneuploidy.

CHROMOSOME CHANGES AND CANCER

Changes in the number and structure of chromosomes are a common feature of cancer cells (Figure 14.12). In some cases, the relationship between a single chromosome change and the development of cancer is not clear. For example, Down syndrome is caused by the presence of an extra copy of chromosome 21. This quantitative change in genetic information is not associated with any known gene mutation. In addition to defects in cardiac structure and in the immune system, children with Down syndrome are 18 to 20 times more likely to develop leukemia than children in the general population. How extra copies of genes on chromosome 21 predispose to cancer is not yet known, but it may be an effect of increasing the dosage of some proto-oncogenes. In a limited number of cases, more direct information is available about the way chromosomal rearrangements are associated with developing and/or maintaining the cancerous condition.

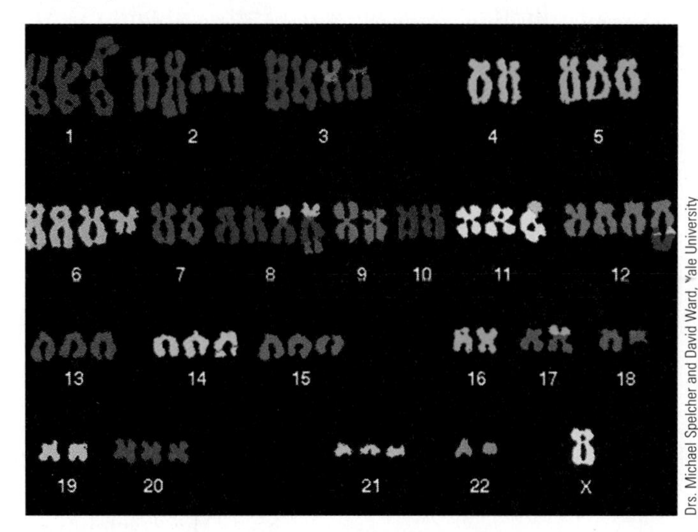

FIGURE 14.12 Chromosome aberrations including translocations, deletions, and aneuploidy result from the genomic instability of cancer cells.

Chromosome Rearrangements and Leukemia

The connection between chromosome rearrangements and cancer is evident in leukemias. In these cancers (involving uncontrolled division of white blood cells), specific chromosome changes are well defined and diagnostic (Table 14.10).

Table 14.10 **Chromosomal Translocation Associated with Human Cancers**

Chromosomal Translocation	Cancer
t(9;22)	Chronic myelogenous leukemia (Philadelphia chromosome)
t(15;17)	Acute promyelocytic leukemia
t(11;19)	Acute monocytic leukemia, acute myelomonocytic leukemia
t(1;9)	Pre-B-cell leukemia
t(8;14),t(8;22),t(2;8)	Burkitt's lymphoma, acute lymphocytic leukemia of the B-cell type
t(8;21)	Acute myelogenous leukemia, acute myeloblastic leukemia
t(11;14)	Chronic lymphocytic leukemia, diffuse lymphoma, multiple myeloma
t(4;18)	Follicular lymphoma
t(4;11)	Acute lymphocytic leukemia
t(11;14)(p13;q13)	Acute lymphocytic leukemia

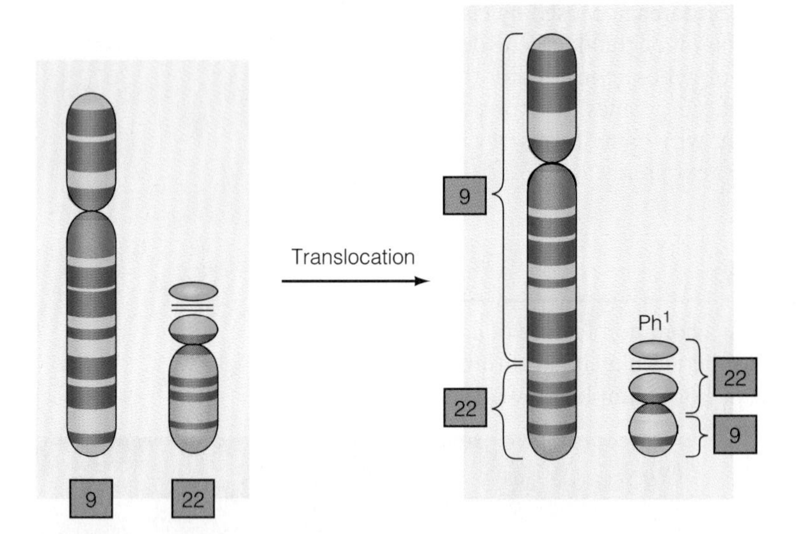

FIGURE 14.13 A reciprocal translocation between chromosomes 9 and 22 results in the formation of a chromosome involved in chronic myelogenous leukemia (CML).

Philadelphia chromosome An abnormal chromosome produced by translocation of parts of the long arms of chromosomes 9 and 22.

One of the best established links between cancer and a chromosomal aberration is the translocation between chromosome 9 and chromosome 22 in chronic myelogenous leukemia (CML) (Figure 14.13). Originally called the **Philadelphia chromosome** (the city in which it was first found), this discovery by Janet Rowley of the University of Chicago was the first example of a chromosome translocation accompanying a human disease.

Other cancers, including acute myeloblastic leukemia, Burkitt lymphoma, and multiple myeloma, are associated with specific translocations (Table 14.10). The finding that certain forms of cancer are consistently associated with specific chromosomal abnormalities suggests that these aberrations are related to the development of the cancer. There is strong evidence that chromosome rearrangements are not byproducts of malignancy but are important steps in the development of certain cancers. The genetic and molecular basis for this role is now becoming clear as the field of cancer cytogenetics merges with the molecular biology of oncogenes.

Translocations, Hybrid Genes, and Leukemias

As oncogenes were identified, cytogeneticists systematically mapped the chromosomal locations of the normal versions of these genes (the proto-oncogenes). It soon became clear that many of these genes are located at or very close to the break points of chromosomal translocations involved with specific forms of leukemia. In fact, chromosome breaks can alter the expression of proto-oncogenes and bring about the development of cancer.

In CML the translocation between chromosomes 9 and 22 creates a hybrid gene. The oncogene c-*abl* is at the breakpoint on chromosome 9, and the *bcr* gene is at the breakpoint on chromosome 22. The translocation produces a hybrid gene with *bcr* sequences at the beginning of the gene and c-*abl* sequences at the end of the gene (Figure 14.14). The hybrid gene is transcribed and translated to produce a fusion protein. The hybrid protein acts as a signal for cell division, causing CML.

Designing New Cancer Drugs

Cancer therapy uses radiation and chemicals to target and kill rapidly dividing cells in the body. While cancer cells are rapidly dividing, so are cells in bone marrow (making red blood cells), in the intestine (replacing worn away cells), and many other tissues. All these cells are destroyed or damaged along with cancer cells during radiation treatment or chemotherapy, often with serious side effects for the patient. The fact that only cancerous CML cells contain the hybrid Bcr-Abl protein offers an opportunity to develop a chemotherapy drug that targets only the cancer cells. The hybrid protein folds to form a pocket for binding of ATP, a molecular energy source. ATP binding is required for Bcr-Abl protein activity. Using this information, a molecule, called STI-571 was designed to fit into the Bcr-Abl pocket, to prevent ATP from entering and to make the hybrid protein inactive (Figure 14.15). In clinical trials, more than 96% of the CML patients treated with this drug went into remission and showed a dramatic reduction in white blood cells carrying the Philadelphia chromosome.

The success of this approach to cancer treatment has changed the way in which anticancer drugs are developed. In the past, such drugs were found by screening hundreds or thousands of chemicals for their ability to slow or stop the growth of

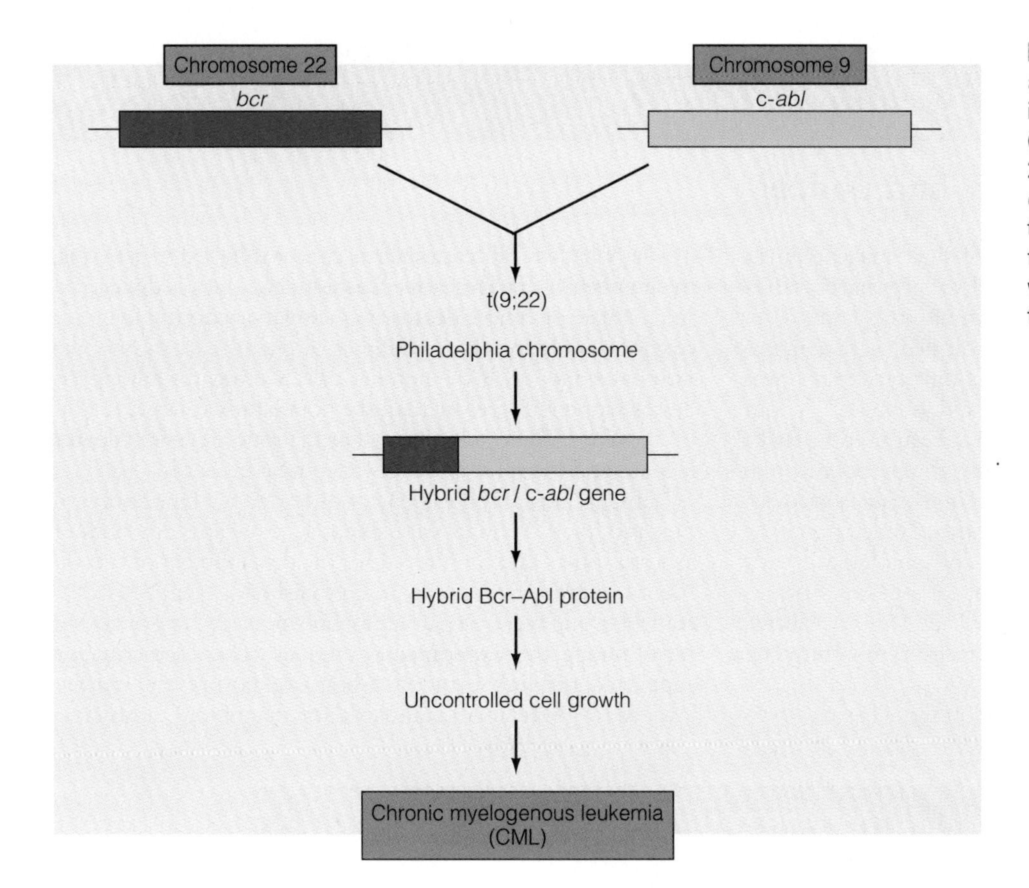

FIGURE 14.14 Gene fusion associated with the 9:22 translocation in chronic myelogenous leukemia (CML). The *bcr* gene on chromosome 22 is fused to the c-*abl* proto-oncogene on chromosome 9. The hybrid gene is transcribed, and the resulting Bcr-Abl fusion protein stimulates cell division in white blood cells. Overproduction of these cells results in CML.

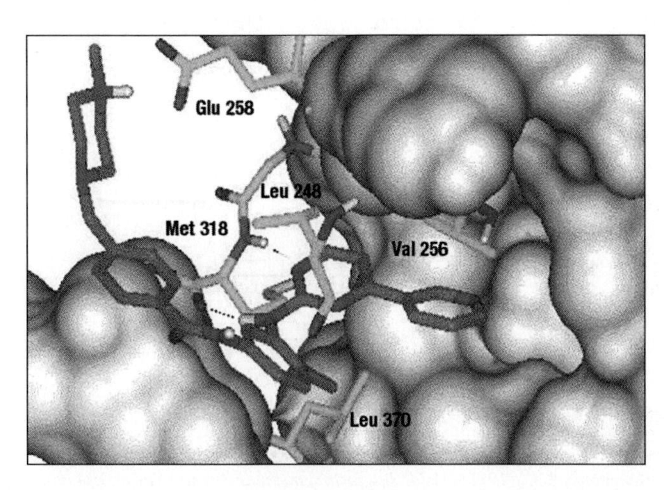

FIGURE 14.15 A molecular model of the Bcr-Abl hybrid protein showing the drug STI-571 in the ATP binding site. Binding of the drug blocks ATP binding and keeps the protein inactive, preventing cell division. (From Buchdunger, E. et al. (2001). Bcr-Abl inhibition as a modality of CML therapeutics, *Biochimica et Biophysica. Acta 1551,* m11-m18, Fig 5A, p. M16. Copyright 2001, reprinted with permission from Elsevier Sciences.)

cancer cells. With an understanding of the molecular events linked to cancer, it is now possible to design drugs for treatment of specific cancers without the side effects of other treatments.

CANCER AND THE ENVIRONMENT

The relationship between environmental factors and cancer has been studied for more than 50 years. During that time, the development of sophisticated methods for gathering and analyzing data has provided solid evidence for the relationship between the environment and cancer. **Epidemiology** is the study of factors that control the presence or absence of a disease. It is an indirect and inferential science that provides correlations between environmental factors and the existence of a disease,

Epidemiology The study of the factors that control the presence, absence, or frequency of a disease.

Concepts and Controversies

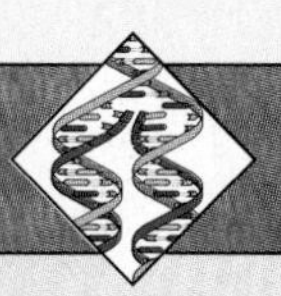

Epidemiology, Asbestos, and Cancer

Asbestos is a fibrous material known since ancient times. The emperor Charlemagne is reported to have had a tablecloth made of asbestos that was thrown into the flames after meals and emerged from the flames unburned, to the amazement of his guests. In more modern times, asbestos was widely used in manufactured goods and is present in our homes, schools, and automobiles. It has been used in brake linings, ceiling tiles, wallboard, textiles, ironing boards, and kitchen gloves. Almost everyone in the United States has been exposed to asbestos in one form or another. To determine the possible role of asbestos in cancer, an epidemiological study compared the cause of death in a group of asbestos workers with an age- and sex-matched group of individuals selected from the general population. Some data from this study are shown in the following table.

A total of 444 deaths were recorded in the asbestos workers, and 301 deaths occurred in the control group. In determining whether a disease such as cancer is linked to asbestos, the number of cancer deaths in asbestos workers is divided by the number of cancer deaths in a similar-sized sample of the control population. If the number of deaths is the same, the ratio is about 1.0. If the number of cancer deaths among asbestos workers is greater than the population at large, the ratio is higher. The results show that the ratio of cancer deaths is 3.86, and the ratio of deaths from lung cancer is 7.62.

This circumstantial evidence was tested in a laboratory on rats and mice. Animals were exposed to various amounts of asbestos fibers and monitored for the development of cancer. A control group was not exposed to asbestos. Cancers in the control group were used as a baseline measurement of cancer rates among the experimental animals. These experiments supported the link between asbestos exposure and cancer, leading to government standards for maximum permissible exposure to asbestos fibers.

Number of Deaths, 1943–1973

Cause of Death	Expected	Observed	Ratio of Obs/Exp
Cancer, all sites	51	198	3.86
Lung cancer	12	89	7.62
Pleural mesothelioma	0	10	
Peritoneal mesothelioma	0	25	
Stomach cancer	5	18	3.53
Colon-rectal cancer	8	22	2.93
Asbestosis	0	37	
All other causes	249	209	0.84
Total deaths, all cases	301	444	1.48

Source: Selicoff, I. J., & Hammond, E. C. (1975). Major risk factors in environmental cancer. In Fraumeni, J. F. (ed.), Persons at High Risk of Cancer. New York: Academic Press, pp. 467–483.

such as cancer. These correlations provide working hypotheses that must be confirmed in laboratory experiments on animal models and then in carefully controlled clinical trials with humans (see Concepts and Controversies: Epidemiology, Asbestos, and Cancer).

Epidemiology and Cancer Links to Environmental Factors

Typically, an epidemiological cancer study measures the incidence of cancer in several different populations. If a statistically significant difference is found, further studies seek to identify factors correlated with this difference. There are geographic variations in cancer rates that are presumably related to environmental factors (Table 14.11). The rates of many forms of cancer in the United States are related to our physical surroundings, personal behavior, or both. Estimates indicate that at least 50% of all cancer can be attributed to environmental factors.

Occupational Hazards and Cancer Risk

A relationship between an occupation and cancer was first noted in 1775 by the English physician Percival Potts, who recorded that London chimney sweeps had a high rate of scrotal cancer, presumably caused by exposure to soot and coal tars. Coal tars and over 500,000 other chemicals are in commercial and industrial use in the United States. Less than 1% of these have been adequately tested for their ability to cause cancer. Occupational exposure to some chemicals is known to cause cancer, but because of the long lag between the time of exposure and the onset of cancer, identification of those at risk is often difficult. One study estimates that occupational exposure to a small number of chemicals used in industrial processes may account for 18% to 38% of all cancer cases in the next few decades. Among these materials is vinyl chloride.

Vinyl chloride is used to make many plastic items, and more than 8 billion pounds is produced each year. Polyvinyl chloride (PVC) is used to make floor tile, bottles, food wrap, and pipes. In laboratory rats, vinyl chloride causes a rare form of liver cancer at a dose of 50 parts per million (ppm). At the time (1970), the acceptable level of occupational exposure was 500 ppm. At that level, workers in vinyl chloride plants developed the same form of liver cancer as rats exposed to vinyl chloride. In 1974, exposure levels were reduced to 50 ppm, and in 1975 they were reduced to 1 ppm. The Food and Drug Administration also banned the use of PVC in beverage containers. Although workers in vinyl chloride plants receive the highest levels of exposure, fin-

Table 14.11 Age-Adjusted Cancer Death Rates per 100,000 Population, 1990–1993

	All Sites	
Country	**Male**	**Female**
United States	165.3 (27)*	111.1 (18)
Australia	158.5 (28)	100.2 (20)
Austria	171.6 (20)	105.6 (16)
Denmark	178.7 (17)	138.1 (1)
Germany	177.3 (18)	108.2 (11)
Hungary	258.7 (1)	135.2 (2)
Japan	149.8 (32)	75.2 (43)
Latvia	206.1 (6)	98.7 (23)
Mauritius	85.4 (47)	63.8 (46)
Mexico	81.6 (48)	77.6 (41)
Poland	204.2 (8)	107.6 (13)
Romania	140.2 (36)	84.5 (38)
Slovenia	203.9 (9)	108.0 (12)
Switzerland	167.2 (24)	96.5 (26)
Trinidad, Tobaco	120.0 (42)	91.4 (31)
United Kingdom	179.1 (16)	124.6 (5)

**Number in parentheses refers to rank order.*

Guest Essay: Research and Applications

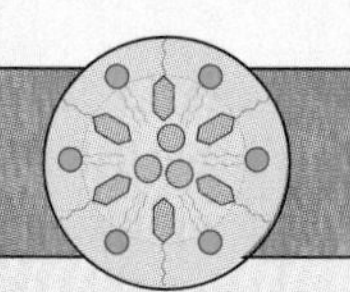

A Journey Through Science

Bruce Ames

As a youngster, I was always interested in biology and chemistry and read the books left lying around the house by my father, who was chairman of a high school chemistry department and later supervisor of science for all of the New York City public schools. During the summers, my sisters and I explored the natural world and collected animals at our family's summer cabin on a lake in the Adirondack Mountains. Throughout my childhood, I read voraciously and would come back from the library every week with a whole stack of books. I attended the Bronx High School of Science, where I conducted my first scientific experiments, studying the effect of plant hormones on the growth of tomato root tips. Motivated by my experience with research, I enrolled at Cornell University to study chemistry and biology. After graduating, I headed west to study at the California Institute of Technology. In the early 1950s, there was an exceptional group of faculty and students at Cal Tech, many of whom were learning about genes by studying biochemical genetics. I joined the laboratory of H. K. Mitchell and began using mutants of the bread mold *Neurospora* to work out the biosynthetic pathway of the amino acid histidine.

After receiving my Ph.D. at Cal Tech, I moved to the National Institutes of Health at Bethesda, Maryland, to do research on gene regulation, using mutant strains of the bacterium *Salmonella*. By 1964, I was married and had two children, a daughter, Sofia, and a son, Matteo. Sometime in that same year, I happened to read the list of ingredients on a box of potato chips and began to think about all the new synthetic chemicals being used, and wondered if they might cause genetic damage to human cells.

To test the ability of chemicals to cause mutation, I devised a simple test, using strains of *Salmonella*. Chemicals that cause mutations in bacteria may cause mutation in human genes. This test showed that a high percentage of the known cancer-causing substances cause mutations. Now, the *Salmonella* test, widely known as the Ames test, is used by more than 3,000 industries and laboratories as a first step in identifying and weeding out chemicals that might cause cancer.

In 1968, I joined the faculty at the University of California at Berkeley, where my students and I developed genetically engineered bacterial strains that can be used to identify what types of DNA changes are caused by mutagens. My colleague Lois Gold and I have assembled a database on the results of animal cancer tests. We have concluded that there is no epidemic of cancer caused by synthetic chemicals, which account for less than 1% of human cancer. Several factors, including tobacco (about one-third of cancer) and diet, have been identified as the major contributors to cancer in the United States. The quarter of the population that has the lowest dietary intake of fruits and vegetables has twice the cancer rate for most types of cancer compared to the highest quarter. These results strongly suggest that a large portion of cancer deaths can be avoided by using knowledge at hand to modify lifestyles.

My current research is on how to delay the decay of mitochondria with age. Mitochondria are the power plants in every cell that make ATP; their decay is a major contributor to aging. My other main interest is the prevention of cancer. The best ways to prevent cancer and delay aging are to decrease smoking and to improve diet. We have been advocating having people take a daily multivitamin/mineral pill as insurance, in addition to eating their five portions of fruits and vegetables a day. We and others are showing that deficiency of particular vitamins and minerals, many of which come from fruits and vegetables, breaks human chromosomes, and thus acts as a radiation mimic.

Bruce Ames is a professor of biochemistry and molecular biology at the University of California, Berkeley. He has been the international leader in the field of mutagenesis and genetic toxicology for more than 30 years. His work has had a major impact on, and changed the direction of, basic and applied research on mutation, cancer, and aging. He earned his B.A. from Cornell University and his Ph.D. from the California Institute of Technology. He is a recipient of the National Medal of Science.

ished plastics always contain entrapped vinyl chloride gas in amounts that can be significant. The effects of this exposure on the general population are yet to be assessed.

Environmental Factors and Cancer

The American Cancer Society estimates that 85% of the lung cancer cases in men and 75% of the cases in women are related to smoking. Smoking produces cancers of the oral cavity, larynx, esophagus, and lungs and accounts for 30% of all cancer deaths. Most of these cancers have very low survival rates. Lung cancer, for example, has a 5-year survival rate of 13%. Cancer risks associated with tobacco are not limited to smoking; the use of snuff or chewing tobacco carries a 50-fold increased risk of oral cancer.

About 800,000 new cases of skin cancer are reported in the United States every year, almost all related to ultraviolet light from the sun or tanning lamps. Skin cancer cases are increasing rapidly in the population, presumably as a result of an increase in outdoor recreation. Epidemiological surveys show that lightly pigmented people are at much higher risk for skin cancer than heavily pigmented individuals. This supports the idea that genetic characteristics can affect the susceptibility of individuals or subpopulations to environmental agents that cause a specific form of cancer.

Summary

Cancers Are Malignant Tumors

1. Cancers are malignant tumors. The primary risk factor for cancer is age. Heritable predispositions to cancer usually show a dominant pattern of inheritance.

Mutations in Specific Genes Can Predispose to Cancer

2. Cancers that result from heritable predispositions have been used to develop models in which cancer is a multistep process, requiring a minimum of two mutations within a cell to produce cancer. According to this model, cancer cells are clonal descendants from a single mutant cell.

Tumor Suppressor Genes, Oncogenes, and the Cell Cycle

3. The study of two classes of genes, tumor suppressor genes and oncogenes, has established the relationship between cancer, the regulation of cell growth and division, and the cell cycle. The discovery of tumor suppressor genes that normally act to inhibit cell division has provided insight into the regulation of the cell cycle. These gene products act at control points in the cell cycle at G1/S or G2/M. Deletion or inactivation of these products causes cells to continuously divide.

Oncogenes and Cancer

4. The discovery of an oncogene in the Rous sarcoma virus led to the identification of other oncogenes and their normal cellular counterparts, proto-oncogenes. Proto-oncogenes regulate cell growth and are converted to oncogenes by mutation. Because all human tumors show alterations in oncogenes, it is possible that a limited number of molecular pathways lead to the development of cancer.

A Genetic Model for Cancer: Colon Cancer

5. Cancer is a multistep process that requires a number of specific mutations. Colon cancer has been studied to provide insight into the number and order of steps involved in transforming normal cells into cancer cells. Two pathways to colon cancer illustrate that some genes are gatekeepers, controlling the cell cycle, whereas others are caretakers, repairing DNA damage to prevent genomic instability.

Chromosome Changes and Cancer

6. Other human disorders, including Down syndrome, are associated with high rates of cancer. This predisposition may result from the presence of an initial mutation or genetic imbalance that moves cells closer to a cancerous state. Other cancers, including leukemia are caused by translocation events, some of which create hybrid genes that activate cell division.

Cancer and the Environment

7. It is now apparent that many cancers are environmentally induced. Occupational exposure to minerals and chemicals poses a cancer risk to workers in a number of industries. The widespread dissemination of these materials poses an undefined but potentially large risk to the general population. Social behavior contributes to approximately 50% of all cancer cases in the United States, most, if not all, of which are preventable.

Case Studies

CASE 1

Julie was a 23-year-old examined in the family cancer risk assessment clinic because of her family history of breast cancer. Julie had been going to her local breast center to be treated for fibrocystic breast disease. At her last mammogram, she requested prophylactic mastectomies. Julie came to the counseling session with a maternal aunt who had prophylactic mastectomies several years earlier. The genetic counseling session explored the reasons that had prompted Julie's request for surgery and reviewed her family history. Three of Julie's six maternal aunts had breast cancer in their late 30s or early 40s. One maternal first cousin had also been diagnosed with breast cancer in her 30s. Julie's mother was in her 50s and had no history of cancer.

Discussion about the family revealed that Julie and her aunt believed that the only way not be diagnosed with cancer was to have their breasts removed. Julie went on to say that her mother had prophylactic mastectomies at age 34, and the aunt attending the session had prophylactic mastectomies last year, when she turned 40. The aunt told stories of caring for her dying sisters. She also stated, "It would be the same as a death sentence not to do this." The women in Julie's family believed that the only way to avoid this disease was to have their healthy breasts removed before cancer developed. This led the remaining aunts to seek surgery, and now Julie was being encouraged to take this step.

The genetic counselor explained that one of the maternal aunts could take a genetic test to determine if an altered gene is contributing to the development of breast cancer in the family. The counselor explained that if a mutation is present in her aunt, Julie could be tested for this same mutation. If present, prophylactic mastectomies might be a reasonable option. However, if Julie did not inherit the mutation, her risk for breast cancer would be the same as for the general population (or 10%), and prophylactic mastectomies would be unnecessary. One of Julie's aunts with cancer was tested, and it was found that she carries a mutation for a predisposition to breast cancer. Julie and her mother were tested for this same mutation and neither of them had the mutation. Julie continues to be followed by the local breast center for her fibrocystic breast disease; however, she has decided not to have prophylactic mastectomies.

1. What if Julie's family had a different gene that causes breast cancer, and no test was available? If you were her, would you get the prophylactic mastectomies?
2. Is fibrocystic breast disease a precursor to cancer?
3. Is prophylactic mastectomy guaranteed to prevent breast cancer? Explain.

CASE 2

Mike was referred for genetic counseling because he was concerned about his extensive family history of colon cancer. His family history, outlined here, is highly suggestive of an inherited form of colon cancer, known as hereditary nonpolyposis colon cancer (HNPCC). This is an autosomal dominant predisposition to colon cancer, and those who carry the altered gene have a 75% chance of developing colon cancer by age 65. Mike was extensively counseled about the inheritance of this condition, the associated cancers, and the possibility of genetic testing (on an affected family member). Mike's aunt elected to be tested for one of the genes that may be altered in this condition, and it was found that she has an altered *MSH2* gene. Other family members are in the process of being tested for this mutation.

1. Seventy-five percent of people who carry the altered gene will get colon cancer by age 65. This is an example of incomplete penetrance. What could cause this?
2. Once a family member is tested for the gene, is it hard for other family members to remain unaware of their own fate, even if they did not want this information? How could family dynamics help (or hurt) this situation?
3. Is colon cancer treatable? What are the common treatments, and how effective are they?

Questions and Problems

Cancers Are Malignant Tumors

1. Theodore Boveri predicted that malignancies would often be associated with chromosomal mutation. What lines of evidence substantiate this prediction?
2. Distinguish between a carcinogen and a mutagen.
3. Benign tumors:
   a. are noncancerous growths that do not spread to other tissues.
   b. do not contain mutations.
   c. are malignant and clonal in origin.
   d. metastasize to other tissues.
   e. none of the above.
4. What does it mean to have a malignant tumor?
5. Metastasis refers to the process in which:
   a. tumor cells die.
   b. tumor cells detach and move to secondary sites.
   c. cancer does not spread to other tissues.
   d. tumors become benign.
   e. cancer can be cured.

Mutations in Specific Genes Predispose to Cancer

6. Cancer is now viewed as a disease that develops in stages. What are the stages that this statement is referring to?
   a. Nonmalignant tumors become malignant.
   b. Proto-oncogenes become tumor suppressor genes.
   c. Younger people get cancer more than older people.
   d. Small numbers of individual mutational events exist that can be separated by long periods of time.
   e. None of the above.
7. It is often the case that a predisposition to certain forms of cancer is inherited. An example is familial retinoblastoma. What does it mean to have inherited an increased probability to acquire a certain form of cancer? What subsequent event(s) must occur?

Tumor Suppressor Genes, Oncogenes, and the Cell Cycle

8. A proto-oncogene is a gene that:
   a. normally causes cancer.
   b. normally suppresses tumor formation.
   c. normally functions to promote cell division.
   d. is involved in forming only benign tumors.
   e. is only expressed in blood cells.
9. What is the difference between an oncogene and a tumor suppressor gene?
10. Distinguish between dominant inheritance and recessive inheritance in retinoblastoma.
11. Describe the likelihood of developing bilateral (both eyes affected) retinoblastoma in the inherited versus the sporadic form of the disease.
12. Parents of a 1-year-old boy are concerned that their son may be susceptible to retinoblastoma because the father's brother had bilateral retinoblastoma. Both parents are normal (no retinoblastoma). They have their son tested for the *RB* gene and find that he has inherited a mutant allele. The father is then tested and also found to carry a mutant allele.
    a. Why didn't the father develop retinoblastoma?
    b. What is the chance that the couple will have another child carrying the mutant allele?
    c. Is there a benefit to knowing their son may develop retinoblastoma?
13. The search for the *BRCA1* breast cancer gene discussed in this chapter was widely publicized in the media (for example, *Newsweek,* Dec. 6, 1993). Describe the steps taken by Mary Clare King and her colleagues to clone this gene. How long did this process take?

Oncogenes

14. Viral oncogenes are often transduced copies of normal cellular genes—they become oncogenic when their gene product or its regulation is subtly altered. What are the roles of cellular proto-oncogenes, and how is this role consistent with their implication in oncogenesis?
15. Which of the following mutations will result in cancer?
    a. homozygous recessive point mutations in a tumor suppressor gene coding for a nonfunctional protein
    b. dominant mutation in a tumor suppressor gene in which the normal protein product is overexpressed
    c. homozygous recessive mutation in which there is a deletion in the coding region of a proto-oncogene, leaving it nonfunctional
    d. dominant mutation in a proto-oncogene in which the normal protein product is overexpressed
16. In DNA repair, what are caretaker genes? How do they work?
17. The following is a description of a family with a history of inherited breast cancer. Betty (grandmother) does not carry the gene. Don, her husband, does. Don's mother and sister had breast cancer. One of Betty and Don's daughters (Sarah) has breast cancer, the other (Karen) does not. Sarah's daughters are in their 30s. Dawn, 33, has breast cancer; Debbie, 31, does not. Debbie is wondering if she will get the disease because she looks like her mother. Dawn is wondering if her 2-year-old daughter (Nicole) will get the disease.
    a. Draw a pedigree indicating affected individuals, and identify all individuals.
    b. What is the most likely mode of inheritance of this trait?
    c. What are Don's genotype and phenotype?

d. What is the genotype of the unaffected women (Betty and Karen)?

e. An RFLP marker has been found that maps very close to the gene. Given the following RFLP data for chromosomes 4 and 17, which chromosome does this gene map to?

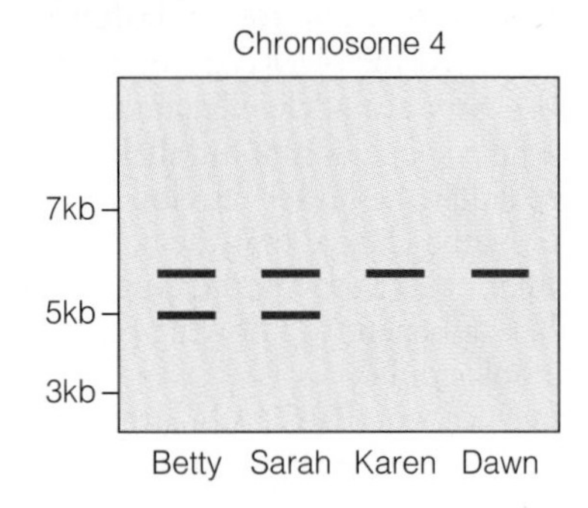

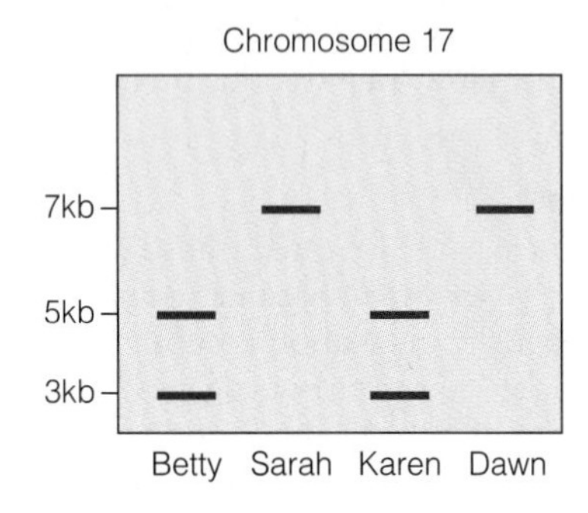

f. Using the same RFLP marker, Debbie and Nicole were tested. The results are shown in the following figure. Based on their genotypes, are either of them at increased risk for breast cancer?

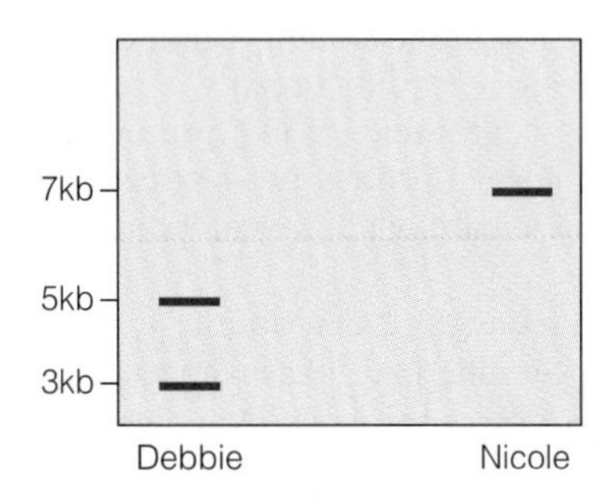

18. You are in charge of a new gene therapy clinic. Two cases have been referred to you for review and possible therapy.

Case 1. A mutation in the promoter of an oncogene causes the gene to make too much of its normal product, a receptor protein that promotes cell division. The uncontrolled cell division has caused cancer.

Case 2. A mutation in an exon of a tumor suppressor gene makes this gene nonfunctional. The product of this gene normally suppresses cell division. The mutant gene cannot suppress cell division and has led to cancer.

Given the current state of knowledge, in which case is gene therapy a viable option? Why?

19. Explain how the *APC* gene starts the progress towards colon cancer.

Chromosome Changes and Cancer

20. Can you postulate a reason or reasons why children with Down syndrome are 20 times more likely to develop leukemia than children in the general population?

21. What mutational event is typically associated with Burkitt lymphoma? Which chromosomes are involved in this tumor?

Environmental Causes of Cancer

22. What are some factors that epidemiologists have associated with a relatively high risk of developing cancer?

23. Smoking cigarettes has been shown to be associated with the development of lung cancer. However, a direct correlation between how many cigarettes one smokes and the onset of lung cancer does not exist. A heavy smoker may not develop lung cancer, although a light smoker may develop the disease. Explain why this may be.

24. Discuss the relevance of epidemiologic and experimental evidence in recent governmental decisions to regulate exposure to asbestos in the environment.

25. Studies have shown that there are significant differences in cancer rates among different ethnic groups. For example, the Japanese have very high rates of colon cancer, but very low rates of breast cancer. It has also been demonstrated that when members of low-risk ethnic groups move to high-risk areas, their cancer risks rise to that of the high-risk area. For example, Japanese who live in the United States, where the risk of breast cancer is high, have higher rates of breast cancer than Japanese who live in Japan. What are some of the possible explanations for this interesting phenomenon? What factors may explain why the Japanese have higher rates of colon cancer than other ethnic groups?

Internet Activities

The following activities use the resources of the World Wide Web to enhance the topics covered in this chapter. To investigate the topics described below, log on to the book's home page (www.brookscole.com/biology_d) and follow the *Student Resource* link to your subject and text.

1. *The "Cancer Gene" TP53 (Li-Fraumeni Syndrome).* The tumor suppressor gene TP53 (also known as p53) helps improve the rate of DNA repair; mutations in this gene are implicated in the onset of many cancers. At the *TP53 Cancer GeneWeb* Web site (http://www.cancerindex.org/geneweb/TB53.htm#summary), you can explore links to many different sources of information on TP53. From this Web site, under the "Other Related Resources" heading, you can link to the Weizmann Institute's p53 home page (http://bioinformatics.weizmann.ac.il/hotmolecbase/entries/p53.htm) and read about the expression, cellular function, and involvement in disease of this critical gene. Why are cancers that include TP53 mutations particularly difficult to treat?
2. *Family Histories of Breast Cancer.* Breast cancer screening followed by prophylactic (preventive) mastectomy has become an option for some women with strong family histories of breast cancer. At Lawrence Berkeley National Laboratory's *ELSI* Web site (http://www.lbl.gov/Education/ELSI/), link to the "Breast Cancer Screening" page. Within the main page, read about breast cancer in the introduction, and then scroll down to "What Would You Do?" and consider how you would answer the questions posed in the various classroom scenarios.
3. *Cancer Case Studies.* The *Genetics of Cancer* Web site (http://www.intouchlive.com/home/frames.htm?http://www.intouchlive.com/cancergenetics&3) provides access to a number of case studies including family history information and pedigrees and then poses personal and ethical questions. Select one of the cases, evaluate it, and answer the questions.

For Further Reading

Alberts, D. S., Lipkin, M., & Levin, B. (1996). Genetic screening for colorectal cancer and intervention. *Int. J. Cancer, 69,* 62–63.

Ames, B., Gold, L. S., & Willett, W. C. (1995). The causes and prevention of cancer. *Proc. Nat. Acad. Sci. U. S. A., 92,* 5258–5265.

Ames, B., Magraw, R., & Gold, L. (1990). Ranking possible cancer hazards. *Science, 236,* 71–80.

Brown, M. A. (1997). Tumor suppressor genes and human cancer. *Adv. Genet., 36,* 45–135.

Burt, R., Bishop, D. T., Cannon, L. A., Dowdle, M. A., Lee, R. G., & Skolnick, M. H. (1985). Dominant inheritance of adenomatous polyps and colorectal cancer. *N. Engl. J. Med., 312,* 1540–1544.

Call, K. M., Glaser, T., Ito, C. Y., Buckler, A. J., Pelletier, J., Haber, D. A., Rose, E. A., Kral, A., Yeger, H., Lewis, W. H., Jones, C., & Housman, D. E. (1990). Isolation and characterization of a zinc finger polypeptide gene at the human chromosome 11 Wilms tumor locus. *Cell, 61,* 509–520.

Cancer facts and figures. (2001). New York: American Cancer Society.

Committee on Diet, Nutrition and Cancer, Assembly of Life Sciences, National Research Council. (1982). *Diet, nutrition, and cancer.* Washington, D.C.: National Academy Press.

Compagni, A., & Christofori, G. (2000). Recent advances in research on multistage tumorigenesis. *Br. J. Cancer, 83,* 1–5.

Croce, C., & Klein, G. (1985, March). Chromosome translocations and human cancer. *Sci. Am., 252,* 54–60.

Fearon, E. R. (1997). Human cancer syndromes: clues to the origin and nature of cancer. *Science, 278,* 1043–1050.

Fearon, E. R., & Vogelstein, B. (1990). A genetic model for colorectal tumorigenesis. *Cell, 61,* 759–767.

Hamm, R. D. (1990). Occupational cancer in the oncogene era. *Br. J. Ind. Med., 47,* 217–220.

Kinzler, K. W., & Vogelstein, B. (1997). Gatekeepers and caretakers. *Nature, 386,* 761–763.

Malkin, D., & Knoppers, B. M. (1996). Genetic predisposition to cancer—issues to consider. *Semin. Cancer Biol., 7,* 49–53.

Marshall, E. (1993). Search for a killer: Focus shifts from fat to hormones. *Science, 259,* 618–621.

Perrera, F. P. (1996). Molecular epidemiology: Insights into cancer susceptibility, risk assessment and prevention. *J. Natl. Cancer Inst., 88,* 496–509.

Rous, P. (1911). Transmission of a malignant new growth by means of a cell-free filtrate. *JAMA, 56,* 1981.

Ruoslahti, E. (1996, September). How cancer spreads. *Sci. Am., 275,* 72–77.

Russell, P. (1998). Checkpoints on the road to mitosis. *Trends Biochem. Sci., 10,* 399–402.

Stahl, A., Levy, N., Wadzynska, T., Sussan, J. M., Jourdan-Fonta, D., & Saracco, J. B. (1994). The genetics of retinoblastoma. *Ann. Genet., 37,* 172–178.

Sugimura, T. (1990). Cancer prevention: Underlying principles and practical proposals. *Basic Life Sci., 52,* 225–232.

Vasen, H. F. (1994). What is hereditary nonpolyposis colon cancer (HNPCC)? *Anticancer Res., 14,* 1613–1615.

Weinberg, R. A. (1985). The action of oncogenes in the cytoplasm and nucleus. *Science, 203,* 770–776.

Weinberg, R. A. (1996, September). How cancer arises. *Sci. Am., 275,* 62–70.

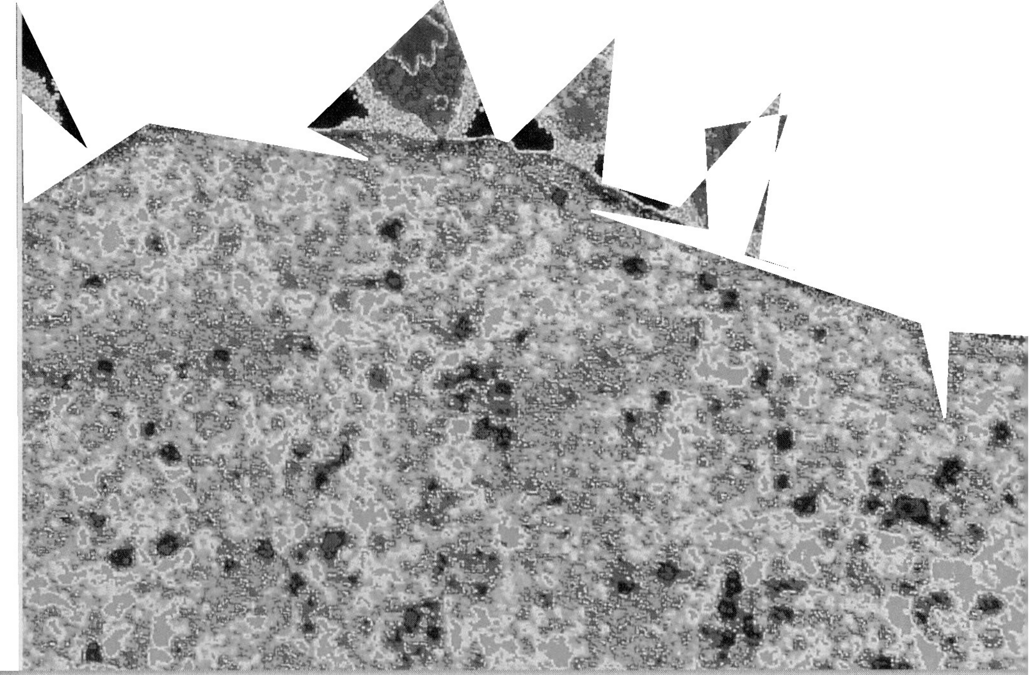

© J.L. Carson/Custom Medical Stock Photo

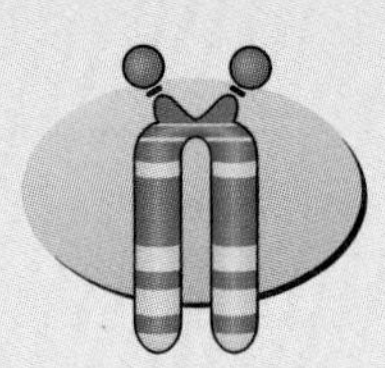

Genetics of the Immune System

Chapter Outline

Approximately 2,500 years ago, a mysterious plague swept through Athens, Greece, killing thousands of residents. The historian Thucydides wrote about the epidemic and observed that anyone who survived the disease could care for the sick without becoming ill a second time. In ancient China, physicians saw that people who had smallpox did not get the disease again. Records from eighth-century China record partially successful attempts to transfer resistance to uninfected individuals by injecting them with fluid from the sores of smallpox victims. The first safe and successful method of transferring resistance to a disease was developed approximately one thousand years later by Edward Jenner, an English physician. He noted that anyone who was sick with cowpox became resistant to smallpox. To see if he could transfer resistance to smallpox, Jenner inoculated a boy with dried scrapings from a cowpox patient and followed the inoculation a few weeks later with an injection from a smallpox patient. The boy did not develop smallpox, and the method, called vaccination (from vaccus, *the Latin word for "cow"), became an effective tool in controlling this disease. The World Health Organization led a worldwide vaccination campaign against smallpox and eradicated the disease by 1980. Because the virus that causes smallpox cannot reproduce outside the body of an infected individual, the disease has disappeared. The only official remaining samples of the virus are stored in research facilities in the United States and Russia.*

Vaccines are used to prevent a number of diseases, including diphtheria, whooping cough, measles, and hepatitis B. The study of the body's resistance to disease is a young and developing science, and new diseases still arise to remind us of our limited knowledge. In the early 1970s, women began to develop toxic shock syndrome, and by 1980 several thousand cases and more than 25 deaths had been reported. Investigation determined that certain brands of highly absorbent tampons led to vaginal infections by strains of Staphylococcus aureus. *These bacteria secrete a toxin that produces fever, rash, low blood pressure, and, in some cases, death.*

Other new diseases remain a mystery. In 1981, acquired immunodeficiency syndrome (AIDS) appeared in the United States. Affected individuals exhibit rare forms of cancer, pneumonia, and other infections. In all cases, there is a complete and irreversible breakdown of the immune system, associated with infection by the human immunodeficiency virus (HIV). This breakdown allows the development of infections, one or more of which proves fatal.

THE IMMUNE SYSTEM DEFENDS THE BODY AGAINST INFECTION

Infections caused by viruses, bacteria, and fungi remain a health problem in industrialized nations and are a major cause of death in developing nations. The responses of the body to infection include a nonspecific response and a specific set of responses known as the immune reaction.

Nonspecific responses are designed (1) to block entry of disease-causing agents into the body and (2) to block the spread of infectious agents if they get into the body. The immune system has two responses to invading agents: antibody-mediated immunity and cell-mediated immunity. Each is highly specific and has two stages—a primary response and a later-developing secondary response. In addition, the immune system is responsible for the success or failure of blood transfusions and organ transplants. In this chapter, we examine the cells of the immune system and their mobilization in an immune response. We also consider how the immune system determines blood groups and mother–fetus incompatibility. The immune system also plays a role in organ transplants and in determining risk factors for a wide range of diseases. Finally, we describe a number of disorders of the immune system, including how AIDS acts to cripple the immune response of infected individuals.

Monocytes White blood cells that participate in the inflammatory response.

Inflammatory response The body's reaction to invading microorganisms.

THE INFLAMMATORY RESPONSE IS A GENERAL REACTION

The skin is a barrier to infectious agents, such as viruses and bacteria, and prevents them from entering the body. The outer surface of the skin is home to bacteria, fungi, and even mites, but they cannot penetrate the protective layers of dead skin cells to cause infection. If microorganisms are able to penetrate the skin or the cells lining the respiratory, digestive, or urinary systems, an inflammatory reaction develops (Figure 15.1). Cells infected by microorganisms release chemical signals, including histamine. These signals increase blood flow in the affected area (that is why the area around a cut or scrape gets red and warm). The increased heat creates an unfavorable environment for microorganism growth, mobilizes white blood cells, and raises the metabolic rate in nearby cells. These reactions promote healing. In addition, white blood cells migrate to the area in response to the chemical signals to engulf and destroy the invading microorganisms.

If infection persists, capillaries in the infected area become leaky and plasma flows into the injured tissue, causing it to swell. Clotting factors in the plasma trigger a cascade of small blood clots that seal off the injured area, preventing the escape of invading organisms. Finally, the area becomes targeted by **monocytes,** which clean up dead viruses, bacteria, or fungi and dispose of dead cells and debris.

This chain of events, beginning with the release of chemical signals and ending with the cleanup by monocytes, is the **inflammatory response.** This response is an active defense mechanism that the body

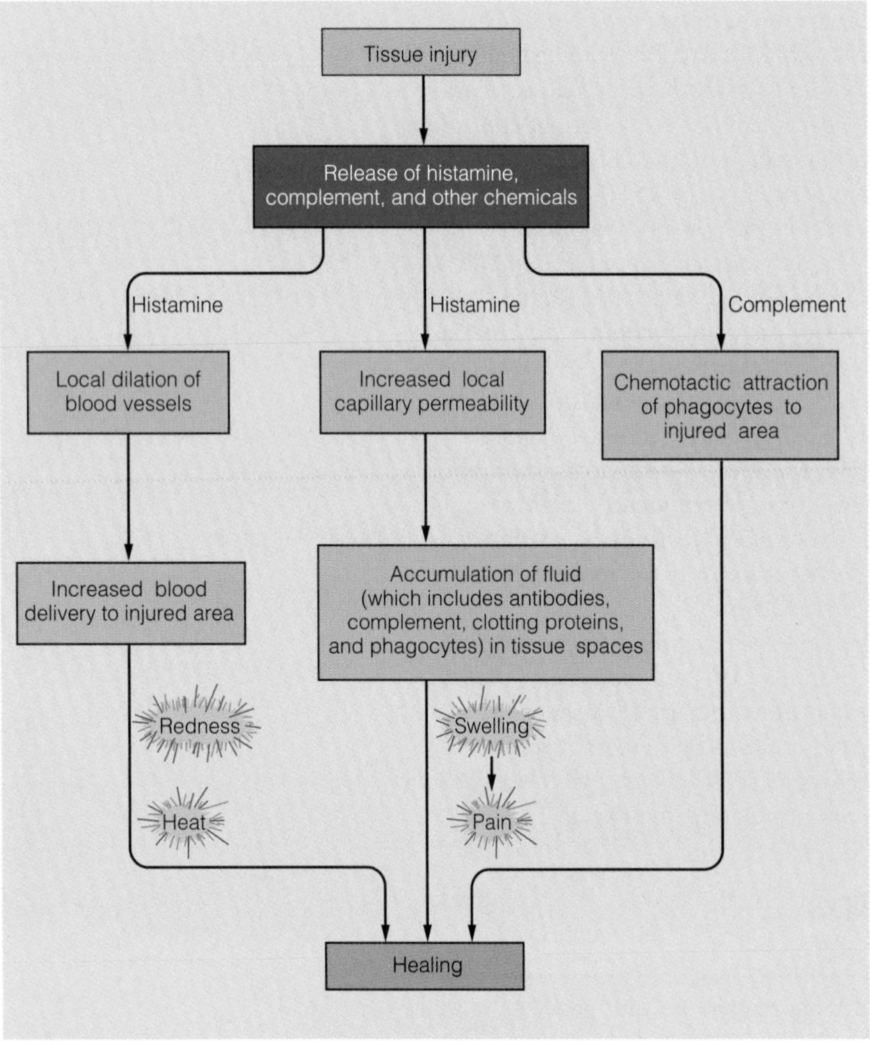

FIGURE 15.1 Tissue injury or infection generates an inflammatory response. The responses generate four signs of inflammation: redness, swelling, heat, and pain.

employs to resist infection. This reaction is usually enough to stop the spread of infection. If this system fails, another more powerful system—the immune response—is called into action.

Genetics and Inflammatory Diseases

The inner cell layer of the intestine is a barrier that prevents bacteria in the digestive system from crossing into the body. Failure to monitor or properly respond to bacteria crossing this barrier results in inflammatory bowel diseases. Inflammatory bowel diseases are genetically complex and involve the interaction of environmental factors with genetically predisposed individuals. Ulcerative colitis (MIM/OMIM 191390) and Crohn disease (MIM/OMIM 266600) are two forms of inflammatory bowel disease caused by malfunctions in the immune system. Crohn disease occurs with a frequency of 1 in 1,000 individuals, mostly young adults. The frequency of this disorder has greatly increased over the last 50 years, presumably as the result of unknown environmental factors. A genetic predisposition to Crohn disease maps to chromosome 16. The gene for this predisposition has been identified and cloned using recombinant DNA techniques. The gene, *NOD2*, encodes a receptor found on the surface of monocytes and other cells of the immune system. The receptor detects the presence of molecules on the surface of bacterial cells. Once activated, the receptor signals a protein in the nucleus to begin the inflammatory response. In Crohn disease, the protein encoded by the mutant allele is defective and causes an abnormal inflammatory response that damages the intestinal wall. Carrying the mutant allele of *NOD2* confers only a predisposition; unknown environmental factors and other genes are probably involved in this disorder.

The Complement System Kills Microorganisms Directly

The **complement system** is a chemical defense system that supplements the inflammatory response, kills microorganisms directly, and works with the immune response (◗ Figure 15.2). Its name derives from the way it complements the action of

Complement system A chemical defense system that kills microorganisms directly, supplements the inflammatory response, and works with (complements) the immune system.

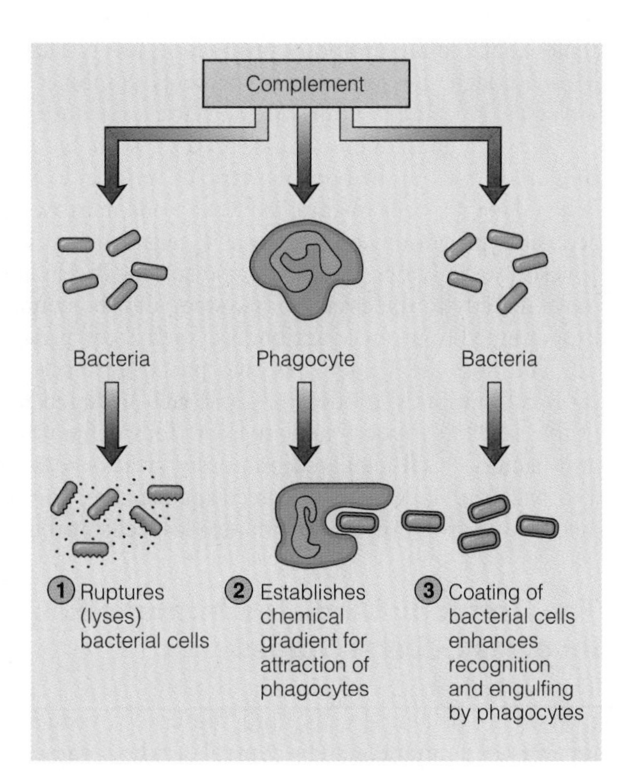

◗ **FIGURE 15.2** The complement system has several actions, all associated with defending the body against infection.

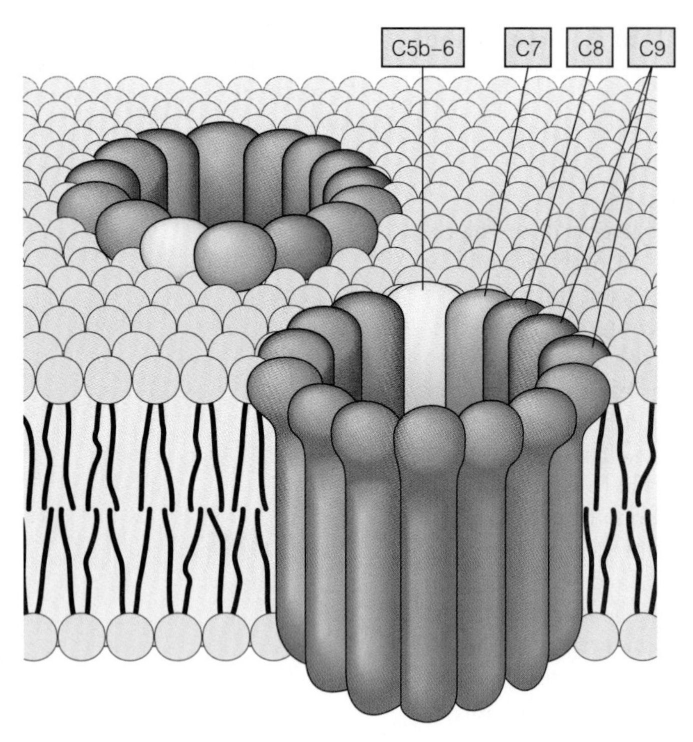

▶ **FIGURE 15.3** The membrane attack complex (MAC). The complement proteins insert into and become part of the plasma membrane of an invading microorganism. Once in the membrane, the complement proteins form a pore, causing water to flow into and burst the cell.

the immune system. Complement proteins are synthesized in the liver and circulate in the bloodstream as inactive precursors. At the site of infection, the first protein in the series (C1) activates the second (C2), and so forth, in a cascade of activation. The final five components (C5 to C9) form a large, cylindrical multiprotein, called the membrane-attack complex (MAC). The MAC embeds itself in the plasma membrane of an invading microorganism, creating a pore (▶ Figure 15.3). Fluid flows into the cell in response to an osmotic gradient, eventually bursting the cell.

In addition to directly destroying microorganisms, some complement proteins guide phagocytes to the site of infection. Other components aid the immune response by binding to the outer surface of microorganisms and marking them for destruction.

THE IMMUNE RESPONSE IS A SPECIFIC DEFENSE AGAINST INFECTION

The immune system generates a response that neutralizes and/or destroys viruses, bacteria, fungi, and cancer cells. The immune system is more effective than the nonspecific defense system and has a memory component that remembers previous encounters with infectious agents. Immunological memory allows a rapid, massive response to a second exposure to a foreign substance.

An Overview of the Immune Response

Immunity results from the production of proteins called **antibodies.** Antibodies bind to foreign molecules and microorganisms and inactivate them in several ways. Molecules that initiate antibody production are called **antigens** (*anti*body *gen*erators). Most antigens are proteins or proteins combined with polysaccharides, but *any* molecule, regardless of its source, that causes antibody production is an antigen.

The immune response is mediated by white blood cells called **lymphocytes.** These cells are formed in the bone marrow by mitotic division of **stem cells** (▶ Figure 15.4). When daughter cells migrate from the bone marrow to the thymus gland, they are irreversibly programmed to become **T cells.** Mature T cells circulate in the blood and concentrate in lymph nodes and the spleen. **B cells** mature in the bone marrow and move directly to the circulatory system and the lymph system. B cells are genetically programmed to produce antibodies. Each B cell produces only one type of antibody.

The immune system has two parts, **antibody-mediated immunity,** regulated by B cell antibody production, and **cell-mediated immunity,** controlled by T cells (Table 15.1). Antibody-mediated reactions defend the body against invading viruses and bacteria. Cell-mediated immunity attacks cells of the body infected by viruses and bacteria. T cells also protect against infection by parasites, fungi, and protozoans. One group of T cells can also kill cells of the body if they become cancerous.

Antibodies A class of proteins produced by B cells that bind to foreign molecules (antigens) and inactivate them.

Antigens Molecules carried or produced by microorganisms that initiate antibody production.

Lymphocytes White blood cells that originate in bone marrow and mediate the immune response.

Stem cells Cells in bone marrow that produce lymphocytes by mitotic division.

T cells A type of lymphocyte that undergoes maturation in the thymus and mediates cellular immunity.

B cells A type of lymphocyte that matures in the bone marrow and mediates antibody-directed immunity.

Antibody-mediated immunity Immune reaction that protects primarily against invading viruses and bacteria by antibodies produced by plasma cells.

Cell-mediated immunity Immune reaction mediated by T cells directed against body cells that have been infected by viruses or bacteria.

The Antibody-Mediated Immune Response Involves Several Stages

The antibody-mediated immune response has several stages: antigen detection, activation of helper T cells, and antibody production by B cells. A specific cell type of the immune system controls each of these steps. White blood cells, called

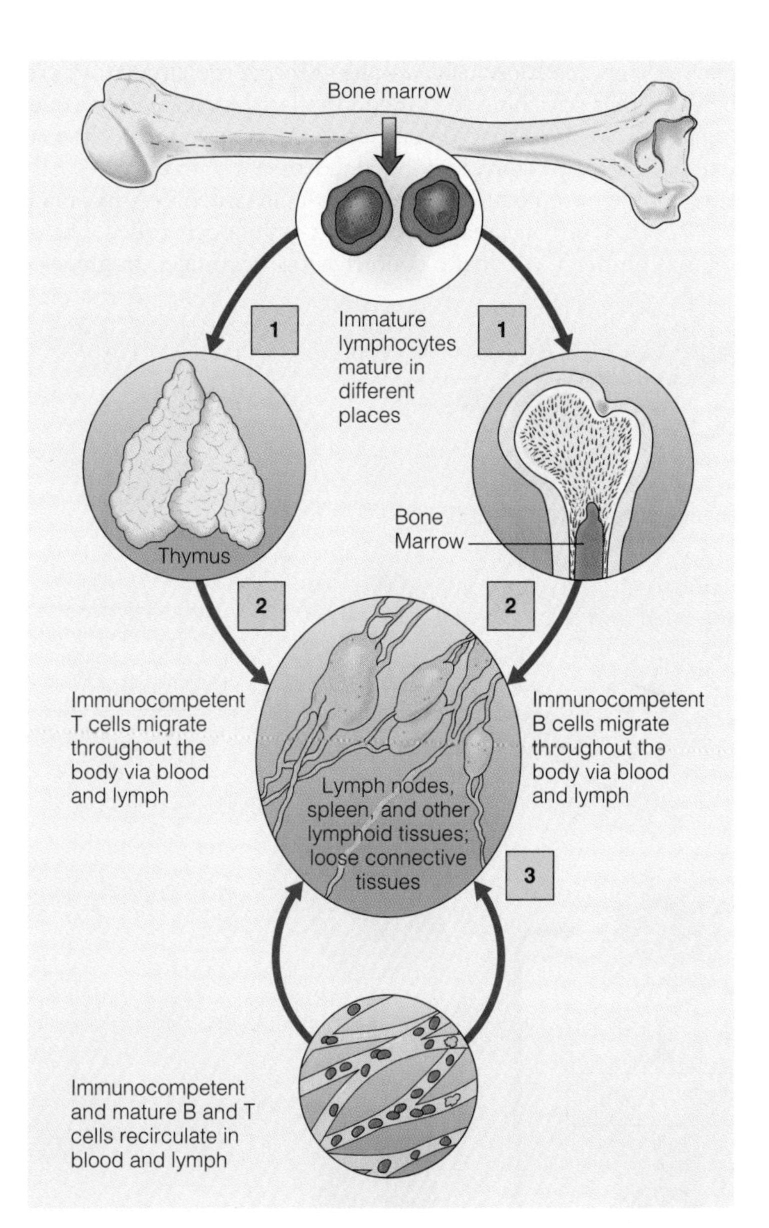

FIGURE 15.4 Immature lymphocytes (lymphoblasts) produced in bone marrow that travel to the thymus for maturation become T cells. Those that remain and mature in bone marrow become B cells. Once mature, T and B lymphocytes migrate through the body in the circulatory system as part of the immune system.

Table 15.1 Comparison of Antibody-Mediated and Cell-Mediated Immunity

Antibody-Mediated	Cell-Mediated
Principal cellular agent is the B cell. B cell responds to bacteria, bacterial toxins, and some viruses.	Principal cellular agent is the T cell. T cells respond to cancer cells, virally infected cells, single-celled fungi, parasites, and foreign cells in an organ transplant.
When activated, B cells form memory cells and plasma cells, which produce antibodies to these antigens.	When activated, T cells differentiate into memory cells, cytotoxic cells, suppressor cells, and helper cells; cytotoxic T cells attack the antigen directly.

Macrophages Large white blood cells derived from monocytes that engulf antigens and present them to T cells, activating the immune response.

Helper T cell A lymphocyte that stimulates production of antibodies by B cells when an antigen is present.

macrophages, continuously wander through the circulatory system and the spaces between cells searching for foreign (nonself) antigenic molecules, viruses, or microorganisms. When a macrophage encounters an antigen, the macrophage engulfs it, internalizes it (Figure 15.5), and destroys it with enzymes. Small fragments of the antigen move to the outer surface of the macrophage's plasma membrane.

This macrophage may encounter a lymphocyte called a **helper T cell.** Surface receptors on the T cell make contact with the antigen fragment on the macrophage.

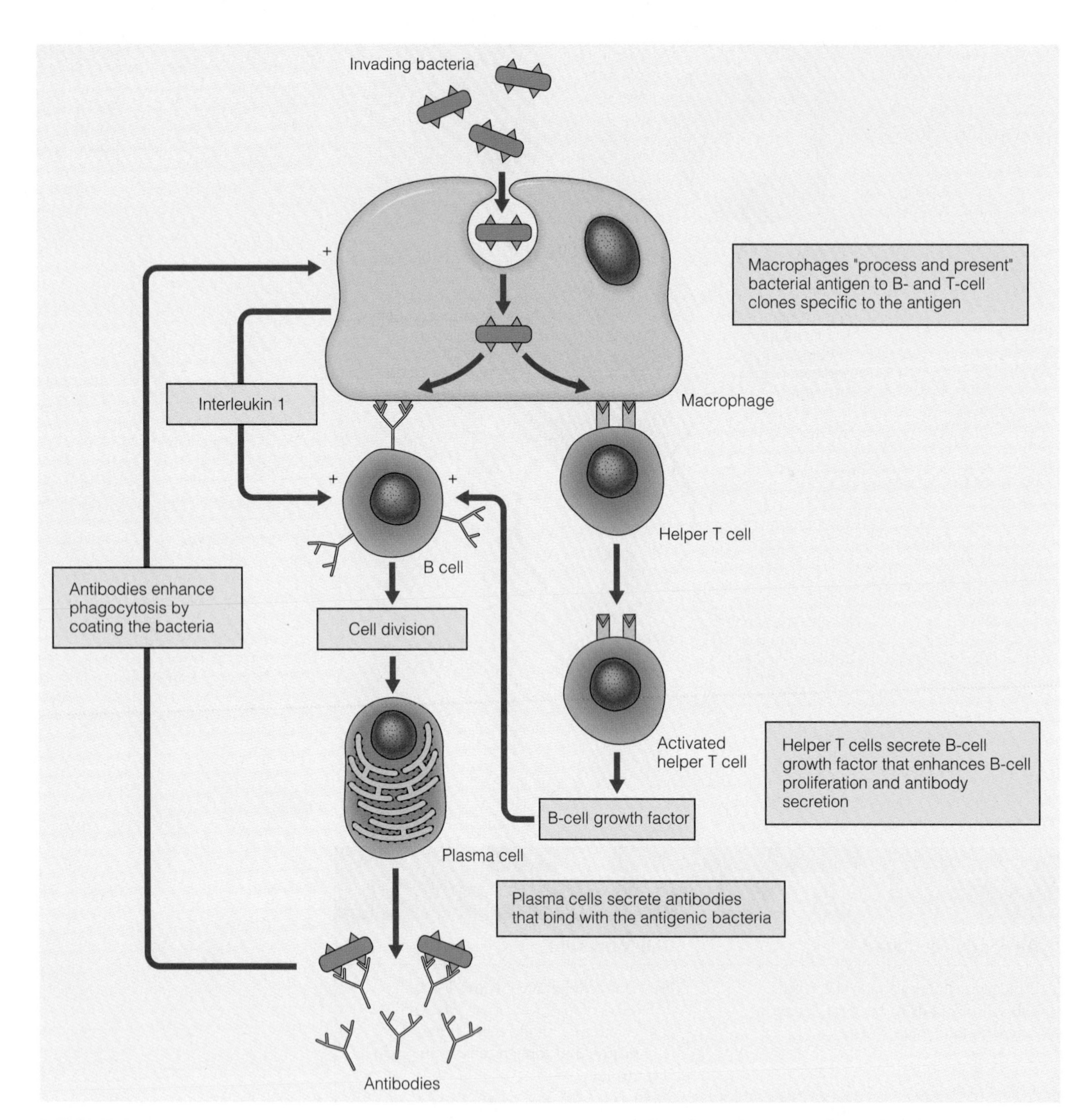

FIGURE 15.5 Macrophages engulf invading microorganisms and process their antigens for presentation to the helper T cells of the immune system. The activated T cell, in turn, stimulates division of the B cell, which produces antibodies against the antigen presented by the macrophage. Daughter B cells become differentiated as plasma cells for the production of large quantities of antibody. Macrophages can also directly stimulate antibody production by interaction with a B cell.

This contact activates the T cell. The activated T cell in turn identifies and activates B cells that synthesize an antibody against the antigen encountered by the T cell. The activated B cells divide and form two types of daughter cells. The first is **plasma cells,** which synthesize and secrete 2,000 to 20,000 antibody molecules *per second* into the bloodstream (Figure 15.6). A second cell type, a **B-memory cell,** also forms at this time. Plasma cells live only a few days, but memory cells have a life span of months or even years. Memory cells are part of the immune memory system and are described in a later section.

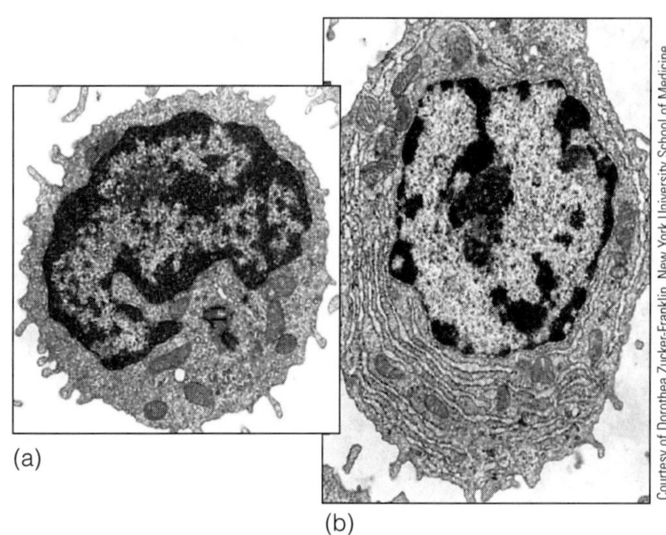

FIGURE 15.6 Electron micrographs of (a) a mature, unactivated B cell and (b) a differentiated plasma cell (an activated B cell).

Antibodies Are Molecular Weapons Against Antigens

Antibodies are Y-shaped protein molecules that bind to specific antigens in a lock-and-key manner to form an antigen–antibody complex (Figure 15.7). Antibodies are secreted by plasma cells and circulate in the blood and lymph. Some types of antibodies are attached to the surface of B cells. Antibodies belong to a class of proteins known as **immunoglobulins.**

There are five classes of immunoglobulins: IgG, IgA, IgM, IgD, and IgE. Each class has a unique structure, size, and function (Table 15.2). In general, antibody molecules have two identical long polypeptides (H chains) and two identical short polypeptides (L chains). The chains are held together by chemical bonds.

Antibody structure is related to its functions: (1) recognize and bind antigens and (2) inactivate the antigen. At one end of the antibody is an antigen-combining site formed by the ends of the H and L chains. This site recognizes and binds to a part of the antigen called the **antigenic determinant.** The formation of an antigen–antibody complex leads to the destruction of an antigen in several ways (Figure 15.8).

Plasma cells Cells, produced by mitotic division of B cells, that synthesize and secrete antibodies.

B-memory cell A long-lived B cell produced after exposure to an antigen that plays an important role in secondary immunity.

Immunoglobulins The five classes of proteins to which antibodies belong.

Antigenic determinant The site on an antigen to which an antibody binds, forming an antigen–antibody complex.

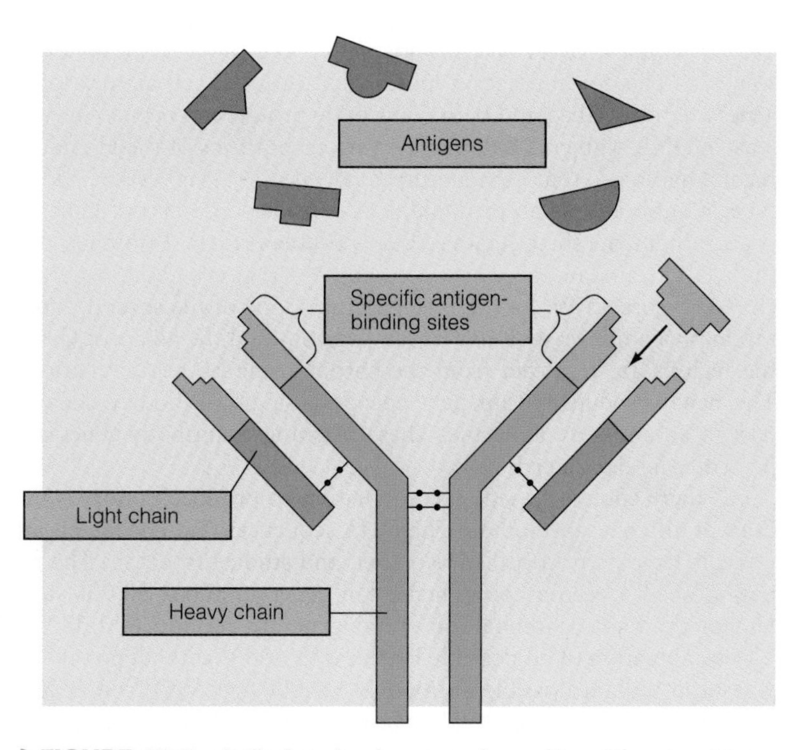

FIGURE 15.7 Antibody molecules are made up of two different proteins (an H chain and an L chain). The molecule is Y-shaped and forms a specific antigen-binding site at the ends.

Table 15.2 Types and Functions of the Immunoglobulins

Class	Location and Function
IgD	Present on surface of many B cells, but function uncertain; may be a surface receptor for B cells; plays a role in activating B cells.
IgM	Found on surface of B cells and in plasma; acts as a B-cell surface receptor for antigens secreted early in primary response; powerful agglutinating agent.
IgG	Most abundant immunoglobulin in the blood plasma; produced during primary and secondary response; can pass through the placenta, entering fetal bloodstream, thus providing protection to fetus.
IgA	Produced by plasma cells in the digestive, respiratory, and urinary systems, where it protects the surface linings by preventing attachment of bacteria to surfaces of epithelial cells; also present in tears and breast milk; protects lining of digestive, respiratory, and urinary systems.
IgE	Produced by plasma cells in skin, tonsils, and the digestive and respiratory systems, overproduction is responsible for allergic reactions, including hay fever and asthma.

Antibody Genes Undergo Recombination in B Cells

Humans can produce billions of different antibody molecules, each of which can bind to a different antigen. Because there are billions of such combinations, it is impossible that each antibody molecule is encoded directly in the genome; there simply is not enough DNA in the human genome to encode hundreds of millions or billions of antibodies.

The vast number of different antibodies are produced as a result of genetic recombination in three clusters of antibody genes: the H-chain genes on chromosome 14, the kappa L genes on chromosome 2, and the lambda light genes on chromosome 22. This recombination takes place during B-cell maturation before antibody genes are transcribed and before antibody production begins. In each antibody gene cluster, DNA segments that encode various portions of the H and L chains undergo recombination so that each mature B cell encodes, synthesizes, and secretes only one type of antibody. As an example, let's consider one class of light chain genes. This gene set contains three regions: V-L (variable-leader), J (joining), and C (constant). In B cell precursors, there are 70 to 300 V-L regions, about 6 J regions, and 1 C region (▶ Figure 15.9). As the B cell matures, one of the several hundred V-L regions randomly combines with one of the J regions and the adjacent C region. The rest of the regions are removed from the chromosome by recombination and destroyed. This newly produced fusion gene encodes a single antibody L chain that will become part of an antibody molecule. These rearranged antibody genes are stable and are passed to all daughter B cells.

Is there enough genetic recombination to produce hundreds of millions or billions of different antibodies? The DNA sequences that encode H- and L-chain components have been identified, isolated, and studied in detail. The random recombination of all H-chain components can generate about 30,000 different H chains. Similar random recombination events generate about 3,600 different light chains. The combination of all possible H chains (30,000) with all possible L chains (3,600) gives 108 million possible antibodies (30,000 × 3,600) from a few hundred components. Other events in B-cell maturation expand the number of possible H and L chains, allowing production of billions of possible antibody combinations from a few hundred segments at three loci.

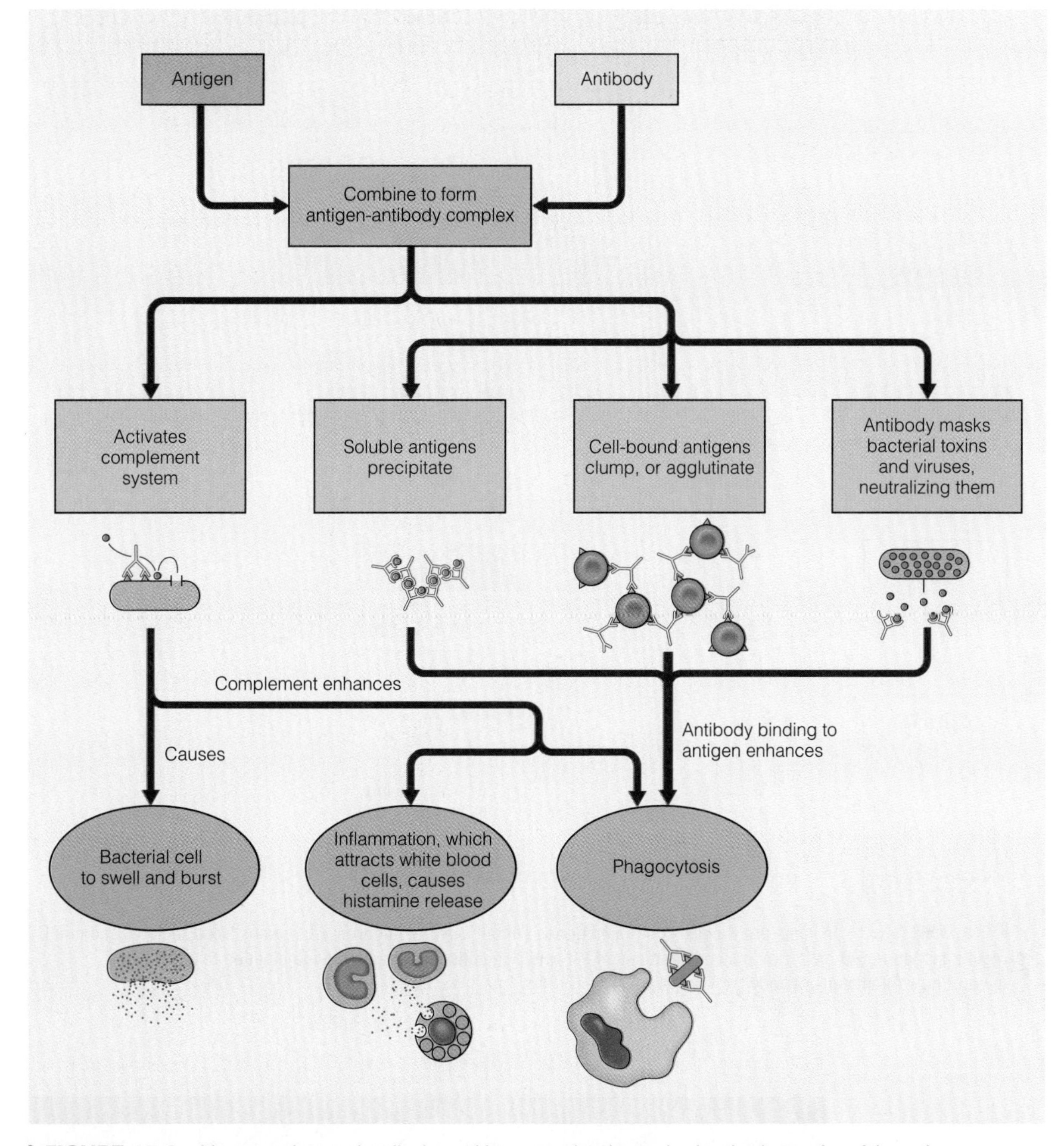

FIGURE 15.8 After an antigen and antibody combine, several pathways lead to the destruction of the antigen.

T Cells Mediate the Cellular Immune Response

There are several types of T cells in the immune system (Table 15.3). Helper T cells, described earlier, activate B cells to produce antibodies. **Suppressor T cells** slow down and stop the immune response and act as an "off" switch for the immune system. A third type, the **killer T cells,** finds and destroys cells of the body that are infected with a virus, bacteria, or other infectious agents (Figure 15.10). If a cell becomes infected with a virus, viral proteins will appear on its surface. These foreign antigens are recognized by receptors on the surface of a killer T cell. The T cell attaches to the infected cell and secretes a protein that punches holes in the plasma membrane of the infected cell. The cytoplasmic contents of the infected cell leak out through these holes, and the infected cell dies and is removed by phagocytes.

Suppressor T cells T cells that slow or stop the immune response of B cells and other T cells.

Killer T cells T cells that destroy body cells infected by viruses or bacteria. These cells can also directly attack viruses, bacteria, cancer cells, and cells of transplanted organs.

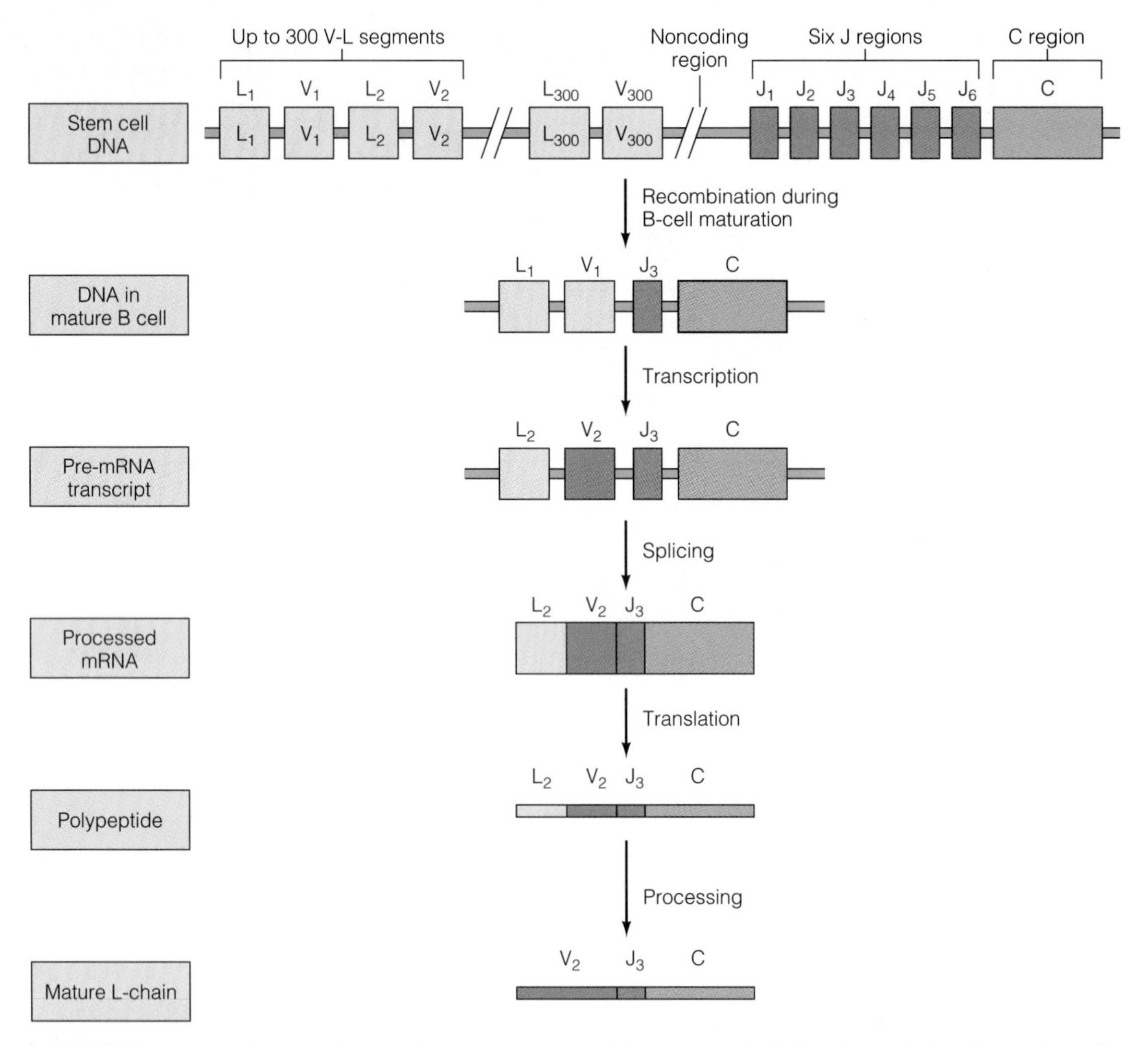

FIGURE 15.9 In immature B cells, light chain genes contain several hundred V-L regions, six J regions, and one C region. In maturing B cells, random combinations of V-L and J regions are fused by recombination to a C region, generating a functional antibody L-chain gene.

Table 15.3 Summary of T Cells

Cell Type	Action
Killer T cells	Destroy body cells infected by viruses and attack and kill bacteria, fungi, parasites, and cancer cells.
Helper T cells	Produce a growth factor that stimulates B-cell proliferation and differentiation and also stimulates antibody production by plasma cells; enhance activity of cytotoxic T cells.
Suppressor T cells	May inhibit immune reaction by decreasing B- and T-cell activity and B- and T-cell division.
Memory T cells	Remain in body awaiting reintroduction of antigen, when they proliferate and differentiate into cytotoxic T cells, helper T cells, suppressor T cells, and additional memory cells.

© Jean Claude Revi/Phototake

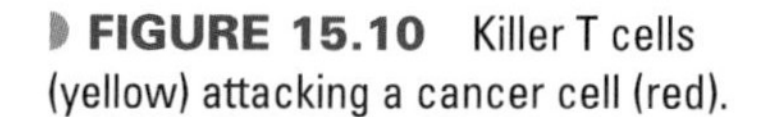

FIGURE 15.10 Killer T cells (yellow) attacking a cancer cell (red).

Cytotoxic T cells also kill cells of transplanted organs if they recognize them as foreign. The nonspecific and specific reactions of the immune system are summarized in Table 15.4.

The Immune System Has a Memory Function

As described in the opening section of the chapter, ancient writers observed that exposure to certain diseases made people resistant to second infections by the same disease. B and T memory cells that are produced in the first infection control this resistance, called **secondary immunity.** A second exposure to the same antigen results in an immediate, large-scale production of antibodies and killer T cells. Because of

Secondary immunity Resistance to an antigen the second time it appears because of T and B memory cells. This second response is faster, is larger, and lasts longer than the first.

Table 15.4 **Nonspecific and Specific Immune Responses to Bacterial Invasion**

Nonspecific Immune Mechanisms	Specific Immune Mechanisms
INFLAMMATION	Processing and presenting of bacterial antigen by macrophages
Engulfment of invading bacteria by resident tissue macrophages	Proliferation and differentiation of activated B-cell clone into plasma cells and memory cells
Histamine-induced vascular responses to increase blood flow to area, bringing in additional immune cells	Secretion by plasma cells of customized antibodies, which specifically bind to invading bacteria
Walling off of invaded area by fibrin clot	Enhancement by helper T cells, which have been activated by the same bacterial antigen processed and presented to them by macrophages
Migration of neutrophils and monocytes/macrophages to the area to engulf and destroy foreign invaders and to remove cellular debris	Binding of antibodies to invading bacteria and activation of mechanisms that lead to their destruction
Secretion by phagocytic cells of chemical mediators, which enhance both nonspecific and specific immune responses	Activation of lethal complement system
NONSPECIFIC ACTIVATION OF THE COMPLEMENT SYSTEM	Stimulation of killer cells, which directly lyse bacteria
Formation of hole-punching membrane attack complex that lyses bacterial cells	Persistence of memory cells capable of responding more rapidly and more forcefully should the same bacterial strain be encountered again
Enhancement of many steps of inflammation	

the presence of the memory cells, the second reaction is faster, more massive, and lasts longer than the primary immune response.

The secondary immune response is the reason that we can be vaccinated against infectious diseases. A **vaccine** stimulates the production of memory cells against a disease-causing agent. A vaccine is really a weakened, disease-causing antigen, given orally or by injection, that provokes a primary immune response and the production of memory cells. Often, a second dose is administered to elicit a secondary response that raises, or "boosts," the number of memory cells (which is why such shots are called booster shots).

Vaccines are made from killed or weakened strains (called attenuated strains) of disease-causing agents that stimulate the immune system but do not produce life-threatening symptoms of the disease. Recombinant DNA methods are now used to prepare vaccines against a number of diseases that affect humans and farm animals. The irony is that although vaccination is a proven and effective means of controlling a wide range of infectious diseases, fewer people in the United States are choosing to be vaccinated.

Vaccine A preparation containing dead or weakened pathogens that elicits an immune response when injected into the body.

Blood type One of the classes into which blood can be separated based on the presence or absence of certain antigens.

Hemolytic disease of the newborn (HDN) A condition of immunological incompatibility between mother and fetus that occurs when the mother is Rh^- and the fetus is Rh^+.

BLOOD TYPES ARE DETERMINED BY CELL-SURFACE ANTIGENS

Antigens on the surface of blood cells determine compatibility in blood transfusions. There are about 30 known antigens on blood cells; each of these constitutes a blood group or **blood type.** For successful transfusions, certain critical antigens of the donor and recipient must be identical. If transfused red blood cells do not have matching surface antigens, the recipient's immune system will produce antibodies against this antigen, clumping the transfused cells. The clumped blood cells block circulation in capillaries and other small blood vessels, with severe and often fatal results. In transfusions, two blood groups are of major significance: the ABO system and the Rh blood group.

ABO Blood Typing Allows Safe Blood Transfusions

ABO blood types are determined by a gene *I* (for isoagglutinin) that encodes an enzyme that alters a cell-surface protein. This gene has three alleles, I^A I^B, and I^O, often written as *A, B,* and *O*. The *A* and *B* alleles each produce a slightly different version of the enzyme, and the *O* allele produces no gene product. Individuals with type A blood have *A* antigen on their red blood cells, so they do not produce antibodies against this cell surface marker. However, people with type A blood have antibodies against the antigen encoded by the *B* allele (Table 15.5). Those with type B blood carry the *B* antigen on their red cells and have antibodies against the *A* anti-

Table 15.5 Summary of ABO Blood Types

Blood Type	Antigens on Plasma Membranes of RBCs	Antibodies in Blood	Safe to Transfuse To	Safe to Transfuse From
A	A	Anti-B	A, AB	A, O
B	B	Anti-A	B, AB	B, O
AB	A + B	none	AB	A, B, AB, O
O	–	Anti-A Anti-B	A, B, AB, O	O

SIDEBAR

Genetically Engineered Blood

The long search for an effective man-made substitute for blood that is both easy to use and free from contamination may be coming to a close. Companies are working to develop a blood substitute that will satisfy the need for the estimated 13 million units of blood transfused in the United States each year. That translates into a domestic market of approximately $2 billion and a global market as large as $8 billion a year.

Commercial research is aimed at producing blood substitutes. One approach to producing blood substitutes involves genetically engineered hemoglobin. Genes that control hemoglobin production are inserted into a bacterium and become incorporated into the bacterium's genetic material. During growth, *Escherichia coli* produces hemoglobin, which is purified and packaged for commercial sale. However, this approach has yet to be perfected in laboratory experiments.

The obvious advantages to genetically engineered hemoglobin is that it can have the efficiency of natal blood; it can be manufactured in unlimited quantities through genetically altered bacteria; and it would not contain contaminants, especially viruses such as HIV and hepatitis.

gen. If you have type AB blood, both antigens are present on the surface of red cells, and no antibodies against *A* or *B* are made. Those with type O blood have neither antigen but do have antibodies against both the *A* and *B* antigen.

Because AB individuals have no antibodies against *A* or *B,* they can receive a transfusion of blood of any type. Type O individuals have neither antigen and can donate blood to anyone, even though their plasma contains antibodies against *A* and *B;* after transfusion, the concentration of these antibodies is too low to cause problems.

When transfusions are made between incompatible blood types, several problems arise. Figure 15.11 shows the cascade of reactions that follows transfusion of someone who is type A with type B blood. Antibodies to the *B* antigen are in the blood of the recipient. These bind to the transfused red blood cells, causing them to clump. The clumped cells restrict blood flow in capillaries, reducing oxygen delivery. Lysis of red blood cells releases large amounts of hemoglobin into the blood. The hemoglobin forms deposits in the kidneys that block the tubules of the kidney and often cause kidney failure.

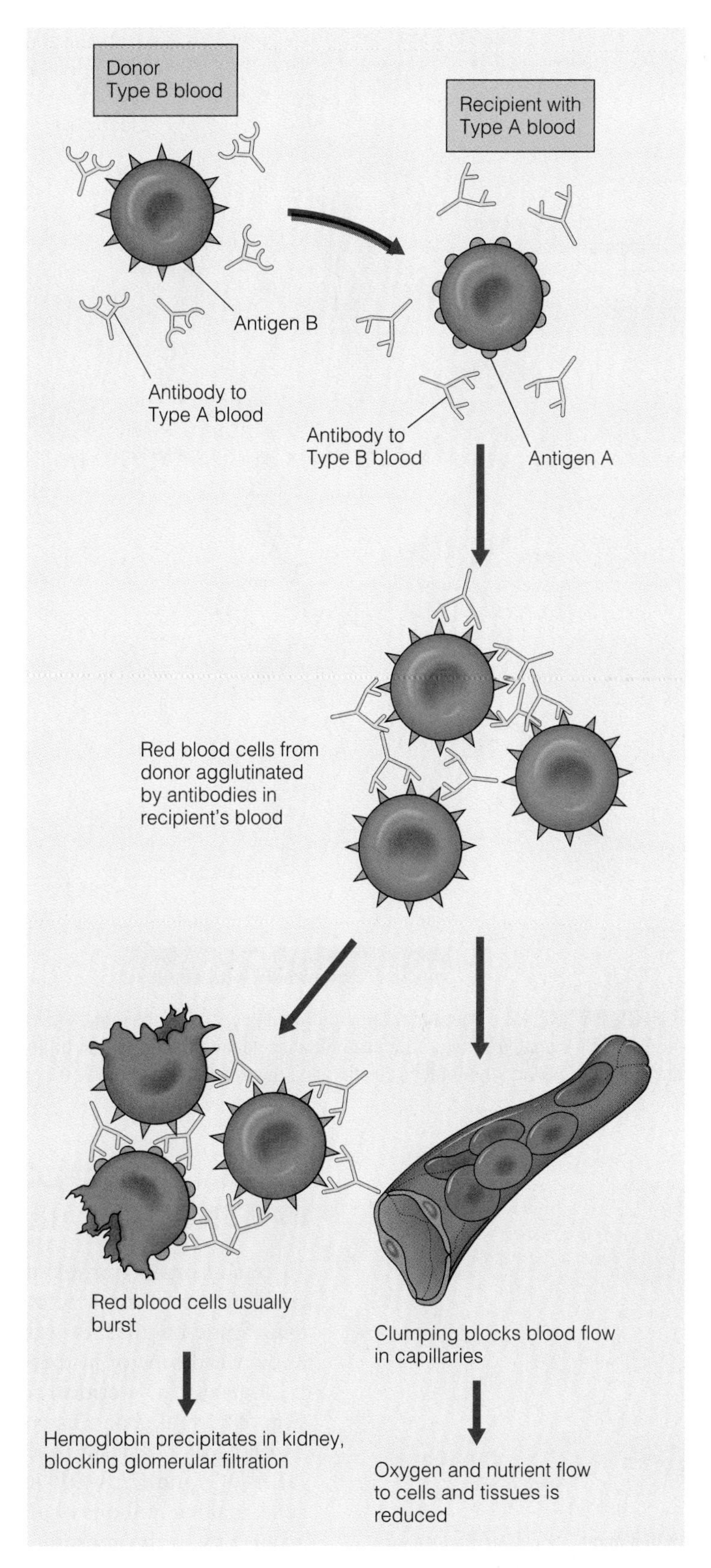

FIGURE 15.11 A transfusion reaction that results when type B blood is transfused into a recipient who has type A blood.

Rh Blood Types Can Cause Immune Reactions Between Mother and Fetus

The Rh blood group (named for the rhesus monkey, in which it was discovered) includes those who can make the *Rh* antigen (Rh positive, Rh^+) and those who cannot make the antigen (Rh negative, Rh^-).

The Rh blood group is a major concern when there is incompatibility between mother and fetus, a condition known as **hemolytic disease of the newborn (HDN).** This occurs most often when the mother is Rh^- and the fetus is Rh^+ (Figure 15.12). If Rh^+ blood from the fetus enters the Rh^- maternal circulation, the mother's immune system makes antibodies against the Rh antigen. Mingling of Rh^+ blood with that of the Rh^- mother commonly occurs during birth, so the first Rh^+ child is often unaffected. However, the maternal circulation now contains antibodies against the Rh antigen, and a subsequent Rh^+ fetus evokes a response from the maternal immune system. Massive amounts of antibodies cross the placenta in late stages of pregnancy and destroy the fetus' red blood cells (Figure 15.12), resulting in HDN.

To prevent HDN, Rh^- mothers are given an Rh-antibody preparation during the first pregnancy in which the fetus is Rh^+ and in all subsequent pregnancies involving an Rh^+ fetus. The injected Rh-antibodies move through the maternal circulatory system and destroy any Rh^+ fetal cells that may have entered the mother's circulation. To be effective, this antibody must be administered before the maternal immune system can make its own antibodies against the Rh antigen.

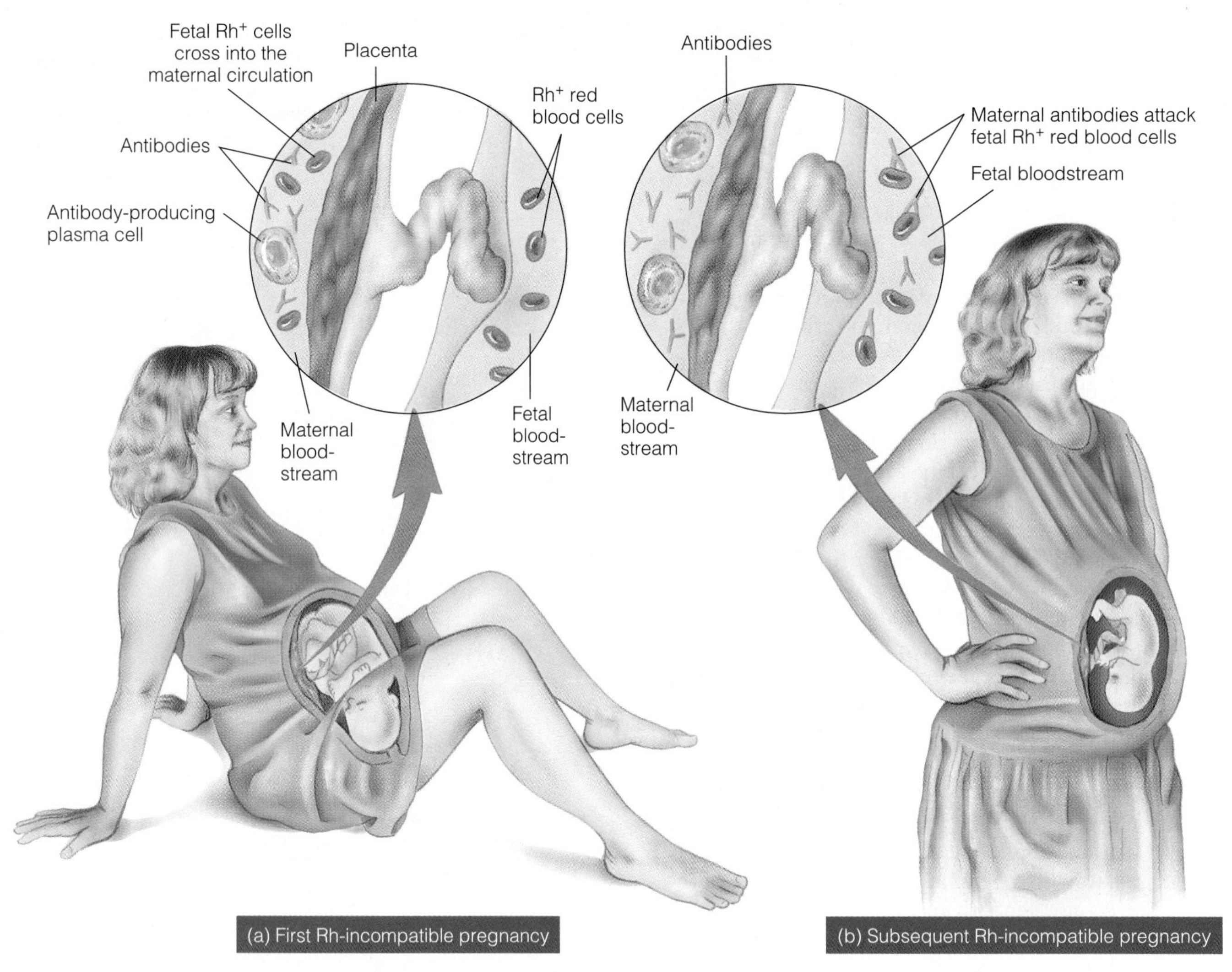

FIGURE 15.12 The Rh factor and pregnancy. (a) Rh-positive cells from the fetus can enter the maternal circulation at birth. If the mother is Rh-negative, she produces antibodies against the Rh factor. (b) In a subsequent pregnancy, if the fetus is Rh-positive, the maternal antibodies cross into the fetal circulation and destroy fetal red blood cells, producing hemolytic disease of the newborn (HDN).

ORGAN TRANSPLANTS MUST BE IMMUNOLOGICALLY MATCHED

Successful organ transplants and skin grafts depend on matches between the histocompatibility antigens of the donor and recipient. These antigens are cell surface proteins found on all cells in the body. In humans, a cluster of genes on chromosome 6, known as the major histocompatibility complex (MHC), produces these antigens. The HLA genes play a critical role in the outcome of transplants. The HLA complex consists of several gene clusters. One group, called class I, consists of *HLA-A, HLA-B,* and *HLA-C.* Adjacent to this is a cluster called class II, which consists of *HLA-DR, HLA-DQ,* and *HLA-DP.* A large number of alleles have been identified for each *HLA* gene, making millions of allele combinations possible. The array of HLA alleles on a given copy of chromosome 6 is known as a **haplotype.** Because each of us carries two copies of chromosome 6, we each have two HLA haplotypes (Figure 15.13).

Haplotype A cluster of closely linked genes or markers that are inherited together. In the immune system, the HLA alleles on chromosome 6 are a haplotype.

Because so many allele combinations are possible, it is rare that any two individuals have a perfect HLA match. The exceptions are identical twins, who have identical HLA haplotypes, and siblings, who have a 25% chance of being matched. In the example shown in Figure 15.13, each child receives one haplotype from each

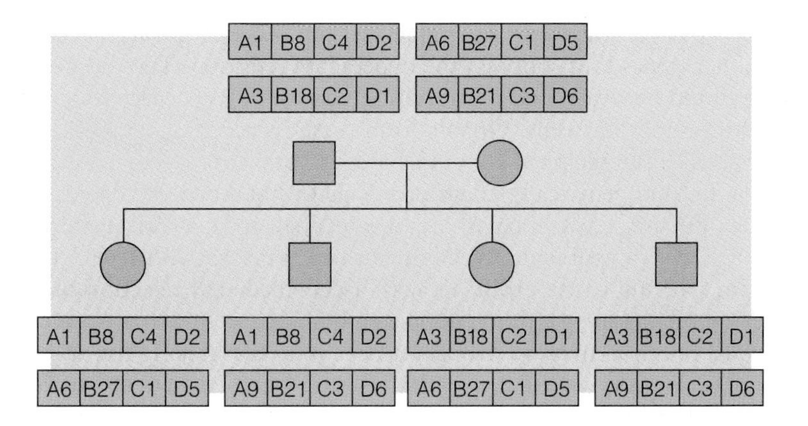

FIGURE 15.13 The transmission of HLA haplotypes. In this simplified diagram, each haplotype contains four genes, each of which encodes a different antigen.

parent. As a result, four new haplotype combinations are represented in the children. (Thus siblings have a one in four chance of having the same haplotypes.)

Successful Transplants Depend on HLA Matching

Successful organ transplants depend largely on matching HLA haplotypes between donor and recipient. Because there are so many HLA alleles, the best chance for a match is usually between related individuals, with identical twins having a 100% match. The order of preference for organ and tissue donors among relatives is identical twin, sibling, parent, and unrelated donor. Among unrelated donors and recipients, the chances for a successful match are only 1 in 100,000 to 1 in 200,000. Because the frequency of HLA alleles differs widely among racial and ethnic groups, matches across racial and ethnic lines are often more difficult. When HLA types are matched, the survival of transplanted organs is dramatically improved. Survival rates for matched and unmatched kidney transplants over a 4-year period are shown in Figure 15.14.

Animal–Human Organ Transplants

In the United States, about 18,000 organs are transplanted each year, but about 40,000 qualified patients are on waiting lists. Each year, almost 3,000 people on the waiting list die before receiving a transplant and another 100,000 die even *before* they are placed on the waiting list. Although the demand for organ transplants is rising, the number of donated organs is growing very slowly. Experts estimate that more than 50,000 lives would be saved each year if enough organs were available.

Xenotransplant Organ transplants between species.

One way to increase the supply of organs is to use animal donors for transplants. Animal–human transplants (called **xenotransplants**) have been attempted many times, but with little success. Two important biological problems are related to xenotransplants: (1) complement-mediated rejection and (2) T-cell mediated rejection. In the first, cell-surface proteins (products of the HLA system) act as antigens that are very different across species. When an animal organ (e.g., from a pig) is transplanted into a human, the cell-surface proteins on the pig organ are so different that they trigger an immediate and massive immune response, known as hyperacute rejection. This reaction, mediated by the complement system, usually destroys the transplanted organ within hours. The second problem is the same as in human organ transplants: how to suppress the cellular immune (T-cell mediated) rejection of the transplant.

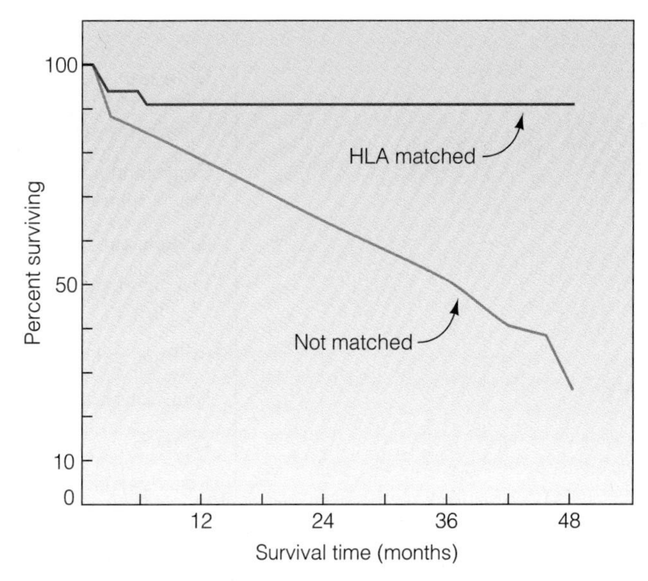

FIGURE 15.14 The outcome of kidney transplants with (upper curve) and without (lower curve) HLA matching.

To overcome the first problem, several research groups have isolated and cloned human genes that suppress the hyperacute rejection. These genes were injected into fertilized pig eggs, and the resulting transgenic pigs carry human-recognition antigens on all their cells. Organs from these transgenic pigs should appear as human organs to the recipient's immune system, preventing a hyperacute rejection. Transplants from genetically engineered pigs to monkey hosts have been successful, but the ultimate step will be an organ transplant from a transgenic pig to a human.

Even if hyperacute rejection can be suppressed, however, transplanting pig organs will still cause problems with T-cell–mediated rejection of the transplant. Because transplants from pig donors to humans occur across species, the tendency toward rejection may be stronger and require the lifelong use of immunosuppressive drugs. These powerful drugs may be toxic when taken over a period of years or may weaken the immune system, paving the way for continuing rounds of infections. To deal with this problem, it may be necessary to transplant bone marrow from the donor pig to the human recipient. The resulting pig–human immune system (called a chimeric immune system) would recognize the pig organ as "self" and still retain normal human immunity. As far-fetched as this may sound, animal experiments using this approach have been successful in preventing rejection more than 2 years after transplantation without the use of immunosuppressive drugs.

As recently as 10 years ago, the possibility of animal–human organ transplants seemed like a remote possibility, more suited to science fiction than to medical fact. However, the advances described here make it likely that xenotransplants to humans will be attempted in the next few years. Although animal organ donors will probably become common in the near future, guidelines for transgenic donors and problems with immunosuppressive drugs and immune tolerance remains to be worked out.

The HLA System and Disease Associations

Ankylosing spondylitis An autoimmune disease that produces an arthritic condition of the spine; associated with HLA allele B27.

In studying the distribution of HLA alleles in the population, researchers have discovered a relationship between certain HLA alleles and specific diseases. For example, more than 90% of those who have the connective tissue disease **ankylosing spondylitis** (MIM/OMIM 106300) carry the HLA-B27 allele. This disease causes

Table 15.6 **HLA Alleles and Disease**

Disease	HLA Allele	Frequency in Patients %	Frequency in General Population %	Relative Risk Factor
Ankylosing spondylitis	B27	>90	8	>100
Congenital adrenal hyperplasia	B47	9	0.6	15
Goodpasture syndrome	DR2	88	32	16
Juvenile rheumatoid arthritis	DR5	50	16	5
Multiple sclerosis	DR2	59	26	4
Pernicious anemia	DR5	25	6	5
Psoriasis	B17	38	8	6
Reiter's syndrome	B27	75	8	50
Rheumatoid arthritis	DR4	70	28	6
Systemic lupus erythematosis (SLE)	DR3	50	25	3

chronic inflammation of joints and the spine, leading to fusion of the joints between vertebrae. Evidence suggests that B27 class I antigens are involved in promoting the joint inflammation characteristic of this disease. Other diseases also show an association with certain HLA alleles; some of these are listed in Table 15.6.

Several ideas have been proposed to explain the relationship between HLA alleles and disease, but so far none has proven valid. It appears that most of these disorders result from a combination of HLA and non-HLA effects with environmental activation, possibly by infection. For example, ankylosing spondylitis may be triggered in B27 patients by an infection, and the bacterium *Klebsiella* has been implicated in the onset of this disease.

DISORDERS OF THE IMMUNE SYSTEM

We are able to resist infectious disease because we have an immune system. Unfortunately, failures in the immune system can result in abnormal or even absent immune responses. The consequences of these failures can range from mild inconvenience to systemic failure and death. In this section, we briefly catalog some ways in which the immune system can fail.

Overreaction in the Immune System Causes Allergies

Allergies result when the immune system overreacts to weak antigens that do not provoke an immune response in most people (Figure 15.15). These weak antigens, called **allergens,** include a wide range of substances: house dust, pollen, cat dander, certain foods, and even medicines such as penicillin. It is estimated that up to 10% of the U.S. population suffers from at least one allergy. Typically, allergic reactions develop after a first exposure to an allergen. The allergen causes B cells to make IgE antibodies, instead of IgG antibodies. The IgE antibodies attach to mast cells in tissues, including those of the nose and respiratory system.

Allergens Antigens that provoke an inappropriate immune response.

In a second exposure, the allergen binds to the IgE and the mast cells release histamine, triggering an inflammatory response that results in fluid accumulation, tissue swelling, and mucus secretion. This reaction is severe in some individuals, and histamine is released into the circulatory system, causing a life-threatening decrease in blood pressure and constriction of airways in the lungs. This reaction, called **anaphylaxis** or anaphylactic shock, most often occurs after exposure to the antibiotic penicillin or to the venom in bee or wasp stings. Prompt treatment of anaphylaxis with antihistamines, epinephrine, and steroids can reverse the reaction. As the name suggests, antihistamines block the action of histamine. Epinephrine opens the airways and constricts blood vessels, raising blood pressure. Steroids, such as prednisone, inhibit the inflammatory response. Some people who have a history of severe reaction to insect stings carry these drugs with them in a kit (Concepts and Controversies: Why Bee Stings Can Be Fatal).

Anaphylaxis A severe allergic response in which histamine is released into the circulatory system.

Autoimmune Reactions Cause the Immune System to Attack the Body

One of the most elegant properties of the immune system is its capacity to distinguish self from nonself. During development, the immune system "learns" not to react against cells of the body. In some disorders, this immune tolerance breaks down, and the immune system attacks and kills cells and tissues in the body. Juvenile diabetes, also known as insulin-dependent diabetes (IDDM) (MIM/OMIM 222100), is an autoimmune disease. Clusters of cells in the pancreas produce insulin, a hormone that lowers blood sugar levels. In IDDM, the immune system attacks and kills the insulin-producing cells, resulting in diabetes and the need for insulin injections to control blood sugar levels.

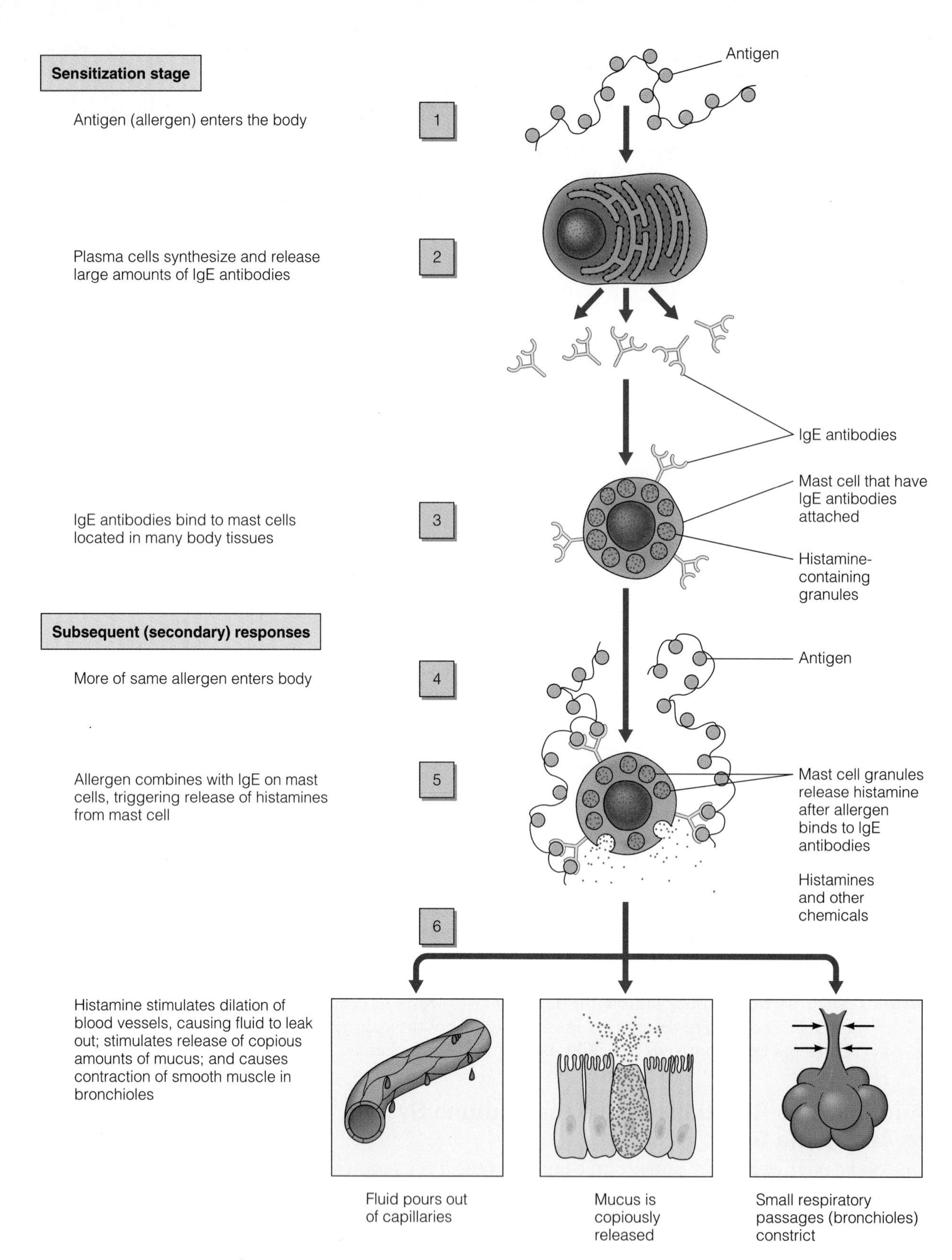

FIGURE 15.15 The steps in an allergic reaction.

X-linked agammaglobulinemia (XLA) A rare, sex-linked, recessive trait characterized by the total absence of immunoglobulins and B cells.

Concepts and Controversies

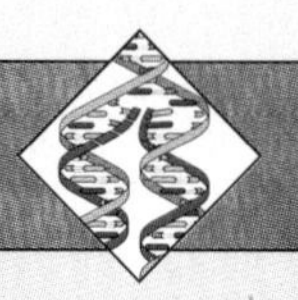

Why Bee Stings Can Be Fatal

In most cases an allergy is an inconvenience, not a life-threatening condition. Typically, the response to contact with an allergen involves localized itching, swelling, and reddening of the skin, often in the form of hives. Inhaled allergens usually result in a localized response accompanied by itching and watery eyes, runny nose, and constricted bronchial tubes. The most common form of allergy involving inhaled allergens is hay fever. In the United States, more than 20 million people are affected.

In response to injected allergen (such as venom from a bee sting) or certain drugs (such as penicillin), the reaction can be systemic rather than localized and result in anaphylactic shock. Anaphylaxis is usually a two-stage process. In the first stage, using a bee sting as an example, the venom enters the body and IgE antibodies are made in response, causing sensitization to future stings. It is not clear why some people become sensitized and others do not. If a person has become sensitized, a second exposure to bee venom can provoke a life-threatening allergic reaction. When a bee stings a sensitized individual, the allergen in the venom enters the body and binds to immunoglobulin E antibodies made in response to previous exposure. Within 1 to 15 minutes of exposure, the IgE antibodies activate mast cells. The stimulated mast cells release large amounts of histamines and chemotactic factors, which attract other white blood cells as part of the inflammatory response. In addition, the mast cells release prostaglandins and the slow-reacting substance of anaphylaxis (SRS-A). SRS-A is more than 100 times more powerful than histamine and prostaglandins in eliciting an allergic reaction and intensifies the response to the allergen.

Release of SRS-A, histamine, and other factors into the bloodstream causes a systemic reaction. In response, the bronchial tubes constrict, closing the airways, and fluids pass from the tissues into the lungs, making breathing difficult. Blood vessels dilate, causing blood pressure to drop, and plasma escapes into the tissues, causing shock. Heart arrhythmias and cardiac shock can develop and cause death within 1 to 2 minutes after the onset of symptoms. Immediate treatment with epinephrine, antihistamines, and steroids is effective in reversing the symptoms.

In the case of hypersensitivity to insect stings, desensitization can be used as immunotherapy. This consists of injecting small quantities of the allergen, gradually increasing the dose. Exposure over time to small doses of the allergen causes the production of IgG antibodies against the allergen rather than IgE antibodies. When desensitization is complete, subsequent exposure to the allergen causes the more numerous IgG antibodies to bind to the mast cells, preventing the binding of IgE antibodies and the resulting anaphylactic reaction.

Other forms of autoimmunity, such as systemic lupus erythematosus (SLE) (MIM/OMIM 152700), attack and slowly destroy major organ systems. Some autoimmune disorders are listed in Table 15.7.

Table 15.7 Some Autoimmune Diseases

Some Autoimmune Diseases
Addison's disease
Autoimmune hemolytic anemia
Diabetes mellitus—insulin-dependent
Graves' disease
Membranous glomerulonephritis
Multiple sclerosis
Myasthenia gravis
Polymyositis
Rheumatoid arthritis
Scleroderma
Sjögren's syndrome
Systemic lupus erythematosus

Genetic Disorders Can Impair the Immune System

In 1952 Ogden Bruton, a physician at Walter Reed Army Hospital, described the first immunodeficiency disease. He examined a young boy who had suffered at least 20 serious infections in the preceding 5 years. Blood tests showed that the child had no antibodies. Other patients with similar problems were soon discovered. Affected individuals were usually boys who were highly susceptible to bacterial infections. In all cases, either B cells were completely absent, or the B cells were immature and unable to produce antibodies. Without functional B cells, no antibodies can be produced. On the other hand, these individuals usually have nearly normal levels of T cells. In other words, antibody-mediated immunity is absent, but cellular immunity is normal. This heritable disorder, called **X-linked agammaglobulinemia (XLA)** (MIM/OMIM 300300), usually appears 5 to 6 months after birth when maternal antibodies disappear and the infant's B-cell population normally begins to produce antibodies. Patients with this disorder are highly susceptible to pneumonia and streptococcal infections and pass from one life-threatening infection to another.

Individuals with XLA lack mature B cells but have normal populations of B-cell precursors, indicating that the defect is in maturation of B cells. The *XLA* gene was mapped to Xq21.3-Xq22 and encodes an enzyme that transmits signals from the

▶ **FIGURE 15.16** David, the "boy in the bubble," the oldest known survivor of severe combined immunodeficiency disease.

© Baylor College of Medicine/Peter Arnold Inc.

Severe combined immunodeficiency disease (SCID) A genetic disorder in which affected individuals have no immune response; both the cell-mediated and antibody-mediated responses are missing.

Acquired immunodeficiency syndrome (AIDS) A collection of disorders that develop as a result of infection with the human immunodeficiency virus (HIV).

cell's environment into the cytoplasm. Chemical signals from outside the cell help trigger B-cell maturation, and the gene product that is defective in XLA plays a critical role in this signaling process. Understanding the role of this protein in B-cell development may permit the use of gene therapy to treat this disorder.

A rare genetic disorder causes a complete absence of *both* antibody-mediated and cell-mediated immune responses. This condition is called **severe combined immunodeficiency (SCID)** (MIM/OMIM 102700, 600802, and others). Affected individuals have recurring and severe infections and usually die at an early age from seemingly minor infections. One of the longest known survivors of this condition was David, the "boy in the bubble," who died at 12 years of age after being isolated in a sterile plastic bubble for all but the last 15 days of his life (▶ Figure 15.16).

One form of SCID is caused by a deficiency of the enzyme adenosine deaminase (ADA). A small group of children affected with ADA-deficient SCID (MIM/OMIM 102700) is currently undergoing gene therapy to give them a normal copy of the gene. Their genetically modified white blood cells are returned to their circulatory systems by transfusion. Expression of the normal *ADA* gene in these cells stimulates development of functional T and B cells and at least partially restores a functional immune system. The recombinant DNA techniques used in gene therapy are reviewed in Chapter 13.

▶ **FIGURE 15.17** Particles of human immunodeficiency virus (HIV).

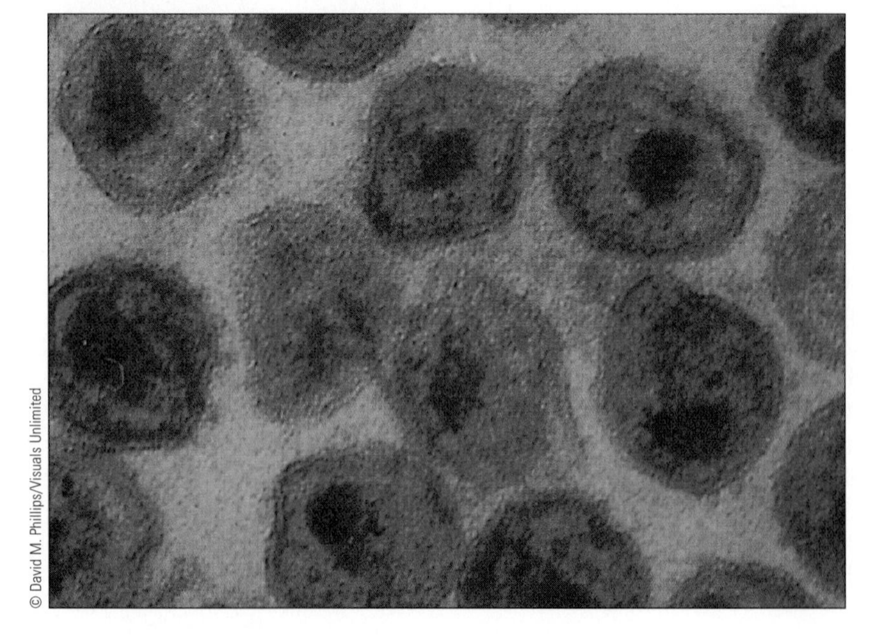

© David M. Phillips/Visuals Unlimited

HIV Attacks the Immune System

The immunodeficiency disorder currently receiving the most attention is **acquired immunodeficiency syndrome (AIDS).** AIDS is a collection of disorders that develop as a result of infection with HIV (▶ Figure 15.17). HIV is a retrovirus with three components:

Guest Essay: Research and Applications

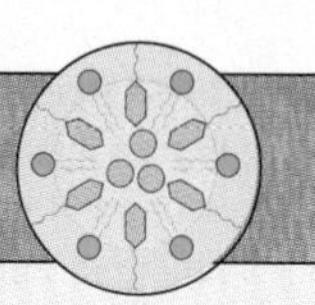

Medicine—A Scientific Safari

M. Michael Glovsky

What motivates a person to become a physician or a scientist? Chance, the desire for adventure, and the opportunity to discover the unknown are important ingredients. Let me explain.

I grew up in a small town in central Massachusetts and attended Tufts University. Chemistry was a favorite subject. I learned that it was possible to question why certain chemical reactions occurred and that several different approaches could be used to solve problems. My desire to work in a rewarding and stimulating profession was also important. I then entered Tufts Medical School. Most medical students are overwhelmed by the amount of knowledge that is available. I was not an exception. Yet, after the initial shock, coupled with long hours of study, I adapted to the need to absorb the most important information in anatomy, physiology, and biochemistry. During my second year, I became intrigued by problems in immunology. How is the fetus able to tolerate the foreign environment of the mother's uterus? Why do habitual abortions occur? I had an opportunity to explore these two questions in the laboratory of Dr. Wadi Bardawell, an obstetrical pathologist. In the final two years of medical school, patient contact and the reality of the consequences of disease were opportunities to integrate knowledge gained in the basic sciences.

After graduation from medical school, I spent almost four years learning about complement and immunoglobulins. I was fortunate to have as mentors Dr. Elmer Becker at the Walter Reed Army Institute of Research and Hugh Fudenberg, M.D., at the University of California, San Francisco. My initial research addressed what chemicals could block the inflammatory reactions in the skin when complement and immunoglobulins interact.

Because of my interest in immunology, I became an allergist. Can chemicals derived from the structure of IgE, the allergy antibody, interfere in the allergic response? We were fortunate to have as collaborators Drs. Bergitt Helm and Hannah Gould from King's College, London. They provided recombinant proteins synthesized from the known structure of the heavy chain regions of the IgE myeloma proteins. Together, we were able to show that the IgE recombinant fragment (301–376) could block ragweed allergens and grass allergen reactions in human skin and in blood basophils. Further studies were performed to pinpoint the binding site of the allergy antibody, IgE, that binds to human basophils.

More recently we have studied substances in air pollution that cause the nose to run and the bronchial tubes to constrict. We are exploring whether latex, an ingredient in natural rubber, produces allergenic proteins that are important in asthma. Together with Dr. Ann Miguel and Dr. Glenn Cass at Caltech, we have found latex allergen in air samples and roadside dust samples. Latex allergen, present in natural rubber gloves, balloons, and tire dust, is one of the most potent allergens of the last 10 years. We are studying whether latex allergy is important in the increasing symptoms of asthma in the last decade.

Medical science has evolved to provide sophisticated tools, especially in molecular biology, to address important questions and seek relevant answers. Biological science is a safari through the wilderness. Every path leads to adventure and exploration with both frustration and rewards. But the trail is always interesting and provocative.

M. Michael Glovsky is director of the Asthma and Allergy Center at Huntington Memorial Hospital in Pasadena, California. He received a B.S. degree from Tufts University in 1957 and an M.D. degree from Tufts Medical School in 1962.

(1) a protein coat (which encloses the other two components), (2) RNA molecules (the genetic material), and (3) an enzyme called reverse transcriptase. The viral particle is enclosed in a coat derived from a T-cell plasma membrane. The virus selectively infects and kills T4 helper cells of the immune system. Inside a cell, reverse transcriptase transcribes the RNA into a DNA molecule, and the viral DNA is inserted into a human chromosome, where it can remain for months or years.

Later, when the infected T cell is called upon to participate in an immune response, the viral genes are activated. Viral RNA and proteins are made, and new viral particles are formed. These bud off the surface of the T cell, rupturing and killing the cell and setting off a new round of T-cell infection (◗ Figure 15.18). Over the course of HIV infection, the number of helper T4 cells gradually decreases. (Recall

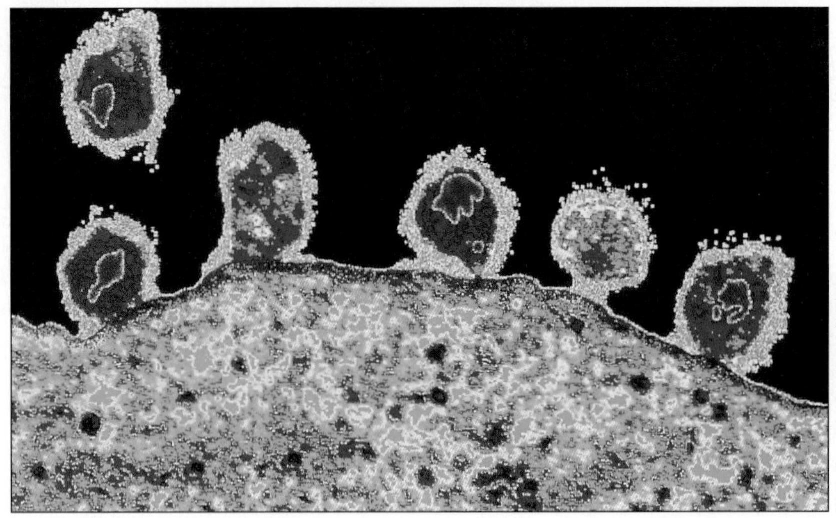

© J.L. Carson/Custom Medical Stock Photo

FIGURE 15.18 A colored electron micrograph showing human immunodeficiency virus (HIV) particles budding off the surface of a helper T-lymphocyte.

that these cells act as the master "on" switch for the immune system.) As the T4 cell population falls, the ability to mount an immune response decreases. The result is increased susceptibility to infection and increased risk of certain forms of cancer. The eventual outcome is premature death brought about by any of a number of diseases that overwhelm the body and its compromised immune system.

HIV is transmitted from infected to uninfected individuals through body fluids, including blood, semen, vaginal secretions, and breast milk. The virus cannot live for more than 1 to 2 hours outside the body and cannot be transmitted by food, water, or casual contact.

Summary

The Immune System Defends the Body Against Infection

1. The immune system protects the body against infection by a graded series of responses that attack and inactivate foreign molecules and organisms.

The Inflammatory Response Is a General Reaction

2. The lowest level of response to infection involves a nonspecific, local, inflammatory response. This response is mediated by cells of the immune system and isolates and kills invading microorganisms. Genetic control of this response is abnormal in inflammatory diseases, including ulcerative colitis and Crohn's disease.

The Immune Response Is a Specific Defense Against Infection

3. The immune system has two components: antibody-mediated immunity, which is regulated by B cells and antibody production, and cell-mediated immunity, which is controlled by T cells. The primary function of antibody-mediated reactions is to defend the body against invading viruses and bacteria. Cell-mediated immunity is directed against cells of the body that have been infected by agents such as viruses and bacteria.

Blood Types Are Determined by Cell Surface Antigens

4. The presence or absence of certain antigens on the surface of blood cells is the basis of blood transfusions and blood types. Two blood groups are of major significance: the ABO system and the Rh blood group. Matching ABO blood types is important in blood transfusions. In some cases, mother–fetus incompatibility in the Rh system can cause maternal antigens to destroy red blood cells of the fetus, resulting in hemolytic disease in the newborn.

Organ Transplants Must Be Immunologically Matched

5. The success of organ transplants and skin grafts depends on matching histocompatibility antigens found on the surface of all cells in the body. In humans, the antigens produced by a group of genes on chromosome 6 (known as the MHC complex) play a critical role in the outcome of transplants.

Disorders of the Immune System

6. Allergies are the result of immunological hypersensitivity to weak antigens that do not provoke an immune response in most people. These weak antigens, known as allergens, include a wide range of substances: house dust, pollen, cat hair, certain foods, and even medicines, such as penicillin. Acquired immunodeficiency syndrome (AIDS) is a collection of disorders that develop as a result of infection with a retrovirus known as the human immunodeficiency virus (HIV). The virus selectively infects and kills the T4 helper cells of the immune system.

CASE 1

Mary and John Smith went for genetic counseling because John did not believe that their newborn son was his. Mary and John wanted blood tests to help rule out the possibility that someone other than John was the father of this baby. The counselor explained that ABO blood typing could give some preliminary indications about possible paternity. Its use is limited, however, because there are only four possible ABO blood types, and the vast majority of people in any population have only two of those types (A and O). This means that a man may have a blood type consistent with paternity and still not be the father of the tested child. Because the allele for type O can be masked by the genes for A or B, inheritance of blood type can be unclear. More modern genetic tests, such as DNA typing, would lead to a more reliable conclusion regarding paternity.

Mary's blood was tested and identified as type O. John's blood was tested and identified as type O. Based on these two parental combinations of blood types, the only possible blood type that their son could be is type O. The baby was tested and his blood type was, indeed, type O.

1. Can any absolute conclusions be drawn based upon the results of these blood tests? Why or why not?
2. If not, was it worth doing the test in the first place?
3. Why do you think DNA testing would be more reliable than blood testing for this purpose?

And the Father is

If the Mother is	A	B	AB	O
A	A or O	A, B, AB or O	A, B, or AB	A or O
B	A, B, AB, or O	B or O	A, B, or AB	B or O
AB	A, B, or AB	A, B, or AB	A, B, or AB	A or B
O	A or O	B or O	A or B	O

The Child must be

CASE 2

The Joneses were referred to a clinical geneticist because their 6-month-old daughter was failing to grow adequately and was having recurrent infections. The geneticist took a detailed family history (which was uninformative) and medical history of their daughter. He discovered that their daughter had a history of several ear infections against which antibiotics had no effect, she had difficulty gaining weight (failure to thrive), and had an extensive history of yeast infection (thrush) in her mouth. The geneticist did a simple blood test to check their daughter's white blood count and determined that she had severe combined immunodeficiency (SCID).

The geneticist explained that SCID is an immune deficiency that causes a marked susceptibility to infections. The defining characteristic is usually a severe defect in both the T- and B-lymphocyte systems. This usually results in one or more infections within the first few months of life. These infections are usually serious and may even be life threatening. They may include pneumonia, meningitis, or bloodstream infections. Based on the family history, it was possible that their daughter inherited an altered gene from each of them and therefore was homozygous for the gene that causes SCID. Each time the Joneses had a child there would be a 25% chance that the child would have SCID. Prenatal testing is available to determine whether the developing fetus has SCID.

1. Genetic testing was performed and it showed that both parents were heterozygous carriers of a mutant allele of adenosine deaminase (ADA) and that the daughter is homozygous for this mutation. Are there any treatment options available for ADA-deficient SCID?
2. If the Joneses wanted to have another child but wanted to be certain that the child would not have SCID, what types of reproductive options do you think they might have?

Questions and Problems

The Inflammatory Response Is a General Reaction

1. What causes the area around a cut or a scrape to become warm? What is the role of this heat in the inflammatory response?
2. The complement system supplements the inflammatory response by directly killing microorganisms. Describe the life cycle of the complement proteins, from their synthesis in the liver to their activity at the site of an infection.

The Immune Response Is a Specific Defense Against Infection

3. Name the class of molecules that includes antibodies, and name the five groups that make up this class.
4. Discuss the roles of the three types of T cells: helper cells, suppressor cells, and killer cells.
5. Compare the general inflammatory response, the complement system, and the specific immune response.
6. Distinguish between antibody-mediated and cell-mediated immunity. What components are involved in each?
7. The molecular weight of IgG is 150,000 kd. Assuming that the two heavy chains are equivalent, the two light chains are equivalent, and the molecular weights of the light chains are half the molecular weight of the heavy chains, what are the molecular weights of each individual subunit?
8. Identify the components of cellular immunity, and define their roles in the immune response.
9. Describe the rationale for vaccines as a form of preventive medicine.
10. Researchers have been having a difficult time developing a vaccine against a certain pathogenic virus due to the lack of a weakened strain. They turn to you because of your wide knowledge of recombinant DNA technology and the immune system. How could you vaccinate someone against the virus, using a cloned gene from the virus that encodes a cell-surface protein?
11. It is often helpful to draw a complicated pathway in the form of a flow chart to visualize the multiple steps and the ways in which the steps are connected to each other. Draw the antibody-mediated immune response pathway that acts in response to an invading virus.
12. Describe the genetic basis of antibody diversity.
13. In cystic fibrosis gene therapy, scientists propose the use of viral vectors to deliver normal genes to cells in the lungs. What immunological risks are involved in this procedure?

Blood Types Are Determined by Cell-Surface Antigens

14. A man has the genotype I^AI^A, and his wife is I^BI^B. If their son needed an emergency blood transfusion, would either parent be able to be a donor? Why or why not?
15. Why can someone with blood type AB receive blood of any type? Why can a blood type O individual donate blood to anyone?
16. What is more important to match during blood transfusions: the antibodies of the donor or the antigens of the donor/recipient?
17. The following data were presented to a court during a paternity suit: (1) the infant is a universal donor for blood transfusions, (2) the mother bears antibodies against the B antigen only, and (3) the alleged father is a universal recipient in blood transfusions.
    a. Can you identify the ABO genotypes of the three individuals?
    b. Can the court draw any conclusions?
18. A patient of yours has just undergone shoulder surgery and is now experiencing kidney failure for no apparent reason. You check his chart and find that his blood is type B, but he has been mistakenly transfused with type A. Describe why he is experiencing kidney failure.
19. Assume that a single gene having alleles that show complete-dominance relationships at the phenotypic level controls the Rh character. An Rh^+ father and an Rh^- mother have eight boys and eight girls, all Rh^+.
    a. What are the Rh genotypes of the parents?
    b. Should they have been concerned about hemolytic disease of the newborn?
20. How is Rh incompatibility involved in hemolytic disease of the newborn? Is the mother Rh^+ or Rh^-? Is the fetus Rh^+ or Rh^-? Why is a second child that is Rh^+ more susceptible to attack from the mother's immune system?

Organ Transplants Must Be Immunologically Matched

21. What mode of inheritance has been observed for the HLA system in humans?
22. A burn victim receives a skin graft from her brother; however, her body rejects the graft a few weeks later. The procedure is attempted again, but this time the graft is rejected in a few days. Explain why the graft was rejected the first time and why it was rejected faster the second time.
23. In the human HLA system there are 23 *HLA-A* alleles, 47 for *HLA-B*, 8 for *HLA-C*, 14 for *HLA-DR*, 3 for *HLA-DQ*, and 6 for *HLA-DP*. How many different human HLA genotypes are possible?
24. In the near future, pig organs may be used for organ transplants. How are researchers attempting to prevent rejection of the pig organs by human recipients?
25. A couple has a young child who needs a bone marrow transplant. They propose that preimplantation screening be done on several embryos fertilized *in vitro* to find a match for their child.
    a. What do they need to match in this transplant procedure?

b. The couple proposes that the matching embryo be transplanted to the mother's uterus and serve as a bone marrow donor when old enough. What are the ethical issues involved in this proposal?

26. A couple have the following HLA genotypes: Male A12, B36, C7, DR14, DQ2, DP5/A9, B27, C3, DR14, DQ3, and DP1; Female A8, B15, C2, DR5, DQ2, DP3/A2, B20, C3, DR8, DQ1, and DP3. What is the probability that an offspring will suffer from ankylosing spondylitis if:
    a. the child is male and has HLA-A9?
    b. the child is female and has HLA-A9?
    c. the child is male and has HLA-DP1?
    d. the child is male and has HLA-C7?

Disorders of the Immune System

27. Why are allergens called "weak" antigens?
28. Antihistamines are used as antiallergy drugs. How do these drugs work to relieve allergy symptoms?
29. Autoimmune disorders involve a breakdown of an essential property of the immune system. What is it? How does this breakdown cause juvenile diabetes?
30. A young boy, who has suffered from over a dozen viral and bacterial infections in the past 2 years, comes to your office for an examination. You examine the boy and determine by testing that he has no circulating antibodies. What syndrome does he have, and what are its characteristics? What component of the two-part immune system is nonfunctional?
31. AIDS is an immunodeficiency syndrome. In the flow chart you drew for question 11, describe where AIDS sufferers are deficient. Why can't our immune systems fight off this disease?
32. An individual has an immunodeficiency that prevents helper T cells from recognizing the surface antigens presented by macrophages. As a result, the helper T cells are not activated, and they, in turn, fail to activate the appropriate B cells. At this point, is it certain that the viral infection will continue unchecked?

Internet Activities

The following activities use the resources of the World Wide Web to enhance the topics covered in this chapter. To investigate the topics described below, log on to the book's home page (www.brookscole.com/biology_d) and follow the *Student Resources* link to your subject and text.

1. *Immune System Function.* At the *CellsAlive!* Web site, access the "Antibody" link (http://www.cellsalive.com/antibody.htm) for a beautiful overview of antibody structure, production, and function.
2. *HIV/AIDS and the Immune System.* The University of Arizona's *Biology Project HIV 2001* (http://www.biology.arizona.edu/immunology/activities/AIDS2001/main.html) allows you to run a simulation of the spread of HIV through a population or work through a tutorial on HIV/AIDS and the immune system. The tutorial includes an overview of immune system function.
3. *Autoimmune Diseases.* The National Library of Medicine maintains the *Medline Plus Autoimmune Diseases* Web page (http://www.nlm.nih.gov/medlineplus/autoimmunediseasesgeneral.html). Here you can find links to various resources on autoimmune diseases such as lupus, multiple sclerosis (MS), and rheumatoid arthritis. Scroll down to the "Anatomy/Physiology" link for a good immune system tutorial from the National Cancer Institute or down to the "Diagnosis/Symptoms" link to find out how the standard ANA or antinuclear antibody test used in the diagnosis of many autoimmune diseases works.
4. *Which Immune Disorders Have Genetic Bases?* Because the normal functioning of the immune response in humans requires the delicate interplay of B cells, T cells, and phagocytic cells, as well as the actions of several types of immunoglobulins and cytokines, it is easy to see that some genetic disorders are likely to be recognized as related to absent of abnormal immune function. The National Institute of Allergy and Infectious Disease maintains a fact sheet on "Primary Immune Deficiencies" (http://www.niaid.nih.gov/factsheets/pid.htm). Which of the disorders described have a demonstrated genetic link, and which have no described clear pattern of inheritance?

For Further Reading

Allan, J. S. (1996). Xenotransplantation at a crossroads: Prevention versus progress. *Nat. Med., 2,* 18–21.

Bach, F. H., Winkler, H., Wrighton, C. J., Robson, S. C., Stuhlmeier, K., & Ferran, C. (1996). Xenotransplantation: A possible solution to the shortage of donor organs. *Transplant. Proc., 28,* 416–417.

Barre-Sinoussi, F. (1996). HIV as the cause of AIDS. *Lancet, 348,* 31–35.

Barrington, R., Zhang, M., Fischer, M. & Carroll, M. C. (2001). The role of complement in inflammation and adaptive immunity. *Immunol. Rev., 180,* 5–15.

Buckley, R. (1992). Immunodeficiency diseases. *JAMA, 268,* 2797–2806.

Cook, G. S., & Hill, A. V. (2001). Genetics of susceptibility to infectious disease. *Nat. Rev. Genet., 2,* 967–977.

Dierich, M. P., Stoiber, H., & Clivio, A. (1996). A complementary AIDS vaccine. *Nat. Med., 2,* 153–155.

Farzaneh-Far, A., Rudd, J., & Wessberg, P. L. (2001). Inflammatory mechanisms. *Br. Med. Bull., 59,* 55–68.

Fischer, G. F. (2000). Molecular genetics of HLA. *Vox Sang., 78,* (suppl. 2), 261–264.

Just, J. J. (1995). Genetic predisposition to HIV-1 infection and acquired immunodeficiency syndrome: A review of the literature examining association with HLA. *Hum. Immunol., 44,* 156–159.

Lokki, M. L., & Colten, H. R. (1995). Genetic deficiency of complement. *Ann. Med., 27,* 451–459.

Morgan, B. P. (1996). Intervention in the complement system: A therapeutic strategy in inflammation. *Biochem. Soc. Trans., 24,* 224–229.

Puck, J. (1993). X-linked immunodeficiencies. *Adv. Hum. Genet., 21,* 107–144.

Quinn, T. C. (1996). Global burden of the HIV pandemic. *Lancet, 348,* 99–106.

Rapaport, F. T. (1995). Skin transplantation—experimental basis for the study of human histocompatibility. *Transplant. Proc., 27,* 2205–2210.

Sprent, J., & Tough, D. F. (2001). T cell death and memory. *Science, 293,* 245–248.

Tomlinson, I. P., & Bodmer, W. F. (1995). The HLA system and the analysis of multifactorial genetic disease. *Trends Genet., 11,* 493–498.

Tonegawa, S. (1985, October). The molecules of the immune system. *Sci. Am., 253,* 123–131.

Vladutiu, A. (1993). The severe combined immunodeficient (SCID) mouse as a model for the study of autoimmune diseases. *Clin. Exp. Immunol., 93,* 1–8.

Yamamoto, F. (1995). Molecular genetics of the ABO histo-blood group system. *Vox Sang., 69,* 1–7.

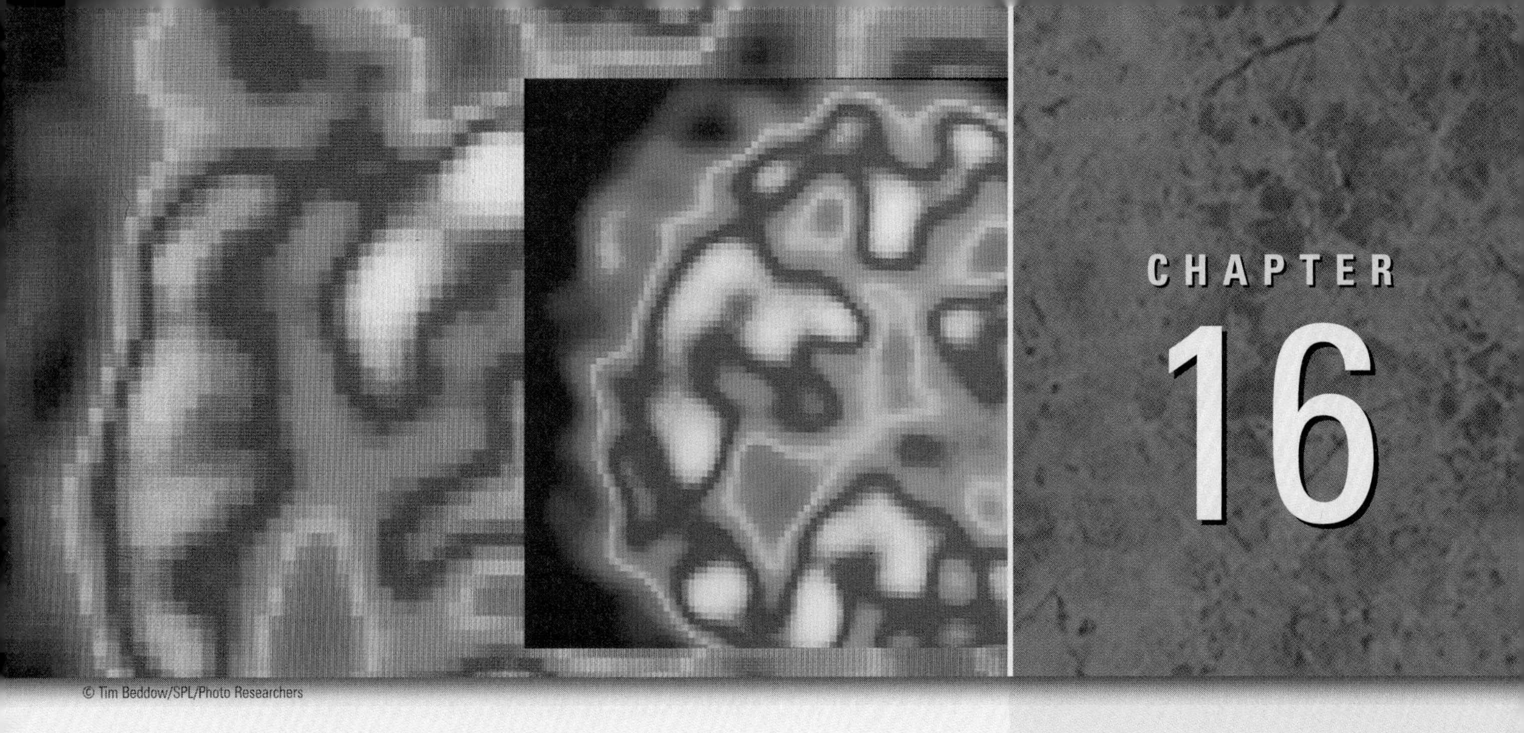

Genetics of Behavior

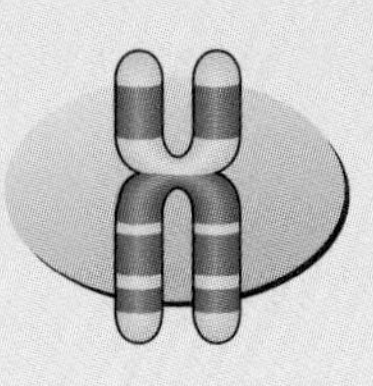

Chapter Outline

In the late 1850s, an 8-year-old boy and his father, a physician, drove along a road on eastern Long Island. They met two women walking along the road, and this chance encounter had a profound effect on the young boy. Years later, he would recall that meeting:

> *I recall it as vividly as though it had occurred but yesterday. It made a most enduring impression upon my boyish mind, an impression every detail of which I recall today, an impression which was the very first impulse to my choosing chorea as my virgin contribution to medical lore. Driving with my father through a wooded road leading from East Hampton to Amagansett, we suddenly came upon two women, mother and daughter, both tall, thin, almost cadaverous, both bowing, twisting, grimacing. I stared in wonderment, almost in fear. What could it mean? My father paused to speak to them and we passed on.*

The young boy, George Huntington, went on to study medicine at Columbia University. In 1872, one year after his graduation, he published an account of the disorder he had witnessed as a boy, which became known as Huntington's chorea, or, as it is called today, Huntington disease. In the paper, his summary of the condition is a model of brevity and clarity:

> *This disorder has three characteristics: 1. Its hereditary nature, 2. A tendency to insanity and suicide, 3. Its manifesting itself as a grave disease only in adult life.*

Huntington's description of the inheritance of the disorder is consistent with an autosomal dominant trait. Having described the condition, Huntington turned his attention elsewhere, and little more was accomplished in the next 88 years until the gene for Huntington disease was mapped to chromosome 4 using recombinant DNA techniques.

Conditions such as Huntington disease (HD) are models for genetic disorders that affect behavior. The pattern of autosomal dominant inheritance is well defined and clear-cut. The gene for HD was one of the first to be mapped using restriction fragment length polymorphisms (RFLPs), which demonstrated the power of molecular techniques in analyzing human behavior. More recently, positional cloning, one of the newer genetic techniques of recombinant DNA technology, was used to isolate the gene. The molecular basis of mutation in the HD gene represents a new class of mutations that affect the nervous system. From all of this, it would seem that by following the trail used in studying Huntington disease, researchers could identify, map, and isolate many genes affecting human behavior.

Unfortunately, searches for single genes (like the HD gene) that control behavior may have only limited success. One of the most difficult problems in human behavioral genetics is defining the phenotype. Although Huntington disease has a well-defined phenotype and progression, many behavioral traits do not. In addition, the phenotype of some conditions, such as schizophrenia (well defined as a medical condition), can be genetically heterogeneous and may actually include several genetic disorders, each of which has a similar phenotype. Finally, many behavioral traits are multifactorial; the phenotype is determined by several genes and environmental interactions, and no single gene has a major effect.

To understand the issues in behavior genetics and how decisions about phenotypic definitions, genetic models of behavior, and methods influence both the speed and the outcome of this research, this chapter begins with a discussion of the genetic models and methods used in studying human behavior. Then we briefly consider animal models, for which single-gene effects on behavior have been well documented. We discuss single genes that affect human behavior through their effect on the nervous system and then consider more complex traits and those that have the greatest social impact. The chapter ends with a summary of the current state of human behavior genetics and the ethical, legal, and social implications of this research.

MODELS, METHODS, AND PHENOTYPES IN STUDYING BEHAVIOR

Pedigree analysis, family studies, adoption, and twin studies suggest that many parts of human behavior are under genetic control. However, behaviors with a genetic component are likely to be controlled by several genes, to interact with other genes, and to be influenced by environmental components. In fact, most behaviors are *not* inherited as single-gene traits, demonstrating the need for genetic models that can explain observed patterns of inheritance. To a large extent, the model proposed to explain the inheritance of a trait determines the methods used to analyze its pattern of inheritance and the techniques to be used in mapping and isolating the gene or genes responsible for the trait's characteristic phenotype (discussed later).

Genetic Models of Inheritance and Behavior

Several models for genetic effects on behavior have been proposed (Table 16.1). The simplest model is a single gene, dominant or recessive, that affects a well-defined behavior. Several behavioral traits—including Huntington disease, Lesch-Nyhan syndrome, fragile-X syndrome, and others—are described by such a model (Figure 16.1). Multiple-gene models are also possible. The simplest of these is a polygenic additive model in which two or more genes contribute equally in an additive manner to the phenotype. This model has been proposed (along with others) to explain

Table 16.1 Models for Genetic Analysis of Behavior

Model	Description
Single gene	One gene controls a defined behavior
Polygenic trait	Additive model that has two or more genes One or more major genes with other genes contributing to phenotype
Multiple genes	Interaction of alleles at different loci generates a unique phenotype

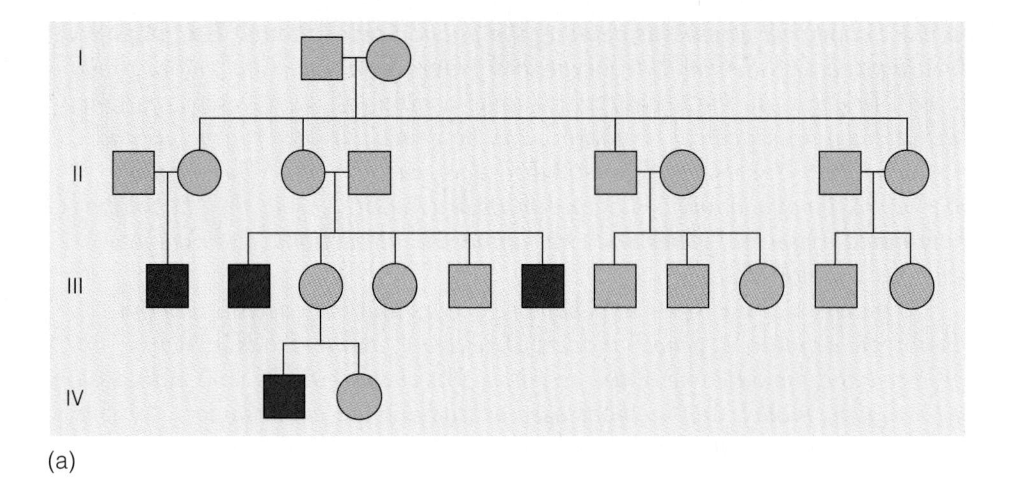

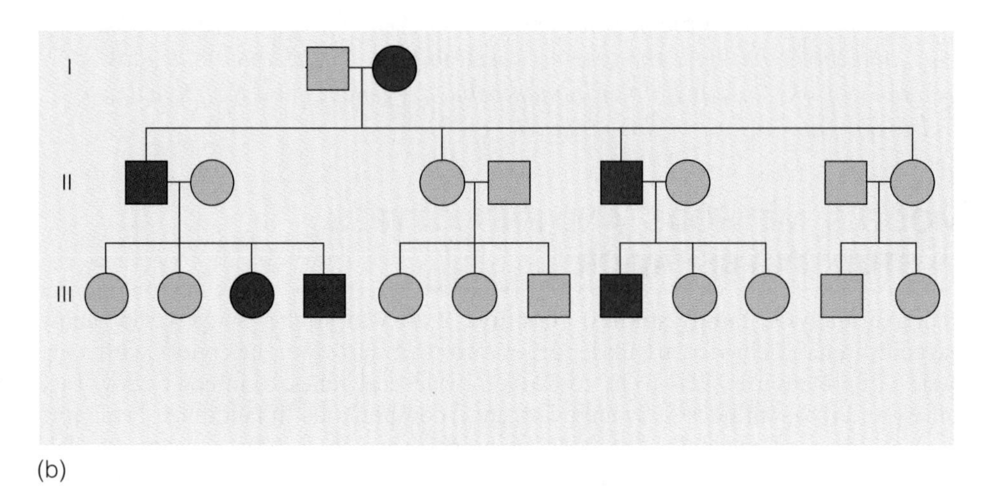

FIGURE 16.1 Many genetic disorders inherited as sex-linked recessive traits (a) or as autosomal dominant traits (b) have behavioral phenotypes.

schizophrenia (the inheritance of additive polygenic traits was considered in Chapter 5). Multigene models can also include situations in which one or more genes have a major effect, while other genes make smaller contributions to the phenotype. Still another multigene model involves epistasis, a form of gene interaction in which an allele of one gene masks the expression of a second gene. This form of gene interaction has been well documented in experimental genetics, although it has not yet been shown to operate in human behavioral traits.

In each of these models, the environment can significantly affect the phenotype, and the study of behavior must take this into account (see Concepts and Controversies: Is Going to Medical School a Genetic Trait?). To assess the role of the environment in the phenotype, geneticists must use methods that measure the genetic and environmental contributions to a trait.

Methods of Studying Behavior Genetics

For the most part, methods for studying behavior genetics follow the pattern used for study of other human traits. If a single gene is proposed, pedigree analysis and linkage studies, including the use of RFLP markers and other methods of recombinant DNA technology, are the most appropriate methods. However, because many behaviors are polygenic, twin studies play a prominent role in human behavior genetics. Concordance and heritability values based on twin studies have established a genetic link to mental illnesses (manic depression and schizophrenia) and to behavioral traits (sexual preference and alcoholism). The results of such studies must be interpreted with caution because there are limitations inherent in interpreting

Concepts and Controversies

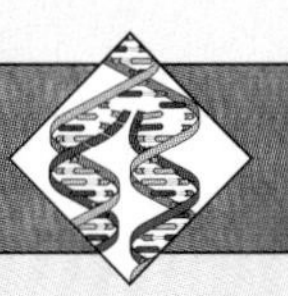

Is Going to Medical School a Genetic Trait?

Many behavioral traits have a familial pattern of inheritance. This observation, along with twin studies and adoption studies, reveals a genetic component in complex behavioral disorders, such as bipolar illness and schizophrenia. In most cases, these phenotypes are not inherited as simple Mendelian traits. Researchers are then faced with selecting a model to describe how a behavioral trait is inherited. Using this model, further choices select the methods used in genetic analysis of the trait. A common strategy is to find a family in which the behavior appears to be inherited as a recessive or an incompletely penetrant dominant trait controlled by a single gene. One or more molecular markers (such as RFLPs) are used in linkage analysis to identify the chromosome that carries the gene that controls the trait.

If researchers are looking for a single gene when the trait is controlled by two or more genes or by genes that interact with environmental factors, then the work may produce negative results, even though preliminary findings can be encouraging. Reports of loci for bipolar illness on chromosome 11 and the X chromosome were based on single-gene models, but after initial successes, it was found that these reports were flawed. Overall, regions on 14 chromosomes have been proposed as candidates for genes that control bipolar illness, but none have been substantiated.

To illustrate some of the pitfalls associated with model selection in behavior genetics, one study selected attendance at medical school as a behavioral phenotype and attempted to determine if the distribution of this trait in families is consistent with a genetic model. This study surveyed 249 first- and second-year medical students. Thirteen percent of these students had first-degree relatives who had also attended medical school, compared with 0.22% of individuals selected from the general population with such relatives. Thus the overall risk factor among first-degree relatives for medical school attendance was 61 times higher for medical students than for the general population, indicating a strong familial pattern. To determine whether this trait was inherited in a Mendelian manner, researchers used standard statistical analysis, which supported familial inheritance and rejected the model of no inheritance. Analysis of the pedigrees most strongly supported a simple recessive mode of inheritance, although other models, including polygenic inheritance, were not excluded. Using a further set of statistical tests, the researchers concluded that the recessive mode of inheritance was just at the border of statistical acceptance.

Similar results are often found in studies of other behavioral traits, and it is usually argued that another, larger study would confirm the results, in this case that attendance at medical school is a recessive Mendelian trait. Although it is true that genetic factors may partly determine whether one will attend medical school, it is unlikely that a single recessive gene controls this decision, regardless of the support of such a conclusion by this family study and segregation analysis of the results.

The authors of this study were not serious in their claims that a decision to attend medical school is a genetic trait, nor was it their intention to cast doubts on the methods used in the genetic analysis of behavior. Rather, it was intended to point out the folly of accepting simple explanations for complex behavioral traits.

heritability (see Chapter 5), and these studies often use small sample sizes, in which minor variations can have a disproportionately large effect on the outcome.

To overcome these problems, geneticists are adapting twin studies as a genetic tool to study behavior. One innovation studies the children of twins to confirm the existence of genes predisposing to a certain behavior. Twin studies are also being coupled to recombinant DNA techniques to search for behavior genes, and this combination may prove to be a powerful method for identifying such genes.

Phenotypes: How Is Behavior Defined?

A second restriction on progress in human behavior genetics is the choice of a consistent phenotype as the basis for study. The phenotypic definition of a behavior must be precise enough to distinguish the behavior from all other behaviors and from the behavior of the control group. However, the definition cannot be so narrow that it excludes some variations of the behavior. Recall that gene mapping uses the phenotype as a guide, and so starting with an accurate description of the phenotype is very important.

For some mental illnesses, clinical definitions are provided in the *Diagnostic and Statistical Manual of Mental Disorders* of the American Psychiatric Association. For other behaviors, phenotypes are poorly defined and may not provide clues to the underlying biochemical and molecular basis of the behavior. For example, alcoholism can be defined as the development of characteristic deviant behaviors associated with excessive consumption of alcohol. Is this definition explicit enough to be useful as a phenotype in genetic analysis? Is there too much room for interpreting what is deviant behavior or what is excessive consumption? As we will see, whether the behavioral phenotype is narrowly or broadly defined can affect the outcome of the genetic analysis and even the model of inheritance for the trait.

The Nervous System Is the Focus of Behavior Genetics

In Chapter 10, we discussed the role of genes in metabolism. Mutations that disrupt metabolic pathways or interfere with the synthesis of essential gene products can influence the function of cells and in turn produce an altered phenotype. If the affected cells are part of the nervous system, alterations in behavior may be part of the phenotype. In fact, some, and perhaps many, genetic disorders affect cells in the nervous system that in turn affect behavior. In phenylketonuria (PKU), for example, brain cells are damaged by excess levels of phenylalanine, causing mental retardation and other behavioral deficits.

For much of behavior genetics, the focus is on the structure and function of the nervous system. This emphasis is reinforced by the finding that many disorders with a behavioral phenotype, including Huntington disease, Alzheimer disease, and Charcot-Marie-Tooth disease, alter the structure of the brain and nervous system. Other behavior disorders such as bipolar illness and schizophrenia may be disorders of function rather than structure.

ANIMAL MODELS: THE SEARCH FOR BEHAVIOR GENES

One way to study the genetics of human behavior is to ask whether behavior genetics can be studied in model systems. Results from experimental organisms can then be used to study human behavior. Several approaches are used to study behavior in animals. One method uses two closely related species or two strains of the same species to detect variations in behavioral phenotypes. Genetic crosses are used to establish whether these variations are inherited and, if so, to determine the pattern of inheritance. More recently, the effects of single genes on animal behavior have been studied. In some cases, these studies have led to isolating and cloning genes that affect behavior. In the following sections, we describe some examples of behavior genetic studies of experimental organisms.

Open-Field Behavior in Mice

Beginning in the mid-1930s, the emotional and exploratory behaviors of mice were tested by studying open-field behavior. When mice are introduced into a brightly lit environment, some mice actively explore the new area, whereas others are apprehensive, do not move about, and have elevated rates of urination and defecation (Figure 16.2). This behavior pattern is under genetic control, because strains exhibiting both types of behavior have been established.

To test the genetic components of this behavior, beginning in the 1960s, studies used an enclosed, illuminated box whose floor is marked into squares. Counting the mouse's movements in different squares tests exploration, and emotion is quantified by counting the number of defecations. The BALB/cJ strain, homozygous for a recessive albino allele, shows low exploratory behavior and is

FIGURE 16.2 An open-field trial. Movements are automatically recorded as the mouse moves across the open field.

highly emotional. The C57BL/6j strain has normal pigmentation, is active in exploration, and shows low levels of emotional behavior.

If these two strains are crossed and the offspring are interbred, each generation beyond the F1 includes both albino and normally pigmented mice. When tested for open-field behavior, pigmented mice behaved like the C57 parental line, showing active exploration and low levels of emotional behavior. The albino mice behaved like the BALB parental line and showed low exploratory activity and high levels of emotional behavior, indicating that the albino gene affects behavior and pigmentation. The overall results show that open-field behavior is a polygenic trait.

Learning in *Drosophila* Is Genetically Controlled

The fruit fly, *Drosophila,* offers several advantages for the study of behavior genetics, including the existence of behavioral mutants (Table 16.2). For learning, flies are presented with a series of odors, one of which is accompanied by an electric shock.

Table 16.2 Some Behavior Mutants of Drosophila

Category	Mutation	Phenotype
Learning	*dunce*	Cannot learn conditioned response.
	turnip	Impaired in learning conditioned response.
	rutabaga	Impaired in several types of learning and memory.
Sexual Behavior	*fruitless*	Males court each other.
	savoir-faire	Males unsuccessful in courtship.
	coitus-interruptus	Males stop copulation prematurely.
Motor Behavior	*flightless*	Lacks coordination for flying.
	sluggish	Moves slowly.
	wings up	Holds wings perpendicular to body.

Flies learn to avoid the odor associated with the shock. Mutant screens have identified a number of genes that influence this learning ability, including *dunce, turnip,* and *rutabaga.*

These single-gene mutants exert their effect through an intracellular signaling system that involves a molecule called cyclic AMP (cAMP). Inside cells of the nervous system, cAMP sets off a cascade of biochemical reactions that controls gene transcription and the responses associated with learning. Cyclic AMP is produced by the enzyme adenyl cyclase (◗ Figure 16.3); the normal *rutabaga* (*rut*) gene encodes this enzyme. In mutant homozygotes, there is no adenyl cyclase and no production of cAMP, and flies cannot learn the relationship between odor and electric shock. The *turnip* gene encodes a protein that activates adenyl cyclase, and the *dunce* gene controls the pathway by which adenyl cyclase is recycled. The clustering of these mutations in the cAMP pathways provides strong evidence for the involvement of cAMP in learning. Experiments in other organisms support this finding and imply that some aspects of learning and memory in humans may involve cyclic nucleotides.

Transgenic Animals Are Used as Models of Human Neurodegenerative Disorders

In addition to studying behavior in animal models using mutants and inbred strains, researchers are creating animal models of nervous system disorders by constructing transgenic animals. Recall from Chapter 13 that transgenic animals are produced by transferring genes from one species to another.

Let's briefly look at how transgenic animals are used in one group of genetic disorders of the nervous system. Neurodegenerative disorders are a group of progressive and fatal diseases. Some of these disorders, such as Alzheimer disease (AD), amyotrophic lateral sclerosis (ALS), and Parkinson disease (PD) occur sporadically or from inherited mutations. Others, such as Huntington disease (HD) and spinocerebellar ataxias have only a genetic cause. Transgenic animal models can be constructed only after a specific gene for a disorder has been identified and isolated. The use of transgenic models allows research on the molecular and cellular mechanisms of the disorder and on the development and testing of drugs for the treatment of these disorders.

For example, about 10% of all ALS cases are inherited as an autosomal dominant trait (MIM/OMIM 105400). Affected individuals have progressive weakness and muscle atrophy with occasional paralysis caused by degeneration of nerve cells that connect with muscles. Some of these affected individuals have a mutation in the *SOD1* gene on chromosome 21. Mutations cause the SOD1 protein to become toxic. Transgenic mice carrying a mutant human *SOD1* gene develop muscle weakness and atrophy similar to that seen in affected humans (◗ Figure 16.4). These mice are used to study how the mutant SOD1 protein selectively damages some nerve cells, but leaves others untouched, and to study the effects of drugs designed to treat this disorder.

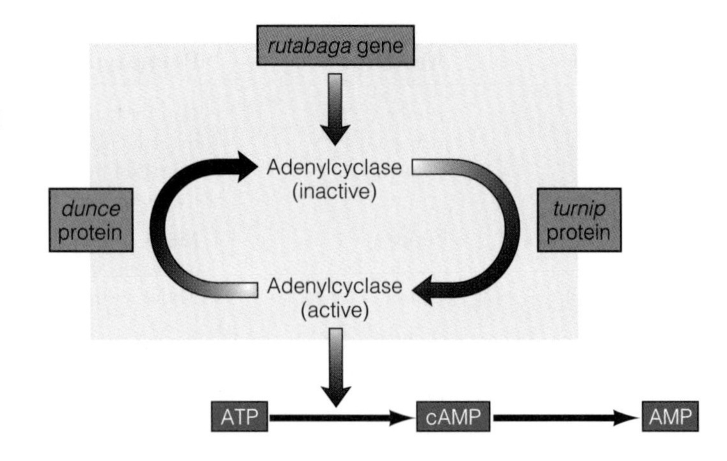

◗ **FIGURE 16.3** The metabolic pathway that involves production of cyclic AMP (cAMP) is involved in learning. In this pathway, ATP is converted into cAMP by the enzyme adenyl cyclase. cAMP is active in signal transduction and then is converted into AMP. In *Drosophila,* the *rutabaga* gene encodes adenyl cyclase. The enzyme is inactive when first produced and is activated by the action of a protein encoded by the *turnip* gene. When no longer needed, a gene product encoded by the *dunce* gene inactivates the enzyme. When any of these genes are mutant in *Drosophila,* the flies have difficulty learning.

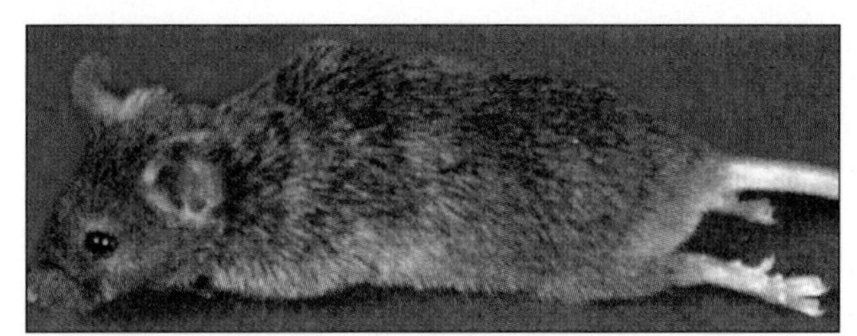
Courtesy of Dr. Donald Price, Johns Hopkins School of Medicine

FIGURE 16.4 A transgenic mouse carrying a mutation in the *SOD-1* gene, which causes paralysis of the limbs. In humans, this mutation causes amyotrophic lateral sclerosis (ALS), a neurodegenerative disease. The mutant mouse serves as model for this disease, allowing researchers to explore the mechanism of the disease and to design therapies to treat humans affected with ALS. (Courtesy of Dr. Donald Price, Johns Hopkins School of Medicine. From Price, D. L. et al, Genetic neurodegenerative diseases: the human illness and transgenic model. Reprinted with permission from *Science 282,* 1079–1083. © 1998 American Association for the Advancement of Science.)

Although mice are widely used in transgenic research, human genes transferred to *Drosophila* are also used as models of human neurodegenerative diseases. Flies carrying mutant human genes for HD and spinocerebellar ataxia 3 have been constructed and are used to study how the mutant proteins kill nerve cells and to identify genes or chemicals that can slow or prevent the loss of cells.

SINGLE GENES AFFECT THE NERVOUS SYSTEM AND BEHAVIOR

In this section we discuss several single-gene defects that have specific effects on the development, structure, and/or function of the nervous system and that consequently affect behavior. Following this we discuss more complex interactions between the genotype and behavior, in which the number and functional roles of genes are not well understood and in which effects on the nervous system may be more subtle.

Huntington Disease Is a Model for Neurodegenerative Disorders

Huntington disease (HD) (MIM/OMIM 143100) is a useful model for single-gene disorders that affect the nervous system and have a behavioral phenotype. HD is an adult-onset neurodegenerative disorder inherited as an autosomal dominant trait. It affects about 1 in 10,000 individuals in Europe and the United States. HD was one of the first disorders to be mapped using recombinant DNA techniques (see Chapter 13). The mutation causing the disorder involves expansion of a trinucleotide repeat (this topic is covered in Chapter 11), and the disorder shows anticipation (also covered in Chapter 11). Predictive genetic testing and transgenic animal models are available, and the condition is being treated by transplantation of fetal nerve cells.

Huntington disease An autosomal dominant disorder associated with progressive neural degeneration and dementia. Adult onset is followed by death 10 to 15 years after symptoms appear.

As described in the opening section of this chapter, the phenotype of HD usually begins in midadult life as involuntary muscular movements and jerky motions of the arms, legs, and torso. As it progresses, personality changes, agitated behavior, and dementia occur. Most affected individuals die within 10 to 15 years after the onset of symptoms.

The gene for HD maps to the short arm of chromosome 4 (4p16.3). Mutant alleles have increased copies of CAG triplet repeats in the gene. HD is one of eight known neurodegenerative disorders caused by the expansion of a CAG trinucleotide repeat (see Chapter 11). In all cases, the mutation leads to an increase in the number of copies of the amino acid glutamine inserted into the gene product. This increase, called a polyglutamine expansion, causes the protein to become toxic and kill nerve cells. Individuals with less than 35 copies of the CAG repeat in the *HD* gene do not develop the disease, and those with 35 to 39 repeats may or may not develop HD. However, those with 40 to 60 repeats will develop HD as adults. People with more than 60 repeats will develop HD before age 20.

Anticipation is a pattern of successive earlier age of disease onset within a pedigree (see Chapter 11). In HD, anticipation is associated with expansion of the number of CAG repeats as the HD gene is passed from parent to offspring. Expansion

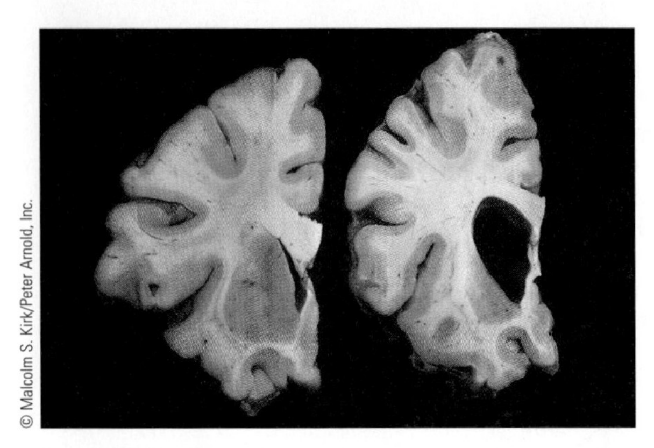

FIGURE 16.5 Section of brain with Huntington disease (left) and normal brain (right).

of paternal copies of CAG is more likely to produce an earlier age of onset, and juvenile cases of HD are almost always related to paternal transmission. The reasons for this are unclear.

Brain autopsies of HD victims show damage to several specific brain regions, including the striatum and the cerebral cortex. In affected regions, cells fill with cytoplasmic and nuclear clusters of the mutant protein and degenerate and die (Figure 16.5). Involuntary movements and progressive personality changes accompany the degeneration and death of these neurons (neurons are cells of the nervous system). The *HD* gene encodes a large protein, huntingtin (Htt). In adult brains, Htt stimulates the production of a protein, BDNF, necessary for the survival of cells in the striatum. Mutant Htt causes a decrease in BDNF production, and cells of the striatum degenerate and die. Because the HD mutation causes the gene product to lose its function, HD can be regarded as a loss-of-function mutation. However, mutation in the HD gene also causes the altered Htt protein to become toxic. Whether its toxic effects are caused by intracellular aggregation, cleavage fragments of the protein, or the accumulation of other proteins in the Htt aggregates is not clear. This effect of the mutant HD protein can be regarded as a gain-of-function mutation.

Strains of transgenic mice that carry and express the mutant human *HD* gene have been developed. In these mice, the mutant *HD* gene is active in the brain and other organs. Phenotypically, these mice show progressive behavioral changes and an increasing loss of muscular control. The brains of affected transgenic mice show the same changes as affected humans: accumulation of Htt clusters leading to degeneration and loss of cells in the striatum and cerebrum (Figure 16.6). These mice are being used to study the early events in Htt accumulation and to develop treatments that work in presymptomatic stages to prevent cell death.

HD is caused by a mutation resulting in degeneration and death of cells in the striatum. Experiments in animals showed that transplantation of fetal nerve cells into the striatum partially restores nerve connections, muscle control, and behavior. Investigators have started treating HD patients with transplants of fetal striatal cells to determine whether the transplanted cells would survive, make connections to other cells, and lead to improvements in muscle control and intellectual functions. Results from the first round of transplants are encouraging, adding HD to the list of disorders that can be treated with such transplants. This success, however, adds to the debate about fetal stem cells, and the direction of stem cell research in this country.

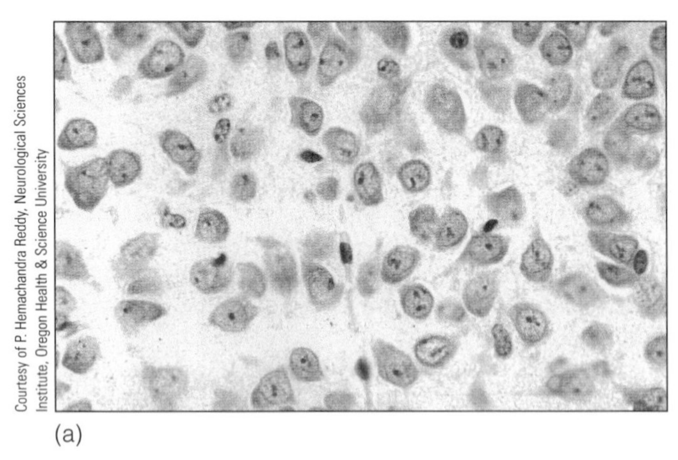

(a)

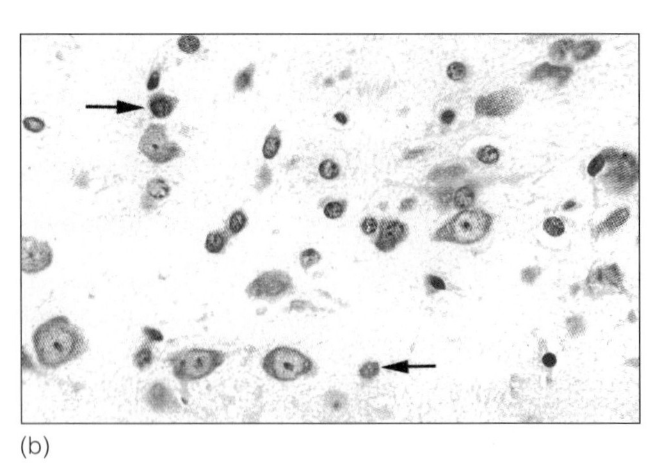

(b)

FIGURE 16.6 Loss of brain cells in a transgenic Huntington disease mouse. (a) Section of normal mouse brain striatum showing densely packed neurons. (b) Section of the striatum from an HD89 mouse showing extensive loss of neurons that accompany this disease. (Courtesy of P. Hemachandra Reddy, Neurological Sciences Institute, Oregon Health & Science University; from Reddy, P. H. et al. (1999). Transgenic mice expressing mutated full-length HD cDNA: a paradigm for locomotor changes and selective neuronal loss in Huntington's disease. *Proc. Roy. Soc. London B, 354,* 1035–1045.)

Exploring the Link Between Language and Brain Development

For over forty years, linguists, psychologists, and geneticists have unsuccessfully argued over the relationship between language and genetics. The linguist Noam Chomsky observed that all children learn to speak without much instruction. Because language seems to be an inherent trait, he proposed that it might have a genetic component. The finding that specific language deficits are a familial trait and can be separated from other conditions such as autism, deafness, and mental retardation supported Chomsky's idea. However, because there was no clear-cut pattern of inheritance, any genetic link to language seemed weak.

About 10 years ago, a large, multigenerational family, the KE family, came to the attention of researchers. In this family, a very specific speech and language disorder is inherited as an autosomal dominant trait (◗ Figure 16.7). Affected members cannot correctly identify language sounds and have difficulty understanding sentences. They also have problems in making language sounds, and it is almost impossible to understand their speech.

With the cooperation of the family, investigators were able to map the disorder to a small region on the long arm of chromosome 7 and named the unknown gene in this region *SPCHI* (*SPEECH1*) (MIM/OMIM 602081). Recently, a child, CS, who is not related to the KE family, was found to have the same speech deficit as the KE family. CS carries a translocation involving the long arm of chromosome 7 in the *SPCH1* region. This allowed researchers to identify the gene, now called *FOXP2* (MIM/OMIM 606354), because it is a member of a previously identified gene family.

Affected members of the KE family have a single nucleotide change in the *FOXP2* gene, in which a G is replaced with an A. This change results in an amino acid substitution in the protein, presumably altering the protein's function and resulting in the language deficit.

FOXP2 is a member of a family of genes that encode transcription factors. Transcription factors are proteins that switch on genes or gene sets, often at specific stages

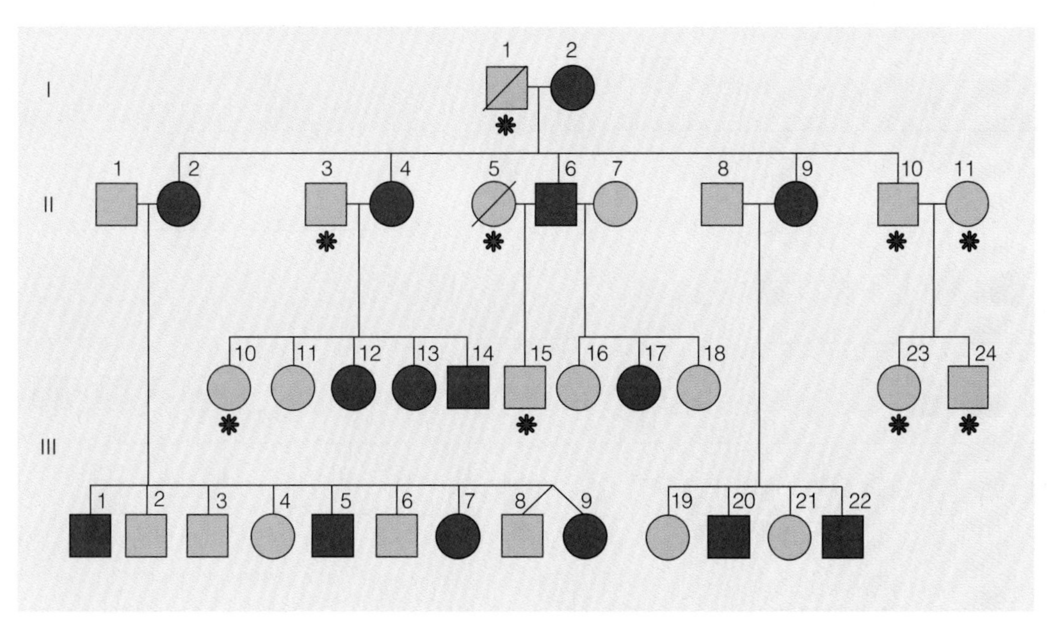

◗ **FIGURE 16.7** Pedigree of a family in which some members are affected with a severe speech and language disorder (solid symbols represent affected members). Asterisks mark individuals who were not analyzed. The pattern of inheritance is consistent with an autosomal dominant trait. The gene for this disorder maps to the long arm of chromosome 7. (From Lai, C. et al. (2000). The *SPCH1* region on human 7q31: genomic characterization of the critical interval and localization of translocations associated with speech and language disorder. *American Journal of Human Genetics, 67,* 357–368, Figure 1, p. 358. Adapted with permission from University of Chicago Press.)

of development. The FOXP2 protein is very active in fetal brains. It may be that affected members have a 50% reduction in the amount of this protein at a critical stage of brain development, leading to an abnormality of language development.

Future work on *FOXP2* may help us understand how the brain understands and processes language and to develop therapies to treat language disorders. In addition, comparing the action of *FOXP2* in the developing brains of chimpanzees and other primates may help us understand how language evolved and what separates us from our fellow primates.

SINGLE GENES CONTROL AGGRESSIVE BEHAVIOR AND BRAIN METABOLISM

In 1993 a new form of X-linked mild mental retardation was identified in a large European family. All affected males showed forms of aggressive and often violent behavior. Gene mapping and biochemical studies indicate that this condition exhibits a direct link between a single-gene defect and a phenotype characterized by aggressive and/or violent behavior (◗ Figure 16.8). In particular, eight males with mild or borderline mental retardation showed a characteristic pattern of aggressive and often violent behavior triggered by anger, fear, or frustration. The behavioral phenotype varied widely in levels of violence and time but included acts of attempted rape, arson, stabbings, and exhibitionism.

Mapping a Gene for Aggression

Using RFLP markers, the gene for this behavior was mapped to the short arm of the X chromosome in region Xp11.23-11.4. A gene in this region encodes an enzyme called monoamine oxidase type A (MAOA) that breaks down a neurotransmitter (Table 16.3). Neurotransmitters are chemical signals that carry nerve impulses across

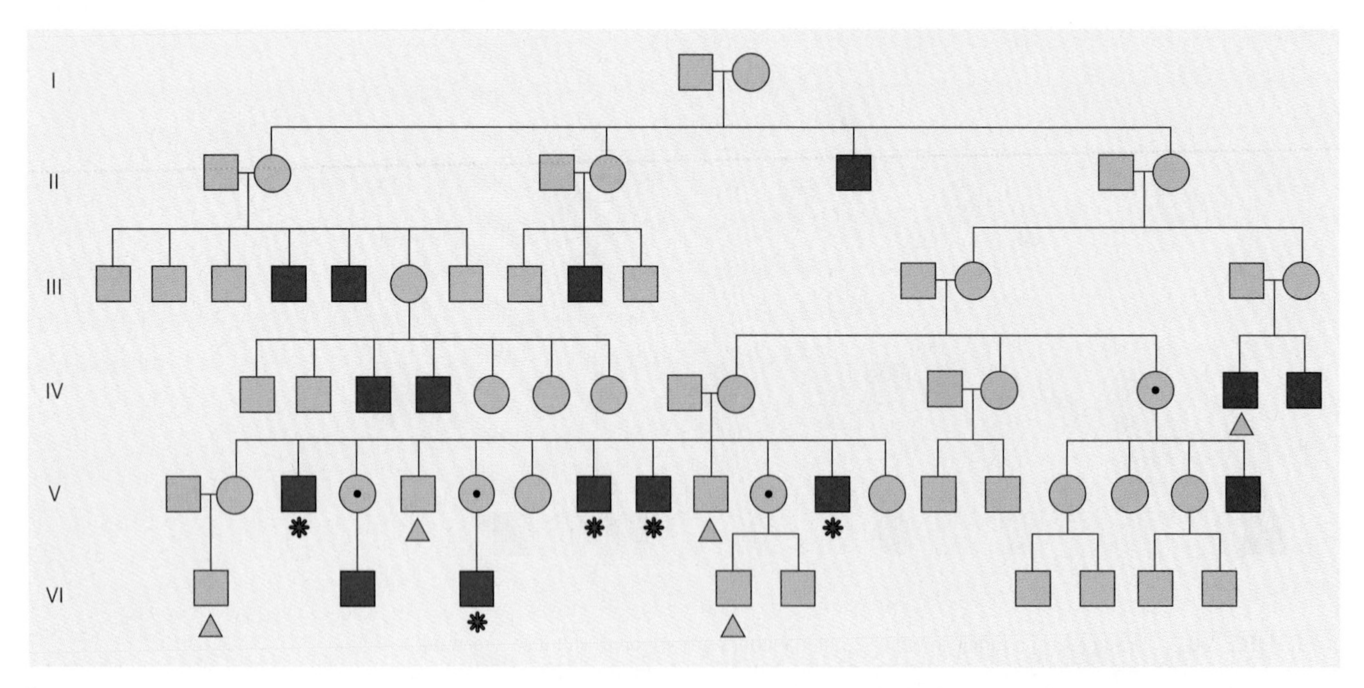

◗ **FIGURE 16.8** Cosegregation of mental retardation, aggressive behavior, and a mutation in the monoamine oxidase type A (*MAOA*) gene. Affected males are indicated by the filled symbols. Symbols marked with an asterisk represent males known to carry a mutation of the *MAOA* gene; those marked with a triangle are known to carry the normal allele. Symbols marked with a dot inside represent females known to be heterozygous carriers.

synapses in the brain and nervous system (◗ Figure 16.9). Failure to break down these chemical signals rapidly can disrupt the normal function of the nervous system.

The urine of the eight affected individuals contains abnormal levels of chemicals produced by MAOA. The researchers concluded that the eight affected males carried a mutation in the gene that encodes this enzyme and that lack of MAOA activity is associated with their behavior pattern.

A follow-up study analyzed the *MAOA* gene (MIM/OMIM 309850) in five of the eight affected individuals and showed that all five carry a mutation that encodes a nonfunctional gene product. This mutation was also found in two female heterozygotes and is not found in any unaffected males in this pedigree. Loss of MAOA activity in affected males prevents normal breakdown of certain neurotransmitters, reflected in elevated levels of toxic compounds in the urine.

Because it is difficult to relate a phenotype such as aggression (for example, what exactly constitutes aggression?) to a specific genotype, further work is needed to determine whether mutations of *MAOA* are associated with altered behavior in other families and in animal model systems. In addition, the interaction of this dis-

Table 16.3

Some Common Neurotransmitters

Acetylcholine
Dopamine
Norepinephrine
Epinephrine
Serotonin
Histamine
Glycine
Glutamate
Gamma-Aminobutyric acid (GABA)

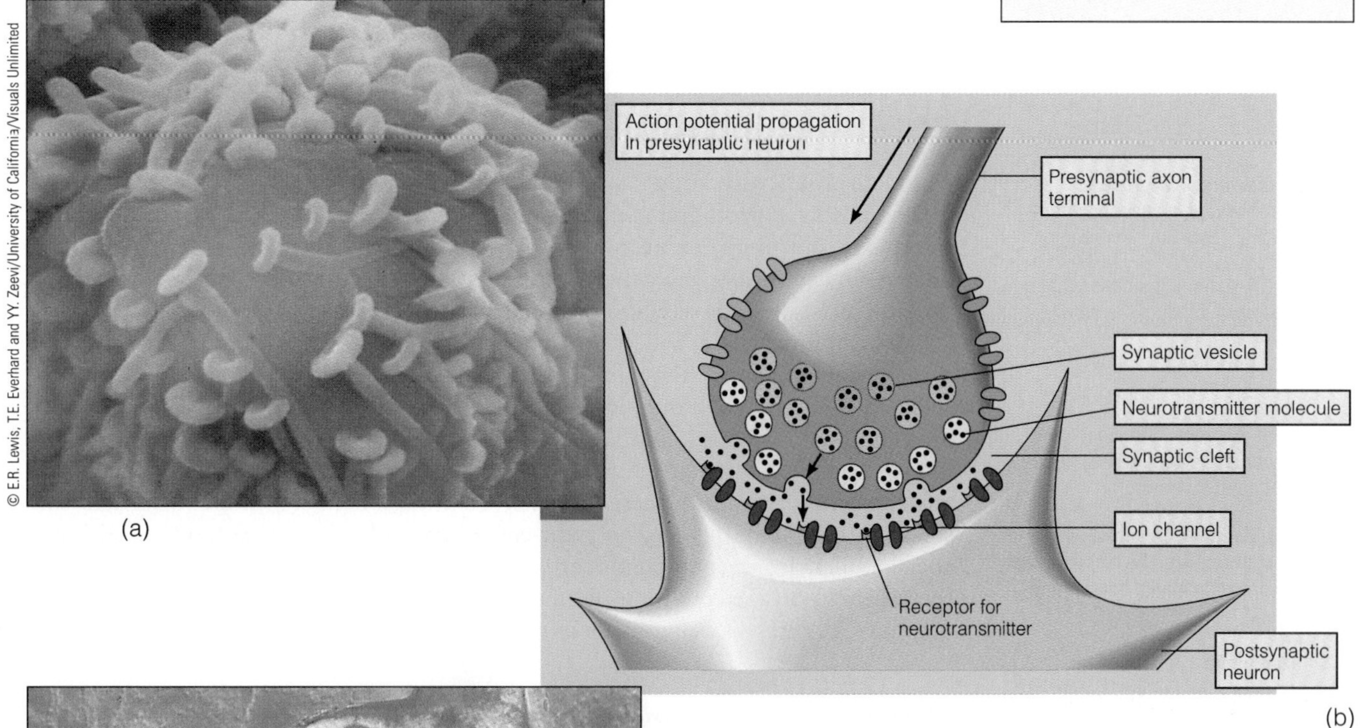

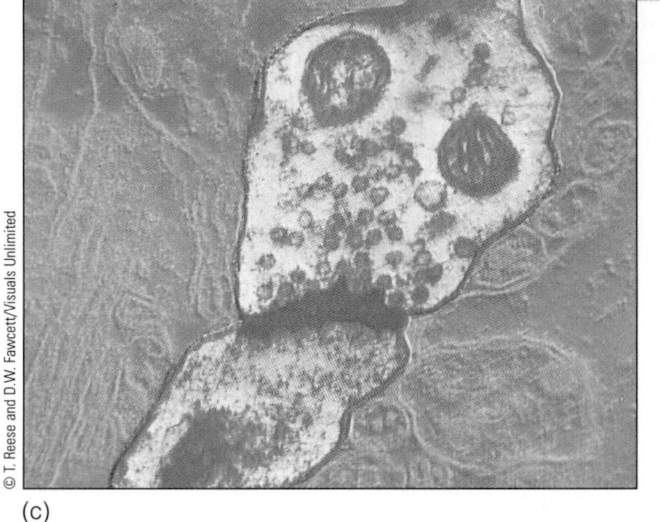

◗ **FIGURE 16.9** The synapse and synaptic transmission. (a) A scanning electron micrograph of the endings (called terminal boutons) of an axon in contact with another cell. (b) In this diagram, a nerve impulse that arrives at the end of an axon triggers the release of a chemical neurotransmitter from storage in synaptic vesicles. The neurotransmitters diffuse across the synapse and bind to receptors on the membrane of the cell on the other side of the synapse (postsynaptic neuron), where they trigger another nerve impulse. (c) A transmission electron micrograph of a synapse. The cell above the synapse contains synaptic vesicles that have stored neurotransmitters.

order with external factors such as diet, drugs, and environmental stress remain to be established. However, the identification of a specific mutation associated with this behavior pattern is an important discovery and suggests that biochemical or pharmacological treatment for this disorder may be possible.

Problems With Single-Gene Models for Behavioral Traits

Although recombinant DNA markers have been successful in identifying, isolating, and cloning single genes that affect behavior, in other cases it has produced erroneous results. In 1987, a DNA linkage study mapped a gene for bipolar illness to a region of chromosome 11. Later, individuals from the study group who did not carry the suspect copy of chromosome 11 developed the illness, indicating that the gene was not on this chromosome. Another study reported linkage between DNA markers on chromosome 5 and schizophrenia; however, the linkage was later found to be coincidental, or, at best, could only explain the disorder in a small, isolated population. These early failures to find single genes that control these disorders led to the reevaluation of single-gene models for many behavior traits and to the development of alternative models for complex traits, as described in the next section.

THE GENETICS OF MOOD DISORDERS AND SCHIZOPHRENIA

Mood disorders A group of behavior disorders associated with manic and/or depressive syndromes.

Mood A sustained emotion that influences perception of the world.

Affect Pertaining to emotion or feelings.

Unipolar disorder An emotional disorder characterized by prolonged periods of deep depression.

Bipolar disorder An emotional disorder characterized by mood swings that vary between manic activity and depression.

Schizophrenia A behavioral disorder characterized by disordered thought processes and withdrawal from reality. Genetic and environmental factors are involved in this disease.

Mood disorders, also known as affective disorders, are conditions in which affected individuals have profound emotional disturbances. **Moods** are sustained emotions; **affects** are short-term expressions of emotion. Individuals with affective disorders have periods of prolonged depression (**unipolar disorder**) or cycles of depression alternating with periods of elation (**bipolar disorder**).

Schizophrenia is a collection of mental disorders characterized by psychotic symptoms, delusions, thought disorders, and hallucinations, often called the schizoid spectrum. Schizoid individuals experience disordered thinking, inappropriate emotional responses, and social deterioration. Mood disorders and schizophrenia are complex, often are difficult to diagnose, and have genetic components and environmental triggers. There are genetic components to mood disorders and schizophrenia, but there is no clear-cut pattern of inheritance, and the roles of social and environmental factors are unclear. Nonetheless, some information about the genetics of these conditions is emerging, and, despite setbacks in identifying single genes, progress is being made in generating genetic models of these disorders.

Mood Disorders: Unipolar and Bipolar Illnesses

The lifetime risk for a clinically identifiable mood disorder in the U.S. population is 8% to 9%. Depression (unipolar illness) is the most common mood disorder. It is more common in females (about a 2:1 ratio), usually begins in the fourth or fifth decade of life, and is often long lasting or a recurring condition. Depression is associated with weight loss, insomnia, poor concentration, irritability and anxiety, and lack of interest in surrounding events.

About 1% of the U.S. population suffers from bipolar illness. The age of onset occurs during adolescence or the second and third decades of life, and males and females are at equal risk for this condition. In bipolar illness, periods of manic activity alternate with depression. Manic phases are characterized by hyperactivity, acceleration of thought processes, low attention span, creativity, feelings of elation or power, and risk-taking behavior.

Family, twin, and adoption studies have linked bipolar illness to genetics. These studies show concordance of 60% for monozygotic (MZ) twins and 14% for dizygotic (DZ) twins. Adoption studies also indicate that genetic factors are involved in this disorder. Studies have also documented the risk to first-degree relatives of those with bipolar illness (Figure 16.10). The fact that concordance in MZ twins is not 100% suggests that environmental factors (such as stress) interact with genetic risk. As discussed earlier, several attempts have been made to map genes for bipolar illnesses by using a single-gene model but were unsuccessful. The failure to find linkage between genetic markers and single genes for bipolar illness does not undermine the role of genes in this condition, but means that new strategies of linkage analysis are required to identify the genes involved.

New strategies, using genomics, have centered on several approaches used in a worldwide effort to screen all human chromosomes for genes that control bipolar illness. Association studies are one such approach. If there is an association between a disease gene and nearby markers, this combination should occur more often in people with bipolar illness than in control populations. Association studies use DNA markers but follow the inheritance of the marker and the disorder (bipolar illness in this case) in unrelated individuals affected with the disorder. The idea is to identify portions of the genome that are more common in affected individuals than in those without the trait. Association studies have identified regions on chromosomes 4, 12, 15, 18, and 21 as candidates that may contain genes for bipolar illness (Figure 16.11). Other chromosomes,

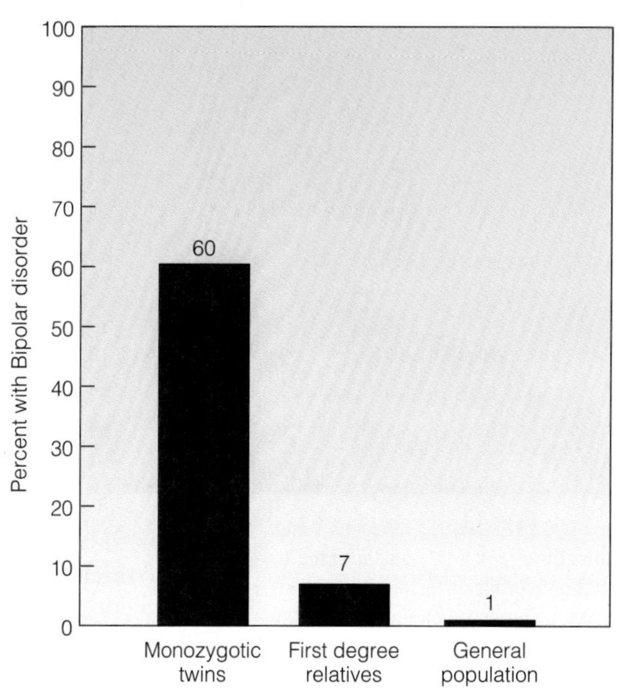

FIGURE 16.10 The frequency of bipolar illness in members of monozygotic twin pairs and in first-degree relatives of affected individuals indicates that genetic factors are involved in this disease. (From Craddock, N. et al. (1995). Mathematical limits of multilocus models: the genetic transmission of bipolar disorder. *American Journal of Human Genetics, 57,* 690–702. Reprinted with permission of University of Chicago Press.)

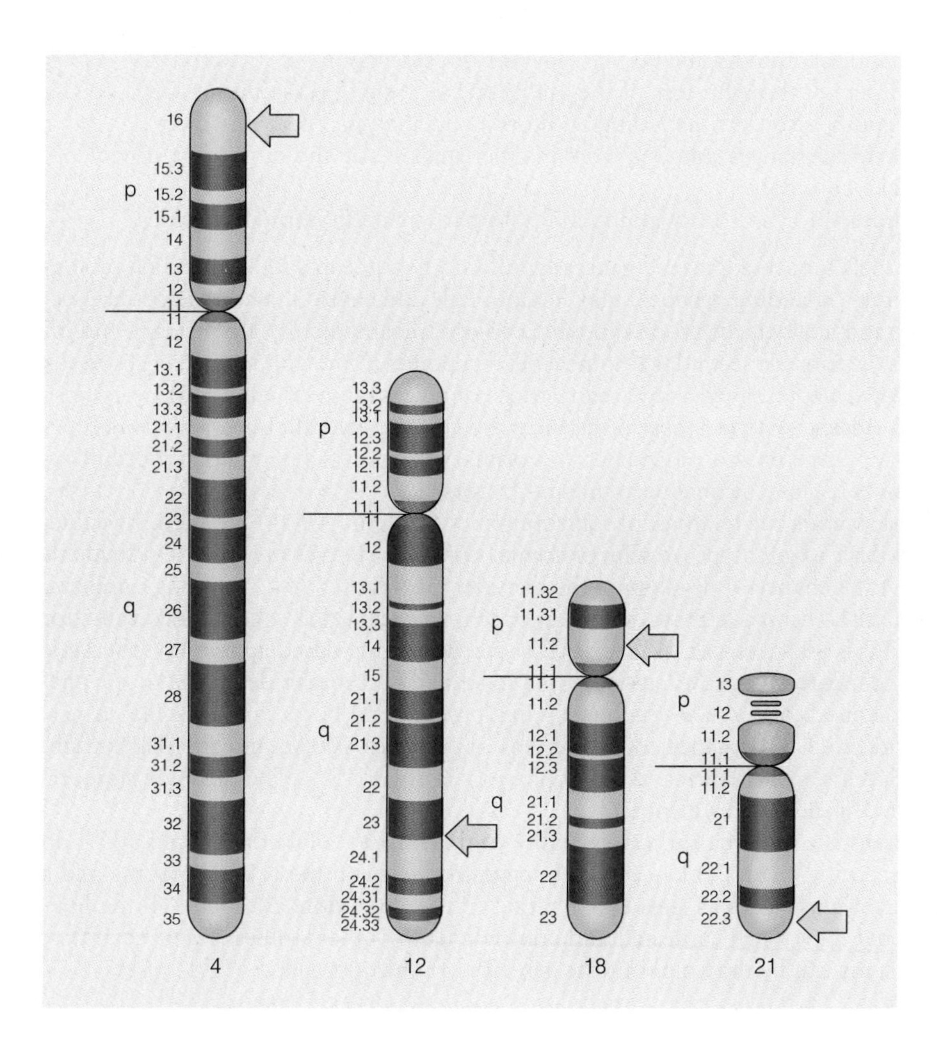

FIGURE 16.11 Arrows indicate chromosome regions initially reported to be linked to bipolar illness, and in which a second round of studies has confirmed linkage. These regions have a high probability of containing one or more genes associated with bipolar illness. (From Potash, J.B. and DePaulo, J.R. (2000). Searching high and low: a review of the genetics of bipolar disorder. *Bipolar Disorders, 2,* 8–26, Figure 3, p. 14.)

FIGURE 16.12 Virginia Woolf, the author and poet, was affected by manic depression. Like others, she often commented on the relationship between creativity and her illness.

including 7, 10, and 13, may also carry genes associated with this disorder. Now that chromosome regions have been identified, they are being closely studied to search for specific genes responsible for this disorder.

In addition to its elusive genetic nature, bipolar illness remains a fascinating behavioral disturbance because of its close association with creativity (see Concepts and Controversies: The Link Between Madness and Genius). Many great artists, authors, and poets have been afflicted with bipolar illness (Figure 16.12). The thought patterns of the creative mind parallel those of the manic stage of bipolar illness. In her book *Touched with Fire,* Kay Jamison explores the relationship among genetics, neuroscience, and the lives and temperaments of creative individuals, including Byron, Van Gogh, Poe, and Virginia Woolf.

Schizophrenia Has a Complex Phenotype

Schizophrenia is a mental illness affecting about 1% of the population (about 2.5 million people in the United States are affected). The disorder usually appears in late adolescence or early adult life. It has been estimated that half of all hospitalized mentally ill and mentally retarded individuals are schizophrenic.

Schizophrenia is a disorder of thought rather than of mood. Diagnosis is often difficult, and there is notable disagreement on the definition of schizophrenia because it has no single distinguishing feature and causes no characteristic brain pathology. Some features of the disorder include

- Psychotic symptoms, including delusions of persecution
- Disorders of thought; loss of the ability to use logic in reasoning
- Perceptual disorders, including auditory hallucinations (hearing voices)
- Behavioral changes, ranging from mannerisms of gait and movement to violent attacks on others
- Withdrawal from reality and inability to participate in normal activities

Models for schizophrenia generally fall into two groups: those in which biological factors (including genetics) play a major role and environmental factors are secondary and, conversely, models in which environmental factors are primary and biological factors are secondary. Some evidence points to metabolic differences in the brains of schizophrenics compared to those of normal individuals (Figure 16.13), but it is unclear whether these differences are genetic. Overall, however, the best evidence supports the role of genetics as a primary factor in schizophrenia, but full expression is dependent on environmental factors.

Risk factors for relatives of schizophrenics (Figure 16.14) are high, revealing the influence of genotype on schizophrenia. Overall, relatives of affected individuals have a 15% chance of developing the disorder (as opposed to 1% among unrelated individuals). Using a narrow definition of schizophrenia, the concordance value for MZ twins is 46%, versus 14% for DZ twins. MZ twins raised apart show the same level of concordance as MZ twins raised together. If a broader definition of the phenotype is used, one which combines schizophrenia and borderline or schizoid personalities, the concordance for MZ twins approaches 100%, and the risk for siblings, parents, and offspring of schizophrenics is about 45%. This strongly supports the role of genes in this disorder.

How schizophrenia (or susceptibility to this illness) is inherited is unknown. In the past, pedigree and linkage studies suggested genes on the X chromosome and a number of autosomes as sites contributing to this condition. Although all the linkage studies have been contested and contradicted by other studies, several tentative conclusions can be drawn about the genetics of schizophrenia.

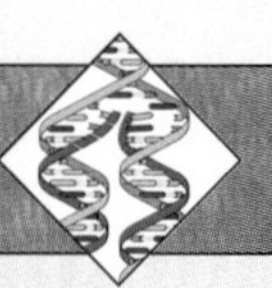

Concepts and Controversies

The Link Between Madness and Genius

As long ago as the 4th century B.C. Aristotle observed that talented philosophers, poets, and artists tend to have mental problems. This idea has been incorporated into the popular wisdom of our culture, as illustrated by statements such as "There is a thin line between genius and madness." In recent years, a substantial amount of evidence has accumulated that distinguished artists, poets, authors, and composers suffer from mood disorders, particularly depression and bipolar illness, at rates 10 to 30 times more frequent than the general population. Based on these results, some researchers have proposed that there is a link between bipolar disorder and creativity. Several books, including *Touched with Fire* (Kay Jamison), *The Price of Greatness* (Arnold Ludwig), and *The Broken Brain* (Nancy Andreasen), have explored this link.

As described by Jamison, the poet Alfred Lord Tennyson and members of his family were affected by mental instability, unstable moods, and insanity. Alfred was affected by life long bouts of depression, as were two of his brothers, his sister, his father, two uncles, an aunt, and his grandfather. Along with bipolar illness went a consuming passion for poetry and verse. Although Alfred is the best known member of his family, his brothers published volumes of poetry and each won prestigious awards for translating ancient Greek poems and epics. Alfred's siblings and his aunt also wrote verse, and one of his nieces became a poet and playwright, illustrating the link between creativity and bipolar illness.

We now know that different regions of the brain are affected in the depressive and the manic stages of bipolar illness. In addition, those with bipolar illness have a unique pattern of metabolism and blood flow in the prefrontal cortex, the part of the brain associated with intellect. It is thought that nerve cell connections ("wiring") may be different in certain regions of the brains of bipolar individuals and that the transition from mania to depression may stimulate mental activity and creativity.

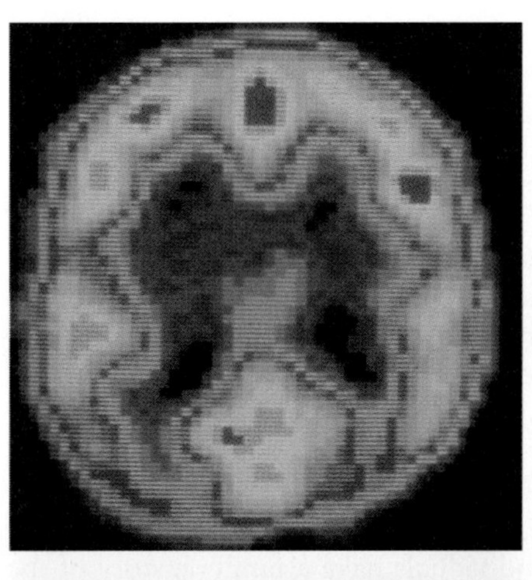
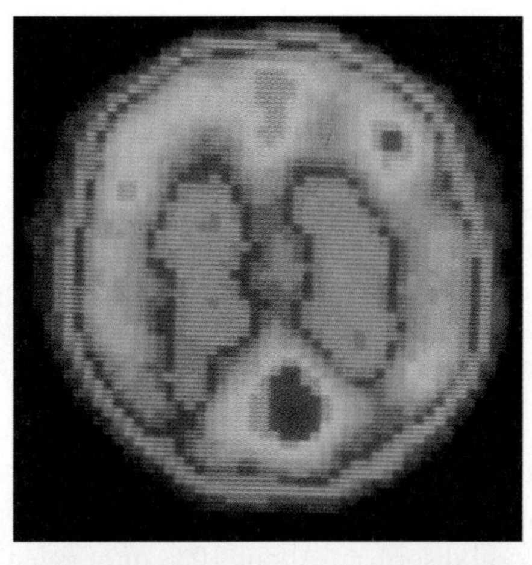
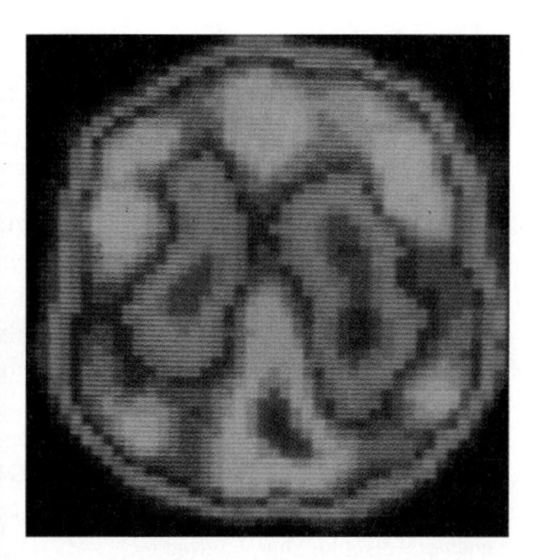
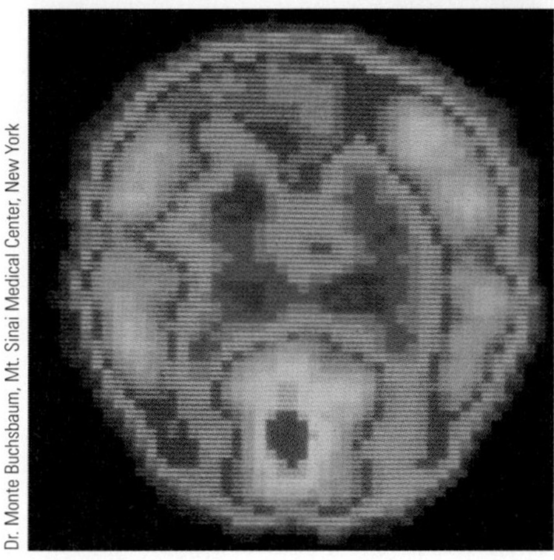
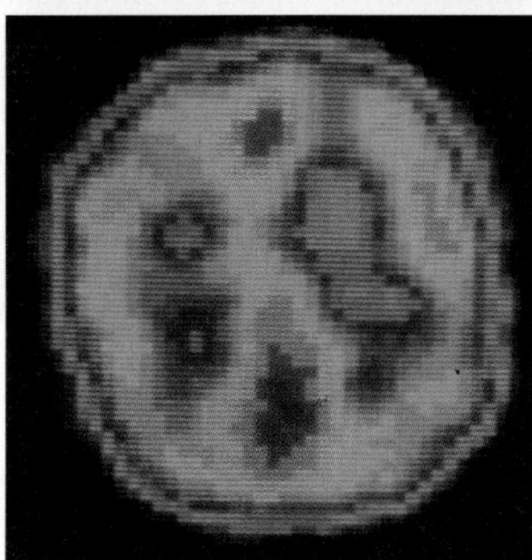

Dr. Monte Buchsbaum, Mt. Sinai Medical Center, New York

FIGURE 16.13 Brain metabolism in a set of monozygotic quadruplets, all of whom suffer from varying degrees of schizophrenia. These scans of glucose utilization by brain cells are visualized by positron emission tomography (PET scan). They show low metabolic rates in the frontal lobe (top of each scan) compared to nonschizophrenics (top left image). The frontal lobe is where cognitive ability resides.

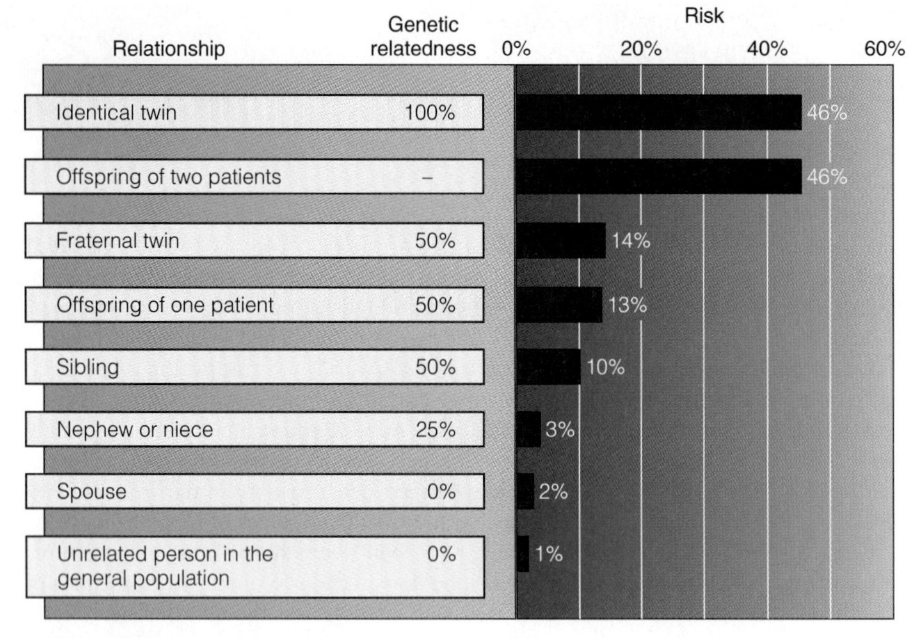

FIGURE 16.14 The lifetime risk for schizophrenia varies with the degree of the relationship to an affected individual. The observed risks are more compatible with a multifactorial mode of transmission than a single-gene or polygenic mode of inheritance.

A polygenic model, in which a single gene makes a major contribution, is consistent with results from family studies, concordance in MZ twins, and incidence of the disorder in the general population. Linkage studies seem more valuable than association studies for identifying genes related to schizophrenia. A group of research laboratories has formed the Schizophrenia Collaborative Linkage Group to conduct genome scans for major genes that contribute to schizophrenia and bipolar illness. In one study, 452 microsatellite markers covering all regions of the genome were used to study individuals in 54 pedigrees, each of which have multiple cases of schizophrenia. To date, linkages between schizophrenia and loci on chromosomes 3, 8, 13, and 22 have been identified.

Another model of schizophrenia studies the neurochemical events in nerve impulse transmission in normal and schizophrenic individuals and in animal model systems. By disrupting specific groups of neurons using chemical or genetic methods, researchers hope to understand how alterations in neurotransmitter levels and interactions result in the behavioral profile seen in this disorder.

GENETICS AND SOCIAL BEHAVIOR

Human geneticists have long been interested in behavior that takes place in a social context, that is, behavior resulting from interactions between and among individuals. This behavior is often complex, and evidence indicates that such behavior involves multifactorial inheritance. Several traits that affect different aspects of social behavior are discussed in the following sections.

Tourette Syndrome Affects Speech and Behavior

Tourette syndrome (GTS) A behavioral disorder characterized by motor and vocal tics and inappropriate language. Genetic components are suggested by family studies that show increased risk for relatives of affected individuals.

Tourette syndrome (GTS) (MIM/OMIM 137580) is characterized by motor and behavioral disorders. About 10% of affected individuals have a family history of the condition. Males are affected more frequently than females (3:1), and onset is usually between 2 and 14 years of age. GTS is characterized by episodes of motor and vocal tics that can progress to more complex behaviors involving a series of grunts and barking noises. Vocal tics include outbursts of profane and vulgar language and parrot-like repetition of words spoken by others. Because there is so much variation in expression, the incidence of the condition is unknown, but some researchers suggest that the disorder may be very common.

Family studies reveal that relatives of affected individuals are at significantly greater risk for GTS than relatives of unaffected controls. The inheritance of GTS is complex and involves a major gene and a number of minor genes, as well as environmental contributions.

Identifying genes associated with GTS is proving difficult. Recent efforts have focused on genetically isolated populations, such as the Afrikaners of South Africa. Genome scans using DNA markers on samples from affected and unaffected members of the Afrikaner population have identified regions on five autosomes that may contain genes associated with GTS. These regions are on chromosomes 2p, 8q, 11q, 20q, and 21q. Further work using data from the Human Genome Project will identify specific genes in these regions that are candidates for GTS and hopefully lead to new methods of diagnosis and treatment for this disorder.

Alzheimer Disease Has Genetic and Nongenetic Components

The symptoms of **Alzheimer disease** (**AD**) begin with loss of memory, progressive dementia, and disturbances of speech, motor activity, and recognition. There is an ongoing degeneration of personality and intellect, and, eventually, affected individuals are unable to care for themselves. Ten percent of the U.S. population older than 65 years has AD, and the disorder affects 50% of those older than 80 years. The cost of treatment and care for AD is more than $80 billion.

Alzheimer disease (AD) A heterogeneous condition associated with the development of brain lesions, personality changes, and degeneration of intellect. Genetic forms are associated with loci on chromosomes 14, 19, and 21.

Brain lesions (Figure 16.15) accompany the progression of AD. The lesions are formed by a protein fragment, amyloid beta-protein, which accumulates outside cells in aggregates known as senile plaques. These plaques cause the degeneration and death of nearby neurons, affecting selected regions of the brain (Figure 16.16). Formation of senile plaques is not specific to AD; almost everyone who lives beyond the age of 80 will have such lesions. The difference between normal aging of the brain and AD appears to be the number of such plaques (greatly increased in AD) and the rate of accumulation (earlier and faster in AD).

The genetics of AD is complex. Less than 50% of all cases can be traced to genetic causes, indicating that the environment plays a significant role in this disorder. We first examine the genetic evidence and then discuss some of the proposed environmental factors associated with AD.

The gene that encodes the amyloid beta-protein, *AD1* (MIM/OMIM 104300), is located on the long arm of chromosome 21. Mutations of this gene are responsible for an early-onset form of AD, inherited as an autosomal dominant trait. A second form of AD, *AD2* (MIM/OMIM 107741), is associated with an allele (*APOE*4*) of the *APOE* gene on chromosome 19. ApoE is a protein involved with cholesterol metabolism, transport, and storage. The *APOE* gene has three alleles (*E*2*, *E*3*, and *E*4*). Those who carry one or two copies of the *E*4* allele have an earlier onset of AD, but these alleles do not affect the progress of the disease. Thus, this allele does not cause AD but does influence the age of onset.

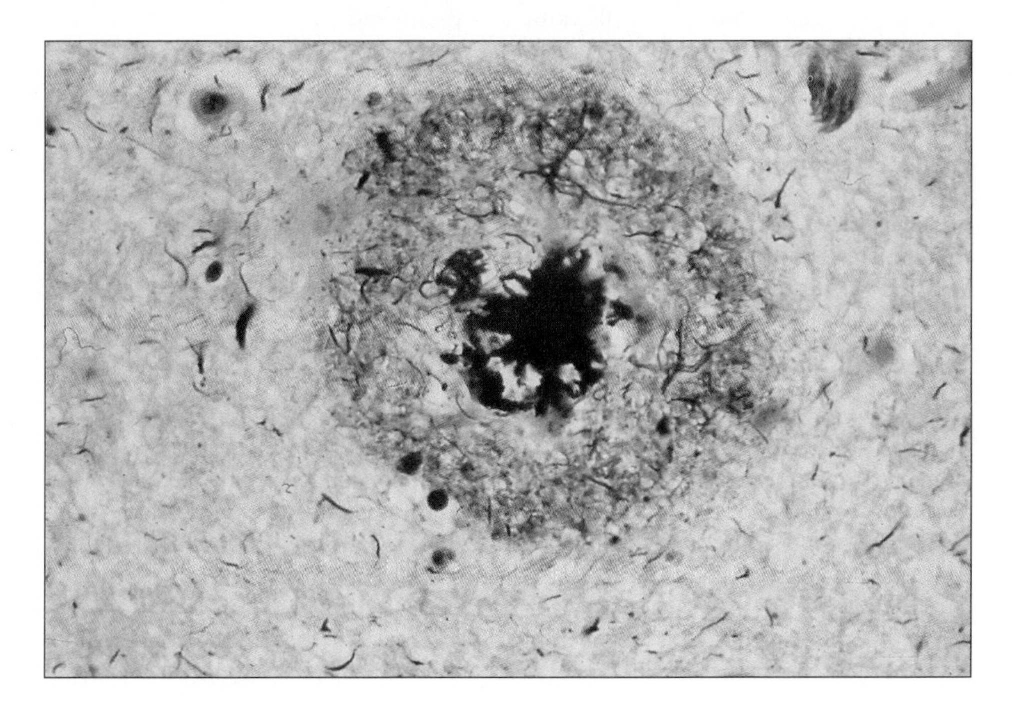

FIGURE 16.15 A lesion called a plaque in the brain of an individual with Alzheimer disease. A ring of degenerating nerve cells surrounds the deposit of protein.

FIGURE 16.16 Location of brain lesions in Alzheimer disease. Plaques are most heavily concentrated in the amygdala and hippocampus. These brain regions are part of the limbic system.

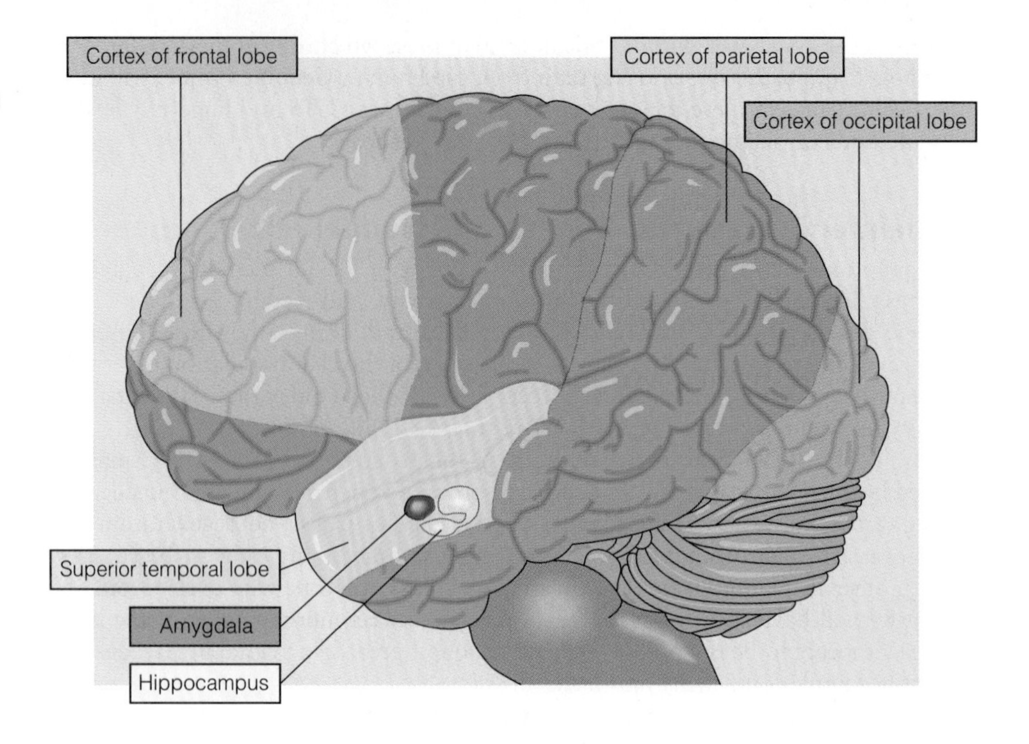

A third form, *AD3*, is caused by a mutation in a membrane protein encoded by a gene on chromosome 14 (MIM/OMIM 104311), and *AD4* (MIM/OMIM 600759) is caused by a mutation in a gene on chromosome 1 that also encodes a membrane protein. There is evidence for AD genes on other chromosomes, and mitochondrial DNA polymorphisms may also play a role in susceptibility to this disease.

It is not clear whether all cases of AD have a genetic basis. Other genes that have not been identified may play a role in AD, or environmental factors may play a role in initiating this disorder. Once β-amyloid accumulation begins, several factors can influence the rate at which the disease progresses. These include

- Free radical production stimulated by β-amyloid accumulation
- Calcium uptake into nerve cells caused by β-amyloid
- β-Amyloid toxicity to nearby nerve cells

In summary, we know that AD has several genetic causes, and mutations in any of several genes can produce the AD phenotype. In addition, the fact that many cases of AD cannot be traced to a genetic source may indicate that there is more than one cause for AD. In addition, there are several factors that influence the rate at which the disease progresses, and the role of nongenetic influences and their mechanisms continue to be investigated to define the risk factors and rate of progression for this debilitating condition.

Alcoholism Has Several Components

As a behavioral disorder, excessive alcohol consumption (MIM/OMIM 103780) has two important components. First, drinking in excess over a long time causes damage to the nervous system and other organ systems. Over time, the accumulation of damage results in altered behavior, hallucinations, and loss of memory. The second component involves the behavior patterns that lead to alcohol abuse and a loss of the ability to function in social settings, the workplace, and the home.

It is estimated that 75% of the adult U.S. population consumes alcohol. About 10% of these adults are classified as alcoholics, and the male to female ratio is about

4:1. From the genetic standpoint, alcoholism is most likely a genetically influenced, multifactorial (genetic and environmental) disorder. The role of genetic factors in alcoholism is indicated by a number of findings:

- In mice, experiments indicate that alcohol preference can be selected for; some strains of mice choose 75% alcohol over water, whereas other strains shun all alcohol.
- There is a 25% to 50% risk of alcoholism in sons and brothers of alcoholic men.
- There is a 55% concordance for alcoholism in MZ twins, and a 28% rate in same-sex DZ twins.
- Sons adopted by alcoholic men show a rate of alcoholism closer to their biological fathers than their adoptive fathers.

Genes that influence alcoholism have not been identified. Segregation analysis in families with alcoholic members has produced evidence against the Mendelian inheritance of a single major gene and for multifactorial inheritance involving several genes. Other researchers have adopted a single-gene model and find an association between an allele of a gene that encodes a neurotransmitter receptor protein (called D2) and alcoholism. This evidence is based on the finding that in brain tissue, the *A1* allele of the *D2* gene was found in 69% of the samples from severe alcoholics but in only 20% of the samples from nonalcoholics, implying that the *A1* allele is involved in alcoholism.

Subsequent linkage and association studies on the *A1* allele have not supported the idea that this allele is involved in alcoholism. Taken together, the available studies have failed to show any relationship between abnormal neurotransmitter metabolism or receptor function and alcoholic behavior.

The search for genetic factors in alcoholism illustrates the problem of selecting the proper genetic model to analyze behavioral traits. Segregation and linkage studies indicate that there is no single gene for alcoholism. If a multifactorial model involving a number of genes, each with a small additive effect, is invoked, the problem becomes more complicated. How do you prove or disprove that a given gene contributes, say, 10% to the behavioral phenotype? At present, the only method would involve studying thousands of individuals to find such effects.

Is Sexual Orientation a Multifactorial Trait?

Most people are heterosexual and prefer the opposite sex, but a fraction of the population is homosexual and prefers sexual activity with members of the same sex. These variations in sexual behavior have been recorded since ancient times, but biological models for these behaviors have been proposed only recently.

Twin studies and adoption studies have investigated the role of genetics in sexual orientation. One twin study involved 56 MZ twins, 54 DZ twins, and 57 genetically unrelated adopted brothers. The concordance for homosexuality was 52% for MZ twins, 22% for DZ twins, and 11% for the unrelated adopted siblings. Overall heritability ranged from 31% to 74%. Another study investigating homosexual behavior in women employed 115 twin pairs and 32 genetically unrelated adopted sisters. In this study, heritability ranged from 27% to 76%.

The results from these and other studies indicate that homosexual behavior has a strong genetic component. These studies have been challenged on the grounds that the results can be affected by the phrasing of the interview questions and by the methods used to recruit participants and that the phenotype is self-described. But the average heritability estimates for sexual preference from these studies parallel those from the Minnesota Twin Project, a long-term study of MZ twins separated at birth and reared apart. Further twin studies are needed to determine whether the heritability values are accurate. If confirmed, the studies to date indicate that homosexual behavior is a multifactorial trait that involves several genes and unidentified environmental components.

Guest Essay: Research and Applications

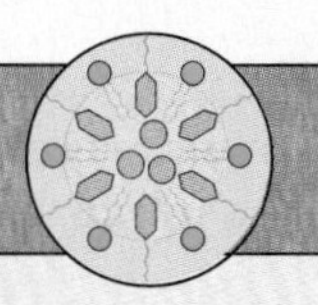

Neurogenetics: From Mutants to Molecules

Barry S. Ganetzky

From the time I was young, I was fascinated by living things and enjoyed reading books on natural history. The introductory biology courses I took as an undergraduate at the University of Illinois-Chicago solidified my interests, especially in the areas of genetics and molecular biology. I was fascinated by learning the molecular basis of life. However, I still had little understanding of biology as a profession.

In my junior year, I began an honor's research project. My mentor was a new, young professor (a certain Michael R. Cummings) who was studying development of eggs and embryos in *Drosophila* (fruit flies). This was my first real exposure to research and to what academic life was all about. That I could actually make a career out of applying my imagination and creativity to discover answers to nature's secrets seemed almost too good to be true. I still feel pretty much the same way. Although my original project was supposed to last only 10 weeks, I worked with Dr. Cummings for the next 2 years and this experience guided my future course.

I pursued my Ph.D. in genetics at the University of Washington with the late Larry Sandler, one of the great *Drosophila* geneticists as well as a gifted and inspiring teacher. He had helped elevate *Drosophila* genetics to a fine art, and under his guidance I acquired these skills.

As a postdoctoral fellow with Seymour Benzer, I became interested in applying genetic techniques to elucidate the molecular basis of neuronal function. It was known that ion channel proteins play key roles in generating nerve impulses, but little was known about their molecular structure or how they worked. To identify genes that encode these proteins, I screened for *Drosophila* mutants that were defective in neuronal activity.

The trick was finding the right mutants. I decided to focus on mutants that became paralyzed when exposed to elevated temperatures. In a stroke of luck, one of the first such mutants I found caused a complete block of action potentials, demonstrating that our approach worked. After I took a faculty position at the University of Wisconsin, my students and I went on to isolate many more such mutants and to clone the corresponding genes. A number of these genes turned out, in fact, to encode ion channel proteins, and our studies provided the first molecular insights into their structure and function. Interestingly, many heritable neurological disorders in humans are now also known to be caused by mutations in ion channel genes. We were able to identify one such human gene because of its close similarity to a gene we discovered originally in *Drosophila*. Genome sequence analysis has now revealed that at least 50% of all human disease genes have closely related counterparts in *Drosophila*. Thus, *Drosophila* is playing an increasingly important role as a model system for understanding human disease. We still study our paralytic mutants, which continue to provide us with novel insights into neuronal function and dysfunction. For example, we recently discovered that some of our mutations cause neurodegeneration in flies, opening up new avenues for investigating the molecular basis of this pathology, which is also a major human health problem. It has been very gratifying to me that studies I pursued primarily because they were interesting and fun have turned out to be of medical relevance and importance as well.

BARRY S. GANETZKY received a B.S. in biology from the University of Illinois-Chicago and a Ph.D. in genetics from the University of Washington. He has been a faculty member in the Laboratory of Genetics at the University of Wisconsin, Madison, since 1979, where he is the Steenbock Professor of Biological Sciences. After spending more than half his life working with fruit flies, he still derives great pleasure from discovering new mutations that have interesting and unusual phenotypes.

Several years ago, a group of U.S. researchers used RFLP markers to study male homosexual behavior and found linkage between one subtype of homosexuality and markers on the long arm of the X chromosome (Figure 16.17). In this study, a two-step approach was employed. First, family histories were collected from 114 homosexual males. Pedigree analysis was performed on 76 randomly selected individuals from this group, using interviews with male relatives to ascertain sexual preference. The results indicated the possibility of X-linked inheritance (MIM/OMIM 306995).

A further pedigree analysis was conducted using 38 families in which there were two homosexual brothers, based on the idea that this might show a stronger trend for X-linked inheritance because there were two homosexual siblings in the same family. The results show a stronger trend toward X-linked inheritance and an absence of paternal transmission.

Using information about traits and relatedness from the pedigree analysis, the second part of the study employed DNA markers to determine whether an X-linked locus or loci were associated with maternally transmitted male homosexual behavior in the 38 families that had two homosexual brothers. Linkage was detected to RFLP markers from the distal region of the long arm in the Xq28 region. These markers were present in two-thirds of the 32 pairs of homosexual brothers and in about one-fourth of the heterosexual brothers.

In a subsequent study of 52 homosexual brothers, a Canadian research team failed to find linkage between the Xq28 markers and homosexuality. Their analysis could not rule out the possibility of a gene with a minor contribution to this behavior but could exclude a gene with a major contribution. More work is needed to confirm the linkage relationship and to search the region for a locus or loci that affects sexual orientation.

Two important factors related to the U.S. study should be mentioned. First, seven sets of homosexual brothers did not share all of the markers in the Xq28 region, and about one-fourth of all heterosexual brothers inherited the markers but did not display homosexual behavior. This indicates that genetic heterogeneity or nongenetic factors are significant in this behavior. Second, the study cannot estimate what fraction of homosexual behavior might be related to the Xq28 region or whether this region influences lesbian sexual behavior. In spite of its preliminary nature and (to some) its controversial conclusions, this work applies genome-wide screening with molecular markers to study the role of genes in one type of male sexual behavior and represents a model for future studies in this area of behavior genetics.

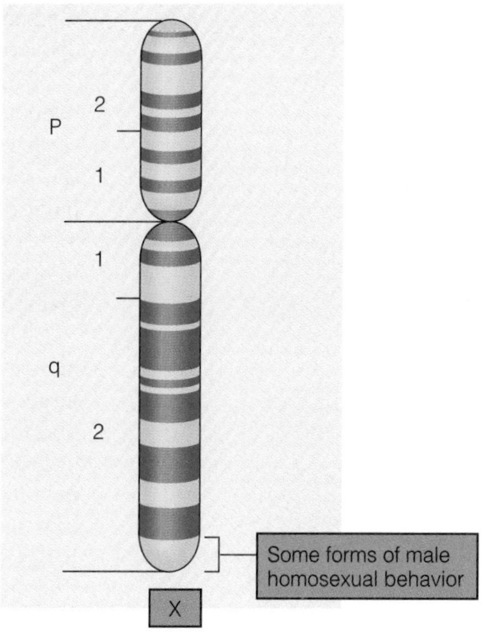

p
2
1
1
q
2
Some forms of male homosexual behavior
X

FIGURE 16.17 Region of the X chromosome found by linkage analysis to be associated with one form of male homosexual behavior.

SUMMING UP: THE CURRENT STATUS OF HUMAN BEHAVIOR GENETICS

In reviewing the current state of human behavior genetics, several elements are apparent. Almost all studies of complex human behavior have provided only indirect and correlative evidence for the role of specific genes. Segregation studies and heritability estimates indicate that most behaviors are complex traits involving several genes. Searches for single-gene effects have proven unsuccessful to date, and initial reports of single genes that control bipolar illness, schizophrenia, and alcoholism have been retracted or remain unconfirmed.

The multifactorial nature of behavioral traits means that methods for identifying genes with small, incremental effects must be developed. Although twin and adoption studies have been valuable in behavior genetics, these studies typically involve small numbers of individuals. For example, fewer than 300 pairs of MZ twins raised apart have been identified worldwide. When traits involve multiple genes, confirmation results can require detailed examination of thousands of individuals in hundreds of families. This process is necessarily slow and labor-intensive.

Recent successes in finding genes involved in other complex traits may point the way to identifying genes associated with behavior. Chromosome regions shared by those with a genetic disorder are larger than originally thought, and fewer markers are needed to trace the coinheritance of a trait and its associated markers, making work easier and faster. Using a small number of markers, along with information from the Human Genome Project and newer methods of data analysis, researchers have identified genes for susceptibility to inflammatory bowel disease (Crohn disease) and insulin-dependent diabetes. Previously, the linkage results for these diseases showed the same inconsistency as the results for behavior disorders such as bipolar illness and schizophrenia. Putting this set of techniques to work in identifying genes for behavior along with more refined definitions of phenotype may be successful in dissecting the genetic components of mental illness and other behavioral phenotypes.

However, success in identifying susceptibility genes for behavior should not overshadow the fact that the environment plays a significant role in behavior. As confirmation of the role of genes in behavior becomes available, investigations into the role of environmental factors cannot be neglected. The history of human behavior genetics in the eugenics movement of the early part of this century provides a lesson in the consequences of overemphasizing the role of genetics in behavior. Attempts to provide single-gene explanations for complex behavior inhibited the growth of human genetics as a discipline.

The identification of genes affecting behavior may lead to improvements in diagnosis and treatment of behavior disorders but also has implications for society at large. As discussed in Chapter 13, the Human Genome Project has raised questions about the way genetic information will be disseminated and used and who will have access to this information and under what conditions. These same concerns need to be addressed for genes that affect behavior. If genes for alcoholism or homosexuality can be identified, will this information be used to predict an individual's future behavioral patterns? Who should have access to this information, and what can be done to prevent health care discrimination in employment or insurance?

Many behavioral phenotypes, such as Huntington or Alzheimer disease, are clearly regarded as abnormal. Few would argue against the development of treatments for intervening in and perhaps preventing these conditions. When do behavior phenotypes move from being abnormal to being variants? If there is a connection between bipolar illness and creativity, to what extent should this condition be treated? If genes that influence sexual orientation are identified, will this behavior be regarded as a variant or as a condition that should be treated and/or prevented?

Although research can provide information about the biological factors that play a role in determining human behavior, it cannot provide answers to questions of social policy. Social policy and laws have to be formulated by using information from research.

Summary

Models, Methods, and Phenotypes in Studying Behavior

1. Many forms of behavior represent complex phenotypes. The methods used to study inheritance of behavior encompass classical methods of linkage and pedigree analysis, newer methods of recombinant DNA analysis, and new combinations of techniques, such as twin studies combined with molecular methods. Refined definitions of behavior phenotypes are also being used in the genetic analysis of behavior.

Animal Models: The Search for Behavior Genes

2. Results from work on experimental animals indicate that behavior is under genetic control and have provided estimates of heritability. The molecular basis of single-gene effects in some forms of behavior has been identified and provides useful models to study gene action and behavior. Transgenic animals carry mutant copies of human genes and are studied to understand the action of the mutant alleles and to develop drugs for treatment of these conditions.

Single Gene Effects on Human Behavior

3. Several single-gene effects on human behavior are known. Most of these affect the development, structure, or function of the nervous system and consequently affect behavior. Huntington disease serves as a model for neurodegenerative disorders. Language and brain development are linked by genes that encode transcription factors.

Single Genes Control Aggressive Behavior and Brain Metabolism

4. Most forms of mental retardation are genetically complex multifactorial disorders. One form of X-linked retardation, associated with aggressive behavior, is linked to abnormal metabolism of a neurotransmitter (a chemical that transfers nerve impulses from cell to cell).

The Genetics of Mood Disorders and Schizophrenia

5. Bipolar illness and schizophrenia are common behavior disorders, each affecting about 1% of the population. Simple models of single-gene inheritance for these disorders have not been supported by extensive studies of affected families, and more complex forms of multifactorial inheritance seem likely.

Genetics and Social Behavior

6. Multifactorial traits that affect behavior include Tourette syndrome, Alzheimer disease, alcoholism, and sexual orientation. Twin studies combined with molecular markers have identified a region of the X chromosome that may affect one form of homosexual behavior. Others have been unable to confirm this link, leaving open the question of genetic control of sexual choice.

Summing Up: The Current Status of Human Behavior Genetics

7. The evidence for genetic control of complex behaviors is indirect. Although some progress has been made in showing linkage between certain chromosome regions and disorders including bipolar illness and schizophrenia, no genes contributing to these or other behaviors such as alcohol abuse or sexual preference have been discovered. New ways of studying linkage, coupled with new methods of analysis of linkage data and information from the Human Genome Project may lead to rapid identification of genes involved in behavior.

Case Studies

CASE 1

Rachel asked to see a genetic counselor because she was concerned about developing schizophrenia. Her mother and maternal grandmother both had schizophrenia and were institutionalized for most of their adult lives. Rachel's three maternal aunts are all in their 60s and have not shown any signs of this disease. Rachel's father is alive and healthy, and his family history does not suggest any behavioral or genetic conditions. The genetic counselor discussed the multifactorial nature of schizophrenia and explained that there have been many candidate genes identified that may be mutated in individuals with this condition. However, a genetic test is not available for presymptomatic testing. The counselor explained that based on Rachel's family history and her relatedness to the individuals who have schizophrenia, her risk of developing it is approximately 13%. If an altered gene is in the family and her mother carries this gene, Rachel would have a 50% chance of inheriting it.

1. Why do you think it has been so difficult to identify genes underlying schizophrenia?
2. If a test were available that could tell you whether you were likely to develop a disorder such as schizophrenia later in life, would you submit yourself to the test? Why or why not?

CASE 2

A genetic counselor was called to the pediatric ward of the hospital for a consultation. Her patient, an 8-year-old boy, was having a "temper tantrum" and was biting his own fingers and toes. The nurse called after she noticed that he was a patient of a clinical geneticist at another institution. The counselor reviewed the boy's chart and noted a history of growth retardation and self-mutilation since the age of 3. His movements were very "jerky," and he was banging his head against the bedpost. The nurses were having a very difficult time controlling him. The counselor immediately recognized these symptoms as part of a genetic disorder known as Lesch-Nyhan syndrome.

Lesch-Nyhan syndrome is an X-linked recessive condition (Xq26) caused by mutation in the hypoxanthine phosphoribosyltransferase gene, and it affects about 1 in 100,000 males. Symptoms usually begin between the ages of 3 and 6 months. Prenatal testing is available, but there is no treatment for Lesch-Nyhan, and most affected individuals die by the second decade of life.

1. If you were the parent of a child with Lesch-Nyhan syndrome and you heard about an experimental gene therapy technique that had shown some promise in treating the disease but which also had significant associated risks, would you attempt to enroll your child in a clinical trial of the technique? Explain.
2. Lesch-Nyhan syndrome is quite rare (1 in 100,000 males), but its effects are devastating. Would you support an effort to screen every developing fetus for this disorder? Explain.

Questions and Problems

Models, Methods, and Phenotypes in Studying Behavior

1. What are the major differences in the methods used to study the behavior genetics of single-gene traits versus polygenic traits?
2. In human behavior genetics, why is it important for the trait under study to be defined accurately?
3. One of the models for behavioral traits in humans involves a form of interaction known as epistasis. In a simplified example involving two genes, the expression of one gene affects the expression of the other. How might this interaction work, and what patterns of inheritance might be shown?

Animal Models: The Search for Behavior Genes

4. What are the advantages of using *Drosophila* for the study of behavior genetics? Can this organism serve as a model for human behavior genetics? Why or why not?
5. You are a researcher studying an autosomal dominant neurodegenerative disorder. You have cloned the gene underlying the disorder and found that it encodes an enzyme that is overexpressed in neurons of individuals suffering from the disorder. To better understand how this enzyme causes neurodegeneration in humans, you make a strain of transgenic *Drosophila* whose nerve cells overexpress the enzyme.
   a. How might you use these transgenic flies to try to gain insight into the disease or to identify drugs that might be useful in the treatment of the disease?
   b. Can you think of any potential limitations of this approach?

Single Genes Affect the Nervous System and Behavior

6. What type of mutation causes Huntington disease? How does this mutation result in neurodegeneration?
7. Perfect pitch is the ability to identify a note when it is sounded. In a study of this behavior, perfect pitch was found to predominate in females (24 out of 35 in one group). In one group of seven families, individuals in each family had perfect pitch. In two of these families, the affected individuals included a parent and a child. In another group of three families, three or more members (up to five) of each had perfect pitch, and in all three families, two generations were involved. Given this information, what, if any, conclusions can you draw as to whether this behavioral trait might be genetic? How would you test your conclusion? What further evidence would be needed to confirm your conclusion?
8. Opposite to perfect pitch is tone deafness, the inability to identify musical notes. In one study, a bimodal distribution in populations was found, with frequent segregation in families and sibling pairs. The author of the study concluded that the trait might be dominant. In a family study, segregation analysis suggested an autosomal dominant inheritance of tone deafness with imperfect penetrance. One of the pedigrees is presented here. On the basis of the results, do you agree with this conclusion? Could perfect pitch and tone deafness be alleles of a gene for musical ability?

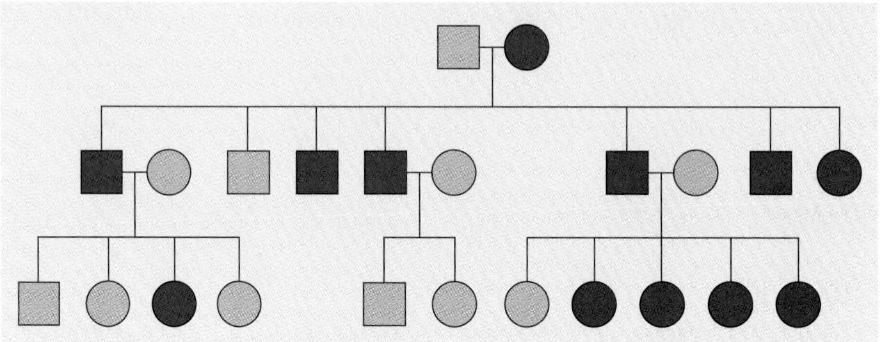

Single Genes Control Aggressive Behavior and Brain Metabolism

9. Name three genes whose mutation leads to an altered behavioral phenotype. Briefly describe the normal function of the mutated gene as well as the altered phenotype.
10. Mutations in the gene encoding monoamine oxidase type A (MAOA) have been linked to aggressive and sometimes violent behavior. Based on this finding, it is conceivable that a genetic test could be developed that could potentially identify individuals likely to exhibit such behaviors. Do you think such a test would be a good idea? What would some of the ethical and societal implications of such a test be?

The Genetics of Mood Disorders and Schizophrenia

11. The two main affective disorders are bipolar illness and schizophrenia. What are the essential differences and similarities between these disorders?
12. A pedigree analysis was performed on the family of a man with schizophrenia. Based on the known concordance statistics, would his MZ twin be at high risk for the disease? Would the twin's risk decrease if he were raised in a different environment than his schizophrenic brother?
13. You are a researcher studying bipolar disorder. Your RFLP data shows linkage between a marker on

chromosome 7 and bipolar illness. Later in the study, you find that a number of individuals lack this RFLP marker but still develop the disease. Does this mean that bipolar disorder is not genetic?

14. A region on chromosome 6 has been linked to schizophrenia, but researchers have not yet found a gene. Explain this linkage and why linkage does not necessarily locate a gene.

Genetics and Social Behavior

15. Of the following findings, which does *not* support the idea that alcoholism is genetic?
    a. Some strains of mice select alcohol over water 75% of the time, whereas others shun alcohol.
    b. The concordance value is 55% for MZ twins and 28% for DZ twins.
    c. Biological sons of alcoholic men who have been adopted have a rate of alcoholism more like their adoptive fathers.
    d. There is a 20% to 25% risk of alcoholism in sons of alcoholic men.
    e. None of the above.
16. A woman diagnosed with Alzheimer disease wants to know the probability of her children inheriting this disorder. Explain to her the complications of determining heritability for this disease.
17. What types of studies have been used to suggest that sexual orientation has a genetic component?
18. In July of 1996, *The Independent,* a popular newspaper published in London, England, reported a study conducted by Dr. Aikarakudy Alias, a psychiatrist who had been working on the relationship between body hair and intelligence for 22 years. Dr. Alias told the 8th Congress of the Association of European Psychiatrists that hairy chests are more likely to be found among the most intelligent and highly educated than in the general population. According to this new research, excessive body hair could also mean higher intelligence. Is correlating body hair with intelligence a valid method for studying the genetics of intelligence? Why or why not? What other factors contribute to intelligence? Is it logical to assume that individuals with little or no body hair are consistently less intelligent than their hairy counterparts? What type of study could be done to prove or disprove this idea?
19. In a long-term study of over 100 pairs of MZ and DZ twins separated shortly after birth and reared apart, one of the conclusions was that "general intelligence or IQ is strongly affected by genetic factors." The study concluded that about 70% of the variation in IQ is due to genetic variability (review the concept of heritability in Chapter 5). Discuss this conclusion, and include in your answer the relationship between IQ and intelligence and to what extent these conclusions can be generalized. In evaluating the study's conclusion, what would you like to know about the twins?

Internet Activities

The following activities use the resources of the World Wide Web to enhance the topics covered in this chapter. To investigate the topics described below, log on to the book's home page (www.brookscole.com/biology_d) and follow the *Student Resources* link to your subject and text.

1. *The Genetics of Personality.* The *Personality Research Site* (http://www.personalityresearch.org/) presents an overview of scientific research programs in personality psychology. Follow the "Behavior Genetics" link. The study of human behavior and behavioral disorders is complex and must account for both environmental and genetic influences. What types of studies do researchers use to attempt to tease out these differences?

For Further Reading

Aldhous, P. (1992). The promise and pitfalls of molecular genetics. *Science, 257,* 164–165.

Aston, C., & Hill, S. (1990). Segregation analysis of alcoholism in families ascertained through a pair of male alcoholics. *Am. J. Hum. Genet., 46,* 879–887.

Blum, K., Noble, E., Sheridan, P., Montgomery, A., Ritchie, T., Jagadeeswaran, P., Nogami, H., Briggs, A., & Cohn, J. (1990). Allelic association of human dopamine D(2) receptor gene in alcoholism. *JAMA, 263,* 2055–2060.

Bolos, A., Dean, M., Lucas-Derse, S., Ramsburg, M., Brown, G., & Goldman, D. (1990). Population and pedigree studies reveal a lack of association between the dopamine D(2) receptor gene and alcoholism. *JAMA, 264,* 3156–3160.

Brunner, H., Nelen, M., Breakfield, X., Ropers, H., & van Oost, B. (1993). Abnormal behavior associated with a point mutation in the structural gene for monoamine oxidase A. *Science, 262,* 578–580.

Brunner, H., Nelen, M., vanZandvoort, P., Abeling, N., van Gennip, A., Walters, E., Kulper, M., Ropers, H., & van Dost, B. (1993). X-linked borderline mental retardation with prominent behavioral disturbance: Phenotype, genetic localization, and evidence for disturbed monoamine metabolism. *Am. J. Hum. Genet., 52,* 1032–1039.

Comings, D.E. (2001). Clinical and molecular genetics of ADHD and Tourette syndrome: Two related polygenic disorders. *Ann. N.Y. Acad. Sci., 931,* 50–83.

Coryell, W., Endicott, J., Keller, M., Andreasen, N., Grove, W., Hirschfield, R., & Scheftner, W. (1989). Bipolar affective disorder and high achievement: A familial association. *Am. J. Psychiatry, 146,* 983–988.

Devore, E., & Cloninger, C. (1989). The genetics of alcoholism. *Ann. Rev. Genet., 23,* 19–36.

Fortini, M. E., and Bonini, N. M. (2000). Modeling human neurodegenerative diseases in *Drosophila;* on a wing and a prayer. *Trends Genet., 16,* 161–167.

Grahame, N.J. (2000). Selected lines and inbred strains: tools in the hunt for genes involved in alcoholism. *Alcohol Res. and Health, 24,* 159–163.

Hamer, D., Hu, S., Magnuson, V., Hu, N., & Pattatucci, A. M. (1993). A linkage between DNA markers on the X chromosome and male sexual orientation. *Science, 261,* 321–327.

Hu, S., Pattatucci, A. M., Patterson, C., Li, L., Fulker, D. W., Cherney, S. S., Krugylak, L., & Hamer, D. H. (1995). Linkage between sexual orientation and chromosome Xq28 in males but not in females. *Nat. Genet., 11,* 248–256.

Jamison, K. (1993). *Touched with fire: Manic depressive illness and the artistic temperament.* New York: Free Press.

Kendler, K., Heath, A., Neale, M., Kessler, R., & Eaves, L. (1992). A population-based twin study of alcoholism in women. *JAMA, 268,* 1877–1882.

Lai, C. S. L., Fisher, S. E., Jurst, J. A., Vargha-Khadem, F., and Monaco, A. P. (2001). A forkhead-domain gene is mutated in a severe speech and language disorder. *Nature, 413,* 519–523.

Lawrence, S., Keats, B., & Morton, N. (1992). The AD1 locus in familial Alzheimer disease. *Ann. Hum. Genet., 56,* 295–301.

LeVay, S. (1991). A difference in hypothalamic structure between heterosexual and homosexual men. *Science, 253,* 1034–1037.

LeVay, S., and Hamer, D. H. (1994). Evidence for a biological influence in male homosexuality. *Sci. Am. 270,* 44–49.

Potash, J. B., and DePaulo, J. R. Jr., (2000). Searching high and low: A review of the genetics of bipolar disorder. *Bipolar Disorders 2,* 8–26.

Powledge, T. (1993). The genetic fabric of human behavior. *Bioscience, 43,* 362–367.

Powledge, T. (1993). The inheritance of behavior in twins. *Bioscience, 43,* 420–424.

Reddy, P. H., Charles, V., Williams, M., Miller, G., Whetsell, W. O. Jr., and Tagle, D. A. (1999). Transgenic mice expressing mutated full-length HD cDNA: A paradigm for locomotor changes and selective neuronal loss in Huntington's disease. *Phil. Trans. R. Soc. Lond. B, 354,* 1035–1045.

Risch, N., & Botstein, D. (1996). A manic depressive history. *Nat. Genet., 12,* 351–353.

Tivol, E. A., Shalish, C., Schuback, D. E., Hsu, Y. P., & Breakefield, V. O. (1996). Mutational analysis of the human *MAOA* gene. *Am. J. Med. Genet., 67,* 92–97.

Wickelgren, I. (1999). Discovery of 'gay gene' questioned. *Science, 284,* 571.

© Tim Davis/Tony Stone Images

Genes in Populations

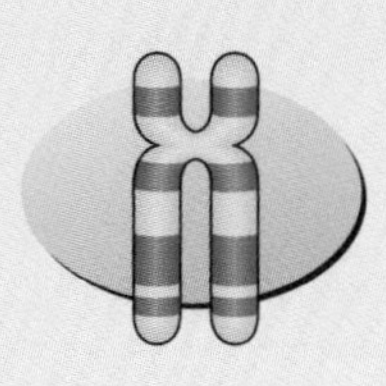

Chapter Outline

Arguments in which a father and son take opposite sides are neither new nor unusual. All offspring occasionally disagree with their parents. These arguments rarely result in ideas that remain controversial some 175 years later, however. An argument between Thomas Malthus and his father began in the years following the French Revolution. This event, like the American Revolution that preceded it, was hailed by many (including the elder Malthus) as the beginning of a new era for humanity. According to his view, throwing off the repressions of monarchy presented an opportunity for unlimited progress in social, political, and economic areas. The younger Malthus disagreed and argued that even if all social impediments were removed, there were fundamental natural constraints that would limit progress. These limits would always result in poverty and misery as part of the human condition. He wrote an essay about his ideas, published in 1798 as "The Essay on the Principle of Population as It Affects the Future Improvement of Society." Later versions were expanded and published as a book and were part of the 1824 supplement to the Encyclopedia Britannica.

In his essay, Malthus observed that populations grow geometrically, with the human population doubling in size every 25 years or so (▶ Figure 17.1). However, resources such as living space and food are more limited and grow more slowly. As a result, population size will rapidly outgrow the environment's ability to support the population. When this point is reached, factors such as war, disease, and starvation begin to limit population growth by increasing the death rate. Voluntary constraints such as delayed marriage, celibacy, and birth control can help limit population growth and bring about a reduction in human suffering and an improvement in living conditions, but sexual passion and human nature cause most people to ignore these constraints. According to the younger Malthus, the result is the existence of unrelieved poverty and marginal living conditions, even in the most prosperous of nations.

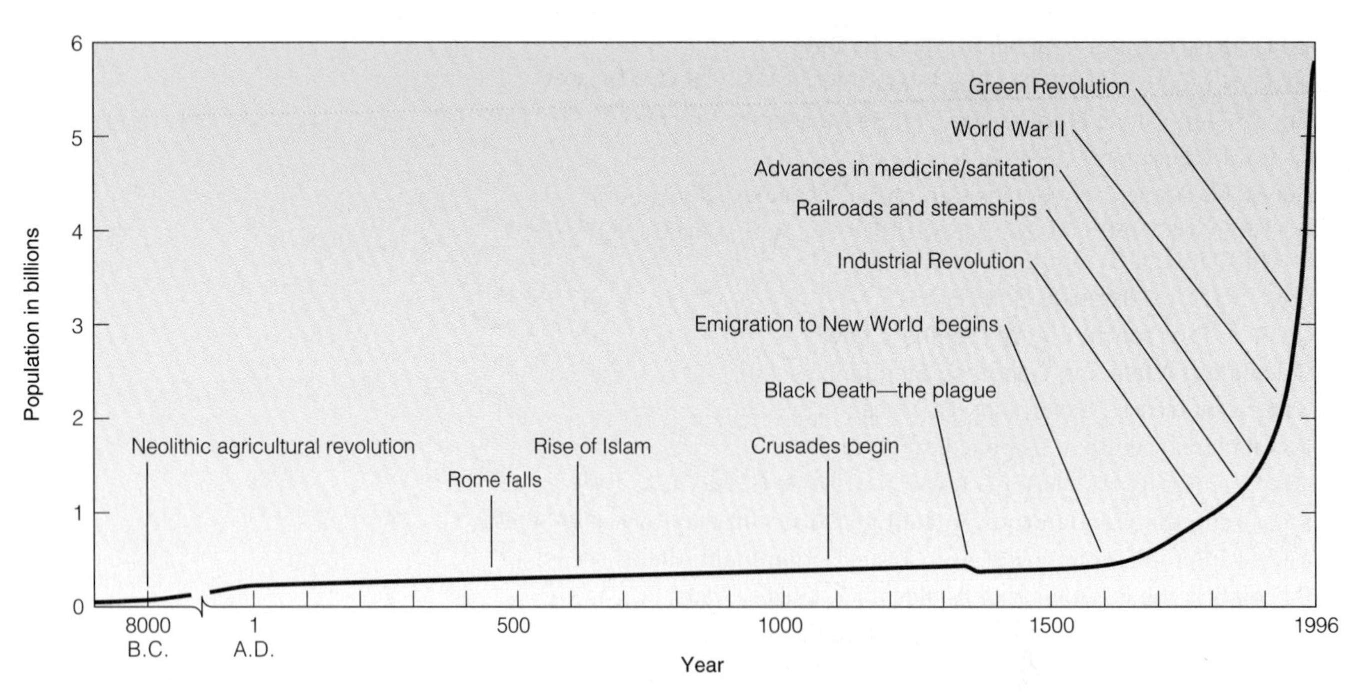

▶ FIGURE 17.1 Exponential growth of the human population. It took 1,800 years to reach 1 billion, but only another 130 years to reach 2 billion and another 45 years to reach 4 billion. In 2000, the population reached 6 billion.

In the nineteenth century, the writings of Malthus were used to argue that social reform and welfare were useless, because poverty was the result of natural law not social inequities. In the twentieth and twenty-first centuries, the debate over Malthus continues. Since the 1960s, population growth has again been regarded as a global threat that could override technological progress, leading to a lower standard of living and an increase in poverty and social problems. Movements calling for zero population growth and universal policies of birth control arose in the United States and other Western countries. More recently, China instituted a policy of one couple, one child. Many developing nations, however, regard these policies as political attempts to limit their growth and economic expansion and, in the extreme, as policies designed to result in extinction of smaller, poorer countries.

Malthus correctly foresaw the dangers of uncontrolled population growth, and he was one of the first to deal with the dynamics of populations. His work on the interaction between a population and its environment influenced both Wallace and Darwin, who incorporated Malthus's ideas into the theory of natural selection.

In the decades after Mendel's work became widely known, it became obvious that phenotypic variations seen in populations have a genetic basis and that the genetic structure of populations is part of the evolutionary process. Today the study of population genetics is closely tied to the study of evolution. Populations represent collections of alleles, and evolution is the result of changes in allele frequencies.

In this chapter we consider the population as a genetic unit and examine its organization, methods of measuring allele frequencies, and the ways in which the genetic structure of the population directly affects the incidence of human genetic disorders. We begin with a definition of populations, their subunits, and the concept of populations as reservoirs of genetic diversity. We then consider how allele frequencies can be measured in populations and how this information is used to answer practical questions about the frequency of disorders and heterozygotes in a population.

THE POPULATION AS A GENETIC RESERVOIR

Humans are distributed over most of the land surfaces of the Earth (Figure 17.2). As such, our species is subdivided into locally interbreeding units known as **populations,** or **demes.** Like individuals, populations are dynamic: They have life histories that include birth, growth, and response to the environment. Like individuals, populations can age and eventually die. Populations can be described by measures such as age structure (Figure 17.3), geography, birth and death rates, and allele frequencies.

Populations (demes) Local groups of organisms belonging to a single species, sharing a common gene pool.

Populations are more genetically diverse than individuals. For example, no single individual can have blood types A, B, and O. Only a group of individuals can carry all three blood types. The set of genetic information carried by a population is known as its **gene pool.** For a given gene, such as *I*, the gene for ABO blood type, the pool includes all the *A*, *B*, and *O* alleles in the population. Zygotes produced by one generation represent samples selected from the gene pool to form the next generation. The gene pool of a new generation is descended from the parental generation, but for several reasons, including chance, the gene pool of the new generation may have different allele frequencies than the parental pool. Over time, changes in allele frequency can cause changes in phenotype frequency. The long-term effect of changes in allele frequency is evolutionary change.

Gene pool The set of genetic information carried by the members of a sexually reproducing population.

C. Mayhew & R. Simmon (NASA/GSFC), NOAA/NGDC, DMSP Digital Archive

FIGURE 17.2 Human population density. As this satellite view of the world at night shows, humans are not randomly distributed across the land areas of the world but are clustered into discrete populations.

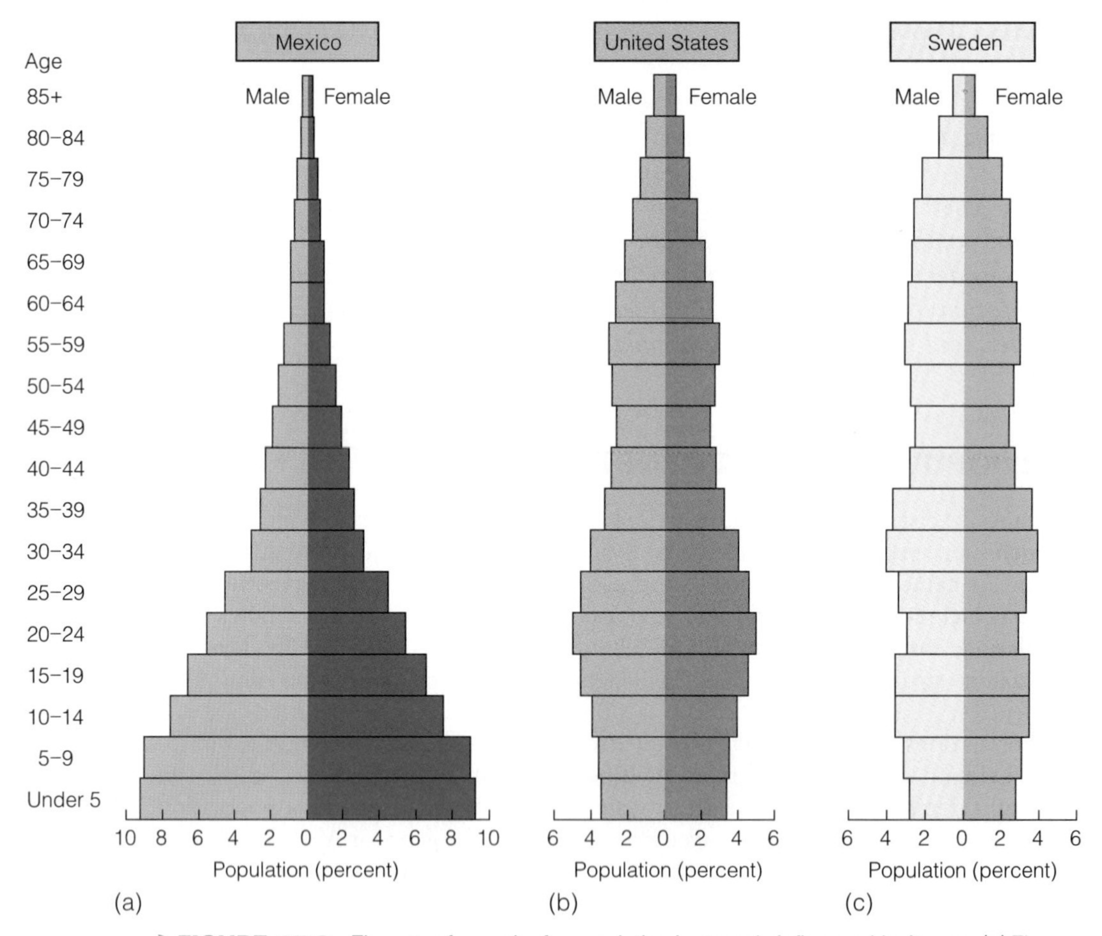

FIGURE 17.3 The rate of growth of a population is strongly influenced by its age. (a) The population of Mexico is pyramid-shaped and has more individuals who have yet to reproduce. This type of population has the potential for explosive growth. (b) The population structure of the United States is tapered at the bottom and has more individuals who have already reproduced. (c) The population of Sweden is of roughly constant size across all age groups and is a stable population.

HOW CAN WE MEASURE ALLELE FREQUENCIES IN POPULATIONS?

Populations can remain stable in size or can drastically expand or contract. In an extreme situation, a population can go extinct. Many factors influence population size, including birth rate, disease, migration, and adaptation to environmental factors such as climate. As these factors change, the genetic structure of a population can also change (see Concepts and Controversies: The Thrifty Genotype).

In studying the genetic structure of populations, geneticists measure **allele frequencies** and study the population over several generations to determine whether allele frequencies are stable.

Allele frequency The frequency with which alleles of a given gene are present in a population.

In our discussion, the term *allele frequency* means the frequency with which alleles of a given gene are present in a population. As several following examples will show, allele frequencies are not the same as genotype frequencies.

We cannot always directly determine allele frequencies, because we can see phenotypes but not genotypes. If, however, we examine the frequencies of codominant alleles, phenotypes are equivalent to genotypes, and we can determine allele frequencies directly. The MN blood group in humans is an example of a codominant system. In this case the gene *L* (located on chromosome 4) has two alleles, L^M and L^N, responsible for the M and N blood types, respectively. Each allele controls the synthesis of a gene product on the surface of red blood cells independently of the other allele. Thus individuals may be type M (L^M/L^M), type N (L^N/L^N), or type MN (L^M/L^N). The genotypes, blood types, and immunological reactions of the MN blood groups are shown in Table 17.1.

Codominant Allele Frequencies Can Be Measured Directly

The frequency of codominant alleles can be determined simply by counting how many copies of each allele are present in a population. For example, suppose that blood typing reveals that a population of 100 individuals contains 54 MM homozygotes, 26 MN heterozygotes, and 20 NN homozygotes. The 54 MM individuals carry 108 copies of the *M* allele (54 individuals, each of whom carries 2 *M* alleles). The 26 MN heterozygotes carry an additional 26 *M* alleles, so the population has a total of 134 *M* alleles (108 + 26 = 134). Each member of the population carries two copies of the *L* gene, for a total of 200 alleles (100 individuals, each of whom has 2 alleles = 200). The frequency of the *M* allele is 134/200, or 0.67 (67%). The frequency of the *N* allele can be calculated by counting 40 *N* alleles in the *NN* homozygotes (20 individuals, each of whom has 2 *N* alleles) and an additional 26 *N* alleles in the *MN* heterozygotes, a total of 66 copies of the *N* allele. The frequency of the *N* allele in the population is 66/200, or 0.33 (33%). This method of calculating gene frequencies in codominant populations is summarized in Table 17.2, and the frequencies of *M* and *N* alleles in several human populations are listed in Table 17.3.

Table 17.1 MN Blood Groups

Genotype	Blood Type	Antigens Present	Antibody Reactions
L^ML^M	M	M	Anti-M
L^ML^N	MN	M, N	Anti-M, Anti-N
L^NL^N	N	N	Anti-N

Concepts and Controversies

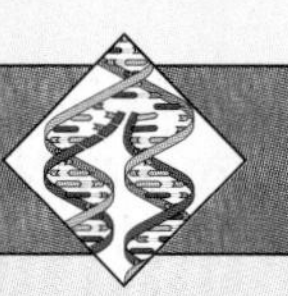

The Thrifty Genotype

The Pima Indians of the American Southwest have one of the highest rates of adult diabetes of any population in the United States. Between 42% and 66% of the Pima older than 35 years are diabetic. This disorder, accompanied by obesity, developed in the Pima only in this century and became recognized as a serious health problem only after 1950. Diabetes is usually followed by blindness, kidney failure, and heart disease.

How is it that an inherited condition with so many deleterious effects can be present in a population at such a high frequency? James Neel of the University of Michigan speculated on this question. He observed that the frequency of diabetes is very low in hunter-gatherer societies, such as the Pima were before the twentieth century. He also noted that females prone to diabetes become sexually mature at an earlier age. In addition, susceptible females give birth to larger-than-average babies (increased birth weight is linked to increased survival). He postulated that these characteristics might provide a reproductive advantage to diabetics and account for the high frequency of this disorder.

In addition, Neel proposed that in the feast-or-famine diet of hunter-gatherers, the diabetic represents a "thrifty" genotype that is more efficient in converting food into energy. In diabetics, more insulin is released after each meal, and more glucose is metabolized. According to Neel, as the age of the individual increases, these repeated cascades of insulin release produce a counterreaction that releases an antagonist to stop insulin action. In the hunter-gatherer culture, this overloading occurs infrequently, and adult-onset diabetes does not fully develop.

We now know that there is no antagonist as envisioned by Neel, but as he reviewed the concept in 1982, he noted that "although incorrect in detail, it may have been correct in principle." It appears that those with adult-onset diabetes release more insulin after a meal than those without diabetes. This quick response prevents loss of blood glucose through the kidneys and is an effective adaptation in hunter-gatherer societies. When the Pima diet converted to one high in refined carbohydrates, the release of extra insulin occurred over and over again, eventually causing a reduction in the number of cell-surface insulin receptors. As a result, although insulin is present, it is ineffective in metabolizing glucose, and the symptoms of diabetes appear.

The adult-onset diabetic genotype may represent one that was well adapted to the environment of the hunter-gatherer societies that prevailed for hundreds of thousands of years. In the case of the Pima Indians, environmental conditions changed dramatically in less than a century, and the genotype is now at a distinct disadvantage in an environment where carbohydrate-rich foods are freely available. This idea is interesting and if confirmed would be an example of natural selection in action.

Table 17.2 Determining Allele Frequencies for Codominant Genes by Counting Alleles

Genotype	MM	MN	NN	Total
Number of individuals	54	26	20	100
Number of L^M alleles	108	26	0	134
Number of L^N alleles	0	26	40	66
Total	108	52	40	200

Frequency of L^M in population: 134/200 = 0.67 = 67%

Frequency of L^N in population: 66/200 = 0.33 = 33%

Recessive Allele Frequencies Cannot Be Measured Directly

In genes with codominant alleles, there is a direct relationship between phenotype and genotype. However, if one allele is recessive, there is no direct relationship between phenotype and genotype because heterozygotes and homozygotes for the dominant allele have identical phenotypes. In this situation, we cannot measure allele frequencies by counting, because we don't know how many people in the population have a homozygous dominant genotype and how many are heterozygotes. However, a math-

Table 17.3 Frequencies of L^M and L^N Alleles in Various Populations

Population	Genotype Frequency (%)			Allele Frequency	
	MM	MN	NN	L^M	L^N
U.S. Indians	60.00	35.12	4.88	0.776	0.224
U.S. Blacks	28.42	49.64	21.94	0.532	0.468
U.S. Whites	29.16	49.38	21.26	0.540	0.460
Eskimos (Greenland)	83.48	15.64	0.88	0.913	0.087

ematical formula can be used to determine allele frequencies when one or more alleles are recessive and a number of conditions (described later) are met. This method, developed independently by Godfrey Hardy and Wilhelm Weinberg, is known as the **Hardy-Weinberg Law.**

Hardy-Weinberg Law The statement that allele and genotype frequencies remain constant from generation to generation when the population meets certain assumptions.

THE HARDY-WEINBERG LAW: MEASURING ALLELE AND GENOTYPE FREQUENCIES

After Mendel's work became widely known, there was intense debate about whether the principles of Mendelian inheritance applies to humans. One of the first genetic traits identified in humans was a dominant allele, brachydactyly (MIM/OMIM 112500). Because the phenotypic ratio of dominant traits is 3:1 in children of heterozygotes (1 *AA*, 2 *Aa*, 1 *aa*), it was thought that over time, the phenotypic ratio for this dominant allele would become 3:1 in the human population. However, as critics of Mendel argued, only a small fraction of the population has brachydactyly, and therefore although the principles of Mendelian inheritance are valid for plants and other animals, they do not explain inheritance in humans.

An English mathematician, Godfrey Hardy, and a German physician, Wilhelm Weinberg, independently recognized that this reasoning was false. The argument about brachydactyly failed to recognize that there is a difference between *how a trait is inherited* (in this case a dominant trait with a 3:1 ratio) and the *frequency* of the dominant and recessive alleles in the population. Hardy and Weinberg each developed a simple mathematical model to estimate allele frequency in a population and to describe how alleles combine to form genotypes.

Assumptions for the Hardy-Weinberg Law

The model developed by Hardy and Weinberg is based on a number of assumptions:

- The population is large. In practical terms, this means that the population is large enough that there are no errors in measuring allele frequencies.
- No genotype is better than any other; that is, all genotypes have equal ability to survive and reproduce.
- Mating in the population is random.
- Other factors that change allele frequency, such as mutation and migration, are absent or rare events and can be ignored.

Calculating Allele and Genotype Frequencies

Let's see how the model works in a population carrying an autosomal gene with two alleles, *A* and *a*. In this example, the frequency of the dominant allele *A* in gametes

SIDEBAR

Selective Breeding Gone Bad

Purebred dogs are the result of selective breeding over many generations, and worldwide, there are now more than 300 recognized breeds and varieties of dogs. Selective breeding to produce dogs with desired traits, such as long noses and closely set eyes in collies, or the low, sloping hindlegs of German Shepherds, can have unintended side effects. About 70% of all collies suffer from hereditary eye problems, and more than 60% of German Shepherds are at risk for hip dysplasia. It is estimated that 25% of all purebred dogs have a serious genetic disorder.

The high level of genetic disorders in purebred dogs is a direct result of selective breeding. Over time, selective breeding has decreased genetic variability and increased animals homozygous for recessive genetic disorders.

Outbreeding is a simple genetic solution to this problem. In the world of dogs, this means mixed-breed dogs, or mutts. For example, crosses between collies and German Shepherds or between Labrador Retrievers and German Shepherds combine the looks and temperaments of both breeds but reduce the risk of offspring that have genetic disorders.

		Sperm	
		$A(p)$	$a(q)$
Eggs	$A(p)$	AA (p^2)	Aa (pq)
	$a(q)$	Aa (pq)	aa (q^2)

FIGURE 17.4 The frequency of the dominant and recessive alleles in the gametes of the parental generation determines the frequency of the alleles and the genotypes of the next generation.

is represented by p, and the frequency of the recessive allele a in gametes is represented by q. Because the sum of p and q represents 100% of the alleles for that gene in the population, $p + q = 1$. A Punnett square can be used to predict the genotypes produced by the random combination of these gametes (Figure 17.4).

In combining gametes, the chance that both the egg and sperm will carry the A allele is $p \times p = p^2$. The chance that the gametes will carry different alleles is $(p \times q) + (p \times q) = 2pq$. The chance that both gametes carry recessive alleles is $q \times q = q^2$. Although p^2 represents the chance that both gametes will carry an A allele, p^2 also represents the frequency of the homozygous AA genotype in the new generation. In the same way, $2pq$ is a measure of heterozygote (Aa) frequency, and q^2 represents the frequency of homozygous recessive (aa) individuals. In other words, the distribution of genotypes in the next generation can be expressed as

$$p^2 + 2pq + q^2 = 1$$

where 1 represents 100% of the genotypes in the new generation. This equation formulates the Hardy-Weinberg Law, which states that both allele and genotype frequencies will remain constant from generation to generation in a large, interbreeding population in which mating is random and there is no selection, migration, or mutation.

The formula can be used to calculate allele frequencies (A and a in our example) and the frequency of the various genotypes in a population. To show how the model works, let's begin with a population for which we already know the frequency of the alleles. Suppose we have a large, randomly mating population in which the frequency of an autosomal dominant allele A is 60% and the frequency of the recessive allele a is 40%. This means that $p = 0.6$ and $q = 0.4$. Because A and a are the only two alleles, the sum of $p + q$ equals 100% of the alleles:

$$p\ (0.6) + q\ (0.4) = 1$$

In this population, 60% of the gametes carry the dominant allele A, and 40% carry the recessive allele a. The distribution of genotypes in the next generation is shown in Figure 17.5.

In the new generation, 36% ($p^2 = 0.6 \times 0.6$) of the offspring will have the genotype AA, 48% ($2\,pq = 2[0.6 \times 0.4]$) will be Aa, and 16% ($q^2 = 0.4 \times 0.4$) will have the genotype aa.

We can also use the Hardy-Weinberg equation to calculate the frequency of the A and a alleles in the new generation. The frequency of A is

$$p^2 + 1/2\ (2pq)$$

$$0.36 + 1/2\ (0.48)$$

$$0.36 + 0.24 = 0.60 = 60\%$$

For the recessive allele a, the frequency is

$$q^2 + 1/2\ (2pq)$$

$$0.16 + 1/2\ (0.48)$$

$$0.16 + 0.24 = 0.40 = 40\%$$

Because $p + q = 1$, we could have calculated the value for the recessive allele a by subtraction:

$$p + q = 1$$

$$q = 1 - p$$

$$q = 1 - 0.60$$

$$q = 0.40 = 40\%$$

		Sperm	
		$A(p = 0.6)$	$a(q = 0.4)$
Eggs	A ($p = 0.6$)	AA ($p^2 = 0.36$)	Aa ($pq = 0.24$)
	a ($q = 0.4$)	Aa ($pq = 0.24$)	aa ($q^2 = 0.16$)

FIGURE 17.5 The frequency of alleles and genotypes in a new generation in which the alleles in the parental generation are present at a frequency of 0.6 for the dominant allele (A) and 0.4 for the recessive allele (a).

Web-Integrated Toolbox

Population Genetics

Studying allele frequencies and genotype frequencies in members of a population reveals the genetic structure of populations. Population genetics has many applications in human genetics, and is used to estimate how many individuals in a population may be heterozygous carriers of a recessive genetic disorder and the differences in allele frequencies found in various ethnic groups, and in isolated populations. Population genetics is also used to determine the geographic distribution of alleles for genetic disorders, and to estimate the age of mutant alleles.

Population genetics uses a simple mathematical relationship called the Hardy-Weinberg Law to measure allele frequencies and genotype frequencies in populations. Use of the Hardy-Weinberg equation also provides us with an insight into the forces of evolution, such as natural selection, that change allele frequencies from generation to generation, causing populations to evolve over time. This provides a picture of a population's evolutionary history, as well as clues to what forces have shaped the gene pool of the population. Overall, population genetics can help tell us who we are and where we came from.

In addition to providing clues about our species' evolutionary history, the Hardy-Weinberg equation can be used to predict how well different populations will respond to infectious diseases, such as HIV/AIDS, typhus, and malaria.

The book's home page contains exercises and links that will help you understand how the Hardy-Weinberg equation works, and how it is used to determine both allele frequencies and genotype frequencies in a population. Learning how to use the Hardy-Weinberg equation will give you the skills to look into a population's history, assess its current situation, and make some predictions about its future.

Populations Can Be in Genetic Equilibrium

In the previous example, the frequencies of *A* and *a* in the new generation are the same as in the parent's generation. If allele frequencies for a gene remain constant from generation to generation, the population is in **genetic equilibrium** for that allele. This doesn't mean the population is in equilibrium for all alleles. On the contrary, if mutation, selection, or migration are operating on other genotypes, the frequency of other alleles may change from one generation to the next.

Genetic equilibrium The situation in which the allele frequency for a given gene remains constant from generation to generation.

The existence of a genetic equilibrium in a population illustrates why dominant alleles do not increase in frequency as new generations are produced. If the allele for brachydactyly is in equilibrium, it will not increase in the population and reach a 3:1 frequency but instead will be maintained at constant frequency from generation to generation.

In addition, genetic equilibrium also helps maintain genetic variability in the population. In the previous example, at equilibrium we can be assured that 60% of the alleles for gene *A* will be dominant (*A*) and 40% will be recessive (*a*) in generation after generation. The presence and maintenance of genetic variability is important to the process of evolution.

USING THE HARDY-WEINBERG LAW IN HUMAN GENETICS

The Hardy-Weinberg Law is one of the foundations of population genetics and has many applications in human genetics and human evolution. We consider only a few of its uses, primarily those that apply measuring allele and genotype frequencies.

Measuring the Frequency of Autosomal Dominant and Recessive Alleles

In most autosomal traits the homozygous dominant and heterozygous genotypes have the same phenotype. If the trait is inherited recessively, we can calculate the frequency of the recessive allele in the population. We begin by counting the number of homozygous recessive individuals in the population. For example, cystic fibrosis is an autosomal recessive trait, and homozygous recessive individuals can be identified by their distinctive phenotype. Suppose that 1 in 2,500 individuals in a population is affected with cystic fibrosis. These individuals have the genotype *aa*. According to the Hardy-Weinberg equation, the frequency of this genotype in the population is equal to q^2. The frequency of the *a* allele in this population is therefore equal to the square root of q^2:

$$q^2 = 1/2500 = 0.0004$$

$$q = \sqrt{0.0004}$$

$$q = 0.02 = 1/50$$

Once we know that the frequency of the *a* allele is 0.02 (2%), we can calculate the frequency of the dominant allele *A* by subtraction:

$$p + q = 1$$

$$p = 1 - q$$

$$p = 1 - 0.02$$

$$p = 0.98 = 98\%$$

In this population, 98% of the alleles for gene *A* are dominant (*A*), and 2% are recessive (*a*). This method can be used to calculate the allele and genotype frequencies for any dominant or recessive trait.

Measuring the Frequency of Autosomal Codominant Alleles

As outlined earlier, allele and genotype frequencies for autosomal codominant traits can be determined directly from the phenotype because each genotype gives rise to a distinctive phenotype, as in the MN blood groups. Allele and genotype frequencies can be determined for this pattern of inheritance by counting individuals in the population. Further, there is no need to make assumptions about Hardy-Weinberg conditions because the genotype and phenotype frequencies are identical.

Calculating the Frequency of Alleles for X-Linked Traits

One of our underlying assumptions in estimating the allele frequency for the autosomal recessive trait that controls cystic fibrosis was that the frequency of *A* and *a* is the same in sperm and eggs. That is, 98% of the sperm and 98% of the eggs in this population should carry the dominant allele *A*, and 2% of the sperm and 2% of the eggs should carry the recessive allele *a*. This situation does not hold true for X-linked traits. Human females carry two X chromosomes and have two copies of all X-linked genes. Males have only one X chromosome and are hemizygous for all genes on the X chromosome. Thus, X-linked genes are not distributed equally in the population: Females (and their gametes) carry two-thirds of the alleles, and males (and their gametes) carry one-third of the alleles. As we will see in the following, the Hardy-Weinberg equation can be used to calculate genotype frequencies in females for recessive X-linked traits. However, because males are hemizygous for all traits on the X chromosome, the allele frequency for recessive X-linked traits equals the number of males with the mutant phenotype. For example, in the United States, about 8% of males are colorblind. Therefore, the frequency of the color blindness allele in this population is 0.08 ($q = 0.08$).

Because females carry two X chromosomes, genotypic frequencies for X-linked recessive traits in females can be calculated using the Hardy-Weinberg equation. If male color blindness has a frequency of 8% ($q = 0.08$), we expect color blindness in females to have a frequency of q^2, or 0.0064 (0.64%). With an allele frequency of 0.08, in a population of 10,000 males, 800 would be colorblind, but in a population of 10,000 females, only 64 would be colorblind. This example reemphasizes the fact that males are at much higher risk for deleterious traits carried on the X chromosome. The relative values for the frequency of X-linked recessive traits in males and females are listed in Table 17.4.

Table 17.4

Frequency of X-Linked Recessive Traits in Males and Females

Males	Females
1/10	1/100
1/100	1/10,000
1/1,000	1/1,000,000
1/10,000	1/100,000,000

The Frequency of Multiple Alleles Can Be Calculated

Until now we have discussed allele frequencies in genes that have only two alleles. For other genes more than two alleles can be present in the population. In ABO blood types, three alleles of the isoagglutinin locus (*I*) are present in the population. The alleles *A* and *B* are codominant, and both are dominant to O. This system has six possible genotypic combinations:

$$AA,\ AO,\ BB,\ BO,\ AB,\ OO$$

Homozygous *AA* and heterozygous *AO* individuals have the same phenotype (type A blood), as do *BB* and *BO* individuals (type B blood). This dominance relationship among the alleles results in four phenotypic combinations, known as blood types A, B, AB, and O.

The Hardy-Weinberg Law can be used to calculate both the allele and genotype frequencies for this three-allele system by adding another term to the equation. For the three blood group alleles

$$p(A) + q(B) + r(O) = 1$$

In other words, when you add together the frequencies of the *A*, *B*, and *O* alleles, you have accounted for 100% of the alleles for this gene in the population. The genotypic frequencies are given by the equation

$$(p + q + r)^2 = 1$$

Allele frequencies for *A*, *B*, and *O* can be calculated from the phenotypic frequencies in a population if random mating is assumed.

Once we know the frequency of the *A*, *B*, and *O* alleles for a given population, we can then calculate the genotypic and phenotypic frequencies for all combinations of these alleles. The genotypic combinations can be calculated by an expansion of the Hardy-Weinberg equation:

$$p^2\ (AA) + 2pq\ (AB) + 2pr\ (AO) + q^2\ (BB) + 2qr\ (BO) + r^2\ (OO) = 1$$

That is, the frequency of the *AA* genotype is predicted to be p^2, the *AB* genotype would be $2pq$, and so forth. The frequencies for the *A*, *B*, and *O* alleles in different populations in the world are listed in Table 17.5. The geographic distribution of ABO alleles is shown in ◗ Figure 17.6. By using the equations shown previously and the values in Table 17.5, we can calculate the genotypic and phenotypic frequencies for the populations shown in Figure 17.6 or for any population in which we know the allele frequencies.

Estimating the Frequency of Heterozygotes in a Population

Most human genetic disorders are inherited as recessive traits. In most cases, a recessive disorder appears in a family by the mating of two heterozygotes. For many reasons, it is important to know the population frequency of heterozygotes carrying a deleterious recessive allele. Calculating the frequency of heterozygotes is an important application of the Hardy-Weinberg Law, because most disease-causing alleles are carried by heterozygotes. To calculate the frequency of heterozygous carriers for recessive traits, we begin by counting the number of homozygous recessive individuals (all of whom show the recessive phenotype) in the population. For example, cystic fibrosis, an autosomal recessive disorder has a frequency of 1 in 2,500 among white Americans. (The disease is much rarer among American blacks and Asians.)

Table 17.5 Frequency of ABO Alleles in Various Populations

Population	Frequency		
	A(p)	B(q)	O(r)
Armenians	36.0	10.4	53.6
Basques	25.5	—	74.5
Eskimos	35.5	4.6	59.9
Belgium	27.0	5.9	67.1
Denmark	29.4	7.7	62.9
Greece	22.9	8.2	68.9
Poland	25.9	14.0	60.1
Russia (Urals)	29.5	19.5	51.0
Russia (Siberia)	13.0	25.1	61.9
Russia (Tadzhikistan)	21.1	37.1	41.7
Sri Lanka (Sinhalese)	14.0	15.2	70.8
India (Assum)	19.2	11.1	69.7
India (Madras)	16.5	20.5	63.0
China (Hong Kong)	19.1	19.1	61.8
Japan	26.2	18.3	55.5
Nigeria (Ibo)	13.2	9.5	77.3
Nigeria (Yoruba)	13.8	14.6	71.6
Upper Volta	14.8	18.2	67.0
Kenya	17.2	14.0	68.8

Homozygous recessive individuals can be identified by a set of clinical symptoms (to review the symptoms, see Chapter 4). The frequency of the homozygous recessive genotype is 1 in 2,500, or 0.0004, and is represented by q^2. As before, the frequency of the recessive allele q in the population can be calculated as

$$q = \sqrt{q^2}$$
$$q = \sqrt{0.0004}$$
$$q = 0.02 = 2\%$$

Because $p + q = 1$, we can calculate the frequency of the dominant allele p:

$$p = 1 - q$$
$$p = 1 - 0.02$$
$$p = 0.98 = 98\%$$

Knowing the allele frequencies, we can use the Hardy-Weinberg equation to calculate genotype frequencies. Recall that in the Hardy-Weinberg equation, $2pq$ is the frequency for the heterozygous genotype. Using the values we have calculated for p and q, we can determine the frequency of heterozygotes as follows:

$$2pq = 2(0.98 \times 0.02)$$
$$2pq = 2(0.0196)$$
$$\text{Heterozygote frequency} = 0.039 = 3.9\%$$

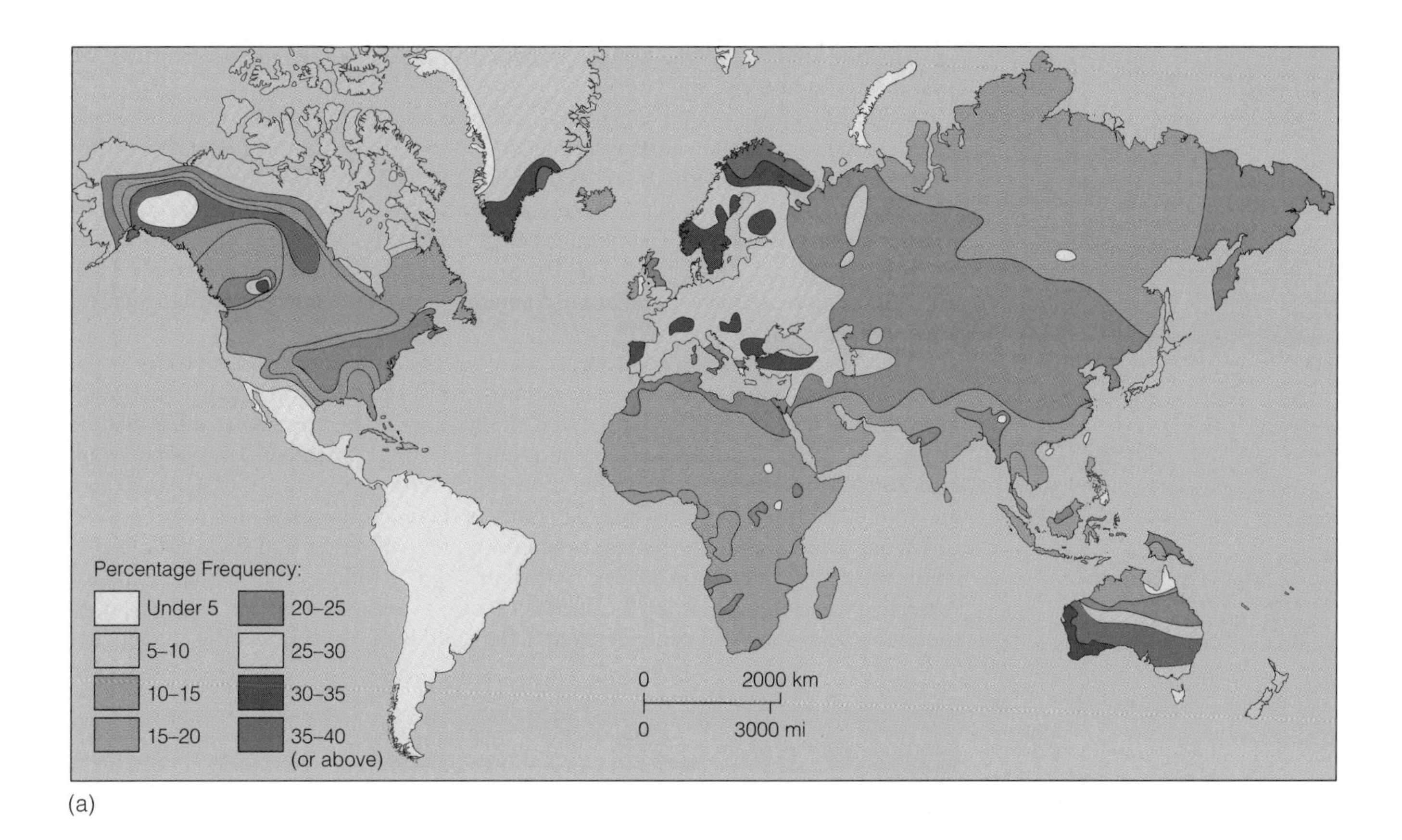

(a)

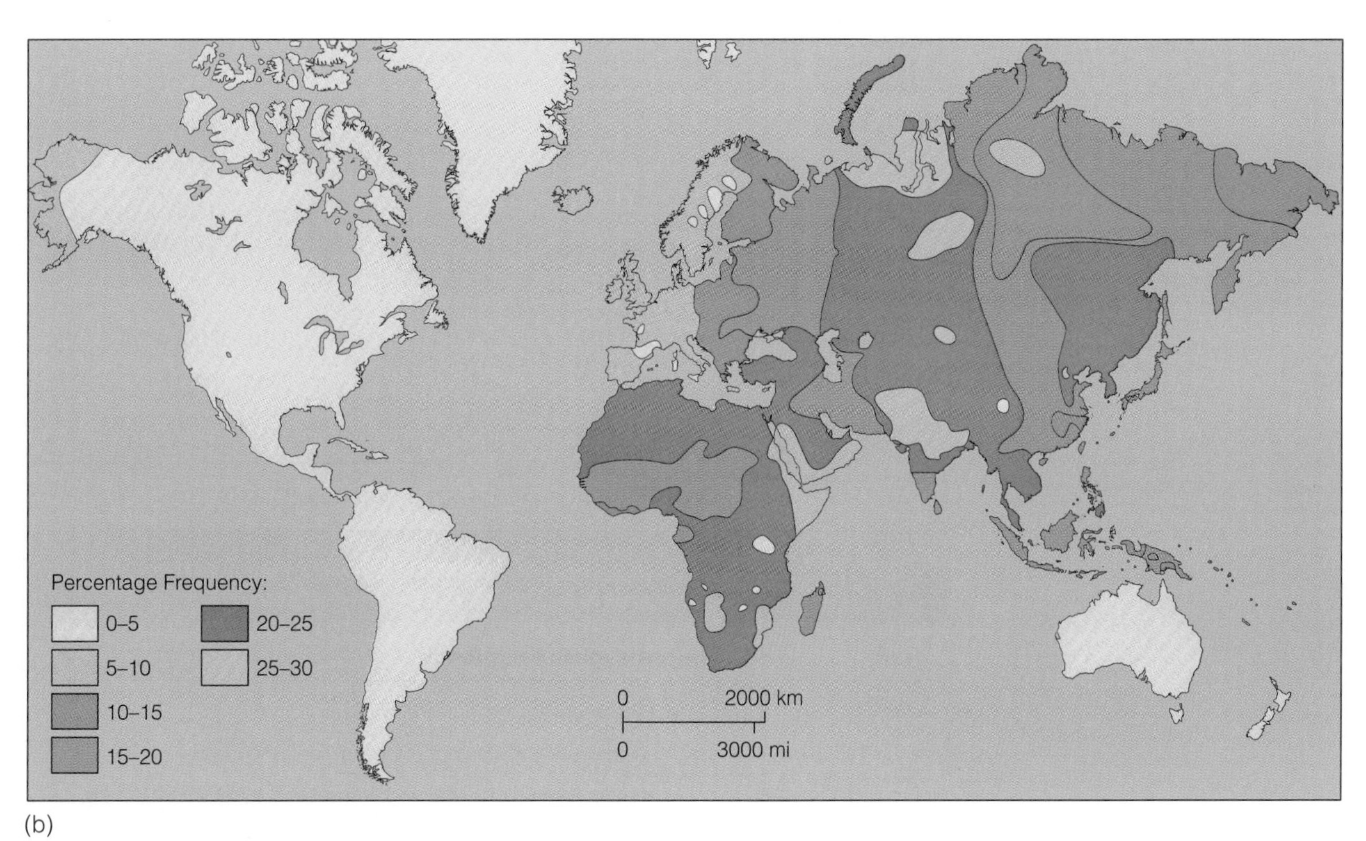

(b)

FIGURE 17.6 The distribution of alleles in the ABO system. (a) The distribution of the *A* allele in the indigenous population of the world (before 1600 A.D.). (b) Distribution of the *B* allele.

This means that 3.9%, or approximately 1 in 25, white Americans carry the gene for cystic fibrosis.

Sickle cell anemia is an autosomal recessive disorder affecting 1 in 500 black Americans. From the phenotype frequency we can calculate the frequency of homozygous recessive individuals (q^2) as 0.002. Using the Hardy-Weinberg equation, we can calculate that 8.5% of black Americans, or 1 in every 12, are heterozygous carriers for sickle cell anemia. The frequencies of heterozygous carriers of recessive alleles with a frequency range from 1 in 10 to 1 in 10,000,000 are listed in Table 17.6. The heterozygote frequencies for some common human autosomal recessive traits are listed in Table 17.7.

Many people are surprised to learn that heterozygotes for recessive traits are so common in the population. If a genetic disorder is relatively rare (say 1 in 10,000 individuals), they generally assume that the number of heterozygotes must also be rather low. In fact, if a disorder is found in 1 in 10,000 members of a population, 1 in 50 (2%) members of the population is a heterozygote (Table 17.6), and there are about 200 times as many heterozygotes as there are homozygotes.

What are the chances that two heterozygotes will mate and have an affected child? We can calculate the answer to this question as follows: The chance that two heterozygotes will mate is $1/50 \times 1/50 = 1/2,500$. Because they are heterozygotes, the chance that they will produce an affected child is one in four. So the chance that

Table 17.6 Heterozygote Frequencies for Recessive Traits

Frequency of Homozygous Recessives (q^2)	Frequency of Heterozygous Individuals ($2pq$)
1/100	1/5.5
1/500	1/12
1/1,000	1/16
1/2,500	1/25
1/5,000	1/36
1/10,000	1/50
1/20,000	1/71
1/100,000	1/158
1/1,000,000	1/500
1/10,000,000	1/1,582

Table 17.7 Frequency of Heterozygotes for Some Recessive Traits (U.S. Population)

Trait	Heterozygote Frequency
Cystic fibrosis	1/22 whites; much lower in blacks, Asians
Sickle cell anemia	1/12 blacks; much lower in most whites and in Asians
Tay-Sachs disease	1/30 among descendants of Eastern European Jews; 1/350 among others of European descent
Phenylketonuria	1/55 among whites; much lower in blacks and those of Asian descent
Albinism	1/10,000 in Northern Ireland; 1/67,800 in British Columbia

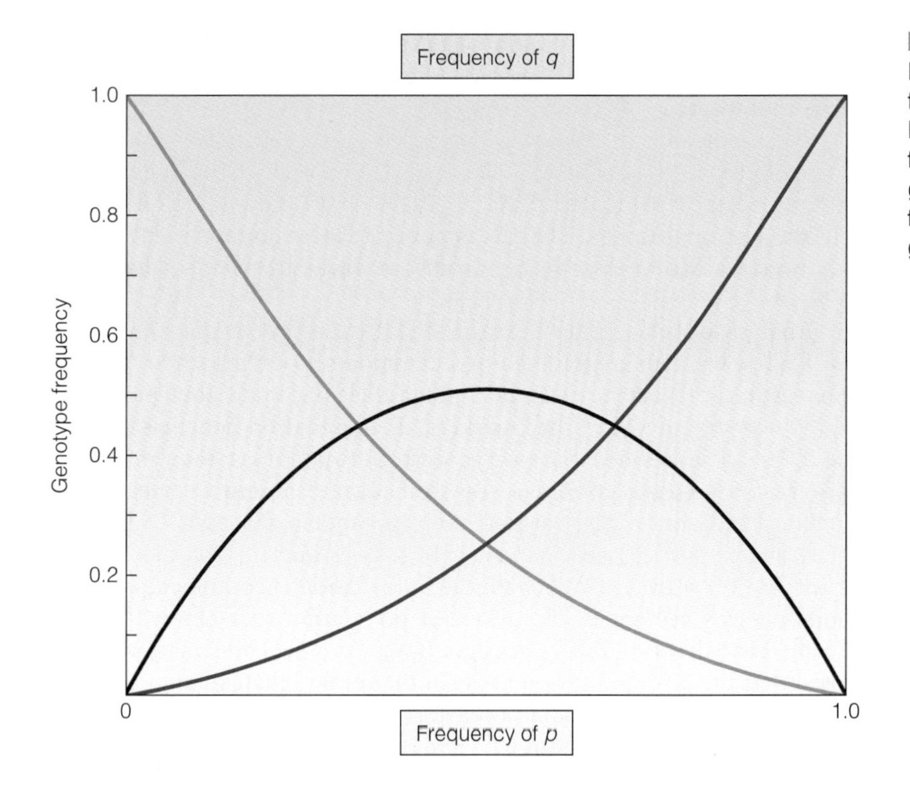

FIGURE 17.7 The relationship between allelic and genotypic frequency in a population that is in Hardy–Weinberg equilibrium. As the frequencies of the homozygous genotypes (p^2 and q^2) decline, the frequency of the heterozygote genotype ($2pq$) rises.

they will mate and produce an affected child is 1/2,500 × 1/4 = 1/10,000. In other words, if the disorder is present in 1 of every 10,000 individuals, 1 in 50 individuals must be a heterozygous carrier of the recessive allele.

Once the frequency of either allele is known, we can calculate the frequency of the homozygous dominant genotypes and the heterozygotes. Remember that the frequency of the genotypes depends on the allele frequency. The relationship between allele and genotype frequency is shown in Figure 17.7. As the frequencies of p and q move away from zero, the percentage of heterozygotes in the population increases rapidly. This again illustrates the point that in disorders such as cystic fibrosis or sickle cell anemia, the majority of the recessive alleles in a population are carried by heterozygotes, in whom the deleterious effects of the allele are not expressed.

POPULATION GENETICS CAN HELP DETERMINE THE AGE AND ORIGIN OF HUMAN GENETIC DISEASES

Measurement of allele frequencies in populations often reveals geographic patterns of distribution. These patterns, in combination with other data, including frequency measurements for closely linked alleles and information from other fields such as demographics, history, and archeology, can be used to reconstruct the history of a genetic disease. Let's look at the distribution of two mutant alleles that cause cystic fibrosis and what they can tell us about the disease and about human history.

Cystic fibrosis is an autosomal recessive disorder frequently found in populations of European origin. The *CFTR* gene, on chromosome 7, encodes a protein that becomes embedded in the plasma membrane of the cell and regulates the flow of chloride ions into and out of the cell. Affected individuals have chronic pulmonary problems, digestive and nutritional abnormalities, and other symptoms (see Chapter 4 for a review of the phenotype). Several hundred different mutations have been identified in the *CFTR* gene (review *CFTR* mutations in

Chapter 11). The most common *CFTR* mutation is a deletion of phenylalanine at position 508, called *Δ508*. The second most common mutation, *G542X,* is a nonsense mutation.

The frequency of the *Δ508* mutation has been determined in populations across Europe (Table 17.8). The frequency of *Δ508* is highest in northwest Europe, where it represents 90% to 100% of all mutant *CFTR* alleles (Denmark and the Faeroe Islands). The frequency gradually diminishes along a gradient to its lowest values in southeast Europe (in Turkey), where it has a frequency of 27% (◗ Figure 17.8).

Analysis of other genes near the CFTR locus on chromosome 7 show that the *Δ508* allele has remained linked to a common set of markers in northwest Europe, whereas in southeast regions, the mutant allele has undergone more recombination and is now found with other markers. To population geneticists, this means that the *Δ508* allele has been in southeastern Europe longer than in the northern and western parts of the continent. By studying the mutation rate of nearby microsatellite DNA sequences (variable-number tandem repeats [VNTRs], also used in DNA fingerprinting [see Chapter 13]), it is estimated that the *Δ508* mutation is about 50,000 years old. When all the information is combined, it appears that this mutation was present in the ancestral population that entered Europe from the Middle East about 40,000 years ago. Based on other information, it has been proposed that the *Δ508* mutation arose in the Iranian Plateau about 50,000 years ago and spread from there by migration to Europe.

About 2.5% of all mutant *CFTR* alleles are a nonsense mutation called *G542X.* This allele is the second most common *CFTR* mutation. Distribution of this allele shows geographic variation, but it has a very different pattern than the *Δ508* allele. The highest frequencies of the *G542X* allele are found in areas around the Mediterranean Sea, including Spain, Italy, Turkey, Tunisia, Sicily, and in the Canary Islands. These areas correspond to sites settled by the Phoenicians, who originated in the Middle East and spread from Carthage (Tunisia) to Western Europe and the Canary Islands approximately 2,500 to 3,000 years ago, and the *G542X* mutation may have been carried to these sites by this ancient seafaring nation.

Table 17.8 Geographic Distribution of the *Δ508* Allele in Europe

Country	Incidence of CF	Town/Region	Number of CF Chromosomes Studied	Frequency of *Δ508*
Denmark	1/4700	Copenhagen	586	0.87
		Faroe Islands	24	1.0
Great Britain	1/2400	Aberdeen	116	0.80
Germany	1/3300	Munich	448	0.70
Czechoslovakia	1/2600	Prague (Czech)	456	0.68
		Bratislava (Slovak)	76	0.58
Poland	1/6000	Warsaw	168	0.47
Italy		Naples	154	0.50
Romania	—	—	36	0.30
Turkey	—	Istanbul	30	0.27

From: *Lucotte, G., Hazout, S., and De Braekeleer, M. (1995). Complete map of cystic fibrosis mutation DF508 frequencies in Western Europe and correlation between mutation frequencies and incidence of disease.* Human Biology, *67, 797–803, with permission.*

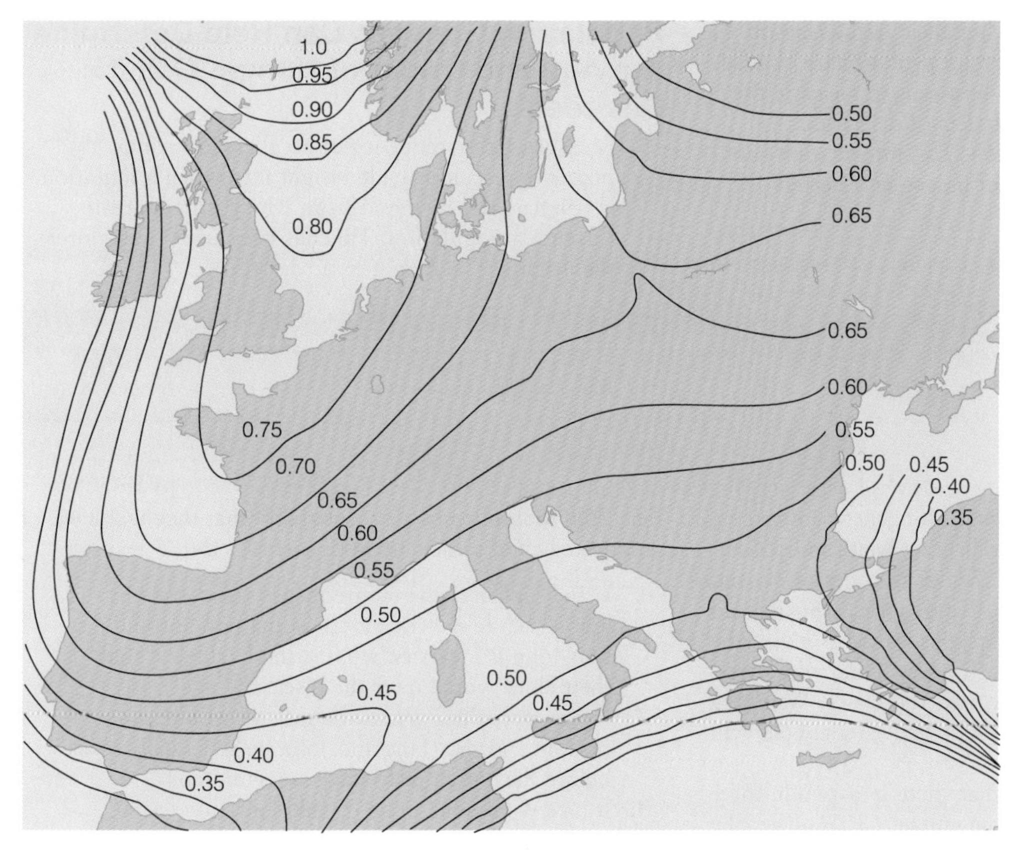

FIGURE 17.8 A map of Europe showing the distribution of the *Δ508* allele. There is a frequency gradient from northwest to southeast. The highest values are in northwest Europe (100% in the Faeroe Islands) and are lowest in Turkey (27%). Combined with other information from demographics and genetics, these results indicate that the mutant allele entered Europe from the southeast and spread to the northwest and that it is older than the original populations who entered Europe about 40,000 years ago. (From Lucotte, G., Hazout, S., and De Braekeleer, M. (1995). Complete map of cystic fibrosis mutation DF508 frequencies in Western Europe and correlation between mutation frequencies and incidence of disease. *Human Biology, 67,* 797–803, Figure 1, p. 800. Reprinted with permission.)

Summary

The Population as a Genetic Reservoir

1. In the early decades of the twentieth century, genes were recognized as the agents that cause phenotypic variations, giving rise to the field of population genetics. After the mathematical and theoretical basis of this field was established, experimentalists began to study allele frequencies in populations rather than in the offspring of a single mating. This work has produced the basis for our understanding of evolution.

How Can We Measure Allele Frequencies in Populations?

2. In some cases, including codominant alleles, allele frequency can be measured directly by counting phenotypes, because in these cases phenotypes are equivalent to genotypes. In other cases, the Hardy-Weinberg Law provides a means of measuring allele frequencies within populations.

The Hardy-Weinberg Law: Measuring Allele and Genotype Frequencies

3. The Hardy-Weinberg equation assumes that the population is large and randomly interbreeding and that factors such as mutation, migration, and selection are absent. The presence of equilibrium in a population explains why dominant alleles do not replace their recessive alleles. In equilibrium populations, the Hardy-Weinberg equation can be used to measure allele and genotype frequencies from generation to generation.

Using the Hardy-Weinberg Law in Human Genetics

4. The Hardy-Weinberg Law can also be used to estimate the frequency of autosomal and X-linked alleles in a population. It can also be used to detect when allele frequencies are shifting in the population. The

conditions that lead to changing allele frequencies in a population are those that produce evolutionary change. One of the law's most common uses is to measure the frequency of heterozygous carriers of deleterious recessive alleles in a population. This information can be used to calculate the risk of having an affected child.

Population Genetics Can Help Determine the Age and Origin of Human Genetic Disorders

5. Measuring the frequency of an allele in various populations can provide insight into when a mutation originated and how and from where it has spread through a population. This has been used to explore the origins of human genetic disorders.

CASE 1

Jane, a healthy woman, was referred for genetic counseling because she had two siblings, a brother Matt and a sister Edna, with cystic fibrosis who died at the ages of 32 and 16, respectively. Jane has a husband, John, who has no family history of cystic fibrosis. Jane wants to know the probability that she and John will have a child with cystic fibrosis. The genetic counselor used the Hardy-Weinberg model to calculate the probability that this couple will have an affected child.

The counselor explained that there is a two-in-three chance that she is a carrier for the mutant *CFTR* allele. She used a Punnett square to illustrate this. The probability that John is a carrier is equal to the population carrier frequency ($2pq$). The probability that John and Jane will have a child who has cystic fibrosis equals the probability that Jane is a carrier (2/3) multiplied by the probability that John is a carrier ($2pq$) multiplied by the probability that they will have an affected child if they are both carriers (1/4).

1. Using the heterozygote frequency for cystic fibrosis among white Americans to estimate the probability that John is a carrier, what is the likelihood that their child would have the disease?
2. If you were their genetic counselor, would you recommend that Jane and John be genetically tested before they attempt to have any children?
3. It is now possible to use preimplantation testing, which involves *in vitro* fertilization plus genetic testing of the embryo before implantation, to ensure that a heterozygous couple has a child free of cystic fibrosis. Do you see any ethical problems or potential future dangers associated with this technology?

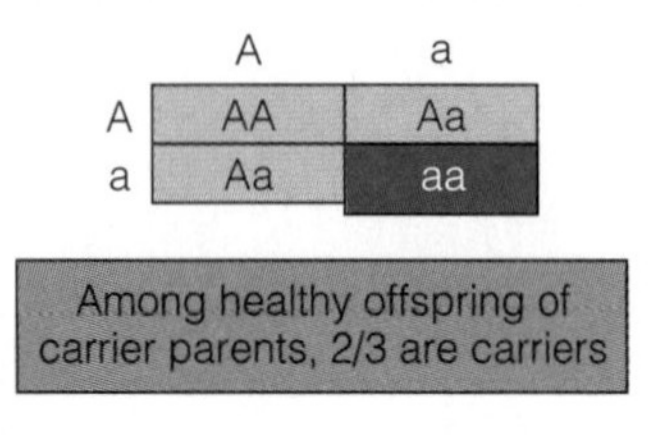

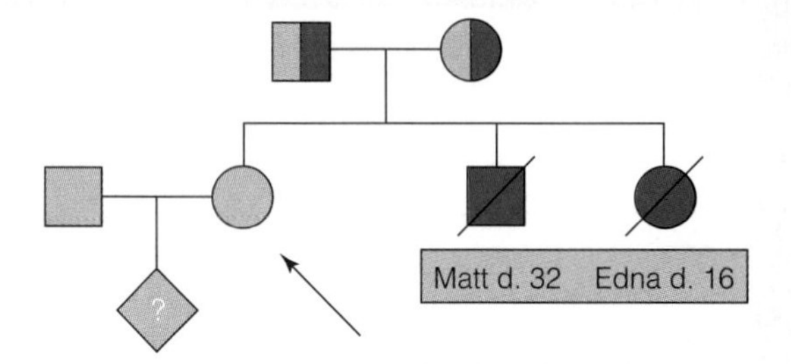

CASE 2

A population geneticist sampled 200 people on a small tropical island to determine the allelic frequency of the MN blood group. He knows that this single locus (*L*) has two codominant alleles, *M* and *N*. Thus three phenotypes, M, MN, and N, correspond to three genotypes, MM, MN, and NN, respectively. Among the 200 individuals sampled, 72 are M, 104 are MN, and 24 are N. The frequencies of the respective genotypes are 0.36 (MM), 0.52 (MN), and 0.12 (NN). The allelic frequencies can be precisely determined as follows:

This sample consists of 200 (N) individuals and 400 (2*N*) alleles. Each M individual carries 2 *M* alleles ($2 \times 72 = 144$); each MN carrier has 1 *M* allele ($1 \times 104 = 104$). This sample contains $144 + 104 = 248$ *M* alleles (out of the 400 alleles in the sample). Therefore, the frequency of *M* is $248/400 = 0.62$ and by subtraction, the frequency of the *N* allele $= 1 - 0.62 = 0.38$.

1. You compared the allele frequencies on this island with those of the mainland population from which the island inhabitants are descendants, and you found that the frequency of the *M* allele is much higher on the mainland than it is on the island. What could account for this difference?
2. You recently visited the same island 20 years after the initial testing and found that the *M* allele frequency was now much closer to the frequency on the mainland. What might have happened in the meantime to cause this change?

Questions and Problems

The Population as a Genetic Reservoir

1. Draw a graph showing the difference between a population that grows at a geometric rate and one that is only permitted to grow at a constant arithmetic rate. Be sure to label each axis.
2. What are some of the natural constraints referred to by Malthus that keep the size of human populations in check?
3. Explain why a population carries more genetic diversity than an individual carries.

How Can We Measure Allele Frequencies in Populations?

4. Define the following terms:
   a. population
   b. gene pool
   c. allele frequency
   d. genotype frequency
5. Why are codominant alleles ideal for studies of allele frequencies in a population?
6. Explain the connection between changes in population allele frequencies and evolution, and relate this to the observations made by Darwin and Wallace concerning natural selection.
7. Do you think populations can evolve without changes in allele frequencies?
8. Design an experiment to determine if a population is evolving.
9. Which pairs of populations in Table 17.3 appear to be most closely related in terms of their allelic frequencies at the MN locus? What scenario would you propose to account for these relationships?

The Hardy-Weinberg Law: Measuring Allele and Genotype Frequencies

10. What are four assumptions of the Hardy-Weinberg Law?
11. Drawing on your newly acquired understanding of the Hardy-Weinberg equilibrium law, point out why the following statement is erroneous: "Because most of the people in Sweden have blond hair and blue eyes, the genes for blond hair and blue eyes must be dominant in that population."
12. In a population where the females have the allelic frequencies $A = 0.35$ and $a = 0.65$ and the frequencies for males are $A = 0.1$ and $a = 0.9$, how many generations will it take to reach Hardy-Weinberg equilibrium for both the allelic and genotypic frequencies? Assume random mating, and show the allelic and genotypic frequencies for each generation.

Using the Hardy-Weinberg Law in Human Genetics

13. Suppose you are monitoring the allelic and genotypic frequencies of the MN blood group locus in a small human population. You find that for 1-year-old children the genotypic frequencies are MM = 0.25, MN = 0.5, and NN = 0.25, whereas the genotypic frequencies for adults are MM = 0.3, MN = 0.4, and NN = 0.3.
    a. Compute the *M* and *N* allele frequencies for the 1-year-olds and adults.
    b. Are the allele frequencies in equilibrium in this population?
    c. Are the genotypic frequencies in equilibrium?
14. Using Table 17.6, determine the frequencies of p and q that result in the greatest proportion of heterozygotes in a population.
15. In a given population, the frequencies of the four phenotypic classes of the ABO blood groups are found to be A = 0.33, B = 0.33, AB = 0.18, and O = 0.16. What is the frequency of the *O* allele?
16. If a trait determined by an autosomal recessive allele occurs at a frequency of 0.25 in a population, what are the allelic frequencies? Assume Hardy-Weinberg equilibrium, and use *A* and *a* to symbolize the dominant and recessive alleles, respectively.
17. Five percent of the males of a population express an X-linked recessive trait.
    a. What are the frequencies of the dominant and recessive alleles?
    b. What are the genotypic frequencies for the males and females in the population?
18. For each population in Table 17.3, use the given allelic frequencies to determine if the genotypic frequencies are in general agreement with the Hardy-Weinberg law.

Population Genetics Can Help Determine the Age and Origin of Human Genetic Disorders

19. The *Δ508* mutation in the *CFTR* gene is frequently linked to a particular set of neighboring markers in northwest Europe, whereas the relationship between this mutation and its neighboring markers is much less consistent in southeast Europe. What does this tell us about the age of this mutation in these two regions?

Internet Activities

The following activities use the resources of the World Wide Web to enhance the topics covered in this chapter. To investigate the topics described below, log on to the book's home page (www.brookscole.com/biology_d) and follow the *Student Resources* link to your subject and text.

1. *Comparing DNA Sequences.* GenBank is the National Institute of Health's (NIH) database of all known nucleotide and protein sequences, including supporting bibliographic and biological data. In the next two activities you will use two features of GenBank, the Entrez system (to search for a DNA sequence) and BLAST (to find similar sequences in GenBank).
   a. Begin this activity by searching the nucleotide database of *NCBI Entrez* (http://www.ncbi.nlm.nih.gov/Entrez). Enter the gene designation "glycosyltransferase" (an enzyme that places blood group antigens on red blood cell surfaces) in the query block. Select "retrieve" to find all data in GenBank on this particular enzyme. Select one of the documents related to this gene in humans, and then retrieve the sequence in FASTA format. Highlight a portion of the sequence (300–500 bases), and copy it to the clipboard.
   b. Access *BLAST* (http://www.ncbi.nlm.nih.gov/BLAST/). Enter the sequence you copied to the clipboard by pasting it into the query box. Submit the query. How many different species have similar sequences to the one you selected? What percentage homology do these sequences have? Are there other human sequences that are identical or similar to the sequence you selected?
2. *Exploring the Hardy-Weinberg Equilibrium Equation.* At the Access Excellence *Activities Exchange* site (http://www.accessexcellence.org/AE/), there are several Hardy-Weinberg–related exercises. To see how selection can affect a population's allele frequencies, try the *Fishy Frequencies* activity (http://www.accessexcellence.org/AE/AEPC/WWC/1994/fishfreq.html). This exercise can be run alone or as part of a group—and you get to eat fish crackers as you work!

For Further Reading

Barrantes, R., Smouse, P. E., Mohrenweiser, H. W., Gershowitz, H., Azofeifa, J., Arias, T. D., & Neel, J. V. (1990). Microevolution in lower Central America: Genetic characterization of the Chibcha-speaking groups of Costa Rica and Panama, and a consensus taxonomy based on genetic and linguistic affinity. *Am. J. Hum. Genet., 46,* 63–84.

Bodmer, W. F., & Cavelli-Sforza, L. L. (1976). *Genetics, evolution and man.* San Francisco: Freeman.

Chakraborty, R., Smouse, P. E., & Neel, J. V. (1988). Population amalgamation and genetic variation: Observations on artificially agglomerated tribal populations of Central and South America. *Am. J. Hum. Genet., 43,* 709–725.

Constans, J. (1988). DNA and protein polymorphism: Application to anthropology and human genetics. *Anthropol. Anz., 46,* 97–117.

Deevey, E. S., Jr. (1960, September). The human population. *Sci. Am., 203,* 194–205.

Feldman, M. W., & Christiansen, F. B. (1985). *Population genetics.* Palo Alto, CA: Blackwell Scientific.

Friedlaender, J. S. (1975). *Patterns of human variation.* Cambridge, MA: Harvard University Press.

Mettler, L. E., Gregg, T. G., & Schaffer, H. E. (1988). *Population genetics and evolution* (2nd ed.). Englewood Cliffs, NJ: Prentice Hall.

Neel, J. V. (1978). The population structure of an Amerindian tribe, the Yanomama. *Ann. Rev. Genet., 12,* 365–413.

Relethford, J. H. (1985). Isolation by distance, linguistic similarity and the genetic structure on Bougainville Island. *Am. J. Physiol. Anthropology, 66,* 317–326.

Romeo, G., Devoto, M., & Galietta, L. J. (1989). Why is the cystic fibrosis gene so frequent? *Hum. Genet., 84,* 1–5.

Spiess, E. B. (1989). *Genes in populations* (2nd ed.). New York: Wiley.

Woo, S. L. (1989). Molecular basis and population genetics of phenylketonuria. *Biochemistry, 28,* 1–7.

© Paul Hanny/Liaison Agency

Human Diversity and Evolution

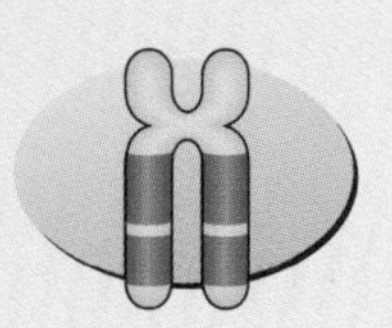

Chapter Outline

Two humanlike creatures, one larger, one smaller, walked across a muddy field of volcanic ash in a region of what is now Tanzania, Africa. As it does now, the landscape consisted of small lakes, patches of vegetation, and a volcano. The pair passed through the area during or just after a rain and shortly after an eruption had covered the ground with a layer of volcanic ash. Along the way, the smaller one stopped, turned away to look at something, and then resumed walking. Soon after these creatures passed, several new layers of volcanic ash covered their footprints. This unremarkable event was later to provide important clues about several aspects of human evolution.

Three and a half million years later, the footprints, exposed by erosion, were discovered by a team of fossil hunters led by Mary Leakey. Altogether, more than 50 prints covering a distance of approximately one hundred feet were recovered and preserved as casts.

The prints show that ancient primates, known as hominids, had feet with well-developed arches, heels, and toes. More importantly, these prints reveal that two of the key events in human evolution, upright posture and walking, evolved more than 3 million years ago. Other fossil evidence suggests that these small, primitive hominids had apelike faces with receding foreheads and small brains. However, they had humanlike pelvis and leg anatomy, allowing them to stand upright and walk much like modern humans. Upright posture and bipedal locomotion freed their hands for important tasks, such as carrying food or young.

In this chapter, we examine the evidence about how and where our species, Homo sapiens, *arose from primate ancestors through a series of evolutionary adaptations over millions of years. The broad outlines of this process are now well established. New discoveries and interpretations of the evidence are constantly refining our knowledge of the nature and sequence of the historical events of our own species. Although much of the evidence consists of bits of fossilized bones dug out of rocks, there is a direct and thrilling connection between our ancestors and ourselves when we view humanlike footprints from 3 million years ago.*

MEASURING GENETIC DIVERSITY IN HUMAN POPULATIONS

Understanding our evolutionary history depends on identifying factors that lead to variations in allele frequencies between populations and the way that these variations are acted upon by natural selection. In the following sections, we explore how genetic variation is produced and the role of natural selection and culture as a force in changing allele frequencies.

For studies of genetic variation in populations, blood types, such as ABO, are ideal. Everyone has a blood type, and it is easily analyzed from a few drops of blood. Soon after their discovery, detailed maps of the geographic distribution of blood-type alleles became available (▶ Figure 18.1).

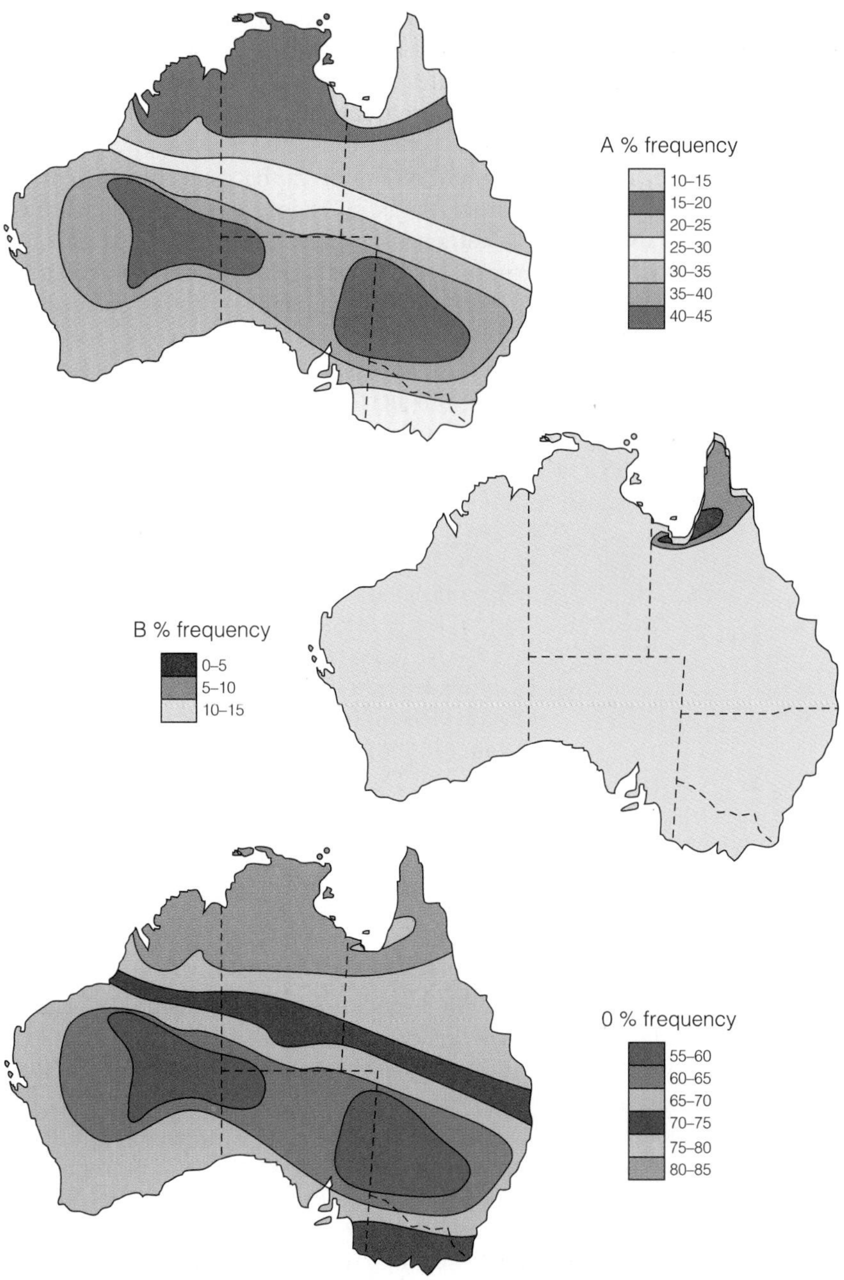

FIGURE 18.1 A reconstruction of the geographic distribution of the ABO alleles in the prehistoric population of Australia.

ABO blood types are genetic **polymorphisms** because two or more distinct alleles of the gene are present in the population at frequencies greater than 1%. The study of polymorphisms tells us something about how much genetic variation exists in a population. Some polymorphisms are responsible for genetic disorders (such as cystic fibrosis or sickle cell anemia) and others (such as the ABO blood types) are of interest only as genetic markers.

Polymorphism The occurrence of two or more genotypes in a population in frequencies that cannot be accounted for by mutation alone.

Mutation Generates New Alleles but Has Little Impact on Allele Frequency

As discussed in Chapter 17, the gene pool is reshuffled each generation to produce the genotypes of the offspring. In the process, genetic variation (new combinations of alleles) is produced by recombination and Mendelian assortment. However, these events do not produce any new alleles. Mutation is the ultimate source of all new alleles and is the origin of all genetic variability.

We discussed mechanisms and rates of mutation in Chapter 11. In this section, we consider the effect of mutation on allele frequencies in a population. If the mutation rate for a gene is known, we can use the Hardy-Weinberg Law to calculate the change in allele frequency resulting from new mutations in that gene in each generation. Let's use the dominant trait achondroplasia as an example. Statues and murals indicate that this form of dwarfism was known in Egypt more than 2,500 years ago, or about 125 generations (allowing 20 years per generation) ago. If mutant achondroplasia alleles were introduced into the gene pool by mutation in each generation beginning 2,500 years ago, how much would the frequency of achondroplasia change over this time period? Should we expect a higher frequency of achondroplasia among residents of an ancient city, such as Cairo, than among residents of a recently established city like Houston?

For this calculation we assume that initially, only homozygous recessive individuals with the genotype *dd* (normal stature) were present in the population and that mutation has added new mutant (*D*) alleles to each generation at the rate of 1×10^{-5}. The change in allele frequency over time that results from this rate of mutation is shown in Figure 18.2. To change the frequency of the recessive allele (*d*) from 1.0 (100%) to 0.5 (50%) at this rate of mutation will require 70,000 generations, or 1.4 million years. Thus, the frequency of achondroplasia need not be any higher in Houston than in Cairo. Our conclusion in this case is that mutation alone has a minimal impact on the genetic variability present in a population.

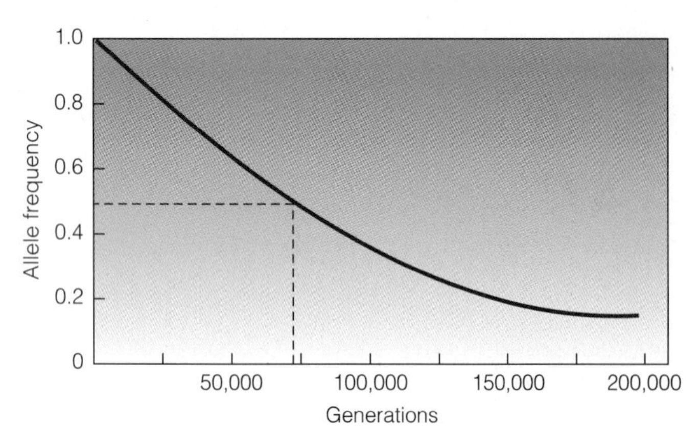

FIGURE 18.2 The rate of replacement of a recessive allele *a* by the dominant allele *A* by mutation alone. Even though the initial rate of replacement is high, it will take about 70,000 generations or 1.4 million years to drive the frequency of the allele *a* from 1.0 to 0.5.

Genetic Drift Can Change Allele Frequencies

Occasionally populations start with a small number of individuals, or founders. Alleles carried by the founders, whether they are advantageous or detrimental, become established in the new population. These events take place simply by chance and are known as **founder effects.**

Founder effects Allele frequencies established by chance in a population that is started by a small number of individuals (perhaps only a fertilized female).

Genetic drift The random fluctuations of allele frequencies from generation to generation that take place in small populations.

Random changes in allele frequency that occur from generation to generation in small populations are examples of **genetic drift.** In addition to founder effects, genetic drift can occur in small populations by drastic reductions in population size. These reductions, called population bottlenecks, are often caused by natural disasters. In extreme cases, drift can lead to the elimination of one allele from all members of the population. Small interbreeding groups on isolated islands often provide examples of genetic drift.

The island of Tristan da Cuhna (Figure 18.3) is located in the southern Atlantic Ocean, 2,900 km (about 1,800 mi) from Capetown, South Africa, and 3,200 km (about 2,000 mi) from Rio de Janeiro, Brazil. British troops were stationed here in 1816 to prevent Napoleon from escaping his exile on the island of St. Helena. After Napoleon's death, Corporal William Glass, his wife, and two daughters received permission to remain after the British army withdrew. Others joined Glass at intervals, and the development of the isolated and highly inbred population that formed here can be traced with great accuracy.

The genotypes of the few hundred residents have been studied for over 40 years to see how isolation and inbreeding affect the island's gene pool. As might be expected, one effect of inbreeding is an increase in homozygosity for recessive traits. Table 18.1 lists some genetic markers for which all the islanders tested are homozygous.

Because of a founder effect, traits carried by one or a small number of early settlers often end up in a large fraction of their descendants. On Tristan, a deformity

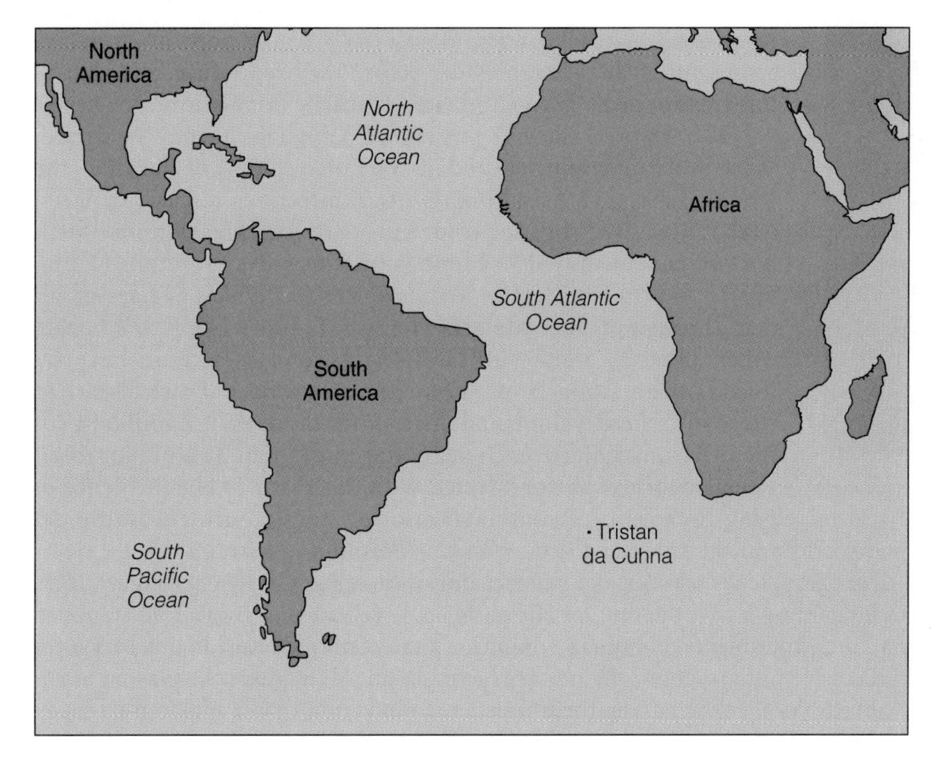

FIGURE 18.3 Location of the island of Tristan de Cuhna, first discovered by a Portuguese admiral in 1506.

Table 18.1

Homozygous Markers among Tristan Residents

Transferrins
Phosphoglucomutase
6-Phosphogluconate dehydrogenase
Adenylate kinase
Hemoglobin A variants
Carbonic anhydrase (2 forms)
Isocitrate dehydrogenase
Glutathione peroxide
Peptidase A, B, C, D

SOURCE: *Data from T. Jenkins, P. Beighton, & A. G. Steinberg, (1985) Ann. Hum. Biol., 12, 363–371,* **Table 2.**

of the fifth finger, known as **clinodactyly** (MIM/OMIM 112700), is present at a high frequency. This autosomal dominant trait is especially prominent in members of the Glass family, the first permanent residents of the island. The high frequency of this trait in the present island population can be explained by its presence in one of the original colonists.

Clinodactyly An autosomal dominant trait that produces a bent finger.

This brief example illustrates how genetic drift can be responsible for changing allele frequencies in populations that are isolated, inbred, and stable for long periods. Most human populations, however, do not live on remote islands and are not subject to prolonged isolation and inbreeding. Yet there are differences in the distribution and frequency of alleles among populations, indicating that founder effects and drift are not the only factors that can change allele frequencies.

Natural Selection Acts on Variation in Populations

Wallace and Darwin recognized that some members of a population are better adapted to the environment than others. These better-adapted individuals have an increased chance of leaving more offspring than those with other genotypes. The ability of a given genotype to survive and to reproduce is a measure of its **fitness.** Fitter genotypes are better at survival and reproduction and make a larger contribution to the gene pool of the next generation than less-fit genotypes.

Fitness A measure of the relative survival and reproductive success of a given individual or genotype.

In time this reproductive difference leads to changes in allele frequencies within the population. The differential reproduction of fitter genotypes is **natural selection.** The relationship between the allele for sickle cell anemia and malaria is an example of how natural selection changes allele frequencies. Sickle cell anemia is an autosomal recessive condition associated with a mutant form of hemoglobin. Affected individuals have a wide range of clinical symptoms (see Chapter 4 for a review of the symptoms). Although many untreated homozygotes die in childhood, the mutant allele is present in very high frequencies in certain populations. In some West African countries, 20% of the population may be heterozygous for this trait, and along

Natural selection The differential reproduction shown by some members of a population that is the result of differences in fitness.

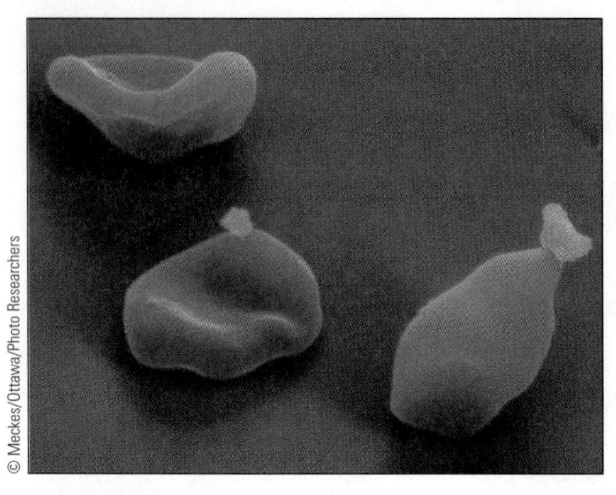

▶ **FIGURE 18.4** *Plasmodium* parasites (yellow) attacking and infecting red blood cells. Infection by *Plasmodium* causes malaria.

rivers such as the Gambia, almost 40% of the population is heterozygous. If homozygotes die before they reproduce, why hasn't this mutant allele been eliminated from the population?

The mutant allele is present at a high frequency in certain West African countries and in certain regions of Europe, the Middle East, and Asia because of an infectious disease, malaria. Malaria, caused by a protozoan parasite, *Plasmodium falciparum,* is transmitted to humans by infected mosquitoes (▶ Figure 18.4). Once infected, victims suffer recurring episodes of illness throughout life. Malaria victims are more likely to contract other diseases, often with fatal results, and as a result have reduced fitness. Malaria may seem like an exotic and rare disease to those in more developed countries, but more than 2 million people die from malaria each year, and more than 300 million individuals worldwide are infected with this disease. The incidence of malaria is increasing because of a combination of population growth in the developing world and the spread of drug-resistant *Plasmodium* strains.

The geographic distributions of malaria and sickle cell are shown in ▶ Figure 18.5. Research has shown that the mutant allele for sickle cell anemia confers resistance to malaria, and experiments on human volunteers have confirmed this. In heterozygotes and in recessive homozygotes, the mutant hemoglobin (HbS) alters the plasma membrane of red blood cells, making them resistant to infection by the malarial parasite. This resistance makes heterozygotes fitter than those with homozygous normal genotypes. Those with the homozygous recessive genotype are also resistant to malaria but have sickle cell anemia and are less fit than the other two genotypes. In this case, selection favors the survival and differential reproduction of heterozygotes.

Because of their fitness, heterozygotes leave more offspring than the other genotypes, and the HbS allele is spread through the population and maintained at a high frequency.

▶ **FIGURE 18.5** (a) The distribution of sickle cell anemia in the Old World. (b) The distribution of malaria in the same region.

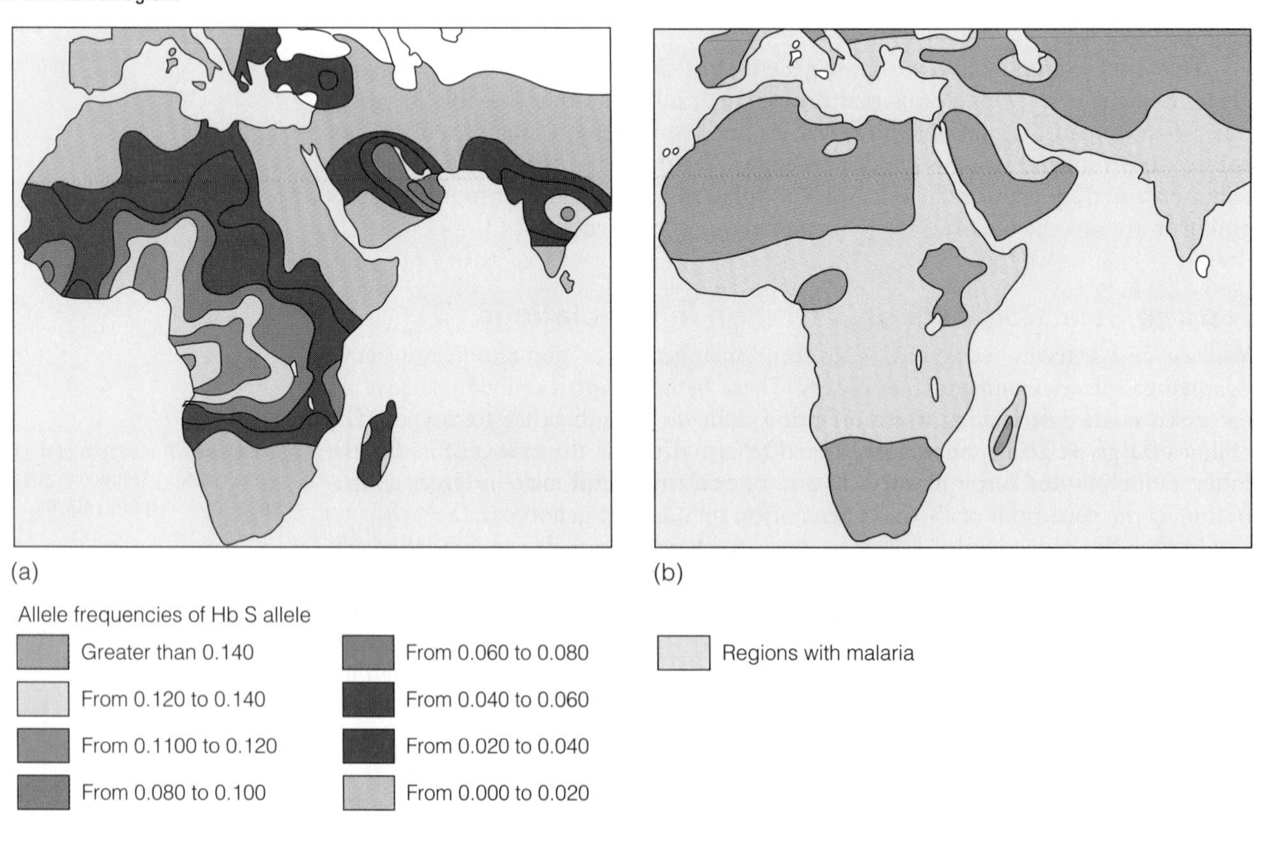

Selection Can Rapidly Change Allele Frequencies

The effects of selection can be seen in geographically separated subpopulations that originated from a single ancestral population. The emigration of Jews from what is now Israel over the last 2,500 years has formed subpopulations in Europe, North Africa, the Middle East, and Asia. The frequency of a glucose-6-phosphate dehydrogenase (*G6PD*) deficiency allele (MIM/OMIM 305900) is now very different among these subpopulations (◗ Figure 18.6). The frequency ranges from near zero in Central Europe (Ashkenazi) to around 70% in the Middle East (Kurdish Jews).

Two explanations for these differences in allele frequency are possible: founder effects and selection. The distribution of the *G6PD* deficiency allele and malaria are very similar. This distribution and the rapid change in allele frequency that occurred in only 100 to 125 generations indicates that selection is the cause for the differences in allele frequency. As with sickle cell anemia, malaria is the selective force. Malarial parasites reproduce well in cells with G6PD. As a result, G6PD-deficient homozygous females and hemizygous males are resistant to malaria. If we assume that the ancestral population in Israel had a low frequency of the *G6PD* deficiency allele, selection has dramatically changed its frequency in a relatively short time, illustrating the power of natural selection.

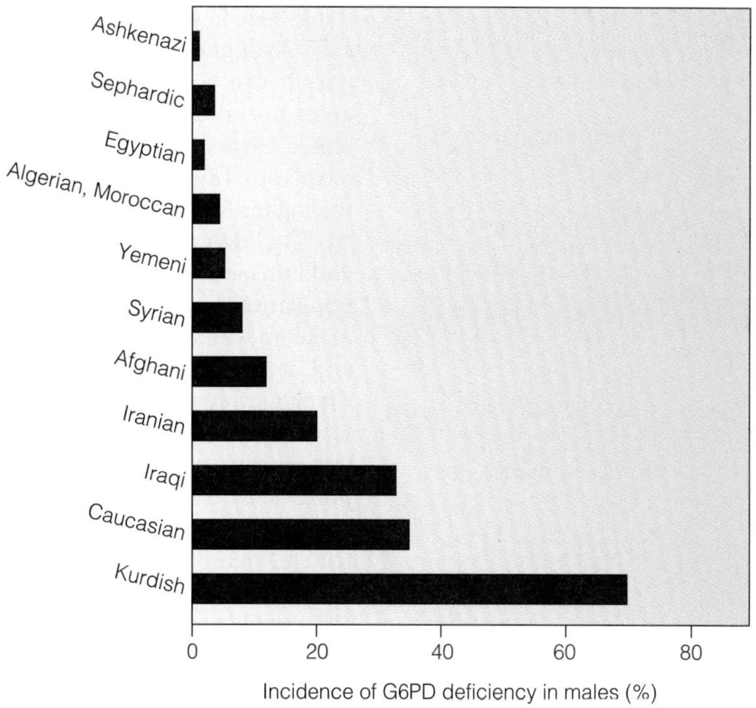

◗ FIGURE 18.6 Distribution of glucose-6-phosphate dehydrogenase (G6PD) deficiency in various Jewish populations. Because intermarriage with native populations is rare, the differences in frequency are attributed to selection. G6PD deficiency confers resistance to malaria, and the frequency of the allele is highest in regions where malaria is endemic.

NATURAL SELECTION AFFECTS THE FREQUENCY OF GENETIC DISORDERS

Many genetic disorders are disabling or fatal, so why are they so common? In other words, what keeps natural selection from eliminating the deleterious alleles responsible for these disorders? In analyzing the frequency and population distribution of human genetic disorders, it is clear that there is no single answer. One conclusion, drawn from the Hardy-Weinberg Law, is that rare lethal or deleterious recessive alleles survive because the vast majority of them are carried in the heterozygous condition and thus are hidden in the gene pool. Other factors, however, can cause the differential distribution of alleles in human populations, and several of these are discussed in the following.

Almost all affected individuals with Duchenne muscular dystrophy (DMD) (MIM/OMIM 310200) die without reproducing. If this is the case, eventually the mutant allele should be eliminated from the population. However, because the mutation rate for *DMD* is high (perhaps as high as 1×10^{-4}), mutation replaces the *DMD* alleles lost when affected individuals die before reproducing. Thus the frequency of the *DMD* allele in a population is a balance between alleles introduced by mutation and those removed by the death of affected individuals.

Selection can cause detrimental alleles to have high frequencies in large, well-established populations. Heterozygote advantage in sickle cell anemia is an example. For other genetic disorders with a high frequency of the disease allele, the selective advantage of heterozygotes may be less obvious or perhaps no longer exists.

Tay-Sachs disease (MIM/OMIM 272800) is an autosomal recessive disorder that is fatal in early childhood. Although this is a rare disease in most populations,

Ashkenazi Jews (those who live in or have ancestors from Eastern Europe) have a ten-fold higher incidence of the disease. In these populations, heterozygote frequency can be as high as 11%. There is indirect evidence that Tay-Sachs heterozygotes are more resistant to tuberculosis, a disease endemic to cities and towns, where most of the European Jews lived. As in sickle cell anemia, the death of homozygous Tay-Sachs individuals is the genetic price paid by the population to allow the higher fitness and survival of the more numerous heterozygotes.

The high incidence of cystic fibrosis (CF) (MIM/OMIM 219700) in European and European-derived populations has been a more difficult case to explain. In these populations, cystic fibrosis has an incidence of about 1 in 2,000, and heterozygotes make up 4% to 5% of the population. Untreated cases are lethal in early adulthood, and, until recently, almost all cases resulted from matings between heterozygotes. The high frequency of the mutant allele in some populations is explained by the defects in the mutant CFTR protein. At the cellular level, CF impairs chloride ion transport in secretory cells, and heterozygotes have a reduced level of chloride transport. Reduced chloride transport may make heterozygotes more resistant to certain infectious diseases. In many parts of the world, as in the past, bacterial diarrhea contributes significantly to infant mortality. Toxins produced in bacterial diarrhea cause an oversecretion of chloride ions, and it has been proposed that CF heterozygotes are more likely to survive illnesses that kill by electrolyte depletion and dehydration. Evidence indicates that CF heterozygotes may be resistant to one such disease, typhoid fever.

These examples illustrate that several factors contribute to disease allele frequency and that each disease must be analyzed individually. In some cases mutant alleles are maintained by mutation. In other cases migration and founder effects increase the frequency of deleterious alleles. In still other cases, natural selection favors heterozygous carriers, and affected homozygotes bear the burden of conferring advantages on other genotypes.

HUMAN ACTIVITY CAN CHANGE ALLELE FREQUENCIES

In addition to natural selection, human activities can influence the frequency and distribution of alleles. Collectively, these activities are defined as culture. Social customs and technology are powerful forces that dictate and shape human mating patterns and can change allele frequencies. For example, the invention of technology for long-distance travel in the sixteenth century led to worldwide movement of populations, causing a redistribution of alleles. Social constraints that limit mate selection to those who have some common bonds (language, religion, and economic status) inhibit random mating and can change allele frequencies. In this section we consider how two human activities, migration and mate selection, bring about such changes.

Migration Reduces Genetic Variation Between Populations

Distribution of the *B* allele of the ABO system (◗ Figure 18.7) shows the large-scale effect of migration. The highest frequency of the *B* allele is in indigenous populations of Central Asia. Several waves of migration from Central Asia into Europe produced a gradient (called a cline) that runs from east to west. The highest frequencies in this gradient are in Eastern Europe and the lowest in Southwest Europe.

Other studies have uncovered genetic relics of previous patterns of human migration. A large-scale study of 19 loci and 63 alleles carried out at more than 3,000 locations in Europe identified 33 boundaries of abrupt changes in allele frequency (◗ Figure 18.8). The sharp change in allele frequencies across each boundary suggests that little mixing between the neighboring populations has occurred. Twenty-two of the boundaries are physical barriers (mountains or ocean) across which there is little genetic exchange. The remaining 11 are not physical

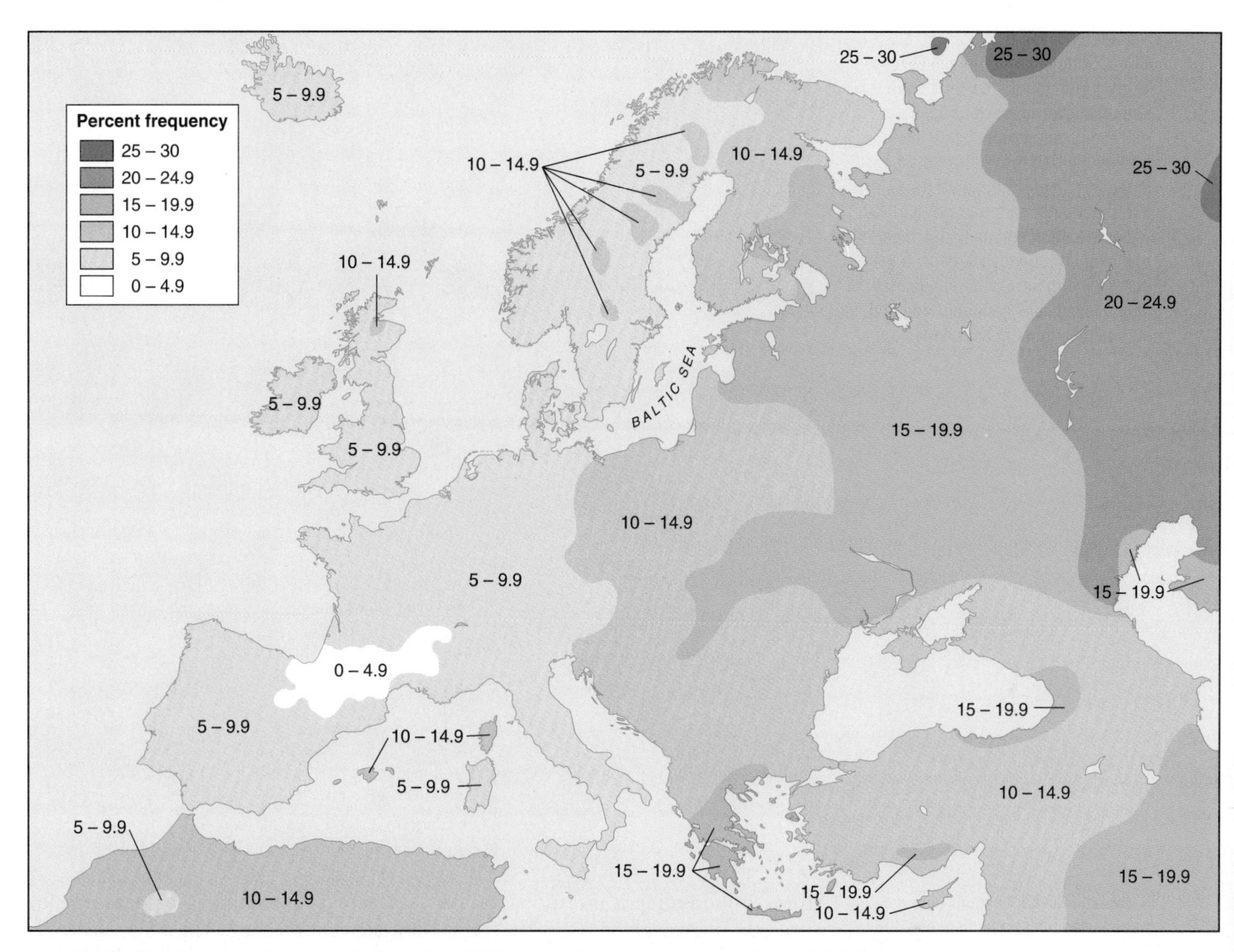

FIGURE 18.7 The gradient of the *B* allele of the ABO locus is due to waves of migration from Asia into Europe.

barriers; instead, they represent linguistic barriers that separate populations. (Some of the 22 physical boundaries also have linguistic barriers.)

These studies suggest that the genetic structure of European populations results from adaptation to local environments and reflects the diverse origins of populations that arrived by migration. The results also suggest that language is an effective barrier to gene flow and can be a selective force that maintains genetic differences between populations. Although European populations have been in contact for thousands of years, there has been little breakdown of allelic differences in some regions.

Mate Selection Is Usually Nonrandom

One assumption of the Hardy-Weinberg Law is that mating in a population is at random. This ideal condition rarely occurs in human populations, partly because of geographic proximity and partly because of social structure and cultural factors. One type of nonrandom mating is **assortative mating.** Cultural forces including a common language, physical characteristics, economic status, and religion are often important in mate selection. These factors can influence allele frequency.

Inbreeding, another form of nonrandom mating, involves mating between close relatives, who share common genes because of common ancestry. The most extreme

Assortative mating Reproduction in which mate selection is not random but instead is based on physical, cultural, or religious grounds.

Inbreeding Production of offspring by related parents.

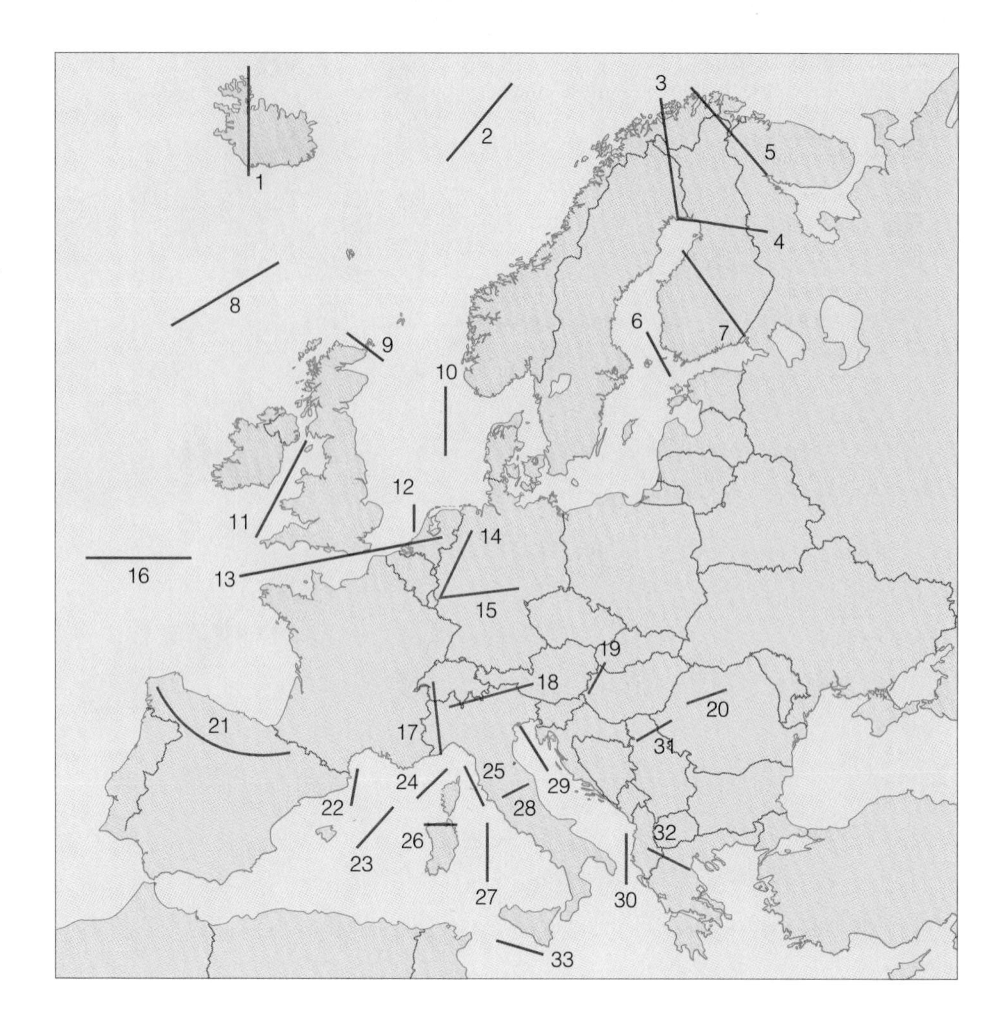

FIGURE 18.8
Genetic/linguistic barriers in Europe. Thirty-three genetic barriers have been discovered, each of which represents an abrupt shift in allele frequencies. Of these, 31 correspond to modern linguistic boundaries. The two that do not match are #1 and #32. The correlation between language and allele frequency suggests that culture in the form of language may play a major role in establishing and maintaining genetic boundaries.

Incest Sexual relations between parents and children or between brothers and sisters.

Consanguineous matings Matings between two individuals who share a common ancestor in the preceding two or three generations.

form of inbreeding is **incest**, mating between parents and children or between siblings. In almost all societies, incest is culturally unacceptable. In some societies, however, incest or inbreeding was common among royalty, including the Ptolemaic dynasties of Egypt (Figure 18.9). Such matings increase homozygosity and decrease heterozygosity in the offspring. Frequencies of some autosomal recessive conditions among offspring of **consanguineous matings** are shown in Table 18.2.

For many centuries, matings between cousins were common in many cultures, including those of Western Europe (Figure 18.10a). As population mobility has increased, the frequency of these matings has decreased, contributing to more random matings (Figure 18.10b). Many states in the United States have laws regulating the degree of consanguinity permitted in marriages, but these laws are based on social or religious customs rather than on genetics.

Culture Is a Force in Altering Allele Frequencies

Collective human activity in the form of cultural practices can act as a selective force to alter the frequency of genotypes and alleles. The interaction of culture, environment, and allele frequency associated with the herding of dairy animals and lactose digestion is an example.

Lactose is the principal sugar in milk (human milk is 7% lactose) and is a ready energy source. In infants, the enzyme lactase converts lactose into glucose and galactose, sugars that are easily absorbed by the intestine. Lactase production in most humans slows at the time of weaning, and lactase production stops as children grow into adults (review the biochemistry of lactose breakdown in Chapter 10). Adults

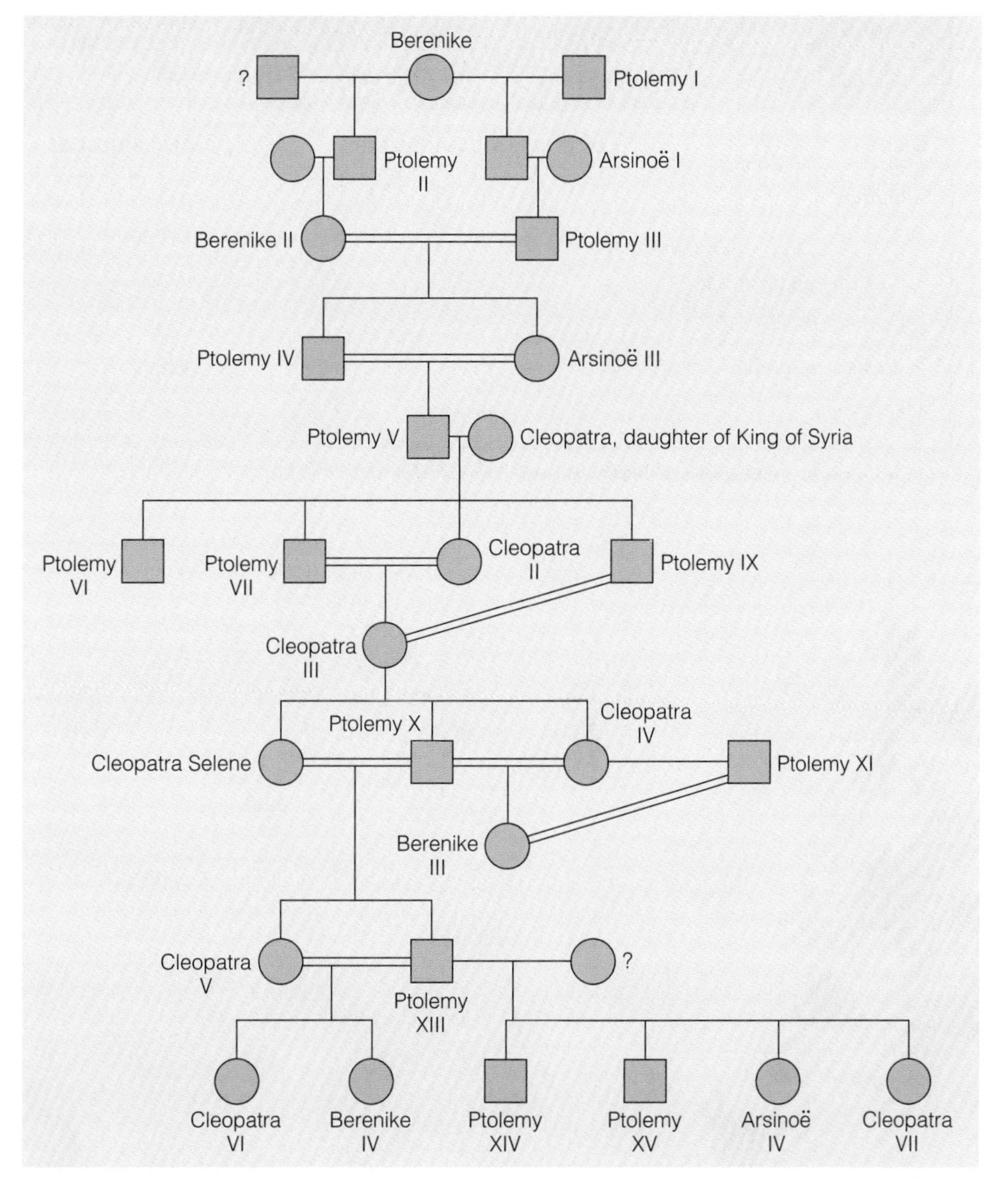

FIGURE 18.9 Pedigree of the Ptolemaic dynasty of Egypt. Incest and consanguineous matings (indicated by the double horizontal lines) were common. Cleopatra VII, the great Cleopatra, is known for her romantic involvement with Julius Caesar and Marc Antony. She had one son by Caesar and three children by Antony.

with low lactase levels cannot metabolize lactose (MIM/OMIM 223100), and have bad reactions to milk or dairy products. After eating foods that contain lactose, they develop gas, cramps, diarrhea, and nausea. In these individuals, undigested lactose passes from the small intestine into the colon, where it is metabolized by bacteria, resulting in gas and diarrhea. However, in some human populations, lactase is produced throughout adulthood; these individuals are called lactose absorbers (LA). Genetic evidence indicates that adult lactose utilization (and the adult production of lactase) is inherited as an autosomal dominant trait. Across different populations, the frequency of the lactose absorption allele varies from 0.0% to 100% (Table 18.3).

Why does the frequency of this allele vary so widely across populations? The answer appears to be that cultural practices are a selective force. According to this hypothesis, the human species originally was lactose-intolerant as adults, as are all other land mammals. As dairy herding developed in some populations, adult LA had

Table 18.2 Percentage of First-Cousin Parents in Children Who Have Recessive Genetic Diseases

Disease	% of First-Cousin Parents	MIM/OMIM Number
Albinism	10	203100
PKU	10	261600
Xeroderma pigmentosum	26	278700
Alkaptonuria	33	203500
Ichthyosis congenita	40	242300
Microcephaly	54	215200

SOURCE: *S. Reed (1980)* Counseling in Medical Genetics *(3rd ed.). New York: Alan Liss, p. 77, Table 10.1.*

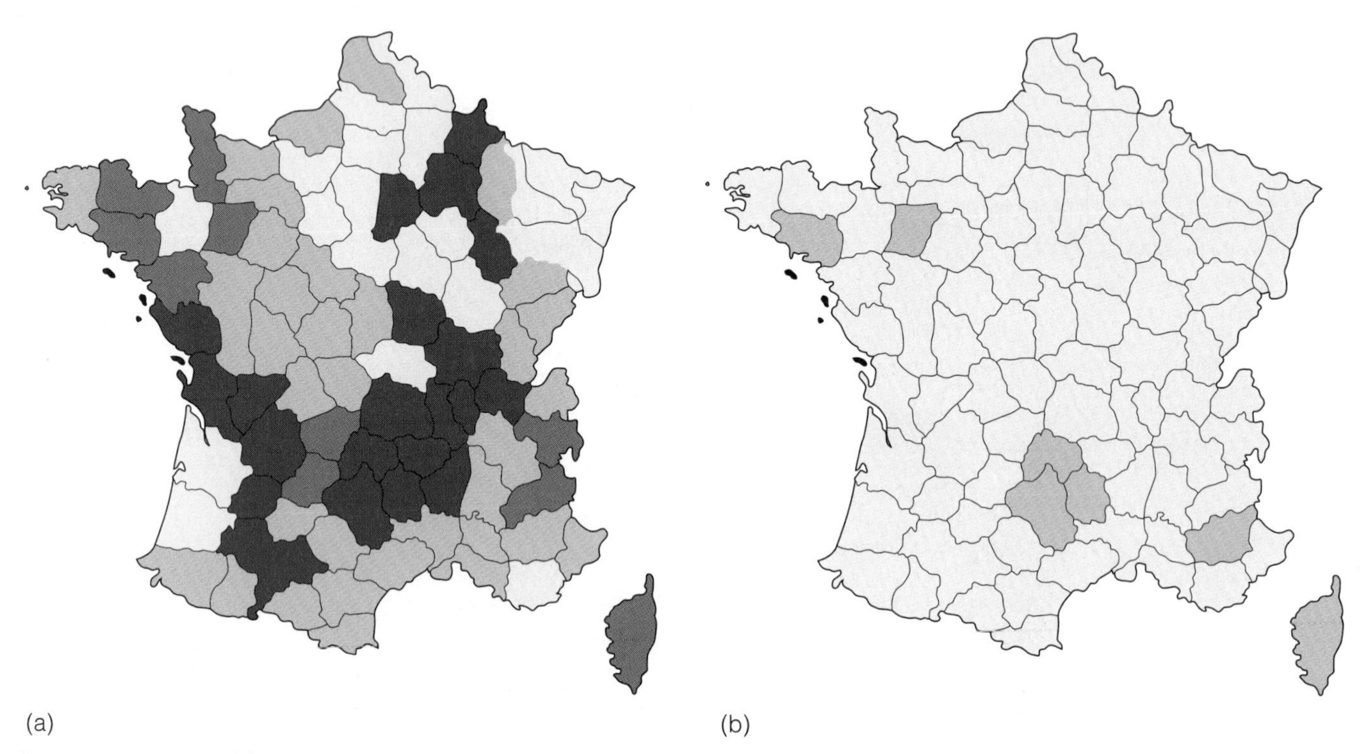

FIGURE 18.10 (a) Relative levels of consanguinity in France between 1926 and 1930. Darker shades represent higher levels of inbreeding. (b) Relative levels of consanguinity in France between 1956 and 1958.

the selective advantage of being able to derive nutrition from milk. This improved their chances of survival and success in leaving offspring. As a result, the cultural practice of maintaining dairy herds was the selective factor that provided a fitness advantage for the LA genotype.

Population surveys provide a strong correlation between a society's history of dairy herding and the frequency of LA. The Tuareg of North Africa are nomadic herders who have been in the Central Sahara region of Africa for 2,000 to 3,000 years (Figure 18.11). Other food is often unavailable, and adult consumption of several liters of milk per day is common. There is a high frequency of the *LA* allele in this population. In fact, all populations with high LA frequencies (60% to 100% is a high frequency) have a history of dairy herding that can be traced back for over 1,000 years. In contrast, populations in the tropical forest belt of Africa, where sleeping sickness carried by the tsetse fly prevents dairy herding, the frequency of lactose absorption is

Table 18.3 Lactose Absorption in Various Populations

Population	Percent Lactose Absorbers (LA)
Eskimos—Greenland	15
!Kung—Africa	2.5
Tuareg—Africa	85
Bantu—Africa	0
Arabs—Saudi Arabia	86
Sephardic Jews—North Africa	38
Danes—Europe	98
Czechs—Europe	100
U.S.—Asians	3
Aborigines	15
U.S.—Blacks	25

FIGURE 18.11 The Tuareg people of Africa inhabit the central region and the southern edge of the Sahara Desert.

low (0% to 20%). Taken together, the evidence from genetics, anthropology, and geography supports the idea that variation in the frequency of lactose absorption in present human populations is derived from cultural practices acting as a selective force on allele frequencies. The role of culture as a selective force makes humans unique among land mammals in having populations with a high frequency of adult LA.

GENETIC VARIATION IN HUMAN POPULATIONS

Evidence indicates that a great deal of genetic variation is present in the human genome. The process of mutation has introduced all this variation. Natural selection and drift are the primary mechanisms by which alleles spread through local population groups.

The Spread of Polymorphisms

One way to monitor the spread of alleles through a population is to study polymorphisms that show a geographical gradient. This distribution may result from migratory patterns, selection, or a combination of factors. An interesting polymorphic gradient is the distribution of blond or tawny hair among Australian aborigines (◗ Figure 18.12). The allele for this codominantly inherited phenotype apparently originated in the west-central desert region (◗ Figure 18.13) and spread from there to much of the west coast. The selective advantage, if any, for this hair color allele is not known. It is possible that tawny hair is only a phenotypic byproduct of a gene that controls a biochemical trait with more direct selective significance. The aborigines live in a harsh desert environment, and this allele may contribute to an increase in fitness or may interact with other genes to increase fitness. On the other hand, perhaps those with tawny hair are more often selected as mates.

Photo courtesy of Peter Weigand

◗ **FIGURE 18.12** An Australian aborigine near Kolomburu in Western Australia who has a polymorphic hair color.

Measuring Gene Flow Between Populations

Anthropologists and geneticists have had a long-standing interest in estimating gene flow between populations to reconstruct the origin and history of hybrid populations formed when European and non-European populations come into contact. The best-

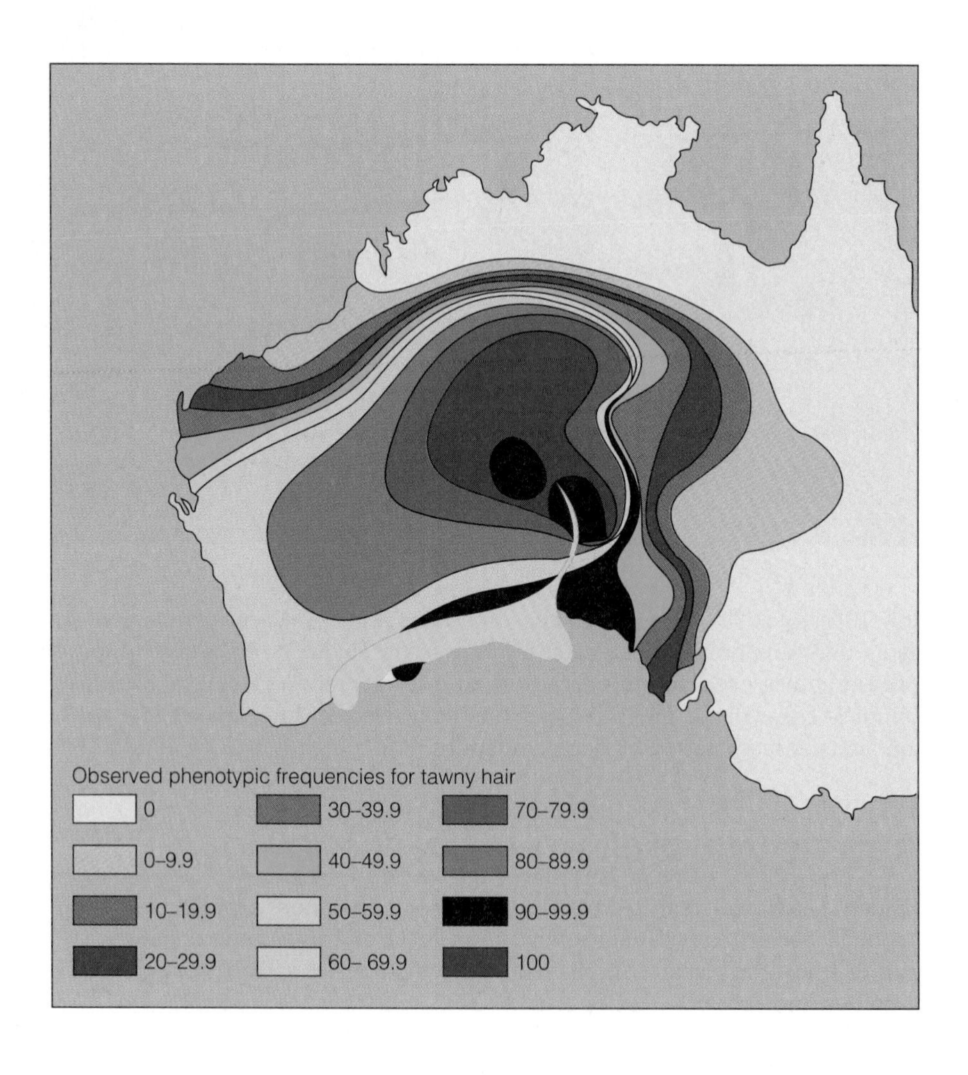

◗ **FIGURE 18.13** The frequency of tawny hair color in Australian aborigines forms a gradient from a center in the west-central desert region.

documented case is gene flow into the American black population from Europeans, but other populations have also been studied.

Most of the black population in the United States originated in West Africa, and the majority of the white population arrived from Europe. The Duffy blood group has three alleles: *FY**A*, *FY**B*, and *FY**O* (MIM/OMIM 110700). The frequency of the *FY**O* allele is close to 100% in West African populations, whereas in Europeans, this allele has a frequency close to zero. Almost all Europeans carry *FY**A* or *FY**B*, and these alleles are very rare in African populations. By measuring the frequencies of the *FY**A* and *FY**B* alleles in the U.S. black population, we can estimate of how much genetic mixing has occurred between these populations over the last 300 years. The frequency of the *FY**A* allele among black populations in West Africa and in several locations in the United States is shown in ◗ Figure 18.14. Using this as an average gene, we can calculate that about 20% of the genes in the black population in some northern cities are derived from Europeans.

Other studies that use restriction fragment polymorphisms (RFLPs) in the gene encoding the Duffy gene product give similar results. Studies using other genes reinforce this conclusion. For example, the frequency of phenylketonuria (PKU) (MIM/OMIM 261600) in U.S. blacks has been estimated as 1 in 50,000, about one-third of that of the U.S. white population. Using a cloned copy of the PKU gene and cloned molecular markers, such as RFLP sites, the origins of the PKU mutation and the phenylalanine hydroxylase (PAH) gene in U.S. blacks have been studied. Results suggest that about 20% of the PAH genes in U.S. blacks originated from a Caucasian population, whereas the rest are likely to be of West African origin.

Using 15 genes to examine 52 alleles in U.S. blacks from the Pittsburgh region (including 18 unique alleles of African origin), another study found that the proportion of European genes in this black population is approximately 25%. These studies are all consistent with the idea that members of the U.S. black population derive approximately 20% of their genes from Europeans.

However, not all contact between genetically distinct populations leads to a reduction in genetic differences. Using a combination of genetic and historical methods, researchers investigated the origin and extent of European admixture in the Gila River American Indian community of central Arizona. Results show a Euro-

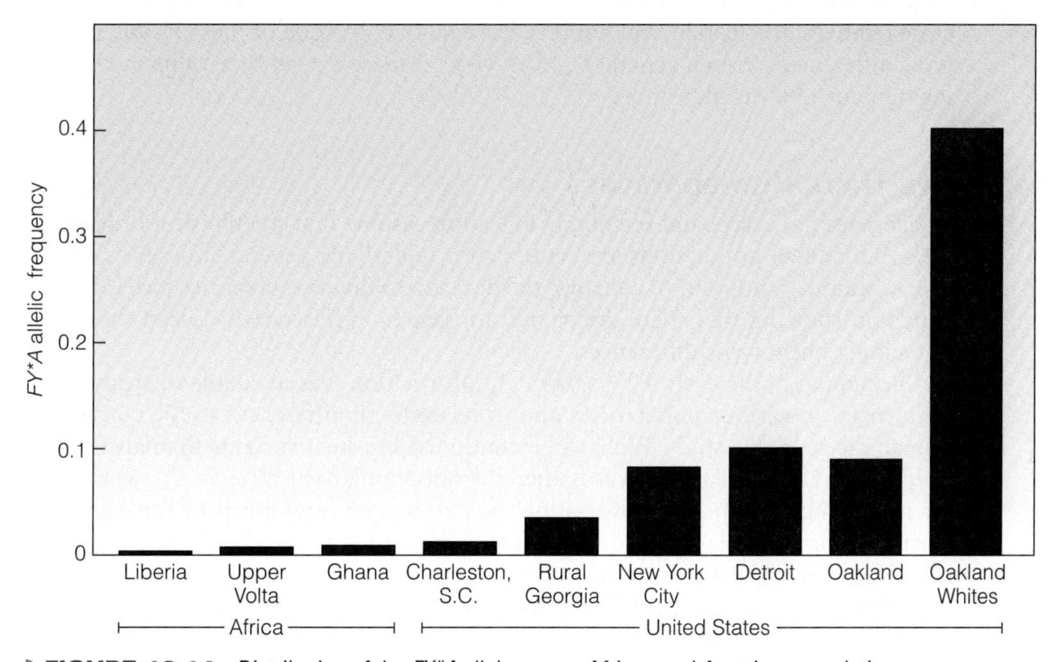

◗ **FIGURE 18.14** Distribution of the *FY**A* allele among African and American populations.

pean admixture of 5.4%, whereas the demographic data indicate an admixture of 5.9%. The first contact between Europeans and the Gila River community occurred in 1694. Using 20 years as the estimate for one generation, approximately 15 generations of contact have occurred. Because gene flow is approximately 0.4% per year, the Gila River American Indian community has retained almost 95% of its native gene pool even after close contact with other gene pools for 300 years.

WHAT ARE RACES?

Race A genotypically distinct group within a species.

In the nineteenth century, biologists used the term **race** to describe groups of individuals within a species that were phenotypically different from other groups in that species. For example, biologists of that time might find two-spotted beetles and four-spotted beetles in the same population and call them separate races, even though some two-spotted and four-spotted beetles might be siblings.

Races as Subspecies

As population genetics emerged as a separate field of study in the early twentieth century, the phenotypic concept of race was replaced by a genetic definition. Geneticists regard races as populations with significant differences in allele frequencies when compared to other populations. To the biologist studying nonhuman organisms, races represent groups that are genetically differentiated to the point at which they are regarded as subspecies; members of different subspecies can still interbreed and therefore have not yet formed separate species.

How does this happen? How do races (subspecies) form? Let's examine a simple case. Suppose a population with a single gene pool becomes geographically separated into two isolated populations, splitting the gene pool and the population in two. Over time, these isolated populations respond to selection in their separate environments and their gene pools accumulate differences. If this process continues over time, the populations can accumulate enough genetic differences to be classified as geographic races or subspecies.

This concept of geographic race does not clarify the difficulty of defining what we mean by the term race as it applies to humans. Clearly there are phenotypic differences among humans. Residents of the Kalahari Desert are rarely mistaken for close relatives of Aleutian seal hunters, for example. In spite of these visible phenotypic differences, from a genetic point of view, is there a need or a value in classifying humans into racial groups?

Are There Human Races?

In one sense, it is easy and seemingly logical to assume that the physical differences we see in human populations are evidence for underlying genetic differences. However, from the standpoint of genetics, the decision to divide our species into races depends on showing that there are significant genetic differences between the races, not simply phenotypic differences.

Beginning in the early 1970s, enough information was available to study allele frequencies in various populations and to relate these differences to the concept of human races. In one study, Richard Lewontin used protein variants to analyze allele frequencies at 15 loci. Data was gathered from populations all over the world. After measuring the amount of variation, he determined how much of this variation exists within a population and how much is present between populations. He also asked how much genetic variation exists among what was then considered to be conventional racial groups. His results show that 85% of the detected variation is present within populations and that less than 7% of the variation is present between racial groups. Studies of other protein markers confirmed these findings. These results, however, had little impact on the idea that the human species is divided into racial groups distinguished by large biological differences.

More recently, using DNA markers, geneticists have continued to search for genetic variation between and among populations. One study analyzed the distribution of variation within populations and among populations using 109 DNA markers (microsatellite alleles and RFLPs). The differences were analyzed for members of the same population, between and among populations on the same continent, and among four or five different geographic groups. Again, as in the protein studies, most of the genetic variation was within groups (about 85%), and only about 10% of the variation was among groups on different continents (equivalent to racial groups).

Taken together, the work on proteins and DNA indicates that most of the genetic variation in the species is present within human populations and that there is little variation among populations, including those classified as different racial groups (◗ Figure 18.15). In keeping with the genetic definition of race, if humans are to be divided into racial groups, large-scale genetic differences should occur along sharp boundaries. If such genetic differences exist, they have not yet been observed. On the other hand, is it possible that the small amount of genetic variation observed among groups is enough to warrant the classification of humans into racial groups? Almost all geneticists would answer no to that question and agree that presently there is no genetic basis for subdividing our species into racial groups.

What then is the genetic basis for the stereotypic view of race as differences in skin color and hair color? Most of these phenotypic differences are due to alleles of *MCIR* (MIM/OMIM 155555), a gene that controls the balance of pigment production and storage. Nucleotide sequencing of this gene from different populations reveals high levels of nucleotide diversity. This variation may be the result of selection for different alleles in different environments, correlated with latitude and exposure to ultraviolet radiation.

Human Genetic Variation: Summing Up

In earlier sections of the chapter we examined genetic differences over geographic distances in the form of clines (ABO blood types, HbS, G6PD) and genetic borders that coincide with linguistic borders. How can these findings be reconciled with the idea that as a species we have a highly homogeneous gene pool?

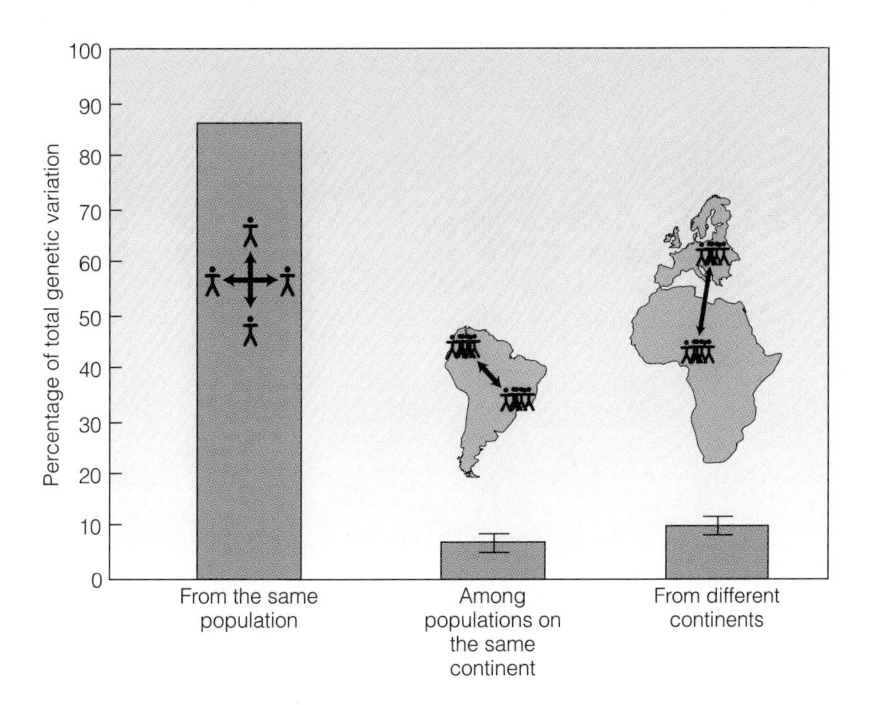

◗ **FIGURE 18.15** Distribution of human genetic variation. More than 80% of all variation is among individuals of the same population. Only about 10% of all variation is between populations from different continents (roughly equivalent to racial groups). From Owens, K., and King, Mary Clairè. (1999). Genomic views of human history. *Science, 286,* 451–453, Figure 1, p. 452. Copyright 1999 American Association for the Advancement of Science. Reprinted with permission.)

The patterns found in clines or gradients of allele frequencies represent responses to forces such as migration or natural selection. These patterns can be detected only when allele frequencies for one or a small number of genes are analyzed. When a large number of alleles are examined (using either protein markers or DNA markers), it is clear that natural selection does not have a substantial effect on the rest of the genome. As a result, there may be large differences in the frequency of one or two alleles among populations, but when many loci are considered, few differences are found.

One other conclusion can be drawn from studying genetic diversity across human populations. If diversity is analyzed by the size of populations, ranging from small villages to large nations and continents, a large fraction of the genetic variation in our species is present in small populations. This is consistent with the idea that our species has undergone a recent expansion from a small, common population. This insight plays an important part in our current understanding of how, when, and where our species originated. The following sections explore the evidence that surrounds the evolution of our species.

PRIMATE EVOLUTION AND HUMAN ORIGINS

The process of species formation is difficult to observe, mainly because of the timescale over which it operates. In the case of human evolution, the events have been reconstructed from the study of fossils and DNA variations. Primate evolution from the early Miocene Epoch (20 to 25 million years ago) to the appearance of modern man some 100,000 years ago has been reconstructed mainly from the fossil record. Primates that are ancestors to both humans and apes are found as fossils from the Miocene Epoch (Figure 18.16). These **hominoids** originated in Africa between 4 and 8 million years ago and underwent a series of rapid evolutionary steps, giving rise to a diverse array of species. This diversity, coupled with the fragmentary fossil record, makes it difficult to draw conclusions about phylogenetic relationships among the early hominoids and to identify the ancestral line leading to modern **hominids.**

Hominoids Members of the primate superfamily Hominoidea, including the gibbons, great apes, and humans.

Hominids Member of the family Hominidae, which includes bipedal primates such as *Homo sapiens.*

Humanlike Hominids Appeared About 4 Million Years Ago

Reconstructions from the fossil record and molecular studies of higher primates indicate that humans are more closely related to chimpanzees than to other great apes

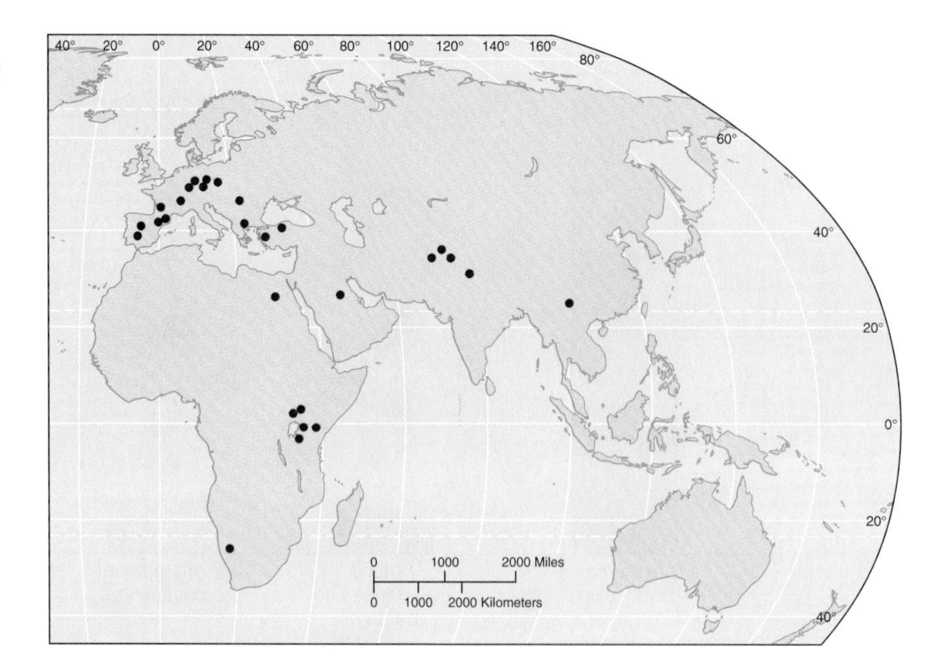

FIGURE 18.16 Hominid fossil sites from the Miocene (20 to 25 million years ago).

such as the gorilla and the orangutan. Fossil evidence places the most recent common ancestor of chimps and humans as living about 5 to 6 million years ago. Fossils spanning the next 3 to 4 million years indicate that five main groups of humanlike primates lived in Africa: *Kenyanthropus, Ardipithecus, Australopithecus, Paranthropus,* and our own genus, *Homo*. At least 13 species are represented in these five groups. There is general agreement that *Paranthropus* represents an evolutionary dead end and is not an ancestor to *Homo*. Fossils from the other three groups contain an overlapping mixture of human and nonhuman features that make it impossible at present to establish which of these groups is our ancestor. Evolutionary trees constructed from fossils of these species show a highly branched structure that creates more questions than answers (◗ Figure 18.17).

Let's briefly examine some of the features of these fossils. *Australopithecus afarensis* first appeared 3.6 to 4.0 million years ago. This species, which lived in East Africa, had a combination of apelike and humanlike characteristics. Members of this species walked upright on two legs (they were the creatures described in the opening vignette of the chapter), but the body proportions were apelike, with short legs and relatively long arms (◗ Figure 18.18). The arm bones of australopithecines are long and curved like those of chimpanzees but have humanlike elbows that could not support body weight while knuckle walking (which is how chimpanzees and gorillas move around). This combination of features is well suited for a species that spends time climbing in trees but walks on two legs across the ground. The australopithecines have apelike skulls (◗ Figure 18.19) and small brains; receding faces; and large, canine teeth.

The species *Kenyanthropus platyops*, which lived around a lake in western Kenya between 3.2 and 3.5 million years ago, is known from a skull and some den-

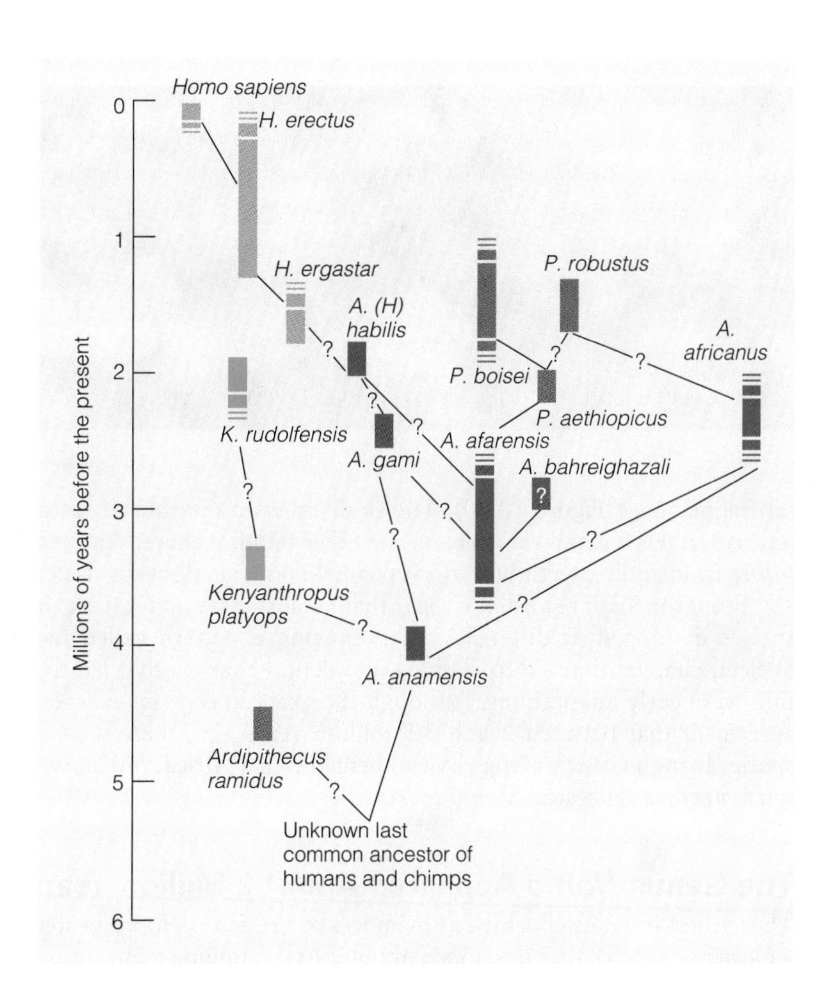

◗ **FIGURE 18.17** One of the possible patterns of evolutionary relationships among the five major groups of hominids: *Ardipithecus* (purple), *Australopithecus* (blue), *Paranthropus* (red), *Kenyanthropus* (green), and *Homo* (orange). Question marks indicate uncertain relationships. The horizontal bars represent uncertainty about time spans. (From Lieberman, D.E. (2001). Another face in our family tree. *Nature, 410,* 419–420, Figure 2, p. 420.) Adapted with permission from *Nature.* Copyright 2001 Macmillan Magazines Limited.)

▶ **FIGURE 18.18** A reconstruction of a group of *Australopithecus afarensis* gathering and eating fruits and seeds.

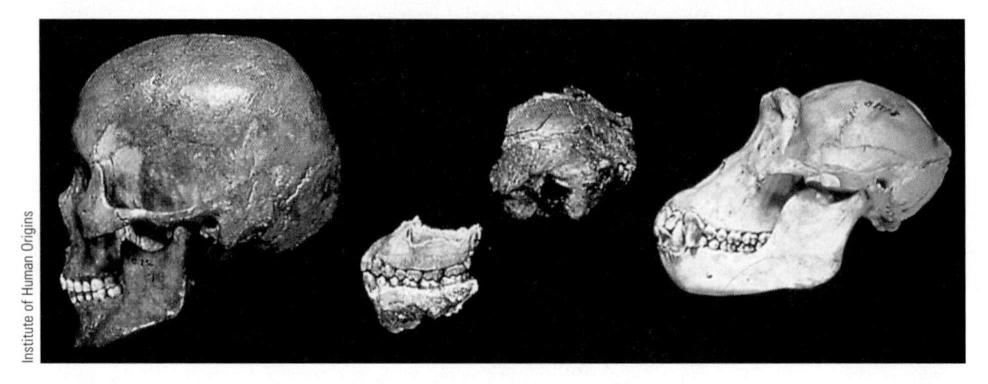

▶ **FIGURE 18.19** Comparison of the skulls of modern humans (left), *Australopithecus* fragments (center), and chimpanzee (right).

tal fragments (▶ Figure 18.20). The skull shows a mixture of features. Like chimpanzees, it has a small ear hole; on the other hand, it shares features with *Australopithecus*, including teeth with thick enamel and a small brain size.

From the fossil record, it is clear that evolutionary adaptations in these hominid species developed at different rates, showing a pattern called mosaic evolution. Skeletal changes that led to walking and dental changes that led to diet diversification were early adaptations. Although the exact lineage is unclear, there is general agreement that between 2 and 3.5 million years ago, there were several hominid species living in overlapping environments in East Africa, one of which gave rise to our immediate ancestor, *Homo erectus*.

The Genus *Homo* Appeared About 2 Million Years Ago

The earliest humans classified as members of the genus *Homo* probably appeared in a cluster of species that developed about 2 to 2.5 million years ago. This group con-

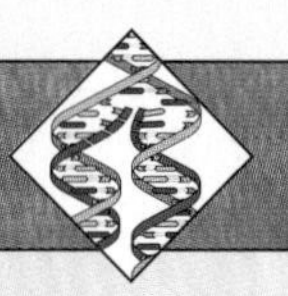

Concepts and Controversies

Tool Time: Did *Homo erectus* Use Killer Frisbees?

The appearance of *Homo erectus* about 2 million years ago was a turning point in human evolution. Tooth wear patterns indicate that meat was an important component of *H. erectus'* diet. This species was also the first hominid to move out of Africa into Europe and Asia, exploiting new habitats as resources. Evidence indicates that these and other changes were accompanied by technological innovation in the form of new tools. The use of tools predates *H. erectus* by about a million years, but these earlier tools (called Oldowan tools) were small, were used mainly for chopping, and remained unchanged over a span covering a million years.

The tools of *H. erectus* (called Acheulian tools) were large, had two cutting faces, and included hand axes, cleavers, and picks (see art). The typical tool kit also included scrapers and trimmers for producing and sharpening new tools. Acheulian tools were gradually refined but changed little in the million years they were used. Few new tools were added to the basic tool kit. These tools disappeared about 200,000 years ago, at a time when toolmaking entered a period of technological innovation and refinement that marks the Middle Paleolithic (Middle Stone Age).

Of all the tools in the kit of *H. erectus*, the possible uses of the hand ax have remained a subject of speculation. Modern-day anthropologists have learned how to make such tools and have used them as small axes for chopping or as heavy-duty knives for slicing animal hides or skinning carcasses. Examination of fossil hand axes by electron microscopy indicates that these tools may have been used on many materials including hide, meat, bone, and even wood. One form of the hand ax is ovoid and has a pointed end (see art). Their size and shape has led to the suggestion that they may have been thrown like Frisbees into a herd of small animals, stunning or wounding one, which could then be overtaken and killed. Flight tests of fossils and replicas support the idea of killer Frisbees, but this use is not widely accepted by anthropologists and remains among the most speculative proposals for the function of hand axes. Whatever their uses, the tools of *H. erectus* were associated with dramatic changes in behavior, including diet, migration, systematic hunting, the use of fire, and the establishment of home bases or camps. The role of new technology in promoting or supporting these new behaviors remains an area of intense investigation in paleoanthropology.

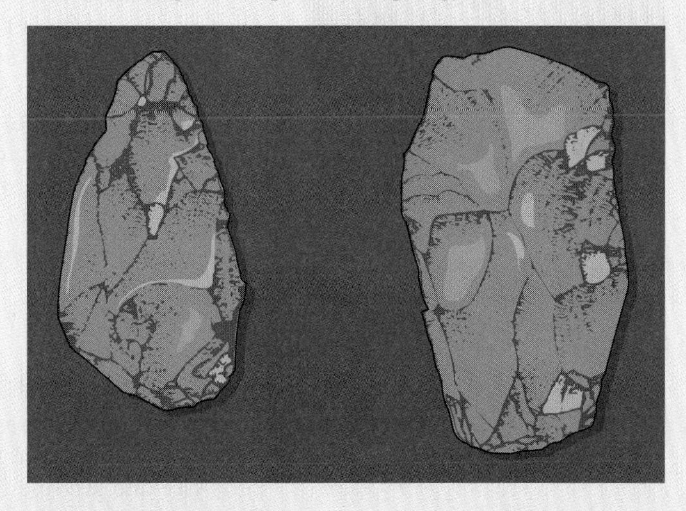

National Museums of Kenya, Nairobi, Kenya

FIGURE 18.20 (Left) A recently discovered skull of *Kenyanthropus platyops.* (Right) A skull from another species of *Kenyanthropus.* (From Lieberman, D. E. (2001). Another face in our family tree. *Nature 410,* 419–420.)

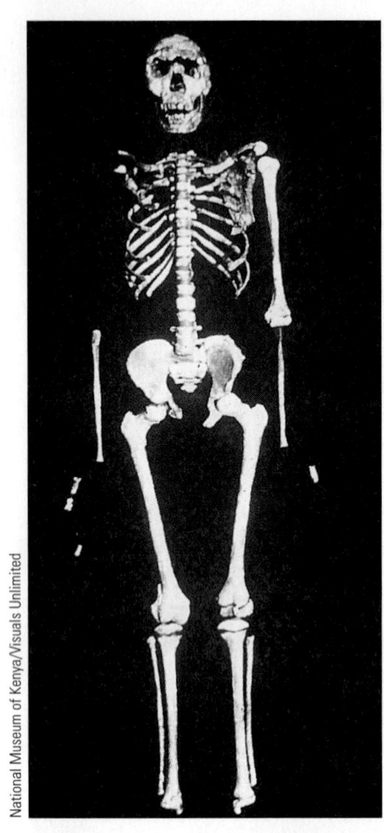

▸ **FIGURE 18.21** A skeleton of *Homo erectus.* This is the most complete skeleton of *H. erectus* found to date and is of a boy who lived some 1.6 million years ago.

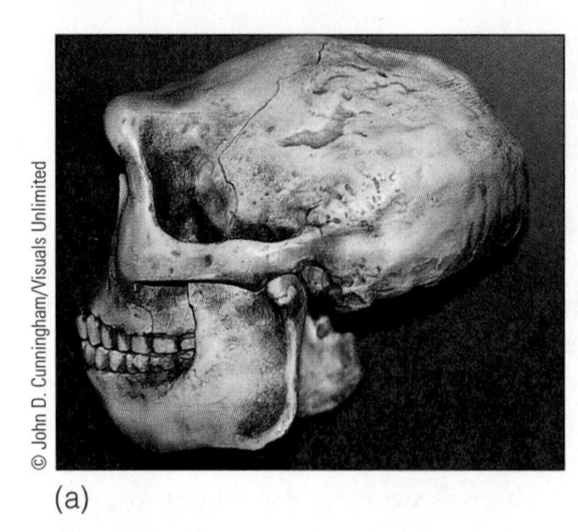

(a)

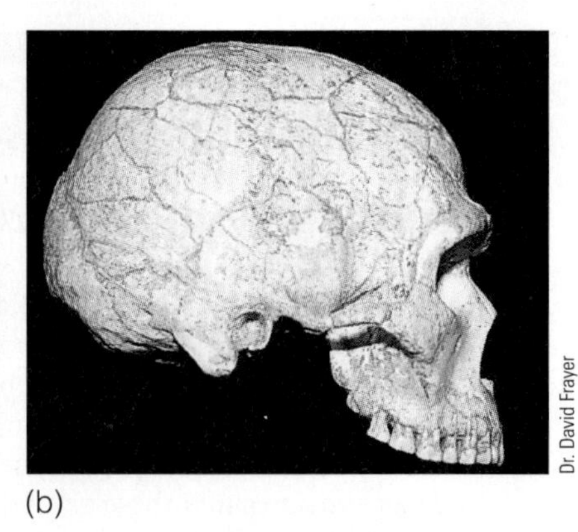

(b)

▸ **FIGURE 18.22** Skulls of (a) *Homo erectus* and (b) *Homo sapiens.* The skull of *H. erectus* is low and pointed in the rear, whereas the *H. sapiens* skull is high and dome-shaped.

tains several species. Members of this species cluster, which arose about 2 million years ago, had a brain size about 20% larger than that of the australopithecines. They also had differently shaped skulls and teeth distinct from those of australopithecines. At least one and probably several species of the genus *Homo* lived alongside the australopithecines in East Africa for about a million years, and other members of the genus *Homo* lived in South Africa at the same time. After this period, the australopithecines became extinct. Sometime during this period, a species that became our immediate ancestor arose within the *Homo* line. This hominid, called *Homo erectus* (▸ Figure 18.21), represents an important turning point in human evolution (see Concepts and Controversies: Tool Time: Did *Homo erectus* Use Killer Frisbees?)

Homo erectus Originated in Africa

African fossils of *H. erectus* date to about 1.8 million years ago, but the species may actually be much older. Although *H. erectus* originated in Africa, members migrated into Asia and Europe. Fossils of *H. erectus* have been recovered from Indonesia (Java man), China (Peking man), and sites in North Africa. The recent dating of Indonesian fossils of *H. erectus* to about 1.8 million years ago indicates that, as a species, *H. erectus* is probably older than 2 million years.

Physically *H. erectus* is different from earlier members of the genus *Homo* in several respects: increased brain size, a flatter face, and prominent brow ridges. There are physical similarities between *H. erectus* and modern humans, including height and walking patterns, but there are also some significant differences. In some populations of *H. erectus,* the skull is pointed in the back, not dome-shaped as in *Homo sapiens* (▸ Figure 18.22). In addition, *H. erectus* had a receding chin, a prominent brow ridge, and some differences in teeth.

THE APPEARANCE AND SPREAD OF *HOMO SAPIENS*

Tracing the origins of our species has become a multidisciplinary task, using the tools and methods of anthropology, paleontology, and archeology; the techniques of genetics and recombinant DNA technology; and more recently, satellite mapping from space. These methods are being used to reconstruct the origins and ancestry of populations of *H. sapiens* and to determine how and when our species originated and became dispersed across the globe.

Two Theories Differ on How and Where *Homo sapiens* Originated

From the evidence provided by fossils and artifacts, it is clear that populations of *H. erectus* moved out of Africa 1 to 2 million years ago and spread through parts of the Middle East and Asia. What currently divides anthropologists and evolutionary biologists is the question of how and where *H. sapiens* originated. In a general sense, there are two opposing views about the origin of modern humans. One idea (often called the out-of-Africa hypothesis) argues that after *H. erectus* moved out of Africa, populations that remained behind continued to evolve and gave rise to *H. sapiens* about 200,000 years ago. From this single source in Africa, modern humans migrated to all parts of the world, displacing and driving into extinction members of our related species, *H. erectus*. According to this model, modern human populations are all derived from a single speciation event that took place in a restricted region within Africa. As a result, the human populations of today should show a high degree of genetic relatedness (which they do).

This model of human evolution was based on evidence that members of African populations have the greatest amount of genetic diversity, measured by differences in mitochondrial DNA nucleotide sequence. Members of non-African populations show much less diversity. The underlying assumption in these studies is that these mutational changes accumulate at a constant rate, providing a "molecular clock" that can be calibrated by studying the fossil record.

Studies of mitochondrial DNA reveal a single ancestral mitochondrial lineage for our species that originated in Africa. Calculations using the molecular clock indicate that our species originated about 200,000 years ago from an African population that might have been made up of about 10,000 individuals.

The second idea about the origin of our species is called the multiregional hypothesis. According to this idea, after populations of *H. erectus* spread from Africa over the Middle East and Asia, *H. sapiens* developed as the result of an interbreeding network descended from the original colonizing populations of *H. erectus*. The evidence to support this model is derived from a combination of genetic and fossil evidence. The fossil record shows a gradual transition from archaic to modern humans that took place at multiple sites outside of Africa. In this model, *H. erectus* became gradually transformed into *H. sapiens,* instead of being replaced by *H. sapiens*.

The two opposing ideas can be summarized as follows: one favors speciation and replacement (the out-of-Africa model), and the other favors evolution and transition within a single species (the multiregional model). These ideas are hotly debated and have received a great deal of attention in the press and other media. These alternative explanations for the appearance of modern humans show that scientists can reach different conclusions about the same problem.

The accuracy of the molecular clock and the method used to construct mitochondrial phylogenetic trees has been called into question. However, studies of genetic variation in nuclear genes support a single point of origin and a timescale consistent with the out-of-Africa hypothesis. In addition, the distribution of RFLPs on the Y chromosome (which is passed from father to son) is consistent with the origin of our species at a single site about 200,000 to 270,000 years ago. In sum, the genetic evidence supports the out-of-Africa model across studies of mitochondrial DNA, microsatellite data, autosomal genes, and Y chromosomes. In spite of this, the issue over the origins of *H. sapiens* has not been resolved for several reasons: Some genetic evidence is difficult to reconcile with the out-of-Africa model, the fossil record is difficult to interpret, and human population dynamics are very complex. Although we have considered only two possible origins for our species, further work may produce additional ideas. Each will have to be evaluated by using the available information, by acquiring new information, and perhaps by applying new techniques.

SIDEBAR

Stone Age Fabrics

When did humans begin weaving fabrics? This is a difficult question to answer because plant fibers decay and leave little for archaeologists to discover. Twisted fibers found in Israel about 19,000 years old are the earliest indications that humans used cord or fibers. Recently, pottery excavated from a site in the Czech republic between 1952 and 1972 has provided evidence that textile production in Central Europe was well developed some 27,000 years ago.

Researchers who examined pottery fragments from the site spotted fabric imprints, pressed into the clay of the vessel while it was still wet. These impressions were made by loosely woven cloth or flexible basketry. The texture of the imprints suggests that the material might have been made using a loom. Because of the sophisticated weave of the material, archaeologists suggest that weaving and fabric use had a beginning much earlier than 27,000 years ago and must have been used in cultures before the development of agriculture.

The Spread of Humans Across the World

As summarized previously, there is strong evidence to support the idea that *H. sapiens* originated in Africa and spread from there to other parts of the world (▶ Figure 18.23). Based on the molecular evidence and some fossil evidence, it appears that modern *H. sapiens* originated in Africa some 130,000 to 170,000 years ago and that a small subset of this population emigrated from Africa about 137,000 years ago. There may have been one primary migration or several from a base in northeast Africa. The emigrants carried a subset of the variation present in the African population, consistent with the finding that present-day non-African populations have a small amount of genetic variation.

Modern forms of *H. sapiens* spread through Central Asia some 50,000 to 70,000 years ago and into Southeast Asia and Australia about 40,000 to 60,000 years ago. *H. sapiens* moved into Europe some 40,000 to 50,000 years ago, displacing the Neanderthals who lived there from about 100,000 years ago to about 30,000 years ago. Recent analysis of DNA from the bones of Neanderthal skeletons indicates that they were not the ancestors to European populations of *H. sapiens,* further supporting the out-of-Africa hypothesis.

Genetic data and recent archeological findings indicate that North and South America were populated by three or four waves of migration that occurred 15,000 to 30,000 years ago. Migrations from Asia across the Bering Sea are well supported by archeological and genetic findings, but Asia may not have been the only source of the first Americans. Some skeletal remains, such as Kennewick man and the Spirit Cave mummy, have features that more closely resemble Europeans than Asians. Evidence from a mitochondrial DNA variant called haplotype X, found only in Europeans, as well as a reinterpretation of stone tool technology, all make it seem likely that Europeans migrated to North America more than 10,000 years ago. Although a model with migrations from two sources explains most of the data available, there are other issues that remain to be resolved. Nonetheless, it is clear that genetic analysis of present-day populations coupled with anthropology, archaeology, and linguistics can provide a powerful tool for reconstructing the history of our species.

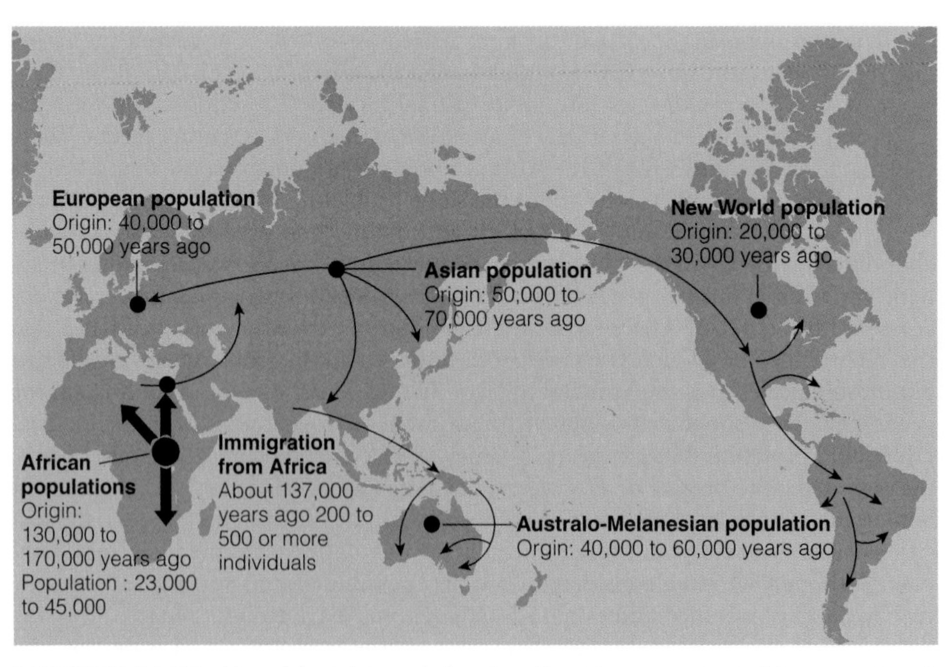

▶ **FIGURE 18.23** The origin and spread of modern *H. sapiens* reconstructed from genetic and fossil evidence.

Summary

Measuring Genetic Diversity in Human Populations

1. Studies indicate that human populations carry a large amount of genetic diversity. Natural selection acts on genetic diversity in populations to drive the process of evolution. All genetic variants originate by mutation, but mutation is an insignificant force in bringing about changes in allele frequency. Other forces, including genetic drift, act on the genetic variation in the gene pool and are responsible for changing the frequency of alleles in the population. Drift is a random process that acts in small, isolated populations to change allele frequency from generation to generation. Examples include island populations or those separated from general population by socioreligious practices.

Natural Selection and the Frequency of Genetic Disorders

2. Selection increases the reproductive success of fitter genotypes. As these individuals make a disproportionate contribution to the gene pool of succeeding generations, genotypes change. The differential reproduction of fitter genotypes is known as natural selection. Darwin and Wallace identified selection as the primary force in evolution that leads to evolutionary divergence and the formation of new species. The high frequency of genetic disorders in some populations is the result of selection that often confers increased fitness on the heterozygote.

Human Activity as a Force in Changing Allele Frequencies

3. Humans have developed a set of adaptive mechanisms collectively known as culture. Culture has a profound effect on the interaction of the gene pool with the environment and can also act as a force that brings about changes in allele frequencies. Although both plant and animal species employ strategies of dispersal and migration, human technology makes it possible for humans to migrate to any place on Earth within a short time. The net effect of this short migratory time is the reduction of allele frequency differences among populations. In some instances, migration into small, relatively isolated populations can rapidly generate changes in allele frequencies.

Genetic Variation in Human Populations

4. Patterns of geographic distribution of polymorphisms can be used to assess the spread of new alleles through populations. The frequency of tawny hair in Australian aborigines is an example of such a polymorphism.

The Question of Race

5. The biological concept of race evolved from emphasis on phenotypic differences to genotypic differences. Information from variations in proteins, microsatellites, and nuclear genes shows that most human genetic variation is present within populations rather than between populations. For this reason, there is no genetic basis for dividing our species into races.

Primate Evolution and Human Origins

6. The fossil record from the Miocene Epoch can be used to trace the evolution of hominoids and hominids, primates that gave rise to the present human species. The incomplete nature of the fossil record makes it difficult to construct a phylogeny, but now techniques of molecular biology are being used to clarify the origin of humans.

The Appearance and Spread of *Homo sapiens*

7. A combination of linguistics, archaeology, anthropology, and genetics is being used to reconstruct the dispersal of human populations across the globe. The evidence available suggests that North and South America were populated by waves of migration sometime during the last 15,000 to 30,000 years. In Europe, the alignment of genetic and linguistic barriers suggests that culture, in the form of language, can be a force in establishing and maintaining genetic differences among populations.

Case Studies

CASE 1

Natural selection alters genotypic frequencies by increasing or decreasing fitness (that is, differential fertility or mortality). There are several examples of selection associated with human genetic disorders. Sickle cell anemia and other abnormal hemoglobins are the best examples of selection in humans. Carriers of the sickle and other hemoglobin mutations are more resistant to malaria than either homozygous class. Therefore, in areas where malaria is endemic, carriers are less likely to die of malaria and will have proportionally more offspring than homozygotes, thus passing on more genes. Balancing selection may also have influenced carrier frequencies for more "common" recessive diseases, such as cystic fibrosis in Europeans and Tay-Sachs in the Ashkenazi Jewish population, but the selective agent is not known for certain.

Selection may favor homozygotes over heterozygotes, resulting in an unstable polymorphism. One example is selection against heterozygous fetuses when an Rh-negative mother carries an Rh-positive (heterozygous) fetus. This should result in a gradual elimination of the Rh-negative allele. However, the high frequency of the Rh-negative allele in so many populations suggests that other unknown factors may maintain the Rh-negative allele in human populations.

1. If you suspected that heterozygous carriers of a particular disease gene had a selective advantage in resisting a type of infection, how would you go about testing this hypothesis?
2. If allele frequencies in the hemoglobin gene are influenced by sickle cell anemia on one hand, and by resistance to malaria on the other, what factors may cause a change in these allele frequencies over time?

Questions and Problems

Measuring Genetic Diversity in Human Populations

1. Distinguish between mutations and polymorphisms. How are polymorphisms used in the study of genetic variation?
2. Why is it that mutation, acting alone, has little effect on gene frequency?
3. Do the differences we see in the ABO blood group polymorphisms represent adaptive changes, or do they reflect some other process not important to the fitness of populations in which they occur?
4. Successful adaptation is defined by:
   a. evolving new traits.
   b. producing many offspring.
   c. increasing fitness.
   d. moving to a new location.
5. What is the relationship between founder effects and genetic drift?
6. How would a drastic reduction in a population's size affect the population's gene pool?
7. The major factor causing deviations from Hardy-Weinberg equilibrium is:
   a. selection.
   b. nonrandom mating.
   c. mutation.
   d. migration.
   e. early death.
8. A specific mutation in the *BRCA1* gene has been estimated to be present in approximately 1% of Ashkenazi Jewish women of Eastern European descent. This specific alteration, 185delAG, is found about three times more often in this ethnic group than the combined frequency of the other 125 mutations found to date. It is believed that the mutation is the result of a founder effect from many centuries ago. Explain the founder principle.
9. The theory of natural selection has been popularly summarized as "survival of the fittest." Is this an accurate description of natural selection? Why or why not?

Natural Selection Affects the Frequency of Genetic Disorders

10. Will a recessive allele that is lethal in the homozygous condition ever be removed from a large population by natural selection?
11. How are genetic polymorphisms maintained in a population?

Human Activity Can Change Allele Frequencies

12. What are examples of nonrandom mating, and how does this behavior affect genetic equilibrium?
13. How do you think the development of culture affected the process of human evolution? Is our present culture affecting selection? Can you give specific examples?
14. Do you think that our species is still evolving or are we shielded from natural selection by civilization? Is it

possible that misapplications of technology will end up exposing our species to more rather than less natural selection (consider the history of antibiotics)?

What Is Race?

15. a. Provide a genetic definition of race.
    b. Using this definition, can modern humans be divided into races? Why or why not?

Primate Evolution and Human Origins

16. You are a paleoanthropologist searching for remnants of the earliest human ancestors and are examining some fossils you found in a geological layer dating to about 4 million years ago. What types of characteristics would you look for in the fossils that should resemble modern humans? What features would you expect to be more apelike?

The Appearance and Spread of *Homo sapiens*

17. a. Briefly describe the two major theories discussed in this chapter about the origin of modern humans.
    b. Which of these two theories would predict a closer relationship for the various modern human populations?
    c. Which of the two theories is best supported by the genetic evidence?

Internet Activities

The following activities use the resources of the World Wide Web to enhance the topics covered in this chapter. To investigate the topics described below, log on to the book's home page (www.brookscole.com/biology_d) and follow the *Student Resources* link to your subject and text.

1. *DNA, Archeology, and Human History.* Read the article "Scientists Rough Out Humanity's 50,000 Year Old History" (http://www.nytimes.com/learning/general/featured_articles/001115wednesday.html) at the *New York Times Learning Network* site. When did the earliest migrations of modern humans into Europe seem to have occurred? What percentage of the current population of Europe seems to be descended from those earliest immigrants? What kind of DNA evidence was used in this analysis?
2. *Human Evolution Overview.* Review Jim Foley's Web site at the *Talk Origins Archive* (http://www.talkorigins.org/faqs/homs/) for a good overview of hominid evolution, descriptions, and photographs of important fossil finds that currently shape our views on hominid evolution, and rebuttals of creationists' arguments regarding evolutionary theory.
   a. Compare and contrast the scientific evidence and hypotheses with the creationists' viewpoints on the australopithecines, *Homo erectus,* and the Neanderthals. Read the overview article of creationists' arguments. For more sites on both sides of the argument, click on the "Links" icon at the bottom of the home page. Can you summarize the differences between the creationist and evolutionist points of view? Which arguments do you find more persuasive? Does the debate itself have merit?
   b. Click on the *Evolution* link (http://www.talkorigins.org/origins/faqs-evolution.html) to get an evolutionist's perspective on some of the reasons for confusion about evolution between scientists and nonscientists. What do you think of the ideas presented here?
3. *Solving a Genetic Mystery.* Access the Cold Spring Harbor DNA Learning Center's *Mystery of the Romanovs* page (http://vector.cshl.org/bioforms/guide.pl?b_id=1&p_id=1). Polymorphisms in mitochondrial DNA can provide information about genetic relatedness as well as diversity and species origins. You can solve the mystery of the Romanovs by comparing single nucleotide polymorphisms (SNPs) in the mitochondrial DNA sequence.

For Further Reading

Ayala, F. (1984). Molecular polymorphism: How much is there, and why is there so much? *Dev. Genet., 4,* 379–391.

Bahn, P. (1993). 50,000 year old Americans of Pedra Furada. *Nature, 362,* 114–115.

Barbujani, G., & Sokol, R. (1990). Zones of sharp genetic change in Europe are also linguistic boundaries. *Proc. Nat. Acad. Sci. U S A, 87,* 1816–1819.

Bodmer, W. F., & Cavelli-Sforza, L. L. (1976). *Genetics, evolution and man.* San Francisco: Freeman.

Cavalli-Sforza, L., Menzonni, P., & Piazza, A. (1993). Demic expansion and human evolution. *Science, 259,* 639–646.

Durham, W. (1991). *Coevolution: Genes, culture and human diversity.* Stanford, CA: Stanford University Press.

Gould, S. J. (1982). Darwinism and the expansion of evolutionary theory. *Science, 216,* 380–387.

Gyllensten, U. B., & Erlich, H. A. (1989). Ancient roots for polymorphism at the HLA-DQ alpha locus in primates. *Proc. Nat. Acad. Sci. U S A, 86,* 9986–9990.

Kennedy, K. (1976). *Human variation in space and time.* Dubuque, Iowa: Brown.

Lewontin, R. (1974). *The genetic basis of evolutionary change.* New York: Columbia University Press.

Little, B. B., & Malina, R. M. (1989). Genetic drift and natural selection in an isolated Zapotec-speaking community in the valley of Oaxaca, Southern Mexico. *Hum. Hered., 39,* 99–106.

Mayr, E. (1963). *Animal species and evolution.* Cambridge, MA: Harvard University Press.

Molnar, S. (1983). *Human variation.* Englewood Cliffs, NJ: Prentice Hall.

Relethford, J. H. (1988). Heterogeneity of long-distance migration in studies of genetic structure. *Ann. Hum. Biol., 15,* 55–63.

Roberts, D. F. (1988). Migration and genetic change. Raymond Pearl lecture 1987. *Hum. Biol., 60,* 521–539.

Romero, G., Devoto, M., & Galietta, L. J. (1989). Why is the cystic fibrosis gene so frequent? *Hum. Genet., 84,* 1–5.

Schanfield, M. (1992). Immunoglobulin allotypes (GM and KM) indicate multiple founding populations of native Americans: Evidence of at least four migrations to the New World. *Hum. Biol., 64,* 381–402.

Semino, O., Torrino, A., Scozzari, R., Brega, A., De Benedictis, G., & Santachiara-Benerecetti, A. S. (1989). Mitochondrial DNA polymorphisms in Italy: III. Population data from Sicily: A possible quantitation of maternal African ancestry. *Ann. Hum. Genet., 53,* 193–202.

Shields, G., Schmiechen, A., Frazier, B., Redd, A., Voevoda, M., Reed, J., & Ward, R. (1993). mtDNA sequences suggest a recent evolutionary divergence for Beringian and northern North American populations. *Am. J. Hum. Genet., 53,* 549–562.

Smith, J. M. (ed.). (1982). *Evolution now: A century after Darwin.* San Francisco: Freeman.

Stanton, W. (1960). *The leopard's spots.* Chicago: University of Chicago Press.

Torroni, A., Schurr, T., Cabell, M., Brown, M., Neel, J., Larsen, M., Smith, D., Vullo, C., & Wallace, D. (1993). Asian affinities and continental radiation of the four founding native American mtDNAs. *Am. J. Hum. Genet., 53,* 563–590.

Torroni, A., Sukernik, R., Schurr, T., Starikovskaya, Y., Cabell, M., Crawford, M., Comuzzie, A., & Wallace, D. (1993). mtDNA variation of aboriginal Siberians reveals distinct genetic affinities with native Americans. *Am. J. Hum. Genet., 53,* 591–608.

Towne, B., & Hulse, F. S. (1990). Generational changes in skin color variation among Habbani Yemeni Jews. *Hum. Biol., 62,* 85–100.

Wallace, D., & Torroni, A. (1992). American Indian prehistory as written in the mitochondrial DNA: A review. *Hum. Biol., 64,* 403–416.

Yunis, J. J., & Prakash, O. (1982). The origin of man: A chromosomal pictorial legacy. *Science, 215,* 1525–1530.

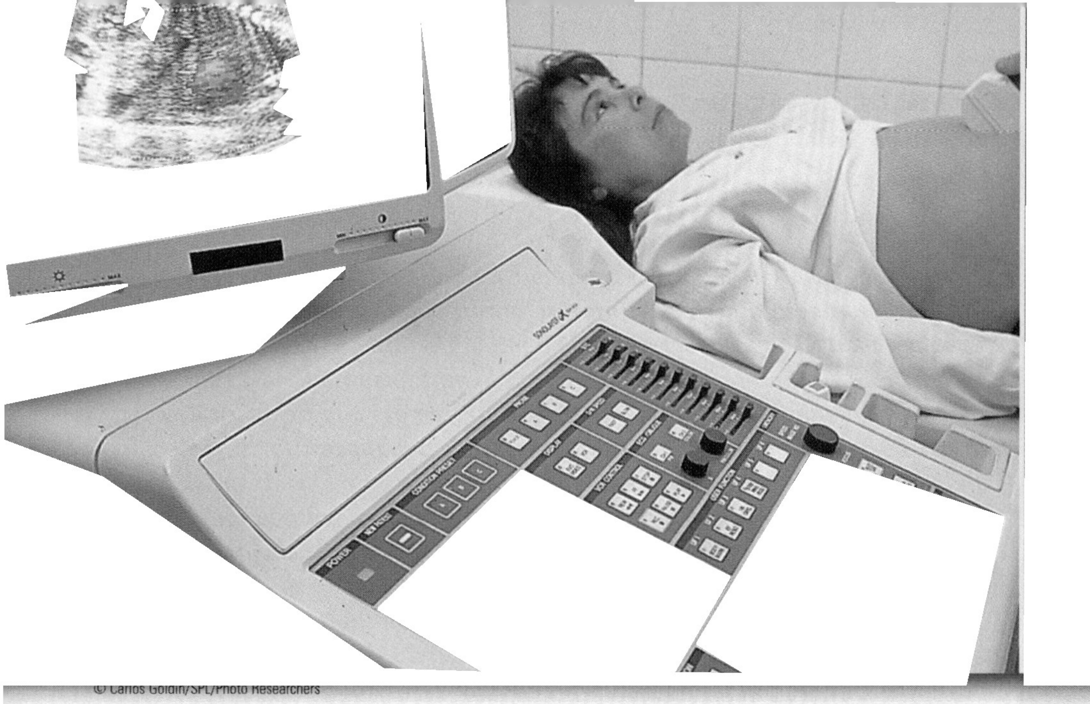

Genetic Testing, Gene Therapy, and Counseling

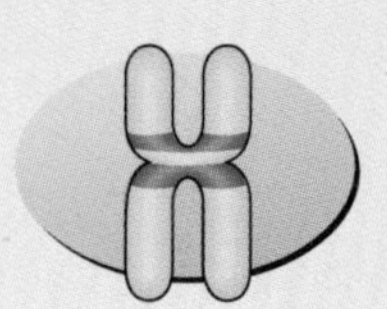

Chapter Outline

Concepts and Controversies The Business of Making Babies

How Does Genetic Counseling Work?

Future Directions

Huntington disease is inherited as an autosomal dominant trait. It is characterized by a gradual loss of motor coordination, degenerative personality changes, and progressive dementia. The disease develops slowly and leads to death within 15 years of onset. Symptoms usually develop around the age of 40, beginning with involuntary limb tremors and behavioral changes. The gene for Huntington disease has been mapped (to chromosome 4) and cloned and can be used to identify those at risk for this disorder.

Suppose that your uncle, age 49, has recently been diagnosed with Huntington disease. His physician calls you to inform you of the diagnosis and indicates that you have a 25% chance of developing the disease. You are 25 years old, married, and have one child. Should you undergo testing to determine whether you are at risk? Can you live with the outcome if the results indicate that you will develop the disease? Do you want your children to be informed of your condition? Do you want them to be tested? Do you want them to know if they will develop the disease? Should you have any more children? What if you were not tested, but your child was tested and was found to have the Huntington disease gene? You are the only parent who could have transmitted the mutant allele. You now have information about your genotype that you may not have wanted. Does this mean that if you do not wish to be tested, you can prevent your child from being tested?

This scenario and the questions it raises are not hypothetical. Although Huntington disease affects only a small fraction of the population (3 or 4 out of 100,000), it is now possible to identify those who will develop this devastating disease. Helping individuals to understand the implications of genetic testing and to make informed decisions about actions to be taken are goals of genetic counseling.

In this chapter we survey genetic screening and testing and explore the rationale, methods, and potential for its use and misuse. We also consider the role of the genetic counselor in informing and educating those who undergo genetic screening. With the growth in genetic technology, these fields will have a great impact on our own lives and personal decisions and on those of our family members and friends.

GENETIC TESTING AND GENETIC SCREENING

Genetic testing The use of methods to determine if someone has a genetic disorder, will develop one, or is a carrier.

Genetic screening The systematic search for individuals who have certain genotypes.

Before we explore the rationale, methods, and issues surrounding genetic testing, let's explore the differences between genetic testing and genetic screening. **Genetic testing** uses a variety of techniques to determine if someone has a certain genotype. The term covers testing those who have or may carry a genetic disease, those at risk of producing a genetically defective child, and those who may have a genetic susceptibility to environmental agents. **Genetic screening** is carried out on populations, members of which may have a low incidence and a low risk for a particular genetic disorder (Figure 19.1). Genetic testing is most often a matter of choice, whereas genetic screening is often a matter of law. In this chapter we will review testing and screening.

Several outcomes of genetic testing should be considered. First, if someone is identified as having or being at risk for a genetic disorder, it often leads to the discovery of other affected or at-risk individuals within the same family. Second, testing often identifies individuals who will develop genetic disorders later in adult life. When the genetic defect diagnosed is traumatic and fatal, such as Huntington disease, this knowledge often has serious personal and social effects. Third, the results of genetic testing often have a direct impact on the offspring of the person who was tested. We will discuss these issues in a later section of the chapter.

There are several types of genetic testing:

- Carrier identification is used on members of families or cultural groups with a history of a genetic disorder, such as sickle cell anemia or cystic fibrosis.
- Prenatal diagnosis of a genetic condition is performed for a developing fetus, such as testing for Down syndrome
- Late-onset testing identifies individuals who are susceptible to conditions such as cancer or who will develop adult-onset genetic disorders such as Huntington disease or polycystic kidney disease (PCKD).

Carrier screening programs are often directed at phenotypically normal members of groups with a high frequency of an autosomal recessive or X-linked recessive disease. Some genetic disorders, including Tay-Sachs disease and sickle cell anemia, have been screened for on a large scale. The development of programs for these two diseases has been made possible by three factors:

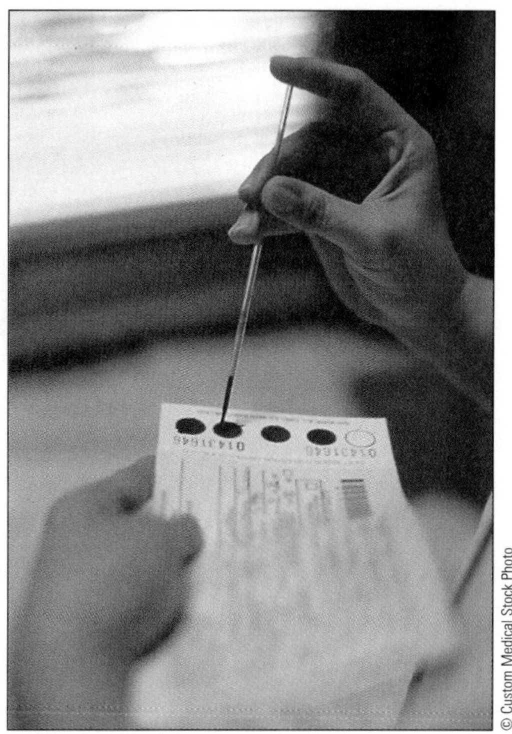

FIGURE 19.1 In a Guthrie test for phenylketonuria (PKU), paper disks that contain a drop of blood are placed on the surface of a bacterial culture that will grow only in the presence of phenylalanine. If an infant has PKU, rapid growth occurs around the disk and is visible as a halo.

- The diseases occur mainly in defined populations. Tay-Sachs carriers are found most frequently among Jews of East European origin, and sickle cell carriers are most common in U.S. blacks of West African origin.
- Carrier detection for these disorders is inexpensive and rapid.
- Prenatal testing for these disorders gives couples at risk the option of having only unaffected children.

In the following section, we will examine two carrier screening programs.

Carrier Screening for Tay-Sachs Disease

Tay-Sachs disease (MIM/OMIM 272800) is an autosomal recessive trait with a frequency of 1 in 360,000. As we discussed in Chapter 2, this is a disease of lysosome function and leads to mental retardation, blindness, and death by the age of 3 or 4. In Jews of Eastern European ancestry (Ashkenazi Jews), the frequency is almost 100 times higher (1 in 4,800 births) than in the general population (Figure 19.2). In the 1970s, carrier-detection programs were set up to identify heterozygotes. More than 300,000 individuals were tested in the first 10 years of screening. Of these, 268 couples were identified in which both members were carriers and had not yet had an affected child. The programs were coupled with counseling sessions that provided education about the risks of having an affected child, the availability of prenatal testing, and reproductive options. None of these programs is mandatory, although some states have laws that require informing couples that testing for Tay-Sachs disease is available. In the 1970s there were 50 to 100 Tay-Sachs births annually in the United States. Because of carrier-testing programs, there are now fewer than 10 such births each year.

Carrier Screening for Sickle Cell Trait

Sickle cell anemia (MIM/OMIM 141900) is an autosomal recessive condition that differentially affects Americans with family origins in West Africa and in the lowlands of the Mediterranean Sea, including Sicily, Italy, Greece, Lebanon, and Israel. Blood tests and recombinant DNA-based tests can identify heterozygous carriers and affected homozygotes. In 1971 Connecticut instituted a program of testing black schoolchildren in grades 7 to 12, with parental consent, for **sickle cell trait,** a term used to designate

Sickle cell trait The symptoms shown by those heterozygous for sickle cell anemia.

▶ **FIGURE 19.2** In Jews of Eastern European ancestry, the frequency of Tach-Sachs disease is almost 100 times higher (1 in 4,800 births) than in the general population.

heterozygous carriers. In 1972 Congress passed the National Sickle Cell Anemia Control Act, part of which was designed to establish carrier-detection programs. Some of the programs established as a result of this legislation were compulsory, requiring that black children be tested before attending school; others required testing before obtaining a marriage license. Professional football players were tested, as were cadets at the Air Force Academy in Colorado Springs, where heterozygotes were excluded from enrollment. The exclusion was based on the assumption that heterozygotes might undergo sickling of red blood cells at high altitudes under reduced oxygen concentrations. This policy was reversed in 1981 under the threat of a lawsuit. Other individuals who tested as positive heterozygotes were reportedly turned down for insurance and employment, even though carriers do not have any inherent health problems.

Some of the programs for sickle cell carrier detection were criticized for laxity in confidentiality of records and the failure to provide counseling to those identified as heterozygotes. In the late 1970s, many of the sickle cell testing programs were cut back or reorganized, and currently only a few states offer sickle cell testing, although more than 40 states screen all newborns to detect sickle cell homozygotes.

PRENATAL AND PRESYMPTOMATIC GENETIC TESTING

Genetic testing determines the genetic status of those suspected to be at high risk for a particular inherited condition. Prenatal testing detects genetic disorders and birth defects in the fetus. More than 200 single-gene disorders can be diagnosed prenatally (Table 19.1). In most cases the disorders are rare, and genetic testing is usually done only when there is a family history or another indication that warrants testing. For some conditions, such as Tay-Sachs disease or sickle cell anemia, parents can be tested to determine if either is a carrier. If the tests for both parents are positive, the fetus has a 25% chance of being affected. In such cases prenatal testing can determine whether the fetus is a recessive homozygote afflicted with the disorder. Similarly, if the mother is a carrier for certain deleterious, X-linked genetic disorders, testing can be offered.

For other genetic conditions, such as Down syndrome (trisomy 21), analysis of the fetal chromosomes is the most direct way to detect an affected fetus. Testing for

Table 19.1 Some Metabolic Diseases and Birth Defects That Can Be Diagnosed by Prenatal Testing	
Acatalasemia	Mannosidosis
Adrenogenital syndrome	Maple syrup urine disease
Chédiak-Higashi syndrome	Marfan syndrome
Citrullinemia	Muscular dystrophy, X-linked
Cystathioninuria	Niemann-Pick disease
Cystic fibrosis	Oroticaciduria
Fabry disease	Progeria
Fucosidosis	Sandhoff disease
Galactosemia	Spina bifida
Gaucher disease	Tay-Sachs disease
G6PD deficiency	Thalassemia
Homocystinuria	Werner syndrome
I-Cell disease	Xeroderma pigmentosum
Lesch-Nyhan syndrome	

Down syndrome is usually carried out not because of a familial history of genetic disease or detection of heterozygotes in the parents, but because of advanced maternal age. Because the risk of Down syndrome increases dramatically with maternal age (see Chapter 6), cytogenetic testing is recommended for all pregnant females older than 35 years of age.

Samples or images for prenatal testing can be obtained in several ways, including amniocentesis, chorionic villus sampling (CVS), ultrasonography, and fetoscopy. The fluids and cells obtained for testing can be analyzed by several techniques, including cytogenetics, biochemistry, and recombinant DNA technology. Images obtained by ultrasonography and fetoscopy can be used to diagnose conditions such as neural tube defects, cardiac abnormalities, and malformations of the limbs, associated with chromosomal aberrations.

Because recombinant DNA technology analyzes the genome directly, it is the most specific and sensitive method currently available. The accuracy, sensitivity, and ease with which recombinant DNA technology can be used to assemble a profile of the genetic diseases and susceptibilities carried by an individual have raised a number of legal and ethical issues that have yet to be resolved.

PRENATAL TESTING FOR GENETIC DISORDERS

Using recombinant DNA techniques, it is possible to test prenatally for the presence of mutant genes. In addition, both children and adults can be tested for genetic disorders before symptoms appear (some genetic disorders appear only later in life). Prenatal testing can detect the presence of a genetic disorder in an embryo or fetus. The use of recombinant DNA techniques extends the use of amniocentesis and CVS in prenatal testing (see Chapter 6 for a discussion of amniocentesis and CVS). Rather than testing for chromosomal or biochemical abnormalities, linking amniocentesis or CVS with recombinant DNA technology allows the direct testing of the genotype of an embryo or fetus. We will use the example of sickle cell anemia to illustrate how recombinant DNA methods are used in prenatal testing.

Prenatal Testing for Sickle Cell Anemia

As described in Chapter 4, sickle cell anemia is a recessive trait caused by a mutant form of the oxygen-carrying protein, β-globin. The β-globin gene is not active before birth, so the condition cannot be prenatally diagnosed by analyzing fetal hemoglobin, but it can be diagnosed using recombinant DNA techniques.

The mutation that causes sickle cell anemia destroys a restriction-enzyme cutting site, changing the length and number of DNA fragments generated by a specific enzyme (◗ Figure 19.3). By collecting fetal cells with amniocentesis and analyzing restriction fragments from the fetal DNA, the genotypes of normal homozygous, heterozygous, and affected homozygous individuals can be visualized directly on a Southern blot. Prenatal diagnosis can be performed by amniocentesis after the fifteenth week of development or by CVS in the eighth week of development. At present, dozens of genetic disorders can be detected prenatally using recombinant DNA techniques, and the list is growing steadily.

Testing Embryonic Blastomeres

In preimplantation testing, human eggs are fertilized *in vitro* and allowed to develop in a culture dish for several days. By this time, the fertilized egg has divided to form an early embryo. For testing, one of the six to eight embryonic cells (called blastomeres) is removed by microdissection. DNA is extracted from this cell, amplified by polymerase chain reaction (PCR), and tested for the presence of a genetic disorder in the embryo (◗ Figure 19.4). Blastomere testing is used for the most common form of cystic fibrosis, muscular dystrophy, Lesch-Nyhan syndrome, and hemophilia.

Removing a single cell from an early-stage embryo does not damage the embryo or alter its potential for development, and embryos not affected with a genetic disorder are implanted for development and birth. As the genes for more genetic dis-

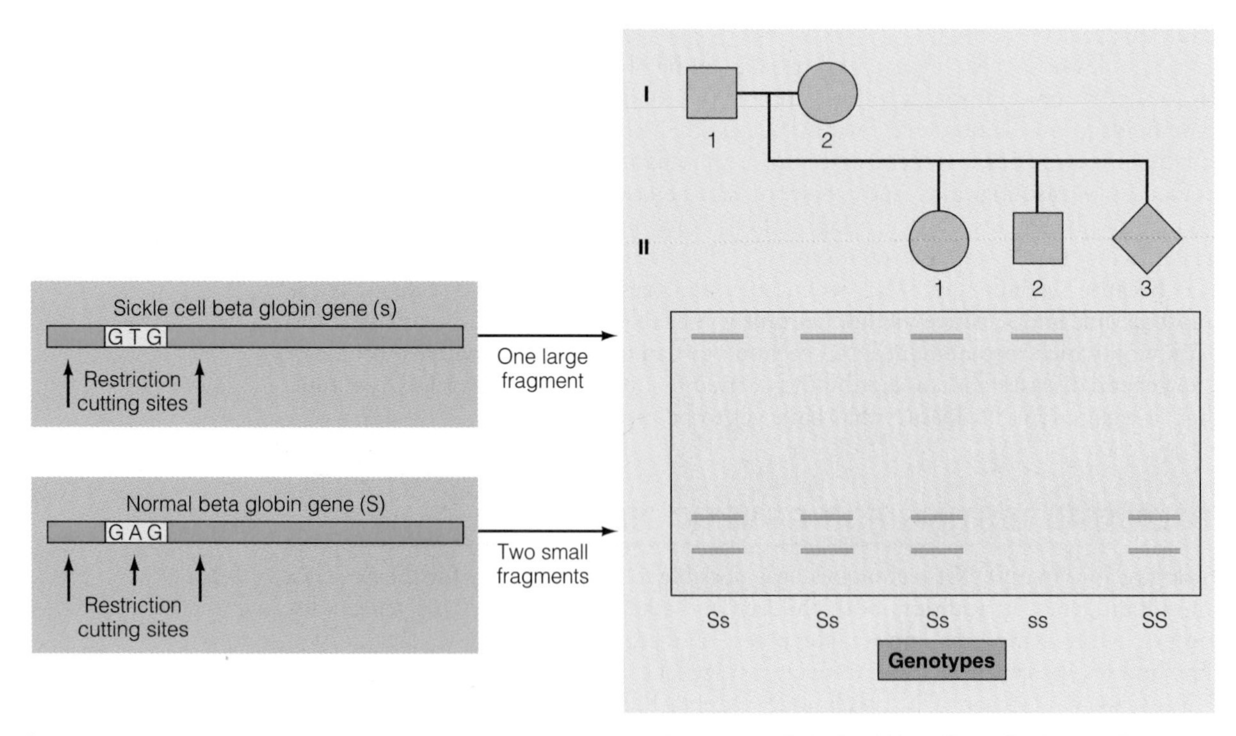

◗ **FIGURE 19.3** Prenatal diagnosis for sickle cell anemia by Southern blot analysis. In sickle cell anemia, the mutation destroys a cutting site for a restriction enzyme. As a result, the number and size of restriction enzyme fragments are altered. Because the mutant allele has a distinctive band pattern, the genotypes of family members can be read directly from the blot. The parents (I-1, I-2) are heterozygotes. The first child (II-1) is a heterozygous carrier; the second child (II-2) is affected with sickle cell anemia; and the unborn child (II-3) is homozygous for the normal alleles and therefore will be unaffected and will not be a carrier.

orders are identified and sequenced, it will be possible to test for an increasing number of these diseases.

In addition to genetic disorders, prenatal testing can detect defects of embryonic development. Among these are defects in the formation of the neural tube, a structure that arises in the first 2 months of development. One such defect, spina bifida, is a condition in which the spinal column is open or partially open. Neural tube defects can be diagnosed accurately by testing the amniotic fluid for elevated levels of alpha fetoprotein. In about 80% of cases, alpha fetoprotein levels in the maternal blood serum are also elevated. A test using maternal blood can identify mothers for whom further tests, such as amniocentesis, are recommended. Other methods of obtaining information about the fetus, including ultrasonography and fetoscopy, use visual images to examine the fetus and detect birth defects.

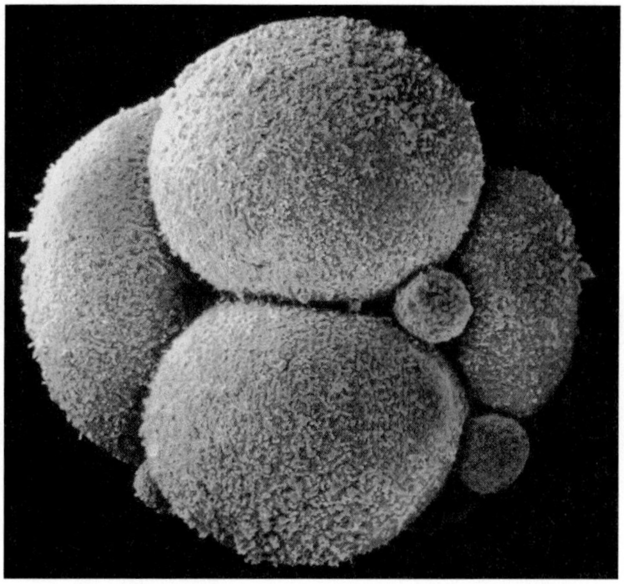

FIGURE 19.4 In blastomeric testing, a single cell is surgically removed from an early embryo, and the DNA from this cell is used to test for a genetic disease.

Ultrasonography

Ultrasonography is based on sonar technology developed for military use during World War II. For prenatal diagnosis, a transducer is placed on the abdomen over the enlarged uterus (Figure 19.5). The probe emits pulses of ultrasonic energy, and as the ultrasound strikes the surface of the fetus, some waves are reflected and return to the transducer. These reflected waves are electronically converted to images and displayed on a screen (Figure 19.6).

Ultrasound can help diagnose multiple pregnancy; determine fetal sex; and identify neural tube defects, other central nervous system defects, skeletal disorders, limb malformations, and congenital heart defects.

Fetoscopy

Fetoscopy directly visualizes the fetus using a fiberoptic endoscope. For fetoscopy, a hollow needle is inserted through the abdominal wall into the amniotic cavity, and the fiberoptic cable is threaded through the needle. The image is transmitted to a

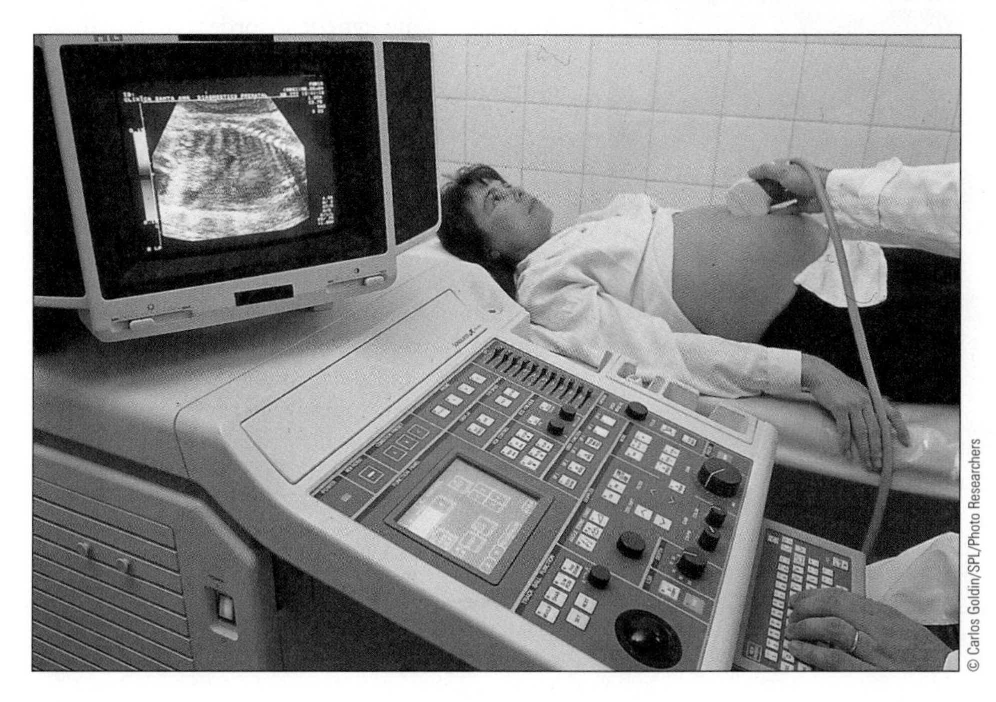

FIGURE 19.5 A pregnant woman is undergoing an ultrasound examination.

▸ **FIGURE 19.6** Baby Lavery as seen in ultrasound at 20 weeks.

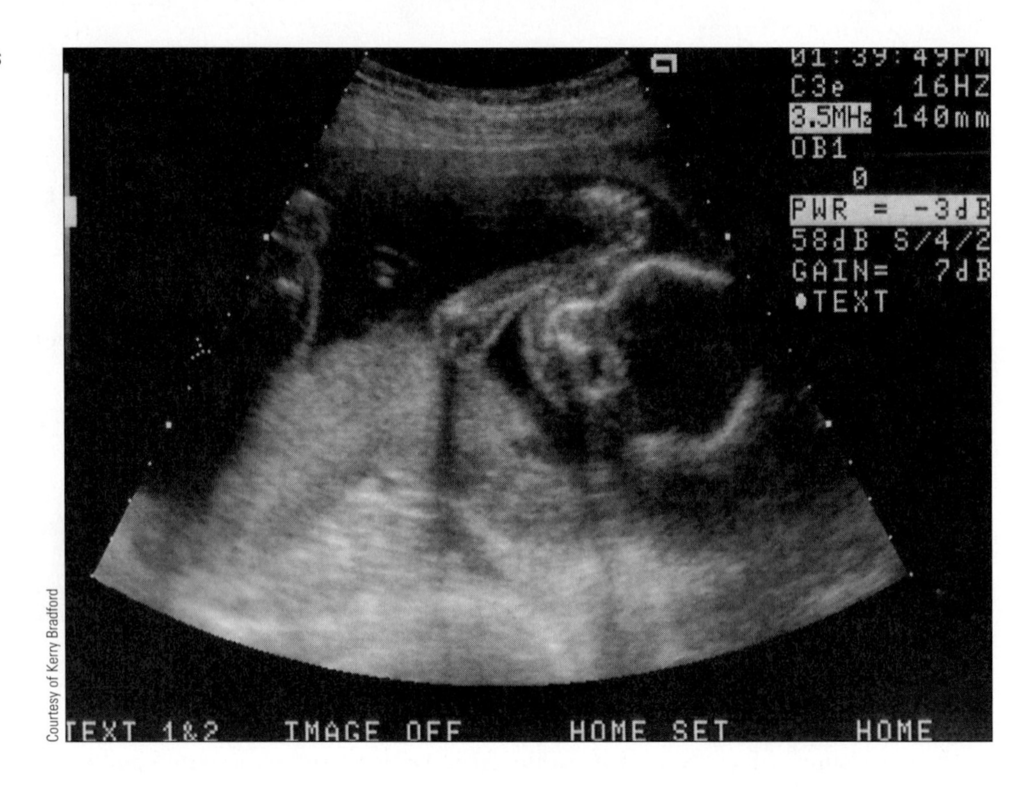

Courtesy of Kerry Bradford

video screen and can be viewed and recorded from the screen (▸ Figure 19.7). The technique is most useful when a disorder cannot be diagnosed by cytogenetic or biochemical methods. Fetoscopy can also be used to obtain samples of fetal blood, allowing diagnosis of genetic diseases such as hemophilia and certain forms of thalassemia. Fetoscopy poses a danger to the fetus, however, and there is a 2% to 5% chance of spontaneous abortion in using this method.

▸ **FIGURE 19.7** The hand and face of a 9-week-old fetus seen by fetoscopy.

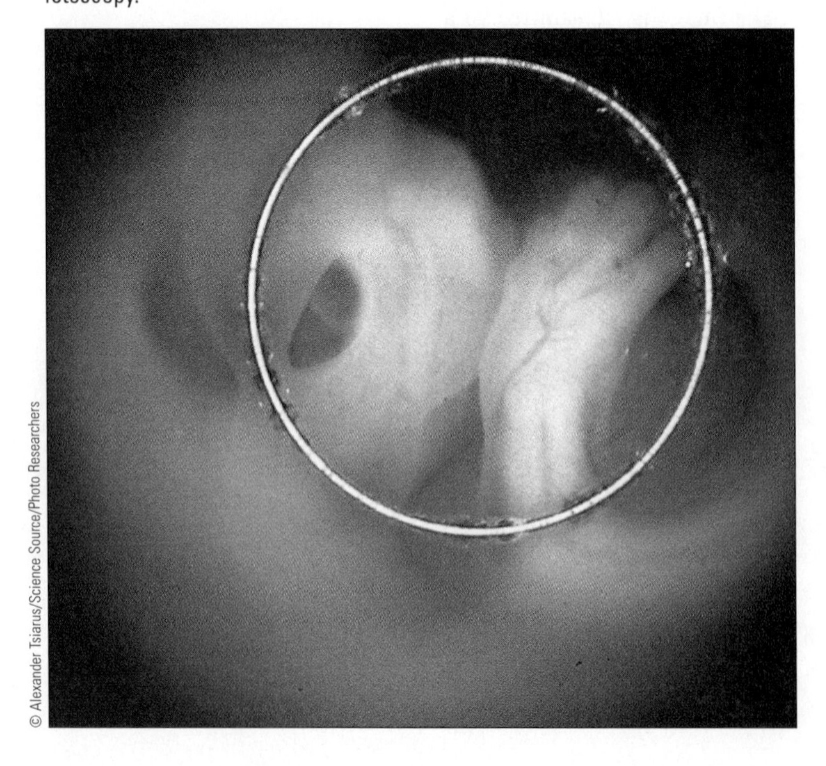

© Alexander Tsiarus/Science Source/Photo Researchers

Risks and Problems Associated with Prenatal Testing

Although many genetic disorders and birth defects can be detected by prenatal testing, the technique has some limitations. These include measurable risks to the mother and fetus, including infection, hemorrhage, fetal injury, and spontaneous abortion. Keep in mind, however, that there is a 3.5% overall risk of bearing a child with a birth defect, genetic condition, or mental retardation.

Conventional strategies for prenatal testing will not always detect the majority of certain defects. For example, amniocentesis is recommended for all mothers over the age of 35 years to test for Down syndrome. (In reality, only a small percentage of pregnant women older than 35 are tested by amniocentesis.) However, about 80% of all Down syndrome births are to mothers younger than the age of 35. This distribution of Down syndrome births reflects differences in the number of pregnancies in women younger than and older than the age of

35. Younger mothers may have 80% of the Down syndrome children, but they also have 93% of all children. Older women have about 7% of all children but 20% of the Down syndrome births, emphasizing once again the relationship between maternal age and increased risk of Down syndrome.

In the case of neural tube defects, approximately 90% of all affected infants are born to parents who have no family history of such conditions. Thus, testing couples who have had an affected child will have little effect on the overall rate of prenatal detection for this birth defect. On the other hand, screening of all pregnant females is not cost effective or possible, given the limited number of prenatal screening clinics.

PRESYMPTOMATIC TESTING FOR GENETIC DISORDERS

Presymptomatic testing involves detecting genetic disorders that become apparent only later in life. The phenotypes of many dominant traits, including polycystic kidney disease (PCKD) (MIM/OMIM 173900) are expressed in adulthood. PCKD usually appears between 35 and 50 years of age and is characterized by the formation of kidney cysts, which grow and gradually destroy the organ. Treatment includes kidney dialysis and transplantation of a normal kidney, but many affected individuals die prematurely. Because PCKD is a dominant trait, anyone heterozygous or homozygous for the gene will be affected. The PCKD gene maps to chromosome 16, and markers can be used to determine which family members are most likely to develop this disorder. Testing can be done prenatally or at any age before (or after) the condition appears.

DNA Chips and Genetic Testing

The recombinant DNA methods described earlier can test for a single genetic disorder or several alleles of a mutant gene. A new technology employing DNA chips (◗ Figure 19.8) allows screening for all known mutations in a disease gene and can be used to screen for mutations in thousands of genes at a time.

DNA chips employ semiconductor technology and are made using the same equipment as computer chips. DNA chips are glass, one-half-inch–by–one-half-inch squares. The chip is divided into fields (smaller squares), each the width of a human hair. Each field contains linker molecules to which synthetic DNA probes of about 20 nucleotides are attached. In a row of fields, each probe differs by one nucleotide. Thus, for the first nucleotide in the probe, a set of four fields (one for each nucleotide) is needed. With a probe that is 20 nucleotides long, 160,000 fields (20^4) are needed to give all possible combinations of nucleotides. Current chips hold 400,000 fields, but chips with 1.6 million fields are now in development, and even larger chips are planned.

For testing, DNA is extracted from a blood sample and cut with one or more restriction enzymes. The DNA fragments are tagged with a fluorescent dye, melted into single strands, and pumped into the chip. Fragments with sequences that exactly match probe sequences will bind, and those that do not match are washed from the chip. A laser scanner reads the chip and detects the pattern of fluorescence, which is analyzed by a linked software program. The results are presented in the form of a nucleotide sequence for the DNA tested. This sequence can be compared with the normal sequence to detect any mutations.

Chips are now used to screen for mutations in the *p53* gene, which is mutated in 60% of all cancers, and to screen for mutations in *BRCA1*, the gene that predisposes women to breast cancer. In the near future, chips will be loaded with gene sets for heart diseases, diabetes, Alzheimer disease, and other genetically defined disease subtypes. Eventually, chips may be used to test for mutations in the entire human gene set, giving individuals a genetic profile of all the mutations they carry and evaluating the risks they have for developing a genetic disorder or for transmitting a disorder to their children.

FIGURE 19.8 Probes are used to detect DNA mutations and monitor gene expression. Each chip holds between 50,000 and 500,000 probes. The nucleic acids to be analyzed are made single-stranded, tagged with fluorescent dyes, and hybridized to the probes on the chips. Unbound molecules are washed off the chip, and a laser scans the chip and detects the fluorescence that signals hybridization with a probe.

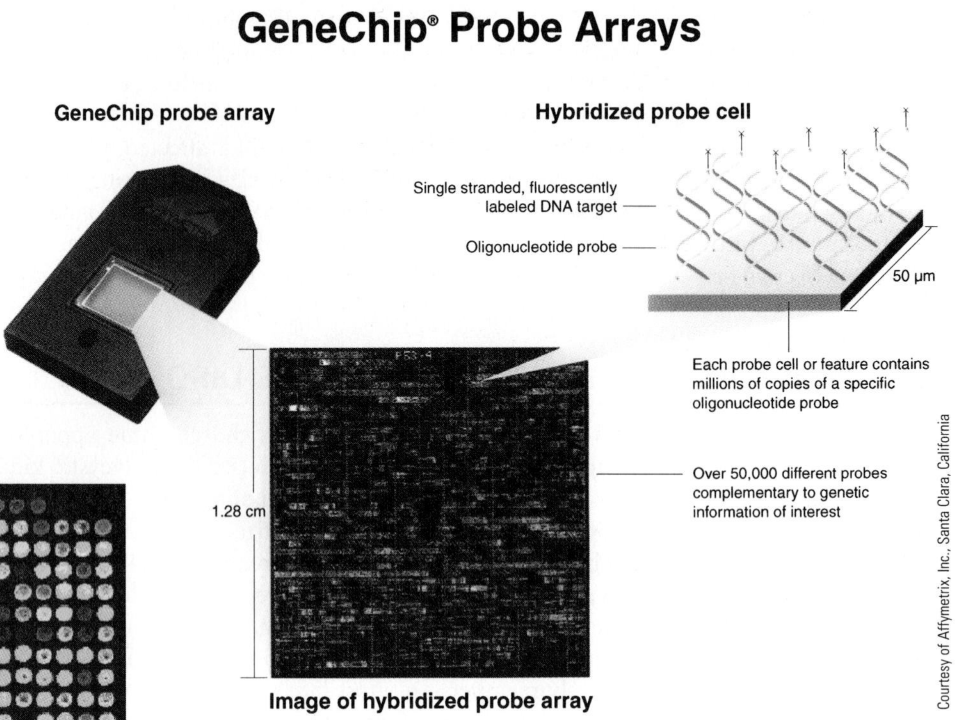

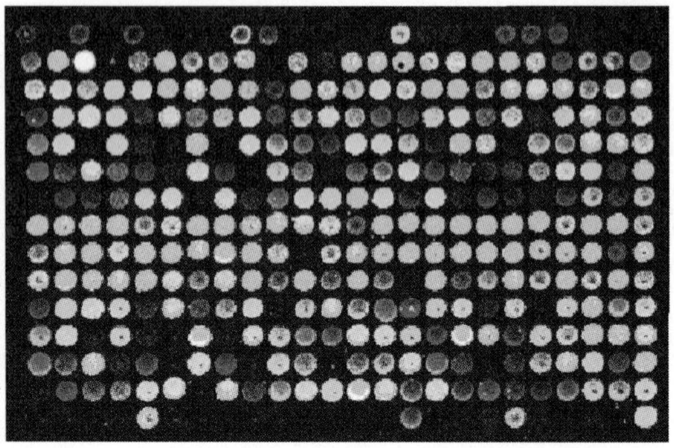

Courtesy of Tyson Clark, Dept. of Molecular, Cellular and Developmental Biology, University of California Santa Cruz

Courtesy of Affymetrix, Inc., Santa Clara, California

GENE THERAPY PROMISES TO CORRECT MANY DISORDERS

Recombinant DNA technology has made it possible to diagnose genetic disorders and to treat them by transferring normal copies of genes into cells that carry defective copies. There are several methods for transferring cloned genes into human cells, including viral vectors, chemical methods to transfer DNA across the cell membrane, and physical methods such as microinjection or fusion of cells with vesicles that carry cloned DNA sequences.

Viral vectors, especially retroviruses, are used for gene therapy. These vectors have been genetically modified, and some viral genes are removed, allowing a human gene to be inserted (◗ Figure 19.9). Human gene therapy began in 1990, when a human gene for the enzyme adenosine deaminase (ADA) was inserted into a retrovirus and then transferred into the white blood cells of a young girl, Ashanti DeSilva, who had a form of severe combined immunodeficiency disease (SCID) (MIM/OMIM 102700). She had no functional immune system and was prone to infections, many of which can be fatal. The normal *ADA* gene, inserted into her white blood cells, encodes an enzyme that allows cells of the immune system to mature properly. As a result, she now has a functional immune system and is leading a normal life.

Gene Therapy: Setbacks and Starting Over

In the early to mid-1990s, gene therapy trials were started for several genetic disorders, including cystic fibrosis and familial hypercholesterolemia. Over a 10-year period, more than 4,000 people underwent gene transfer. Unfortunately, these trials were largely failures and led to a loss of confidence in gene therapy. Hopes for gene therapy plummeted even further in September, 1999, when Jesse Gelsinger died during gene therapy. His death was triggered by a massive immune response to the vector, a modified adenovirus (adenoviruses cause colds and respiratory infections).

Efforts are now directed at developing new vectors that are less visible to the immune system and that transfer genes to target cells with a higher efficiency. Some successes in animal models and human transfers are encouraging researchers and clinicians to continue work on gene therapy. In 2000 a French team used a modified first-generation retroviral vector to successfully treat an X-linked form of SCID, and other workers, using a new, second-generation vector, have had initial successes in treating hemophilia B.

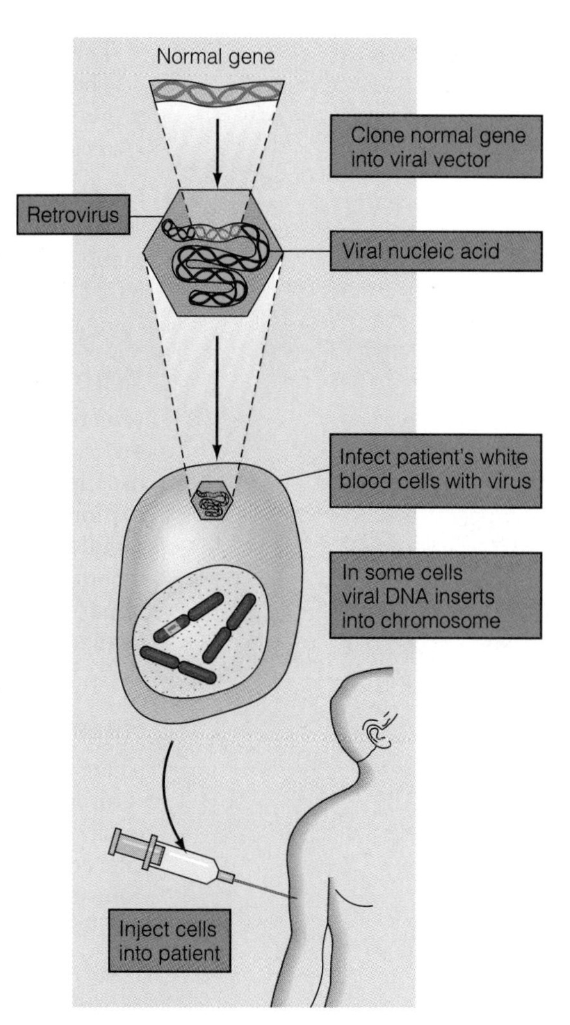

FIGURE 19.9 The present method of gene therapy uses a virus as a vector to insert a normal copy of a gene into the white blood cells of a patient who has a genetic disorder. The normal gene becomes active, and the cells are reinserted into the affected individual, curing the genetic disorder. Because white blood cells die after a few months, the procedure has to be repeated regularly. In the future, it is hoped that transferring a normal gene into the mitotically active cells of the bone marrow will make gene therapy a one-time procedure.

Ethical Issues Related to Gene Therapy

Gene therapy is carried out under an established set of ethical and medical guidelines. All patients are volunteers, gene transfer is started after several reviews, and the trials are monitored to protect the patients' interests. Newer guidelines instituted after Gelsinger's death strengthen these protections and coordinate the role of government agencies that regulate gene therapy. Other ethical concerns have not yet been resolved, as described next.

At present, gene therapy uses somatic cells as targets for transferred genes. This form of gene therapy is called **somatic gene therapy.** In somatic therapy, gene transfer involves only the target tissue and only one person is treated (only after obtaining informed consent and permission for the treatment). Two other forms of gene therapy are not yet used, mainly because ethical issues have not been clarified. One of these is **germ-line therapy,** in which germ cells (gametes and the cells that give rise to them) are targets for gene transfer. In germ-line therapy, the transferred gene would be present in all cells of the individual produced from the genetically altered gamete, including his or her germ cells. As a result, members of future generations are affected by this gene transfer, without their consent. Does anyone have the right to genetically modify others without their consent? Can we make this decision for members of future generations? These and other ethical concerns are not resolved, and germ-line therapy is currently prohibited.

The second form of gene therapy, **enhancement gene therapy,** raises even more ethical concerns. If we discover genes that control a desirable trait, such as intelligence or athletic ability, should we use them to enhance someone's intellectual ability or athletic skills? For now, the consensus is that we should not use gene transfer for such purposes. As the number of genes identified in the Human Genome Project grows, this issue is likely to be revisited. The debate over the use of germ-line therapy and enhancement therapy has long-term consequences for us as a species.

Somatic gene therapy Gene transfer to somatic target cells to correct a genetic disorder.

Germ-line gene therapy Gene transfer to gametes or the cells that produce them. Transfers the gene to all cells in the next generation, including germ cells.

Enhancement gene therapy Gene transfer to enhance traits, such as intelligence or athletic ability, not to treat genetic disorders.

Genetic counseling A process of communication that deals with the occurrence or risk that a genetic disorder will occur in a family.

GENETIC COUNSELING

Genetic counseling is a process of communication that deals with the occurrence or risk for occurrence of a genetic disorder in a family. Counseling involves one or more appropriately trained persons who help an individual or family to understand each of the following:

- The medical facts, including the diagnosis, probable course of the disorder, and the available treatment and management

- The way heredity contributes to the disorder and the risk of recurrence
- The alternatives for dealing with the risk of recurrence
- How to adjust to the disorder in an affected family member or to the risk of recurrence

Genetic counselors achieve these goals in a nondirective way. Genetic counselors provide all of the information available to and desired by an individual or family, so that the person or family can make the decisions most suitable to them based on their own cultural, religious, and moral beliefs.

Who Are Genetic Counselors?

Genetic counselors are health care professionals with specialized graduate training and experience in the areas of medical genetics, psychology, and counseling. They usually work as members of a multidisciplinary health care team and offer information and support to families that have relatives with genetic conditions and to families who may be at risk for a variety of inherited conditions. Genetic counselors identify families at risk, investigate the problem in the family, interpret information about the disorder, analyze inheritance patterns and risk of recurrence, and review available options with the family (▶ Figure 19.10).

Reasons to Seek Genetic Counseling

There are many reasons that people seek genetic counseling. The most typical case is an individual or family with a history of a genetic disorder, cancer, birth defect, or developmental disability. Women older than 35 years and individuals from specific ethnic groups in which particular genetic conditions occur more frequently are counseled to learn of their increased risk for genetic or chromosomal disorders and the diagnostic testing that is available. Counseling is especially recommended for the following individuals or families:

- Women who are pregnant, or are planning to become pregnant, after age 35
- Couples who already have a child with mental retardation, an inherited disorder, or a birth defect
- Couples who would like testing or more information about genetic defects that occur more frequently in their ethnic group

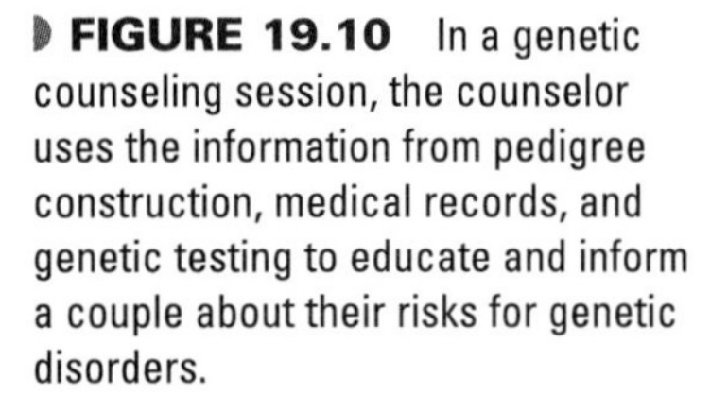

▶ FIGURE 19.10 In a genetic counseling session, the counselor uses the information from pedigree construction, medical records, and genetic testing to educate and inform a couple about their risks for genetic disorders.

Concepts and Controversies

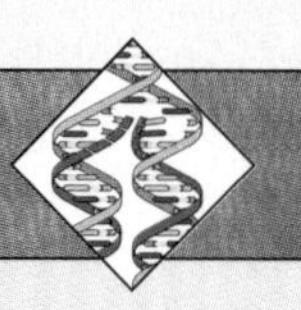

The Business of Making Babies

Recombinant DNA methods are revolutionizing the fields of genetic testing and genetic counseling. New technology has also made the business of human fertilization a part of private enterprise. One in six couples in the United States (over 3 million couples of childbearing age) is classified as infertile, and most of these couples want to have children. The first successful *in vitro* fertilization (IVF) was accomplished in 1981 at the Medical College of Virginia at Norfolk. Since then, more than 150 hospitals and clinics that use these and other techniques have opened. Many of these clinics are associated with university medical centers, but others are operated as freestanding businesses. Some are public companies that have sold stock to raise start-up money or to cover operating costs. It is estimated that a capitalization of about $1 million is required to start an IVF clinic and that 50 to 60 fertilization attempts per month are necessary for the venture to be profitable. Each IVF attempt costs between $5,000 and $10,000, and several attempts (four to six) are usually required for success. Because these costs are not usually covered by insurance, IVF is a major expense for couples who want children.

In IVF, egg maturation is induced with drugs, and the mature eggs are recovered by laparoscopy (a small incision is made in the abdominal wall, and a fiberoptic device is used to recover the eggs). In an alternative procedure, an ultrasonically guided needle is inserted into the vagina and moved up to the ovary to remove the eggs. The eggs are fertilized in a dish (*in vitro* literally means "in glass"). After the fertilized egg begins to develop, it is implanted into the woman's uterus.

If extra eggs are recovered, they are fertilized, and the resulting embryos are frozen in liquid nitrogen. This eliminates the need to retrieve eggs every month for fertilization. If fertilization is successful, the extra embryos can remain in storage for implantation at a later time, or they can be donated to another couple.

Several companies, such as IVF Australia, are open in several locations in the United States. Because of the high start-up costs and expertise required, it is possible that the field will be dominated by a small number of companies through franchising agreements. Some investment analysts predict that IVF will grow into a $6 billion annual business. In parallel, a genetic testing and screening industry is beginning to emerge that offers tests in the areas of prenatal diagnosis, newborn screening, carrier screening, adult-onset screening, and forensic testing.

- Couples who are first cousins or other close blood relatives
- Individuals who are concerned that their jobs, lifestyle, or medical history may pose a risk to a pregnancy, including exposure to radiation, medications, chemicals, infection, or drugs
- Women who have had two or more miscarriages or babies who died in infancy
- Couples whose infant has a genetic disease diagnosed by routine newborn screening
- Those who have, or are concerned that they might have, an inherited disorder or birth defect
- Pregnant women who have been told that, based on ultrasound tests or blood tests for alpha fetoprotein, their pregnancy may be at increased risk for complications or birth defects

How Does Genetic Counseling Work?

Most people go for counseling after a prenatal test or after the birth of a child with a genetic condition. The counselor usually begins by constructing a detailed family and medical history, or pedigree.

Prenatal screening and cytogenetic or biochemical tests can be used along with the pedigree to help determine risk of occurrence or recurrence. The counselor uses as much information as possible to establish whether the trait is genetically determined.

If it is found that a condition in a family is genetically determined, the counselor constructs a risk assessment for the couple. In this process, the counselor uses all of the information available to explain the risk of having another child affected with the condition or to explain the risk that the individual who is being counseled will be affected with the condition. Conditions that are considered high risk include dominant conditions (50% risk if one parent is heterozygous), simple autosomal recessive conditions (25% when both parents are heterozygotes), and certain chromosomal translocations. Often conditions are difficult to assess because they involve polygenic traits or conditions that have high mutation rates (like neurofibromatosis).

Genetic counselors explain basic concepts of biology and inheritance to all couples. This helps them understand how genes, proteins, or cell-surface antigens are related to the defects seen in their child or family. The counselor provides information that allows informed decision making about future reproductive choices. Reproductive alternatives, such as adoption, artificial insemination, *in vitro* fertilization, egg donation, and surrogate motherhood are options that the counselor presents to the couple.

Future Directions

As the Human Genome Project accelerates the number of genetic disorders that can be detected by heterozygote and prenatal screening and as these techniques become more available, the role of the genetic counselor will become more important. The Human Genome Project is changing the focus of genetic counseling from reproductive risks to adult-onset conditions, such as cancer, PCKD, and Huntington disease. Although counseling sessions address reproductive risks for these conditions, the primary focus is on the individual being counseled. The areas addressed include the risk of inheriting the gene, the potential severity of the condition, and the age of onset.

Summary

Genetic Testing and Genetic Screening

1. Genetic testing is the search for individuals who have a particular genotype. Genetic screening tests general populations that may have a low frequency for a disorder. Carrier screening is the search for heterozygotes who may be at risk of producing a defective child. Large-scale carrier detection programs have been conducted for two autosomal recessive diseases that affect discrete population segments: Tay-Sachs disease and sickle cell anemia.

Prenatal and Presymptomatic Genetic Testing

2. Several techniques are used in prenatal diagnosis, including amniocentesis, chorionic villus sampling (CVS), ultrasonography, and fetoscopy. Prenatal screening can also detect chromosome abnormalities, such as Down syndrome, and birth defects, such as spina bifida, that may have a genetic component.

Prenatal Testing for Genetic Disorders

3. Recombinant DNA-based prenatal testing can detect genetic disorders such as sickle cell anemia that cannot otherwise be detected before birth. Testing can also be done on embryos following *in vitro* fertilization and before implantation. Methods for visualizing the fetus can be used to check for visible genetic disorders and developmental problems.

Presymptomatic Testing for Genetic Disorders

4. Recombinant DNA methods can be used to identify those who will develop late-onset genetic disorders such as polycystic kidney disease (PCKD) or Huntington disease. Testing for a wide range of genetic disorders is possible using DNA chips, which can hold thousands of genes.

Gene Therapy Promises to Correct Many Disorders

5. Gene therapy transfers a normal copy of a gene into target cells of individuals carrying a mutant allele. After initial successes, gene therapy suffered several setbacks, including the death of a participant. Ethical issues over the use of germ-line therapy and enhancement therapy are not resolved, and these therapies are not used.

Genetic Counseling

6. Genetic counseling involves developing an accurate assessment of a family history to determine the risk of genetic disease. In many cases this is done after the birth of a child affected with a genetic disorder, but in other cases counseling is entirely retrospective. Decisions about whether to have additional children, to undergo abortion, or even to marry are always left to those being counseled.

Case Studies

CASE 1

Trudy was a 33-year-old woman who went with her husband, Jeremy, for genetic counseling. Trudy has had three miscarriages. The couple has a 2-year-old daughter who is in good health and is developing normally. Chromosomal analysis was done on the products of conception of the last miscarriage and was found to be 46, XY. The last miscarriage occurred in January 1999. Peripheral blood samples for both parents were taken at the time and sent to the laboratory. Trudy's chromosomes were 46, XX and Jeremy's were 46, XY, t(6;18) (q21;q23). Jeremy appears to have a balanced translocation between chromosome number 6 and 18. There is no family history of stillbirths, neonatal death, infertility, mental retardation, or birth defects. Jeremy's parents both died in their 70s from heart disease, and he is unaware of any pregnancy losses experienced by his parents or siblings.

The recurrence risks associated with a balanced translocation between chromosome 6 and 18 were discussed in detail. The counselor used illustrations to demonstrate the approximate 50% risk of unbalanced gametes; the other 50% of the gametes result in either normal or balanced karyotypes. The family was informed that empirical risk for unbalanced conceptions is significantly less than the 50% relative risk. Prenatal diagnostic procedures were described, including amniocentesis and chorionic villus sampling. The benefits, risks, and limitations of each were described.

The couple indicated a desire to pursue another pregnancy and were interested in proceeding with an amniocentesis.

1. Draw each of the possible combinations of chromosomes 6 and 18 that could be present in Jeremy's gametes, showing how there is an approximately 50% chance that they are either normal, or balanced, and a 50% chance that they are unbalanced.
2. Trudy became pregnant again, and an amniocentesis showed that the fetus received the balanced translocation from her father. Is she likely to have any health problems because of this translocation? Will it affect her in any way?

CASE 2

Genetic counselors frequently use empirical risk estimates for multifactorial conditions or in situations in which Mendelian principles do not apply. Let's assume that a couple comes for genetic counseling with no other positive family history except for one child affected with a disorder that could be either multifactorial or chromosomal. The couple wants to know the risk of having another affected child. If the child is affected with spina bifida, the empirical risk to a subsequent child is approximately 4%. If the child has Down syndrome, the empirical risk of recurrence would be approximately 1% if the karyotype were trisomy 21, but it might be substantially higher if one of the parents were a carrier of a Robertsonian translocation involving chromosome 21.

1. For what types of disorders would a genetic counselor most likely have to rely on an empirical risk estimate?
2. If you had a child with spina bifida and you knew that the risk of having another child with the same disorder was 4%, would you consider having another child? Would you consider relying on any available reproductive technologies?

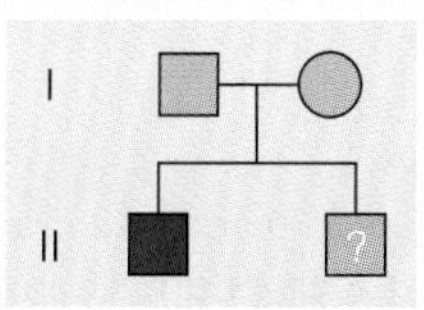

Questions and Problems

Genetic Testing and Genetic Screening

1. What is the difference between genetic testing and genetic screening?
2. The reaction to screening for Tay-Sachs disease and sickle cell anemia offers an interesting contrast in the institution and administration of genetic screening programs. Cystic fibrosis is an autosomal disease that mainly affects the white population, and 1 in 20 whites are heterozygotes. Now that the gene has been mapped to chromosome 7, assume that restriction fragment length polymorphism (RFLP) markers are available to diagnose heterozygotes. Should a genetic screening program for cystic fibrosis be instituted? Should the federal government fund this? Should the program be voluntary or mandatory and why?

Prenatal Testing for Genetic Disorders

3. The measurement of alpha fetoprotein levels is used to diagnose neural tube defects. For every 1,000 such tests, approximately 50 positive cases will be detected. However, up to 20 (40%) of these cases may be false positives. In a false positive, the alpha fetoprotein level is elevated, but the child has no neural tube defect. Your patient has undergone testing of the maternal blood for alpha fetoprotein, and the results are positive. She wants to abort a defective child but not a normal one. What are your recommendations?
4. You are a governmental science policy advisor, and you learn about a new technique that is being developed that promises to accurately predict IQ based on a particular combination of genetic markers. You also learn that this technique may potentially be applied to preimplantation testing, so that parents would be able to not only select an embryo for implantation that is free of a genetic disorder but also select one that is likely to be relatively smart. What policy recommendations would you make concerning this technology? Do you think parents should have the right to choose any characteristic of their children, or should preimplantation be limited to ensuring that embryos are free of genetic disorders? Should guidelines be imposed to regulate this process, or should it even be banned?

Presymptomatic Testing for Genetic Disorders

5. List the types of genetic testing covered in this chapter, and briefly summarize the unique characteristics of each type.

Gene Therapy Promises to Correct Many Disorders

6. Gene therapy involves:
   a. the introduction of recombinant proteins into individuals.
   b. cloning human genes into plants.
   c. the introduction of a normal gene into an individual carrying a mutant copy.
   d. DNA fingerprinting.
   e. none of the above.
7. In selecting target cells to receive a transferred gene in gene therapy, what factors do you think would have to be taken into account?
8. The prospect of using gene therapy to alleviate genetic conditions is still a vision of the future. Gene therapy for adenosine deaminase deficiency has proven to be quite promising, but many obstacles remain to be overcome. Currently, the correction of human genetic defects is done using retroviruses as vectors. For this purpose, viral genes are removed from the retroviral genome, creating a vector capable of transferring human structural genes into sites on human chromosomes within target tissue cells. Do you see any potential problems with inserting pieces of a retroviral genome into humans? If so, are there ways to combat or prevent these problems?
9. Is gene transfer a form of eugenics? Is it advantageous to use gene transfer to eliminate some genetic disorders? Can this and other technology be used to influence the evolution of our species? Should there be guidelines for the use of genetic technology to control its application to human evolution? Who should create and enforce these guidelines?

Genetic Counseling

10. A couple who wishes to have children visits you, a genetic counselor. There is a history of a deleterious, recessive trait in males in the woman's family but not in the man's family. The couple is convinced that because his family shows no history of this genetic disease, they are at no risk of having affected children. What steps would you take to assess this situation and educate this couple?
11. A couple has had a child born with neurofibromatosis. They come to your genetic counseling office for help. After taking an extensive family history, you determine that there is no history of this disease on either side of the family. The couple wants to have another child and wants to be advised about risks of having another child with neurofibromatosis. What advice do you give them?
12. The initial step in the process of genetic counseling is to construct a family pedigree and analyze it for inheritance patterns. What are the major differences between autosomal recessive and autosomal dominant conditions in pedigree analysis?
13. An interesting polymorphism in the human population has to do with the ability to roll one's tongue (curl up the sides of the tongue to make a trough). Some people can do this trick, and others cannot. Therefore it is an example of a dimorphism. Its significance is a mystery. In one family, a boy was unable to roll his tongue, but to his great dismay his sister could. Furthermore, both his parents were rollers, and so were both grandfathers and one paternal uncle and one paternal aunt. One paternal aunt, one paternal uncle, and one maternal uncle could not. Draw the pedigree for this family, clearly defining your symbols and deduce the genotypes of as many individuals as possible.
14. Phenylketonuria (PKU) is a human hereditary disease that prevents the body from processing the amino acid phenylalanine, which is contained in dietary protein. Symptoms of PKU present in early infancy, and if it remains untreated, it leads to severe mental retardation. PKU is caused by a recessive allele with simple Mendelian inheritance. A couple intends to have children but seeks genetic counseling because the woman has a sister with PKU and the man has a brother with PKU. There are no other known cases in their families. They ask you, the genetic counselor, to determine the probabilities that their first child will have PKU. What is this probability?
15. You are a genetic counselor and your patient has asked to be tested to determine if she carries a gene that

predisposes her to early-onset cancer. If your patient has this gene, there is a 50/50 chance that all of her siblings inherited this gene and there is also a 50/50 chance that it will be passed on to their offspring. Your patient is very concerned about confidentiality and does not want anyone in her family to know she is being tested, including her identical twin sister. Your patient is tested and found to carry an altered gene that gives her an 85% lifetime risk of developing breast cancer and a 60% lifetime risk of developing ovarian cancer. At the result-disclosure session, she once again reiterates that she does not want anyone in her family to know her test results.

   a. Knowing that a familial mutation is occurring in this family, what would be your next course of action in this case?

   b. Is it your duty to contact members of this family despite the request of your patient? Where do your obligations lie—with your patient or with the patient's family? Would it be inappropriate to try to convince the patient to share her results with her family members?

16. A young woman (proband) and her partner are referred for prenatal genetic counseling because the woman has a family history of sickle cell anemia. The proband has sickle cell trait (*Ss*), and her partner is not a carrier nor does he have sickle cell anemia (*SS*). Prenatal testing indicates that the fetus is affected with sickle cell anemia (*ss*). The results of this and other tests indicate that the only way the fetus could have sickle cell disease is if the woman's partner is not the father of the fetus. The couple is at the appointment seeking their test results.

   a. How would you handle this scenario? Should you have contacted the proband beforehand to explain the results and the implications of these results?

   b. Is it appropriate to keep this information from the partner because he believes he is the father of the baby? What other problems do you see with this case?

17. You are a genetic counselor and have been asked to review a proposal to screen local high school students for cystic fibrosis (CF) carrier status. The investigator's protocol states that the test will be done by extracting cells from inside the mouth with a cotton swab from all Caucasian students. Before testing, students will be given a booklet about CF and about the test and asked to sign a consent form. Results will be distributed to students in sealed envelopes with a toll-free number to call if they have questions about the test results. Of the various objections that you might raise with this screening protocol, the LEAST compelling of the following is that:

   a. non-Caucasian students will not be able to be tested if they wish to be.

   b. the method of informing the students of results does not adequately protect privacy or provide for appropriate follow-up counseling.

   c. offering the screening in the classroom may unduly pressure students into being tested or stigmatize them.

   d. the students' parents must be involved in providing consent.

Internet Activities

The following activities use the resources of the World Wide Web to enhance the topics covered in this chapter. To investigate the topics described below, log on to the book's home page (www.brookscole.com/biology_d) and follow the *Student Resources* link to your subject and text.

1. *Overview and History of Genetic Counseling.* At the *Access Excellence: Classics Collection* site (http://www.accessexcellence.org/AE/AEC/CC/), click on the link to the article "Genetic Counseling: Coping with the Impact of Human Disease." This article gives an overview of the history of genetic counseling and how genetic counseling is used today. How does the use of genetic information by eugenicists early in the twentieth century compare to the use of genetic information by genetic counselors today? (For review, you may want to refer to Chapter 1, where eugenics was first discussed.) What kinds of ethical questions and issues may arise as a result of genetic counseling?
2. *Genetic Counseling Resources.* The New York Online Access to Health program (*NOAH*) (http://www.noah-health.org/english/illness/genetic_diseases/geneticdis.html) has an excellent home page on genetic disorders and genetic counseling. This site is a good place to start if you or someone in your family has any concerns about genetic disorders.
3. *Ethical, Legal and Social Issues.* The U.S. Department of Energy maintains the *Human Genome Project Information* site (http://www.ornl.gov/TechResources/Human_Genome/home.html), which is a valuable source of information on many aspects of genetic issues.

   a. At the site, click on "Ethical, Social and Legal Issues." Link to "ELSI research," then scroll down to choose "The Genetic Privacy Act and Commentary." Read parts A and B. Do you think that adequate safeguards have been established to protect individuals' genetic information? What, if anything, would you like to see added?

   b. We are often curious about dead celebrities or major political or historical figures. If a person, such as Abraham Lincoln, has been dead for over one hundred years, should samples of hair or other body substances be tested to see if the individual carried any of the genetic markers for diseases such as Marfan syndrome or manic depression? Why or why not? Should descendants or other living relatives have the right to prevent such testing, in part to protect their own privacy?

4. *More Questions About Genetics Issues.* The *Human Genome Project Information* site (http://www.ornl.gov/TechResources/Human_Genome/home.html) also has links to other resources that address issues raised by genetic research. Two such publications include "To Know Ourselves" and "Your Genes, Your Choices"; both publications can be requested in hardcopy or downloaded.

For Further Exploration: Try the karyotyping activity in Chapter 6 to see how some screening for genetic disease can be done.

For Further Reading

Agency for Health Care Policy and Research. Sickle Cell Disease Panel. Sickle cell disease: Comprehensive screening and management in newborns and infants. Rockville, MD, Public Health Service, Department of Health and Human Services, April, 1993.

Billings, P., & Beckwith, J. (1992). Genetic testing in the workplace: A view from the USA. *Trends Genet., 8,* 198–202.

Calabrese, E. J. (1986). Ecogenetics: Historical foundation and current status. *J. Occup. Med., 28,* 1096–1102.

Dagenais, D., Courville, L., & Dagenais, M. (1985). A cost-benefit analysis for the Quebec Network of Genetic Medicine. *Soc. Sci. Med., 20,* 601–607.

Draper, E. (1991). *Risky business: Genetic testing and exclusionary practices in the hazardous workplace.* New York: Cambridge University Press.

Emery, A. E. H., & Pullen, I. (1984). *Psychological aspects of genetic counseling.* New York: Academic.

Fuhrmann, W., & Vogel, F. (1983). *Genetic counseling* (3rd ed.). New York: Springer-Verlag.

Gibbs, R. A., & Caskey, C. T. (1989). The application of recombinant DNA technology for genetic probing in epidemiology. *Ann. Rev. Pub. Health, 10,* 27–48.

Hodgson, S. V., & Bobrow, M. (1989). Carrier detection and prenatal diagnosis in Duchenne and Becker muscular dystrophy. *Br. Med. Bull., 45,* 719–744.

Jinks, D. C., Minter, M., Tarver, D. A., Vanderford, M., Hejtmancik, J. F., & McCabe, E. R. B. (1989). Molecular genetic diagnosis of sickle cell disease using dried blood specimens on blotters used for newborn screening. *Hum. Genet., 81,* 363–366.

Kolata, G. (1986). Genetic screening raises questions for employers and insurers. *Science, 232,* 317–319.

Modell, B., & Kuliev, A. (1993). A scientific basis for cost-benefit analysis of genetics services. *Trends Genet., 9,* 46–52.

Murray, R. F. (1986). Tests of so-called genetic susceptibility. *J. Occup. Med., 28,* 1103–1107.

Rowley, P. (1984). Genetic screening: Marvel or menace? *Science, 225,* 138–144.

Sandovnick, A., & Baird, P. (1985). Reproductive counseling for sclerosis patients. *Am. J. Med. Genet., 20,* 349–354.

Selva, J., Leonard, C., Albert, M., Auger, J., & David, G. (1986). Genetic screening for artificial insemination by donor (AID). *Clin. Genet., 29,* 389–396.

Sommer, S. S., Cassady, J. D., Sobell, J. L., & Bottema, C. D. (1989). A novel method for detecting point mutations or polymorphisms and its application to population screening for carriers of phenylketonuria. *Mayo Clin. Proc., 64,* 1361–1372.

U.S. Congress, Office of Technology Assessment, Cystic Fibrosis and DNA Tests: Implications of Carrier Screening, OTA-BA-532. Washington, DC: U.S. Government Printing Office, August, 1992.

Uzych, L. (1986). Genetic testing and exclusionary practices in the workplace. *J. Public Health Policy, Spring 1986,* 37–57.

Wapner, R. J., & Jackson, L. (1988). Chorionic villus sampling. *Clin. Obstet. Gynecol., 31,* 328–344.

Williams, C., Weber, L., Williamson, R., & Hjelm, M. (1988). Guthrie spots for DNA-based carrier testing in cystic fibrosis. *Lancet, ii,* 693.

Probability

Mendel's use of mathematics to analyze the results of his experiments is frequently overlooked as an important contribution to biology. His application of mathematical reasoning to the analysis of data helped transform biology from an observational and descriptive science into a quantitative and experimental one. When Mendel carried out his experiments, statistics and statistical methods were not highly developed or widely used. In analyzing the results of his crosses, Mendel converted the numbers of individuals with particular genotypes or phenotypes into ratios. From these ratios, he was able to deduce the mechanisms of inheritance.

As we now know, the ratios Mendel observed are the result of random segregation and assortment of genes into gametes during meiosis and their union at fertilization in random combinations. This randomness provides an element of chance in the outcome and prevents us from making exact predictions. In counting pea seeds in the F2, we may expect three-fourths of the seeds to be yellow, but we cannot be absolutely certain that the first seed in an unopened pod will be yellow. The rules of probability can, however, help us guess how often such an event will take place.

DEFINING PROBABILITY

Most people have an innate sense of probability that seems part of common sense. For example, almost everyone would agree with the idea that a January snowfall is more probable in Minneapolis than in Miami. Other aspects of probability also seem obvious. When a coin is flipped, the probable outcomes are heads or tails. In the birth of a child, we expect the outcome to be a boy or a girl.

Unfortunately, the use of intuition alone in matters of probability is not always reliable. For example, what would you say is the probability that in a crowd of 20 people, two individuals share the same birth date? Considering that there are 365 days in the year (excluding leap year), intuition may say that it is not very likely. In fact, in a group of 20, there is almost an even chance that two people share the same birthday. The probability of a shared birthday for groups of various sizes is as follows: for a group of 23, the probability is 51%; for a group of 30, it is 71%; for 40, it is 89%; and for 50, there is a 97% chance that two people will share the same birthday. We will not explore the mathematical reasoning behind this probability, but it is based on the fact that if one person can have any of the 365 days for his or her birthday, the second person can have any of the remaining 364 days, the third person can have any of the remaining 363 days, and so on.

From the preceding example, it should be clear that, to be useful in genetics and in science, probability must be expressed in more quantitative terms. The use of a quantitative approach to probability allows us to assign a numerical value to the probability that a given event will occur and prevents us from leaping to conclusions about the possible outcome of genetic crosses.

QUANTIFYING PROBABILITY

In quantifying probability, let's begin at the limits. If an event is certain to occur, it has a probability of 1; if the event is certain not to occur, then the probability is 0. In genetics as in most other areas, we usually deal with events between these limits and have a mixture of probabilities. Although it is certain that we will all die (a probability of 1), when we die is less certain and therefore must be assigned a probability somewhere between 0 and 1. Insurance companies spend a great deal of time and effort in attempting to determine such probabilities, although we may prefer not to think about them.

In general terms, we can express the probability (p) of an event as the proportion of times that such an event occurs (r) out of the number of times that the event can occur (n):

$$p = r/n$$

In other words, if an event occurs r times in n trials, the probability that the event will take place is r/n. This probability is somewhere between the limits of 0 and 1. If we toss a coin, it may land with heads up or tails up. The probability that it will land with heads up is

$$p = r/n = 1/2$$

Likewise, the probability of a child being a boy or a girl is 1/2. Other events have different probabilities. In a pair of dice, each die has six faces. When a die is thrown (1 throw), the probability of any of the faces (6 faces) being up is

$$p = r/n = 1/6$$

In a deck of 52 cards, the probability of drawing any given card (the ace of spades for example) is

$$p = r/n = 1/52$$

In roulette, the wheel contains the numbers 1 to 36 plus 0 and 00. The probability of the ball landing on any number is therefore

$$p = r/n = 1/38$$

In a monohybrid cross, the probability that an offspring of the self-fertilized F1 pea plant will have a dominant phenotype is

$$p = 3/4$$

In considering probability, we must consider not only the probability of one type of outcome but also the probability of other outcomes. If an event has a probability of p, the probability of an alternative outcome is $q = 1 - p$. In other words, the sum of the probability of p and q equals 1. In the preceding examples, the probability of drawing an ace of spades is 1/52; the probability of drawing another card is 51/52, which when added to 1/52 equals 1. In the monohybrid cross, the probability of a dominant phenotype is 3/4, and the probability of a recessive phenotype is 1/4. Because we are certain that the F2 will have either a dominant or recessive phenotype, adding the probabilities of both phenotypes ($3/4 + 1/4 = 1$) covers all the possible phenotypic combinations.

COMBINING PROBABILITIES

Two rules of probability are useful in analyzing genetics problems. The first, called the product rule, is used when we wish to calculate the probability of two or more independent events occurring at the same time. The second, called the sum rule, is used when two or more events are mutually exclusive or are alternative events.

In the product rule, we are asking what is the probability that event A and event B will occur together. This rule can be summarized as follows: The probability that independent events will occur together is the product of their independent probabilities.

When a coin is tossed, the probability that it will be heads is 1/2, and the chance that it will be tails is 1/2. If we toss a coin four times and it turns up heads each time, the probability that it will turn up tails on the fifth try is still 1/2. In other words, chance has no memory. The probability that two heterozygotes will have a child with cystic fibrosis is 1/4. This does not mean that if their first child has cystic fibrosis, they can be assured of having three unaffected children. It means that for each child, there is a 1 in 4 chance that it will have cystic fibrosis, no matter whether they have 1 child or 20 children. If they have four unaffected children, the chance that their fifth child will have cystic fibrosis is still 1/4.

If, on the other hand, we want to ask what is the probability that we can toss a coin four times and get heads each time, we use the product rule. Because each coin toss is an independent event, the probability of getting heads four times out of four is $1/2 \times 1/2 \times 1/2 \times 1/2 = (1/2)^4 = 1/16$. Similarly, the probability that heterozygous parents will have four children affected with cystic fibrosis is $1/4 \times 1/4 \times 1/4 \times 1/4 = 1/256$.

In using the sum rule, we are asking how often one or the other of two mutually exclusive events can occur. For example, in rolling a die, what is the probability of a three or a five coming up? Because on a single throw only one number can come up, it is impossible to get both numbers on a single throw. The probability of a three is 1/6, and the probability of a five is 1/6. If we want to know what is the probability of either a three *or* a five coming up on a single throw, we add the individual probabilities:

$$1/6 + 1/6 = 2/6 = 1/3$$

In considering the possible genotypes of children whose parents are both heterozygotes for cystic fibrosis, what is the probability that a child will have either one or two copies of the dominant allele? The probability of being homozygous dominant (two copies) is 1/4, and the probability of being heterozygous (one copy) is 1/2. Therefore, the probability of being either heterozygous or homozygous dominant is $1/2 + 1/4 = 3/4$. This is in fact the proportion of individuals with the dominant phenotype seen in the F2 of a monohybrid cross.

In applying probability to the analysis of genetic problems, first determine whether you want to know the probabilities of event A *and* event B or the probability of event A *or* event B. If you want A *and* B, use the product rule and multiply the probabilities of A and B. If you want the probability of A *or* B, use the sum rule and add the probability of event A to the probability of event B. For example, the frequency of albinism is about 1/10,000, and the frequency of cystic fibrosis is about 1/2,000. If we want to know the probability of having both albinism and cystic fibrosis, we multiply the probabilities:

$$1/10{,}000 \times 1/2{,}000 = 1/20{,}000{,}000$$

If we want to know the probability of having either albinism or cystic fibrosis, we add the probabilities:

$$1/10{,}000 + 1/2{,}000 = 6/10{,}000 = 1/1{,}666$$

As you can see, the probabilities are very different and reflect whether we are asking that both events occur or that one or the other event will occur.

Answers to Selected Questions and Problems

Chapter 2

1. e
5. d
8. Cells undergo a cycle of events involving growth, DNA replication, and division. Daughter cells undergo the same series of events. During S phase, DNA synthesis and chromosome replication occurs. During M phase, mitosis takes place.
9. a, e
11. Meiosis II, the division responsible for the separation of sister chromatids would no longer be necessary. Meiosis I, wherein homologues segregate, would still be required.
19. Cell cycle genes can turn cell division on and off. If a gene normally promotes cell division, it can mutate to cause too much cell division. If a gene normally turns off cell division, it can mutate so that it can no longer repress cell division.
20.

	Mitosis	Meiosis
Number of daughter cells produced	2	4
Number of chromosomes per daughter cell	2n	n
Number of cell divisions	1	2
Do chromosomes pair? (Y/N)	N	Y
Does crossing over occur? (Y/N)	N	Y
Can the daughter cells divide again? (Y/N)	Y	N
Do the chromosomes replicate before division? (Y/N)	Y	Y
Type of cell produced	SOMATIC	GAMETE

21.

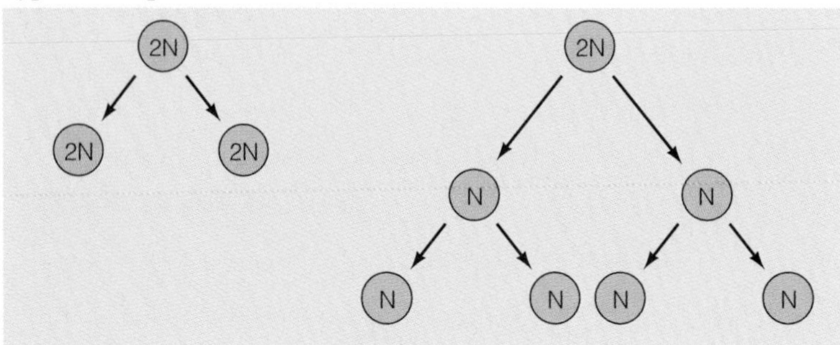

25. a. mitosis; b. meiosis I; c. meiosis II
28. Meiotic anaphase I: no centromere division, chromosomes consisting of two sister chromatids are migrating; meiotic anaphase II: centromere division, the separating sister chromatids are migrating. Meiotic anaphase II more closely resembles mitotic anaphase by the two criteria cited above.

Chapter 3

1. a. A gene is the fundamental unit of heredity. The gene encodes a specific gene product (e.g., a pigment involved in determining eye color). Alleles are alternate forms of a gene that may cause various phenotypic effects. For example, there may be a blue eye color allele, a brown eye color allele, and a green eye color allele of a gene. Locus is the position of a gene on a chromosome. In a normal situation, all alleles of a gene would have the same locus.
   b. Genotype refers to the genetic constitution of the individual (*AaBb* or *aabb*). Notice that the genotype always includes at least two letters, each

representing one allele of a gene pair in a diploid organism. A gamete would contain only one allele of each gene because of its haploid state (*Ab* or *ab*). Phenotype refers to an observable trait. For example, *Aa* (the genotype) will cause a normal pigmentation (the phenotype) in an individual, whereas *aa* will cause albinism.

c. Dominance refers to a trait that is expressed in the heterozygous condition. Therefore, only one copy of a dominant allele needs to be present to express the phenotype. Recessiveness refers to a trait that is not expressed in the heterozygous condition. It is masked by the dominant allele. To express a recessive trait, two copies of the recessive allele must be present in the individual.

d. Complete dominance occurs when a dominant allele completely masks the expression of a recessive allele. For example, in pea plants, yellow seed color is dominant to green. In a heterozygous state, the phenotype of the seeds is yellow. This is the same phenotype seen in seeds homozygous for the yellow allele.

Incomplete dominance occurs when the phenotype of the heterozygote is intermediate between the two homozygotes. For example, in *Mirabilis*, a red flower crossed with a white flower will give a pink flower.

2. Genotypes: a, c, e; phenotypes b, d

a. 1/2 A, 1/2 a
all A
all a

11. a. All F1 plants will be long stemmed.

b. Let S = long stemmed and s = short stemmed. The long-stemmed P1 genotype is *SS*, the short-stemmed P1 genotype is *ss*. The long-stemmed F1 genotype is *Ss*.

c. Approximately 225 long stemmed and 75 short stemmed.

d. The expected genotypic ratio is: 1 *SS*:2 *Ss*:1 *ss*

13. a. $\frac{1}{2}$ A_B_, $\frac{1}{2}$ A_bb

b. $\frac{1}{4}$ A_B_, $\frac{1}{4}$ A_bb, $\frac{1}{4}$ aaB_, $\frac{1}{4}$ aabb

c. $\frac{9}{16}$ A_B_, $\frac{3}{16}$ A_bb, $\frac{3}{16}$ aaB_, $\frac{1}{16}$ aabb

15. Punnett Square:

	AB	Ab	aB	ab
AB	AABB	AABb	AaBB	AaBb
Ab	AABb	AAbb	AaBb	Aabb
aB	AaBB	AaBb	aaBB	aaBb
ab	AaBb	Aabb	aaBb	aabb

phenotypes:
A_B_ $\frac{9}{16}$
A_bb $\frac{3}{16}$
aaB_ $\frac{3}{16}$
aabb $\frac{1}{16}$

genotypes:
AABB $\frac{1}{16}$
AABb $\frac{2}{16}$
AAbb $\frac{1}{16}$
AaBB $\frac{2}{16}$
AaBb $\frac{4}{16}$
Aabb $\frac{2}{16}$
aaBB $\frac{1}{16}$
aaBb $\frac{2}{16}$
aabb $\frac{1}{16}$

Fork-line:

phenotypes:

$\frac{3}{4}$ A
- $\frac{3}{4}$ B → $\frac{9}{16}$ A_B_
- $\frac{1}{4}$ b → $\frac{3}{16}$ A_bb

$\frac{1}{4}$ a
- $\frac{3}{4}$ B → $\frac{3}{16}$ aaB_
- $\frac{1}{4}$ b → $\frac{1}{16}$ aabb

genotypes:

$\frac{1}{4}$ AA
- $\frac{1}{4}$ BB → $\frac{1}{16}$ AABB
- $\frac{2}{4}$ Bb → $\frac{2}{16}$ AABb
- $\frac{1}{4}$ bb → $\frac{1}{16}$ AAbb

$\frac{2}{4}$ Aa
- $\frac{1}{4}$ BB → $\frac{2}{16}$ AaBB
- $\frac{2}{4}$ Bb → $\frac{4}{16}$ AaBb
- $\frac{1}{4}$ bb → $\frac{2}{16}$ Aabb

$\frac{1}{4}$ aa
- $\frac{1}{4}$ Bb → $\frac{1}{16}$ aaBB
- $\frac{2}{4}$ Bb → $\frac{2}{16}$ aaBb
- $\frac{1}{4}$ bb → $\frac{1}{16}$ aabb

16. a. Both are 3:1

b. 9:3:3:1

c. Swollen is dominant to pinched, yellow is dominant to green.

d. Let P = swollen and p = pinched; C = yellow and c = green. Then: P_1 = $PPcc \times ppCC$ or $PPCC \times ppcc$. F1 = $PpCc$

18. a. Let S = smooth and s = wrinkled and Y = yellow and y = green. The parents are $SSYY \times ssyy$. The F1 offspring are *SsYy*.

b. Using the symbols from the problem above, the smooth, yellow parent is *SsYy*. The genotypes of the F1 are *SsYy*, *Ssyy*, *ssYy* and *ssyy*.

22. $\frac{3}{4}$ for $A \times \frac{1}{2}$ for $b \times 1$ for $C = \frac{3}{8}$ for A, b, C

24. The P1 generation is $FF \times ff$. The F1 generation is *Ff*. The mode of inheritance is incomplete dominance.

27. Because neither species produces progeny resembling a parent, simple dominance is ruled out. The species producing pink-flowered progeny from red and white (or very pale yellow) suggests incomplete dominance as a mode of inheritance. However, in the second species, the production of orange-colored progeny cannot be explained in this fashion. Orange would result from an equal production of red and yellow; instead in this case, codominance is suggested, with one parent producing bright red flowers and the other producing pale yellow flowers.

32. During meiotic prophase I, the replicated chromosomes synapse or pair with their homologues. These paired chromosomes align themselves on the equatorial plate during metaphase I. During anaphase I it is the homologues (each containing two chromatids) that separate from each other. There is no preordained orientation for this process—it is equally likely that a maternal or a paternal homologue will migrate to a given pole. This provides the basis for the law of random segregation. Independent assortment results from the fact that the polarity of one set of homologues has absolutely no influence on the orientation of a second set of homologues. For example, if the maternal homologue of chromosome 1 migrated to a certain pole, it will have no bearing on whether the maternal or paternal homologue of chromosome 2 migrates to that same pole.

Chapter 4

2. d
3. a. female
   b. yes
   c. 3 siblings, the proband is the youngest child
5. Autosomal dominant with incomplete penetrance
7. a.

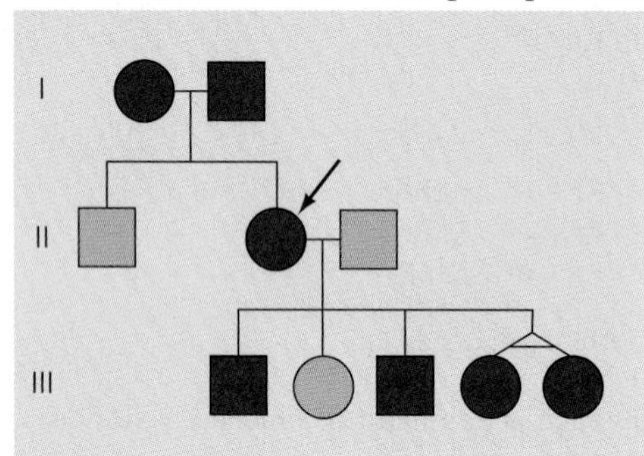

   b. The mode of inheritance is consistent with an autosomal dominant trait. Both of the proband's parents are affected. If this trait were recessive, all of their children would have to be affected (*aa* × *aa* can only produce *aa* offspring). As we see in this pedigree, the brother of the proband is not affected indicating that this is a dominant trait. His genotype is *aa*, the proband's genotype is *AA* or *Aa*, and both parents' genotype is most likely *Aa*.
   c. Because the proband's husband is unaffected, he is *aa*.
9. a. This pedigree is consistent with autosomal recessive inheritance.
   b. If inheritance is autosomal recessive, the individual in question is heterozygous.
10.

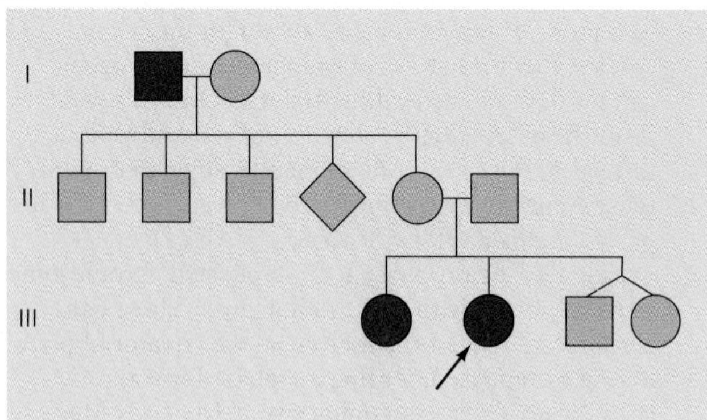

14. Due to the rarity of the disease, we assume the paternal grandfather is heterozygous for the gene responsible for Huntington disease. His son has a 1/2 chance of possessing the deleterious allele. In turn, should the father carry the HD allele, his son would have a 1/2 chance of inheriting it. Therefore, at present, the child has a $1/2 \times 1/2 = 1/4$ chance of having inherited the HD allele.
16. a. 50% chance for sons, 25% of all children
    b. 50% for daughters, 25% of all children
19. In the autosomal case, the parents are Aa and Aa where the disease is recessive. They can have a son or daughter that is affected (*aa*) or unaffected (*AA* or *Aa*). In the X-linked situation, the parents are $X^AX^a \times X^AY$ in a recessive case. The children would be X^AX^a and X^aY. Because an unaffected daughter and affected son are possible in each case, this limited information is not enough to determine the inheritance pattern.
20. X-linked dominant
21. X-linked recessive
23. Mitochondria contain DNA carrying genetic information, and are maternally inherited.
27. 20% of 90 = 18
34.

```
        4 map units          8 map units
C-------------------A-------------------B
```

Chapter 5

2. a. Height in pea plants is determined by a single pair of genes with dominant and recessive alleles. Height in humans is determined by polygenes.
   b. For traits determined by polygenes, the offspring of matings between extreme phenotypes show a tendency to regress toward the mean phenotype in the population.
5. a. F1 genotype = $A'AB'B$, phenotype = height of 6 ft.
   b. $A'AB'B \times A'AB'B$

↓

Genotypes	Phenotypes
$A'A'B'B'$	7 ft
$A'A'B'B$	6 ft 6 in.
$A'A'BB$	6 ft
$A'AB'B'$	6 ft 6 in.
$A'AB'B$	6 ft
$A'ABB$	5 ft 6 in.
$AAB'B'$	6 ft
$AAB'B$	5 ft 6 in.
$AABB$	5 ft

6. In the case of polygenes, the expression of the trait depends on the interactions of many genes, each of which contributes a small effect to the expression of the trait. Thus, the differences between genotypes often are not clearly distinguishable. In the case of monogenic determination of a trait, the alleles of a single locus have major effects on the expression of the trait, and the differences between genotypes are usually easily discerned.
7. Liability is caused by a number of genes acting in an additive fashion to produce the defect. If exposed to certain environmental conditions, the person above the threshold will most likely develop the disorder. The person below the threshold is not predisposed to the disorder and will most likely remain normal.

11. Relatives are used because the proportion of genes held in common by relatives is known.
13. No. Dizygotic twins arise from two separate fertilized eggs. Only monozygotic twins can be Siamese, since they originate from the same fertilized egg and are genetically identical.
14. b
18. a. The study included only men who were able to pass a physical exam that eliminated markedly obese individuals, and so the conclusions cannot be generalized beyond the group of men inducted into the armed forces.
    b. To design a better study, include MZ and DZ twin men and women, maybe even children. Include a cross section of various populations (ethnic groups, socioeconomic groups, weight classifications, etc.) Control the diet so that it remains a constant. Another approach is to study MZ and DZ twins that were reared apart (and presumably in different environments), or adopted and natural children who where raised in the same household (same environment). There are other possible answers.
22. First, intelligence is difficult to measure. Also, for such a complex trait, many genes and a significant environmental component are likely to be involved.
24. The heritability difference observed between the racial groups for this trait cannot be compared because heritability measures variation within one population at the time of the study. Heritability cannot be used to estimate genetic variation between populations.

Chapter 6

1. a. Chemical treatment of chromosomes resulting in unique banding patterns
   b. Q banding with quinacrine and G banding with Giemsa
6. Advanced maternal age, previous aneuploid child, presence of a chromosomal rearrangement, presence of a known genetic disorder in the family history.
8. Triploidy
13. The embryo will be tetraploid. Inhibition of centromere division results in nondisjunction of an entire chromosome set. After cytoplasmic division, some cytoplasm is lost in an inviable product lacking genetic material and the embryo develops from the tetraploid product.
18. Condition 2 is most likely lethal. This condition involves a chromosomal aberration, trisomy. This has the potential for interfering with the action of all genes on the trisomic chromosome. Condition 1 involves an autosomal dominant lesion to a single gene, which is more likely to be tolerated by the organism.
22. Turner syndrome (45,X) is monosomy for the X chromosome. A paternal nondisjunction event could contribute a gamete lacking a sex chromosome to result in Turner syndrome. The complementary gamete would contain both X and Y chromosomes. This gamete would contribute to Kleinfelter syndrome (47,XXY).
23.

Loss of a chromosome segment	deletion
Extra copies of a chromosome segment	duplication
Reversal in the order of a chromosome segment	inversion
Movement of a chromosome segment to another, nonhomologous chromosome	translocation

25. In theory, the chances are 1/3.
26. Two or three possibilities should be considered. The child could be monosomic for the relevant chromosome. The child has the paternal copy carrying the allele for albinism (father is heterozygous) and a nondisjunction event resulted in failure to receive a chromosomal copy from the homozygous mother. The second possibility is that the maternal chromosome carries a small deletion, allowing the albinism to be expressed. The third possibility is that the child represents a new mutation, inheriting the albino allele and having the other by mutation. Because monosomy is lethal, either the second or third possibility seems likely.

Chapter 7

2. There are significant economic and social consequences associated with FAS, including the costs of surgery for facial reconstruction, treatment of learning disorders and mental retardation, and caring for institutionalized individuals. Prevention depends on the education of pregnant women and the early treatment of pregnant women with alcohol dependencies. Other answers are possible.
4. RU486 is controversial because it doesn't prevent fertilization. It prevents implantation of an embryo. To some, this is the same as abortion because it is killing a potential life.
10. d
11. Female
13. A mutation causing the loss of the SRY gene, testosterone, or testosterone receptor gene function. Also a defect in the conversion from testosterone to DHT can cause the female external phenotype until puberty.
17. Pattern baldness acts as an autosomal dominant trait in males and an autosomal recessive trait in females.
20. Random inactivation in females, so the genes from both X chromosomes are active in the body as a whole.

Chapter 8

2. Proteins are found in the nucleus. Proteins are complex molecules composed of 20 different amino acids; nucleic acids are composed of only four different nucleotides. Cells contain hundreds or thousands of different proteins; only two main types of nucleic acids.
4. Protease destroyed any small amounts of protein contaminants in the transforming extract. Similarly, treatment with RNAse destroyed any RNA present in the mixture.

6. The process is transformation, discovered by Frederick Griffith. The P bacteria contain genetic information that is still functional even though the cell has been heat killed. However, it needs a live recipient host cell to accept its genetic information. When heat-killed P and live D bacteria are injected together, the dead P bacteria can transfer its genetic information into the live D bacteria. The D bacteria are then transformed into P bacteria and can now cause polkadots.

10. Chargaff's Rule: A = T and C = G
    If A = 27%, then T must equal 27%
    If G = 23%, then C must equal 23%
    Base composition:
    A = 27%
    T = 27%
    C = 23%
    G = 23%
    100%

12. b and e

13. b

15. c

20.

	DNA	RNA
a. Number of chains:	2	1
b. Bases used:	A,C,G,T	A,C,G,U
c. Sugar used:	deoxyribose	ribose
d. Function:	blueprint of genetic information	transfer of genetic information from nucleus to cytoplasm

23. a

Chapter 9

2. Replication is the process of making DNA from a DNA template. Transcription makes RNA from a DNA template, and translation makes an amino acid chain (a polypeptide) from an mRNA template. Replication and transcription happen in the nucleus, and translation occurs in the cytoplasm.

3. There would be 4^4 or 256 possible amino acids encoded.

7. b

8. 1. removal of introns: to generate a contiguous coding sequence that can make an amino acid chain
   2. addition of the 5′ cap: ribosome binding
   3. addition of the 3′ polyA tail: mRNA stability

12. Codons are triplets of bases on an mRNA molecule. Anticodons are triplets of bases on a tRNA molecule, and are complementary in sequence to the nucleotides in codons.

13. Answer: 25% Total length: 10 kb
    Coding region: 2.5 kb

16.

tRNA:	UAC	UCU	CGA	GGC
mRNA:	AUG	AGA	GCU	CCG
DNA:	TAC	TCT	CGA	GGC-sense strand
protein:	met	arg	ala	pro

Hydrogen bonds present in the DNA: 31
7 GC pairs × 3 = 21
5 AT pairs × 2 = 10

24. a. No
    b. Yes

Chapter 10

3. c

4. a. Buildup of substance A, no substance C
   b. Buildup of substance B, no substance C
   c. Buildup of substance B, as long as A is not limiting factor
   d. 1/2 the amount of C

5. a. Yes. Each will carry the normal gene for the other enzyme (individual 1 will be mutant for enzyme 1 but normal for enzyme 2. This is because enzyme 1 and 2 are encoded by two different genes)
   b. Let D = dominant mutation in enzyme 1, let normal allele = d
   Let A = dominant mutation in enzyme 2, let normal allele = a
   Ddaa × *ddAa*
   Offspring: *DdAa* mutation in enzyme 1 and 2, A buildup, no C
   Ddaa mutation in enzyme 1, A buildup, no C
   ddAa mutation in enzyme 2, B buildup, no C
   ddaa no mutation, normal
   Ratio would be 1:2:1 for substance B buildup, no C:substance A buildup, no C:normal

6. Alleles for enzyme 1: *A* (dominant, 50% activity); *a* (recessive, 0% activity). Alleles for enzyme 2: *B* (dominant, 50% activity); *b* (recessive, 0% activity).

	Enzyme 1	Enzyme 2	Compound		
			A	B	C
1*AABB*	100	100	N	N	N
2*AaBB*	50	100	N	N	N
4*AaBb*	50	50	N	N	N
2*AABb*	100	50	N	N	N
1*Aabb*	100	0	N	B	L
2*Aabb*	50	0	N	B	L
1*aaBB*	0	100	B	L	L
2*aaBb*	0	50	B	L	L
1*aabb*	0	50	B	L	L

N, normal; B, buildup; L, less.

12. b

17. No, because individuals who are GD/GD show 50% activity. The g allele reduces activity by 50% so heterozygotes appear normal. It is not until the level of activity falls below 50% that the mutant phenotype is observed.

18. Let H = the mutant allele for hypercholesterolemia
    Let h = the normal allele

Answer: *HH* heart attack as early as the age of 2, definite heart disease by age 20, death in most cases by 30. No functional LDL receptors produced.
Hh heart attack in early 30s. Half the number of functional receptors are present and twice the normal levels of LDL.

hh normal. Both copies of the gene are normal and can produce functional LDL receptors.

23. It would cause a frame shift mutation very early in the protein. Most likely, the protein would lose all of its functional capacity.
26. Drugs usually act on proteins. Different people have different forms of proteins. Different proteins are inherited as different alleles of a gene.
27. People have different abilities to smell and taste chemical compounds such as phenylthiocarbamide (PTC); some people are unable to smell skunk odors; different reactions to succinylcholine, a muscle relaxant, and to primaquine, an antimalarial drug. Others are sensitive to the pesticide parathion.

Chapter 11

1. A measure of the occurrence of mutations per individual per generation.
2. 245,000 births represent 490,000 copies of the achondroplasia gene, because each child carries two copies of the gene. The mutation rate is therefore 10/490,000 or 2×10^{-5} per generation.
7. Muscular dystrophy is an X-linked disorder. A son receives an X chromosome from his mother and a Y chromosome from his father. In this case, the mother was a heterozygous carrier of muscular dystrophy, and passed the mutant gene to her son. The father's exposure to chemicals in the work place is unrelated to this condition in his son.
10. Missense-same
    Nonsense-shorter
    Sense-longer
16. **a.** See pedigree below
19. Our bodies can correct base pairing errors made during DNA replication. We can also repair UV damaged DNA after thymine dimers form.

Chapter 12

2. d
4. b
8. They leave compatible ends that can be pasted (ligated) to other DNA cut with the same enzyme.
9. *Eco*R1: 2 kb, 11.5 kb, 10 kb
   *Hind*III/*Pst*1: 7 kb, 3 kb, 7.5 kb, 6 kb
   *Eco*R1/*Hind*III/*Pst*1: 2 kb, 5 kb, 3 kb, 3.5 kb, 4 kb, 6 kb
13. b
15. a
20. DNA derived from individuals with sickle cell anemia will lack one fragment contained in the DNA from normal individuals. In addition, there will be a large (uncleaved) fragment not seen in normal DNA.

Chapter 13

2. RFLPs are inherited as codominant alleles. Useful RFLPs have at least two, but not too many alleles, some of which should be fairly common in the population being studied. For linkage, the RFLP should be not only on the same chromosome as the gene being studied, but should be close enough to the gene so that crossovers are infrequent.
5. d
7. **a.** 1/100 × 1/500 = 1/50,000 individuals with this combination of alleles are present in the population.
   b. This is not very convincing, because in a large city, say with a population of 3 million, there will be approximately 60 individuals with this profile
   c. The lab should test two or more additional loci to lower the probability of another person having this profile to a much lower number, such as 1 in 50 million or more.
   d. This answer is an opinion and a point of discussion.
10. c
16. Bacteria have been used as a relatively easy and inexpensive way to make human proteins. However, they sometimes cannot make the proper modifications to proteins that are needed for normal function. Eukaryotic hosts can overcome this problem. Sheep and tobacco plants are examples of organisms currently in use.
17. There are several potential problems. For example with gene transfer, are there harmful side effects with human proteins grown in animals or plants? Do genetically engineered crops have any short- or long-term harmful effects to humans or the environment? Extensive testing and a rigorous approval processes should accompany any recombinant DNA technology before it is put to use.

See Chapter 11, answer 16.a

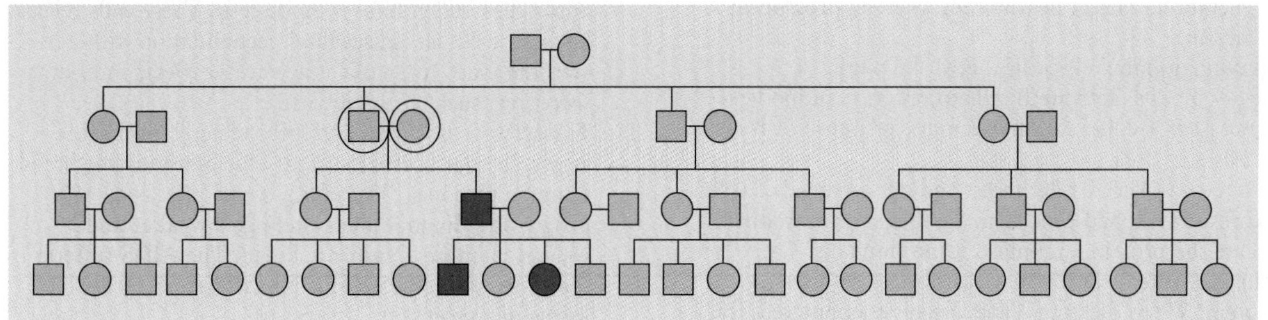

Chapter 14

3. a
6. d
9. An oncogene is a mutated proto-oncogene that promotes uncontrolled cell division that leads to cancer. A tumor suppressor gene is a normal gene that stops cell division when it is not needed. If mutated, a tumor suppressor gene can lose its function and no longer be able to control cell division. The result is too much cell division.
10. The inheritance of predisposition is dominant because only one mutant allele causes the predisposition to retinoblastoma. However, the second allele must also be mutated in at least one eye cell to produce the disease. Therefore, the expression of retinoblastoma is recessive.
15. Conditions a and d would produce cancer. The loss of function of a tumor suppressor gene would allow cell growth to go unchecked. The overexpression of a proto-oncogene would promote more cell division than normal.
19. Mutations in *APC* form hundreds or thousands of benign tumors. If the right mutations occur in one of these tumors, the benign growths can progress to cancer. The large number of benign tumors makes the chance of acquiring other mutations likely.
21. c-*myc* lies at the breakpoint of a translocation involving chromosome 8 and either chromosome 14, 22, or 2. The translocation places the *myc* gene in an altered chromosomal milieu and thus disrupts its normal expression. Altered expression of c-*myc* is thought to be necessary for the production of Burkitt's lymphoma.
25. Diet is suspected as the cause in both cases. When Japanese move to the United States and adopt an American diet, the rate of breast cancer goes up, but the rate of colon cancer goes down. The reverse is also the case: the Japanese diet in Japan predisposes to colon cancer, but not to breast cancer.

Chapter 15

1. **a.** Microorganisms that penetrate the skin infect cells, which then release chemical signals such as histamine. This causes an increased blood flow into the area, resulting in an increase in temperature.
   b. The heat serves to inhibit microorganism growth, to mobilize white blood cells, and to raise the metabolic rate in nearby cells, thereby promoting healing.
3. Immunoglobulins: IgG, IgA, IgM, IgD, IgE
4. Helper T cells: activate B cells to produce antibodies
   Suppressor T cells: stop the immune response of B and T cells
   Cytotoxic (killer) T cells: target and destroy infected cells
10. Express the cloned gene to make the protein product, isolate the protein, and inject it into humans. The immune system should make antibodies to that protein. When the actual live virus is encountered, the immune system will have circulating antibodies and T cells that will recognize the protein (antigen) on the surface of the virus.
16. The antigens of the donor/recipient are more important. The antigen of the donor will be rejected if the recipient does not have the same antigen. The antigen of the recipient determines which antibodies can be produced. For example, a blood type A individual will make B antibodies if exposed to the B.
20. **a.** The mother is Rh^-. She will produce antibodies against the Rh antigen if her fetus is Rh^+. This happens when blood from the fetus enters the maternal circulation.
    b. The mother already has circulating antibodies against the Rh protein from her first Rh^+ child. She can mount a greater immune response against the second Rh^+ child by generating a large number of antibodies.
24. One approach is to clone the human gene that suppresses hyperacute rejection and inject it into pig embryos. The hope is that the pig's cells will express this human protein on the cell surface. The transplanted organ may then be recognized by the human recipient as "self." In addition, transplants of bone marrow from donor pigs into human recipients may help in preventing rejection mediated by T cells. This dual bone marrow system will recognize the pig's organ as "self" but will retain the normal human immunity.
25. **a.** They need to test one cell of each 8-cell embryo for an ABO and Rh blood type match and also an HLA complex match. If a match exists, they will implant the embryo(s) into the mother and hope that pregnancy occurs. When the baby is born, bone marrow will be taken and transplanted into the existing child.
    b. Ethically, it is difficult to imagine having a child for the primary purpose of being a bone marrow donor. It may demean the value of the life of the new child. Also, what happens to the embryos that are not a match to the couple's existing child? These embryos are completely healthy; they simply have the wrong blood type and histocompatibility complex. On the other hand, if the couple will love and provide for this new child, it may be a wonderful experience that the new child has the opportunity to save the life of his/her sibling.
28. Antihistamines block the production or action of histamine. The allergen causes a release of IgE antibodies which bind to mast cells. These cells release histamine which causes fluid accumulation, tissue swelling (such as swollen airways or eyes), and mucus secretion (such as a runny nose).
31. The HIV virus infects and kills helper T4 cells, the very cells that normally trigger the antibody-mediated immune response. Therefore, as the infection progresses, the immune system gets weaker and weaker as more T cells are killed. The AIDS sufferer is then susceptible to various infections and certain forms of cancer.

Chapter 16

2. The definition must be precise enough to distinguish the behavior from other, similar behaviors and from the behavior of the control group. The definition of the behavior can significantly affect the results of the genetic analysis, and even the mode of inheritance of the trait.
4. *Drosophila* has many advantages for the study of behavior. Mutagenesis and screening for behavior mutants allows the recovery of mutations that affect many forms of behavior. The ability to perform genetic crosses and recover large numbers of progeny over a short period of time also enhances the genetic analysis of behavior. This organism can serve as a model for human behavior, because cells of the nervous system in both *Drosophila* and humans use similar mechanisms to transmit impulses and store information.
6. Huntington disease is caused by expansion of a CAG trinucleotide repeat within the HD gene. Expansion of the repeat causes an increase in the number of glutamines in the encoded protein, causing the protein to become toxic to neurons. In regions of the nervous system expressing the mutant protein, cells fill with clusters of the protein, degenerate, and die.
10. It may be argued that using such a test to identify potentially violent individuals would allow them to be given appropriate therapy (such as drugs to increase MAOA activity) before they harm anyone. On the other hand, the use of such a test would result in individuals being labeled as aggressive or violent not based on their behavior, but on their genotype. This could have significant consequences in their work and personal lives.
13. No, it means that there are probably other genes involved or the environment plays a significant role. The linkage to chromosome 7 is still valid, and the next goal would be to find the gene on chromosome 7 that is linked to manic depression. Also, finding the other genes involved is important. A researcher may find that a subset of manic depressive individuals has a defect in the gene on chromosome 7 and another subset in a gene on chromosome 12. Both defects can contribute to the same disease.
16. The heritability of Alzheimer disease, a multifactorial disorder, cannot be established because of interactions between genetic and environmental factors. Less than 50% of Alzheimer cases can be attributed to genetic causes, indicating that the environment plays a large role in the development of this disease. Other nongenetic factors may involve aluminum and prions.
18. Probably not, because correlations must eventually be related to a causal relationship, in this case, between body hair and intelligence. Many factors contribute to intelligence, including environmental factors. Lacking an explanation for the relationship between body hair and intelligence, the assumption is unwarranted. Testing would depend on the definition of intelligence used in the study, and may involve IQ testing by individuals who know nothing about the person's body hair. Alternatively, testing could be done for g, a measure of cognitive ability in blind testing, where the presence of the subject's body hair is unknown to the test administrator.

Chapter 17

1.

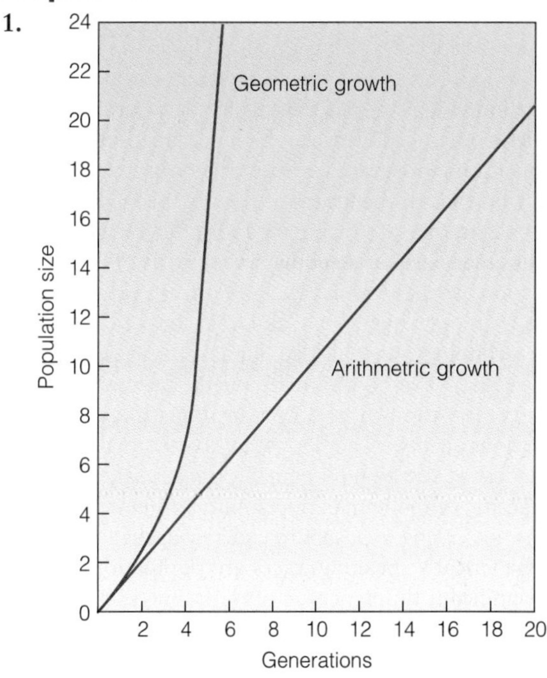

4. a. Population: local groups of individuals occupying a given space at a given time.
   b. Gene pool: the set of genetic information carried by a population.
   c. Allele frequency: the frequency of occurrence of particular alleles in the gene pool of a population.
   d. Genotype frequency: the frequency of occurrence of particular genotypes among the individuals of a population.
11. The frequency of an allele in a population has no relationship to its mode of inheritance. For example, a dominant allele may exist at a very low frequency in a population, and cannot ultimately overtake a recessive allele in frequency.
13. a. Children: $M = 0.5$, $N = 0.5$. Adults: $M = 0.5$, $N = 0.5$.
    b. Yes. Allelic frequencies are unchanged.
    c. No. The genotypic frequencies are changing within each generation.
19. It means that there has been more recombination between the marker and the surrounding DNA in southeast Europe, which indicates that the mutation has been in this region longer than it has been in northwest Europe.

Chapter 18

1. All polymorphisms are mutations. When a mutation resulting in a particular allele is present in at least 1%

of the population, it is termed a polymorphism. This percentage is chosen because it is clearly higher than can be accounted for by mutation alone. Because such genetic variations cannot be accounted for by mutation alone, other forces such as natural selection must be active. The study of such variation is important in understanding the process of evolution.

9. No. It is not a particularly accurate description. Natural selection depends not just on an ability to survive, but also to reproduce. It is the differential reproduction of some individuals that is the essence of natural selection.
11. Genetic polymorphisms can be maintained in populations for several reasons: they can make different contributions to the fitness of the organisms in which they occur, they are maintained in different frequencies in different parts of a subdivided population by natural selection, or they are in Hardy-Weinberg equilibrium.
14. This is an open question. Culture, in the form of society and technology, has shielded humans from many forms of selective forces in the environment, but has also created new forms of selection for humans and for infectious agents. The abuse of antibiotics may increase the effect of selection on human populations, as will the expansion of the human population into new geographic areas, increasing exposure to endemic agents of disease.
15. **a.** Genetically speaking, races are populations with significant differences in allele frequencies compared to other populations.
    b. No. No systematic differences have been identified in allele frequencies within modern human populations that would justify the use of the term race. Studies into the level of genetic variation within and between populations have consistently found that there is much more variety within each population than between them.
17. **a.** The out-of-Africa hypothesis holds that modern humans appeared in Africa, and then left the continent to replace all of the then-existing hominid populations in the world. The multi-regional hypothesis proposes that, through a network of interbreeding populations, *Homo erectus* gradually evolved into modern *Homo sapiens* in different regions of the world.
    b. The out-of-Africa hypothesis.
    c. Both mitochondrial and nuclear DNA evidence favor the out-of-Africa hypothesis.

Chapter 19

1. Genetic testing determines if an individual has a particular genotype. It is often used to test those at risk of having a genetic disease, such as a developing fetus, a family member of someone affected by a homozygous recessive genetic disorder, or a member of a cultural group with a high rate of a particular genetic disorder.

    Genetic screening looks for the presence of usually rare genetic disorders in members of a population. Genetic screening programs are often mandated by law.
3. The first recommendation would be for another α-fetoprotein test to confirm the initial result. If the second test is still positive or ambiguous, ultrasonography should be used to look directly at the spinal cord of the infant, and amniocentesis could be considered in order to examine α-fetoprotein levels in the amniotic fluid. Finally, if any questions remain, fetoscopy could be considered, although the significant risks associated with this technique would have to be taken into account.
5. Prenatal testing: used to diagnose genetic defects and birth defects in the developing fetus.

    Preimplantation testing: used to test for the presence of a mutant gene in one cell taken from a 6- to 8-cell embryo produced by *in vitro* fertilization. This is done to ensure that the embryo is free of genetic defects prior to implantation.

    Carrier testing: used to identify phenotypically normal individuals who carry deleterious recessive traits. Typically performed on family members of an affected individual or in members of a cultural group with a relatively high level of the disease.

    Presymptomatic testing: used to test for the presence of genetic mutations that do not produce an effect until later in life. If a disease-causing mutation is detected, then a preventive treatment may be initiated.
14. If we let the allele causing PKU be *p* and the normal allele be *P*, then the sister and the brother of the woman and the man, respectively must have been *pp*. In order to produce these affected individuals, all four grandparents must have been heterozygous normal. The pedigree can be summarized as:

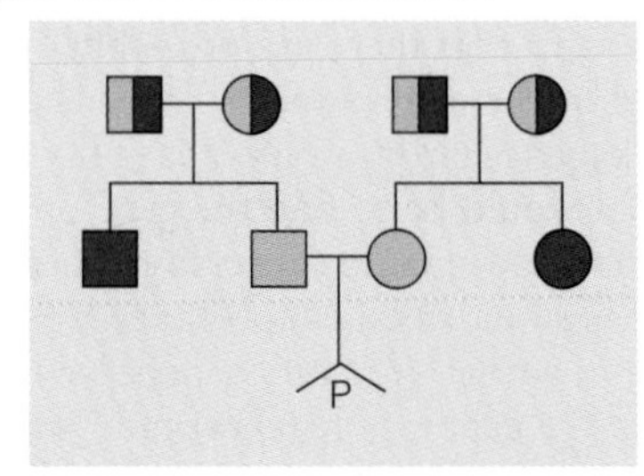

The only way the man and woman can have a PKU child is if both of them are heterozygotes, because they do not have the disease. Both the grandparental matings are simple Mendelian monohybrid crosses expected to produce progeny in the following proportions:

¼ — PP — Normal
½ — Pp — Normal
¼ — pp — PKU

We know that the man and the woman are normal, so the probability of either being a heterozygote is 2/3, because within the *P_* classes, 2/3 are *Pp* and 1/3 are *PP*. The probability of both the man and the woman being heterozygotes is $2/3 \times 2/3 = 4/9$. If they are both heterozygous, then one fourth of their children would have PKU. This means that the probability that their first child will have PKU is 1/4, and the total probability of their being heterozygous and of their first child having PKU is $4/9 \times 1/4 = 4/36$ or $1/9$.

16. **a.** The story of nonpaternity during family genetic counseling is familiar to genetic counselors. When deciding how to approach this type of unexpected findings, counselors need to weigh the benefits and harms of nondisclosure against that of disclosure. The first considerations include the relevance of the information to the patient's situation and the consequences of the findings. The 1983 President's Commission recommends that patients be advised before testing that unexpected information may be revealed.

b. It is reasonable for the counselor to call the woman beforehand and explain the results and the implications of the findings. Given the sensitivity of this information, the long-term effect on the couple's relationship may be dramatic, and disclosure may do more harm than good.

Glossary

Acquired immunodeficiency syndrome (AIDS) A collection of disorders that develop as a result of infection with the human immunodeficiency virus (HIV).

Acrocentric chromosome A chromosome whose centromere is placed very close to, but not at, one end.

Adenine and guanine Nitrogen-containing purine bases found in nucleic acids.

Affect Pertaining to emotion or feelings.

Alkaptonuria An autosomal recessive trait with altered metabolism of homogentisic acid. Affected individuals do not produce the enzyme to metabolize this acid.

Allele One of the possible alternative forms of a gene, usually distinguished from other alleles by its phenotypic effects.

Allele frequency The frequency with which alleles of a given gene are present in a population.

Allelic expansion Increase in gene size caused by an increase in the number of trinucleotide sequences.

Allergens Antigens that provoke an inappropriate immune response.

Alpha thalassemia Genetic disorder associated with an imbalance in the ratio of alpha and beta globin caused by reduced or absent synthesis of alpha globin.

Alzheimer disease A heterogeneous condition associated with the development of brain lesions, personality changes, and degeneration of intellect. Genetic forms are associated with loci on chromosomes 14, 19, and 21.

Amino group A chemical group (NH_2) found in amino acids and at one end of a polypeptide chain.

Amniocentesis A method of sampling the fluid surrounding the developing fetus by inserting a hollow needle and withdrawing suspended fetal cells and fluid; used in diagnosing fetal genetic and developmental disorders; usually performed in the sixteenth week of pregnancy.

Anaphase A stage in mitosis during which the centromeres split and the daughter chromosomes begin to separate.

Anaphylaxis A severe allergic response in which histamine is released into the circulatory system.

Androgen insensitivity An X-linked genetic trait that causes XY individuals to develop into phenotypic females.

Aneuploidy A chromosome number that is not an exact multiple of the haploid set.

Ankylosing spondylitis An autoimmune disease that produces an arthritic condition of the spine; associated with HLA allele B27.

Antibody A class of proteins produced by B cells that bind to and inactivate foreign molecules (antigens).

Antibody-mediated immunity Immune reaction that protects primarily against invading viruses and bacteria by antibodies produced by plasma cells.

Anticipation Onset of a genetic disorder at earlier ages and with increasing severity in successive generations.

Anticodon A group of three nucleotides in a tRNA molecule that pairs with a complementary sequence (known as a codon) in an mRNA molecule.

Antigenic determinant The site on an antigen to which an antibody binds, forming an antigen-antibody complex.

Antigens Molecules carried or produced by viruses and microorganisms that initiate antibody production.

Assisted reproductive technologies (ART) The collection of techniques used to help infertile couples.

Assortative mating Reproduction in which mate selection is not random but instead is based on physical, cultural, or religious grounds.

Assortment The random distribution of members of homologous chromosomal pairs during meiosis.

Atherosclerosis Arterial disease associated with deposition of plaques on inner surface of blood vessels.

Autosomes Chromosomes other than the sex chromosomes. In humans, chromosomes 1 to 22 are autosomes.

B cells A type of lymphocyte that matures in the bone marrow and mediates antibody-directed immunity.

Background radiation Radiation in the environment that contributes to radiation exposure.

Base analogs A purine or pyrimidine that differs in chemical structure from those normally found in DNA or RNA.

Beta thalassemia Genetic disorder associated with an imbalance in the ratio of alpha and beta globin caused by reduced or absent synthesis of beta globin.

Bioinformatics The use of computers and software to acquire, store, analyze, and visualize the information from genomics.

Bipolar disorder An emotional disorder characterized by mood swings that vary between manic activity and depression.

Blastocyst The developmental stage at which the embryo implants into the uterine wall.

Blood type One of the classes into which blood can be separated based on the presence or absence of certain antigens.

B-memory cells Long-lived B cells produced after exposure to an antigen that play an important role in secondary immunity.

Camptodactyly A dominant human genetic trait that is expressed as immobile, bent, little fingers.

Cap A modified base (guanine nucleotide) attached to the 5′ end of eukaryotic mRNA molecules.

Carboxyl group A chemical group (COOH) found in amino acids and at one end of a polypeptide chain.

Caretaker genes Genes that help maintain the integrity of the genome, for example, DNA repair genes.

Cell cycle The sequence of events that takes place between successive mitotic divisions.

Cell-mediated immunity Immune reaction mediated by T cells directed against body cells that have been infected by viruses or bacteria.

Centromere A region of a chromosome to which microtubule fibers attach during cell division. The location of a centromere gives a chromosome its characteristic shape.

Chiasmata The crossing of nonsister chromatid strands seen in the first meiotic prophase. Chiasmata represent the structural evidence for crossing-over.

Chorion A two-layered structure formed from the trophoblast.

Chorionic villus sampling A method of retrieving fetal chorionic cells by inserting a catheter through the vagina or abdominal wall into the uterus. Used in biochemical and cytogenetic defects in the embryo. Usually performed in the eighth or ninth week of pregnancy.

Chromatid One of the strands of a duplicated chromosome, joined by a single centromere to its sister chromatid.

Chromatin The complex of DNA and proteins that makes up a chromosome.

Chromosomes The threadlike structures in the nucleus that carry genetic information.

Clinodactyly An autosomal dominant trait with a bent finger as a phenotype.

Clone-by-clone method A method of genome sequencing that begins with genetic and physical maps and sequences clones after they have been placed in order.

Clones Genetically identical organisms, cells, or molecules all derived from a common ancestor.

Codominance Full phenotypic expression of both members of a gene pair in the heterozygous condition.

Codon Triplets of nucleotides in mRNA that encode the information for a specific amino acid in a protein.

Color blindness Defective color vision caused by reduction or absence of visual pigments. There are three forms: red, green, and blue blindness.

Complement system A chemical defense system that kills microorganisms directly, supplements the inflammatory response, and works with (complements) the immune system.

Concordance Agreement between traits exhibited by both twins.

Consanguineous mating Matings between two individuals who share a common ancestor in the preceding two or three generations.

Continuous variation Phenotypic characters distributed from one extreme to another in an overlapping, or continuous, fashion.

Correlation coefficient A measure of the degree to which variables vary together.

Covalent bond A chemical bond that results from electron sharing between atoms. Covalent bonds are formed and broken during chemical reactions.

Cri du chat syndrome A deletion of the short arm of chromosome 5 associated with an array of congenital malformations, the most characteristic of which is an infant cry that resembles a meowing cat.

Crossing-over The process of exchanging parts between homologous chromosomes during meiosis, which produces new combinations of genetic information.

C-terminus The end of a polypeptide or protein with a free carboxyl group.

Cystic fibrosis A fatal recessive genetic disorder associated with abnormal secretions of the exocrine glands.

Cytogenetics The branch of genetics that studies the organization and arrangement of genes and chromosomes using the techniques of microscopy.

Cytokinesis The process of cytoplasmic division that accompanies cell division.

Cytosine, thymine, and uracil Nitrogen-containing pyrimidine bases found in nucleic acids.

Cytoskeleton A system of protein microfilaments and microtubules that allows a cell to have a characteristic shape.

Deletion A chromosomal aberration in which a segment of a chromosome is missing.

Deoxyribonucleic acid (DNA) A molecule consisting of antiparallel strands of polynucleotides that is the primary carrier of genetic information.

Deoxyribose and ribose Pentose sugars found in nucleic acids. Deoxyribose is found in DNA, ribose in RNA.

Dermatoglyphics The study of the skin ridges on the fingers, palms, toes, and soles.

Diploid The condition in which each chromosome is represented twice as a member of a homologous pair.

Discontinuous variation Phenotypes that fall into two or more distinct, nonoverlapping classes.

Dispermy Fertilization of haploid egg by two haploid sperm, forming a triploid zygote.

Dizygotic (DZ) twins Twins derived from two separate and nearly simultaneous fertilizations, each involving one egg and one sperm. Such twins share, on average, 50% of their genes.

DNA A helical molecule consisting of two strands of nucleotides that is the primary carrier of genetic information.

DNA fingerprint A pattern of restriction fragments that is unique to an individual.

DNA polymerase An enzyme that catalyzes the synthesis of DNA using a template DNA strand and nucleotides.

DNA restriction fragment A segment of a longer DNA molecule produced by the action of a restriction endonuclease.

DNA sequencing A technique for determining the nucleotide sequence of a fragment of DNA.

Dominant The trait expressed in the F1 (or heterozygous) condition.

Dosage compensation A mechanism that regulates the expression of sex-linked gene products.

Duplication A chromosomal aberration in which a segment of a chromosome is repeated and therefore is present in more than one copy within the chromosome.

Ecogenetics A branch of genetics that studies genetic traits related to the response to environmental substances.

Endoplasmic reticulum (ER) A system of cytoplasmic membranes arranged into sheets and channels that functions in synthesizing and transporting gene products.

Enhancement gene therapy Gene transfer to enhance traits, such as intelligence or athletic ability, not to treat genetic disorders.

Environmental variance The phenotypic variance of a trait in a population that is attributed to differences in the environment.

Epidemiology The study of the factors that control the presence, absence, or frequency of a disease.

Essential amino acids Amino acids that cannot be synthesized in the body and must be supplied in the diet.

Eugenics The attempt to improve the human species by selective breeding.

Eukaryote An organism with a nuclear membrane surrounding the genetic material and has other membrane-bound organelles in the cytoplasm.

Exons DNA sequences that are transcribed, joined to other exons during mRNA processing and translated into the amino acid sequence of a protein.

Expressivity The range of phenotypes resulting from a given genotype.

Familial adenomatous polyposis (FAP) An autosomal dominant trait resulting in the development of polyps, benign growths, in the colon. Polyps often develop into malignant growths and cause cancer of the colon and/or rectum.

Familial hypercholesteremia (FH) Autosomal dominant disorder with defective or absent low-density lipoprotein (LDL) receptors. Affected individuals are at increased risk for cardiovascular disease.

Fertilization The fusion of two gametes to produce a zygote.

Fetal alcohol syndrome A constellation of birth defects caused by maternal alcohol consumption during pregnancy.

Fitness A measure of the relative survival and reproductive success of a given individual or genotype.

Fluorescence *in situ* hybridization (FISH) A method of mapping DNA sequences to metaphase chromosomes by using probes labeled with fluorescent dyes.

Founder effects Allele frequencies established by chance in a population started by a small number of individuals (perhaps only a fertilized female).

Fragile X An X chromosome that carries a nonstaining gap, or break, at band q27; associated with mental retardation in males.

Frameshift mutation Mutational event in which a number of bases (other than multiples of three) are added to or removed from DNA, causing a shift in the codon reading frame.

Galactosemia A heritable trait associated with the inability to metabolize the sugar galactose. If left untreated, high levels of galactose-1-phosphate accumulate, causing cataracts and mental retardation.

Gatekeeper genes Genes that regulate cell growth and passage through the cell cycle, for example, tumor suppressor genes.

Gene The fundamental unit of heredity.

Gene pool The set of genetic information carried by the members of a sexually reproducing population.

General cognitive ability A measure of intelligence that includes verbal and spatial abilities, memory, speed of perception, and reasoning.

Genetic code The sequence of nucleotides that encodes the information for amino acids in a polypeptide chain.

Genetic counseling A process of communication that deals with the occurrence or risk that a genetic disorder will occur in a family.

Genetic drift The random fluctuations of allele frequencies from generation to generation that take place in small populations.

Genetic equilibrium The situation when the allele frequency for a given gene remains constant from generation to generation.

Genetic library In recombinant DNA terminology, a collection of clones that contains all of the genetic information in an individual.

Genetic map The arrangement and distance between genes on a chromosome deduced from studies of genetic recombination.

Genetic screening The systematic search for individuals in a population who have certain genotypes.

Genetic testing The use of methods to determine if someone has a genetic disorder, will develop one, or is a carrier.

Genetic variance The phenotypic variance of a trait in a population that is attributed to genotypic differences.

Genetics The scientific study of heredity.

Genomic imprinting Phenomenon in which the expression of a gene depends on whether it is inherited from the mother or the father. Also known as genetic or parental imprinting.

Genomics The collection of techniques used to clone and sequence a genome.

Genotype The specific genetic constitution of an organism.

Germ-line gene therapy Gene transfer to gametes or the cells that produce them. Transfers the gene to all cells in the next generation, including germ cells.

Golgi apparatus Membranous organelles composed of a series of flattened sacs. Proteins synthesized in the endoplasmic reticulum (ER) are sorted, modified, and packaged in the Golgi.

Haploid The condition in which each chromosome is represented once in an unpaired condition.

Haplotype A cluster of closely linked genes or markers that are inherited together. In the immune system, the HLA alleles on chromosome 6 are a haplotype.

Hardy-Weinberg Law The statement that allele and genotype frequencies remain constant from generation to generation when the population meets certain assumptions.

Helper T cells A lymphocyte that stimulates production of antibodies by B cells when an antigen is present.

Hemizygous gene A gene present on the X chromosome that is expressed in males in both the recessive and dominant condition.

Hemoglobin variants Alpha and beta globins with variant amino acid sequences.

Hemolytic disease of the newborn (HDN) A condition of immunological incompatibility between mother and fetus that occurs when the mother is Rh^- and the fetus is Rh^+.

Hereditarianism The idea that human traits are determined solely by genetic inheritance, ignoring the contribution of the environment.

Hereditary nonpolyposis colon cancer (HNPCC) A form of colon cancer associated with genomic instability of microsatellite DNA sequences. Caused by mutation in DNA repair genes.

Heritability An expression of how much of the observed variation in a phenotype is due to differences in genotype.

Heterozygous Carrying two different alleles for one or more genes.

Histones DNA-binding proteins that help compact and fold DNA into chromosomes.

Hominid A member of the family Hominidae, which includes bipedal primates such as *Homo sapiens.*

Hominoid A member of the primate superfamily Hominoidea, including the gibbons, great apes, and humans.

Homologs Members of a chromosomal pair.

Homozygous Having identical alleles for one or more genes.

Huntington disease A dominant genetic disorder characterized by involuntary movements of the limbs, mental deterioration, and death within 15 years of onset. Symptoms appear between 30 and 50 years of age.

Hydrogen bond A weak chemical bonding force between hydrogen and another atom.

Hypertension Elevated blood pressure, consistently above 140/90 mm Hg.

Hypophosphatemia An X-linked dominant disorder. Those affected have low phosphate levels in blood and skeletal deformities.

Immunoglobulins The five classes of proteins to which antibodies belong.

***In vitro* fertilization (IVF)** A procedure in which gametes are fertilized in a dish in the laboratory, and the resulting zygote is implanted in the uterus for development

Inborn error of metabolism The concept advanced by Archibald Garrod that many genetic traits result from alterations in biochemical pathways.

Inbreeding Production of offspring by related parents.

Incest Sexual relations between parents and children or between brothers and sisters.

Incomplete dominance Expression of a phenotype that is intermediate between those of the parents.

Independent assortment The random distribution of genes into gametes during meiosis.

Inflammatory response The body's reaction to invading microorganisms.

Inner cell mass A cluster of cells in the blastocyst that gives rise to the embryo.

Intelligence quotient (IQ) A score derived from standardized tests that is calculated by dividing the individual's mental age (determined by the test) by his or her chronological age and multiplying the quotient by 100.

Interphase The period of time in the cell cycle between mitotic divisions.

Intracytoplasmic sperm injection (ICSI) Fertilization of an egg by microinjection of a sperm, to overcome defects in sperm motility or fertilization events.

Introns DNA sequences present in some genes that are transcribed but are removed during processing and therefore are not present in mature mRNA.

Inversion A chromosomal aberration in which a chromosomal segment has been rotated 180° from its usual orientation.

Ionizing radiation Radiation that produces ions during interaction with other matter, including molecules in cells.

Karyotype A complete set of chromosomes from a cell that has been photographed during cell division and arranged in a standard sequence.

Killer T cells T cells that destroy body cells infected by viruses or bacteria. These cells can also directly attack viruses, bacteria, cancer cells, and cells of transplanted organs.

Klinefelter syndrome Aneuploidy of the sex chromosomes involving an XXY chromosomal constitution.

Leptin A hormone produced by fat cells that signals the brain and ovary. As fat levels become depleted, secretion of leptin slows and eventually stops.

Linkage A condition in which two or more genes do not show independent assortment. Rather, they tend to be inherited together. Such genes are located on the same chromosome. By measuring the degree of recombination between linked genes, the distance between them can be determined.

Lipoproteins Particles that have protein and phospholipid coats that transport cholesterol and other lipids in the bloodstream.

Locus The position occupied by a gene on a chromosome.

Lod method A probability technique used to determine whether genes are linked.

Lod score The ratio of the probability that two loci are linked to the probability that they are not linked, expressed as a $\log_{10}$. Scores of 3 or more are taken as establishing linkage.

Lymphocytes White blood cells that originate in bone marrow and mediate the immune response.

Lyon hypothesis The proposal that dosage compensation in mammalian females is accomplished by partially and randomly inactivating one of the two X chromosomes.

Lysosomes Membrane-enclosed organelles that contain digestive enzymes.

Macrophages Large white blood cells derived from monocytes that engulf antigens and present them to T cells, activating the immune response.

Mad-cow disease A prion disease of cattle; bovine spongiform encephalopathy (BSE). Can be transmitted to humans.

Marfan syndrome An autosomal dominant genetic disorder affecting the skeletal system, the cardiovascular system, and the eyes.

Meiosis The process of cell division during which one cycle of chromosomal replication is followed by two successive cell divisions to produce four haploid cells.

Metabolism The sum of all biochemical reactions by which cells convert and utilize energy.

Metacentric chromosome A chromosome that has a centrally placed centromere.

Metaphase A stage in mitosis during which the chromosomes move and become arranged near the middle of the cell.

Millirem Each rem is equal to 1000 millirems.

Missense mutation A mutation that causes the substitution of one amino acid for another in a protein.

Mitochondria (singular: mitochondrion) Membrane-bound organelles present in the cytoplasm of all eukaryotic cells that are the sites of energy production within cells.

Mitosis Form of cell division that produces two cells, each of which has the same complement of chromosomes as the parent cell.

Molecular genetics The study of genetic events at the biochemical level.

Molecule A structure composed of two or more atoms held together by chemical bonds.

Monocytes White blood cells that participate in the inflammatory response.

Monohybrid cross A cross between individuals that differ with respect to a single gene pair.

Monosomy A condition in which a member of a chromosomal pair is missing; having one less than the diploid number ($2n - 1$).

Monozygotic (MZ) twins Twins derived from a single fertilization involving one egg and one sperm; such twins are genetically identical.

Mood disorders A group of behavior disorders associated with manic and/or depressive syndromes.

Mood A sustained emotion that influences perception of the world.

mRNA A single stranded complementary copy of the nucleotide sequence in a gene.

Müllerian inhibiting hormone (MIH) A hormone produced by the developing testis that causes the breakdown of the Müllerian ducts in the embryo.

Multifactorial traits Traits that result from the interaction of one or more environmental factors and two or more genes.

Multiple alleles Genes that have more than two alleles have multiple alleles.

Muscular dystrophy A group of genetic diseases associated with progressive degeneration of muscles. Two of these, Duchenne and Becker muscular dystrophy, are inherited as X-linked, allelic, recessive traits.

Mutation rate The number of events that produce mutated alleles per locus per generation.

Natural selection The differential reproduction shown by some members of a population resulting from differences in fitness.

Nitrogen-containing base A purine or pyrimidine that is a component of nucleotides.

Nondisjunction The failure of homologous chromosomes to separate properly during meiosis or mitosis.

Nonsense mutation A mutation that changes an amino acid specifying codon to one of the three termination codons.

N-terminus The end of a polypeptide or protein that has a free amino group.

Nucleolus (plural: nucleoli) A nuclear region that functions in the synthesis of ribosomes.

Nucleosomes A bead-like structure composed of histones wrapped with DNA.

Nucleotide substitutions Mutations involving substitutions of one or more nucleotides in a DNA molecule.

Nucleotides The basic building blocks of DNA and RNA. Each nucleotide consists of a base, a phosphate, and a sugar.

Nucleus The membrane-bounded organelle in eukaryotic cells that contains the chromosomes.

Oncogene A gene that induces or continues uncontrolled cell proliferation.

Oogonia Mitotically active cells that produce primary oocytes.

Ootid The haploid cell produced by meiosis that becomes the functional gamete.

Organelle A cytoplasmic structure with a specialized function.

Palindrome A word, phrase, or sentence that reads identically in both directions. In DNA, a sequence of nucleotides that reads identically on both strands when read in the 5′ to 3′ direction.

Pattern baldness A sex-influenced trait that acts like an autosomal dominant trait in males and an autosomal recessive trait in females.

Pedigree analysis The construction of family trees and their use to follow the transmission of genetic traits in families. It is the basic method of studying the inheritance of traits in humans.

Pedigree construction Use of family history to determine how a trait is inherited and to determine risk factors for family members.

Penetrance The probability that a disease phenotype will appear when a disease-related genotype is present.

Pentose sugar A five-carbon sugar molecule found in nucleic acids.

Peptide bond A covalent chemical link between the carboxyl group of one amino acid and the amino group of another amino acid.

Pharmacogenetics A branch of genetics concerned with the inheritance of differences in the response to drugs.

Phenotype The observable properties of an organism.

Phenylketonuria (PKU) An autosomal recessive disorder of amino acid metabolism that results in mental retardation if untreated.

Philadelphia chromosome An abnormal chromosome produced by translocation of parts of the long arms of chromosomes 9 and 22.

Plasma cells Cells produced by mitotic division of B cells; plasma cells synthesize and secrete antibodies.

Polar body A cell produced in the first or second division in female meiosis that contains little cytoplasm and will not function as a gamete.

Poly-A tail A series of A nucleotides added to the 3′ end of mRNA molecules.

Polygenic trait A phenotype that depends on the action of a number of genes.

Polymerase chain reaction (PCR) A method for amplifying DNA segments that uses cycles of denaturation, annealing to primers, and DNA-polymerase directed DNA synthesis.

Polymorphism The occurrence of two or more genotypes in a population in frequencies that cannot be accounted for by mutation alone.

Polyp A growth attached to the substrate by a small stalk. Commonly found in the nose, rectum, and uterus.

Polypeptide A molecule made of amino acids joined together by peptide bonds.

Polyploidy A chromosomal number that is a multiple of the normal haploid chromosomal set.

Population A local group of organisms belonging to a single species, sharing a common gene pool; also called a deme.

Population genetics The branch of genetics that studies inherited variation in populations of individuals and the forces that alter gene frequency.

Porphyria A genetic disorder inherited as a dominant trait that leads to intermittent attacks of pain and dementia. Symptoms first appear in adulthood.

Positional cloning Identification and cloning of a gene responsible for a genetic disorder that begins with no information about the gene or the function of the gene product.

Prader-Willi syndrome A disorder associated with chromosome 15 characterized by uncontrolled eating and obesity.

Precocious puberty An autosomal dominant trait expressed in a sex-limited fashion. Heterozygous males are affected, but heterozygous females are not.

Primary structure The amino acid sequence in a polypeptide chain.

Prion A protein folded into an infectious conformation that is the cause of several disorders, including Creutzfeldt-Jakob disease and mad-cow disease.

Proband First affected family member who seeks medical attention.

Probe A labeled nucleic acid used to identify a complementary region in a clone or genome.

Product The specific chemical compound that is the result of enzymatic action on a substrate molecule. In biochemical pathways, a compound can serve as the product of one reaction and the substrate for the next reaction.

Prokaryote An organism without a nucleus.

Promoter A region of a DNA molecule to which RNA polymerase binds and initiates transcription.

Prophase A stage in mitosis during which the chromosomes become visible and split longitudinally except at the centromere.

Proto-oncogene A gene that initiates or maintains cell division and may become a cancer gene (oncogene) by mutation.

Pseudogene A nonfunctional gene that is closely related (by DNA sequence) to a functional gene present elsewhere in the genome.

Pseudohermaphroditism An autosomal genetic condition that causes XY individuals to develop the phenotypic sex of females.

Purine A class of double-ringed organic bases found in nucleic acids.

Pyrimidines A class of single-ringed organic bases found in nucleic acids.

Quantitative trait loci (QTLs) Two or more genes that act on a single polygenic trait.

Quaternary structure The structure formed by the interaction of two or more polypeptide chains in a protein.

R group A term used to indicate the position of an unspecified group in a chemical structure.

Race A genotypically distinct group within a species.

Radiation The process by which electromagnetic energy travels through space or a medium such as air.

Recessive The trait unexpressed in the F1 but reexpressed in some members of the F2 generation.

Reciprocal translocation A chromosomal aberration involving the exchange of chromosomal material between two nonhomologous chromosomes. This changes the chromosomal location of the exchanged genes but not the number of genes in the genome.

Recombinant DNA technology Technique for joining DNA from two or more different organisms to produce hybrid, or recombined, DNA molecules.

Recombination The exchange of genetic material between homologous chromosomes. Also known as crossing-over.

Regression to the mean In a polygenic system, the tendency of offspring of parents who have extreme differences in phenotype to exhibit a phenotype that is the average of the two parental phenotypes.

Rem The unit of radiation exposure used to measure radiation damage in humans. It is the amount of ionizing radiation that has the same effect as a standard amount of x-rays.

Restriction enzymes Enzymes that recognize a specific base sequence in a DNA molecule and cleave or nick the DNA at that site.

Restriction fragment length polymorphism (RFLP) Variations in the length of DNA fragments generated by a restriction endonuclease. Inherited codominantly, RFLPs are used as markers for specific chromosomes or genes.

Retinoblastoma A malignant tumor of the eye arising in retinoblasts (embryonic retinal cells that disappear at about 2 years of age). Mature retinal cells do not transform into tumors, so this is a tumor that usually occurs in children. It is associated with a deletion on the long arm of chromosome 13.

Retrovirus Viruses with an RNA genome. During the viral life cycle, the RNA is reverse-transcribed into DNA. The name retrovirus symbolizes this backward order of transcription.

Ribonucleic acid (RNA) A nucleic acid molecule that contains the pyrimidine uracil and the sugar ribose. The several forms of RNA function in gene expression.

Ribosomal RNA (rRNA) RNA molecules that form part of the ribosome.

Ribosomes Cytoplasmic particles composed of two subunits that are the site of protein synthesis.

Robertsonian translocation Breakage in the short arms of acrocentric chromosomes followed by fusion of the long parts into a single chromosome.

Sarcoma A cancer of connective tissue. One type of sarcoma in chickens is caused by the Rous sarcoma retrovirus.

Schizophrenia A behavioral disorder characterized by disordered thought processes and withdrawal from reality. Genetic and environmental factors are involved in this disease.

Secondary immunity Resistance to an antigen the second time it appears, due to T and B memory cells. This second response is faster, larger, and lasts longer than the first.

Secondary oocyte The large cell produced by the first meiotic division.

Secondary structure The pleated or helical structure in a protein molecule that is brought about by the formation of bonds between amino acids.

Segregation The separation of members of a gene pair from each other during gamete formation.

Semiconservative replication A model of DNA replication that provides each daughter molecule with one old strand and one newly synthesized strand. DNA replicates in this fashion.

Sense mutation A mutation that changes a termination codon into one that codes for an amino acid. Such mutations produce elongated proteins.

Severe combined immunodeficiency disease (SCID) A genetic disorder in which affected individuals have no immune response; both the cell-mediated and antibody-mediated responses are missing.

Sex chromosome In humans, the X and Y chromosomes that are involved in sex determination.

Sex ratio The relative proportion of males and females belonging to a specific age group in a population.

Sex-influenced genes Loci that produce a phenotype conditioned by the sex of the individual.

Sex-limited genes Loci that produce a phenotype in only one sex.

Short tandem repeats (STRs) Short nucleotide sequences, repeated in tandem, clustered at many sites in the genome. The number of repeats at homologous loci is variable.

Shotgun cloning A method of genome sequencing that selects clones at random from a genomic library and assembles the sequence using software.

Sickle cell anemia A recessive genetic disorder associated with an abnormal type of hemoglobin, a blood transport protein.

Sickle cell trait The symptoms shown by those heterozygous for sickle cell anemia.

Sister chromatids Two chromatids joined by a common centromere. Each chromatid carries identical genetic information.

Somatic gene therapy Gene transfer to somatic target cells to correct a genetic disorder.

Southern blotting A method for transferring DNA fragments from a gel to a membrane filter, developed by Edward Southern for use in hybridization experiments.

Spermatids The four haploid cells produced by meiotic division of a primary spermatocyte.

Spermatogonia Mitotically active cells in the gonads of males that give rise to primary spermatocytes.

SRY A gene called the sex-determining region of the Y, located near the end of the short arm of the Y chromosome that plays a major role in causing the undifferentiated gonad to develop into a testis.

Start codon A codon present in mRNA that signals the location for translation to begin. The codon AUG functions as an initiator codon.

Stem cells Cells in bone marrow that produce lymphocytes by mitotic division.

Stop codons Codons present in mRNA that signal the end of a growing polypeptide chain. UAA, UGA, and UAG function as stop codons.

Submetacentric chromosome A chromosome whose centromere is placed closer to one end than the other.

Substrate The specific chemical compound acted upon by an enzyme.

Suppressor T cells T cells that slow or stop the immune response of B cells and other T cells.

Synapsis The pairing of homologous chromosomes during prophase I of meiosis.

T cells A type of lymphocyte that undergoes maturation in the thymus and mediates cellular immunity.

Telophase The last stage of mitosis, during which division of the cytoplasm occurs.

Template The single-stranded DNA that serves to specify the nucleotide sequence of a newly synthesized polynucleotide strand.

Teratogen Any physical or chemical agent that brings about an increase in congenital malformations.

Terminator region the nucleotide sequence at the end of a gene that signals the end of transcription.

Tertiary structure The three-dimensional structure of a protein molecule brought about by folding on itself.

Testosterone A steroid hormone produced by the testis; the male sex hormone.

Tetraploidy A chromosomal number that is four times the haploid number, having four copies of all autosomes and four sex chromosomes.

Thalassemia A disorder associated with an imbalance in the production of alpha or beta globin.

Thymine dimer A molecular lesion in which chemical bonds form between a pair of adjacent thymine bases in a DNA molecule.

Tourette syndrome A behavioral disorder characterized by motor and vocal tics and inappropriate language. Genetic components are suggested by family studies that show increased risk for relatives of affected individuals.

Trait Any observable property of an organism.

Transcription Transfer of genetic information from the base sequence of DNA to the base sequence of RNA mediated by RNA synthesis.

Transfer RNA (tRNA) A small RNA molecule that contains a binding site for a specific type of amino acid and a three-base segment known as an anticodon that recognizes a specific base sequence in messenger RNA.

Transformation The process of transferring genetic information between cells by DNA molecules.

Transforming factor The molecular agent of transformation; DNA.

Translation Conversion of information encoded in the nucleotide sequence of an mRNA molecule into the linear sequence of amino acids in a protein.

Translocation A chromosomal aberration in which a chromosomal segment is transferred to another, nonhomologous chromosome.

Transmission genetics The branch of genetics concerned with the mechanisms by which genes are transferred from parent to offspring.

Trinucleotide repeat A form of mutation associated with the expansion in copy number of a nucleotide triplet in or near a gene.

Triploidy A chromosomal number that is three times the haploid number, having three copies of all autosomes and three sex chromosomes.

Trisomy A condition in which one chromosome is present in three copies, whereas all others are diploid; having one more than the diploid number ($2n + 1$).

Trisomy 13 The presence of an extra copy of chromosome 13 that produces a distinct set of congenital abnormalities resulting in Patau syndrome.

Trisomy 18 The presence of an extra copy of chromosome 18 that results in a clinically distinct set of invariably lethal abnormalities known as Edwards syndrome.

Trisomy 21 Aneuploidy involving the presence of an extra copy of chromosome 21, resulting in Down syndrome.

Trophoblast The outer layer of cells in the blastocyst that gives rise to the membranes surrounding the embryo.

Tubal ligation A contraceptive procedure in women in which the oviducts are cut, preventing ova from reaching the uterus.

Tumor suppressor gene A gene encoding a protein that suppresses cell division.

Turner syndrome A monosomy of the X chromosome (45,X) that results in female sterility.

Uniparental disomy A condition in which both copies of a chromosome are inherited from one parent.

Unipolar disorder An emotional disorder characterized by prolonged periods of deep depression.

Vaccine A preparation containing dead or weakened pathogens that elicit an immune response when injected into the body.

Vasectomy A contraceptive procedure in men in which the vas deferens is cut and sealed to prevent the transport of sperm.

Vector A self-replicating DNA molecule that is used to transfer foreign DNA segments between host cells.

Xenotransplant Organ transplants between species.

X-linked The pattern of inheritance that results from genes located on the X chromosome.

X-linked agammaglobulinemia (XLA) A rare, sex-linked, recessive trait characterized by the total absence of immunoglobulins and B cells.

XYY karyotype Aneuploidy of the sex chromosomes involving XYY chromosomal constitution.

Yeast artificial chromosome (YAC) A cloning vector that has telomeres and a centromere that can accommodate large DNA inserts and that uses the eukaryote, yeast, as a host cell.

Y-linked The pattern of inheritance that results from genes located only on the Y chromosome.

Y-linked genes Genes located only on the Y chromosome.

Zygote The diploid cell resulting from the union of a male haploid gamete and a female haploid gamete.

Credits

CHAPTER 1

Opener © David M. Phillips/SPL/Photo Researchers
Figure 1.2 © Alfred Pasieka/SPL/Photo Researchers
Figure 1.3 Portrait by Marcus Alan Vincent
Figure 1.4 © Don Fawcett/Visuals Unlimited
Figure 1.6 Courtesy of Ifti Ahmed
Figure 1.7 Courtesy of Calgene
Figure 1.8 © Corbis
Figure 1.9 American Philosophical Association
Figure 1.10 © Ken Eward Biografx/Photo Researchers
Figure 1.11 © Dr. Yorgos Nikas/SPL/Photo Researchers

CHAPTER 2

Opener © David M. Phillips/Visuals Unlimited
Figure 2.3b © K.G. Murti/Visuals Unlimited
Figure 2.5b © Bill Longcore/Photo Researchers
Figure 2.6a © Biophoto Associates/Science Source/Photo Researchers
Figure 2.6b © Don W. Fawcett/Visuals Unlimited
Figure 2.6c © Don W. Fawcett/Visuals Unlimited
Figure 2.7 © Biophoto Associates/Science Source/Photo Researchers
Figure 2.10 All: Andrew S. Bajer, University of Oregon
Figure 2.12 Photo R.M.N. Musee du-Louvre, Paris, France
Figure 2.13 (both) © David M. Phillips/Visuals Unlimited
Figure 2.14 Printed with permission of Dr. W. Ted Brown
Figure 2.19 © Cabisco/Visuals Unlimited

CHAPTER 3

Opener © Yoav Levy/Phototake
Figure 3.1a © James W. Richardson/Visuals Unlimited
Figure 3.1b © David Sieren/Visuals Unlimited
Figure 3.1c © R. Calentine/Visuals Unlimited
Figure 3.2 © Malcolm Gutter/Visuals Unlimited
Figure 3.14 All © E.R. Degginger/Photo Researchers
Figure 3.15 © Dr. P. Marazzi/SPL/Photo Researchers

CHAPTER 4

Opener © Sinclair Stammers/SPL/Photo Researchers
Figure 4.2 Images by kind permission of Family-Genetix Ltd. © 2002, FamilyGenetix Ltd. All rights reserved
Figure 4.3 © Patricia Barber/Custom Medical Stock Photo
Figure 4.4 Courtesy of Online Mendelian Inheritance in Man, National Center for Biotechnology Information
Figure 4.7 © Jeff Greenberg/Visuals Unlimited
Figure 4.10 Courtesy of B. Carragher, D. Bluemke, R. Josephs, Electron Microscope and Image Laboratory, University of Chicago
Figure 4.11 (both) © Omikron/Photo Researchers
Figure 4.13 © Steven E. Sutton/DUOMO
Figure 4.15 © Biophoto Associates/Photo Researchers
Figure 4.19 (both) © Eastcott/Momatiuk/Photo Researchers
Figure 4.20 © Visuals Unlimited
Figure 4.23 (both) Courtesy of Dr. Poh-San Lai, Department of Paediatrics, National University of Singapore, Lower Kent Ridge Road, Singapore 119074
Figure 4.26 The Granger Collection, New York

CHAPTER 5

Opener © Mark Burnett/David R. Frazier Photolibrary, Inc.
Figure 5.3a © Joe McDonald/Visuals Unlimited
Figure 5.3b © Tim Hauf/Visuals Unlimited
Figure 5.3c © M. Long/Visuals Unlimited
Figure 5.3d © Marek Litman/Visuals Unlimited
Figure 5.3e © Mark D. Cunningham/Visuals Unlimited
Figure 5.12 © Mark C. Burnett/Photo Researchers
Figure 5.14 From http://www.cdc.gov/nchs/images/obsefig1.gif
Figure 5.15 John Sholtis, The Rockefeller University, New York, NY
Figure 5.18a © Cabisco/Visuals Unlimited
Figure 5.18b © W. Ober/Visuals Unlimited

CHAPTER 6

Opener © M. Coleman/Visuals Unlimited
Figure 6.1 Courtesy of Ifti Ahmed
unnumbered photos, page 145 Courtesy of Ifti Ahmed
Figure 6.6 © Saturn Stills/SPL/Photo Researchers
Figure 6.9 Courtesy of Ifti Ahmed
Figure 6.10 Reproduced by permission of *Pediatrics* vol. 74, p. 296, © 1984, Falix et al, *Pediatrics* 74: 296–299, 1984, Figure 29.1
Figure 6.13 Courtesy of Dr. Ira Rosenthal, Dept. of Pediatrics, University of Illinois at Chicago
Figure 6.14a Courtesy of Dr. Ira Rosenthal, Dept. of Pediatrics, University of Illinois at Chicago
Figure 6.14b Courtesy of Ifti Ahmed
Figure 6.15a Photo courtesy of Dr. Irene Uchida, Genetic Services, Oshawa General Hospital, Hamilton, Ontario, Canada
Figure 6.15b © Hattie Young/SPL/Photo Researchers
Figure 6.18a Photo courtesy of Dr. Irene Uchida, Genetic Services, Oshawa General Hospital, Hamilton, Ontario, Canada
Figure 6.18b Courtesy of Dr. Ira Rosenthal, Dept. of Pediatrics, University of Illinois at Chicago
Figure 6.19 From Weiss, E., et al. (1982). Monozygotic twins discordant for Ulrich-Turner syndrome. Am. J. Med. Genet. 13: 389–399
Figure 6.20a © Martin M. Rotker
Figure 6.20b Photo courtesy of Dr. Irene Uchida, Genetic Services, Oshawa General Hospital, Hamilton, Ontario, Canada
Figure 6.21 Courtesy of Ifti Ahmed
Figure 6.28 From Hassold, T. and Hunt, P. (2001). To err (meiotically) is human: the genesis of human aneuploidy. *Nature Reviews Genetics, 2,* 280–291. Table 1, p. 283
Figure 6.31 © Grant R. Sutherland/Visuals Unlimited

CHAPTER 7

Opener © David M. Phillips/Photo Researchers
Figure 7.1 © David M. Phillips/Visuals Unlimited
Figure 7.6 From Lennart Nilsson, A Child is Born, © 1966,1977 Dell Publishing Co., Inc.
Figure 7.8 Photo courtesy of Dr. Marilyn Miller, University of Illinois at Chicago
Figure 7.11 © Sovereign/Phototake
Figure 7.13 © Sandra Wavrick/Photonica
Figure 7.17 From Zourlas, P. et al, 1965, Clinical histologic and cytogenetic findings in male hermaphroditism, *Obstetrics and Gynecology* 25: 768–778, Figure 7
Figure 7.19 (both) © George Wilder/Visuals Unlimited
Figure 7.20 Courtesy of Dr. Martin Feather and Dr. C. Bird
Figure 7.21 (both) Courtesy of Dr. George Brewer, University of Michigan
Figure 7.22 Courtesy of Dr. Neil Brockdorff, Imperial College School of Medicine, London. From Duthie, S. et al, 1999, XIST RNA exhibits a banded localization of the inactive X chromosome and is excluded from autosomal material in cis, *Human Molecular Genetics* 8: 195–204
Figure 7.24 © Leonard Lessin/Peter Arnold

CHAPTER 8

Opener Courtesy of Dmitry Pruss, National Institutes of Health
Figure 8.5 From Franklin, R. and Gosling, R.G. 1953, Molecular configuration in sodium thymonucleate, *Nature* 171: 740–741
Figure 8.10a © H. Swift and D.W. Fawcett/Visuals Unlimited
Figure 8.10b © D.R. Wolstenholme, I.B. Dawid and D.W. Fawcett/Visuals Unlimited

Figure 8.11 Reprinted by permission from American Scientist, 66: 704–711, Olin D. and Olin A., Nucleosomes: the structural quantum in chromosomes
Figure 8.12 From Harrison, C. et al., 1982, High resolution scanning electron microscopy of human metaphase chromosomes, *Journal of Cell Science 56:* 409–422, Figure 3, © The Company of Biologists, Limited
Figure 8.14 Photo by Thomas Cremer. Courtesy of William C. Earnshaw from Lamond, A.I., and Earnshaw, W.C. Structure and function in the nucleus. Reprinted with permission from *Science,* V. 280: 547–553 © American Association for the Advancement of Science

CHAPTER 9

Opener © SPL/Photo Researchers
Figure 9.2 © John D. Cunningham/Visuals Unlimited
Figure 9.12 © S.L. McKnight & Oscar L. Miller/SPL/Photo Researchers

CHAPTER 10

Opener © Walter Reinhart/Phototake
Figure 10.1 Museo del Prado, Madrid, Spain/Giraudon, Paris/SuperStock
Figure 10.12 (both) B. Carragher, D. Bluemke, R. Josephs, from the Electron Microscope & Image Processing Laboratory, University of Chicago
Figure 10.15a © Ray Coleman/Photo Researchers
Figure 10.15b © Ken Brate/Photo Researchers

CHAPTER 11

Opener © Lior Rubin/Peter Arnold, Inc.
Figure 11.11 © Kenneth E. Greer/Visuals Unlimited

CHAPTER 12

Opener © David Parker/SPL/Photo Researchers
Figure 12.2 © E. Webber/Visuals Unlimited
Figure 12.4 © Le Corre-Ribiero/Liaison Agency
Figure 12.5 Photo May 27, 1999 by David Au, courtesy of University of Hawaii
Figure 12.9 © SPL/Custom Medical Stock Photo
Figure 12.12 © Michael Gabridge/Custom Medical Stock Photo
Figure 12.16 © Alfred Pasieka/SPL/Photo Researchers
Figure 12.17 © 1992 NIH. All rights reserved
Figure 12.19 Courtesy of Suzanne McCutcheon

CHAPTER 13

Opener (all) Courtesy of CropTech Corporation
Figure 13.4 Beatty, Barbara, et al., 1999. Chromosomal localization of a phospholipase A2 activating protein, an Ets2 target gene, to 9p21. *Genomics 62:* 529–532.
Figure 13.5 Courtesy of Drs. Jorma Isola & Minna Tanner, Univ. of Tampere, Finland; Järvinen TAH, Tanner M, Bärlund M, Borg A, Isola J; Characterization of topoisomerase II gene amplification and deletion in breast cancer; *Genes, Chromosomes and Cancer 26:* 142–150 1999
Figure 13.7 Courtesy of Affymetrix, Inc., Santa Clara, California
unnumbered photo, page 326 © UPI/Corbis
Figure 13.8 From International Human Genome Sequencing Consortium. (2001). Initial sequencing and analysis of the human genome. *Nature,* 409:860–921. Figure 1, p. 862. Adapted with permission from *Nature.* Copyright 2001 Macmillan Magazines Limited.
Figure 13.10 From Venter, C. et al. (2001). The sequence of the human genome. *Science, 291,* 1304–1351. Figure 15, p. 1335. Copyright 2001 American Association for the Advancement of Science. Reprinted with permission.
Figure 13.11 Courtesy of Sirano Dhe-Paganon and Steven Shoelson
Figure 13.12 Provided by Drs. Tomokazu Amano and Andras Nagy
Figure 13.13 © Phillipe Plailly/Eurelios/SPL/Photo Researchers
Figure 13.14 Courtesy of Dwayne Kirk, Boyce Thompson Institute for Plant Research
Figure 13.15 Agricultural Research Service/USDA
Figure 13.16 Reprinted with permission of the Department of Soil & Crop Sciences, Colorado State University. http://www.colostate.edu/programs/lifesciences/TransgenicCrops/current.html
Figure 13.17 © EURELIOS/Phototake

CHAPTER 14

Opener © Nancy Kedersha/SPL/Photo Researchers
Figure 14.03 © Custom Medical Stock Photo
Figure 14.07 © K.G. Murti/Visuals Unlimited
Figure 14.10 © Benjamin/Custom Medical Stock Photo
Figure 14.12 Drs. Michael Spelcher and David Ward, Yale University
Figure 14.15 From Buchdunger, E et al. (2001). Bcr-Abl inhibition as a modality of CML therapeutics, *Biochimica et Biophysica. Acta 1551,* m11-m18, Fig 5A, p. M16. Copyright 2001, reprinted with permission from Elsevier Sciences.

CHAPTER 15

Opener © J.L. Carson/Custom Medical Stock Photo
Figure 15.6a Courtesy of Dorothea Zucker-Franklin, New York University School of Medicine
Figure 15.10 © Jean Claude Revi/Phototake
Figure 15.16 © Baylor College of Medicine/Peter Arnold Inc.
Figure 15.17 © David M. Phillips/Visuals Unlimited
Figure 15.18 © J.L. Carson/Custom Medical Stock Photo

CHAPTER 16

Opener (both) © Tim Beddow/SPL/Photo Researchers
Figure 16.2 © Columbus Instruments/Visuals Unlimited
Figure 16.4 Courtesy of Dr. Donald Price, Johns Hopkins School of Medicine. From Price, D.L. et al, Genetic neurodegenerative diseases: the human illness and transgenic model. Reprinted with permission from *Science 282:* 1079–1083 © 1998 American Association for the Advancement of Science
Figure 16.5 (both) © Malcolm S. Kirk/Peter Arnold, Inc.
Figure 16.6 Courtesy of P. Hemachandra Reddy, Neurological Sciences Institute, Oregon Health & Science University, from Reddy, P.H. et al, 1999, Transgenic mice expressing mutated full-length HD cDNA: a paradigm for locomotor changes and selective neuronal loss in Huntington's disease. Proceedings of the Royal Society of London B *354,* 1035–1045
Figure 16.7 From Lai, C. et al. (2000). The *SPCH1* region on human 7q31: genomic characterization of the critical interval and localization of translocations associated with speech and language disorder. *American Journal of Human Genetics, 67,* 357–368, Figure 1, p. 358. Adapted with permission from University of Chicago Press.
Figure 16.9a © E.R. Lewis, T.E. Everhard and YY. Zeevi/University of California/Visuals Unlimited
Figure 16.9c © T. Reese and D.W. Fawcett/Visuals Unlimited
Figure 16.10 From Craddock, N. et al. (1995). Mathematical limits of multilocus models: the genetic transmission of bipolar disorder. *American Journal of Human Genetics, 57,* 690–702. Reprinted with permission of University of Chicago Press.
Figure 16.11 From Potash, J.B. and DePaulo, J.R. (2000). Searching high and low: a review of the genetics of bipolar disorder. *Bipolar Disorders, 2,* 8–26, Figure 3, p. 14
Figure 16.12 © Corbis
Figure 16.13 Dr. Monte Buchsbaum, Mt. Sinai Medical Center, New York, New York
Figure 16.15 Dr. Dennis Selkoe, Center for Neurologic Diseases, Harvard Medical School, Boston, Massachusetts

CHAPTER 17

Opener © Tim Davis/Tony Stone Images
Figure 17.2 C. Mayhew & R. Simmon (NASA/GSFC), NOAA/NGDC, DMSP Digital Archive
Figure 17.8 From Lucotte, G. et al. (1995). Complete map of cystic fibrosis mutation DF508 frequencies in Western Europe and correlation between mutation frequencies and incidence of disease. *Human Biology, 67,* 797–803, Figure 1, p. 800

CHAPTER 18

Opener © Paul Hanny/Liaison Agency
Figure 18.4 © Meckes/Ottawa/Photo Researchers
Figure 18.11 © George Holton/Photo Researchers
Figure 18.12 Fr. Peter Weigand, OSB
Figure 18.15 Owens, K., and King, Mary Clairè. (1999). Genomic views of human history. *Science, 286,* 451–453, Figure 1, p. 452. Copyright 1999 American Association for the Advancement of Science. Reprinted with permission.
Figure 18.17 From Lieberman, D.E. (2001). Another face in our family tree. *Nature, 410,* 419–420, Figure 2, p. 420. Adapted with permission from *Nature.* Copyright 2001 Macmillan Magazines Limited.
Figure 18.19 Institute of Human Origins
Figure 18.20 From Lieberman, D.E. 2001 Another face in our family tree. *Nature 410,* 419–420 Credit National Museum of Kenya, Nairobi, Kenya
Figure 18.21 National Museum of Kenya/Visuals Unlimited
Figure 18.22a © John D. Cunningham/Visuals Unlimited
Figure 18.22b Dr. David Frayer

CHAPTER 19

Opener © Carlos Goldin/SPL/Photo Researchers
Figure 19.1 © Custom Medical Stock Photo
Figure 19.2 © Rafael Macis/Photo Researchers
Figure 19.4 © Dr. Yorgos Nikas/SPL/Photo Researchers
Figure 19.5 © Carlos Goldin/SPL/Photo Researchers
Figure 19.6 Courtesy of Kerry Bradford
Figure 19.7 © Alexander Tsiarus/Science Source/Photo Researchers
Figure 19.8a Courtesy of Affymetrix, Inc., Santa Clara, California
Figure 19.8b Courtesy of Tyson Clark, Dept. of Molecular, Cellular and Developmental Biology, University of California Santa Cruz
Figure 19.10 © Martha Cooper/Peter Arnold, Inc.

Index

A

Page numbers in italics indicate illustrations.

I

J

K

L

M

O

P

S

T

U

X